SIDNEY BOROWITZ *New York University*

ARTHUR BEISER

ESSENTIALS OF PHYSICS

A text for students of science and engineering

ADDISON-WESLEY WORLD STUDENT SERIES

ADDISON-WESLEY PUBLISHING COMPANY

Reading, Massachusetts · Palo Alto · London · Sydney · Manila

THIS BOOK IS AN ADDISON-WESLEY WORLD STUDENT SERIES EDITION

FIRST PRINTING 1967

Library of Congress Catalog Card No. 65-19242

Preface

In recent years it has become more and more difficult to present a meaningful first course in physics at the college level for students of science and engineering. The two main reasons for this difficulty are the ever-expanding scope of physics and the fact that considerable changes have occurred in the engineering profession, changes which are in turn reflected in altered curricula. Engineering students must often be exposed early in their educations to subject matter which only a few years ago was considered advanced modern physics. At the same time the more traditional engineering disciplines must not by any means be neglected, and hence the basic subjects of classical physics—mechanics, electricity, thermodynamics, and kinetic theory—are as important as ever in the training of engineers.

A similar evolution has occurred in the education of chemists because of the necessity for more sophisticated approaches to atomic structure and the introduction of such developments as nuclear chemistry. Physics, too, has begun to incorporate material from related fields, such as plasma physics and solid-state theory, which combine classical and modern notions on terms of equal importance. Soon students of the biological sciences will probably also require increased grounding in the whole spectrum of physics.

It is this lack of obsolescence in physics which places such a heavy burden on an elementary course. Merely adding new material to what is being covered in existing courses is impossible: such a compendium would be too gigantic. Nor is it feasible simply to omit whole chunks of the subject in order to suit the limited course time available. Some reworking and rethinking of the basic content of this course is called for. This book represents such an effort.

Our primary aim was to write a book which would be relatively concise but which yet would not neglect any major aspect of the subject. Conciseness has several advantages for the student. There is the psychological advantage of having the necessary material available in one volume which the student feels he can cover in a reasonable time. More important, this same conciseness enables him to perceive the unity and structure of the subject. He can grasp the relationships between various fields of physics while the course is underway, not at some later time.

Compactness has been achieved in a variety of ways. We use the calculus freely, taking advantage of the fact that the present college freshman is better prepared mathematically than his predecessors of even five or ten years ago. At the same time, since we are also aware that many students who take an elementary physics course are taking the calculus concurrently, full use is not made of this subject until the student is well along in his first semester. But we *do* make full use of vector methods, since they are developed in the text itself. Much of the compactness of the book should be credited to our attitude that this is not primarily a text for solitary study. We took it for granted that deft guidance on the part of an instructor would save pages of explanation in the text. At the same time, we have left it to the instructor to choose which topics he wishes to pursue in greater detail.

However, despite the concise treatment, we have developed many aspects of the subject in more depth than is usual in an introductory course. We hoped to provide the student with a strong framework that will need no major alterations when he wishes to expand or flesh it out in his further work in science and engineering. Finally, we have tried to avoid any pedantry in our explanations and proofs; an elaborate proof often serves only to confuse the beginner and obscure the point.

The basic emphasis is on the microscopic properties of matter and the fundamental force laws and conservation theorems common to all branches of physics. Thus the student not particularly interested in atomic and nuclear physics is not ignored, any more than is the student who expects to specialize in these subjects. Applications have been kept to a minimum and are mentioned only when they are needed to illustrate essential ideas. Only the most significant aspects of contemporary physical research are discussed.

No attempt has been made to achieve an artificially constant level of difficulty. Each topic is treated as simply as it is possible to treat it and still be consistent with correctness, which means that more effort is required on the part of the student in some places than in others. The justification for this approach is twofold: first, the student not only does not need to unlearn anything at a later date, but he will also see the relationship between the material covered here and the deeper, more sophisticated treatment he will eventually encounter. Second, students of all abilities and degrees of preparation will find challenges that will stimulate their interest and extend their competence. The exercises, which are essential to the learning process, are also aimed at students with varying degrees of ability.

The book is planned for a three-semester course, but an instructor may, by a judicious selection of material, use it for a one-year course. The further development of some of the topics covered would make it possible for a two-year sequence to be based on the book.

Thanks are due to the various persons who have been kind enough to read and comment on the manuscript during the preparation of the book. We are especially grateful to Professor Howard Carr and to Mr. John B. Lewis for their help.

New York S.B.
Mallorca A.B.
August 1965

Contents

1

The Quantities of Physics

Physics is rooted in measurement. Its great success has come about through a strictly quantitative approach—theory and experiment both must deal with precisely defined quantities capable of being measured. While much of physics seems abstract and distant from what we like to think of as reality, every statement that it makes is ultimately reducible to a measurement, to a "pointer reading." It is for this reason that we begin our study of physics with a discussion, in this chapter and the next, of the quantities found in physics and how they are handled mathematically.

An important aspect of physical quantities is their *scale*: how large they are relative to one another and to things we are familiar with, what is important and what can be neglected when several effects occur simultaneously. The ability to perceive the essentials of any physical situation requires a feeling for distinguishing between dominant quantities and small corrections. This facility generally develops with experience.

1–1 Measurement

When we speak of measuring something, we actually imply two separate operations. The first is the establishment of a *standard unit*; the second is a comparison between the standard unit and the quantity being measured.

A total of four standard units is all that is necessary to express the results of measuring almost any physical quantity that we know of in the universe. To a surprisingly large extent the choice of the fundamental quantities in terms of which everything is to be measured is arbitrary. The accepted ones today are length, mass, time, and electric current. The two chief systems of units in use today both employ the same units—the *second* and the *ampere*, respectively—for the latter two quantities, but have different units for length and mass. The unnecessarily cumbersome *British system* uses the *foot* (abbreviated ft) for length and the *pound* (lb) for weight. The *metric system* uses the *meter* (m) for length and the *kilogram* (kg) for mass. (The important distinction between mass and weight will be discussed in Chapter 5.)

In recent years the standards of length and time have been accurately compared with quantities found in nature whose magnitudes are believed to be invariant. Thus the length of the meter may be specified in terms of one of the characteristic wavelengths of light emitted by a particular isotope of krypton, and the duration of the second may be specified in terms of one of the characteristic frequencies of vibration of the ammonia molecule. In this way every laboratory in the world has direct access to the fundamental standards of measurement.

The specific combination of length, mass, time, and current units required to express one of the diverse quantities found in physics is called the *dimension* of the quantity. To give an example, "work," in the sense that this term is used in physics, has the dimensions of (mass)(length)^2/(time)^2, which would be kg-m^2/sec^2 in the metric system. This combination of units is called the *joule* as an abbreviation, but, like its fellow specialized units, it can be reduced to its equivalent in terms of the primary units. A useful check on the correctness of a calculation or derivation is to see whether the dimensions throughout are consistent; in this way we can learn at once if we have omitted some pertinent quantity or made an incorrect conversion from one size unit to another. Many of the illustrative examples in this book have been worked out with the units given for each important step to show how this is done.

One quantity not normally specified in terms of length, mass, time, and electric current is *temperature*. Temperature, as we shall see, is in a unique category, since it has no meaning without reference to certain statistical properties of matter and therefore cannot be defined in terms of a standard in the same sense as the other fundamental quantities.

The second operation involved in measurement is that of *comparison*. How many standard units are there in whatever it is that we wish to measure? A comparison of a length with the standard meter is, in concept, simple enough. We take the standard meter, or a replica of it, and place it successively along the subject of our measurement as many times as it will fit. If there is not exactly a whole number of meters in our specimen, we then estimate, perhaps with the help of geometry, what fraction of a meter is left over. However, when we attempt to make such a comparison involving one of the other fundamental quantities, we find that the method by which the comparison is made is of great significance. This is particularly true when measuring things either much smaller or much larger than the standard units. In every case we must be careful not to attribute to the result of the measurement more accuracy than it deserves.

1-2 Significant figures

When we look up the mass of the planet Uranus in a reference book, we read that it is 14.6 times the mass of the earth. Does that mean that the mass ratio is actually 14.600000 . . .? The answer, clearly, is no. The astronomer who arrived at the figure 14.6 felt confident that the correct value was closer to 14.6 than it was to either 14.5 or 14.7; his means of comparing the masses of the two planets were not sufficiently accurate to permit him to specify their ratio more exactly.

There exist mathematical techniques, based on the theory of probability, for establishing the degree of confidence we may place in a particular measurement. A number of individual determinations must be made, and their average value calculated. Thus, when the height of a man is measured with a meter stick several times, the values 1.73, 1.76, 1.74, and 1.78 might be obtained. We have no way of assessing the merits of each particular determination and must rely on the *average* of all of them as being the most reliable figure. The average of the four values quoted is 1.7525 m. But since each determination yielded three *significant figures*, we are not justified in presenting the result of the entire measurement as being accurate to five significant figures. The appropriate thing to do here is to specify his height as 1.75 ± 0.02 m, where 1.75 represents the average value and ± 0.02 m represents the error.

The theory of error has certain specific prescriptions for assigning the number of significant figures to be retained in an average value, and for determining the magnitude of the uncertainty in the average. Thus the uncertainty is often expressed in terms of *probable error*. If the 0.02 m above is the probable error (actually many more than four determinations are required in order for the theory to be applicable), then there is a 50% likelihood that any individual measurement will fall between 1.73 and 1.77 m. For our purposes we shall be content merely with an awareness of experimental error and the importance of not attributing a greater number of significant figures to a quantity than is justified.

All numerical statements contain significant figures, which are the essential results of the measurement, plus some means of indicating the size of the number. An elephant weighs 3500 lb. If the scale is accurate to ± 100 lb, only the first two digits are significant. The zeros serve only to indicate the position of the decimal point. If the scale is accurate to ± 10 lb, the first three digits are significant. It is sometimes difficult to tell just how many significant figures are present in a given number.

A common dilemma involving significant figures occurs whenever quantities differing in scale are to be combined in a calculation. A bee of mass 0.000070 kg (*not* 0.00007000 . . . kg) alights on a flower of mass 0.0172 kg. What is the mass that must now be supported by the stem of the flower? Adding the mass of the bee to the mass of the flower yields 0.017270 kg, which suggests more accuracy in the sum than the sum is entitled to. The rule to follow when adding or subtracting quantities of limited accuracy is to round off the answer at the point where any one of them runs out of significant figures. In other words, the *least* accurate quantity determines the accuracy of the sum or difference being computed. Thus we have

$$\begin{array}{rl} 0.0172 & \text{kg} \\ +0.000070 & \\ \hline 0.0173 & \text{kg.} \end{array}$$

Example: An elephant weighing 3500 ± 100 lb has a mahout weighing 182.81 ± 0.01 lb sitting on its head. What is the total weight?

Solution: It is absurd to write 3682.81 ± 100 lb; a proper statement here is 3700 ± 100 lb, since the uncertainty in the elephant's weight far exceeds that in the man's weight.

A similar rule must be followed when quantities of limited accuracy are multiplied or divided by one another. Suppose that we wish to find the average speed of a man who runs 100.00 m in 10.3 sec. It is absurd to express the answer as 9.708737 m/sec. The proper procedure, again, is to retain in the answer only as many significant figures as there are in the least accurate number that goes into the calculation, although one should carry along an additional figure during the arithmetic to avoid errors in rounding-off the final result. Thus the man's speed should be written 9.71 m/sec.

Of course, there are quantities which can be determined theoretically to an indefinite number of significant figures. Such quantities are π, logarithms, trigonometric functions, and so on. As many figures should be used in numbers of this kind as there are in the least accurately known number in the calculation, plus an additional digit to prevent rounding-off errors.

Example: Find the volume of a sphere of radius 6.1 m.

Solution: The volume of a sphere of radius R is

$$V = \tfrac{4}{3}\pi R^3,$$

and so here the required volume is

$$V = 950.29378 \text{ m}^3 = 950 \text{ m}^3$$

if we let $\pi = 3.14$. If we had let $\pi = 3.1$, as many significant figures as there are in R, then

$$V = 938.1881 \text{ m}^3 = 940 \text{ m}^3,$$

a distinct loss in accuracy.

1-3 Exponential notation

It is clear from the above example that the number of significant figures in a given number may readily be ambiguous in the absence of an explicit indication of the uncertainty. The volume of the sphere is not precisely 950 m^3, but is merely closer to 950 m^3 than to either 940 m^3 or 960 m^3. There exists a convenient method for removing such ambiguities, called *exponential notation*, which is widely used in scientific and engineering work. It is helpful not only in providing a measure of the accuracy of a given number but also in abbreviating very large and very small numbers. For instance, the mean distance from the earth to the sun is 149,600,000,000 m, and 0.000,000,000,053 m is the effective radius of the hydrogen atom; both are cumbersome to write and to treat arithmetically.

In exponential notation we write only as many digits as there are significant figures in any quantity, and indicate the magnitude of the quantity by a power of ten. In this notation we would write 950 as 9.5×10^2, since

$$950 = 9.5 \times 100 = 9.5 \times 10^2.$$

Similarly 0.097 would be written 9.7×10^{-2}, since

$$0.097 = 9.7 \times 0.01 = 9.7 \times 10^{-2}.$$

In exponential notation we would write the mean earth–sun distance as 1.4960×10^9 m, and the radius of the hydrogen atom as 5.3×10^{-11} m. Note that the first factor should lie between 1 and 10; that is, 1.4960×10^9 rather than, say, $14{,}960 \times 10^5$ or 0.14960×10^{10}. The reason for this convention is that it facilitates arithmetic. The various powers of ten may be separated out initially, and the remaining computation involves only numbers of the same order of magnitude.

Example: The radius of the earth is 6.37×10^6 m and its mass is 5.975×10^{24} kg. What is its average density?

Solution: Density is mass per unit volume, and so here we must divide the mass m of the earth by its volume $V = \frac{4}{3}\pi R^3$. Hence

$$\begin{aligned} d &= \frac{m}{\frac{4}{3}\pi R^3} \\ &= \frac{3 \times 5.975 \times 10^{24}}{4 \times 3.142 \times (6.37 \times 10^6)^3} \\ &= \frac{3 \times 5.975}{4 \times 3.142 \times (6.37)^3} \times \frac{10^{24}}{(10^6)^3} \\ &= 0.005517 \times 10^6 \text{ kg/m}^3 \\ &= 5.52 \times 10^3 \text{ kg/m}^3. \end{aligned}$$

Note that $(10^6)^3 = 10^{18}$. Only three significant figures belong in the density we have computed, since that is the accuracy of the poorest figure that went into the calculation.

1–4 Mathematics and physics

The laws of physics are expressed most compactly and conveniently as mathematical relationships among quantities appropriate in describing the particular system involved. There is a great distinction between mathematics and physics, however. Mathematics is essentially the study of certain relationships among symbols, with no reference to nature at all. Physics, on the other hand, is concerned exclusively with nature, and uses mathematics only as an aid in obtaining and stating its findings. To the physicist, mathematics is one of many tools, though an important one, and his work consists of much more than manipulating equations.

While all the material in this book could have been presented using no more than the most elementary geometry and algebra, such a procedure would be very wasteful. An equation in physics is an abstraction of reality, and its concise form helps us to grasp the meaning of the things it describes. The more advanced the mathematics we employ, generally speaking, the fewer and briefer the equations. A mathematical technique, for instance integration, may be complicated in itself, but mastering it permits us to absorb a good deal of information less painfully than would be possible otherwise. For this reason more mathematics is used in what follows than perhaps is conventional, and as a result we are able to penetrate a little more quickly into the fundamental principles governing the physical world. The next chapter, in fact, is a discussion of vector analysis, a branch of mathematics especially useful in the study of physics.

Problems

1. Express the following numbers in exponential notation:

6,460,000,000	8400
0.14	0.000,000,007,819
1,000,000	70
0.007,890	351,600
3.81	0.000,013

2. Express the following numbers in exponential notation:

13,008,000,000	800
0.0054	790,000
12.45	0.000,000,890
12,345.09	0.01

3. Express the following numbers in decimal notation:

4.51×10^8	2.0×10^5
7.819×10^2	5.1×10^0
8×10^{-2}	9.56×10^{-5}
1.003×10^{-8}	10^{-10}

4. Express the following numbers in decimal notation:

10^9	6.78×10^5
8.000×10^4	9×10^2
7.416×10^1	9.4×10^{-1}
6.350×10^{-5}	7×10^{-9}

2

Vector Analysis

Before going into the actual subject matter of physics, we must learn how to deal with the various measurable quantities we shall encounter. Often the ordinary arithmetical operations of addition, subtraction, multiplication, and division will be employed; in such cases the quantities treated are called *scalars*. However, there is another, more general kind of quantity in physics called a *vector*, and in order to treat vectors we must master a new kind of arithmetic.

2–1 Scalars

Quantities which are found by experiment to combine with one another according to the laws of ordinary arithmetic are called *scalars*. A mass of 7 kilograms, a time interval of 38 minutes, a length of 4.3 meters are all scalars. There are many other scalar quantities which we shall deal with: density, pressure, energy, work, electric charge, capacitance. Scalars are completely described by a magnitude only. A unit is usually included after the number in order to indicate precisely what is meant. When we add or subtract scalars, we can use only numbers associated with the same units. Thus 7 ft + 4 ft = 11 ft, but 7 ft + 38 min is meaningless. (This is, of course, obvious, but it is easy to make blunders of the latter type unless units are carefully carried along in all calculations.) In multiplying or dividing we can sometimes meaningfully combine quantities associated with different units. For example, we may divide a distance of 72 mi by a time interval of 3 hr to obtain a speed of 24 mi/hr.

2–2 Vectors

Many quantities of importance in physics are not completely specified by a magnitude alone. If we apply a force to some object, it is not sufficient to say only that the force is 27 pounds. What we must also give is the *direction* of the force, for instance 27 pounds vertically upward, if the description is to be complete.

Another example of a vector quantity is displacement. A boat travels 45 mi, a scalar statement. The boat may have gone 45 mi to the north, it may have gone 20 mi west and then 25 mi south, or it may have gone 22.5 mi east and then 22.5 mi west, to end up at its starting point. By adding a direction to a magnitude, for instance by saying that the boat went 45 mi southwest, we have a complete description of its displacement.

It is customary to print the symbols of vector quantities in boldface type, while italics are used both for scalar quantities (such as mass, time, and area) and for the magnitudes of vectors. Thus the magnitude of the vector **A** is A. In handwriting, vectors are represented by placing arrows over their symbols, for instance $\vec{A}$.

In this book we shall meet numerous vectors, all of which share the property of having a direction as well as a magnitude associated with them. When we discuss vector arithmetic, a particular kind of vector, namely a displacement, will serve as the chief example, but it is important to keep in mind that the results apply for *all* vectors.

2–3 Vector addition

If an object is originally at the point O in Fig. 2–1 and is then moved to the point A, its change in position may be represented by the line segment OA, whose arrow indicates the sense of the displacement. This line segment has both a magnitude (its length) and a direction, and it is a vector which we shall designate as **A**. If we wanted to specify the displacement OA without the aid of Fig. 2–1, we should have to state three things: (1) the plane it lies in, here the plane of the paper; (2) the distance l from O to A; and (3) the angle θ between OA and some reference direction, say the x-axis.

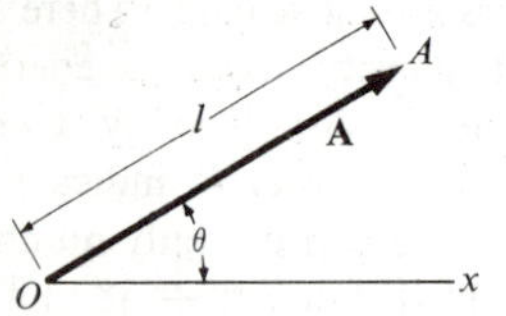

FIG. 2–1. The vector **A** represents a displacement from O to A of length l at an angle of θ with the $+x$-axis.

There are other ways of describing OA. A common one is to have the point O at which the vector starts as the origin of a coordinate system and to give the cartesian coordinates of the point A at which the vector terminates. As shown in Fig. 2–2, the projection of A on the x-axis is A_x, on the y-axis A_y, and on the z-axis A_z. That is, the coordinates of the point A are A_x, A_y, and A_z. Together A_x, A_y, and A_z completely describe the vector **A**.

Sometimes it is convenient to use spherical polar coordinates for designating a vector. Again we let O be the origin and, as in Fig. 2–3, we locate point A by its distance A_r from O, the angle θ which OA makes with the z-axis, and the angle ϕ which the projection of the vector on the xy-plane makes with the x-axis.

No matter what means we employ to describe OA, though, we require three separate quantities. (In the example of the boat given above, it is implied that the boat is on the surface of the earth, providing the third part of the vector description of its displacement.)

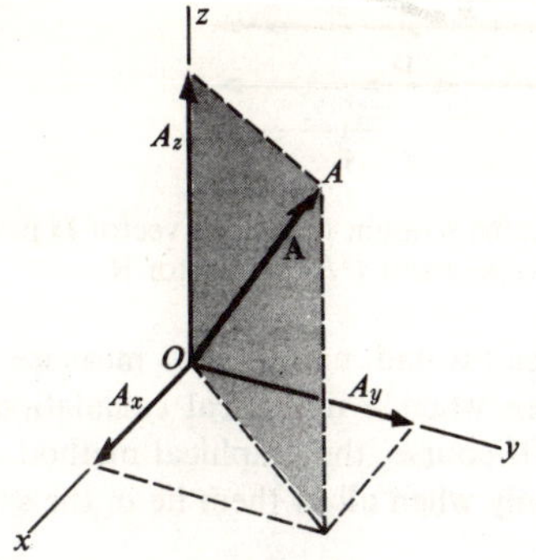

FIG. 2–2. The vector **A** and its components A_x, A_y, and A_z.

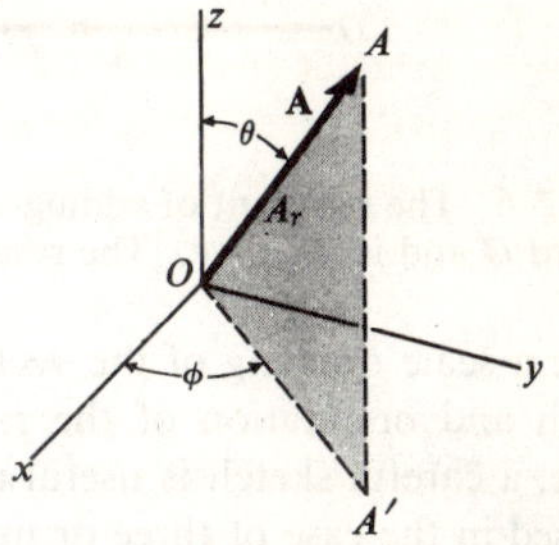

FIG. 2–3. The vector **A** can also be specified by its length A_r, polar angle θ, and azimuthal angle ϕ.

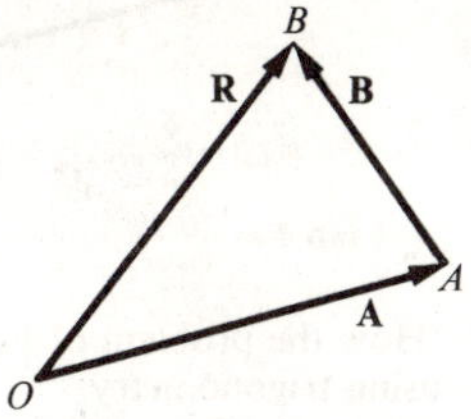

FIG. 2–4. The resultant **R** of a displacement **A** followed by a displacement **B**, illustrating the method of adding vectors.

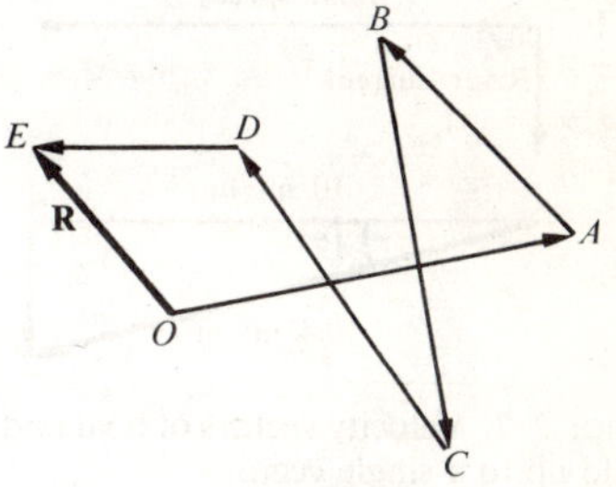

FIG. 2–5. The resultant **R** of the five vectors shown.

Let us now move the object through two successive displacements, OA followed by AB (Fig. 2–4). The net result is the displacement OB, which is defined as the *sum* of OA and AB. In general, if we wish to add the vector **A** to the vector **B**, we draw **A** and at its end draw **B**, as shown. The sum of the two, the *resultant* **R**, is the vector extending from the beginning of **A** to the terminus of **B**.

If more than two vectors are to be added, the same procedure is followed. Each successive vector is drawn starting from the terminus of the preceding one, and the resultant is the vector joining the start of the first vector and the terminus of the last one. This is illustrated in Fig. 2–5. In the special case in which all the vectors are parallel to one another, vector addition reduces to scalar addition, provided that positive and negative signs are assigned to the two directions possible for each vector (Fig. 2–6).

The sum of two vectors can always be found by trigonometry, since all that is involved is the solution of a triangle two of whose sides and the included angle are given. Where no great accuracy is required, vector addition may be performed graphically with the help of a ruler and protractor. The procedure is simply to

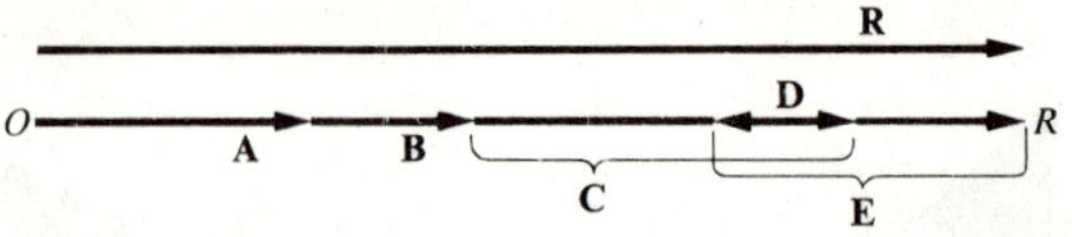

FIG. 2-6. The resultant of adding vectors in the same straight line. The vector **D** points toward *O* and is negative. The resultant is the line segment *OR*, the vector **R**.

make a scale drawing of the vectors placed head-to-tail, and then to measure the length and orientation of the resultant. Even when a numerical calculation is made, a careful sketch is useful as a check. Of course, the graphical method can be used in the case of three or more vectors only when all of them lie in the same plane.

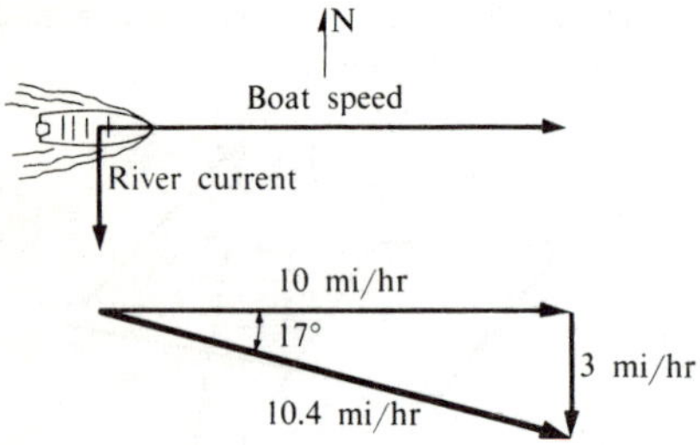

FIG. 2-7. Velocity vectors of boat and river add up to a single vector.

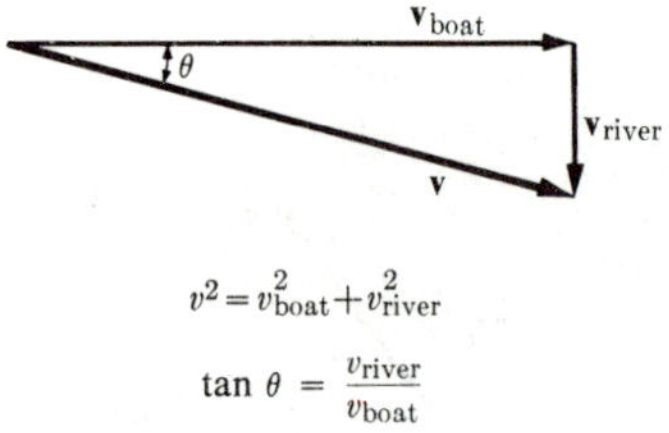

FIG. 2-8. How the problem of Fig. 2-7 is solved by using trigonometry.

Example: A boat is heading east at 10.0 mi/hr in a river flowing south at 3.0 mi/hr. What is the boat's velocity relative to the earth?

Solution: Figure 2-7 shows how the problem is solved by using vectors to represent the boat's velocity and the river's current. The boat is traveling at 10.4 mi/hr in a direction about 17° south of east.

More accurate results may be obtained with the help of trigonometry. Figure 2-8 illustrates how the problem may be solved by using trigonometric methods. Because $\mathbf{v}_{boat}$ is perpendicular to $\mathbf{v}_{river}$, the speed v of the boat relative to the earth may be obtained from the Pythagorean theorem:

$$v^2 = v_{boat}^2 + v_{river}^2,$$

$$v = \sqrt{v_{boat}^2 + v_{river}^2} = \sqrt{100 + 9.0}\ \text{mi/hr} = 10.4\ \text{mi/hr}.$$

The angle θ between $\mathbf{v}$ and $\mathbf{v}_{boat}$ is specified by

$$\tan\theta = \frac{v_{river}}{v_{boat}} = \frac{3.0\ \text{mi/hr}}{10.0\ \text{mi/hr}} = 0.3,$$

$$\theta = 17°.$$

Example: What is the resultant of a vector of magnitude 12 that is at an angle of 30° from the x-axis (angles specified in this manner are normally to be taken counterclockwise from the $+x$-direction), and another of magnitude 8 that is at an angle of 100° from the x-axis?

Solution: The problem is sketched in Fig. 2–9. The magnitude of the resultant is the side of the triangle whose other sides are of lengths 12.0 and 8.0 with an included angle of 110°. From the law of cosines,

$$R = \sqrt{(8.0)^2 + (12.0)^2 - 2 \times (8.0) \times (12.0) \cos 110^\circ}$$
$$= 16.5.$$

The angle ϕ between R and the side of length 12 may be found from the law of sines:

$$\frac{\sin \phi}{8.0} = \frac{\sin 110^\circ}{16.5}, \qquad \phi = 27^\circ.$$

The resultant, therefore, has the magnitude 16.5 and makes an angle of 57° with the x-axis.

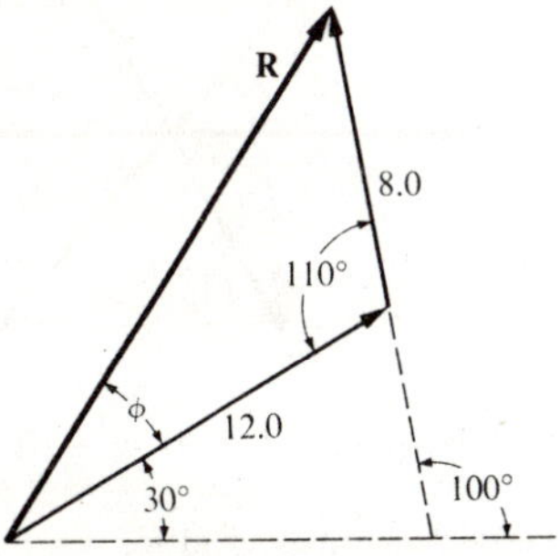

FIGURE 2–9

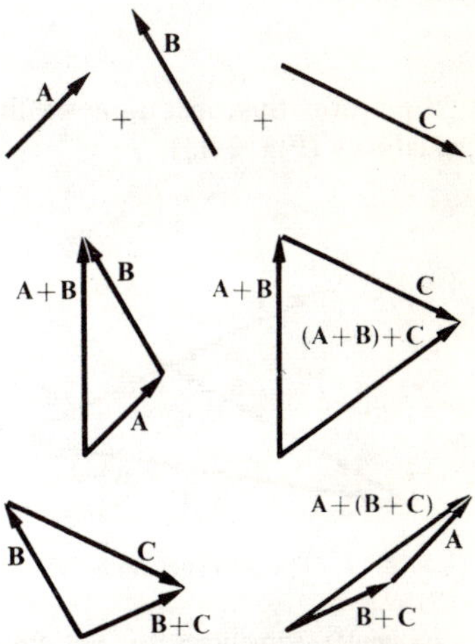

FIG. 2–10. The commutative and associative laws of vector addition.

Although in the addition of vectors it is not, perhaps, so obvious as it is for scalars, vectors may be added in any order. They obey both commutative and associative rules; that is, as shown in Fig. 2–10,

$$\mathbf{A} + \mathbf{B} = \mathbf{B} + \mathbf{A} \quad \text{(commutative law)} \tag{2–1}$$

and

$$(\mathbf{A} + \mathbf{B}) + \mathbf{C} = \mathbf{A} + (\mathbf{B} + \mathbf{C}) \quad \text{(associative law).} \tag{2–2}$$

In adding vectors it is permissible to shift them parallel to their original orientations in order to have them originate at any point required.

2-4 Vector subtraction

The operation of subtracting one vector from another is similar to the corresponding operation in ordinary arithmetic. We first introduce the *zero element*, here the *null vector* **0**. This **0** is a vector having no magnitude, so that it is a point with no direction associated with it. Now we define the negative of a vector by means of the equation

$$(-\mathbf{A}) + \mathbf{A} = \mathbf{0}.$$

That is, the negative of **A** is that vector which, when added to **A**, yields the null vector **0** as the resultant. The negative of a vector has the same magnitude as the original vector and acts along the same line, but it points in the opposite direction.

To subtract the vector **B** from the vector **A**, then, we add vectorially **A** and the negative of **B**:

$$\mathbf{A} - \mathbf{B} = \mathbf{A} + (-\mathbf{B}). \tag{2-3}$$

To perform this operation we first construct −**B** and then add **A** to it in the usual way (Fig. 2-11).

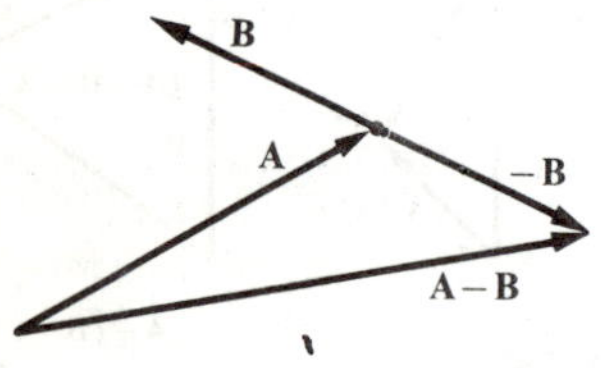

FIG. 2-11. The subtraction of vectors.

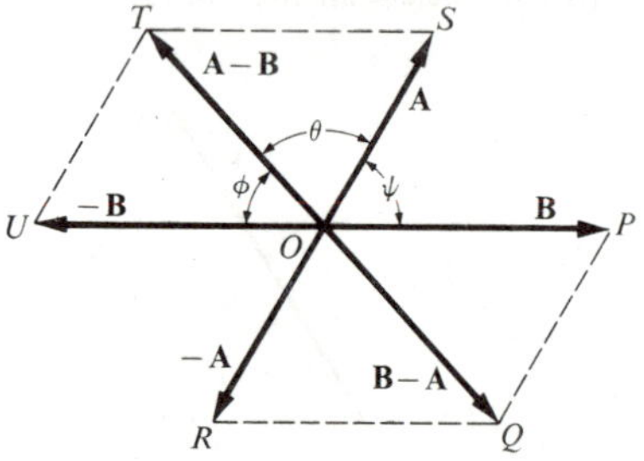

FIG. 2-12. **A** − **B** = −(**B** − **A**).

Example: Sometimes it is not obvious that certain identities valid for scalars are also valid for vectors. For this reason we might enquire whether

$$\mathbf{A} - \mathbf{B} = -(\mathbf{B} - \mathbf{A}),$$

where **A** and **B** are arbitrary vectors.

Solution: We begin by constructing **A** − **B** and **B** − **A** (Fig. 2-12) according to the law of subtraction given in this section. The lengths of the vectors **A** − **B** and **B** − **A** are the same, since the parallelograms *OPQR* and *OSTU* are congruent and the two vectors are the diagonals of these parallelograms. By construction,

$$\theta + \phi + \psi = 180°.$$

Since angles *POQ* and ϕ are vertical angles, they are equal, and *TOQ* is a straight line. Hence

$$\mathbf{A} - \mathbf{B} = -(\mathbf{B} - \mathbf{A}).$$

2–5 Resolution into components

When two vectors are to be added or subtracted, the labor necessary to obtain the result can be large. When more than two are involved, the labor is almost prohibitive unless they are parallel. However, there is a means for simplifying such a calculation by taking advantage of the fact that adding or subtracting vectors along the same line is the same as arithmetical addition or subtraction. What is done is to establish certain fixed directions in space, usually the x-, y-, and z-coordinate axes, and to express each vector we encounter in terms of three vectors along these directions. The breaking down of a vector into an equivalent set of other vectors in fixed directions is called the *resolution* of the vector into its *components.*

The process of resolving a vector into components is not a unique one, since it depends on which axes are chosen as the reference set. Hence any set of cartesian axes may be used, provided that the same one is used throughout the calculation. In fact, any suitably chosen coordinate system, cartesian or not, can be used. This situation is analogous to the lack of uniqueness in breaking up one number into two others whose sum is the same as the original one; $3 + 7$, $5 + 5$, and $6 + 4$ all are equal to 10.

Let us first consider vectors in a plane. In Fig. 2–13 the vector $\mathbf{C}$ is evidently the sum of the vectors $\mathbf{C}_x$ and $\mathbf{C}_y$, which are the components of $\mathbf{C}$. Their magnitudes are

$$C_x = C \cos \theta, \qquad C_y = C \sin \theta,$$

where C is the magnitude of the vector. A vector in a plane may equivalently be specified either by its magnitude and direction or by its components along two perpendicular axes, and we can go from one method to the other by using simple trigonometry.

Example: A weight of 100 lb rests on an inclined plane which makes an angle of 30° with the horizontal (Fig. 2–14). (Weight is a force acting downward perpendicular to the earth's surface.) Resolve the weight into components parallel and perpendicular to the plane itself.

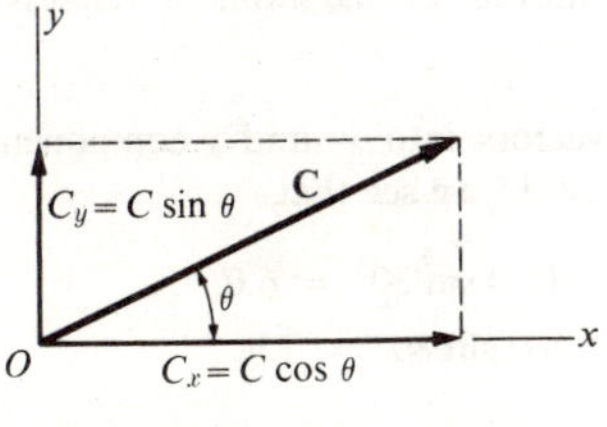

FIG. 2–13. The resolution of the vector $\mathbf{C}$ into x- and y-components.

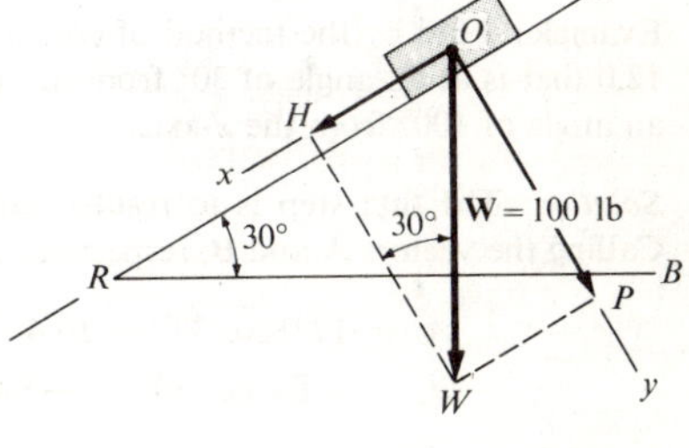

FIG. 2–14. The resolution of a weight on an inclined plane into components parallel and perpendicular to the plane.

Solution: We draw a set of axes parallel and perpendicular to the plane with their origin at O. Next, lines are drawn from the end of **W**, which is the vector representing the weight perpendicular to Ox and Oy. The components we seek are OH and OP. Angle OWH is equal to ARB, which is given as 30°, since the sides of these angles are mutually perpendicular. We therefore have

$$OH = 100 \sin 30° = 50 \text{ lb}, \qquad OP = 100 \cos 30° = 86.6 \text{ lb}.$$

In adding two vectors, we begin by resolving each one into components. Then the components along the same axes are added as though they were ordinary numbers, and the resultant is the vector determined by the new set of components. The resultant may be given in terms of these components, or its magnitude and direction may be computed from them. If we write

$$\mathbf{A} = (A_x, A_y)$$

to express the fact that the vector **A** (in a plane) is completely defined by its components A_x and A_y, and similarly with B, the resultant of $\mathbf{A} + \mathbf{B}$ is in the same notation

$$\mathbf{R} = \mathbf{A} + \mathbf{B} = (A_x, A_y) + (B_x, B_y) = (A_x + B_x, A_y + B_y). \qquad (2\text{–}4)$$

The last part of the above equation is a mathematical description of the process we have described: The resultant is completely defined by components which are the sums of the components of the individual vectors being added.

Equation (2–4) can be generalized in an obvious way to cover more than two vectors in space. Adding the three-dimensional vectors **A**, **B**, and **C** thus yields in this notation

$$\begin{aligned} \mathbf{R} &= \mathbf{A} + \mathbf{B} + \mathbf{C} \\ &= (A_x, A_y, A_z) + (B_x, B_y, B_z) + (C_x, C_y, C_z) \\ &= (A_x + B_x + C_x, A_y + B_y + C_y, A_z + B_z + C_z) \\ &= (R_x, R_y, R_z). \end{aligned}$$

Example: Find by the method of components the resultant of a vector of magnitude 12.0 that is at an angle of 30° from the x-axis and another of magnitude 8.0 that is at an angle of 100° from the x-axis.

Solution: The first step is to resolve each of the vectors into x- and y-components. Calling the vectors **A** and **B**, respectively, from Fig. 2–15 we see that

$$A_x = 12.0 \cos 30° = 10.4, \qquad A_y = 12.0 \sin 30° = 6.0,$$
$$B_x = -8.0 \cos 80° = -1.4, \qquad B_y = 8.0 \sin 80° = 7.9.$$

The components of the resultant **R** are

$$R_x = A_x + B_x = 9.0, \qquad R_y = A_y + B_y = 13.9,$$

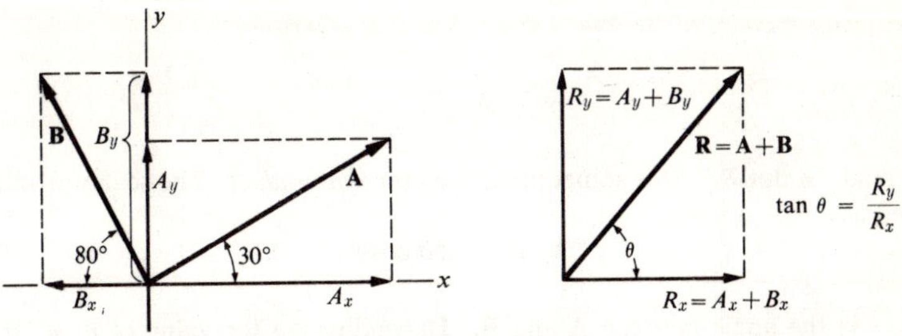

FIG. 2–15. The addition of vectors by the component method.

which specifies **R** completely. The alternate description in terms of magnitude and direction is not difficult to obtain. From the Pythagorean theorem, we have

$$R = \sqrt{R_x^2 + R_y^2} = 16.5,$$

and the angle made by R with the x-axis is determined by

$$\tan\theta = \frac{R_y}{R_x} = 1.54, \qquad \text{so that} \qquad \theta = 57°.$$

In all but the simplest problems, it is preferable to use vector components in addition and subtraction; thus it is well to become familiar with them.

2–6 Scalar product

In the development of ordinary arithmetic, the next step after the introduction of addition and subtraction is multiplication. The process of multiplication is a useful one because problems of repeated addition appear so often in practice. The multiplication in vector analysis which corresponds to scalar multiplication is the multiplication of a vector by a scalar. The vector $\mathbf{A} + \mathbf{A}$ is written $2\mathbf{A}$. More generally $\mathbf{A} + \mathbf{A} + \mathbf{A} + \cdots$ (n times) $= n\mathbf{A}$, where n is some scalar. There are other types of multiplication that occur in vector analysis which involve the product of two vector quantities. There is not so intimate a relationship between addition and multiplication in the case of two vectors as there is in the case of multiplication of scalars. Products of vectors are introduced because they occur so frequently in the analysis of physical problems.

Two methods of combining vectors which have been found to be very useful in treating physical situations are customarily called vector multiplication; in one of them the result is a scalar quantity, and in the other it is a vector quantity. The two are quite distinct, and will be discussed separately. The techniques of vector multiplication do not provide any new information when applied to a particular problem, but by abbreviating the purely mathematical aspects they make possible a clearer perspective of the physical principles involved.

The *scalar product* of the two vectors **A** and **B** is written

$$\mathbf{A} \cdot \mathbf{B},$$

to be read "*A* dot *B*." It is sometimes called the *dot product.* The scalar product is defined as

$$\mathbf{A} \cdot \mathbf{B} = AB\cos\theta, \tag{2–5}$$

where θ is the angle between **A** and **B**. Depending on the value of θ, $\mathbf{A} \cdot \mathbf{B}$ can vary from AB to $-AB$, the former occurring when **A** and **B** are parallel and the latter when they are antiparallel. When **A** and **B** are perpendicular, $\mathbf{A} \cdot \mathbf{B} = 0$, no matter what the magnitudes of **A** and **B** are separately, since in that case $\theta = 90°$ and $\cos\theta = 0$.

There is an interpretation of the scalar product that is often quite helpful. Let us rewrite Eq. (2–5) as

$$\mathbf{A} \cdot \mathbf{B} = (A\cos\theta)B$$

or (2–6)

$$\mathbf{A} \cdot \mathbf{B} = A(B\cos\theta).$$

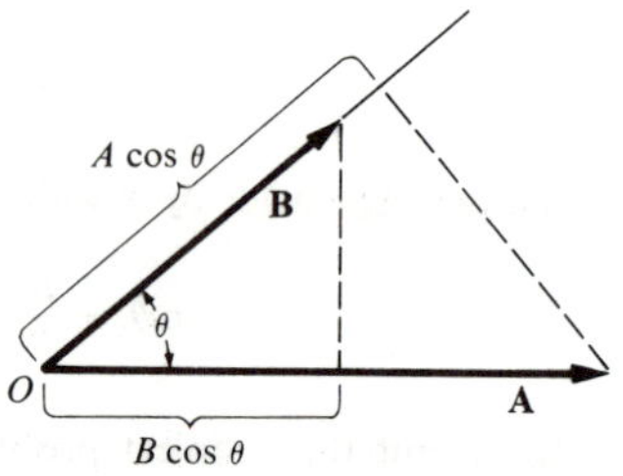

FIG. 2–16. Components for scalar products. $\mathbf{A} \cdot \mathbf{B} = (A\cos\theta)B = A(B\cos\theta)$.

Now $A\cos\theta$ is the component of **A** in the direction of **B** (Fig. 2–16), and similarly $B\cos\theta$ is the component of **B** in the direction of **A**. Hence we can say that the scalar product of two vectors is the component of one of them in the direction of the other, multiplied (in the ordinary sense) by the magnitude of the latter. It makes no difference which of the two forms of Eq. (2–6) is used.

Example: Work is defined as the scalar product of a force and the displacement through which it acts. If a box is moved a distance of 5 ft by means of a horizontal force of 10 lb along an inclined plane which makes an angle of 60° with the horizontal, how much work is done?

Solution: Since $W = \mathbf{F} \cdot \mathbf{d}$,

$$W = Fd\cos\theta = 10\text{ lb} \times 5\text{ ft} \times \cos 60° = 25\text{ ft-lb}.$$

Most of the time we shall deal with vector components, and it is necessary to determine the form of the scalar product in terms of them. Figure 2–17 shows the vectors **A** and **B** that make the angles ϕ and ψ, respectively, with the $+x$-axis. The angle θ included between them is

$$\theta = \psi - \phi,$$

and therefore

$$\cos\theta = \cos\phi\cos\psi + \sin\phi\sin\psi.$$

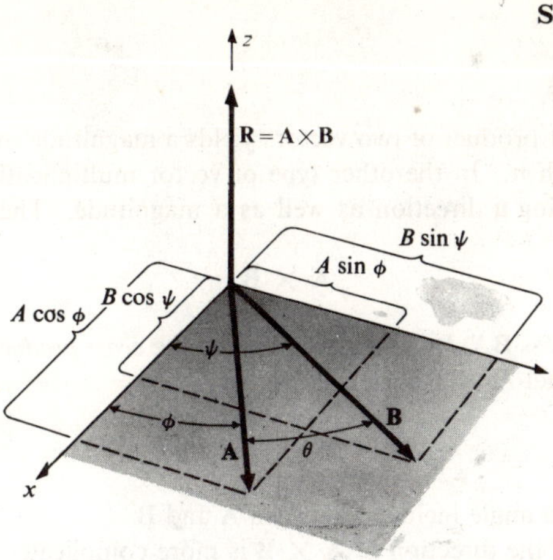

FIG. 2–17. A diagram for determining **A · B** and **A × B** in terms of components.

Hence the scalar product of **A** and **B** becomes

$$\mathbf{A}\cdot\mathbf{B} = AB\cos\theta = AB\cos\phi\cos\psi + AB\sin\phi\sin\psi$$
$$= (A\cos\phi)(B\cos\psi) + (A\sin\phi)(B\sin\psi).$$

However,

$$A\cos\phi = A_x, \qquad A\sin\phi = A_y,$$
$$B\cos\psi = B_x, \qquad B\sin\psi = B_y,$$

which leads to the interesting result that $\mathbf{A}\cdot\mathbf{B} = A_xB_x + A_yB_y$. In three dimensions,

$$\mathbf{A}\cdot\mathbf{B} = A_xB_x + A_yB_y + A_zB_z. \tag{2–7}$$

To obtain the scalar product of two vectors whose components are known, then, we simply multiply each pair of components that lie along the same axis and add the results for the three axes.

Example: What is the angle between two vectors which both originate at (0, 0, 0) and which terminate respectively at (1, 2, 3) and (3, 2, 1)?

Solution: From Eqs. (2–5) and (2–7),

$$\mathbf{A}\cdot\mathbf{B} = AB\cos\theta = A_xB_x + A_yB_y + A_zB_z.$$

Hence

$$\cos\theta = \frac{A_xB_x + A_yB_y + A_zB_z}{AB} = \frac{3\times 1 + 2\times 2 + 1\times 3}{\sqrt{1^2+2^2+3^2}\sqrt{1^2+2^2+3^2}} = \frac{10}{14} = 0.7143,$$

and

$$\theta = 45.4^\circ.$$

2-7 Vector product

Taking the scalar product of two vectors yields a magnitude only, with no direction associated with it. In the other type of vector multiplication the product is itself a vector, having a direction as well as a magnitude. The *vector product* of **A** and **B** is written

$$\mathbf{A} \times \mathbf{B},$$

to be read as "**A** cross **B**." It is sometimes called the *cross product*. The magnitude of the vector product **A** × **B** is defined to be

$$|\mathbf{A} \times \mathbf{B}| = AB \sin \theta, \tag{2-8}$$

where again θ is the angle included between **A** and **B**.

The definition of the direction of **A** × **B** is more complicated to describe. The vectors **A** and **B** together define a plane, as in Fig. 2-18, and **A** × **B** lies perpendicular to this plane, as shown. (When **A** and **B** are in the same direction they do not determine a plane, but if this is the case, the angle between the two vectors is zero and the vector product automatically vanishes.) To determine the sense of the direction of the cross product, that is, whether **A** × **B** is pointing up or down in the drawing, we must imagine the line representing **A** being rotated about the origin through the angle θ until it reaches **B**. The direction in which an ordinary screw (known as a "right-handed" screw) would move under this rotation is the direction of **A** × **B**. If the two vectors lie in the plane of the page, for example, a clockwise rotation of **A** into **B** means that **A** × **B** points down into the paper, while a counterclockwise rotation means that **A** × **B** points upward out of the paper.

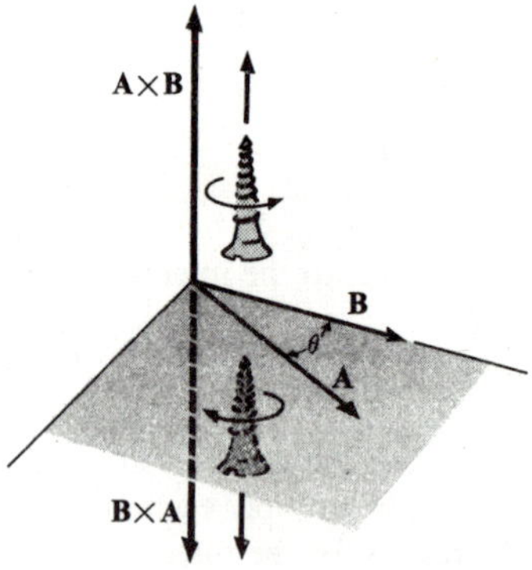

FIG. 2-18. The right-hand-screw rule for vector products.

An important consequence of the above definition is that

$$\mathbf{A} \times \mathbf{B} = -(\mathbf{B} \times \mathbf{A}), \tag{2-9}$$

or, in other words, the vector product is *noncommutative* (Fig. 2-18).

The vector product can also be expressed in terms of the components of **A** and **B**, in which case the direction of **A** × **B** appears automatically. For simplicity we shall again start with **A** and **B** in the xy-plane with the components (A_x, A_y) and (B_x, B_y), respectively, as illustrated in Fig. 2-17. The vector product **R** of **A** and **B** must lie along the z-direction from the results of the previous paragraph, and its magnitude is

$$R = R_z = AB \sin \theta.$$

When we evaluate $\sin\theta$ in terms of ϕ and ψ, we find that

$$\sin\theta = \sin(\psi - \phi) = \sin\psi\cos\phi - \sin\phi\cos\psi,$$

so that

$$R_z = A\cos\phi\, B\sin\psi - A\sin\phi\, B\cos\psi = A_xB_y - A_yB_x. \tag{2–10}$$

Similarly, if **A** and **B** were in the yz-plane, we would find that

$$R_x = A_yB_z - A_zB_y, \tag{2–11}$$

and if they were in the xz-plane,

$$R_y = A_zB_x - A_xB_z. \tag{2–12}$$

Each of Eqs. (2–10) through (2–12) above can be readily remembered by noticing that the subscript of R and the first two subscripts on the right-hand side are cyclic permutations of x, y, and z. These equations hold in general, even though we used a specific example in deriving them. The general result for the vector product in three dimensions obtained from Eqs. (2–10), (2–11), and (2–12) is

$$\mathbf{A} \times \mathbf{B} = (A_yB_z - A_zB_y,\ A_zB_x - A_xB_z,\ A_xB_y - A_yB_x). \tag{2–13}$$

We have not yet justified introducing vector multiplication, which at this point may seem difficult and confusing. The usefulness of these procedures will become apparent in the next few chapters, however, and they are used throughout the book as a means of clarifying the physical content of the theoretical material.

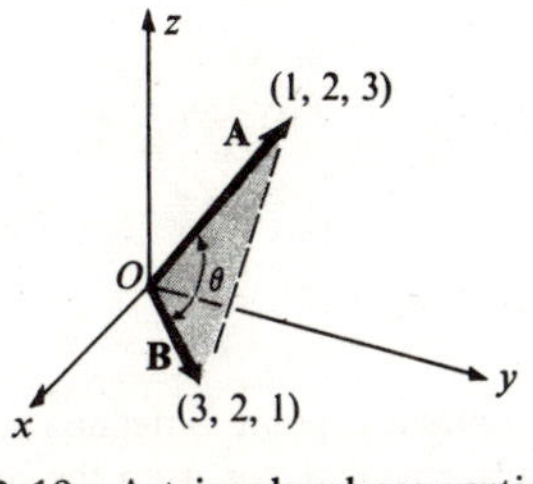

FIG. 2–19. A triangle whose vertices are (0, 0, 0), (1, 2, 3), (3, 2, 1).

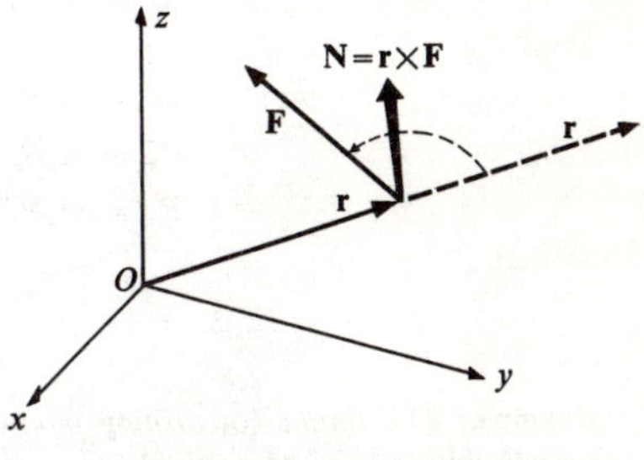

FIG. 2–20. The torque **N** of a force about an axis is equal to **r** × **F**.

Example: Find the area of the triangle contained between the vectors **A** and **B** (Fig. 2–19) which both originate at the origin, and terminate respectively at the points (1, 2, 3) and (3, 2, 1).

Solution: The area of the triangle is

$$S = \tfrac{1}{2}AB\sin\theta = \tfrac{1}{2}|\mathbf{A} \times \mathbf{B}|.$$

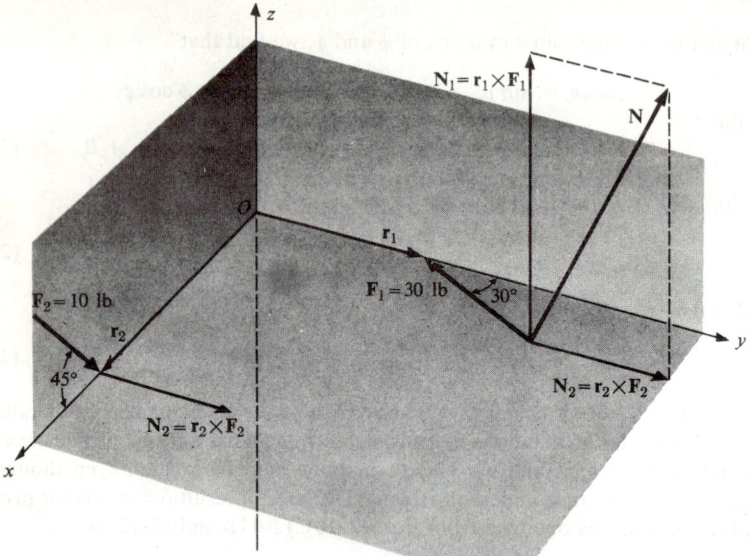

FIG. 2–21. Addition of torques.

Instead of finding the value of θ, we can use the coordinates of the termini of **A** and **B** directly to find the area S, which is

$$S = \tfrac{1}{2}\sqrt{R_x^2 + R_y^2 + R_z^2}.$$

We have

$$R_x = A_yB_z - A_zB_y = -4.0,$$
$$R_y = A_zB_x - A_xB_z = 8.0,$$
$$R_z = A_xB_y - A_yB_x = -3.0,$$

so that

$$S = \tfrac{1}{2}\sqrt{16.0 + 64.0 + 9.0} = 4.7.$$

Example: The *torque* (or *turning moment*) of a force about a point is defined as $\mathbf{N} = \mathbf{r} \times \mathbf{F}$, where **F** is the applied force and **r** is the displacement vector from the specified point at which the force is applied (Fig. 2–20). Calculate the resultant torque about the origin O when a force $\mathbf{F}_1$ in the xy-plane of magnitude 30.0 lb is applied at the point $y = 2.0$ ft, $x = 0$ at an angle of 30° with the y-axis, and another force $\mathbf{F}_2$ in the xz-plane, of magnitude 10.0 lb, is applied at the point $x = 3.0$ ft, $z = 0$ at an angle of 45° to the x-axis (Fig. 2–21).

Solution: The torque about O due to $\mathbf{F}_1$ is along the positive z-axis and has the magnitude

$$N_1 = N_z = r_1F_1 \sin 30° = 2.0 \times 30.0 \times 0.5 = 30.0 \text{ ft-lb}.$$

The torque about O due to the force $\mathbf{F}_2$ is along the positive y-axis and has the magnitude

$$N_2 = N_y = r_2F_2 \sin 45° = 3.0 \times 10.0 \times 0.71 = 21.0 \text{ ft-lb}.$$

The resultant of $\mathbf{N}_1$ and $\mathbf{N}_2$ is evidently in the yz-plane and is shown in Fig. 2–21. By the Pythagorean theorem we find that

$$N = \sqrt{N_y^2 + N_z^2} = \sqrt{(21.0)^2 + (30.0)^2} = \sqrt{1341} = 36.6 \text{ ft-lb}.$$

The angle ϕ between $\mathbf{N}$ and the y-axis is specified by

$$\tan\phi = \frac{N_z}{N_y}, \qquad \text{and so} \qquad \phi = \tan^{-1}\frac{30.0}{21.0} = 55°.$$

2–8 Unit vectors

There is another method of calculating scalar and vector products which, while it requires slightly more elaborate mathematical preliminaries, reduces the labor involved by a large degree.

Let us first recall the meaning of the operation of a scalar multiplying a vector, which we write as

$$\mathbf{B} = c\mathbf{A}, \tag{2–14}$$

where c is the scalar and $\mathbf{A}$ the vector. The result is the vector $\mathbf{B}$, in the same direction as $\mathbf{A}$ but c times longer. That is, we add the vector $\mathbf{A}$ to itself a total of c times in order to obtain $\mathbf{B}$.

Now we take advantage of the fact that a vector $\mathbf{A}$ is completely determined by its components A_x, A_y, A_z and write

$$\mathbf{A} = A_x\mathbf{i} + A_y\mathbf{j} + A_z\mathbf{k}, \tag{2–15}$$

where, $\mathbf{i}, \mathbf{j}, \mathbf{k}$ are *unit vectors* along the coordinate axes x, y, z, respectively, as shown in Fig. 2–22. (A unit vector is a vector whose length is precisely 1. Its function is to specify the direction of the scalar multiplying it.) The sum on the right-hand side of Eq. (2–15) is a vector sum.

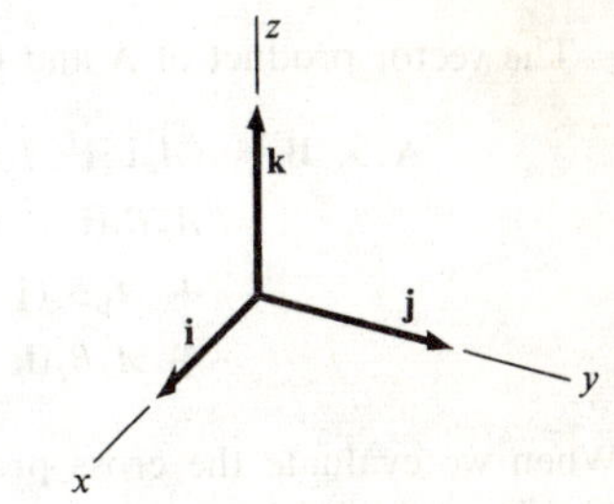

FIG. 2–22. The unit vectors **i, j, k.**

From the definition of the scalar product of the vectors $\mathbf{A}$ and $\mathbf{B}$ as

$$\mathbf{A}\cdot\mathbf{B} = AB\cos\theta,$$

where θ is the angle between $\mathbf{A}$ and $\mathbf{B}$, we can see that

$$\begin{aligned} \mathbf{i}\cdot\mathbf{i} = \mathbf{j}\cdot\mathbf{j} &= \mathbf{k}\cdot\mathbf{k} = 1, \\ \mathbf{i}\cdot\mathbf{j} = \mathbf{i}\cdot\mathbf{k} &= \mathbf{j}\cdot\mathbf{k} = 0, \\ \mathbf{j}\cdot\mathbf{i} = \mathbf{k}\cdot\mathbf{i} &= \mathbf{k}\cdot\mathbf{j} = 0. \end{aligned} \tag{2–16}$$

The second and third of the above sets of equations follow from the fact that **i**, **j**, **k** are perpendicular to one another, so that $\cos\theta = \cos 90° = 0$.

The scalar product of the vectors

$$\mathbf{A} = A_x\mathbf{i} + A_y\mathbf{j} + A_z\mathbf{k}, \qquad \mathbf{B} = B_x\mathbf{i} + B_y\mathbf{j} + B_z\mathbf{k}$$

is

$$\begin{aligned}\mathbf{A}\cdot\mathbf{B} &= (A_x\mathbf{i} + A_y\mathbf{j} + A_z\mathbf{k})\cdot(B_x\mathbf{i} + B_y\mathbf{j} + B_z\mathbf{k})\\ &= A_xB_x(\mathbf{i}\cdot\mathbf{i}) + A_xB_y(\mathbf{i}\cdot\mathbf{j}) + A_xB_z(\mathbf{i}\cdot\mathbf{k})\\ &\quad + A_yB_x(\mathbf{j}\cdot\mathbf{i}) + A_yB_y(\mathbf{j}\cdot\mathbf{j}) + A_yB_z(\mathbf{j}\cdot\mathbf{k})\\ &\quad + A_zB_x(\mathbf{k}\cdot\mathbf{i}) + A_zB_y(\mathbf{k}\cdot\mathbf{j}) + A_zB_z(\mathbf{k}\cdot\mathbf{k}).\end{aligned}$$

Applying Eqs. (2–16) to this result, we find that $\mathbf{A}\cdot\mathbf{B} = A_xB_x + A_yB_y + A_zB_z$, which is identical with Eq. (2–7), which we derived earlier with the help of trigonometry.

In order to compute vector products we begin once more with unit vectors. From

$$\mathbf{A}\times\mathbf{B} = AB\sin\theta$$

we find that

$$\begin{aligned}&\mathbf{i}\times\mathbf{i} = \mathbf{j}\times\mathbf{j} = \mathbf{k}\times\mathbf{k} = 0,\\ &\mathbf{i}\times\mathbf{j} = \mathbf{k}, \qquad \mathbf{j}\times\mathbf{i} = -\mathbf{k},\\ &\mathbf{j}\times\mathbf{k} = \mathbf{i}, \qquad \mathbf{k}\times\mathbf{j} = -\mathbf{i},\\ &\mathbf{k}\times\mathbf{i} = \mathbf{j}, \qquad \mathbf{i}\times\mathbf{k} = -\mathbf{j}.\end{aligned} \tag{2–17}$$

The vector product of **A** and **B** is

$$\begin{aligned}\mathbf{A}\times\mathbf{B} &= (A_x\mathbf{i} + A_y\mathbf{j} + A_z\mathbf{k})\times(B_x\mathbf{i} + B_y\mathbf{j} + B_z\mathbf{k})\\ &= A_xB_x(\mathbf{i}\times\mathbf{i}) + A_xB_y(\mathbf{i}\times\mathbf{j}) + A_xB_z(\mathbf{i}\times\mathbf{k})\\ &\quad + A_yB_x(\mathbf{j}\times\mathbf{i}) + A_yB_y(\mathbf{j}\times\mathbf{j}) + A_yB_z(\mathbf{j}\times\mathbf{k})\\ &\quad + A_zB_x(\mathbf{k}\times\mathbf{i}) + A_zB_y(\mathbf{k}\times\mathbf{j}) + A_zB_z(\mathbf{k}\times\mathbf{k}).\end{aligned}$$

When we evaluate the cross products of the unit vectors with the help of Eqs. (2–17), we find that

$$\begin{aligned}\mathbf{A}\times\mathbf{B} &= A_xB_y\mathbf{k} - A_xB_z\mathbf{j} - A_yB_x\mathbf{k} + A_yB_z\mathbf{i} + A_zB_x\mathbf{j} - A_zB_y\mathbf{i}\\ &= (A_yB_z - A_zB_y)\mathbf{i} + (A_zB_x - A_xB_z)\mathbf{j} + (A_xB_y - A_yB_x)\mathbf{k}.\end{aligned} \tag{2–18}$$

We can write this result alternatively as

$$\begin{aligned}(\mathbf{A}\times\mathbf{B})_x &= A_yB_z - A_zB_y,\\ (\mathbf{A}\times\mathbf{B})_y &= A_zB_x - A_xB_z,\\ (\mathbf{A}\times\mathbf{B})_z &= A_xB_y - A_yB_x,\end{aligned} \tag{2–19}$$

in agreement with Eq. (2–13).

The use of unit vectors simplifies many vector calculations, and, as we shall find later, is helpful in other ways. However, we must keep in mind that every vector is an entity in itself, and need not be identified solely by its components in a particular coordinate system. The choice of the coordinate system in which **i**, **j**, **k** are unit vectors is completely arbitrary, while the vector itself is not.

Example: Calculate

$$\mathbf{B} \cdot (\mathbf{C} \times \mathbf{A})$$

in terms of the components of **A**, **B**, and **C** along some set of cartesian coordinate axes.

Solution: We start by designating the vector $(\mathbf{B} \times \mathbf{C})$ as **D**. The first step is to find $\mathbf{A} \cdot \mathbf{D}$, which is

$$\mathbf{A} \cdot \mathbf{D} = A_x D_x + A_y D_y + A_z D_z.$$

From Eq. (2–19), we obtain

$$\begin{aligned} D_x &= (\mathbf{B} \times \mathbf{C})_x = B_y C_z - B_z C_y, \\ D_y &= (\mathbf{B} \times \mathbf{C})_y = B_z C_x - B_x C_z, \\ D_z &= (\mathbf{B} \times \mathbf{C})_z = B_x C_y - B_y C_x. \end{aligned}$$

Substituting, we have

$$\begin{aligned} \mathbf{A} \cdot (\mathbf{B} \times \mathbf{C}) = \mathbf{A} \cdot \mathbf{D} &= A_x B_y C_z - A_x B_z C_y + A_y B_z C_x \\ &\quad - A_y B_x C_z + A_z B_x C_y - A_z B_y C_x. \end{aligned}$$

From this result it is easy to show that

$$\mathbf{A} \cdot (\mathbf{B} \times \mathbf{C}) = \mathbf{B} \cdot (\mathbf{A} \times \mathbf{C}) = \mathbf{C} \cdot (\mathbf{A} \times \mathbf{B}).$$

Problems

1. In going from one city to another, a car travels 120 mi west, 50 mi north, and 40 mi southeast. How far apart are the two cities?

2. The *United States* is heading due west at 20 mi/hr in the presence of a southeast wind of 15 mi/hr. In what horizontal direction will smoke from the liner's funnel go?

3. A boat heads directly across a river with a speed through the water of 8 mi/hr. The river current is 2 mi/hr. (a) What is the magnitude and direction of the boat's velocity relative to the earth? (b) If the river is 1.5 mi wide, how long does it take the boat to cross to the other bank? (c) When the boat reaches the other bank, how far downstream will it be?

4. The sum of two perpendicular forces is 300 lb. If one of the forces has a magnitude of 120 lb, what is the magnitude of the other?

5. Find the magnitude of the sum of the following displacements: 6 m north and 10 m west, both in a horizontal plane, and 12 m vertically upward.

6. Three forces act in a horizontal plane as follows: **A** has a magnitude of 6 n and points north; **B** has a magnitude of 10 n and points west; and **C** has a magnitude of 8 n and points southeast. Find the magnitude and direction of $\mathbf{A} + \mathbf{B} - \mathbf{C}$.

7. A body moves in a circular path and its velocity is 10 units in a direction perpendicular to the radius of the circle. Find the difference between the velocity vectors when the radius points due north and when it points 30° east of north.

8. The sum of the forces acting at a point is zero. If the magnitudes of the forces are 80, 70, and 50, what is the angle between the forces?

9. The following forces act on a body resting on a level, frictionless surface: 10 n at an angle of 10° clockwise from a reference direction; 40 n at an angle of 60° clockwise from the reference direction; 20 n at an angle of 180° clockwise from the reference direction; and 20 n at an angle of 300° clockwise from the reference direction. What is the magnitude and direction of the net force on the body?

10. Given the three vectors

$$\mathbf{a} = 2\mathbf{i} + 3\mathbf{j} + \mathbf{k},$$
$$\mathbf{b} = \mathbf{i} + 2\mathbf{k},$$
$$\mathbf{c} = \mathbf{i} + \mathbf{j} - 3\mathbf{k},$$

find the directions and magnitudes of

(a) $\mathbf{a} + \mathbf{b} + \mathbf{c}$,
(b) $\mathbf{a} - \mathbf{b} + \mathbf{c}$,
(c) $\mathbf{a} + \mathbf{b} - \mathbf{c}$.

11. A force of 20 lb is applied to the lower end of a window pole that is 70° above the horizontal. How much vertical force is being applied to the window?

12. Find the vertical and horizontal components of a 100-lb force that is directed 40° above the horizontal.

13. A body is moving in a plane at 12 m/sec in a direction 40° clockwise from the $+x$-axis. What are the x- and y-components of its velocity?

14. An airplane is heading southwest when it takes off at an angle of 25° from the horizontal. Its speed is 220 mi/hr. (a) What is the upward component of its speed? (b) What is the component of its speed toward the south?

15. Two particles are moving in a plane; one has the velocity components $v_x = 1$ m/sec, $v_y = 5$ m/sec, and the other has the velocity components $v_x = 4$ m/sec, $v_y = 3$ m/sec. If both started from the same point, what is the angle between their paths?

16. Find the angle between (1, 1, 0) and (1, 0, 1).

17. If $\mathbf{A} \cdot \mathbf{C} = \mathbf{B} \cdot \mathbf{C}$, show that **A** need not be equal to **B**. (The **C**'s cannot necessarily be canceled out.) Under what conditions would **A** be equal to **B**?

18. If **A** and **B** are two vectors, show that the component of **A** in the direction of **B** is

$$\left(\frac{\mathbf{A} \cdot \mathbf{B}}{B^2}\right)\mathbf{B}.$$

19. The three vectors which represent the sides of a triangle satisfy the condition that $\mathbf{A} + \mathbf{B} + \mathbf{C} = 0$. By transposing the vector **C** and taking the scalar product of each side with itself, verify the law of cosines.

20. Prove by vector methods that the sum of the squares of the diagonals of a parallelogram is equal to the sum of the squares of the sides.

21. What is the geometrical meaning of $\mathbf{A} \cdot \mathbf{B} \times \mathbf{C}$? Does this suggest the reason that $\mathbf{A} \cdot \mathbf{B} \times \mathbf{C} = \mathbf{B} \cdot \mathbf{C} \times \mathbf{A} = \mathbf{C} \cdot \mathbf{A} \times \mathbf{B}$?

22. Show that $(\mathbf{A} \times \mathbf{B}) \times \mathbf{C}$ lies in the same plane as the vectors **A** and **B**. Show

also that $\mathbf{A} \times (\mathbf{B} \times \mathbf{C})$ lies in the plane of **B** and **C**. Note that $(\mathbf{A} \times \mathbf{B}) \times \mathbf{C}$ cannot be equal to $\mathbf{A} \times (\mathbf{B} \times \mathbf{C})$ and that the vector cross product is not associative.

23. Prove that

$$\mathbf{A} \times (\mathbf{B} \times \mathbf{C}) = \mathbf{B}(\mathbf{A} \cdot \mathbf{C}) - \mathbf{C}(\mathbf{A} \cdot \mathbf{B}).$$

24. Given two vectors $\mathbf{a} = (1, 2, 3)$ and $\mathbf{b} = (2, -1, 4)$. Calculate the components of $\mathbf{a} \times \mathbf{b}$. Verify that **a** is perpendicular to $\mathbf{a} \times \mathbf{b}$ by calculating the scalar product $\mathbf{a} \cdot \mathbf{a} \times \mathbf{b}$.

25. Show that

$$(\mathbf{A} \times \mathbf{B})^2 + (\mathbf{A} \cdot \mathbf{B})^2$$

is independent of the angle between **A** and **B**. Evaluate this sum when the magnitude of $\mathbf{A} = 4$ and the magnitude of $\mathbf{B} = 3$.

3

Linear Kinematics

Everything in the physical world is in motion, from the elementary particles within atoms to the largest galaxies of stars. Since the goal of physics is an understanding of the nature and behavior of the physical world, the physicist is directly concerned with things in motion in nearly all his work. This concern is not just with the motion itself, of course, but also with its causes and results. Before we can go on to discuss these important aspects of physics we must first consider how motion is described mathematically. The analysis of motion without regard to its origin and effects is called *kinematics*.

The phenomena that fall in the province of kinematics are familiar to everybody: a falling stone, a bullet fired from a rifle, the planets traveling around the sun, a spinning top, an automobile rounding a turn. Phenomena like these are close to our direct experience, and their analysis is correspondingly straightforward. Kinematics is largely the creation of Galileo Galilei (1564–1642). Only after Galileo had clarified the notions of kinematics was the world of physics ready for the work of Isaac Newton, whose inquiry into the causes and results of motion forms the foundation of modern physical thought.

3–1 Average speed

When we say that a body moves, we mean that it is changing its position with respect to something else. A car moves with respect to the road, for example. In order to determine whether a body is moving, then, we must first establish a coordinate system somewhere to which we can refer the position of the body. If the coordinates of the body in this fixed coordinate system change with time, the body is moving with respect to the system. Two different quantities are therefore involved in motion; *length* (the distance between the positions of the body at two different instants of time), and *time* (the elapsed time during which the body's position changed).

Let us begin with motions that take place along a single straight line only; for instance, a train traveling along a straight section of track. Let us begin to measure time when the train passes a particular point, and make note of the total distance from this point that the train has gone at the end of each of a number of time intervals. Our results might be the following:

Time, t	Distance, s
0 sec	0 ft
1	40
2	80
3	120
4	160

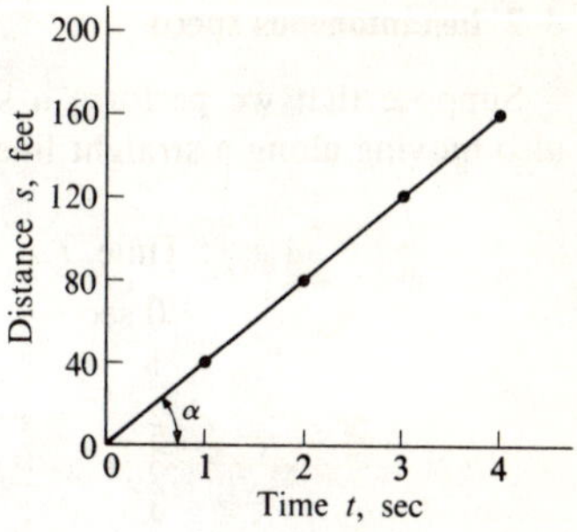

FIG. 3–1. A plot of distance versus time for a body in uniform motion.

These measurements are plotted in Fig. 3–1 as a series of points joined by a smooth curve. Strictly speaking, only the points themselves have meaning; in drawing a straight line through them we have assumed that nothing untoward occurred within each time interval. This kind of approximation is called *interpolation*, and is often (although not always) a perfectly valid procedure. Extending the line beyond the measured points, a process called *extrapolation*, is a more risky affair, since we have no way of knowing whether the train continues on after 4 sec, or comes to a stop or otherwise behaves differently from the way it did before.

We now assume that the distance-versus-time (s, t) curve in Fig. 3–1 accurately represents the motion of the train, and define a property of this motion called *average speed.* The average speed $\bar{v}$ of a body is the distance it covers in a certain period of time divided by that amount of time. That is,

$$\bar{v} = \frac{\Delta s}{\Delta t} = \frac{s_{\text{final}} - s_{\text{initial}}}{t_{\text{final}} - t_{\text{initial}}}. \tag{3–1}$$

By Δs is meant the distance traversed in the time interval Δt. Thus in the interval between 0 and 1 sec the value of $\bar{v}$ is given by

$$\bar{v} = \frac{(40 - 0)\text{ ft}}{(1 - 0)\text{ sec}} = 40\text{ ft/sec.}$$

If we calculate the average speed in a similar manner for the other time intervals in this experiment, we obtain precisely the same figure. This means that the train had a *uniform average speed* during the experiment. This conclusion may also be obtained by inspecting Fig. 3–1, where we note that the (s, t) curve is a straight line. The average speed $\bar{v}$ is a geometric property of the (s, t) curve according to its defining equation, Eq. (3–1): $\bar{v}$ is simply the *slope* of the curve in the interval specified. In Fig. 3–1, $\bar{v} = \tan\alpha$; the slope is of course determined by the scale we have chosen to plot our curve. The unit of average speed is always a distance unit divided by a time unit, such as miles/hour, feet/second, meters/second, etc.

3-2 Instantaneous speed

Suppose that we perform a similar series of measurements on another train, also moving along a straight line, and that our results are:

Time, t	Distance, s
0 sec	0 ft
1	40
2	70
3	90
4	100

Figure 3-2 is a graph of the above data, again with the individual points connected by a curve. Let us apply Eq. (3-1) in order to calculate the average speed of the train in various time intervals. Between 0 and 1 sec we obtain 40 ft/sec for $\bar{v}$, between 1 and 2 sec we obtain 30 ft/sec, between 2 and 3 sec we obtain 20 ft/sec, and in the last interval, between 3 and 4 sec, we obtain 10 ft/sec. The average speed in the entire four-second period is 25 ft/sec. Evidently we are dealing with a more complicated situation, since each average speed that we compute is different from the others.

Geometrically it is not hard to see why $\bar{v}$ should vary so much. The (s, t) curve of Fig. 3-2 is not a straight line, and its slope—which we have identified with the average speed—is not constant. The question arises, then, of finding some quantity to characterize the speed of the train at a particular instant of time, replacing the average speed during an interval which, as we have discovered, has a limited usefulness when the speed is not uniform.

We might wish to know the speed of the train at the particular instant $t = 1$ sec, for example. This speed is called the *instantaneous speed*, because it refers to the rate at which the train is moving at one specific time rather than the average rate over some interval. An indication of the manner in which instantaneous speed, usually denoted by v, may be defined can be obtained from the identification of average speed in an interval as the slope of the chord joining the endpoints of the

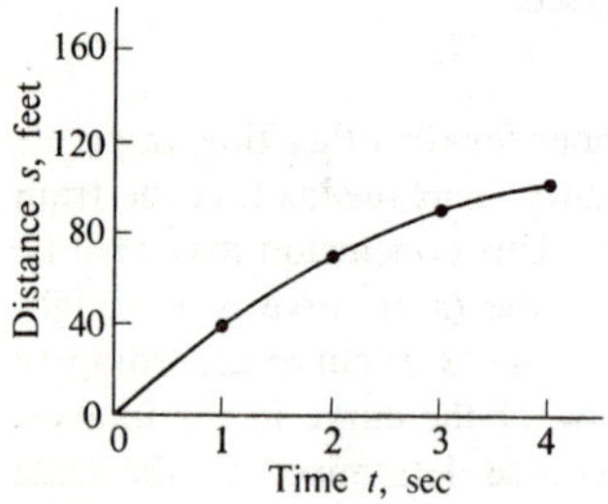

FIG. 3-2. A plot of distance versus time for a body in nonuniform motion.

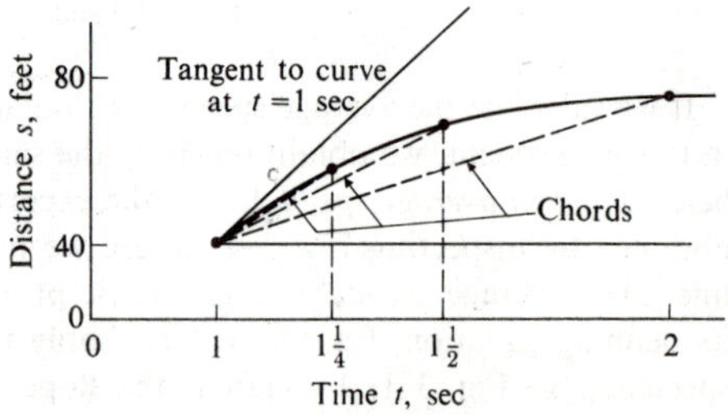

FIG. 3-3. The geometrical description of instantaneous velocity as the tangent to the (s, t) curve.

(s, t) curve in that interval (Fig. 3–3). Let us first calculate the average speed $\bar{v}$ of the train in the interval between 1 and 2 sec. As we saw, this is $\bar{v} = 30$ ft/sec. Now we reduce the time interval so that it extends only from 1 sec to $1\frac{1}{2}$ sec. The average speed $\bar{v}'$ in this briefer interval, interpolating from the graph, is

$$\bar{v}' = \frac{57 - 40}{1\frac{1}{2} - 1} = 34 \text{ ft/sec.}$$

The value of $\bar{v}'$ is closer to the instantaneous speed v of the train at $t = 1$ sec than $\bar{v}$ is. To approach still closer to v we can reduce the interval still further, say from 1 sec to $1\frac{1}{4}$ sec. The average speed is now

$$\bar{v}'' = \frac{49 - 40}{1\frac{1}{4} - 1} = 36 \text{ ft/sec.}$$

With this procedure we can get as near to the value v as we wish.

Essentially the above procedure consists of shrinking the length of the chord whose slope is the average speed $\bar{v}$. In the limit of the chord length approaching 0, the slope of the chord approaches the tangent to the (s, t) curve at the given point as its limit. This is evident from Fig. 3–3. We accordingly define the instantaneous speed v at the time t as the tangent to the (s, t) curve at t. Hence

$$v = \lim_{\Delta t \to 0} \frac{\Delta s}{\Delta t}$$

or, since this limit is simply the derivative taken with respect to time,

$$v = \frac{ds}{dt}. \tag{3–2}$$

How is the instantaneous speed v related to the average speed $\bar{v}$? We rewrite Eq. (3–2) as

$$ds = v\, dt$$

and integrate both sides. Letting the limits of integration be s_{final} and s_{initial} on the left-hand side, and t_{final} and t_{initial} on the right-hand side, we have

$$\int_{s_{\text{initial}}}^{s_{\text{final}}} ds = \int_{t_{\text{initial}}}^{t_{\text{final}}} v\, dt. \tag{3–3}$$

We can perform the integration only if we know the way in which v varies with time. In the case where v *is constant*, we can take v from under the integral sign, and we obtain

$$\int_{s_{\text{initial}}}^{s_{\text{final}}} ds = v \int_{t_{\text{initial}}}^{t_{\text{final}}} dt,$$

$$s_{\text{final}} - s_{\text{initial}} = v(t_{\text{final}} - t_{\text{initial}}),$$

$$v = \frac{s_{\text{final}} - s_{\text{initial}}}{t_{\text{final}} - t_{\text{initial}}}.$$

This result is identical with Eq. (3–1), which defined average speed in an interval. When the instantaneous speed is constant over an interval, then it is precisely the same as the average speed $\bar{v}$ in that interval. However, when the instantaneous speed varies, v and $\bar{v}$ may be different.

3–3 Velocity

The specification of the motion of a body which we have given up to now is adequate only if it is moving in one direction on a straight line. This is clearly insufficient to describe the types of motion with which we are familiar. Even when we are confining ourselves to one-dimensional displacements, it makes a difference whether a body is moving to the right or to the left. These considerations are particularly important if we superpose two motions. Thus if a ball is thrown at a speed of 20 mi/hr on a train moving at the same speed in the same direction as the ball, an observer on the ground will measure a speed of 40 mi/hr for the ball. If the ball is moving in a direction opposite to the direction in which the train is moving, it will appear to be standing still with respect to an observer on the ground.

Clearly we must specify not only a speed associated with a body, but also a direction. This suggests that there is a *vector quantity* describing the motion. This quantity is known as the *velocity* of the body. In the problems we have been considering up to now, since they described one-dimensional motion in one direction, the vector properties of the velocity were not apparent. Under these circumstances, we are considering components of vectors in a particular direction, and they combine as scalars do.

The discussion of the velocity of a body proceeds very much as the discussion of its speed in the previous sections, except that we substitute *displacement* for *distance traveled.* The former is a vector quantity and the latter a scalar. Thus the average velocity in a time interval is:

$$\bar{\mathbf{v}} = \frac{\mathbf{s}_{\text{final}} - \mathbf{s}_{\text{initial}}}{t_{\text{final}} - t_{\text{initial}}}. \tag{3–4}$$

This definition is substantially different from the definition of the average speed.

When we consider more complicated motions, the difference is even more striking. Consider a motion in two dimensions in which a particle is moving in a circle with constant speed; for example, the tip of the minute hand of a watch. A particle on the tip will cover equal distances along the arc of the circle in equal times. Even though its *speed* is constant its *velocity* is constantly changing, not in magnitude but in direction. We shall see later that the changing *velocity* has important physical consequences.

Instantaneous velocity $\mathbf{v}$ is defined in terms of the time derivative of the displacement $\mathbf{s}$ as

$$\mathbf{v} = \frac{d\mathbf{s}}{dt}. \tag{3–5}$$

Even though Eq. (3–5) looks very much like (3–2), it is quite different. In fact we really should take some trouble to define what we mean by the differential $d\mathbf{s}$, the infinitesimal displacement of a vector quantity.

The simplest way to consider Eq. (3–5) is as a shorthand way of writing three equations, one for each of the components of $\mathbf{v}$ in terms of the corresponding components of the infinitesimal displacement. In this way the meaning of Eq. (3–5) is clear in terms of scalar derivatives of each component, and we are familiar with scalar derivatives.

It is interesting to consider the geometrical aspects of Eq. (3–5) also. This is more complicated, and we shall merely illustrate it by the example of the motion of the tip of the minute hand of a clock. Referring to Fig. 3–4, $\mathbf{s}_f$ and $\mathbf{s}_i$ are the displacements of the tip with respect to the origin at two instants infinitesimally separated in time. The figure shows $\mathbf{s}_f - \mathbf{s}_i = \Delta\mathbf{s}$ constructed vectorially; $\mathbf{v}$ is a vector in the direction of $\Delta\mathbf{s}$, and the magnitude is the limit, as Δt becomes smaller, of the fraction $\Delta s/\Delta t$, where Δs is the magnitude of $\Delta\mathbf{s}$. In Chapter 4 we shall show that $\mathbf{v}$ is in the direction of the tangent to the circle and its magnitude is the speed of the tip.

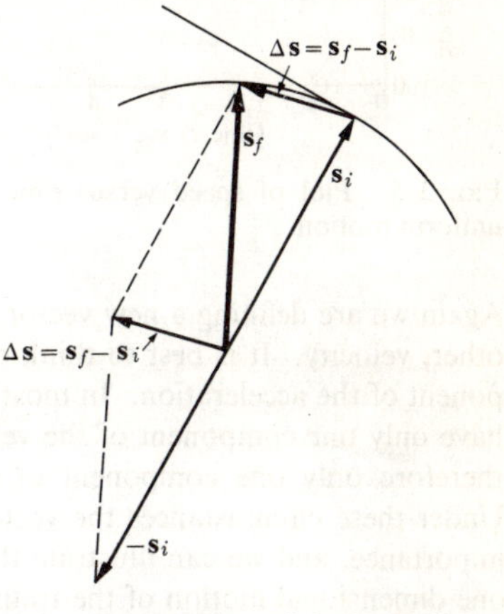

FIG. 3–4. Construction of $\Delta\mathbf{s}$ for uniform circular motion.

One other thing should be mentioned before we temporarily leave the discussion of velocities. We have said that velocity is a vector quantity, implying that when one combines velocities, the results are those given by the laws of combination of vectors stated in Chapter 2. The justification of this remark can only be that the results of this assumption conform to experimental results in kinematics. Whenever we say that a quantity is a vector we are implying the same thing; namely, that this assumption is in accord with the experimental experience of physicists that the laws of combination of this quantity are those of vectors.

3–4 Acceleration

Situations involving constant velocities are rare and usually of little interest. The velocity can change either in magnitude or direction or both, and we require some measure of these changes in order to describe the motion of a body whose velocity varies. Such a measure is called the *acceleration*, and it is defined as

$$\mathbf{a} = \frac{d\mathbf{v}}{dt}\cdot \qquad (3\text{–}6)$$

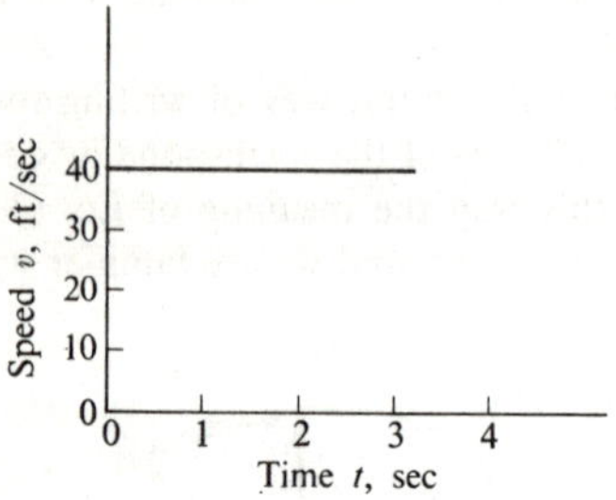

FIG. 3-5. Plot of speed versus time in uniform motion.

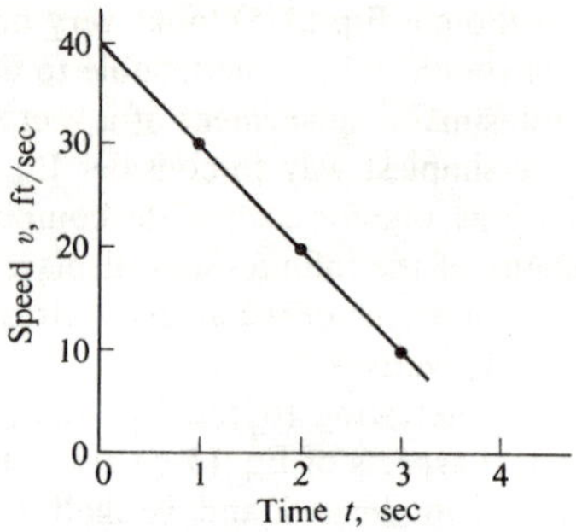

FIG. 3-6. Plot of speed versus time under constant acceleration.

Again we are defining a new vector, acceleration, in terms of the derivative of another, velocity. It is best to think of (3-6) as three equations, one for each component of the acceleration. In most of the applications we shall consider, we shall have only one component of the velocity changing as a function of the time, and therefore only one component of the acceleration will be different from zero. Under these circumstances the vector properties of the acceleration are of minor importance, and we can illustrate the concept by referring to the examples of the one-dimensional motion of the train described at the beginning of this chapter.

Suppose that we take the motion of the train described by the (s, t) curve of Fig. 3-1 and compute from it the instantaneous speed v at every instant of time. Figure 3-5 is a graph of v as a function of time corresponding to the data of Fig. 3-1; it is a horizontal straight line, illustrating the fact that the instantaneous velocity of the train is a constant, and that the acceleration is zero.

The second train we considered, whose displacement varied with time as shown in Fig. 3-2, did not have a constant speed. If we plot its instantaneous speed versus time, as in Fig. 3-6, we have a picture of the manner in which v varied. The straight line means that the speed changed in a uniform way; the acceleration was constant.

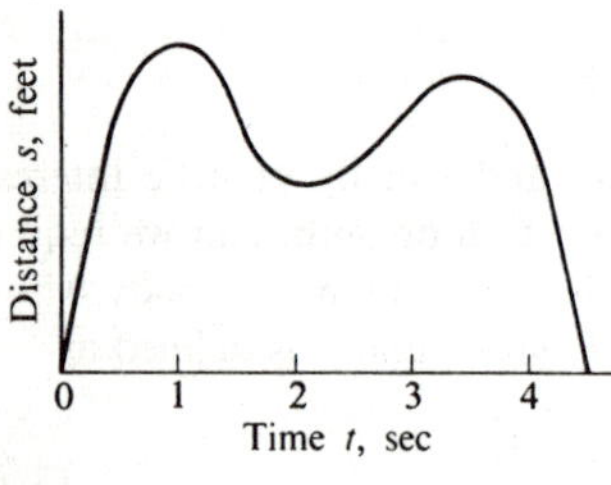

FIG. 3-7. Plot of distance versus time in nonuniform motion.

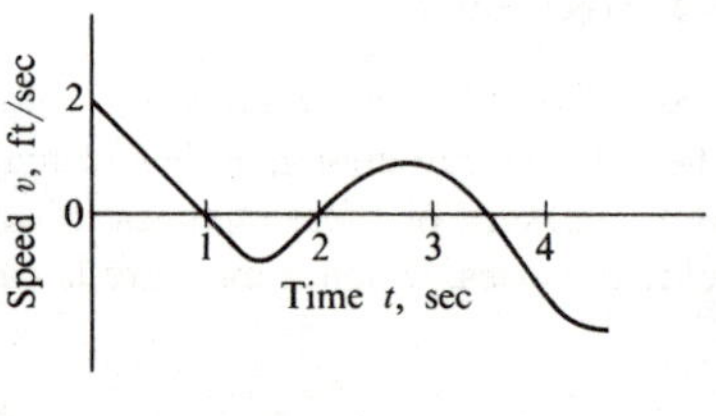

FIG. 3-8. Plot of speed versus time for the nonuniform motion of Fig. 3-7.

A third possibility is that a train may move in such a way that its speed changes nonuniformly. The (s, t) curve of Fig. 3–7 might represent the displacement of such a train as a function of time, and plotting its instantaneous velocity against time results in the irregular curve of Fig. 3–8.

Acceleration is a familiar concept. When an automobile starts moving, its velocity is changing with time, and it is therefore being accelerated. When it reaches its final speed and continues moving at this speed, its acceleration is 0. When it slows down it is again being accelerated, this time negatively.

When the velocity is changing, we have an acceleration. When the velocity is constant the acceleration is zero. For one-dimensional motion, the vector properties of the acceleration are indicated by using a positive or negative sign. The positive sign is used when it is in the direction of positive velocities, the negative sign when it is oppositely directed. Examples given later will illustrate this point. In the following analysis of one-dimensional motion, we shall neglect the vector properties of the acceleration, for convenience.

Most of our work will be restricted to uniform accelerations. In the event of a varying acceleration, it is still possible to consider the *average acceleration* $\bar{a}$ of the body given by

$$\bar{a} = \frac{v_{\text{final}} - v_{\text{initial}}}{t_{\text{final}} - t_{\text{initial}}}. \tag{3–7}$$

The subscripts refer to the beginning and end of the time interval in which the average is taken. For certain types of motion, average acceleration is a convenient quantity, but for others—the oscillations of a pendulum, for example—it is not very meaningful.

Since the instantaneous speed v of a particle is the derivative of the distance traveled with respect to time, $v = ds/dt$, and its instantaneous acceleration is the derivative of its speed with respect to time, $a = dv/dt$, we see that we can express a directly in terms of s and t as the second derivative of s with respect to time:

$$a = \frac{dv}{dt} = \frac{d}{dt}\left(\frac{ds}{dt}\right) = \frac{d^2s}{dt^2}. \tag{3–8}$$

The unit of acceleration is, from its definition, expressed in terms of

$$\frac{\text{Speed}}{\text{Time}} = \frac{\text{distance/time}}{\text{time}} = \frac{\text{distance}}{\text{time}^2}.$$

A body whose speed increases by 10 ft/sec each second would thus have its acceleration expressed as

$$10\,\frac{\text{ft/sec}}{\text{sec}} = 10\,\frac{\text{ft}}{\text{sec}^2}.$$

The time units need not be the same. Thus a common unit of acceleration in the British system is (miles/hour)/second.

3-5 Constant acceleration

When the acceleration is constant, the scalar form of Eq. (3-6) can be integrated easily. We begin by rewriting it

$$dv = a\,dt,$$

and then integrate the left-hand side from the initial speed v_0 to the final speed v, and the right-hand side from the initial time $t = 0$ to the final time t. This procedure yields, with a constant,

$$\int_{v_0}^{v} dv = a\int_0^t dt,$$

$$v - v_0 = at,$$

$$v = v_0 + at. \tag{3-9}$$

Now we replace v by ds/dt in Eq. (3-9) and multiply both sides of the equation by dt:

$$v = \frac{ds}{dt} = v_0 + at, \qquad ds = v_0\,dt + at\,dt.$$

Integrating from the initial position $s = 0$ to the final position s for a total displacement of s, and from the corresponding initial time $t = 0$ to the final time t, we find that

$$\int_0^s ds = v_0\int_0^t dt + a\int_0^t t\,dt,$$

$$s = v_0 t + \tfrac{1}{2}at^2. \tag{3-10}$$

When the body is stationary at the start, $v_0 = 0$, and

$$s = \tfrac{1}{2}at^2 \tag{3-11}$$

gives the distance s it covers in the time t. Thus a uniformly accelerated body starting from rest goes four times as far in 2 sec as it did in 1 sec, nine times as far in 3 sec, and so on.

We can obtain a useful relationship among s, v, v_0, and a without involving t directly. Equation (3-9) can be rewritten as

$$t = \frac{v - v_0}{a}.$$

If we substitute this value of t into Eq. (3-10), we obtain

$$s = v_0\frac{(v - v_0)}{a} + \frac{1}{2}a\frac{(v - v_0)^2}{a^2}$$

$$= \frac{v_0 v}{a} - \frac{v_0^2}{a} + \frac{v^2}{2a} - \frac{v_0 v}{a} + \frac{v_0^2}{2a}$$

$$= \frac{v^2}{2a} - \frac{v_0^2}{2a},$$

so that, multiplying both sides of the equation by $2a$ and rearranging, we have

$$v^2 = v_0^2 + 2as. \tag{3-12}$$

Example: An object has an initial speed of 20 cm/sec and is being accelerated at the rate of 10 cm/sec^2. What is its speed after $\frac{1}{4}$ sec and $\frac{1}{2}$ sec have elapsed? What distance did it travel in $\frac{1}{4}$ sec and $\frac{1}{2}$ sec?

Solution: From Eq. (3–9), after $\frac{1}{4}$ sec

$$v = 10 \times \tfrac{1}{4} + 20 = 22.5 \text{ cm/sec},$$

and after $\frac{1}{2}$ sec

$$v = 10 \times \tfrac{1}{2} + 20 = 25 \text{ cm/sec}.$$

The distance traveled can be found from Eq. (3–10). When $t = \frac{1}{4}$ sec,

$$s = \tfrac{1}{2} \times 10 \times (\tfrac{1}{4})^2 + 20 \times \tfrac{1}{4} = 5\tfrac{5}{16} \text{ cm},$$

and when $t = \frac{1}{2}$ sec,

$$s = \tfrac{1}{2} \times 10 \times (\tfrac{1}{2})^2 + 20 \times \tfrac{1}{2} = 11\tfrac{1}{4} \text{ cm}.$$

3–6 Gravity

The most important example of uniformly accelerated motion in a straight line is that produced by the effect of gravitation on bodies near the earth. We can apply the ideas of this chapter to the motion of falling bodies because of an important observation made by Galileo. He found, after a number of experiments, that all bodies near the earth fall toward the earth with the *same* acceleration. This acceleration, which is 32 ft/sec^2 in British units and 9.8 m/sec^2 (meters per second per second) in metric units, is called the *acceleration of gravity*, denoted by g.

Galileo's work was especially significant because, for the first time, experimental results were accepted despite the fact that they contradicted well-established ideas, in this case the notion that heavy things "should" fall faster than light ones. Modern science owes its success in understanding and utilizing natural phenomena to its reliance on experiment. It may be hard for us today to understand the reluctance of our ancestors to shed their intuitive beliefs in the face of experimental evidence to the contrary; although, if we look carefully, we will see that this unfortunate tendency still persists in the world.

Another aspect of Galileo's work deserves comment. His conclusion that all things fall with the same constant acceleration is only an *idealization* of reality. The actual accelerations with which objects fall depend on many factors: the location on the earth, the size and shape of the object, and the density and state of the atmosphere. For example, an acorn reaches the ground before a leaf does that falls from the same limb at the same time because of the greater effect of buoyancy and air resistance on the leaf. Galileo perceived that the essence of the phenomenon was a constant acceleration downward, with other factors acting merely to cause deviations from the constant value.

We can apply the formulas we derived earlier for motion under constant acceleration to bodies in free fall. It must be kept in mind, of course, that the direction of g is always downward, no matter whether we are dealing with a dropped object or one which is initially thrown upward. A few examples will show the correct procedure.

Example: A stone is dropped from the top of the Empire State Building, which is 1472 ft high. How long does it take to reach the ground, neglecting air resistance? What is its speed when it strikes the ground?

Solution: Since the initial speed is zero, we begin by writing Eq. (3–11),

$$s = \tfrac{1}{2}gt^2.$$

Substituting for s and g, we have

$$1472 \text{ ft} = \frac{32}{2}\frac{\text{ft}}{\text{sec}^2}t^2, \qquad t^2 = 92 \text{ sec}^2, \qquad t = 9.6 \text{ sec}.$$

Knowing the duration of the fall, we find it easy to compute the final speed:

$$v = gt = 32\frac{\text{ft}}{\text{sec}^2} \times 9.6 \text{ sec} = 307\frac{\text{ft}}{\text{sec}}\cdot$$

This is also the speed with which a stone must be thrown upward from the ground if it is to reach the top of the building. To prove this statement, we refer to Eq. (3–12), namely

$$v^2 = v_0^2 + 2as.$$

When the stone is dropped, the acceleration a equals the acceleration of gravity g, and the initial speed v_0 of the stone is zero. Hence

$$v^2 = 0 + 2gs, \qquad v = \sqrt{2gs}.$$

When the stone is thrown upward, on the other hand, $a = -g$ (since the downward acceleration is in the opposite direction to the upward motion), and at the top of its path, $v = 0$. Hence

$$0 = v_0^2 - 2gs, \qquad v_0 = \sqrt{2gs},$$

the same numerical value as before.

Example: An object is thrown vertically upward with a speed of 49 m/sec. How high will it rise? What is its position after it has traveled for 2 sec? 8 sec? 11 sec?

Solution: In this problem it is clear that the initial velocity, which is upward, is in the direction opposite to the acceleration, which is downward. Hence the speed at first decreases to zero, at which time the object is at the top of its path, and then increases as it falls down. Since both acceleration and velocity are vector quantities, but the motion is one-dimensional, it is easy to deal with the situation by defining positive and negative directions. Customarily up is positive and down negative; therefore $v_0 = +49$ m/sec and $a = g = -9.8$ m/sec^2. To solve the first part of the problem,

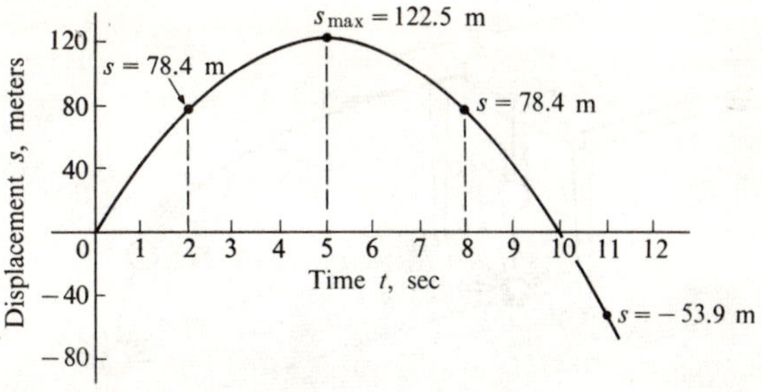

FIG. 3-9. The variation with time of the displacement of a particle thrown upward.

we note that at the top of its path the object has zero velocity. Using Eq. (3-9), we have

$$v = 0 = at + v_0 = -9.8t + 49 \text{ m/sec}$$

and

$$t = 5 \text{ sec}.$$

Equation (3-10) yields for the maximum height

$$s = \tfrac{1}{2}at^2 + v_0t = -\tfrac{1}{2} \times 9.8 \times 5^2 + 49 \times 5 = 122.5 \text{ m}.$$

For the second part of the problem, we again use Eq. (3-10). The results are:

$$t = 2 \text{ sec}: \quad s = -\tfrac{1}{2} \times 9.8 \times 2^2 + 49 \times 2 = 78.4 \text{ m},$$
$$t = 8 \text{ sec}: \quad s = -\tfrac{1}{2} \times 9.8 \times 8^2 + 49 \times 8 = 78.4 \text{ m},$$
$$t = 11 \text{ sec}: \quad s = -\tfrac{1}{2} \times 9.8 \times 11^2 + 49 \times 11 = -53.9 \text{ m}.$$

We note that the position of the object is the same after 8 sec as it was after 2 sec, and that at 11 sec it is negative. Figure 3-9 illustrates the motion. In 2 sec the object has risen to 78.4 m. Three seconds later it is at its maximum height of 122.5 m, and in another 3 sec (making a total of 8 sec since it was thrown), it has fallen down to 78.4 m above the starting point. The motion is symmetrical about the highest point; the speeds at 2 sec and 8 sec are the same, although the velocities are in opposite directions. After 11 sec the object has fallen so far that it is below the position from which it was thrown, as the minus sign indicates. The quantity s is the actual *displacement* of the object, not the total distance it has traveled. For instance, after 11 sec the total distance is 122.5 m up, 122.5 m back down to the starting position, and then 53.9 m farther down, making a total of 298.9 m. The net displacement, however, is −53.9 m.

3-7 Projectile motion

Most moving bodies do not travel in straight lines. Very often, however, the motion lies in a plane, and in such cases the vector properties of displacement **s**, velocity **v**, and acceleration **a** become of great importance.

Suppose that we have two balls at the edge of a table and, as in Fig. 3-10, allow A to drop down while we push B off horizontally with the speed v_x. The experi-

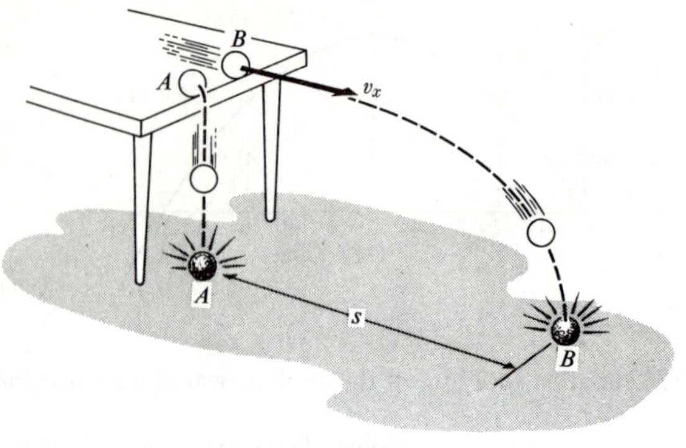

FIG. 3-10. Horizontal and vertical components of velocity are independent.

mental result is that both balls reach the ground at the same instant, showing that the vertical and horizontal motions are independent. This is true regardless of the magnitude of v_x, which in this case remains constant because the acceleration g due to gravity acts only in the vertical direction. (In this example and elsewhere in this chapter it is assumed that there is no air resistance, which would ordinarily provide both horizontal and vertical accelerations.) The vertical component of velocity v_y obeys Eq. (3-9), and is

$$v_y = gt.$$

These velocity components may be added vectorially to give an instantaneous velocity **v** at any time t after the balls are released whose magnitude is

$$v = \sqrt{v_x^2 + v_y^2} = \sqrt{v_x^2 + g^2t^2}.$$

It is not difficult to derive the equation of the path that ball B follows. (Ball A, of course, falls vertically downward.) In the absence of a horizontal acceleration, the horizontal displacement s_x of the ball at the time t is

$$s_x = v_x t,$$

while from Eq. (3-11) we have

$$s_y = \tfrac{1}{2}gt^2.$$

The edge of the table is taken as the origin of coordinates here. Eliminating t from these equations, we obtain

$$s_y = (\tfrac{1}{2}g/v_x^2)s_x^2,$$

which, since the quantities in the parentheses are constant, is a parabola. This conclusion is verified experimentally.

Example: The table of Fig. 3–10 is 1 m high and ball B leaves it at a speed of 5 m/sec. Find the time of fall, how far from the table ball B strikes the floor, and the velocities of A and B when they strike the floor.

Solution: The time of fall, from $s_y = \frac{1}{2}gt^2$, is

$$t = \sqrt{\frac{2 \times 1\ \text{m}}{9.8\ \text{m/sec}^2}} = 0.45\ \text{sec},$$

while the ball reaches the floor at a distance

$$s = 5\,\frac{\text{m}}{\text{sec}} \times 0.45\ \text{sec} = 2.25\ \text{m}$$

away from the table.

The final velocity of A has the components

$$v_{Ax} = 0, \qquad v_{Ay} = gt,$$

while that of B has the components

$$v_{Bx} = v_0, \qquad v_{By} = gt.$$

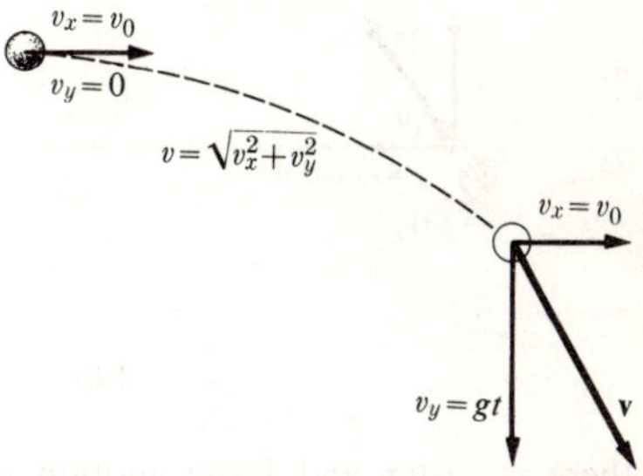

FIG. 3–11. The motion of ball B of Fig. 3–10.

From Fig. 3–11 we see that since v_{Bx} is perpendicular to v_{By}, their vector sum $\mathbf{v}_B$ has the magnitude

$$v_B = \sqrt{v_{Bx}^2 + v_{By}^2} = \sqrt{v_0^2 + (gt)^2}.$$

Inserting the values of v_0 and t for the special case we have been discussing,

$$v_B = \sqrt{(5\ \text{m/sec})^2 + (9.8\ \text{m/sec}^2 \times 0.45\ \text{sec})^2} = 6.7\ \text{m/sec}.$$

The velocity of ball A has the single component v_{Ay}, and so

$$v_A = v_{Ay} = gt = 9.8\,\frac{\text{m}}{\text{sec}^2} \times 0.45\ \text{sec} = 4.4\ \text{m/sec}.$$

It is worth noting that the *vector sum* of v_{Bx} and v_{By} is 6.7 m/sec, while their *algebraic sum* is 9.4 m/sec. The latter number is, of course, completely meaningless, since velocity is a vector quantity and velocity addition must obey the rules of vector addition.

A more general case of projectile motion is exemplified by the flight of a rocket. Let us ignore the curvature of the earth and the frictional resistance of the atmosphere to the passage of the rocket, and assume that the rocket uses up its fuel at a distance from its launching point that is small compared with the total distance it travels. If the initial velocity $\mathbf{v}_0$ of the rocket makes an angle of θ with the ground, we can resolve $\mathbf{v}_0$ into the components

$$v_x = v_0 \cos\theta, \qquad v_y = v_0 \sin\theta.$$

The horizontal velocity component v_x remains constant during the rocket's flight. The vertical component v_y, however, drops to zero as the downward acceleration of gravity takes effect, and then becomes more and more negative (meaning that

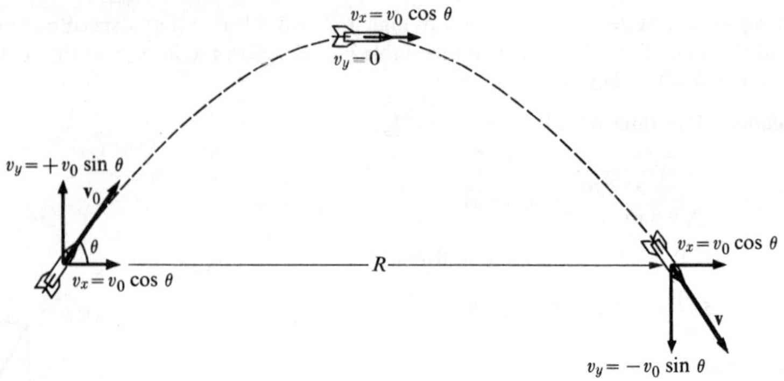

FIG. 3-12. Rocket flight.

there is faster and faster motion downward) until the rocket hits the ground (Fig. 3-12).

We can once again use the formulas of kinematics to discuss the horizontal and vertical aspects of the rocket's motion separately. Let us first calculate the time of flight of the rocket. The rocket will continue to rise until the vertical component of its velocity, given by

$$v_y = v_0 \sin\theta - gt, \tag{3-13}$$

is zero. At this time t,

$$0 = v_0 \sin\theta - gt \quad \text{and} \quad t = \frac{v_0 \sin\theta}{g}.$$

The rocket requires the same period of time to return to the ground, and so its total time of flight T is

$$T = 2t = \frac{2v_0 \sin\theta}{g}. \tag{3-14}$$

We can now find the range R of the rocket; that is, we can see how far from its launching point it will strike the ground. Since the horizontal component v_x of the rocket's velocity is constant,

$$R = v_xT = v_0 \cos\theta \times \frac{2v_0 \sin\theta}{g} = \frac{2v_0^2}{g} \sin\theta \cos\theta. \tag{3-15}$$

Equation (3-15) can be simplified by making use of the trigonometric formula

$$\sin\theta \cos\theta = \tfrac{1}{2} \sin 2\theta.$$

The rocket's range may therefore be written

$$R = \frac{v_0^2}{g} \sin 2\theta. \tag{3-16}$$

This formula gives the range of a rocket (or any other projectile, for that matter, provided it obeys the restrictions given earlier) in terms of its initial speed v_0 and the angle θ at which it is launched. We note that R is a maximum when $\sin 2\theta = 1$, since 1 is the largest value the sine function can have. Since $\sin 90° = 1$, the maximum range occurs when the initial angle θ is 45°. Any other angle, greater or smaller, will result in a shorter range (Fig. 3-13).

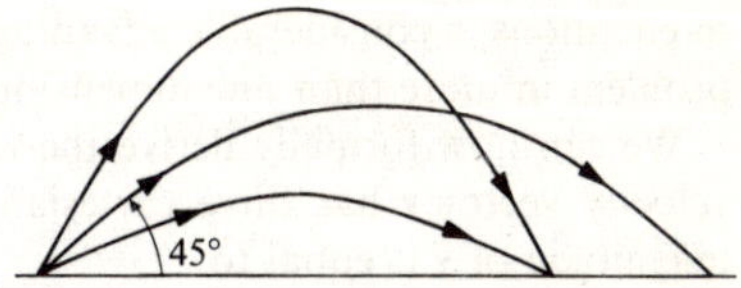

FIG. 3-13. In the absence of air resistance, the maximum range of a projectile occurs when it is fired at an angle of 45°.

Example: What is the minimum initial velocity a rocket must have if it is to be able to reach a target 1000 km (621 miles) away?

Solution: From Eq. (3-16) we obtain

$$v_0 = \sqrt{Rg/\sin 2\theta}.$$

Setting $\theta = 45°$ for maximum range,

$$\begin{aligned} v_0 &= \sqrt{Rg/\sin 90°} = \sqrt{Rg} \\ &= \sqrt{10^3 \text{ km} \times 10^3 \text{ m/km} \times 9.8 \text{ m/sec}^2} \\ &= 3100 \text{ m/sec}. \end{aligned}$$

This is a little less than 2 mi/sec.

3-8 Vector summary

Thus far we have seen how the motion of a body can be conveniently described by the three vectors **s**, **v**, and **a**, which are related by

$$\mathbf{v} = d\mathbf{s}/dt, \qquad \mathbf{a} = d\mathbf{v}/dt = d^2\mathbf{s}/dt^2.$$

In the case of uniformly accelerated motion, these equations can be integrated to give

$$\mathbf{v} = \mathbf{v}_0 + \mathbf{a}t, \qquad \mathbf{s} = \mathbf{v}_0 t + \tfrac{1}{2}\mathbf{a}t^2.$$

Eliminating the time, we obtain

$$\mathbf{v}\cdot\mathbf{v} = \mathbf{v}_0\cdot\mathbf{v}_0 + 2\mathbf{a}\cdot\mathbf{s}$$

in vector notation. The dot products in the above expression represent the scalar products of two vectors.

To generalize from the two dimensions of motion in a plane to the three dimensions of motion in space requires no change in these equations; it is only necessary to give **s**, **v**, and **a** three components each instead of two. Vector notation always permits a very succinct statement of what might otherwise be a complicated set of equations, a considerable advantage when we are trying to understand a physical problem in more than one dimension.

We can now formally derive the relationship between speed and velocity. The velocity vector **v** has three cartesian components, dx/dt, dy/dt, and dz/dt. The magnitude of **v** is equal to

$$v = \sqrt{(dx/dt)^2 + (dy/dt)^2 + (dz/dt)^2} = \frac{\sqrt{(dx)^2 + (dy)^2 + (dz)^2}}{dt}.$$

But $\sqrt{(dx)^2 + (dy)^2 + (dz)^2}$ is just the arc length along the curve on which the body moves. Hence the magnitude of the velocity equals the distance traversed along the actual path divided by the time taken, which is by definition the speed of the body.

Problems

1. A car travels at 60 mi/hr along a straight road for 2 hr and then at 40 mi/hr for the next 3 hr. What is the car's average speed for the entire 5 hr?

2. A car travels at 40 mi/hr along a straight road for 60 mi and then at 50 mi/hr for the next 60 mi. What is the car's average speed for the entire 120 mi?

3. A ball is dropped onto the ground from a height of two feet. It bounces twice and each time returns to its original height. Plot a graph of its velocity versus time.

4. (a) What is the speed with respect to the ground of an airplane whose airspeed is 200 mi/hr when it is bucking a head wind of 56 mi/hr? (b) What distance can it cover relative to the ground in 5 hr?

5. What is the airspeed of an airplane that requires 3 hr and 30 min to go 1200 mi from one city to another when it has a 70 mi/hr tail wind?

6. The speed of light is 3×10^8 m/sec. How many minutes does it take light to reach us from the sun, which is approximately 1.5×10^8 km away?

7. A man 6 ft tall walks away from a point source of light on the ground toward a wall 50 ft from the light. If the man walks at a rate of 3 ft/sec, what is the rate at which the size of his shadow is changing when he is 20 ft from the light?

8. A kite is 100 ft high and 150 ft of weightless cord is out. If the kite moves horizontally at 2 mi/hr, at what rate is the cord being paid out?

9. Two barges are at a fixed distance d from each other in a river whose current velocity is v. The displacement from A to B is in the direction of the current. The barges keep a fixed distance but move with the current. A swimmer swims from A to B and back to A, with a velocity c with respect to the land. Find the time it takes to make the round trip in terms of c, v, and d.

10. Repeat Problem 9 if A and B are so placed that the displacement AB is perpendicular to the river current.

11. A telescope tube is mounted on a flatcar which moves at a speed of 10 m/sec. A raindrop enters the telescope at the top and falls at a constant rate of 50 m/sec.

What angle should the telescope tube make with the horizontal in order for the raindrop to fall through the tube without touching its sides?

12. One end of a ladder 10 ft long rests on a floor and the other on a vertical wall. The lower end is 6 ft from the wall and is slipping at a rate of 1.5 ft/sec. How fast is the other end falling?

13. A car starts from rest and reaches a speed of 50 mi/hr in 20 sec. (a) What was its acceleration (assumed constant) in (mi/hr)/sec? (b) in ft/sec^2?

14. The brakes of a car moving at 50 mi/hr are suddenly applied and the car comes to a complete stop in 4 sec. (a) What was its acceleration (assumed constant) in (mi/hr)/sec? (b) in ft/sec^2?

15. How long would the car of Problem 13 take to reach 80 mi/hr?

16. How long would the car of Problem 14 take to come to a complete stop starting from 80 mi/hr?

17. A car whose initial speed is 20 mi/hr receives an acceleration of 3 (mi/hr)/sec. How far will it go from the point where the acceleration is applied until its speed is 45 mi/hr?

18. The brakes of a particular car are capable of producing an acceleration of -20 ft/sec^2. How far will the car go in the course of slowing down from 90 ft/sec to 30 ft/sec?

19. A stone is dropped from a cliff 64 ft high. How long will it take to fall to the foot of the cliff?

20. A boy throws a ball to a height of 50 ft. How long a period elapses between the time the ball leaves the boy's hand and the moment he catches it on the way down?

21. With what speed will the stone of Problem 19 strike the ground?

22. With what speed did the boy of Problem 20 have to throw the ball?

23. A girl throws a ball vertically upward with a speed of 20 ft/sec from the roof of a building 60 ft high. (a) How long will it take the ball to reach the ground? (b) What will its speed be when it strikes the ground?

24. A girl throws a ball vertically downward with a speed of 20 ft/sec from the roof of a building 60 ft high. (a) How long will it take the ball to reach the ground? (b) What will its speed be when it strikes the ground?

25. An orangutan throws a ball vertically upward at the foot of a cliff 200 ft high while his mate simultaneously drops another ball from the top of the cliff. The two balls collide at an altitude of 100 ft. What was the initial speed of the ball thrown upward?

26. A man in an elevator 100 ft above the ground drops an apple 6 ft above the elevator's floor. With a stopwatch he determines how long the apple takes to fall to the floor. What does he find (a) when the elevator is ascending with an acceleration of 2 ft/sec^2; (b) when it is descending with an acceleration of 2 ft/sec^2; (c) when it is ascending at the constant speed of 8 ft/sec; (d) when it is descending at the constant speed of 8 ft/sec; (e) when the cable has broken and the elevator is descending in free fall?

27. A body is dropped from rest. Find the distance that it traverses during the nth second of its fall.

28. A body moves along a straight path in such a way that its displacement at any time t is given by $s = (4\text{ m/sec}^3)t^3$. (a) Find formulas for the speed and acceleration of the body by successive differentiations of s. (b) Plot graphs of s, v, and a versus t from $t = 0$ to $t = 10$ sec.

29. The motion of a particle along the x-axis is described by the formula $x = A + Bt + Ct^2 + Dt^3$, where $A = 10$ m, $B = 1$ m/sec, $C = -5$ m/sec^2, and $D = 3$ m/sec^3. (a) At what value or values of t is the speed of the particle zero? (b) At what value or values of t is its acceleration zero?

30. A weight suspended by a coil spring oscillates up and down with its vertical coordinate varying with time according to the formula $y = A \cos \omega t$, where A and ω are constants. Show that the acceleration of the weight is directly proportional to y, and in the opposite direction.

31. Find the relations between x and t, v and t, x and t when

(a) $a = 2t$; $x = 0, v = 1$, when $t = 0$.
(b) $a = -2v^2$; $x = 1, v = 5$, when $t = 0$.
(c) $a = -kv$; $x = 0, v = v_0$, when $t = 0$.

32. The velocity of a body is given by the formula

$$v = -\tfrac{1}{5}t^2 + 2t.$$

(a) Find the acceleration as a function of t. (b) Find the distance traversed as a function of t, given that the initial displacement is zero.

33. If the velocity is given by the formula $v^2 = ax^2 + bx + c$, what is the acceleration as a function of x?

34. A ball is thrown horizontally from the roof of a building 60 ft high at a speed of 100 ft/sec. What will the magnitude and direction of the ball's velocity be when it strikes the ground?

35. A bomb is dropped by an airplane in level flight whose speed is 400 mi/hr when it is 12,000 ft above the ground. Where will the bomb strike the ground relative to the location above which it was dropped?

36. A blunderbuss can fire a slug 300 ft vertically upward. (a) What is its maximum horizontal range? (b) With what speeds will the slug strike the ground when fired upward and when fired so as to have maximum range?

37. A football leaves the toe of a punter at an angle of 50° above the horizontal. What was its initial speed if it travels 40 yd?

38. A shell is fired at a speed of 300 m/sec at an angle of 30° above the horizontal. (a) Neglecting friction, how far does it go? (b) How long a time does its flight take?

39. A rifle located at $x = 0$, $y = 0$ is to fire at a target located at the coordinates x, y. What is the minimum muzzle velocity required in terms of x and y?

40. A ball is thrown horizontally toward the north from a rooftop at a speed of 8 m/sec. A 10 m/sec wind is blowing from the east. (a) What is the speed of the ball relative to the ground after 2 sec? (b) What angle does its velocity make relative to the vertical at this time? (c) What angle does its velocity make relative to due north at this time?

41. What is the equation of the trajectory of a particle if the coordinates are given by the parametric equations,

(a) $x = 4t$, $\quad y = 4t - 8t^2$,
(b) $x = \sin t$, $\quad y = \cos 2t$,
(c) $x = A \cos nt$, $\quad y = B \sin nt$.

42. The distance that an object travels as a function of time is given by the formula

$$s = t^3 + 3t^2 + 4 \text{ ft}.$$

What is its instantaneous speed after 1 sec? after 2 sec? What is its average speed during the second second of travel?

43. In climbing up a wall, a caterpillar undergoes a complicated motion in which it ascends 3 ft and slips down 2 ft every 5 min. What is its average speed and average velocity after 5 min?

4

Circular Motion

Bodies in the natural world exhibit a remarkable tendency to travel in curved paths. The orbits of the earth and planets around the sun and of the moon around the earth are very nearly circular, a fact employed by Newton in formulating his law of gravitation. On a microcosmic scale, a useful visualization of the atom invokes electrons circling about a central nucleus, an idea employed by Bohr in devising his theory of the atom. And, of course, circular motion is no novelty in our own experience. We shall find it both interesting and essential for our later work to consider circular motion in some detail.

4–1 Motion about a fixed axis

We shall first consider the motion in a plane of a body such that its distance to a fixed point in the plane is a constant (Fig. 4–1). A line drawn through this fixed point perpendicular to the plane is called the *axis of rotation*. For the moment we shall confine ourselves to situations in which the axis of rotation is fixed in our coordinate system. An example of this type of motion is that of a point on the rim of a wheel as it rotates about its axle.

In principle we could analyze the motion of a point on a wheel in terms of the x- and y-components of its position, velocity, and acceleration. A little reflection, though, reveals that this procedure is unnecessarily complicated. Even though the motion is in two dimensions, a single variable and its time derivatives are enough to specify the motion completely. Since the point is traveling in a circle, by using the polar coordinates r and θ instead of the rectangular coordinates x and y (Fig. 4–1), we can see that θ alone varies with time. A rigid connection between the moving particle and the axis about which it rotates is not even required, provided that r remains constant. An automobile rounding a curve which is an arc of a circle, for instance, may be regarded as in circular motion about a fixed axis. The only parameter involved is the angle θ swept out as the body proceeds on its path.

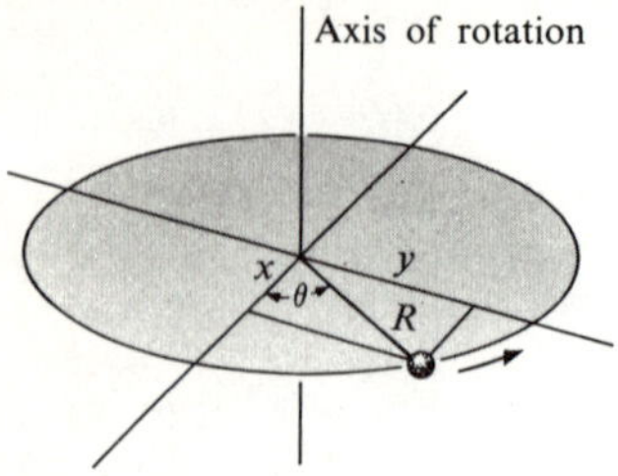

FIG. 4-1. Motion about a fixed axis of rotation.

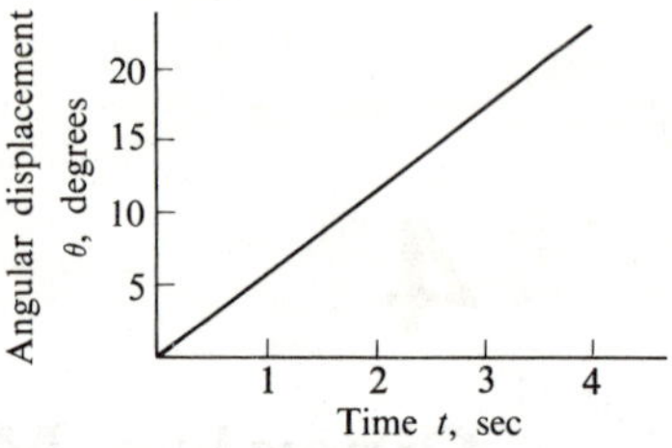

FIG. 4-2. Plot of angular displacement versus time in uniform motion.

When we began studying kinematics at the start of the previous chapter we considered a similar situation, linear motion, where the single parameter used was the linear displacement x of the body. By replacing linear displacement x with angular displacement θ, we can carry over all the concepts developed earlier. Thus if we plot the angular displacement of some particle restricted to move in a circle as a function of time, and find that the resulting curve is a straight line (Fig. 4-2), we can say that we have *uniform angular velocity*. By further analogy with the linear case, we define *average angular velocity*, denoted $\bar{\omega}$, as

$$\bar{\omega} = \frac{\theta_{\text{final}} - \theta_{\text{initial}}}{t_{\text{final}} - t_{\text{initial}}}. \tag{4-1}$$

Equation (4-1) applies whether the angular velocity is uniform or not. It is also convenient to define *instantaneous angular velocity* ω as

$$\omega = d\theta/dt. \tag{4-2}$$

Equations (4-1) and (4-2) give only the magnitude of the angular velocity; we shall postpone discussing the vector properties of angular velocity until Section 4-4.

4-2 Angular measure

What are the units of θ? Since θ is an angle, we may use any convenient angular measure, such as degrees or revolutions. However, the most useful angular unit in physics is the *radian*. An angle in radians is given by the ratio between the arc length s that subtends the angle to the radius R of the circle, as in Fig. 4-3. That is,

$$\theta = s/R. \tag{4-3}$$

One radian therefore corresponds to an angle such that the length of the arc it bounds on a circle equals the radius of the circle. Since the circumference of a circle is $2\pi R$, one revolution is $2\pi R/R$, or 2π radians. The conversion between

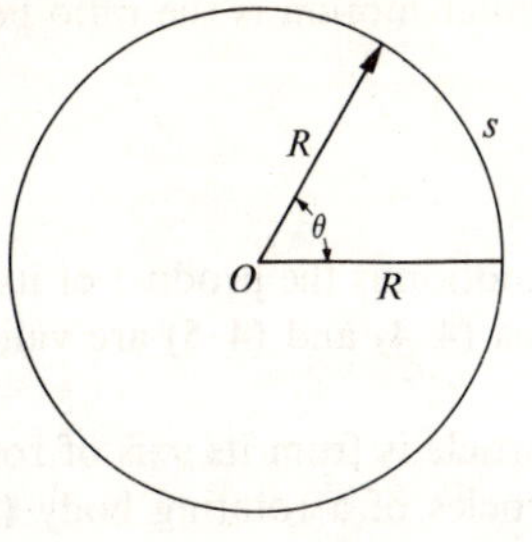

FIG. 4–3. An angle of one radian.

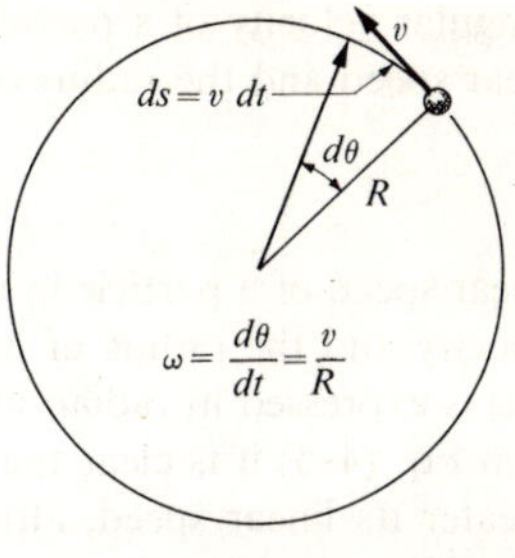

FIG. 4–4. Angular velocity.

degrees and radians follows from

$$1 \text{ revolution} = 2\pi \text{ radians} = 360°,$$

and is

$$1 \text{ radian} = \frac{360°}{2\pi} = 57° \, 17' \, 44'' \cong 57.30°,$$

$$1° = \frac{2\pi}{360} = 0.01745 \text{ radian}.$$

Often it is useful to express angles in radian measure in terms of π itself. For example, an angle of 90° is $\frac{1}{4}$ of a complete circle, and $\frac{1}{4} \times 2\pi = \pi/2$ rad. Of course, this has the same approximate numerical value as $90 \times 0.01745 = 1.571$ rad, since $\pi/2 = 1.571$.

The dimensions of angular velocity are angular displacement divided by time. With $d\theta$ in radians and dt in seconds, the usual units for these quantities, the unit of ω is the rad/sec. Another fairly common unit of angular velocity is the revolution per minute (or rev/min); evidently

$$1 \frac{\text{rev}}{\text{min}} = 1 \frac{\text{rev}}{\text{min}} \times 2\pi \frac{\text{rad}}{\text{rev}} \times \frac{1}{60 \text{ sec/min}} = 0.105 \frac{\text{rad}}{\text{sec}} \cdot$$

Let us consider a particle moving with the uniform speed v in a circle of radius R, as in Fig. 4–4. This particle travels the distance

$$ds = v \, dt$$

in the time dt. The angle through which it moves in that time is

$$d\theta = \frac{ds}{R} = \frac{v \, dt}{R},$$

so that its angular velocity is

$$\omega = \frac{d\theta}{dt} = \frac{v}{R} \cdot \qquad (4\text{–}4)$$

The angular velocity of a particle in uniform circular motion is the ratio between its linear speed and the radius of its path. Also,

$$v = \omega R; \tag{4-5}$$

the linear speed of a particle in uniform circular motion is the product of its angular velocity and the radius of its path. Equations (4–4) and (4–5) are valid only when ω is expressed in radian measure.

From Eq. (4–5) it is clear that the further a particle is from its axis of rotation, the greater its linear speed, although all the particles of a rotating body (except those on the axis) have the same angular velocity. A phonograph record 12 in. in diameter rotates at $33\frac{1}{3}$ rev/min, which is

$$33\tfrac{1}{3}\,\frac{\text{rev}}{\text{min}} \times 0.105\,\frac{\text{rad/sec}}{\text{rev/min}} = 3.50\,\frac{\text{rad}}{\text{sec}}\,.$$

Hence a point 1 in. from the axis has a linear speed of

$$v = \omega R = 3.5\,\frac{\text{rad}}{\text{sec}} \times 1\text{ in} = 3.5\,\frac{\text{in}}{\text{sec}}\,,$$

while a point on the record's rim, where $R = 6$ in., has a speed of

$$v = 3.5\,\frac{\text{rad}}{\text{sec}} \times 6\text{ in} = 21\,\frac{\text{in}}{\text{sec}}\,.$$

4–3 Angular acceleration

When the circular motion of a point is such that its angular velocity is not constant but changes with time, the idea of *angular acceleration* is useful. We define the magnitude of the *instantaneous angular acceleration*, denoted α, by the equation

$$\alpha = \frac{d\omega}{dt} = \frac{d^2\theta}{dt^2}\,. \tag{4-6}$$

(Note that we are still restricting ourselves to the case of a particle moving about a fixed axis in a circular path.) The units of angular acceleration we shall use are rad/sec^2.

Just as we did in the case of linear motion, we may integrate Eqs. (4–2) and (4–6) when α is constant to obtain a useful set of equations:

$$\omega = \alpha t + \omega_0, \tag{4-7}$$

$$\theta = \tfrac{1}{2}\alpha t^2 + \omega_0 t + \theta_0. \tag{4-8}$$

By eliminating t between (4–7) and (4–8), we obtain

$$\omega^2 = \omega_0^2 + 2\alpha\theta. \tag{4-9}$$

4-4 Vector properties of angular velocity

Even when dealing with one-dimensional motion we had to introduce the concept of a vector. There we had to distinguish between the two possible directions of motion, which we could call plus and minus after first specifying which was which unambiguously. In the case of circular motion we have once more to make a distinction, here between clockwise and counterclockwise. Now, however, the way to do this is less obvious. What we shall do is to give a prescription for assigning a direction to an angular velocity which applies in general, whether the axis of rotation is fixed or in motion itself.

The direction associated with the angular velocity vector

$$\boldsymbol{\omega} = d\boldsymbol{\theta}/dt,$$

is the one along which a right-handed screw would travel if it were turned in the direction in which the particle is rotating (Fig. 4-5). This means that it points along the axis of rotation, with the sense given by the right-hand-screw rule. Note the similarity between this specification and the specification of the direction of the vector product.

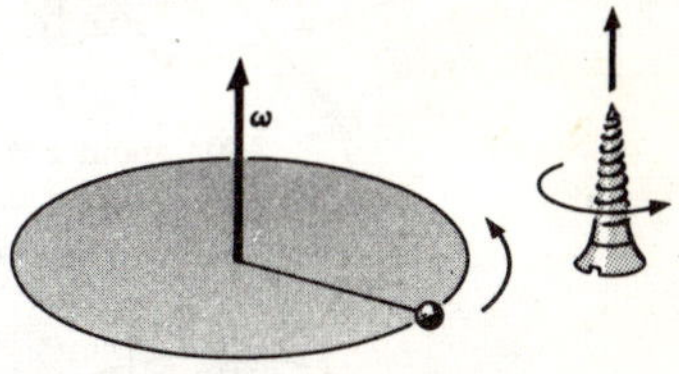

FIG. 4-5. The right-hand-screw rule for the direction of the angular velocity vector.

One might think that simply assigning a direction to a quantity makes it a vector, where by the word vector we mean a quantity which obeys the various combination rules given in Chapter 2. This is not necessarily the case. Suppose that we define a "vector" to represent angular displacement θ in the way we defined a vector to represent angular velocity. Let us have the magnitude of the "vector" equal to the magnitude of the angular displacement, and its direction perpendicular to the plane in which the rotation takes place according to the right-hand-screw rule. Now we take a child's block, as in Fig. 4-6, and turn it first by 90° about the z-axis and then by 90° about the x-axis. The final position of the block is different from what it would be if we first turned it about the x-axis and then about the z-axis! But vectors *must* satisfy the commutative law of addition, $\mathbf{A} + \mathbf{B} = \mathbf{B} + \mathbf{A}$, if they are indeed vectors; in other words, the order of the addition should make no difference. Since this is not true here, the conclusion is that finite rotations are not vectors. However, a similar experiment in which angular *velocities* are added would indicate that they are true vectors, as we assumed.

If angular displacement θ is not a vector, why is its time derivative,

$$\boldsymbol{\omega} = d\boldsymbol{\theta}/dt,$$

a vector? The answer is that, although a finite angular displacement is not a vector, an *infinitesimal* one is. Since $d\boldsymbol{\theta}$ is a vector, $d\boldsymbol{\theta}/dt$ is also. This point is brought up here to emphasize that we must have legitimate grounds for calling a quantity a vector.

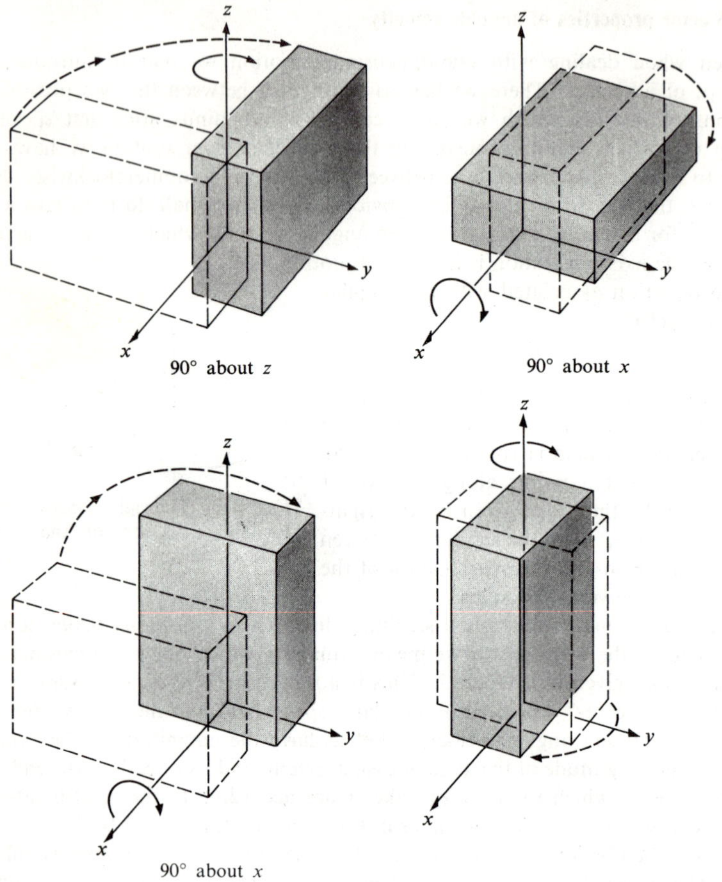

FIG. 4-6. Noncommutativity of finite rotations about z- and x-axes. Hence angular displacement θ is not a vector quantity.

4-5 Vector properties of angular acceleration

Let us now return to angular acceleration and determine the direction associated with it. In general the procedure is to draw the vector describing the angular velocity at the time t and at the time $t + dt$. Subtracting $\boldsymbol{\omega}(t)$ from $\boldsymbol{\omega}(t + dt)$, dividing by dt, and taking the limit as dt approaches zero then gives us $\boldsymbol{\alpha}$.

There are two different cases which must be distinguished. The first, as in Fig. 4-7, involves a fixed axis of rotation, so that only the magnitude of $\boldsymbol{\omega}$ changes. Here $\boldsymbol{\alpha}$ will be in the same direction as $\boldsymbol{\omega}$.

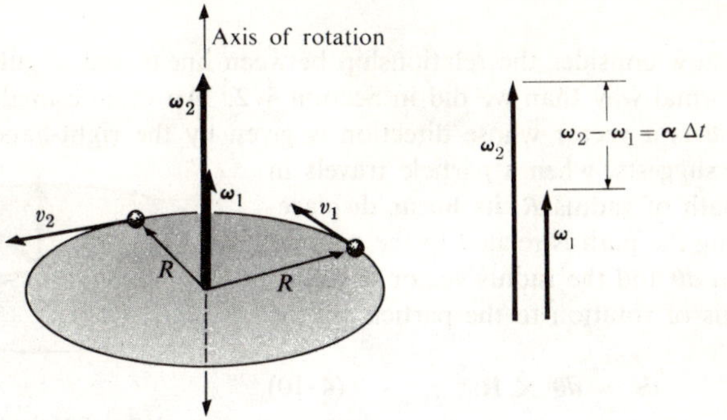

FIG. 4–7. Illustration of changing angular velocity. In a case such as this one, it is the magnitude only of $\boldsymbol{\omega}$ that changes.

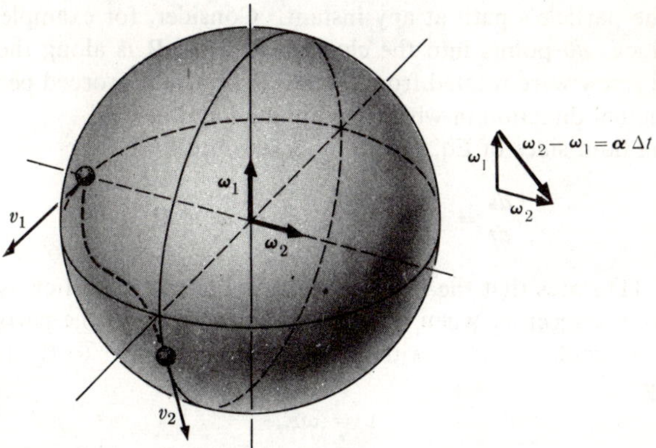

FIG. 4–8. Changing angular velocity. Here the magnitude is constant while the direction varies.

The other possibility is that the magnitude of the angular velocity remains constant but its direction changes (Fig. 4–8). Here the axis is no longer fixed in space. In fact, the plane in which the rotation occurs may even execute an angular motion itself, called *precession.* In the most general type of angular motion, both the magnitude and direction of the angular acceleration vary from point to point in the path of the particle. We shall find it necessary later in this book to consider several specific examples of such complicated forms of motion, but for the present we shall restrict ourselves to fixed axes of rotation.

4-6 Circular and linear motion

We shall now consider the relationship between linear and angular quantities in a more formal way than we did in Section 4-2. An infinitesimal angular displacement $d\boldsymbol{\theta}$ is a vector whose direction is given by the right-hand-screw rule. As Fig. 4-9 suggests, when a particle travels in a circular path of radius R, its linear displacement $d\mathbf{s}$ along the path is related to the angular displacement $d\boldsymbol{\theta}$ and the radius vector $\mathbf{R}$ drawn from the axis of rotation to the particle by the formula

$$d\mathbf{s} = d\boldsymbol{\theta} \times \mathbf{R}. \tag{4-10}$$

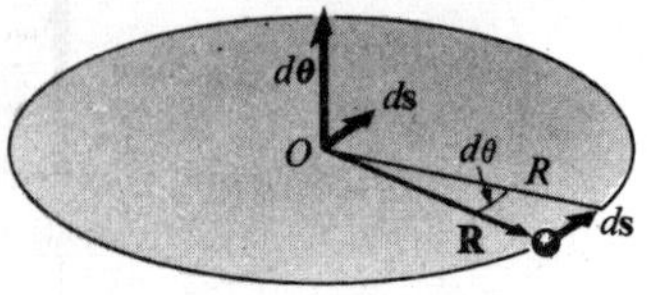

FIG. 4-9. The relationship between the vectors $d\boldsymbol{\theta}$, $d\mathbf{s}$ and $\mathbf{R}$ is $d\mathbf{s} = d\boldsymbol{\theta} \times \mathbf{R}$.

The angular displacement $d\boldsymbol{\theta}$ must be expressed in radians for this equation to hold without an additional numerical factor.

From the definition of vector product it is clear that $d\mathbf{s}$ is always along the tangent to the particle's path at any instant. Consider, for example, the minute hand of a clock: $d\boldsymbol{\theta}$ points into the clock face, while $\mathbf{R}$ is along the hand. If a right-handed screw were rotated from $d\boldsymbol{\theta}$ into $\mathbf{R}$, it would proceed perpendicularly to $\mathbf{R}$ in the actual direction in which the minute hand moves.

By dividing both sides of Eq. (4-10) by dt, we find that

$$\frac{d\mathbf{s}}{dt} = \frac{d\boldsymbol{\theta}}{dt} \times \mathbf{R}, \qquad \mathbf{v} = \boldsymbol{\omega} \times \mathbf{R}. \tag{4-11}$$

Equation (4-11) states that the instantaneous velocity of a particle is always perpendicular to its trajectory when it is in circular motion. If the particle is undergoing uniform circular motion, with $\boldsymbol{\omega}$ always perpendicular to $\mathbf{R}$, the magnitude of $\mathbf{v}$ is simply

$$v = \omega R,$$

which is the same as in Eq. (4-5).

4-7 Centripetal acceleration

As the particle proceeds along its circular orbit, $\mathbf{R}$ continually changes its direction even though $\boldsymbol{\omega}$ may be constant. As a result $\mathbf{v}$ also changes in direction, with its magnitude remaining unchanged if $\boldsymbol{\alpha} = 0$. This means that the particle is being accelerated despite the fact that its angular acceleration may be zero.

In order to compute the acceleration $\mathbf{a}$, we differentiate Eq. (4-11) with respect to time. (A vector product is differentiated exactly as an ordinary product is, except that we must be careful about the order of the factors, since the vector product is not commutative.) We obtain

$$\frac{d\mathbf{v}}{dt} = \frac{d\boldsymbol{\omega}}{dt} \times \mathbf{R} + \boldsymbol{\omega} \times \frac{d\mathbf{R}}{dt}, \qquad \mathbf{a} = \boldsymbol{\alpha} \times \mathbf{R} + \boldsymbol{\omega} \times \frac{d\mathbf{R}}{dt}. \tag{4-12}$$

The first term in Eq. (4–12) represents the contribution to the acceleration that arises from a change in angular velocity. Its direction is evidently tangential to the path of the moving particle, just as the linear velocity is. If there is no angular acceleration—in other words, if the angular velocity (hence linear speed) of the particle is constant—the term $\boldsymbol{\alpha} \times \mathbf{R}$ is equal to 0.

The second term, however, is never zero so long as the particle is moving at all. While it is proceeding about its circular trajectory, $\boldsymbol{\omega}$ at every instant has some definite value, and, because it is constantly changing its position, $\mathbf{R}$ always has a time derivative different from zero. The only occasion when $\boldsymbol{\omega} \times (d\mathbf{R}/dt)$ vanishes is for motion in a straight line, when $\boldsymbol{\omega} = 0$. The quantity $\boldsymbol{\omega} \times (d\mathbf{R}/dt)$ is called *centripetal acceleration.*

The centripetal acceleration can easily be found from Eq. (4–12); with $\alpha = 0$,

$$\mathbf{a}_c = \boldsymbol{\omega} \times \frac{d\mathbf{R}}{dt} = \boldsymbol{\omega} \times \mathbf{v}, \qquad (4\text{–}13)$$

since by definition $\mathbf{v} = d\mathbf{R}/dt$. Equation (4–11) tells us that $\mathbf{v} = \boldsymbol{\omega} \times \mathbf{R}$, which when substituted into (4–13) gives

$$\mathbf{a}_c = \boldsymbol{\omega} \times (\boldsymbol{\omega} \times \mathbf{R}). \qquad (4\text{–}14)$$

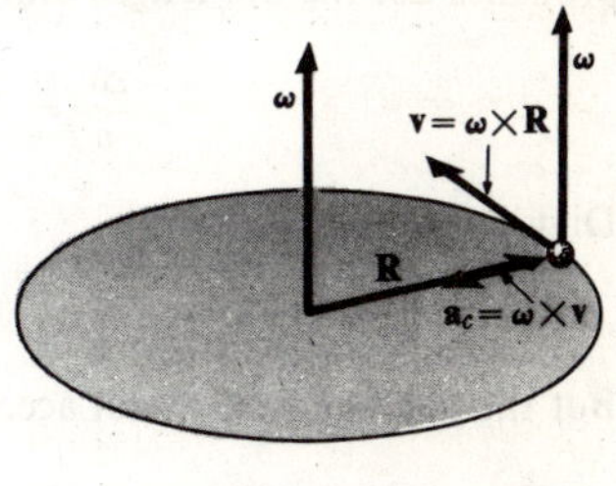

FIG. 4–10. Directional relationships between the various quantities involved in circular motion.

Since the angles between $\boldsymbol{\omega}$ and $\mathbf{R}$, and between $\mathbf{v} = \boldsymbol{\omega} \times \mathbf{R}$ and $\boldsymbol{\omega}$ are 90°, we have immediately that the magnitude of the centripetal acceleration is

$$a_c = \omega^2 R. \qquad (4\text{–}15)$$

The direction can be found by following the definition of a cross product, as illustrated in Fig. 4–10, and is always toward the center of the circle in which the particle moves.

It is possible to obtain an expression for the centripetal acceleration by more conventional means by directly considering the uniform circular motion of a particle. In Fig. 4–11, the velocity vector of such a particle is shown at two nearby points in its path. These vectors have the same magnitude, but their directions differ. The average acceleration the particle experiences between these points is

$$\bar{\mathbf{a}}_c = \frac{\Delta \mathbf{v}}{\Delta t} = \frac{\mathbf{v}_2 - \mathbf{v}_1}{t_2 - t_1}.$$

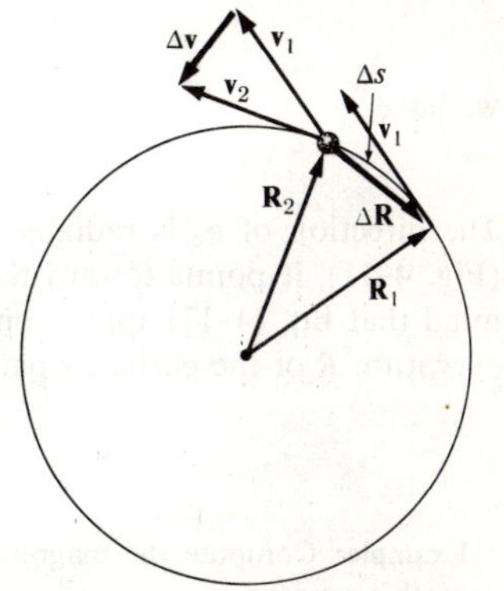

FIG. 4–11. Computing centripetal acceleration. In the limit of an infinitesimal angular displacement, $\Delta\mathbf{v}$ points toward the axis. Hence the direction of the centripetal acceleration $\mathbf{a}_c$ is toward the axis.

The triangles bounded by R_1, ΔR, R_2 and by v_1, Δv, v_2 are similar to one another, since $\mathbf{R}_1 \perp v_1$, $\mathbf{R}_2 \perp v_2$, and $\Delta v \perp \Delta\mathbf{R}$. Because the motion is characterized by a constant speed v and constant radius R (though the *vectors* $\mathbf{v}$ and $\mathbf{R}$ change),

$$R_1 = R_2 = R, \qquad v_1 = v_2 = v,$$

and so, drawing on the similarity between the two triangles of Fig. 4–11, we find that

$$\frac{\Delta v}{v} = \frac{\Delta R}{R}.$$

As the time interval Δt is made smaller, the chord ΔR becomes smaller and approaches Δs, the arc length along the circular path. We may therefore write

$$\frac{\Delta v}{v} = \frac{\Delta s}{R} \qquad \text{or} \qquad \Delta v = \frac{v}{R}\,\Delta s.$$

Dividing by Δt, we have

$$\frac{\Delta v}{\Delta t} = \frac{v}{R}\,\frac{\Delta s}{\Delta t}.$$

But the average centripetal acceleration $\bar{a}_c$ is $\Delta v/\Delta t$, and so

$$\bar{a}_c = \frac{\Delta v}{\Delta t} = \frac{v}{R}\,\frac{\Delta s}{\Delta t}. \tag{4–16}$$

To find the instantaneous centripetal acceleration a_c, we consider the limit of $\bar{a}_c$, the average acceleration, as $\Delta t \to 0$. Because

$$\lim_{\Delta t \to 0} \frac{\Delta s}{\Delta t} = v,$$

we have

$$a_c = v^2/R = \omega^2 R. \tag{4–17}$$

The direction of $\mathbf{a}_c$ is radially inward, since in the limit as $\Delta\mathbf{v}$ approaches zero (Fig. 4–11), it points toward the center of the circle. It is important to keep in mind that Eq. (4–17) applies only to uniform circular motion, with the radius of curvature R of the particle's path constant. In the general case,

$$\mathbf{a}_c = \boldsymbol{\omega} \times \frac{d\mathbf{R}}{dt}.$$

Example: Compute the magnitude of the centripetal acceleration of a point on the earth's equator.

Solution: The radius of the earth is 6.4×10^6 m, and it rotates once each day. One day equals 8.6×10^4 sec. The angular velocity is $\omega = (2\pi/8.6) \times 10^{-4}$ rad/sec, and a_c is given by

$$a_c = \omega^2 R = \frac{4\pi^2}{(8.6)^2} \times 10^{-8} \left(\frac{\text{rad}}{\text{sec}}\right)^2 \times 6.4 \times 10^6\ \text{m} = 3.6 \times 10^{-2}\ \frac{\text{m}}{\text{sec}^2}.$$

If the angular velocity of the particle is not constant, its total acceleration is the *vector* sum of the centripetal acceleration $\mathbf{a}_c$, which points radially inward at the instantaneous center of rotation, and the linear acceleration $\mathbf{a}_T = \boldsymbol{\alpha} \times \mathbf{R}$, which is tangential to the path of the particle at any instant.

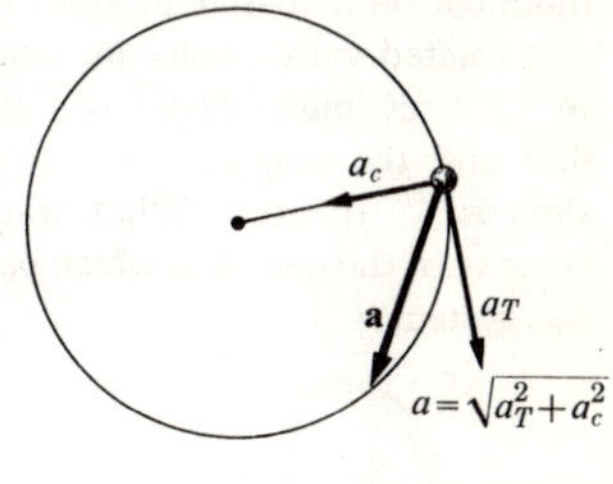

FIG. 4–12. The tangential and centripetal accelerations of a particle in circular motion.

Example: Compute the magnitude of the total linear acceleration of a particle moving in a circle of radius 0.4 m at the instant that its angular velocity is 2 rad/sec and its angular acceleration is 5 rad/sec^2.

Solution: From Eq. (4–12), we have for the particle's tangential and centripetal accelerations

$$a_T = \alpha R = 5 \frac{\text{rad}}{\text{sec}^2} \times 0.4 \text{ m} = 2 \frac{\text{m}}{\text{sec}^2},$$

$$a_c = \omega^2 R = \left(2 \frac{\text{rad}}{\text{sec}}\right)^2 \times 0.4 \text{ m} = 1.6 \frac{\text{m}}{\text{sec}^2}.$$

We must add a_T and a_c vectorially, since they are vector quantities which are in different directions. As shown in Fig. 4–12, the magnitude a of the vector sum $\mathbf{a}$ of $\mathbf{a}_T$ and $\mathbf{a}_c$ is

$$a = \sqrt{a_T^2 + a_c^2} = \sqrt{2^2 + 1.6^2} \frac{\text{m}}{\text{sec}^2} = 2.6 \frac{\text{m}}{\text{sec}^2}.$$

Problems

1. What is the angular velocity in rad/sec of the hour, minute, and second hands of a clock?

2. A grindstone 10 cm in radius is rotating at 1725 rev/min. (a) What is its angular velocity? (b) What is the linear speed of a point on its rim?

3. A drill bit $\frac{1}{4}$ in. in diameter is rotating at 1200 rev/min. (a) What is its angular velocity? (b) What is the linear speed of a point on its circumference?

4. A barrel 2 ft in diameter is rolling with an angular velocity of 5 rad/sec. What is the instantaneous linear velocity of (a) its top, (b) its center, and (c) its bottom with respect to the ground?

5. The propeller of a boat rotates at 1000 rev/min when the speed of the boat is 20 ft/sec. The diameter of the propeller is 18 in. What is the speed of the tip of the propeller?

6. The shaft of a motor rotates at the constant angular velocity of 3000 rev/min. How many radians will it have turned through in $\frac{1}{2}$ min?

7. An automobile makes a *U*-turn in 3 sec. What is its average angular velocity?

8. In some models of an atom, the charge in the atom rotates with an angular velocity of 12.5×10^{15} rad/sec about the nucleus. If the radius of the orbit is 0.5×10^{-8} cm, what is the linear speed of the

charge? What is its centripetal acceleration?

9. In a neutron velocity selector two discs of a neutron-absorbing material are rigidly mounted on a shaft 1 m apart (Fig. 4–13) and rotated with a constant angular speed of 1000 rev/min. There is a slot in each disc and the angular separation between slots is 3″ of arc. What would be the velocity of the neutrons which pass through this system?

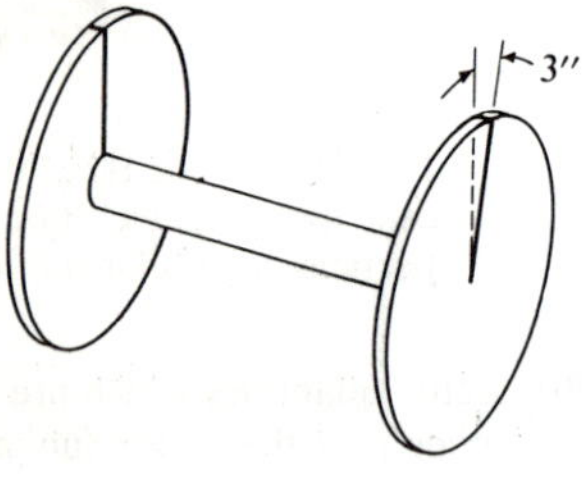

FIGURE 4–13

10. A phonograph record 12 in. in diameter rotates $33\frac{1}{3}$ times per minute. (a) What is the linear speed of a point on its rim in ft/sec? (b) What is the centripetal acceleration of a point on its rim?

11. The minute hand of a large clock is 0.5 m long. (a) What is the linear speed of its tip in m/sec? (b) What is the centripetal acceleration of its tip?

12. A phonograph turntable slows down to a stop from an initial angular velocity of 45 rev/min in 20 sec. (a) What is its angular acceleration? (b) How many turns does it make while slowing down?

13. The speed of a motor increases from 1200 rev/min to 1800 rev/min in 20 sec. What is its angular acceleration? How many revolutions does it make during these 20 sec?

14. A truck undergoes an acceleration of 0.25 m/sec^2. If its wheels are 1 m in diameter, what is their angular acceleration?

15. A wheel starts from rest under the influence of a constant torque and turns through 500 rad in 10 sec. (a) What is its angular acceleration? (b) What is its angular velocity after 10 sec?

16. A grindstone starts from rest and achieves an angular speed of 3000 rev/min in 10 sec. What is its angular acceleration in rad/sec^2?

17. What is the linear speed of a point at the equator on the surface of the earth due to the rotation of the earth? The radius of the earth is 4000 mi.

18. What is the linear speed of the second hand of a watch whose radius is 1 cm?

19. A football is spinning around its long axis with the angular velocity of 10 rad/sec and at the same time it is turning end over end with the angular velocity of 2 rad/sec. What is its resultant angular velocity?

20. The average distance between the earth and the sun is 9.3×10^7 mi. What is the linear speed of the earth in its orbit? What is its centripetal acceleration?

21. The moon is approximately 2.39×10^5 mi from the earth, and requires 27.3 days for each revolution around it. What is the centripetal acceleration of the moon?

22. The angular velocity of a point on a rotating disc whose radius is 10 cm is given in radians/second when t is in sec by

$$\omega = 3t^2 + 2t + 1.$$

(a) What is its angular acceleration when $t = 2$ sec? (b) What is its acceleration when $t = 2$ sec? Direction and magnitude? (c) Through what angle has it turned between $t = 1$ and $t = 2$ sec?

23. The angular displacement of a point on the rim of a wheel whose radius is 1 ft is given in radians when t is in sec by

$$\theta = t^3 + 2t^2 + t + 1.$$

(a) What is its angular velocity when $t = 1$ sec? (b) What is its angular acceleration when $t = 1$ sec? (c) What is its total tangential acceleration when $t =$ 1 sec?

24. The angular velocity of a body points instantaneously along the x-axis. There is a constant angular acceleration only along the z-axis. Describe the motion of the body.

25. Two men are throwing a baseball back and forth on the platform of a merry-go-round whose radius is R feet. Charles is at the center, Ralph is at the rim. The angular velocity of the merry-go-round is ω. If the ball is thrown with a velocity v, (a) along which direction should Charles throw the ball so that it reaches Ralph? (b) How long does the ball take to reach Ralph? (c) Answer (a) and (b) for Ralph's throws to Charles. (d) What is the critical value of ω_c such that for all $\omega > \omega_c$, none of Ralph's throws could reach Charles?

5

Dynamics of Particles

The formulas of kinematics provide us with a way of analyzing the motion of a particle in terms of the quantities we have called linear displacement, velocity, and acceleration, or their angular counterparts. But kinematics goes no further than a description of the motion that takes place. For insight into the circumstances underlying the observed motion, we must turn to the branch of physics known as *dynamics*, whose roots consist of three laws first obtained by Isaac Newton. These laws apply to particles, which we shall take to be material bodies having no extension in space, points with no volume, yet in other respects the same as ordinary bodies. Particles are abstractions, and it would seem that Newton's dynamics could not be applied to bodies in the real world. We shall see in Chapter 8 that, starting with the dynamics of particles, it is possible to derive identical dynamical laws for real extended bodies. In the meantime, we need not be too concerned with the distinction, except that we shall not consider the possibility that our bodies rotate. In this chapter we shall use the words body and particle interchangeably.

5–1 The laws of motion

Almost three centuries ago Isaac Newton (1642–1727) formulated three principles, based on observations he and others had made, which are so fundamental that they have become known as the laws of motion. These laws, like most physical principles, must be approached with an awareness of the processes of selection and idealization inherent in the scientific description of nature.

The first step in analyzing a physical situation is to select those aspects of it which are essential and disregard the others. Thus if we wish to understand the motion of a ball after it has been thrown into the air, we would want to know its initial speed and direction, how high it goes and where it reaches the ground, and so on; we would dismiss as irrelevant the time of day, the color of the ball, and the manner of throwing it.

The second step is to idealize the data we obtain. If we perform careful measurements on the flight of the ball, we would find that its path is almost, but not quite, a parabola. We might reasonably suppose that the deviation from a parabola arises from such factors as air resistance and wind, with which we are not at the moment concerned, and assume that the ball's trajectory in the absence of these factors would be a perfect parabola. In this way we reduce our problem to the basic one of explaining the parabolic path of an object released with a certain initial speed in a certain direction. If we are able to solve this problem, we might then be able to calculate the effects of air resistance and wind to see whether they account for the observed deviations of the ball's path from a parabola. But it would have been difficult or impossible to proceed with the problem of projectile motion unless we had first selected the essential elements of the problem and then idealized them to their simplest form. It is within this framework of selection and idealization that we must approach Newton's three laws of motion.

5–2 The first law of motion

The first law of motion states that

In the absence of a resultant force, a body at rest will remain at rest and a body in motion will continue in motion in a straight line at constant speed.

We note immediately that a new word has been introduced: *force.* We all have an intuitive idea of what a force is, namely a "push" or a "pull." But our own experience is quite limited and not always consistent. We usually think of a force as acting between bodies in direct contact with one another, but gravitational, electric, and magnetic forces, for example, operate without any apparent contact between the origin of the force and the body being affected by it. Furthermore, the relationship between force and motion is not always obvious.

The first law of motion clears up these ambiguities with a simple definition: a force is any influence that can cause a body to be accelerated. The definition of a force is made in terms of a kinematical measurement which we know how to make. If a body moves along a straight path at constant speed or is stationary, no resultant force is acting on it. Of course, several individual forces may be present, but if this is the case their magnitudes and directions are such that they cancel one another out to leave no *resultant* force.

There are two assumptions embodied in the first law of motion, both based on experimental observation. The first states that the effect of many forces acting simultaneously is the same as the sum of the effects of the individual ones. This is an example of a *superposition principle.* It may seem surprising at first that the principle of superposition of forces requires experimental verification, since it is so "obvious." Actually, there are many situations in which superposition principles do not apply, although we might think that they would. An example is a ship powered with three engines; the ship will move faster if the engines are attached to a single propeller than if they are attached to three separate propellers.

The second assumption states that forces are vector quantities, which means that they follow the rules of vector addition when being combined.

The first law of motion has a number of surprising consequences, some of them apparently contradicting common sense. For instance, until Newton's time it was believed that the orbits of the moon and planets were "natural" ones, with no forces necessary to hold these bodies in their orbits. Since there is no visible contact between the heavenly bodies and anything else, no cosmic hand pulling them round and round, the accepted conclusion was that no forces were acting on them. However, the moon and planets travel in curved orbits, so that they are continually accelerated, and by the first law of motion corresponding forces must be present to produce these accelerations. The search for these forces led Newton to the law of gravitation.

A more familiar paradox involving the first law comes up when we drive a car with a uniform speed along a straight, level road. If we turn off the engine, the car slows down, and we conclude that a force must continually be applied in order to maintain the car at a constant velocity. In order to reconcile this observation with the first law of motion, we begin by noting that the first law refers to the *sum* of the forces acting on an object. If there are other forces acting on the car besides the one supplied by the engine such that the grand total adds up to zero, removing one of the forces will produce an acceleration. There are in fact external forces acting on a moving car, frictional forces which tend to resist its motion. Hence when we turn off the engine the resistive forces take over and slow the car down. By pushing harder on the gas pedal, of course, we can cause the force of the engine to exceed the resisting forces, so that the car's speed then increases.

5-3 Newton's second law

The second law is an explicit statement of the relationship between force and motion.

When a resultant force acts on a particle, it will be accelerated in the direction of the force with an acceleration proportional to the magnitude of the force.

The second law provides us with a method for analyzing and comparing forces in terms of the accelerations they produce. Thus a force which causes a particle to have twice the acceleration another force produces must be twice as great. A particle moving to the right but going slower and slower is accelerated to the left; hence there is a force toward the left acting on it. The first law of motion is clearly a special case of the second: When the applied force is zero, the acceleration is also zero.

The first law of motion imputes to material bodies the property of tending to resist changes in their state of rest or uniform motion, a property that is known as *inertia*. We can specify inertia in a precise way with the help of the second law of motion. Since the response of a body to a net force **F** is an acceleration **a** proportional to **F**, we can write

$$\mathbf{F} = k\mathbf{a}, \tag{5-1}$$

where **F** is the symbol for force, **a** the symbol for acceleration, and k a scalar constant of proportionality whose value depends on the particular particle being considered. The larger the value of k, the smaller will be the acceleration produced by a given applied force. Thus k is a measure of the inertia of a body: A large k means a great deal of inertia; small k means little inertia.

Note that Eq. (5–1) is a vector equation; not only does it include all three dimensions in a single formula, but the resulting formula is independent of the particular coordinate system in which $\mathbf{F}$ and $\mathbf{a}$ are specified. If we were to write Eq. (5–1) as three scalar equations, one for each of the rectangular coordinates x, y, and z, they would look very different from the three equations for each of the spherical polar coordinates r, θ, and ϕ. The most useful aspect of vectors is that they permit us to write equations whose form contains no reference to any system of coordinates, leaving us free to choose the most appropriate set for whatever problem we are considering.

5–4 Mass

Equation (5–1) by itself is limited in scope, since it can be applied only to determine force ratios for a particular particle. In order to make it more useful, we must establish how the constant k depends on the inertial behavior of a particle.

The property of a particle that gives rise to inertia is called its *mass*. We know that a cannonball subjected to the same force as a tennis ball will be accelerated less. But we must still ask exactly how mass is to be defined.

Equation (5–1) is of assistance in doing this. It enables us to establish a definition of equal forces: those which give the same particle the same acceleration. Having established what we mean by the same force, we now can determine which masses are equal. Those particles which are accelerated equally with the same force have the same mass. Now we combine two equal masses into a single mass and subject it to the same force. Experimentally, we find that the acceleration is halved. In this way we determine that the second law of motion can be written in the form

$$\mathbf{F} = Km\mathbf{a}, \tag{5–2}$$

where m is the mass of the particle and K is a new constant, independent of the mass, which depends only on the units used to express $\mathbf{F}$, m, and $\mathbf{a}$.

Another method of comparing masses makes use of the fact that all bodies experience the same acceleration g of 9.8 m/sec^2 (or 32 ft/sec^2 in the British system of units) when dropped near the earth's surface. This acceleration indicates that a force acts on every object on the earth, and the second law reveals that this force is proportional to the mass of the object. From Eq. (5–2) we have, writing g for $\mathbf{a}$,

$$F_e = Kmg,$$

where F_e is the force with which the earth apparently pulls downward on the mass m. This force is familiar to all of us, and it is called *weight*. The weight of a body is the gravitational force it experiences by virtue of its mass; mass is therefore different from weight. We shall explore the relationship between mass and weight in more detail later in this chapter, and in Chapter 11.

Let us take a spring and hang from it in succession the masses m_0 and m. If the masses are different, in each case the elongation of the spring is different, as shown in Fig. 5-1. As we have said, the elongation of a spring is proportional to the force on it, and so we have a means for determining the ratio of the weights W_0 and W of the masses m_0 and m. We have that

$$W_0 = Km_0g, \qquad W = Kmg,$$

and so

$$\frac{m}{m_0} = \frac{W}{W_0}.$$

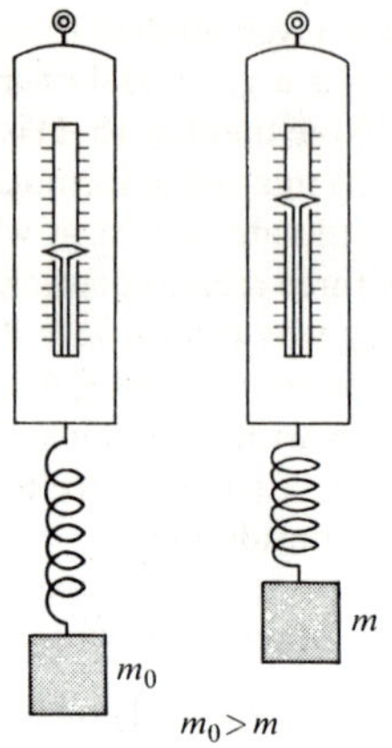

FIG. 5-1. The effect of two different masses on identical springs.

Since the ratio of the weights equals the ratio of the spring elongations, we can again determine m in terms of the standard mass m_0.

These two methods of determining mass are quite different. In the first, in which we measured accelerations, we were dealing with the *inertial* property of mass, which we might think of as the tendency of mass to resist changes in state of motion. In the second, in which we measured the force which the gravitational pull of the earth exerts on mass, we were dealing with the *gravitational* property of mass. On the earth, of course, both inertial and gravitational properties of mass are very much in evidence. If we were to go to interstellar space, however, we would find essentially no gravitational forces acting at all, so that nothing could be thought of as possessing weight there. But objects would still exhibit inertia, and mass determinations could still be made.

It is a very significant empirical fact that the inertial and gravitational masses of a body are proportional to each other. (They are customarily set equal for convenience.) This is by no means "obvious"; a comparable situation is that of the electrical forces exerted by, say, protons and electrons, where electric charge plays a role almost identical to that of mass in gravitational forces, and the intrinsic charges of the proton and electron (as basic a property as gravitational mass) are the same in magnitude although their inertial masses are very different. The proportionality of inertial and gravitational mass is known as the principle of equivalence; it is one of the starting points of Albert Einstein's general theory of relativity.

In Section 5-12, we shall see that Newton's third law of motion gives us an alternative method of establishing a scale of masses.

5-5 Units

Before we can apply Newton's second law of motion to actual problems, we must have an appropriate system of units in which force, mass, and acceleration can be expressed. Although the entire question of units and standards seems trivial the first time it is raised, a clear knowledge of the different systems in use and how they are related is as essential to the physicist and engineer as a clear knowledge of currency systems is essential to the international businessman.

Acceleration has the dimensions of length/time^2, and needs no special unit of its own. The unit of mass depends on what we choose to be our standard mass. The second law expresses force in terms of mass and acceleration, and so we do not need special force units once the mass standards have been established. This is the scheme used in the metric system. In the British system, on the other hand, the unit of force is considered primary, and masses are given in terms of units of force and acceleration.

The metric system of units is based on a standard of length called the *meter* (abbreviated m), which consists of a metal bar with a scratch at each end to fix the precise length, and a standard of mass called the *kilogram* (kg), which consists of a metal cylinder. Once the British system maintained its own length and force standards, but today it also uses the metric standards and defines its units in terms of them. Thus the *foot* (ft) is, by definition, equal to 0.3048 meter. In the British system the unit of force is the *pound* (lb), which represents the weight of a 0.4535-kilogram mass. That is, 1 lb is the force with which the earth attracts a mass of 0.4535 kg. Metric units are invariably used for scientific work, while British units find application in engineering in English-speaking countries.

The unit of time in both systems is the *second* (sec). Time is more difficult to establish a standard for than length or mass. What has been agreed upon is to use the earth's rotation as a clock, with the second defined as 1/86,164 of the time required by the earth to make a complete rotation with respect to the "fixed" stars. Because the length of the day is gradually increasing as a result of tidal friction, and has irregular fluctuations as well, the "standard" day is considered to be the average length which the day had in 1900.

5-6 The metric system

The meter-kilogram-second (mks) system is the one we shall emphasize in this book. The unit of force in the mks system is defined by Newton's second law, Eq. (5-2), with the constant of proportionality K set equal to 1. Hence

$$\mathbf{F} = m\mathbf{a}, \tag{5-3}$$

and the units of force $\mathbf{F}$ are the kg·m/sec^2. This unit has been given the name *newton* (n), so that

$$1 \text{ newton} = 1 \frac{\text{kg}\cdot\text{m}}{\text{sec}^2}. \tag{5-4}$$

The newton is that force which, when applied to a particle having a mass of 1 kg, imparts to it an acceleration of 1 m/sec^2.

Another version of the metric system uses the *centimeter* (cm), the *gram* (gm), and the second (sec) as its basic units. The cgs system, as it is known, is still extensively used, although it is being steadily replaced by the mks system which has many advantages when it comes to treating electricity and magnetism. The conversions between the cgs and mks systems are

$$100 \text{ centimeters} = 1 \text{ meter},$$
$$1000 \text{ grams} = 1 \text{ kilogram}.$$

In the cgs system we also set $K = 1$ in Eq. (5–2), so that once more

$$\mathbf{F} = m\mathbf{a}.$$

The resulting force unit, called the *dyne*, is equal to 1 gm·cm/sec^2, and 10^5 dynes = 1 newton.

Example: A force of 10 n is applied to a 4-kg block that is on a perfectly smooth, level surface. Find the speed of the block and how far it has gone after 6 sec.

Solution: We start from the second law of motion,

$$F = ma.$$

Since we know what F and m are, we can solve for the acceleration a of the block as follows:

$$a = \frac{F}{m} = \frac{10 \text{ n}}{4 \text{ kg}} = 2.5 \frac{\text{m}}{\text{sec}^2}.$$

The second law of motion tells us that the direction of a is the same as that of F. To find the speed of the block after 6 sec, we return to Eq. (3–9) with $v_0 = 0$, so that $v = at$, and obtain

$$v = 2.5 \frac{\text{m}}{\text{sec}^2} \times 6 \text{ sec} = 15 \frac{\text{m}}{\text{sec}}.$$

For the distance the block has traveled in 6 sec we require Eq. (3–11),

$$s = \tfrac{1}{2}at^2,$$

with the result that

$$s = \frac{1}{2} \times 2.5 \frac{\text{m}}{\text{sec}^2} \times (6 \text{ sec})^2 = \frac{1}{2} \times 2.5 \frac{\text{m}}{\text{sec}^2} \times 36 \text{ sec}^2 = 45 \text{ m}.$$

After 6 sec a 4-kg mass acted upon by a 10-n force will have gone 45 m and have a speed of 15 m/sec (Fig. 5–2).

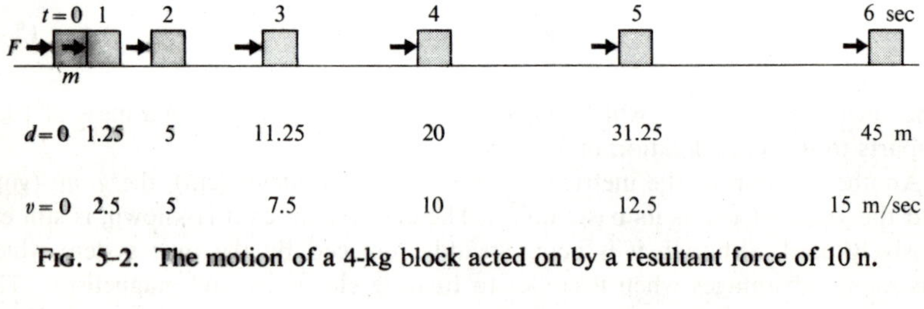

FIG. 5–2. The motion of a 4-kg block acted on by a resultant force of 10 n.

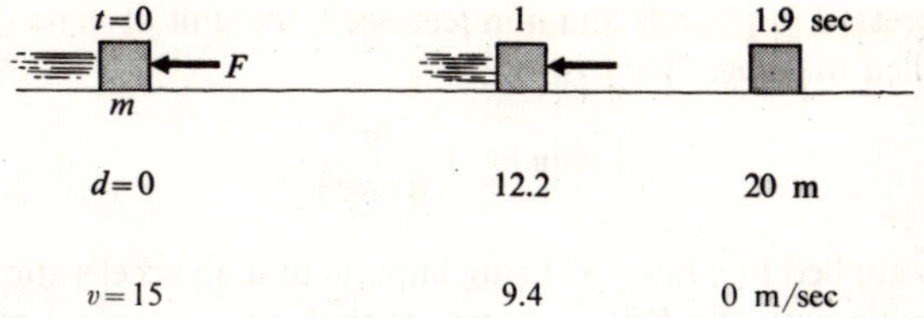

FIG. 5-3. A mass of 4 kg with an initial speed of 15 m/sec comes to a stop in a distance of 20 m when a resistive force of 22.4 n acts on it.

Example: Now let us consider the same block sliding on a level but slightly rough surface without the 10-n force acting on it. The block starts at a speed of 15 m/sec, and gradually slows down and stops in a distance of 20 m. What was the frictional resistive force that acted?

Solution: We start by determining the acceleration of the block. Since we know that its speed has gone from 15 m/sec to 0 m/sec in 20 m, we may use Eq. (3-12):

$$v^2 = v_0^2 + 2as.$$

Here the final speed $v = 0$, the initial speed $v_0 = 15$ m/sec, and the distance $s = 20$ m; hence

$$0 = \left(15 \frac{\text{m}}{\text{sec}}\right)^2 + 2a \times 20\text{ m} = 225 \frac{\text{m}^2}{\text{sec}^2} + 40a\text{ m},$$

and so, solving for the acceleration a, we obtain

$$a = \frac{-225\text{ m}^2/\text{sec}^2}{40\text{ m}} = -5.6 \frac{\text{m}}{\text{sec}^2}\cdot$$

The minus sign indicates a negative acceleration, that is, an acceleration directed opposite to the motion of the block. The acceleration is given as -5.6 m/sec^2 rather than -5.625 m/sec^2 (the value of 225/40) since the initial quantities are given to only two significant figures. When we know the acceleration and mass of the block, it is easy to find the resistive force. We use Eq. (5-3) directly:

$$F = ma = 4\text{ kg} \times (-5.6) \frac{\text{m}}{\text{sec}^2} = -22.4\text{ n}.$$

The minus sign once more indicates that the direction of the force is opposite to that of the block's velocity (Fig. 5-3).

5-7 The British system

In the British system we begin with units for force, length, and time, and from Eq. (5-2) obtain a suitable unit of mass. Again setting $K = 1$, we solve for m, yielding

$$m = \frac{F}{a}\cdot$$

When F is expressed in pounds and a in feet/sec^2, the unit of mass defined by this equation is called the *slug*. That is,

$$1 \text{ slug} = 1 \frac{\text{lb}}{\text{ft/sec}^2}. \tag{5-5}$$

A force of 1 lb applied to a mass of 1 slug imparts to it an acceleration of 1 ft/sec^2.

The big trouble with the British system is that, as a practical matter, *weights* rather than *masses* are specified in actual situations. We speak of 10 tons of concrete and of 5 pounds of potatoes, but never of 100 slugs of coal. The first step in the solution of almost all dynamical problems using the British system is to convert the weight W of the body in question to its mass m in slugs. In order to make this conversion, we recall that all particles in free fall experience the acceleration g of 32.2 ft/sec^2. Since the force on the body that causes it to fall with this acceleration is, by definition, its weight W, we substitute W for $\mathbf{F}$ and g for $\mathbf{a}$ in Eq. (5-3), $\mathbf{F} = m\mathbf{a}$, to give

$$W = mg. \tag{5-6}$$

Hence $m = W/g$ in general, and, in the specific case of the British system of units,

$$m \text{ (slugs)} = \frac{W \text{ (pounds)}}{32.2 \text{ ft/sec}^2}. \tag{5-7}$$

Equation (5-7) is the prescription for converting weights to masses in the British system. Thus the *weight* of a 1-slug mass is 32 lb, and the *mass* of a 1-lb weight is $\frac{1}{32}$ slug.

Example: A particle of weight 4 lb is acted on by a force of 6 lb. What is its acceleration?

Solution: First we convert the mass value from pounds (weight) to slugs. From Eq. (5-7), we have

$$\frac{4 \text{ lb (weight)}}{32 \text{ ft/sec}^2} = \frac{1}{8} \text{ slug}.$$

Now we proceed as before, and find that

$$a = \frac{F}{m} = \frac{6 \text{ lb (force)}}{\frac{1}{8} \text{ slug}} = 48 \text{ ft/sec}^2.$$

Example: A 3000-lb car has an initial speed of 10 mi/hr. How much force is required to accelerate the car to a speed of 50 mi/hr in 9 sec?

Solution: The mass of the car is

$$m = \frac{W}{g} = \frac{3000 \text{ lb}}{32 \text{ ft/sec}^2} = 94 \text{ slugs}.$$

Before we can compute the acceleration involved in going from 10 to 50 mi/hr in 9 sec we must convert the speeds from mi/hr to ft/sec. Using Appendix II, we see that

$$1\ \text{mi/hr} = 1.467\ \text{ft/sec},$$

and therefore, to two significant figures,

$$10\ \text{mi/hr} = 15\ \text{ft/sec} \quad \text{and} \quad 50\ \text{mi/hr} = 73\ \text{ft/sec}.$$

The car's acceleration in the proper units of ft/sec^2 is

$$a = \frac{v_{\text{final}} - v_{\text{initial}}}{t} = \frac{73\ \text{ft/sec} - 15\ \text{ft/sec}}{9\ \text{sec}} = \frac{58\ \text{ft/sec}}{9\ \text{sec}} = 6.4\,\frac{\text{ft}}{\text{sec}^2}\,.$$

Hence the force acting on the car during its acceleration is

$$F = ma = 94\ \text{slugs} \times 6.4\ \text{ft/sec}^2 = 600\ \text{lb}.$$

There is a very basic difference between the metric *absolute* and British *gravitational* systems of units. In a gravitational system force is the standard unit, and mass is computed by taking into account the value of the acceleration due to gravity g. But g is observed to vary somewhat over the earth's surface, implying in such a system that the mass of each body varies in the same way. Since we like to think of mass as an intrinsic property of matter, manifesting itself as inertia whether or not any gravitational force is present, absolute systems in which mass is primary are logically preferable. To avoid confusion when using the British gravitational system, we conventionally assume that g has the constant value 32.2 ft/sec^2, corresponding to 9.80 m/sec^2.

5-8 Applications of Newton's first and second laws

In considering the application of the first and second laws of motion to specific situations, it is convenient to distinguish between problems in *statics* and problems in *dynamics.* In the special case in which a body is not accelerated, the resultant force acting on it must be zero, and we have a problem in statics. Even if motion is present, provided that the velocity is constant, the resultant force is zero and the procedures of statics are applicable. When an acceleration is present the problem is one in dynamics, and Newton's second law must be used in its solution.

Let us begin by considering several problems involving the transmission of forces with ropes. Ropes can convey "pulls" but not "pushes"; they are able to change the direction of a force, even to reverse it, with the help of a pulley, while leaving the magnitude of the force unaltered. In Fig. 5-4, we have a rope attached at one end to a wall with a force **F** applied to the other end. If we consider a thin slice of the rope anywhere along its length, as shown, equal and opposite forces must act across the slice, since it is stationary and the resultant force must there-

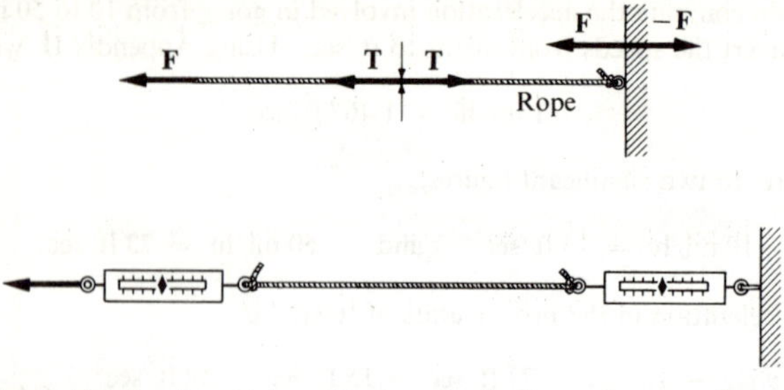

Fig. 5-4. Tension in a string. Both scales indicate the same force, which is equal to the tension.

fore vanish. Either one of these forces is called the *tension* in the rope. At the left-hand end of the rope the force **F** is applied. In the rope this force becomes the tension **T** = **F**. The rope is not accelerated, and therefore the net force acting on it is zero; hence the wall must exert a force −**F** on the end of the rope in order to keep it stationary. At the right-hand end the tension **T** reappears as the force **F**.

To further clarify the meaning of tension, suppose that we connect a spring scale between the left-hand end of the rope and the agency pulling on it, and another between the right-hand end and the wall. Each scale has the same reading, and the value of this reading equals the tension T in the rope. By thinking of tension in this way it is possible to avoid the confusion that sometimes arises in solving problems involving ropes.

Example: A 100-lb weight is attached to the center of a rope, which sags. In the equilibrium position the rope makes an angle of 30° with the horizontal. What is the tension in the rope? (See Fig. 5-5.)

Solution: Since the system is in equilibrium, the sum of the forces acting at the point A must be zero. Thus we have for the horizontal component

$$T_2 \cos 30° - T_1 \cos 30° = 0, \qquad (1)$$
$$T_1 = T_2.$$

The vertical component is found in a similar way to be

$$T_1 \sin 30° + T_2 \sin 30° - 100 = 0, \qquad (2)$$
$$\tfrac{1}{2}(T_1 + T_2) = 100 \text{ lb},$$
$$T_1 = T_2 = 100 \text{ lb}.$$

Example: A particle of mass 50 gm is attached to a string and is being accelerated upward with an acceleration of 10 cm/sec^2. What is the upward force?

Solution: There are two forces acting on the particle. The first is its weight W, which is the downward force exerted by the earth on a 50-gm mass. The second is the upward force we are to find, which is equal to the tension T in the string; T must be sufficiently

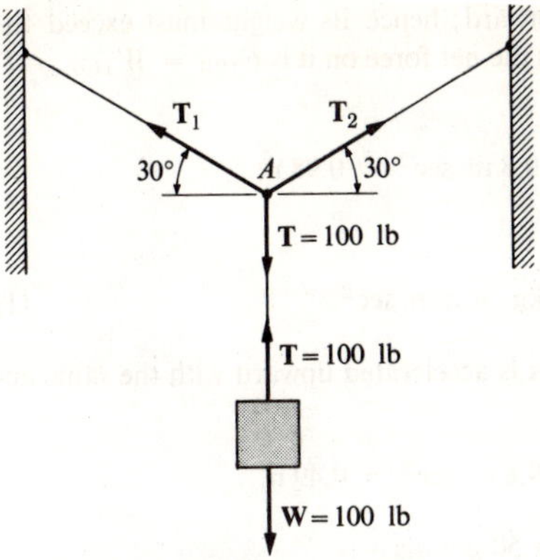

FIG. 5-5. A problem in statics.

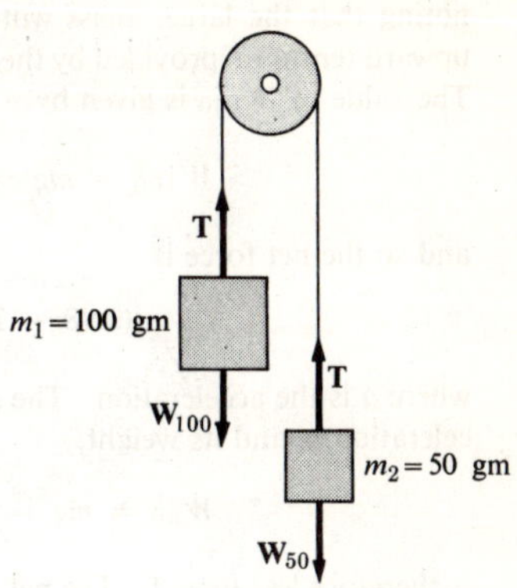

FIG. 5-6. Atwood's machine.

greater than W to produce the observed acceleration. The resultant force is in the direction of the acceleration, and is

$$F = T - W = ma = 0.05 \text{ kg} \times 0.1 \text{ m/sec}^2 = 0.005 \text{ n}.$$

The weight of the body is

$$W = mg = 0.05 \text{ kg} \times 9.8 \text{ m/sec}^2 = 0.49 \text{ n},$$

and so

$$T = 0.495 \text{ n}.$$

Example: A particle of mass 50 gm is attached to a string and is being accelerated downward with an acceleration of 10 cm/sec^2. What is the upward force?

Solution: Since the resultant force F is in the direction of the acceleration, the weight of the particle must exceed the upward force provided by the tension in the string. Therefore

$$F = W - T = ma = 0.005 \text{ n},$$

and since $W = 0.49$, as before, then

$$T = 0.485 \text{ n}.$$

Example: Two particles of masses 50 and 100 gm are joined by a string and placed over a frictionless massless pulley (Fig. 5-6). This is called an Atwood's machine, and it is assumed that the length of the string does not change and that the masses of the string and pulley are so small that they can be neglected. Find the acceleration of each particle.

Solution I: Since the pulley is massless, the tension in the string is constant throughout its length. To solve the problem we consider each particle separately and set the net force acting on it equal to its mass multiplied by its acceleration. We begin by noting that the larger mass will move downward; hence its weight must exceed the upward tension T provided by the rope. Thus the net force on it is $F_{100} = W_{100} - T$. The value of W_{100} is given by

$$W_{100} = mg = 0.1 \text{ kg} \times 9.8 \text{ m/sec}^2 = 0.98 \text{ n},$$

and so the net force is

$$(0.98 - T)\text{ n} = 0.1 \text{ kg} \times a \text{ m/sec}^2, \tag{1}$$

where a is the acceleration. The smaller mass is accelerated upward with the same acceleration a, and its weight,

$$W_{50} = mg = 0.05 \text{ kg} \times 9.8 \text{ m/sec}^2 = 0.49 \text{ n},$$

is therefore less than T. The net force on the 50-gm mass is

$$(T - 0.49)\text{ n} = 0.05 \text{ kg} \times a \text{ m/sec}^2. \tag{2}$$

Solving Eqs. (1) and (2) simultaneously, we find that

$$T = 0.653 \text{ n} \quad \text{and} \quad a = 3.27 \text{ m/sec}^2.$$

Solution II: In the above solution we have considered the problem in detail, and were able to obtain the tension in the rope in addition to the acceleration. However, since all we are asked for is the acceleration, there is a much simpler method of attack. The net force acting on the system of both particles is the difference between the two weights

$$F = W_{100} - W_{50} = 0.49 \text{ n}.$$

The mass that this force acts on is the total mass of the system, 150 gm, and from Newton's second law, $F = ma$, we have

$$0.49 \text{ n} = 0.15 \text{ kg} \times a \text{ m/sec}^2,$$

and thus we obtain

$$a = 3.27 \text{ m/sec}^2,$$

as before. This approach, in which a system as an entity is investigated rather than the component particles separately, is often employed in physics, and we shall make use of it later in the book.

Example: A mass of 100 kg is on a smooth inclined plane which makes an angle of 60° with the horizontal. What is its acceleration down the plane? (See Fig. 5–7.)

Solution: Since we wish to find the acceleration in the direction of the plane, we resolve the force of

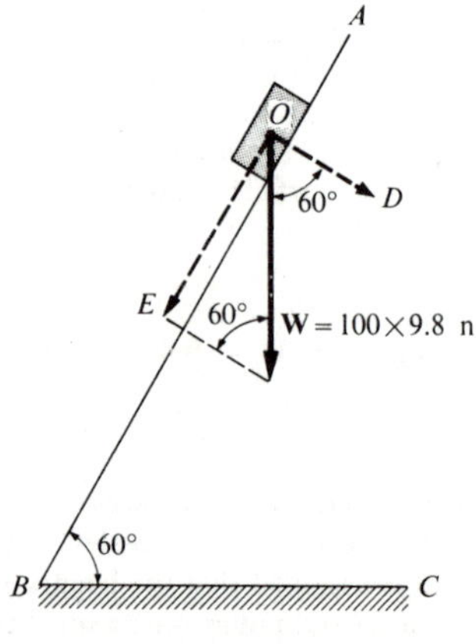

FIG. 5–7. The weight of a body that rests on an inclined plane may be resolved into forces parallel to and perpendicular to the plane.

the weight into two components, one parallel to the plane and one perpendicular to the plane. Since the vector **W** is perpendicular to BC and $OD \perp AB$, the angle DOW is 60°, and the component OE is

$$W \sin 60° = 100 \times 9.8 \times 0.866 = 848.7 \text{ n}.$$

From Newton's second law, $F = ma$, we obtain

$$848.7 = 100a, \qquad a = 8.49 \text{ m/sec}^2.$$

Historically the previous problem was of great importance. Galileo realized that by changing the angle of the plane he could reduce the effective acceleration due to gravity by a factor equal to the sine of the angle that the plane made with the horizontal, which facilitated experiments. Many of his kinematical laws and his statement on the constancy of g derived from inclined-plane experiments.

5-9 Friction

There is a special class of forces, very common in nature, which are grouped under the name of *frictional* forces. We are all familiar with the fact that when we push some object with only a small amount of force, it does not move at all. Since the acceleration is then zero, by Newton's first law the sum of the forces acting on the object is also zero, and the surface upon which the object is resting must therefore be exerting a force on the object in the direction opposite to the one in which we are pushing (Fig. 5-8). This static force is said to be due to friction. We might push harder and finally cause the object to be accelerated. If we measure the acceleration corresponding to the force F, we find that it is less than the amount F/m predicted by Eq. (5-3), which leads us to conclude that some retarding force is still present. The dynamic force exerted by the surface along which the pushed object moves is also an example of friction. Frictional forces act to hamper the motion of bodies in liquids and gases as well as between solids in contact with one another, although we shall confine ourselves to the latter.

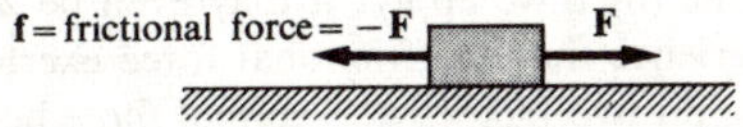

FIG. 5-8. Applied and frictional forces on a stationary body.

Frictional forces are of many kinds, but they all act to impede motion. The effects of friction must be distinguished from those of inertia. The term *inertia* merely refers to the fact that bodies maintain their original states of rest or of motion in the absence of net forces on them; but even a relatively minute force is sufficient to accelerate a body despite its inertia. The term *friction*, on the other hand, refers to actual forces that are exerted in opposition to applied forces.

To understand clearly the characteristic properties of frictional forces, let us consider what happens when we attempt to move a box across a level floor

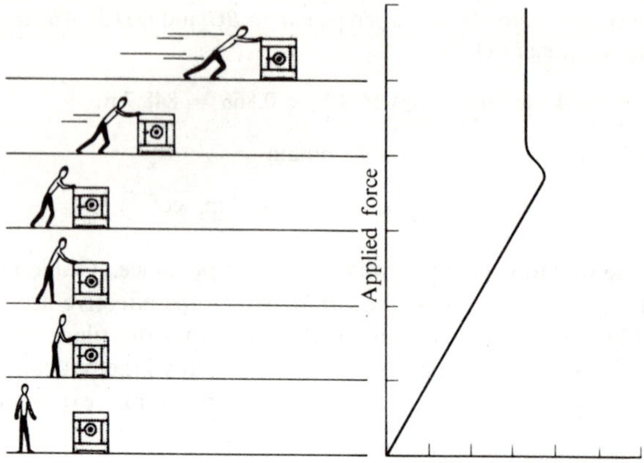

FIG. 5-9. As force is applied to a box on a level floor, the frictional resistive force increases to a certain maximum, decreases somewhat as the box begins to move, and then remains constant.

(Fig. 5-9). At first the box is stationary; no horizontal forces whatever act on it. As we begin to push, the box remains in place, because the floor exerts a force on the bottom of the box which opposes the force we apply. This opposition force is friction, and it arises from the nature of the contact between the floor and box. As we push harder, the frictional force also increases to match our efforts, until finally we are able to exceed the frictional force and begin to move the box. Evidently the opposing frictional force has a maximum value beyond which it cannot go, and when we apply a force greater than this maximum the box will experience a net force. As the box moves under the influence of the net force, the frictional force usually is slightly less than when the box was at rest. Because the net force on the box is the force of our push *minus* the force of friction, it is always less than (or equal to) the force we apply; it may even be zero, as we have seen.

It is a matter of experience that the frictional force exerted by one surface on another surface depends on two factors, the *normal force* holding the surfaces together, and the nature of the surfaces. To a good degree of approximation both these factors can be related to the resulting frictional force F_f by the equation

$$F_f = \mu N, \tag{5-8}$$

where N is the normal force and μ is the *coefficient of friction.* Table 5-1 is a list of coefficients of kinetic friction for several surfaces. Note that the value of F_f computed according to Eq. (5-8) represents a maximum; when the applied force is less than F_f, the frictional force always equals the applied force. Otherwise, since F_f acts in the direction opposite to an applied force, objects would move *backward* when pushed weakly, something that does not, of course, occur.

TABLE 5-1

Approximate coefficients of kinetic friction for various substances in contact*

Materials	μ	Materials	μ
Wood on wood	0.3	Masonry on clay	0.5
Wood on stone	0.4	Rubber tire on dry concrete road	0.7
Metal on metal	0.15		
Metal on wood	0.5	Rubber tire on wet concrete road	0.5
Leather on wood	0.4		

* These values hold only when there is relative motion between the substances; otherwise they are somewhat higher.

When an object is being pushed or pulled horizontally, the normal force N holding it against the surface it is on is simply its weight mg. Thus in such cases,

$$F_f = \mu N = \mu mg.$$

We must apply a force greater than μmg in order to move a body of mass m across a level surface where the coefficient of friction is μ.

In general there are two different coefficients of friction: a static one, which holds for stationary objects, and a kinetic one, which holds for slowly moving objects. Static friction exactly equals the applied force until the latter equals μN, whereupon the object starts to move and the frictional retarding force must be computed from the coefficient of kinetic friction, which is usually smaller than that of static friction. We shall disregard the distinction in what follows so as not to obscure the physical principles involved.

Example: Let us consider a 100-lb wooden crate being pushed across a stone floor with a horizontal force of 45 lb (Fig. 5-10). How long will it take for the crate, starting from rest, to travel 200 ft?

Solution: The first step is to calculate the acceleration of the crate. It is acted on by an applied force F_A of 45 lb and, in the reverse direction, by a frictional force F_f. The latter is

$$F_f = \mu N = 0.4 \times 100 \text{ lb} = 40 \text{ lb},$$

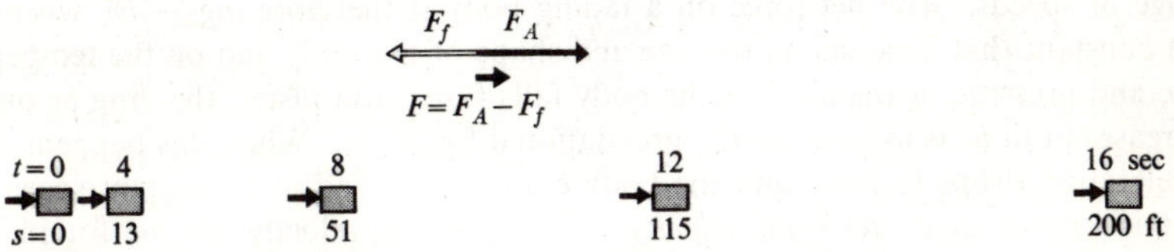

FIG. 5-10. A 100-lb crate travels 200 ft in 16 sec under the influence of a 45-lb applied force and a 40-lb frictional force.

since the coefficient of friction μ is, from Table 5–1, equal to 0.4 here and the normal force is the 100-lb weight of the block. The mass of the block is

$$m = \frac{W}{g} = \frac{100 \text{ lb}}{32 \text{ ft/sec}^2} = 3.1 \text{ slugs.}$$

Inserting these values in the second law of motion, we have

$$F = ma, \qquad F_A - F_f = ma,$$

$$45 \text{ lb} - 40 \text{ lb} = 3.1 \text{ slugs} \times a,$$

so that

$$a = \frac{(45 - 40) \text{ lb}}{3.1 \text{ slugs}} = \frac{5 \text{ lb}}{3.1 \text{ slugs}} = 1.6 \frac{\text{ft}}{\text{sec}^2} \cdot$$

Now that we know the acceleration of the block, we can determine the time required for it to travel 200 ft by rewriting $s = \frac{1}{2}at^2$ in the form $t = \sqrt{2s/a}$. Substituting 200 ft for s and 1.6 ft/sec^2 for a, we find that

$$t = \sqrt{\frac{2 \times 200 \text{ ft}}{1.6 \text{ ft/sec}^2}} = 16 \text{ sec.}$$

Example: A particle of mass m rests on an inclined plane. The coefficient of friction between the plane and particle is μ. What must be the angle of the plane for the particle to be just ready to accelerate down it (Fig. 5–11)?

Solution: We resolve the weight $W = mg$ into components along OE and OF parallel and perpendicular to the plane. Thus $\measuredangle FOG = \measuredangle ABC$, since their sides are mutually perpendicular, and the maximum frictional force $f = \mu N = \mu W \cos \theta$. If the particle is just ready to accelerate down the incline, the component of the weight along OE must equal f; thus

$$W \sin \theta = \mu W \cos \theta \qquad \text{or} \qquad \tan \theta = \mu.$$

Note that since θ can be measured, this constitutes a practical method for determining μ.

The resistive forces that act on a body moving through a fluid usually vary with its speed, sometimes markedly so. A rowboat can tow quite a substantial vessel at low speed, but even the most highly powered warships have maximum speeds less than half that of an automobile. In the case of an object moving through air, the "drag" force turns out to be directly proportional to its speed over a wide range of speeds. The net force on a falling body is therefore $mg - bv$, where b is a constant that depends on the size and shape of the body and on the temperature and pressure of the air. As the body falls faster and faster, the drag bv on it increases until bv is as large as the gravitational force mg. When this happens the acceleration drops to zero, and the body continues to fall at a constant speed v_t that is known as its *terminal velocity*. The terminal velocity can be found by equating the downward force mg with the upward resistive force bv_t:

$$mg = bv_t, \qquad v_t = mg/b. \tag{5–9}$$

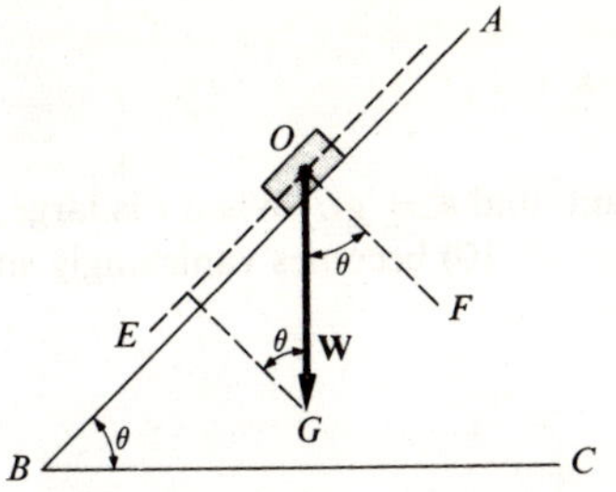

FIG. 5-11. Experiment to determine coefficient of friction of plane *AB*.

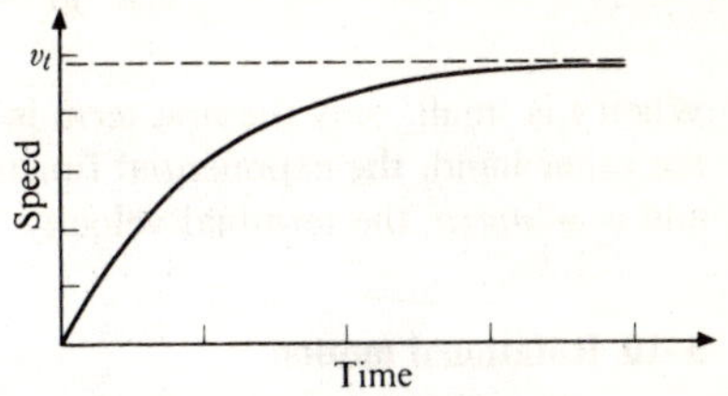

FIGURE 5-12

The terminal velocity of a parachutist is about 14 mi/hr, as compared with a terminal velocity of about 120 mi/hr for a man in free fall; the corresponding values of b are approximately 8 and 1 slug/sec, respectively.

We can determine the exact variation of speed with time for a falling body by starting from the second law of motion. Here

$$F = ma \quad \text{and} \quad mg - bv = m\frac{dv}{dt},$$

or, separating the variables v and t, we have

$$\frac{dv}{(mg/b) - v} = \frac{b}{m}\,dt.$$

Now we integrate the left-hand side from the initial speed 0 to the speed v, and the right-hand side from the initial time 0 to the time t:

$$\int_0^v \frac{dv}{(mg/b) - v} = \frac{b}{m}\int_0^t dt,$$

$$\ln\frac{mg}{b} - \ln\left(\frac{mg}{b} - v\right) = -\ln\frac{(mg/b) - v}{mg/b} = \frac{bt}{m},$$

$$v = \frac{mg}{b}(1 - e^{-bt/m}). \tag{5-10}$$

This result is plotted in Fig. 5-12.

Just after a body is dropped, its speed is low and the air resistance negligible. Hence Eq. (5-10) should reduce to $v = gt$ when t is small, which we can verify by expanding the exponential function in a power series. Since

$$e^x = 1 + x + \frac{x^2}{2!} + \frac{x^3}{3!} + \cdots,$$

we have here

$$v = gt - \frac{bg}{2m}t^2 + \cdots.$$

When t is small, only the first term is significant, and $v = gt$. When t is large, on the other hand, the exponential function in Eq. (5-10) becomes vanishingly small and $v = mg/b$, the terminal velocity.

5-10 Rotational motion

We learned in the previous chapter that, even though a particle moving with constant angular velocity has a constant linear speed, its motion is nevertheless accelerated. The magnitude of the acceleration, called the centripetal acceleration a_c, is

$$a_c = v^2/r,$$

and its direction is always toward the center of the circular path which it describes.

According to the first law of motion, the presence of an acceleration requires the presence of a force, and the second law gives its magnitude and direction. For a body in uniform circular motion, the *centripetal force* F_c which leads to its centripetal acceleration is

$$F_c = ma_c \qquad \text{or} \qquad F_c = \frac{mv^2}{r}. \tag{5-11}$$

The centripetal force F_c points toward the center of curvature of the body's trajectory.

Centripetal forces are quite common, since they must be present whenever rotational motion occurs. Everybody who has whirled a stone at the end of a string is aware that he had to apply an inward force in order to keep the stone traveling in a circle. If the string breaks, the stone will go off along a straight line tangent to the circle in which the string held it. The centripetal force is simply the force required to change the direction of motion of a particle from the straight line which it would normally follow to a curved path. Strings are not the only means of providing centripetal forces. Gravitation provides the centripetal force that keeps the planets moving around the sun and the moon around the earth. Friction between its tires and the road provides the centripetal force needed by an automobile in rounding a curve. Cohesive forces are the centripetal forces that keep a rotating flywheel together.

Example: Calculate the centripetal force required by a 3000-lb car making a turn of radius 100 ft at a speed of 15 mi/hr (Fig. 5-13).

Solution: The car's mass is

$$m = \frac{3000 \text{ lb}}{32 \text{ ft/sec}^2} = 94 \text{ slugs}.$$

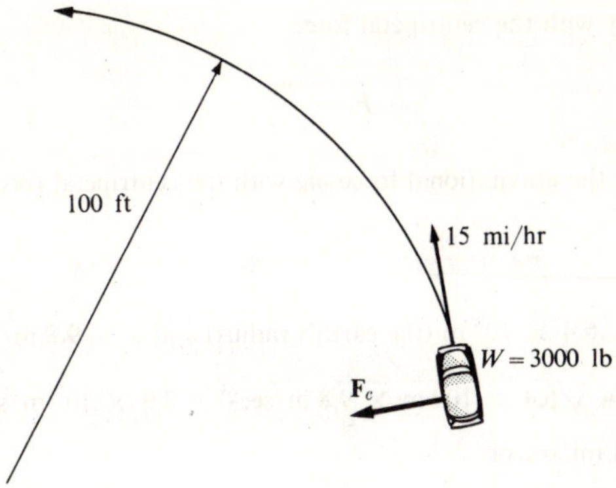

FIGURE 5–13

Since 15 mi/hr = 22 ft/sec (Appendix II), the centripetal force is

$$F_c = \frac{94 \text{ slugs} \times (22 \text{ ft/sec})^2}{100 \text{ ft}} = 455 \text{ lb}.$$

The centripetal force of 455 lb must be provided by the pavement acting on the car's tires through the agency of friction. Let us compute the minimum coefficient of friction μ that must be present if the car is to make the turn without skidding. The frictional force is

$$F_f = \mu N$$

in general. To find μ, we substitute 455 lb for F_f and 3000 lb for the normal force N, which here is the car's weight W. Thus

$$\mu = \frac{F_f}{N} = \frac{F_c}{W} = \frac{455 \text{ lb}}{3000 \text{ lb}} = 0.15$$

to two significant figures, which is readily available under good driving conditions (Table 5–1).

Example: Determine the minimum velocity an earth satellite must have in order to pursue a stable orbit without falling to the ground (Fig. 5–14).

Solution: We neglect the resistance of the atmosphere, which, even at the great altitudes of actual satellites, will ultimately bring them down. Near the earth the gravitational force on a body of mass m is its weight, so that

$$F = mg,$$

where g is the acceleration of gravity. For uniform circular motion about the earth, this force must

FIG. 5–14. The minimum speed an earth satellite must have in order to pursue a stable orbit is 7.9×10^3 m/sec, which is about 18,000 mi/hr.

provide the body with the centripetal force

$$F_c = \frac{mv^2}{r}.$$

Hence, equating the gravitational force mg with the centripetal force F_c,

$$mg = \frac{mv^2}{r}, \qquad v^2 = rg, \qquad v = \sqrt{rg}.$$

Substituting $r = 6.4 \times 10^6$ m (the earth's radius) and $g = 9.8\ \text{m/sec}^2$, we find that

$$v = \sqrt{6.4 \times 10^6\ \text{m} \times 9.8\ \text{m/sec}^2} = 7.9 \times 10^3\ \text{m/sec},$$

which is about 5 mi/sec or

$$5\,\frac{\text{mi}}{\text{sec}} \times 60\,\frac{\text{sec}}{\text{min}} \times 60\,\frac{\text{min}}{\text{hr}} = 18{,}000\,\frac{\text{mi}}{\text{hr}}\,!$$

(The correct value of v is somewhat smaller because g is smaller than $9.8\ \text{m/sec}^2$ at the position of an orbiting satellite.)

5-11 Banked turns

As we saw, the friction between the tires and the road provides a car with the centripetal force needed for it to make a turn. What provides a turning airplane with centripetal force? To answer this question we begin by noting that an airplane flies by virtue of the upward aerodynamic force created by the flow of air past its wings. If the airplane is tilted to one side, the force is still perpendicular to the wings and therefore is tilted as well (Fig. 5-15). Only the vertical component of the aerodynamic force supports the airplane, and its horizontal component is available to furnish centripetal force. To make a turn, an airplane simultaneously tilts (or "banks") in the desired direction and moves its rudder to change its heading. If it only does the latter, it will slip sidewise through the air until eventually air resistance enables it to complete the turn. Since little centripetal force is available in this way, such turns have larger radii than banked turns

From Fig. 5-15 we can determine the proper banking angle. If the aerodynamic lift on the wings is **F**, its horizontal component when the airplane is tilted by the angle θ is $F \sin\theta$. This component furnishes the centripetal force mv^2/r, where m is the airplane's mass, v its speed, and r the turn radius. Thus

$$F \sin\theta = mv^2/r.$$

The vertical component $F\cos\theta$ supports the airplane's weight mg, and so

$$F\cos\theta = mg.$$

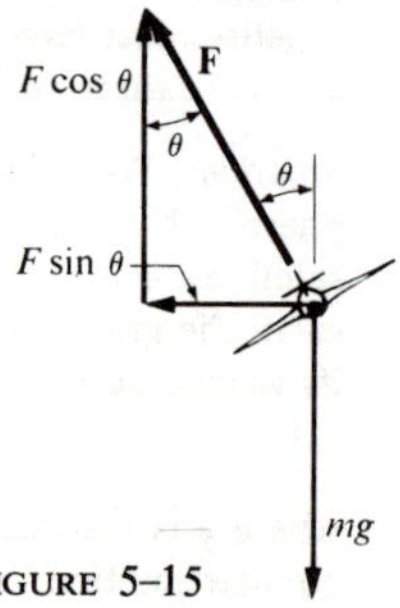

FIGURE 5-15

By dividing the first of the above equations by the second, we obtain

$$\tan\theta = v^2/rg.$$

The tangent of the banking angle θ varies directly with the square of the airplane's speed and inversely with the radius of the turn.

The corresponding calculation of the angle at which a curve in a road should be banked if cars are to stay on the road even when it is slippery (the coefficient of friction $\mu = 0$) yields the same formula as that above. In this case the force on the car is the reaction force supplied by the road, and is perpendicular to the road.

5-12 Rotating coordinate systems

There is another way of looking at the forces acting on a rotating particle. Suppose that the particle is at rest with reference to a laboratory which is rotating, and that the inhabitants of the laboratory are unaware of the rotation. (This is not an unlikely situation; rather a long time passed before man realized that the earth revolves daily on its axis.) From outside the laboratory a stationary observer would say that in order for the particle (Fig. 5-16) to keep rotating so that it does not change its position in the laboratory, a centripetal force mv^2/r must be acting in the direction of the center of rotation O a distance r from the particle. In the diagram the centripetal force is shown supplied by a string attached between O and the particle.

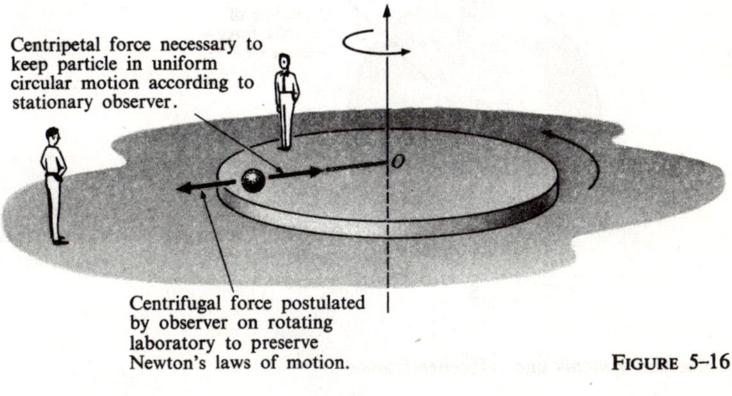

FIGURE 5-16

To a person in the laboratory things seem quite different. He observes that a tension must be maintained in the string in order that the particle stay at rest with respect to him. If the observer in the rotating coordinate system wishes to use Newton's laws of motion, he must postulate the existence of a force equal and opposite to the centripetal force. Only then can he say that a body is at rest with respect to his frame of reference when the sum of the forces acting on it is equal to zero. The force he postulates,

$$F_{c'} = -\frac{mv^2}{r},$$

balances the tension in the string. It is called the *centrifugal force* and always seems to act *away* from the center of rotation. Apparent forces of this kind, which come into play when the observer is himself rotating, are called *inertial forces.* When a passenger in a car is "thrown" to one side when a sharp turn is made, the passenger is in a rotating frame and thinks centrifugal force pushes him outward. The observer on the ground concludes that the thrown passenger moves in a straight-line path because there is no centripetal force to keep him moving in a circular one. In the next section we shall explore what types of coordinate systems do not require the introduction of inertial forces.

If a body is moving with respect to a rotating frame of reference, there is an additional inertial force which comes into play, over and above the centrifugal force. It is called a *Coriolis force* and is equal to $-2m\boldsymbol{\omega} \times \mathbf{v}$, where $\mathbf{v}$ is the velocity in the rotating frame. Thus a body falling near the surface of the earth in the northern hemisphere will have an easterly drift due to the Coriolis force. (See Fig. 5-17.)

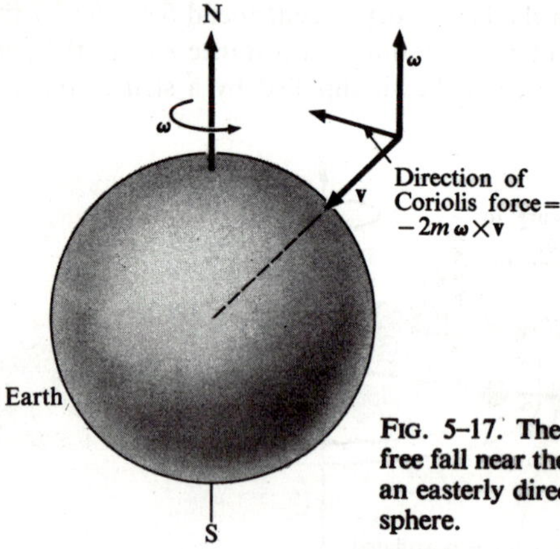

FIG. 5-17. The Coriolis force for a body in free fall near the surface of the earth acts in an easterly direction in the northern hemisphere.

5-13 Inertial systems and reference frames

We have seen that the laws of motion are different in different frames of reference. In a stationary frame, the force in the equation

$$\mathbf{F} = m\mathbf{a} \tag{5-12}$$

is the total external real force; while if the phenomena are observed from a rotating frame, fictitious inertial forces such as centrifugal forces must be introduced in order to account for the apparent accelerations. The choice of a coordinate sys-

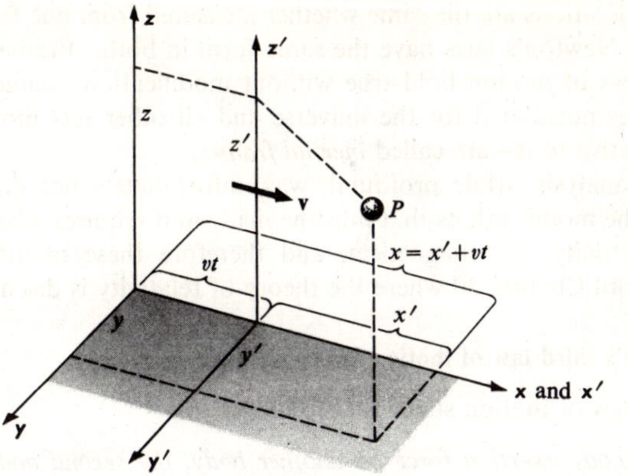

FIG. 5-18. Relation of coordinates in two inertial frames in relative motion in the x-direction.

tem from which to make measurements is obviously of great importance if the results are to be assessed properly.

The problem of finding a coordinate system in which Eq. (5-12) holds exactly as it stands troubled Newton. This system cannot be attached to the earth, since the earth is rotating. But how do we know that the earth rotates without direct reference to a fixed set of coordinates? The answer is that, since inertial forces must be invented to account for the behavior of bodies on the earth, it must be rotating. Any statement about motion implies the existence somewhere of a fixed set of coordinates, and therefore Newton could legitimately say that there must be such a set in the universe in which Eq. (5-12) is valid.

This fixed frame is not the only one in which the laws of motion hold, however. These laws describe forces which can be measured only in terms of the accelerations they produce; hence experiments performed in other frames of reference moving with constant velocities with respect to the fixed frame yield the same laws. Consider the axes x', y', z', moving with the velocity v parallel to the x-axis of the fixed frame x, y, z (Fig. 5-18). The coordinates of a point in the fixed frame are related to its coordinates in the moving frame by the equations

$$x' = x - vt, \qquad y' = y, \qquad z' = z.$$

Taking the second derivative with respect to time on both sides,

$$\frac{d^2x'}{dt^2} = \frac{d^2x}{dt^2}, \qquad \frac{d^2y'}{dt^2} = \frac{d^2y}{dt^2}, \qquad \frac{d^2z'}{dt^2} = \frac{d^2z}{dt^2},$$

so that

$$\mathbf{a}' = \mathbf{a}.$$

Because accelerations are the same whether measured from one frame of reference or the other, Newton's laws have the same form in both. Frames of reference in which the laws of motion hold true without modification—namely, the fixed set of coordinates postulated for the universe and all other sets moving at constant velocities relative to it—are called *inertial frames.*

Newton's analysis, while profound, was unfortunately not quite correct. To understand the modifications that must be made in it requires a background in the ideas of electricity and magnetism, and therefore these modifications will be postponed until Chapter 30 where the theory of relativity is discussed.

5-14 Newton's third law of motion

The third law of motion states that

When one body exerts a force on another body, the second body exerts a force on the first body of the same magnitude but in the opposite direction.

In vector notation, we can write this as

$$\mathbf{F}_{AB} = -\mathbf{F}_{BA}. \tag{5-13}$$

This law is important in analyzing the behavior of systems of particles, and also enables us to compare masses without resorting to calibrated springs. Before going any further, though, we must be sure that we understand the subtleties it contains.

Let us apply the third law, also known as the law of action and reaction, to the situation of a horse pulling a cart. For the sake of simplicity we shall consider both of them as particles, which is reasonable since all we are interested in is their effect on one another. According to the third law, the force the horse exerts on the cart is matched by an equal and opposite force exerted by the cart on the horse. At first glance the forces, being equal and opposite, seem to cancel out and leave no net force at all. Further thought reveals that it is only the force on the cart which determines its motion, and that the force exerted by the cart on the horse affects the horse alone. Hence if the horse is able to apply enough force to overcome the frictional forces present, the cart will accelerate.

In general, the forces of action and reaction cannot balance out, since they act on different objects.

Another example is that of a body near the earth's surface. The earth acts on it with a force causing it to experience the acceleration g of 32 ft/sec^2, or 9.8 m/sec^2. Simultaneously the body exerts the same force on the earth. Hence the ratio of the accelerations produced is inversely proportional to the ratio of the masses of the body and the earth. Because the earth is so much larger, the reaction force on it causes too small an acceleration to be detected.

Suppose that we permit two masses to collide with one another. According to the third law the magnitudes of the forces exerted on them during the collision are the same. From the second law of motion, the ratio of their accelerations, which can be measured, is inversely proportional to their mass ratio. As before, by having one of the colliding bodies be a unit mass, we can compare the other mass with it. Here the comparison is direct, and no spring or balance is needed.

We might conceivably find ourselves on a frozen lake with a perfectly smooth surface. Now we cannot walk because the absence of friction prevents us from exerting a backward force on the ice which would produce a forward force on us. But what we can do is exert a force on some object we may have with us, say a rock. We throw the rock forward by applying a force to it; at the same time the rock is pressing back on us with the identical force but in the opposite direction, and in consequence we find ourselves moving backward (Fig. 5–19).

FIGURE 5–19

We shall learn in Chapter 9 how these notions are expressed in the *law of conservation of linear momentum,* one of the most useful formulations of the laws of motion.

Newton's three laws of motion have been presented here as axioms, more like the axioms of geometry than ideas applicable to a living science. This has been done only to make their meaning clear. The formulation of these laws by Newton actually is a classic example of the application of the scientific method. Newton found that the laws of motion, together with the law of gravitation which we shall consider later, could account for all the planetary data then in existence. He then made some predictions which were subsequently verified by experiment and observation. No discrepancies were uncovered between theory and practice until about sixty years ago, when certain subatomic and large-scale astronomical phenomena apparently contradicted the laws of motion and of gravitation as Newton had formulated them. These laws therefore had to be revised in ways to be described in Chapter 30. For familiar terrestrial phenomena they are still entirely adequate, however, and we shall assume their correctness unless we state otherwise.

Problems

1. A force of 20 n acts on a body whose mass is 4 kg. (a) What is the weight of the body? (b) What is its acceleration?

2. A force of 100 lb acts on a body whose mass is 1.5 slugs. (a) What is the weight of the body? (b) What is its acceleration?

3. A force of 20 n acts on a body whose weight is 8 n. (a) What is the mass of the body? (b) What is its acceleration?

4. A force of 100 lb acts on a body whose weight is 96 lb. (a) What is the mass of the body? (b) What is its acceleration?

5. An elevator having a mass of 1500 kg descends with an acceleration of 1 m/sec^2. What is the tension in its supporting cable?

6. An unloaded truck weighing 4000 lb has a maximum acceleration of 1 ft/sec^2. What is its maximum acceleration when it is carrying a 2000-lb load?

7. An 80-lb kangaroo exerts a constant force on the ground in the first 2 ft of its jump, and rises 6 ft higher. When it carries a baby kangaroo in its pouch, it can rise only $5\frac{1}{2}$ ft higher. What is the weight of the baby kangaroo?

8. When a force equal to its weight is applied to a body, what is its acceleration?

9. A 200-lb man stands on a scale in an elevator. What does the scale read (a) when the elevator is ascending with an acceleration of 3 ft/sec^2; (b) when it is descending with an acceleration of 3 ft/sec^2; (c) when it is ascending at the constant speed of 10 ft/sec; (d) when it is descending at the constant speed of 10 ft/sec; (e) when the cable has broken and it is descending in free fall?

10. A sled whose weight is 5 lb is pulled by a rope which makes an angle of 45° with the horizontal. The coefficient of friction is 0.08. What must be the force exerted on the rope to give the sled an acceleration of 2 ft/sec^2?

11. A wooden crate weighing 240 lb rests on a level wooden floor. What is the minimum force required to move it horizontally?

12. The coefficient of friction between a rubber tire and a dry concrete road is 0.7. (a) What is the maximum acceleration possible for a car on such a road? (b) What is the minimum distance in which a car on such a road can be stopped when it is moving at 60 mi/hr?

13. A 5-kg block resting on a horizontal surface is attached to a 5-kg block that

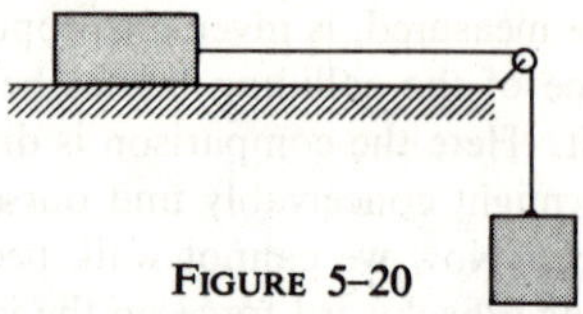

FIGURE 5–20

hangs freely by a string passing over a pulley, an arrangement similar to that shown in Fig. 5–20. (a) If there is no friction between the first block and the surface, what is the block's acceleration? (b) If the coefficient of friction between the first block and the surface is 0.2, what is the block's acceleration?

14. A 10-kg block resting on a horizontal surface is attached to a 5-kg block that hangs freely by a string passing over a pulley, an arrangement similar to that shown in Fig. 5–20. (a) If there is no friction between the first block and the surface, what is the block's acceleration? What is the block's acceleration if the coefficient of friction between the first block and the surface is (b) 0.4? (c) 0.5? (d) 0.6?

15. A 100-lb block resting on a horizontal surface is attached to a 50-lb block that hangs freely by a string passing over a pulley, an arrangement similar to that shown in Fig. 5–20. The coefficient of friction between the first block and the surface is 0.4. (a) How long does it take the blocks to move 20 ft starting from rest? (b) What is their speed at that time?

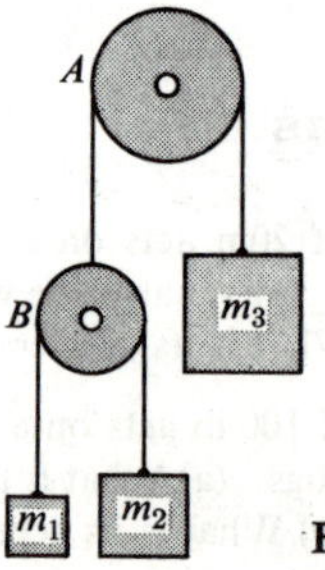

FIGURE 5–21

16. Three masses are suspended by strings on either side of pulleys as shown in Fig. 5–21. The weights of the masses are:

$$m_1 = 1\text{ kg},\ m_2 = 2\text{ kg},\ m_3 = 3\text{ kg}.$$

The pulleys are massless and the strings inextensible. What is the acceleration of each mass?

17. A crate weighing 500 lb is being slid down a ramp which makes an angle of 20° with the horizontal. The coefficient of friction is 0.3. How much force parallel to the plane must be applied to the crate if it is to slide down at constant speed? In which direction must the force be applied?

18. A car is stuck in the mud. To get it out, the driver ties one end of a rope to the car and the other to a tree 100 ft away. He then pulls sideways on the rope at its midpoint. If he exerts a force of 120 lb, how much force is applied to the car when he has pulled the rope 5 ft to one side?

19. A 1-kg bird sits on a telephone wire midway between two poles 50 m apart. The wire, assumed weightless, sags by 10 cm. What is the tension in the wire?

20. Find the tension in AB and BC (Fig. 5–22).

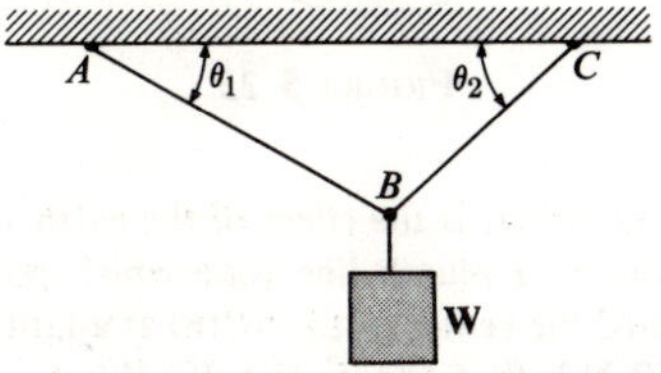

FIGURE 5–22

21. In Fig. 5–23, $W = 50$ lb. What must be the magnitude of **F** so that BC is vertical and the angle $CAB = 60°$?

22. A bucket of mass M is initially filled with water of mass m. The water leaks out at a constant rate, so that the mass of the water remaining in the bucket at any time is given by the formula $m_0 = m(T - t)$, for $t < T$. It is pulled up with a constant force F. What is the velocity of the bucket at the time T?

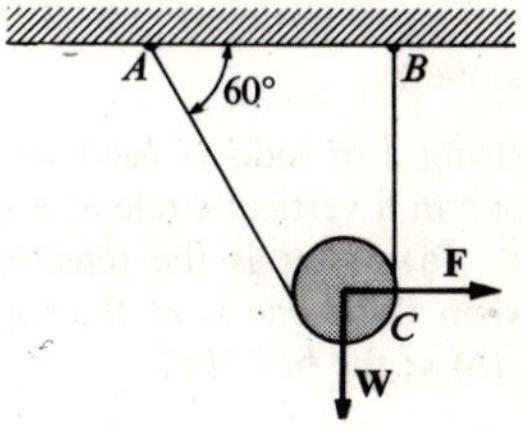

FIGURE 5–23

23. A conical pendulum consists of a mass of 1 kg supported by a rope. It rotates in a horizontal plane with an angular velocity of 4π rad/sec, with the angle of the cone equal to 60°, in a circular path of radius 25 cm (Fig. 5–24). What is the tension in the rope?

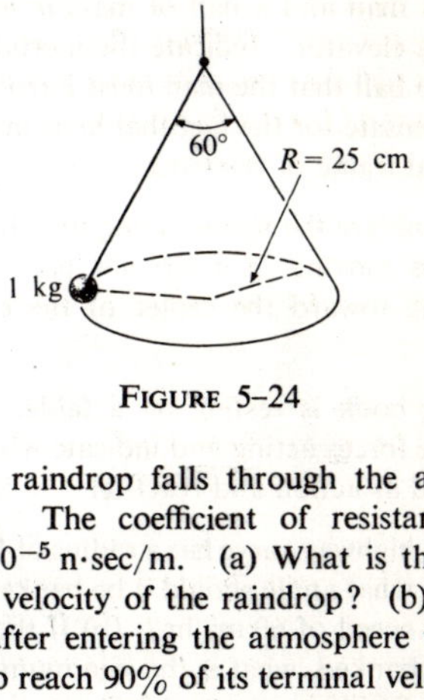

FIGURE 5–24

24. A raindrop falls through the atmosphere. The coefficient of resistance is $\frac{1}{4} \times 10^{-5}$ n·sec/m. (a) What is the terminal velocity of the raindrop? (b) How long after entering the atmosphere will it take to reach 90% of its terminal velocity?

25. What is the centripetal force needed to keep a 3-kg mass moving in a circle of radius 0.5 m at a speed of 8 m/sec?

26. What is the centripetal force needed to keep a 12-lb weight moving in a circle of radius 2 ft at a speed of 10 ft/sec?

27. A certain string 4 ft long breaks when its tension is 40 lb. (a) What is the greatest speed in feet/second at which it can be used to whirl a stone weighing 3 lb? (b) Express this speed in revolutions/second. (Neglect the gravitational pull of the earth on the stone.)

28. A string 1 m long is used to whirl a $\frac{1}{2}$-kg stone in a vertical circle at a speed of 5 m/sec. (a) What is the tension in the string when the stone is at the top of the circle? (b) at the bottom?

29. A string 2 ft long is used to whirl a 5-lb stone in a vertical circle. What must the minimum speed of the stone be if the string is to be just taut when the stone is at the top of the circle?

30. A string 1.5 m long is used to whirl a 1-kg stone in a vertical circle at a speed of 5 m/sec. What is the tension in the string when it is horizontal?

31. A man and a ball of mass m are in a falling elevator. Indicate the inertial force on the ball that the man must introduce to compensate for the fact that he is in a noninertial frame of reference.

32. Analyze the forces acting on a bicyclist who is moving in a circular path and is leaning toward the center of the circular path.

33. A book is resting on a table. Show all the forces acting and indicate which related as action and reaction.

34. A highway curve has a radius of 1000 ft. (a) At what angle should it be banked for a traffic speed of 60 mi/hr? (b) If the curve is not banked, what is the minimum coefficient of friction required between tires and road?

35. An airplane traveling at 300 mi/hr banks at an angle of 45° as it makes a turn. What is the radius of the turn in miles?

36. A car whose speed is 60 mi/hr rounds a curve 600 ft in radius which is properly banked for a speed of 30 mi/hr. Find the minimum coefficient of friction between tires and road that will permit the car to make the turn. (Assume that the car is a particle.)

37. What should the length of the day be if the earth is to rotate fast enough for g to be zero at the equator?

38. A test tube has a sleeping fly resting on the bottom. The tube is dropped and the fly takes off and escapes. Compare the time it takes for the test tube to fall with the speed of another falling tube which has no fly in it.

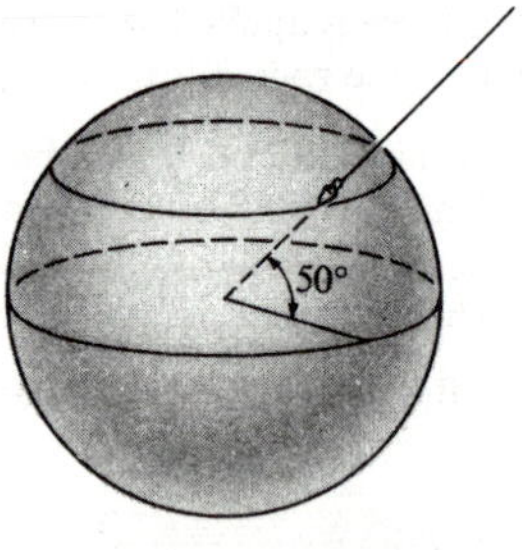

FIGURE 5–25

39. (a) What is the effect of the earth's rotation on a plumb line (one which points toward the center of the earth) at a latitude of 50°N? (See Fig. 5–25.) (b) What is the effect on the point of impact of a body falling from the top of a building? (Give a qualitative answer only; but include the effect of the Coriolis force, if any.)

6

Work and Power

One of the chief distinctions between the physical sciences and nearly all other disciplines is that in the former certain very general conservation theorems have been found to be valid. A conservation theorem states that, no matter what changes a system of some kind undergoes, a certain quantity keeps the same value it had originally.

In our ordinary experiences in life we frequently assume (often incorrectly) that some conservation theorem applies. An example can be found in the training of athletes, mainly distance runners. Many years ago a distance runner would taper off his training before a big meet. It was felt that he would "burn himself out." This really implied that some conservation theorem applies, namely a conservation of strength. His burning himself out meant that there was a limited amount of strength that he had available and that if he were to use it up before a meet he would not perform so well. In more recent years coaches have found that rigorous training up to the date of the meet makes athletes perform better. This would indicate that strength is not among quantities in the body that are conserved, but that it is self-regenerative.

Given one or more conservation theorems applicable to a certain system, we can immediately determine which classes of events can take place in the system and which cannot. Often it is possible to draw some conclusions about the behavior of the particles composing a system without a detailed investigation, basing our analysis simply on the conservation of some particular quantities. The power of this method of approach is exemplified by the great success of physics, a field in which a considerable variety of quantities may, under appropriate circumstances, be conserved: energy, linear momentum, angular momentum, electrical charge, and others, all of which we shall consider in this book.

In the next chapter we shall concern ourselves with conservation of energy. As a preliminary step, however, we must learn about a related quantity, *work*. Work is one of those words whose popular meaning differs from its scientific meaning, and we must be careful to keep the two distinct in our minds.

6-1 Work

Consider a constant force acting through the distance s on a free particle capable of moving in only one direction. From Newton's first law, we have that $F = ma$. We wish to find the final velocity in terms of the force. For this purpose we use Eq. (3-12),

$$v_f^2 = v_i^2 + 2as, \qquad a = \frac{1}{2s}(v_f^2 - v_i^2),$$

where v_i is the particle's initial speed, v_f its final speed, and a its acceleration. Substituting into Newton's first law, we have

$$F = \frac{1}{s}\left(\frac{mv_f^2}{2} - \frac{mv_i^2}{2}\right),$$

$$Fs = \tfrac{1}{2}mv_f^2 - \tfrac{1}{2}mv_i^2. \tag{6-1}$$

Equation (6-1) indicates that if the product Fs is zero, the quantity $\frac{1}{2}mv^2$ for the particle is conserved, since from Eq. (6-1) we would have

$$\tfrac{1}{2}mv_f^2 = \tfrac{1}{2}mv_i^2.$$

In another example, we consider the effect of a force when it is exerted on a free particle near the surface of the earth (Fig. 6-1). We have that $F - mg = ma$. Again

$$a = \frac{1}{2s}(v_f^2 - v_i^2)$$

and

$$Fs = mgs + \tfrac{1}{2}mv_f^2 - \tfrac{1}{2}mv_i^2.$$

From Fig. 6-1, $s = h_f - h_i$, and

$$Fs = mgh_f - mgh_i + \tfrac{1}{2}mv_f^2 - \tfrac{1}{2}mv_i^2$$

$$= (mgh_f + \tfrac{1}{2}mv_f^2) - (mgh_i + \tfrac{1}{2}mv_i^2). \tag{6-2}$$

Now we see that if the product Fs is zero,

$$mgh_f + \tfrac{1}{2}mv_f^2 = mgh_i + \tfrac{1}{2}mv_i^2. \tag{6-3}$$

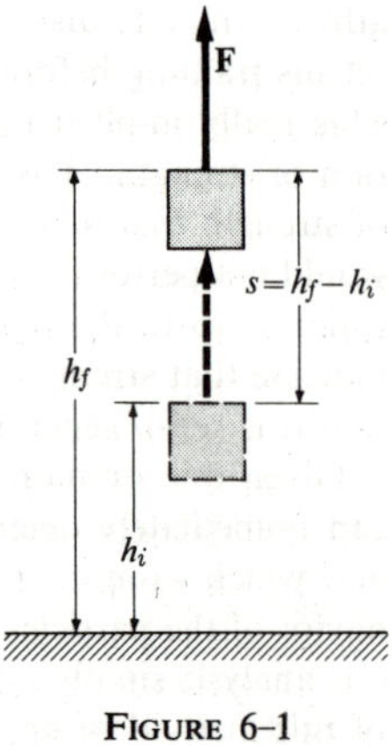

FIGURE 6-1

The quantity $mgh + \frac{1}{2}mv^2$ is conserved, since it is the same at the beginning and at the end of the process. We shall see in the next chapter that $\frac{1}{2}mv^2$ is an important quantity called *kinetic energy*. The quantity mgh is a form of *potential energy*.

These two examples suggest that we examine the quantity Fs, which seems to play an important role in the conservation or nonconservation of energy; the difference in energy in both cases is equal to the product of Fs. This quantity, the product of a force times a displacement of the force, is called *work*:

$$W = Fs. \tag{6–4}$$

Unless the force acts through a distance, no work is done. [A man pushing against a wall would get tired and would think he was doing work even if no displacement occurred. The mechanical work he would be doing according to (6–3) would be zero, however.] One qualification must be made in the definition of work. If we consider the centripetal acceleration for a free particle, even though the centripetal force mv^2/R acts during a displacement, the speed v of the particle, and the $\frac{1}{2}mv^2$ of the particle—namely its kinetic energy—do not change. The reason for this is that the force is perpendicular to the displacement. An arbitrary force can always be resolved into two components, one parallel and one perpendicular to the displacement of the particle, and we see that only forces in the direction of the displacement contribute to the work. We shall consider two- and three-dimensional problems, where the distinction between parallel and perpendicular forces is important, later in the chapter.

The unit in which work is expressed follows from its definition. In the mks system of units, W is given in newton-meters; because of the frequency with which this combination occurs, it has been given the special name of the *joule*. That is,

$$1 \text{ joule} = 1 \text{ newton}\cdot\text{meter}.$$

In the British system work is given in foot-pounds (ft·lb).

Example: A man pushes an automobile with a force of 60 lb for a distance of 200 ft. How much work does he perform?

Solution: From Eq. (6–1), we have

$$W = Fs = 60 \text{ lb} \times 200 \text{ ft} = 1200 \text{ ft}\cdot\text{lb}.$$

It is worth repeating that there must be a displacement in order for work to be done; if the man had pushed against the car with the same force but had been unable to move it, no work would have been done no matter how long he applied the force. Similarly, no work is done if motion occurs without the application of a force in the direction of motion; for instance, when a bowling ball rolls along an alley.

Example: How much work is needed to lift a brick of mass 2 kg a height of 10 m above the ground?

Solution: The force that must be applied in order to lift the brick is that required to overcome the force with which gravity pulls it down. The latter force is the weight of the brick mg, so that the work involved is

$$W = mg \times h,$$

where h is the height to which the brick is lifted. Since m is given as 2 kg and h as 10 m, and we know that $g = 9.8 \text{ m/sec}^2$, then

$$W = 2 \text{ kg} \times 9.8 \text{ m/sec}^2 \times 10 \text{ m} = 196 \text{ joules}.$$

6–2 Power

Very often we wish to know the *rate* at which something is being accomplished. Thus we find that speed—the time rate at which a particle is traveling—is a very useful concept. In the same way the time rate at which a force does work turns out to be a quantity that frequently finds application in physics. This rate is called *power*, symbol P, and is defined by the equation

$$P = dW/dt. \tag{6–5}$$

Equation (6–5) gives the *instantaneous power*, the rate at which work is being done at some particular moment. It is sometimes convenient to speak of *average power* $\overline{P}$, which is the total work ΔW done in some interval of time Δt divided by that interval:

$$\overline{P} = \Delta W/\Delta t. \tag{6–6}$$

Special units have been assigned to power in both systems of units. In the mks system power is expressed in *watts*, where

$$1 \text{ watt} = 1 \text{ joule/sec}.$$

The watt is rather small for most purposes, and is often replaced by a unit 1000 times larger, the *kilowatt*, symbol kw. Thus

$$1 \text{ kilowatt} = 1000 \text{ watts}.$$

In the British system the traditional unit of power is the *horsepower*, symbol hp, where

$$1 \text{ hp} = 550 \text{ ft·lb/sec}.$$

The horsepower is exclusively an engineering unit, while the watt and kilowatt are used in scientific work as well. The relationship between the two kinds of power units is

$$1 \text{ hp} = 746 \text{ watts}.$$

In the event that a constant force is being applied,

$$P = \frac{dW}{dt} = \frac{d}{dt}(Fs) = F\frac{ds}{dt}\cdot$$

Since ds/dt is the velocity v of the particle upon which the force F is acting,

$$P = Fv. \tag{6–7}$$

The rate at which a constant force does work on a particle is equal to the magnitude of the force multiplied by the velocity of the particle. (We are still speaking of one-dimensional motion, with both **F** and **v** in the same direction.)

Example: A certain motorboat requires 160 hp in order to move through the water at the constant speed of 15 mi/hr (22 ft/sec). What is the force with which the water is resisting the motion of the boat at this speed?

Solution: We can solve $P = Fv$ for F, obtaining $F = P/v$. Since 1 hp = 550 ft·lb/sec,

$$P = 160 \text{ hp} \times 550 \frac{\text{ft·lb/sec}}{\text{hp}} = 88{,}000 \frac{\text{ft·lb}}{\text{sec}},$$

and

$$F = \frac{P}{v} = \frac{88{,}000 \text{ ft·lb/sec}}{22 \text{ ft/sec}} = 4{,}000 \text{ lb}.$$

The boat must overcome a retarding force of 2 tons in order to travel at 15 mi/hr.

6-3 Work done by variable forces

In practice we seldom encounter constant forces. In the event that the applied force varies with the displacement x of the particle on which it acts, we can plot F as a function of x, as in Fig. 6-2. Let us divide the total distance through which the particle moves into a large number of intervals Δx long. If Δx is small enough, F is nearly constant within each interval even though it changes from one interval to the next. The work ΔW_i done by the force in propelling the body from the position x_i to the adjacent position $x_i + \Delta x$ is therefore

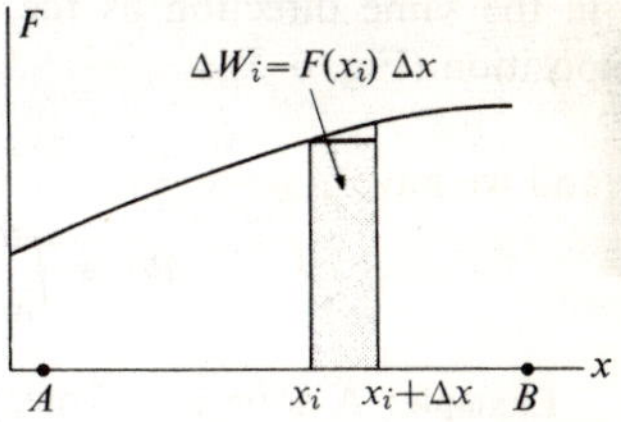

FIG. 6-2. How work is defined when a varying force acts.

$$\Delta W_i = F(x_i)\,\Delta x,$$

where $F(x_i)$ is the average value of the force in the interval between x_i and $x_i + \Delta x$. The total amount of work done in going from A to B is the sum of the amounts of work ΔW_i done in each of the intervals Δx. If there is a total of N of these intervals—in other words, if the distance between A and B has been divided into N parts—then

$$W = \sum_{i=1}^{i=N} \Delta W_i = \sum_{i=1}^{i=N} F(x_i)\,\Delta x. \tag{6-8}$$

Since the force F actually varies continuously in Fig. 6-2, the smaller we make our intervals Δx the more accurate our value for W will be. Finally, as Δx is made smaller and smaller and the number of intervals N approaches infinity, the finite sum of Eq. (6-8) becomes an integral:

$$W = \lim_{N\to\infty} \sum_{i=1}^{i=N} F(x_i)\,\Delta x = \int_A^B F(x)\,dx. \tag{6-9}$$

Equation (6-9) is very helpful if we happen to have an analytical expression for the manner in which the force $F(x)$ varies with distance x. There are two extremely important cases of this sort: (1) when the force varies linearly with the distance away from some particular point, and (2) when the force varies inversely with the square of the distance away from some particular point. An example of case (1) is an ordinary spring. When a spring is either stretched or compressed by an amount x from its normal position, the force that acts to pull or push the spring back again is directly proportional to x. That is, we can write for this force

$$F = -kx, \tag{6-10}$$

where the minus sign indicates that the force is in the direction opposite to the displacement x. [Equation (6-10) is often called *Hooke's law*.] The constant k refers to the stiffness of the spring, or its ability to resist being pulled out or compressed. Thus a spring might have a k-value of 4 lb/ft, meaning that a force of 4 lb must be applied for each foot of compression or extension.

How much work must we perform in order to stretch a spring by the amount L? The force we apply must be directed oppositely to the force of the spring [Eq. (6-10)], which resists any change in the spring. The force that we apply is in the same direction as the displacement x of the spring from its equilibrium position (Fig. 6-3),

$$F(x) = kx,$$

and we have

$$W = \int_A^B F(x)\,dx = \int_0^L kx\,dx = \tfrac{1}{2}kL^2. \tag{6-11}$$

Example: A scale is so constructed that its spring is compressed by $\frac{1}{2}$ in. for every pound of weight placed on it. How much work is done in compressing the spring by a 12-lb weight that is placed on the scale?

Solution: The first step is to determine the value of the spring constant k. From Eq. (6-10), we have

$$k = \frac{F}{x} = \frac{1\ \text{lb}}{\frac{1}{24}\ \text{ft}} = 24\ \text{lb/ft},$$

since $\frac{1}{2}$ in. $= \frac{1}{24}$ ft. Next, we must determine the amount by which the spring is compressed by the 12-lb weight. Again making use of Eq. (6-10), substituting 12 lb for the

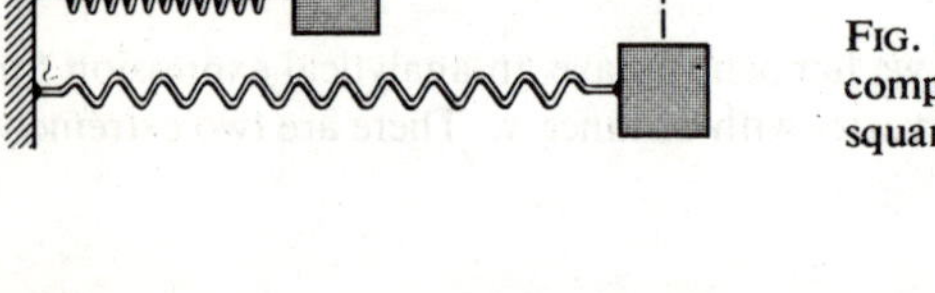

FIG. 6-3. The work required to stretch or compress a spring is proportional to the square of the displacement.

force F and 24 lb/ft for k, we obtain

$$x = \frac{F}{k} = \frac{12\text{ lb}}{24\text{ lb/ft}} = \tfrac{1}{2}\text{ ft}.$$

Now we simply substitute $\frac{1}{2}$ ft for L in Eq. (6–11), and obtain

$$W = \tfrac{1}{2}kL^2 = \frac{24\text{ lb/ft} \times (\frac{1}{2}\text{ ft})^2}{2} = 3\text{ ft·lb}.$$

At first glance it might seem that we could use the definition of work, $W = Fx$, directly in the preceding problem, substituting 12 lb for F and $\frac{1}{2}$ ft for x. The result would be 6 ft·lb, twice the correct value. The reason that we would obtain the wrong answer by proceeding in this way is that we would be neglecting the fact that it is the *resultant force* of the weight pushing down and the spring pushing upward against it that is doing work; the weight is always the same 12 lb, but the upward force of the spring varies from 0 at the start of the motion to 12 lb when it is fully compressed. Because the resultant force varies with the amount of compression, we must perform an integration, as we did in deriving Eq. (6–11).

The second important case of a force which varies with distance is the inverse-square force, which may be written

$$F(x) = k/x^2. \tag{6–12}$$

This force may be either attractive, implying that the body it acts upon is being attracted toward the center of force, or repulsive, meaning that the body is being repelled from the center of force, depending on the sign of k. Examples of inverse-square forces are gravitational forces, which are invariably attractive, and electrostatic ones, which may be either attractive or repulsive, depending on the circumstances.

Let us calculate the work done in carrying a particle from x_1 to x_2 under the influence of an inverse-square force. We have

$$W = \int_A^B F(x)\,dx = k\int_{x_1}^{x_2} \frac{dx}{x^2} = \frac{k}{x_1} - \frac{k}{x_2}. \tag{6–13}$$

Example: The force of gravity with which the earth attracts objects is an inverse-square one. The force is always directed radially inward toward the earth's center, with x measured from that point, so that F and x are parallel. Find the work that must be done by this force when a 100-kg mass from the moon, approximately 3.8×10^8 m away, falls to the earth's surface, approximately 6.4×10^6 m from its center. (Neglect the force exerted by the moon.)

Solution: We must first determine the constant k in Eq. (6–12). A mass of 100 kg means that the corresponding weight, which is the gravitational force on the mass at the earth's surface, is 980 n, and, since $x = 6.4 \times 10^6$ m there,

$$k = Fx^2 = 980\text{ n} \times (6.4 \times 10^6\text{ m})^2 = 4 \times 10^{16}\text{ n·m}^2.$$

Since the force is attractive, k is negative. Next we substitute $x_2 = 6.4 \times 10^6$ m and $x_1 = 3.8 \times 10^8$ m into Eq. (6–13), and find that

$$W = \frac{k}{x_1} - \frac{k}{x_2} = -\frac{4 \times 10^{16}\ \text{n·m}^2}{6.4 \times 10^6\ \text{m}} + \frac{4 \times 10^{16}\ \text{n·m}^2}{3.8 \times 10^8\ \text{m}} = 6.1 \times 10^9\ \text{joules}.$$

This amount of work is equivalent to that done by a 100-hp engine running at full power for 24 hr. (In making this calculation, we have neglected the work needed to overcome the frictional retardation of the earth's atmosphere.)

6–4 Work in two and three dimensions

Let us now remove the restriction that the force and the displacement through which it acts be in the same direction. We shall begin with a situation involving a constant force, and consider first only motions in a plane containing the line of action of the force.

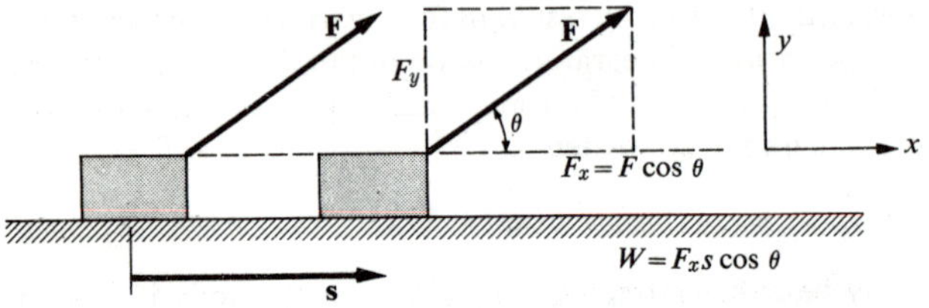

FIG. 6–4. When a force and its displacement are not parallel, the work done is the product of the displacement and the component of the force in the direction of the displacement.

As shown in Fig. 6–4, we take the direction of the displacement as the x-axis. The applied force **F** has a component in the x-direction and one in the y-direction, perpendicular to the displacement. Since we have defined work in one dimension as the product of a force and the displacement through which it acts when the two lie along the same line, it follows that in the present case the work done by **F** will be the product of its component along the x-axis, denoted F_x, multiplied by the displacement x. The y-component of **F**, F_y, does no work since it undergoes no displacement along its line of action. Since $F_x = F\cos\theta$, for the situation shown in Fig. 6–4 we have

$$W = F_x s = Fs\cos\theta.$$

As a general rule the work done by a force **F** is equal to the product of the displacement **s** it undergoes and the component of **F** in the direction of **s**. As we recall from Chapter 2, this is precisely what is described by the scalar ("dot") product, so that we may write

$$W = \mathbf{F} \cdot \mathbf{s}. \qquad (6\text{–}14)$$

Equation (6–14) holds in three dimensions as well as in two. It should be added, as is now clear from its definition as a scalar product, that work is a scalar quantity.

When the applied force is not constant, we proceed exactly as we did in the one-dimensional case. First we break the displacement up into infinitesimal portions $d\mathbf{s}$, in each of which we can consider $\mathbf{F}$ as constant, and then we add up the contributions

$$dW = \mathbf{F} \cdot d\mathbf{s}$$

over the entire displacement to give

$$W = \int_A^B \mathbf{F} \cdot d\mathbf{s}. \tag{6–15}$$

When we actually use Eq. (6–15), it is often necessary to perform three integrations corresponding to the three components of $\mathbf{F}$ and $d\mathbf{s}$. Thus in cartesian coordinates we would write Eq. (6–15) as

$$W = \int_A^B F_x\,dx + \int_A^B F_y\,dy + \int_A^B F_z\,dz. \tag{6–16}$$

Example: A boy pulls a sled of weight 15 lb with a force of 10 lb exerted at an angle of 30° above the horizontal (Fig. 6–5). How much work does he do in moving the sled 100 ft? If it takes the boy 1 min to pull the sled that distance, what is his average power output in horsepower?

Solution: There is no need to integrate here because the force is constant, and so

$$W = \mathbf{F} \cdot \mathbf{s} = Fs\cos\theta$$
$$= 10 \text{ lb} \times 100 \text{ ft} \times \cos 30° = 866 \text{ ft·lb}.$$

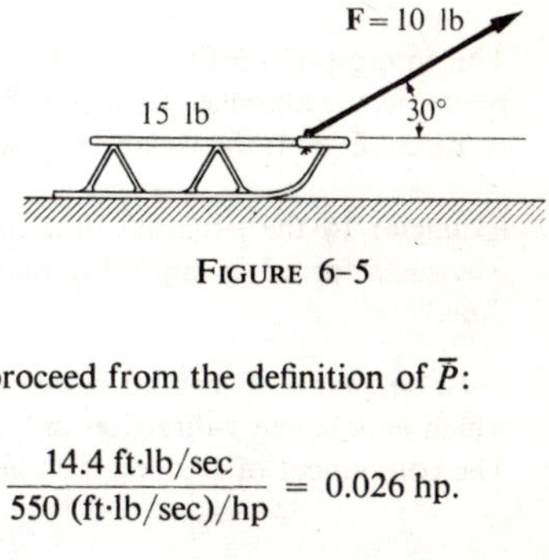

FIGURE 6–5

Note that the mass of the sled is irrelevant for this calculation.

To compute the power output of the boy, we proceed from the definition of $\overline{P}$:

$$\overline{P} = \Delta W/\Delta t = \frac{866 \text{ ft·lb}}{60 \text{ sec}} = 14.4 \text{ ft·lb/sec} = \frac{14.4 \text{ ft·lb/sec}}{550 \text{ (ft·lb/sec)/hp}} = 0.026 \text{ hp}.$$

Example: How much work is done in moving a body of mass m to a height h up a frictionless plane of angle θ with the horizontal? Assume that just enough work is done for the lifting, so that the body has effectively zero velocity at the height h.

Solution: Let us use Eq. (6–16) to solve this problem. If the body is to have zero velocity when it reaches the height h, the only component of the force we must consider is the one which balances the component of the weight along the plane. (A greater force would result in an acceleration and hence a velocity greater than zero.) The

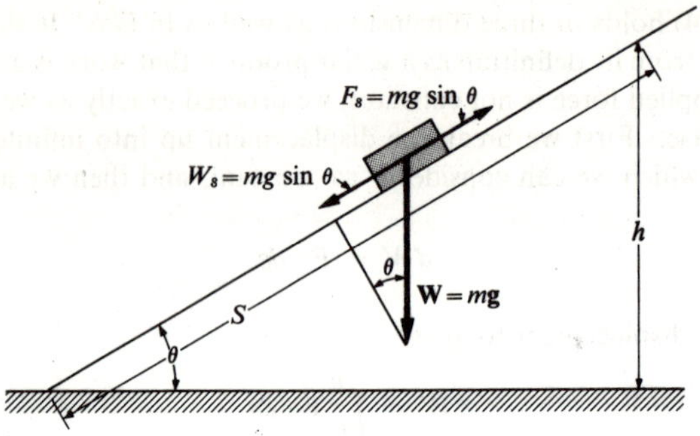

FIG. 6-6. The work which is done in moving a body up a frictionless plane depends on h and not on θ.

other components of the force must also be equal to zero. We call the direction along the plane s; ds is an infinitesimal displacement along the plane. (See Fig. 6-6.) The force in the direction of the displacement, F_s, is

$$F_s = mg \sin\theta \qquad \text{and} \qquad W = \int_0^S F_s\, ds = mgS \sin\theta,$$

where S is the distance the body moves. But $h/S = \sin\theta$, and therefore

$$W = mgh.$$

This interesting result means that the work done does not depend on the angle of the plane (in the absence of friction) but only on the total height through which the weight is lifted. Even if the weight is raised straight up, the same amount of work is involved.

Example: In the situation described in the previous example, let us suppose that a *horizontal* force **F** is applied to move the body up the inclined plane. What is the work done?

Solution: While **F** has no component in the y-direction, both it and the weight mg, which acts in the y-direction only, have components along the plane itself (Fig. 6-7). The component of **F** along the plane, which is

$$F_s = F \cos\theta,$$

must equal in magnitude the component of the weight mg along the plane, which is

$$W_s = mg \sin\theta.$$

The applied horizontal force **F** must be able to supply this force along the plane, and so

$$F = mg \frac{\sin\theta}{\cos\theta} = mg \tan\theta.$$

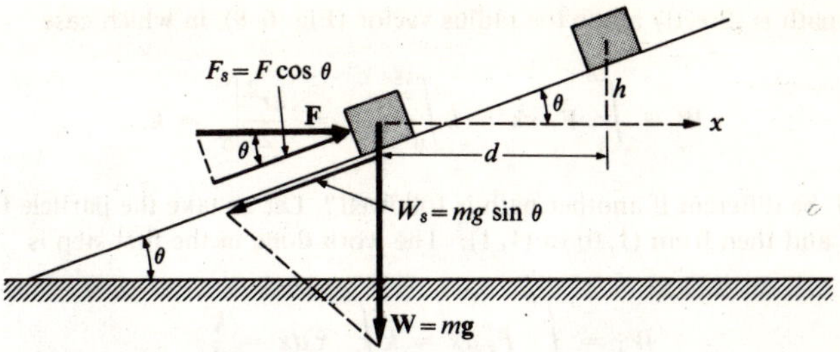

FIG. 6–7. The work done in moving a body up a frictionless plane is the same for a horizontal force as for a force parallel to the plane.

In the course of moving the body up the plane, the force **F** acts through the distance d, and so

$$W = \int_0^d F\,dx = mgd\tan\theta.$$

But d is related to the height of the plane h by the relationship $d = h/\tan\theta$, and thus

$$W = mgh,$$

just as it did in the previous example.

Example: A particle is acted on by a central force in the xy-plane. The direction of the force is toward the origin, and its magnitude is proportional to the distance from the origin to the particle, that is,

$$\mathbf{F}' = -k\mathbf{r}.$$

Calculate the work done in going from (0, 0) to (1, 1) without acceleration.

Solution: The force **F** to be exerted on the particle $= -\mathbf{F}' = k\mathbf{r}$. Here the precise path to be followed by the particle in going from (0, 0) to (1, 1) is not specified. The

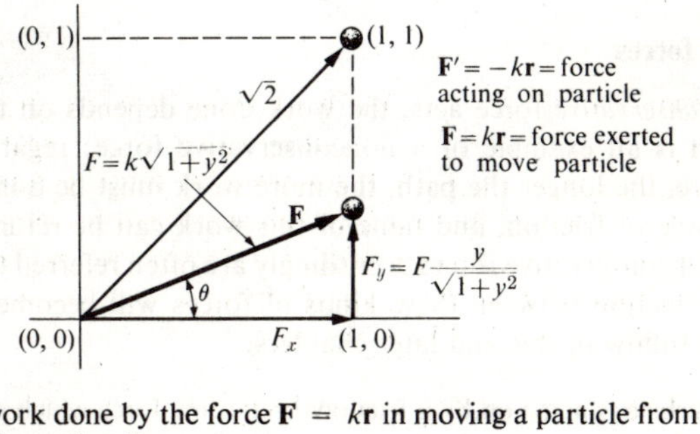

FIG. 6–8. The work done by the force $\mathbf{F} = k\mathbf{r}$ in moving a particle from (0, 0) to (1, 1) is independent of the path taken.

simplest path is directly along the radius vector (Fig. 6-8), in which case

$$W = \int_A^B \mathbf{F} \cdot d\mathbf{s} = k\int_0^{\sqrt{2}} r\,dr = \left.\frac{kr^2}{2}\right|_0^{\sqrt{2}} = k.$$

Will W be different if another path is followed? Let us take the particle from (0, 0) to (1, 0), and then from (1, 0) to (1, 1). The work done in the first step is

$$W_1 = \int_A^B F_x\,dx = k\int_0^1 x\,dx = \frac{k}{2}.$$

For the second step, we note that the magnitude of $\mathbf{F}$ is

$$F = kr = k\sqrt{1^2 + y^2} = k\sqrt{1 + y^2},$$

so that

$$F_y = F\sin\theta = F\frac{y}{\sqrt{1 + y^2}} = ky.$$

Hence

$$W_2 = \int_0^1 F_y\,dy = k\int_0^1 y\,dy = \frac{k}{2},$$

and the total work done in going from (0, 0) to (1, 1) by this path is

$$W = W_1 + W_2 = k.$$

The amount of work done is the same regardless of the path taken, something that is characteristic of what are called *conservative forces.* When only conservative forces are acting, the work involved in moving a body from one place to another depends only on the initial and final positions, and not on the details of the displacement, that is, not on the particular path. In the above example an amount k of work had to be done against the central force to carry it to (1, 1). In going from (1, 1) back to (0, 0), exactly the same work k would be done, this time by the central force.

6-5 Dissipative forces

When a *nonconservative* force acts, the work done depends on the actual path taken. Friction is an example of a nonconservative force: regardless of where the endpoints are, the longer the path, the more work must be done to overcome the retarding force of friction, and none of this work can be returned by reversing the path. Nonconservative forces accordingly are often referred to as *dissipative* forces. The distinction between these kinds of forces will become clearer in the discussions that follow in this and later chapters.

Example: A particle is acted on by a frictional retarding force which obeys Eq. (5-8). The particle is free to move in the xy-plane, as in the previous example. How much work is done in moving the particle from (0, 0) to (1, 1)? Once more assume that the work done is just sufficient to produce the desired displacement, so that the particle is at rest at (1, 1).

Solution: We shall start with the first of the paths used in the previous example. The work done against friction is

$$W = \int_A^B F_f\,dr = \mu N \int_0^{\sqrt{2}} dr = \sqrt{2}\,\mu N.$$

The work done against the retarding frictional force when the particle traverses the second path is

$$W = W_1 + W_2 = \int_0^1 F_f\,dx + \int_0^1 F_f\,dy = 2\mu N,$$

which is greater than the work done in the shorter path! Furthermore, in returning from (1, 1) to (0, 0), work would still have to be done against the retarding frictional force, the amount depending on the length of the path.

Example: A block of wood of mass m is being pushed up the inclined plane of Fig. 6–9. The coefficient of friction between the block and the plane is μ. How much work is required to get the block to the top of the plane?

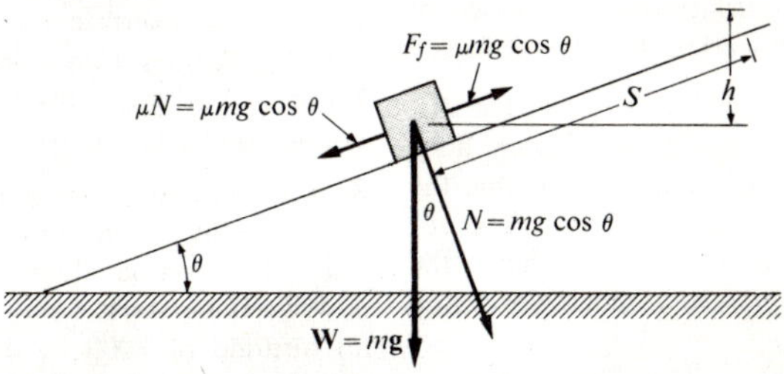

FIG. 6–9. The work done against friction in moving a body up a plane depends on both h and θ.

Solution: The work needed to raise the block in this problem is evidently made up of two parts: (1) the amount mgh involved in raising the block the height h, which we have previously calculated, and (2) an additional amount which represents the work done against friction. The latter may be calculated in the following way. First we find the normal force pushing the block against the plane; this is, as Fig. 6–9 shows, the component of the block's weight mg that is perpendicular to the plane,

$$N = mg\cos\theta.$$

The external force that must be applied to overcome this frictional force is

$$F_f = \mu N = \mu mg \cos\theta.$$

The distance through which the force acts is the length of the plane S, so that the total work done against friction is

$$W_f = F_f S = \mu mgS \cos\theta$$

or, since $S = h/\sin\theta$, we have

$$W_f = mgh(\mu \cot\theta).$$

Thus the total amount of work done in pushing the block to the top of the inclined plane is

$$W = mgh + W_f = mgh(1 + \mu \cot\theta).$$

When $\theta = 90°$ the work reduces to mgh, because in that event the plane is vertical and there is no normal force pushing the block against it, and hence no frictional force.

Problems

1. (a) A force of 130 n is used to lift a 12-kg mass to a height of 8 m. How much work is done by the force? (b) A force of 130 n is used to push a 12-kg mass on a horizontal, frictionless surface for a distance of 8 m. How much work is done by the force?

2. A 20-kg wooden box is pushed a distance of 15 m on a horizontal stone floor by a force just sufficient to overcome the friction between box and floor. The coefficient of friction is 0.4. (a) What is the required force? (b) How much work does the force do?

3. Four thousand joules are used to lift a 30-kg mass. If the mass is at rest before and after its elevation, how high does it go?

4. A boy pulls a sled with a force of 10 lb for 100 ft. The rope attached to the sled is at an angle of 30° above the horizontal. How much work is done?

5. A centripetal force of 18 n is used to keep a 2-kg ball in uniform circular motion at the end of a string 1 m long. How much work does the force do in each revolution of the ball?

6. A horizontal force of 5 n is used to push a box up a ramp 5 m long that is at an angle of 15° above the horizontal. How much work is done?

7. A horse is towing a barge with a rope that makes an angle of 20° with the canal. If the horse exerts a force of 80 n, how much work does it do in moving the barge 1 km (1 km $= 10^3$ m)? If he moves it in one hour, what is the power used?

8. A 160-lb man carrying a 32-lb knapsack climbs to the summit of a 14,000-ft mountain. The average slope of the mountain is 40°. If he starts from a base camp at an altitude of 8000 ft, how much work does he perform in making the ascent?

9. A uniform chain 1 m long of total mass 1 kg lies on a table with $\frac{1}{2}$ of its length dangling over the edge. What is the minimum amount of work required to put the chain back on the table if friction is neglected?

10. Prove that if a body is accelerated from rest by a device delivering constant power, its distance from the starting point is proportional to $t^{3/2}$.

11. An escalator carries passengers from one floor to another 30 ft higher. The escalator is 50 ft long and its linear speed is 2.5 ft/sec. The escalator is designed to

carry 100 passengers per minute, whose average weight is 150 lb. What is the required horsepower of the motor?

12. A 3-lb box is on an inclined plane whose angle is 45°. The coefficient of friction between the plane and the box is 0.15. If the box is released 5 ft up the plane, calculate the velocity at the foot of the incline. How much work was done against friction? How much work was done by gravity?

13. A particle lying in the xy-plane is acted on by a force directed toward the origin whose magnitude is kr, where r (equal to $\sqrt{x^2 + y^2}$) is the distance from the particle to the origin. (a) Find the work that must be done to move the particle from a point $x = 4$, $y = 12$ to the point $x = 4$, $y = 7$ along the radius vector to that point. (b) Find the work that must be done if the particle is first moved to $x = 1$, $y = 12$, and from there to $x = 4$, $y = 12$. (c) Is this particular force conservative or nonconservative?

14. A particle lying in the xy-plane is acted on by a frictional force equal to μN. (a) Find the work that must be done to move the particle from the point $x = 1$, $y = 3$ to the point $x = 4$, $y = 12$ along the radius vector. (b) Find the work that must be done if the particle is first moved to $x = 4$, $y = 3$, and from there to $x = 4$, $y = 12$. (c) Is this particular force conservative or nonconservative?

15. A body is repelled by a central force according to the law $F = kr^3$, where r is its distance from the center. Find the work needed to bring the body from $r = a$ to $r = b$.

16. The equation of the force in newtons on a body as a function of its displacement in meters from the origin is $F = x^2 - 5x + 6$. Find the work done in joules in moving the body from $x = 2$ to $x = 3$ m.

17. The following is the force acting in a displacement:

$0 \leq x < 4$, $F = 3$ n,

$4 < x \leq 7$, $F = -1$ n,

$7 \leq x \leq 10$, $F = [3(x - 7) - 1]$ n.

What is the work done in a displacement from $x = 0$ to $x = 10$?

18. Estimate the horsepower required to move a ski tow on a 30° slope, 1000 ft long. The rope is to move at 5 mi/hr and is designed to move 100 riders, each weighing an average of 150 lb.

7

Energy

Now we turn to the subject of energy—what it is, when it is conserved, and why it is so useful a quantity. We approach this subject by enquiring about what happens to a particle when work is done on it. When the resultant force acting is zero, as in the case of a body being lifted against gravity by a force equal to its weight, the work that is done changes the position of the body. When work is done against friction, again the resultant force is zero (the applied force exactly balances the oppositely directed frictional force), and the work done, as we know from experience, goes into heating up the contacting surfaces. (We shall have more to say about this later on.) When the resultant force acting on the particle is not zero, the consequence of the work done is a change in the particle's velocity.

7–1 Kinetic energy

In this section we shall repeat with more generality some of the introductory remarks in Chapter 6. We know by Newton's second law of motion that when there is a net resultant force acting on a particle, the particle experiences an acceleration in the same direction as that of the force. If the mass of the particle is m, the acceleration $\mathbf{a}$ is determined by

$$\mathbf{F} = m\mathbf{a}.$$

We first consider a force which has a component in the x-direction only. The particle is accelerated in the x-direction. If it moves from a position x_1 to x_2, the work that the force does on the particle is

$$W = \int_{x_1}^{x_2} F_x\, dx.$$

Substituting

$$F_x = ma_x = m\frac{dv_x}{dt},$$

we have

$$W = m\int_{x_1}^{x_2} \frac{dv_x}{dt}\, dx = m\int_{x_1}^{x_2} dv_x \frac{dx}{dt} = m\int_{v_{1x}}^{v_{2x}} v_x\, dv_x.$$

In obtaining the last integral above, we made use of the definition

$$v_x = dx/dt,$$

and, since we now are integrating over velocity v_x instead of displacement x, we replace the limits of x_1 and x_2 by the velocities the particle has at each of those points, which are respectively v_{1x} and v_{2x}.

Integrating, we have

$$W = m\int_{v_{1x}}^{v_{2x}} v_x\,dv_x = \tfrac{1}{2}mv_x^2\Big|_{v_{1x}}^{v_{2x}} = \tfrac{1}{2}mv_{2x}^2 - \tfrac{1}{2}mv_{1x}^2.$$

The work done on the particle by the force F_x appears in the form of the difference between $\frac{1}{2}mv_x^2$ evaluated at the final position x_2, where the particle's velocity is v_{2x}, and at the initial position x_1, where the particle's velocity is v_{1x}.

The calculation of the effect of doing an amount of work W on a particle free to move in three dimensions, again in the absence of gravity and friction, is very similar. We have

$$W = \int \mathbf{F}\cdot d\mathbf{s} = m\int \mathbf{a}\cdot d\mathbf{s} = m\int \frac{d\mathbf{v}}{dt}\cdot d\mathbf{s} = m\int d\mathbf{v}\cdot\frac{d\mathbf{s}}{dt}$$

$$= m\int d\mathbf{v}\cdot\mathbf{v} = \tfrac{1}{2}mv^2\Big|_{v_1}^{v_2} = \tfrac{1}{2}mv_2^2 - \tfrac{1}{2}mv_1^2. \qquad (7\text{–}1)$$

The quantity $\frac{1}{2}mv^2$ is called the *kinetic energy* of a particle of mass m and velocity v. What Eq. (7–1) says is that *the net amount of work done by all forces acting on a particle is equal to the change in its kinetic energy.* The symbol for kinetic energy is T, so that

$$T = \tfrac{1}{2}mv^2, \qquad (7\text{–}2)$$

and, like work, it is a scalar quantity whose unit is the joule in the mks system and the foot-pound in the British system.

To obtain a feeling for the magnitudes of the kinetic energies characteristic of various common motions, we shall compute a few such values. A ball of mass 1 kg (Fig. 7–1a) is thrown with a speed of 5 m/sec. Its kinetic energy is

$$\text{KE} = \frac{1}{2}mv^2 = \frac{1}{2}\times 1\text{ kg}\times\left(5\,\frac{\text{m}}{\text{sec}}\right)^2 = 12.5\text{ joules}.$$

An automobile which weighs 3200 lb (Fig. 7–1b) is traveling at a speed of 15 mi/hr (22 ft/sec). Its kinetic energy is

$$\text{KE} = \frac{1}{2}mv^2 = \frac{1}{2}\frac{w}{g}v^2 = \frac{1}{2}\times\frac{3200\text{ lb}}{32\text{ ft/sec}^2}\times\left(22\,\frac{\text{ft}}{\text{sec}}\right)^2$$

$$= 24{,}200\text{ ft}\cdot\text{lb} = 2.4\times 10^4\text{ ft}\cdot\text{lb}.$$

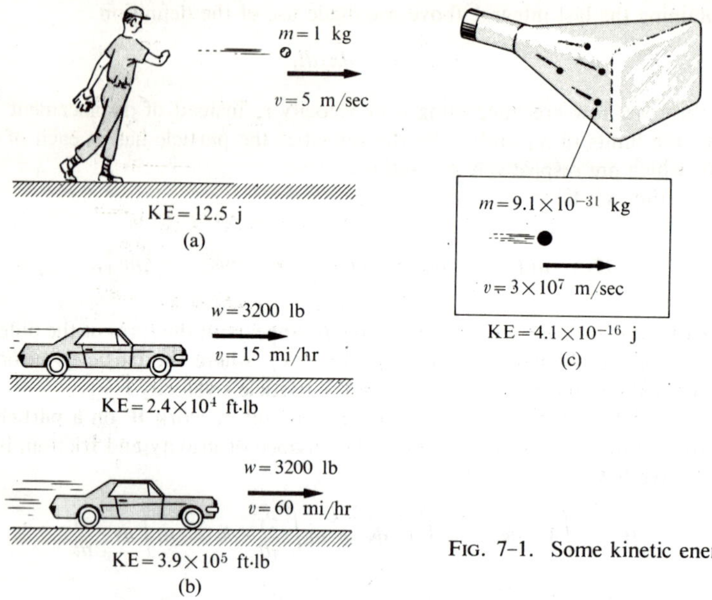

FIG. 7–1. Some kinetic energies.

The same automobile at a speed of 60 mi/hr (88 ft/sec) has a kinetic energy of

$$KE = \frac{1}{2}\frac{w}{g}v^2 = \frac{1}{2} \times \frac{3200 \text{ lb}}{32 \text{ ft/sec}^2} \times \left(88 \frac{\text{ft}}{\text{sec}}\right)^2$$
$$= 387{,}200 \text{ ft·lb} = 3.9 \times 10^5 \text{ ft·lb},$$

which is 16 times greater than at a velocity of 15 mi/hr. The fact that kinetic energy is proportional to the *square* of the velocity is responsible for the severe consequences of automobile accidents at high speeds.

Electrons are the smallest particles present in ordinary matter. The electron mass (provided that the electron speed is not too close to the speed of light, a requirement we shall discuss in Chapter 30) is 9.1×10^{-31} kg. Electrons constitute the beam in a television picture tube (Fig. 7–1c), whose impacts on the tube screen produce tiny flashes of light. Such electrons might have speeds of 3×10^7 m/sec; hence their kinetic energies are

$$KE = \frac{1}{2}mv^2 = \frac{1}{2} \times 9.1 \times 10^{-31} \text{ kg} \times \left(3 \times 10^7 \frac{\text{m}}{\text{sec}}\right)^2$$
$$= 4.1 \times 10^{-16} \text{ joules.}$$

If a particle has kinetic energy, it can do work on other particles by colliding with them. Suppose that particle A collides with another particle, B, which is at rest. Particle A will exert a force on B, and will increase the latter's kinetic energy.

By Newton's third law, B will exert an equal and opposite force on A, and decrease its kinetic energy. If these are the only forces involved, the gain in the kinetic energy of B will be equal to the loss of the kinetic energy of A.

What emerges from such considerations is that the work done on particles must be accounted for in terms of kinetic energy. This kinetic energy can be recovered if the particle which has it does work on another particle, for example by colliding with it. One might say that the *kinetic energy that a particle has is the maximum amount of work that it can do by virtue of the fact that it has a velocity.* In a certain sense, work and energy are interchangeable quantities. When work is done, it appears as energy. The energy can be decreased by permitting the particle to do work on other particles. We can often simplify dynamical problems by making use of this interchangeability.

Example: A particle of mass 20 kg is at rest when a force of 5 n is applied. What is the velocity of the particle after it has gone a distance of 10 m?

Solution: From Eq. (7–1), since the initial velocity is zero, we have

$$W = \tfrac{1}{2}mv^2 = Fs, \qquad v = \sqrt{\frac{2Fs}{m}} = \sqrt{\frac{2 \times 5 \times 10}{20}} = \sqrt{5}\ \text{m/sec.}$$

This same result, of course, can be obtained by the methods we have developed in kinematics and dynamics. For $v_0 = 0$,

$$s = \tfrac{1}{2}at^2 \qquad \text{and} \qquad v = at.$$

Hence $v = \sqrt{2as}$, and since $\mathbf{F} = m\mathbf{a}$,

$$v = \sqrt{2Fs/m} = \sqrt{5}\ \text{m/sec,}$$

as before. Even in this simple example one can see the advantages of using kinetic energy as an aid in solving problems.

Example: A particle of mass m is dropped from rest at a height h above the ground. What is its velocity when it strikes the ground?

Solution: Let us do this problem from the point of view of work and energy. As the particle falls, the work done on it by the earth's gravitational attraction is the force associated with this attraction, the weight mg, multiplied by the distance through which the force acts, the height h. Hence

$$W = mgh.$$

At the ground all this work is in the form of kinetic energy,

$$T = \tfrac{1}{2}mv^2;$$

here the initial velocity of the particle is zero and v is its final velocity. Hence

$$W = T, \qquad mgh = \tfrac{1}{2}mv^2, \qquad v = \sqrt{2gh}.$$

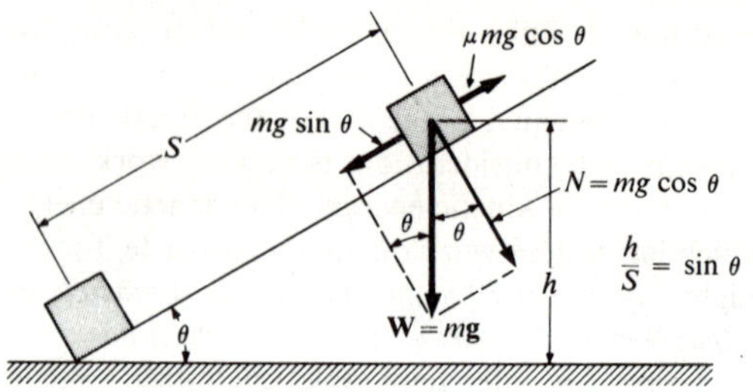

FIGURE 7-2

Example: A block of mass m, starting from rest, slides down the frictionless plane of Fig. 7-2. What is its velocity at the bottom?

Solution: The component of force along the plane F_S is, in terms of the weight mg of the block,

$$F_S = mg \sin \theta.$$

Thus the work done on the block as it slides down is

$$W = F_S S = mgS \sin \theta,$$

and the kinetic energy of the block at the bottom is

$$T = \tfrac{1}{2}mv^2 = W, \qquad \text{and} \qquad v = \sqrt{mgS \sin \theta}.$$

But

$$h = S \sin \theta;$$

thus the velocity of the block is the same as it would be if it were dropped straight down, namely

$$v = \sqrt{2gh}.$$

Example: The block of the previous problem is allowed to slide down a similar inclined plane, but now the coefficient of friction between the block and the plane is μ, where $\mu > 0$. What is the velocity of the block at the bottom of the plane?

Solution: Now we must take into account the frictional retarding force acting on the block as it slides down the plane. This force is

$$F_{fr} = -\mu N = -\mu mg \cos \theta,$$

since the normal force N pressing the block against the plane is $mg \cos \theta$. The minus sign is used because F_{fr} always acts in the direction opposite to that in which the block moves.

The total force $\mathbf{F}$ on the block is the resultant of the frictional force $\mathbf{F}_{fr}$, which acts parallel to the inclined plane, and the gravitational force $\mathbf{F}_{grav}$, which acts vertically downward. We are concerned here with the component of $\mathbf{F}_{grav}$ parallel to the inclined plane, since the block is restricted to move along this plane. This component is, from the diagram,

$$mg \sin \theta.$$

Thus the total force on the block *parallel to the plane* is

$$F_S = mg\sin\theta - \mu mg\cos\theta = mg(\sin\theta - \mu\cos\theta).$$

The work W done by this force as the block slides down the distance S along the plane is equal to the kinetic energy T of the block at the bottom, since it started out from rest. Hence

$$W = T, \qquad F_S S = mg(\sin\theta - \mu\cos\theta)\frac{h}{\sin\theta} = \tfrac{1}{2}mv^2,$$

since S, the length of the plane, is $h/\sin\theta$. Solving for the block's velocity v, we find that

$$v = \sqrt{2gh(1 - \mu\cot\theta)}.$$

When either $\mu = 0$ or $\theta = 90°$, the velocity of the block is the same as the velocity it would have after free fall from a height h, namely $\sqrt{2gh}$, as we might expect. We may also note that $v = 0$ when

$$\mu\cot\theta = 1, \qquad \mu = \tan\theta.$$

If this is the case, the block is barely able to reach the bottom of the plane—in fact, strictly speaking, it requires an infinite time to get there, since its velocity is zero. If the coefficient of friction μ exceeds $\tan\theta$, the velocity is imaginary, which simply expresses the fact that when the frictional retarding force exceeds the gravitational force, the block cannot move at all.

7–2 Potential energy

Thus far we have shown that the work done by the total force acting on a particle is equal to the change in its kinetic energy. We shall now, in this section, separate the total force into an external force and a field force. This will be just a generalization of the situation we discussed at the beginning of Chapter 6, when we considered a body being lifted against the force of gravity. It would be well to keep this example in mind during the general discussion that follows. For a body which is near the surface of the earth, the field force is the force of gravity and the external force is the one we are exerting. The separation is usually made on the basis of the fact that the field force is conservative and the external force is not.

There are many conservative forces with which we shall deal. Central forces as well as gravitational forces near the earth are some examples we have already met. We shall run across many others, especially in our study of electricity and magnetism. For this reason, it is wise to keep the discussion general even though it would be helpful to keep a specific example in mind.

The work done by a conservative force has the important property that the amount of work is independent of the path and depends only on the coordinates of the endpoints of the path. We can write

$$\int_{\mathbf{s}_1}^{\mathbf{s}_2} \mathbf{F}_f \cdot d\mathbf{s} = -(V_2 - V_1), \tag{7–3}$$

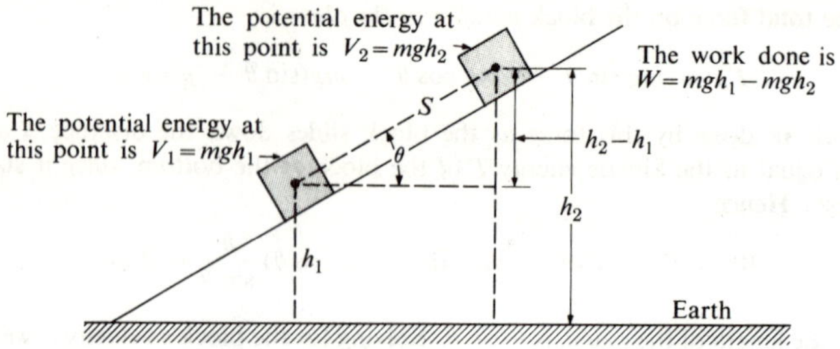

FIG. 7–3. The work done in moving a body up a distance h is independent of how the body is moved if there is no friction.

where we have labeled the conservative force $\mathbf{F}_f$ (field force). Here V is some function of the coordinates: V_2 means that we evaluate V at the position of the upper limit of the integral, $\mathbf{s}_2$, and V_1 is the function evaluated at the lower limit, $\mathbf{s}_1$. The minus sign is part of the convention used in defining the function V.

In previous examples, we have seen that when we consider the gravitational forces near the surface of the earth, the force is independent of whether the body is moved straight down or whether it is moved along a smooth inclined plane. In the first case (Fig. 7–3), the work done by the gravitational force is $mgh = mg(h_1 - h_2) = mgh_1 - mgh_2$. Thus $V = mgh$, $V_1 = mgh_1$ and $V_2 = mgh_2$. If the body moves down a smooth plane (Fig. 7–2), the component of the gravitational force in the direction of the displacement is $mg \sin \theta$. The work done is $mgS \sin \theta$. But $S \sin \theta = h_1 - h_2$, and the work done is $mgh_1 - mgh_2$. Again it is equal to $-(V_2 - V_1)$.

The function $V(\mathbf{r})$ $[V(\mathbf{r}_1) = V_1, V(\mathbf{r}_2) = V_2]$ is called the *potential energy.*

Another example of conservative forces is a central force; that is, one which is directed toward or away from a center of force, and whose magnitude depends only on the distance from the center of force (Fig. 7–4). The magnitude may be written $F(r)$, where r is the distance from the origin of the force to the particle on which it is acting.

To prove that a central force is conservative we consider, as in the figure, the work done in moving a particle from 1 to 2 by a central force $\mathbf{F}(r)$. First consider an infinitesimal element of length ds along the path $1 \rightarrow 2$. The work done by $\mathbf{F}$ in moving along ds from A to B is

$$\mathbf{F} \cdot d\mathbf{s} = F\, ds \cos \theta,$$

where θ is the angle between $\mathbf{F}$ and $d\mathbf{s}$. We draw two concentric spheres whose center is at O through the termini of $d\mathbf{s}$ at A and at B. The force $\mathbf{F} \cdot d\mathbf{s}$ is equal to $F\, dr$, where dr is the distance along a radius vector. This is true since, by the definition of a central force $\mathbf{F}$, it acts along the radius vector, and the projection of $d\mathbf{s}$ in the direction of $\mathbf{F}$ is dr.

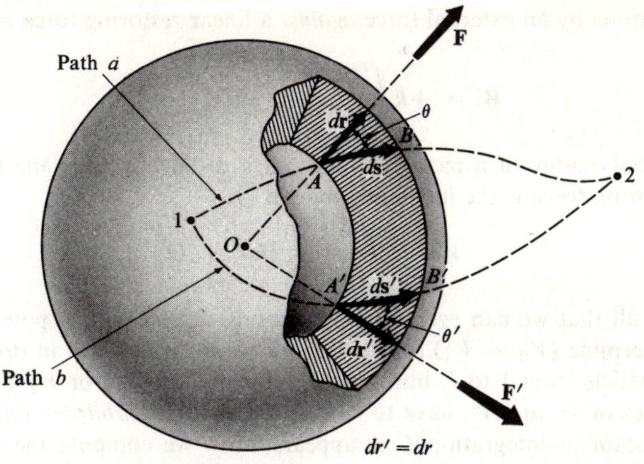

FIG. 7–4. The work done in moving a particle against a central force depends only on the distances of the endpoints of the motion from the center of force O.

Now we consider an alternative path from 1 to 2 (see Fig. 7–4). The spheres cut this path in A' and B'. The work done in going from A' to B' is

$$\mathbf{F}' \cdot d\mathbf{s}' = F' \, dr',$$

by the same argument as in the previous paragraph. Since the force is a central force and since A' is the same distance from O as A, then $F' = F$, $dr' = dr$, and

$$\mathbf{F} \cdot d\mathbf{s} = F \, dr = F' \, dr' = \mathbf{F}' \cdot d\mathbf{s}'.$$

Since every element of length of every possible path from 1 to 2 could be treated the same way, the total work along any two paths is the same. Hence

$$W = \int_1^2 \mathbf{F}_{\text{Path } a} \cdot d\mathbf{s}$$

$$= \int_1^2 \mathbf{F}'_{\text{Path } b} \cdot d\mathbf{s}' = \int_{r_1}^{r_2} F(r) \, dr. \tag{7–4}$$

The integral of Eq. (7–4) depends only on the coordinates of the endpoints of the path and not on the details of the path, and so central forces are conservative.

Two significant central forces are the *linear restoring force*, whose magnitude is

$$F(r) = -kr, \tag{7–5}$$

and the *inverse square force*, whose magnitude is

$$F(r) = k/r^2. \tag{7–6}$$

The work done by an external force *against* a linear restoring force is

$$W = +k\int_{r_1}^{r_2} r\,dr = \tfrac{1}{2}k(r_2^2 - r_1^2). \tag{7-7}$$

The potential energy of a particle at a point a distance r from the origin of the linear restoring force is the indefinite integral

$$V = k\int r\,dr = \tfrac{1}{2}kr^2 + C. \tag{7-8}$$

Because all that we can ever actually measure is a *change* in potential energy, we can determine $(V_2 - V_1)$ from the work we must perform in order to transport the particle from 1 to 2, but not the value of either V_1 or V_2 alone; the absolute values of V_1 and V_2 have to be referred to some *arbitrary reference point*.

The constant of integration C disappears when we compute the difference in potential energy between two points, which is either the work done *on the particle* against the restoring force, in which case the final position has the higher potential energy V, or the work done *by the particle*, in which case the final position has the lower V. Sometimes it is convenient to give the constant C an arbitrary value, for instance the value zero at the center of force, so that $V(0) = 0$. A procedure of this kind enables us to speak loosely of the potential energy of a particle at a particular point, although strictly speaking only a change in V can ever be measured, and hence it alone is significant.

We can readily perform a similar calculation of the work done by an external force against an inverse-square central force of the form

$$F(r) = k/r^2.$$

When the constant k is positive, the force is one which acts to *repel* particles it acts upon *away* from zero; when k is negative, the force acts to *attract* particles it acts upon *toward* zero. The work done on a particle *against* this force is

$$W = -k\int_{r_1}^{r_2} \frac{dr}{r^2} = k\left(\frac{1}{r_1} - \frac{1}{r_2}\right). \tag{7-9}$$

Here the potential energy of a particle a distance r from zero is

$$V = -\frac{k}{r} + C, \tag{7-10}$$

where again the value of C is arbitrary. It is usual to fix C for an inverse-square force so that the potential energy of a particle at infinity is zero; that is, so that $V(\infty) = 0$.

The concept of potential energy is a general one, for it refers to the energy something has as a consequence of its position anywhere. The earth itself, for

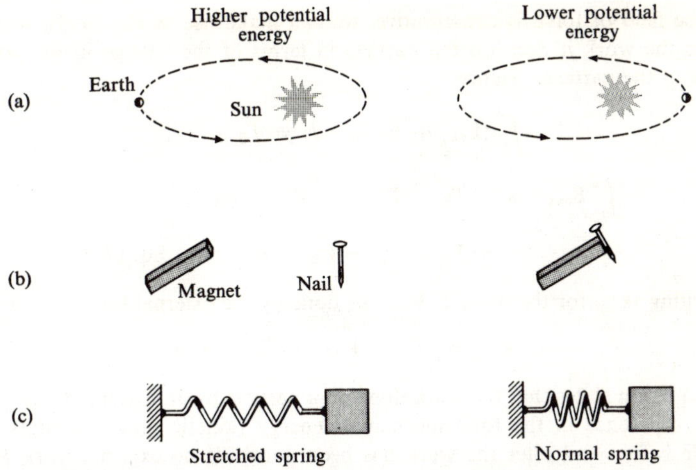

FIG. 7–5. Three examples of potential energy.

instance, has potential energy with respect to the sun, since if its orbital motion were to cease it would fall toward the sun (Fig. 7–5a). An iron nail has potential energy with respect to a nearby magnet, since it will fly to the magnet if released (Fig. 7–5b). An object at the end of a stretched spring has potential energy with respect to its position when the spring has its normal extension, since if let go the object will move as the spring collapses (Fig. 7–5c). In each of these cases the body in question has the potentiality of doing work in its original position, and in each of them the force involved is a conservative one. A body can have potential energy only through the action of a conservative force. Work done against gravity leads to an increase in potential energy; work done against friction does not.

7–3 Conservation of mechanical energy

We have already seen that the work done by all the forces acting on a particle is equal to the difference of the kinetic energy before and after the force acts. If part of the total force is a conservative one, we can derive an additional useful principle by separating the total force into $\mathbf{F}_{\text{ext}} + \mathbf{F}_{\text{field}}$. The force $\mathbf{F}_{\text{field}}$ will later be taken to be the conservative force. That is,

$$\int_1^2 (\mathbf{F}_{\text{ext}} + \mathbf{F}_{\text{field}}) \cdot d\mathbf{s} = T_2 - T_1, \tag{7–11}$$

where T_1 is the kinetic energy of the particle at the start of its motion under the influence of $\mathbf{F}_{\text{ext}} + \mathbf{F}_{\text{field}}$, and T_2 is its kinetic energy at the end of its path. The kinetic energy of the particle may either increase or decrease, depending on whether work is done *on* it or *by* it, respectively.

If the field of force is conservative, we can make use of Eq. (7–3), which expresses the work it does on the particle in terms of the change in the potential energy of the particle. Hence

$$\int_1^2 (\mathbf{F}_{\text{ext}} + \mathbf{F}_{\text{field}}) \cdot d\mathbf{s} = T_2 - T_1, \tag{7–11}$$

$$\int_1^2 \mathbf{F}_{\text{ext}} \cdot d\mathbf{s} = T_2 - T_1 - \int_1^2 \mathbf{F}_{\text{field}} \cdot d\mathbf{s}$$

$$= T_2 - T_1 + V_2 - V_1 \qquad \text{by Eq. (7–3)}$$

or, writing W_{ext} for the work $\int_1^2 \mathbf{F}_{\text{ext}} \cdot d\mathbf{s}$ done by the external force,

$$W_{\text{ext}} = (T_2 + V_2) - (T_1 + V_1). \tag{7–12}$$

This equation states that the work done on a particle by an external force is equal to the difference in the total mechanical energy (kinetic plus potential) of the particle before and after the work has been done. If no external work W_{ext} is performed on the particle, then the quantity $(T + V)$, the particle's total mechanical energy, remains constant in whatever motions the particle undergoes. In the course of such motions, energy can be shifted from kinetic to potential or vice versa, but the total of the two cannot change. The statement that, in a conservative force field, if no external work is done on a particle,

$$T + V = \text{const}, \tag{7–13}$$

is a *conservation theorem* of the kind described at the beginning of Chapter 6.

Example: A brick of mass 2 kg is dropped from a rest position 5 m above the ground. What is its velocity at a height of 3 m above the ground?

Solution: The potential energy V of the brick at a height h above the ground is equal to the work that had to be done in lifting it there, which is mgh. Hence

$$V_1 = mgh_1, \qquad V_2 = mgh_2,$$

where 1 and 2 refer, respectively, to the initial and final heights of the brick. At the initial position the kinetic energy of the block is zero, since it is initially at rest, and $T_1 = 0$. At the 3-m height, however, the brick now has the unknown velocity v, corresponding to a kinetic energy T_2 of $T_2 = \frac{1}{2}mv^2$. According to Eq. (7–7),

$$T_1 + V_1 = T_2 + V_2, \qquad 0 + mgh_1 = \tfrac{1}{2}mv^2 + mgh_2, \qquad \tfrac{1}{2}mv^2 = mgh_1 - mgh_2.$$

The change in potential energy of the brick in falling from h_1 to h_2 is converted into kinetic energy of motion. We note that

$$\tfrac{1}{2}mv^2 = mg(h_1 - h_2),$$

which means that it is the difference in height $(h_1 - h_2)$ that is significant, and not the individual values of h_1 and h_2. This is true only when the magnitude of the force is independent of h, as it is for gravity near the earth's surface. Solving, we obtain

$$v = \sqrt{2g(h_1 - h_2)} = 6.3 \text{ m/sec.}$$

Example: A particle is moving at the end of a spring whose spring constant is k. The particle has a zero velocity when the spring is compressed an amount A. What is the velocity of the particle when the spring is neither compressed nor extended? What is the velocity when the displacement from the center is x?

Solution: The total energy of the particle can be found when it is completely compressed. The energy is all potential, and is equal to $\frac{1}{2}kA^2$. At the midpoint of the motion when the spring is not compressed, the energy is all kinetic, and we have

$$\tfrac{1}{2}mv^2 = \tfrac{1}{2}kA^2, \qquad v = \sqrt{k/m}\,A.$$

When the displacement is x, the energy is partly kinetic and partly potential, and equals $\frac{1}{2}kx^2 + \frac{1}{2}mv^2$. Therefore

$$\tfrac{1}{2}kx^2 + \tfrac{1}{2}mv^2 = \tfrac{1}{2}kA^2, \qquad v = \sqrt{(A^2 - x^2)(k/m)}.$$

7–4 Nonconservative force fields

Not all force fields have the property of being conservative. When work is done on a particle under the influence of a *nonconservative* (or *dissipative*) force field, the amount of work that must be done in taking the particle from one point to another, if there is no change in its kinetic energy, depends not only on the locations of the two endpoints but also on the path taken by the particle in going from one to the other.

An important example of a nonconservative force is friction. As we saw in Section 6–5, the longer the path taken by a block going from one place to another on a horizontal surface when friction is acting, the more work must be done. Let us now consider a block sliding down an inclined plane, where the coefficient of friction between block and plane is μ. At the top of the plane the total mechanical energy of the block is exclusively potential, with $V = mgh$; at the foot of the plane the block has only the kinetic energy $T = \frac{1}{2}mv^2$. We have already computed the value of v, namely

$$v = \sqrt{2gh(1 - \mu \cot \theta)}.$$

Hence the kinetic energy of the block at the foot of the plane is

$$T = \tfrac{1}{2}mv^2 = mgh(1 - \mu \cot \theta) = V(1 - \mu \cot \theta),$$

so that the total mechanical energy at the bottom is less than it was at the top. The missing energy has gone into heat, and is no longer available for direct conversion into mechanical work as V and T are; it has been *dissipated.* We shall leave the precise fate of the dissipated energy until we discuss the kinetic theory of heat and thermodynamics in later chapters.

In nearly all actual macroscopic situations, friction is present in one guise or another. Why then do we bother with the concept of a conservative force field when we almost never encounter one without an accompanying nonconservative field?

There are two separate answers to this question. First, at the atomic and nuclear level there are only conservative forces; conservation of mechanical energy is a valid principle in the atomic domain, and its application facilitates our understanding of phenomena there.

Second, even though conservation of mechanical energy is seldom strictly fulfilled in the macroscopic world, it serves as a limit to the actual events that occur there. If we slide something down an inclined plane, we know that the maximum speed it can have at the bottom is $\sqrt{2gh}$. Friction will make the actual speed somewhat less, but it can never exceed this value if it starts from rest.

When we isolate a mechanical system, the effect of nonconservative forces is to remove energy from the mechanical system. Frictional forces remove it and convert the mechanical energy into heat. Other nonconservative forces convert mechanical energy into other forms of energy which we shall consider later. The important generalization that physicists have made is that *all energy is conserved.* The energy that leaves the mechanical system due to nonconservative forces is accounted for by the increase of the energy of something else. And when a mechanical system gains energy, the external agency responsible for this, when looked at closely, will always be found to have lost some energy. Taking the universe as a whole, the energy within it (properly defined to include the energy associated with mass) remains constant.

Problems

1. A 3200-lb car is moving at 40 mi/hr. What is its kinetic energy?

2. A 0.02-kg bullet has a speed of 500 m/sec. What is its kinetic energy?

3. A 2-kg ball is at rest when a horizontal force of 5 n is applied. In the absence of friction, what is the speed of the ball after it has gone 10 m?

4. A 3-kg stone is lifted to a height of 100 m and then dropped. What is its kinetic energy when it is 50 m from the ground?

5. A man uses a rope and system of pulleys to lift a 160-lb weight to a height of 5 ft. He exerts a force of 45 lb on the rope, and pulls a total of 20 ft of rope through the pulleys in the course of lifting the weight. (a) How much work does the man do? (b) What is the change in the potential energy of the weight? (c) If the answers to (a) and (b) are different, explain.

6. A boy slides down an inclined plane from a starting point 3 m above the ground. His speed at the bottom is 4 m/sec. What percentage of his initial potential energy was dissipated?

7. A force of 100 lb is used to lift an 80-lb weight to a height of 20 ft. There is no friction present. (a) How much work is

done by the force? (b) What is the change in the potential energy of the weight? (c) What is the change in the kinetic energy of the weight?

8. (a) A force of 8 n is used to push a 0.5-kg ball over a horizontal, frictionless table a distance of 3 m. If the ball starts from rest, what is its final kinetic energy? (b) The same force is used to lift the same ball a height of 3 m. If the ball starts from rest, what is its final kinetic energy?

9. At what point in its motion is the kinetic energy of a pendulum bob a maximum? At what point is its potential energy a maximum?

10. At her highest point, a girl on a swing is 8 ft from the ground, while at her lowest point she is 3 ft from the ground. What is her maximum speed?

11. A puck is thrown with an initial speed of 25 ft/sec onto ice. The coefficient of friction is $\mu = 0.0125$. How far will it go before it comes to rest?

12. If an electric car starting from rest coasts 2000 ft down a 2% grade and 1000 ft up a 1% grade, find the coefficient of friction (assumed the same on both grades).

13. The shot-put record is a throw of approximately 65 ft. The shot weighs 16 lb. Assume a trajectory whose initial angle of elevation is 45° and neglect air resistance. What is the initial kinetic energy of the shot?

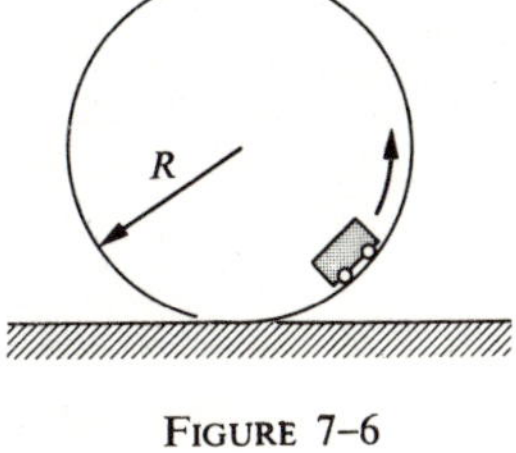

FIGURE 7–6

14. A car is projected into a vertical circular loop of radius R (Fig. 7–6). What is the minimum velocity it should have at the bottom so that it can go around the loop without falling down?

15. In looping a loop of radius R, the car starts from rest at a height h above the top of the loop (Fig. 7–7). Neglecting friction, show that if $h = \frac{1}{2}R$ the car does not leave the track.

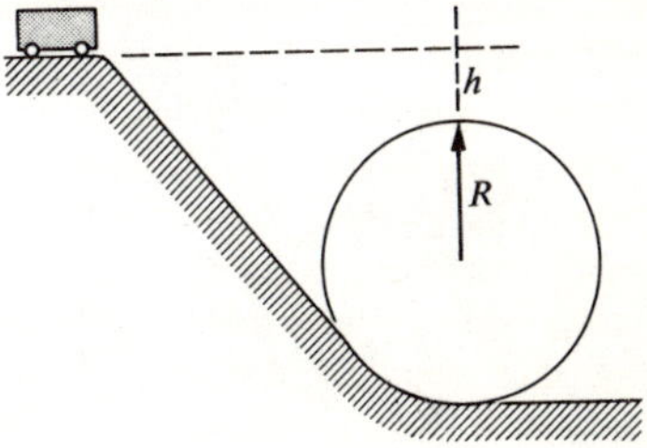

FIGURE 7–7

16. A block whose mass is 5 kg is dropped from a height of 40 m onto a spring whose spring constant is 3,000 n/m. How much will it compress the spring?

17. Show that the tension in a cord tied to a ball moving in a vertical circle at its lowest point exceeds the tension in the cord when the ball is at its highest point by six times the weight of the ball.

18. Repeat Problem 9 of Chapter 6, using the energy method.

19. What is the potential energy function if the force is central and is equal to

(a) $-kr$, (c) e^{-ar}
(b) $-G/r$, (d) $-k/r^7$.

20. The radius of the earth is R. A body is projected vertically from the surface of the earth with an initial velocity v_0. The force of attraction between the body and the earth is

$$F = -\lambda M/r^2,$$

where M is the mass of the body, r is its distance from the center of the earth, and λ is some constant. Show that the height to which it will rise is given by

$$h = \frac{Rv_0^2}{(2\lambda/R) - v_0^2}.$$

What happens when $2\lambda_0/R = v_0^2$? When $v_0^2 > 2\lambda_0/R$?

21. The force of attraction between the earth and the moon is the same as given in Problem 20, with M the mass of the moon. Assume that the moon's orbit about the earth is a circular one, and that the force of attraction supplies the necessary centripetal force. Show that the magnitude of the kinetic energy is $\frac{1}{2}$ the magnitude of the potential energy, and that the total energy is minus $\frac{1}{2}$ of the kinetic energy.

8

Systems of Particles

Whenever we discussed Newton's laws of motion in the past few chapters, it was with the understanding that they applied to abstractions called particles which, strictly speaking, do not exist. We implied, though, that under appropriate circumstances these laws may also be used to understand the behavior of extended bodies composed of a great many particles acting together. In this chapter we shall show how Newton's laws for particles can be applied to extended bodies, provided that we place the proper interpretation on the various quantities. That is, when we employ the second law of motion,

$$\mathbf{F} = m\mathbf{a},$$

with an extended body as its subject, it is not clear at first glance what aspect of the body's behavior the acceleration $\mathbf{a}$ refers to; the body is no longer simply a point mass in space, and it can undergo, in general, deformations and changes in orientation, as well as the shifts in position of a point mass. The problem of interpreting the laws of motion so that they may be employed in the real world as well as in the abstract world, then, is not a trivial one.

To begin with, we require a model of an extended body so that we know exactly what we are talking about. The definition we shall adopt states that an extended body consists of a very large number of particles held together by forces acting among them which obey the third law of motion. The first body we shall treat is one consisting of only three particles, which, though as much of an abstract notion as a single particle, nevertheless allows us to develop certain ideas of wide applicability without undue mathematical complication. What follows, in fact, may be applied to *any* system of particles in addition to those that obey the above definition, no matter what laws govern the interactions of their constituent particles, and no matter whether they are bound together into collective bodies or remain as separate entities.

8-1 System of three particles

All the forces acting on each of the three particles shown in Fig. 8-1 may be divided into two classes: (1) external forces that act from outside the system, and (2) internal forces exerted by the other particles of the system. Consider the forces acting on particle 1. The resultant of all of the external forces is $\mathbf{F}_1^e$. The internal forces are $\mathbf{F}_{12}^i$ and $\mathbf{F}_{13}^i$, exerted respectively by particle 2 and particle 3 on particle 1. Applying the second law of motion to particle 1,

$$\mathbf{F}_1^e + \mathbf{F}_{12}^i + \mathbf{F}_{13}^i = m_1\mathbf{a}_1.$$

For the other two particles we find similar relations:

$$\mathbf{F}_2^e + \mathbf{F}_{21}^i + \mathbf{F}_{23}^i = m_2\mathbf{a}_2$$

and

$$\mathbf{F}_3^e + \mathbf{F}_{31}^i + \mathbf{F}_{32}^i = m_3\mathbf{a}_3.$$

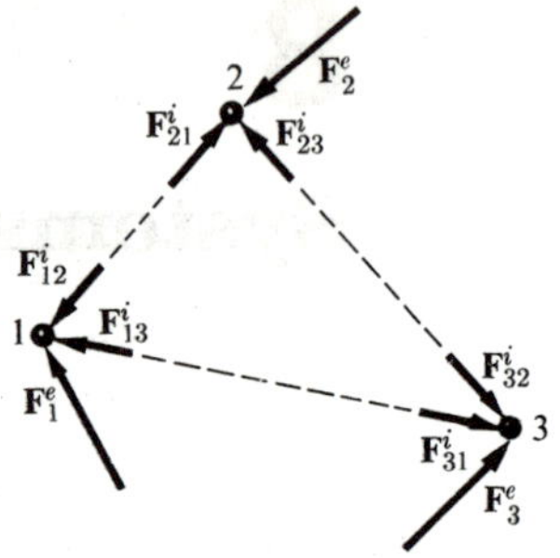

FIG. 8-1. Forces acting on a rigid body consisting of three particles.

Adding these three equations, we have

$$\mathbf{F}_1^e + \mathbf{F}_2^e + \mathbf{F}_3^e + \mathbf{F}_{12}^i + \mathbf{F}_{21}^i + \mathbf{F}_{13}^i + \mathbf{F}_{31}^i + \mathbf{F}_{23}^i + \mathbf{F}_{32}^i = m_1\mathbf{a}_1 + m_2\mathbf{a}_2 + m_3\mathbf{a}_3. \tag{8-1}$$

At this point we make use of the fact that the various internal forces all obey Newton's third law. This means that the force exerted by one particle on another is equal in magnitude and opposite in direction to the force exerted by the latter on the former, so that

$$\mathbf{F}_{12}^i = -\mathbf{F}_{21}^i, \qquad \mathbf{F}_{13}^i = -\mathbf{F}_{31}^i, \qquad \mathbf{F}_{23}^i = -\mathbf{F}_{32}^i. \tag{8-2}$$

Combining Eqs. (8-1) and (8-2) yields

$$\mathbf{F}_1^e + \mathbf{F}_2^e + \mathbf{F}_3^e = \mathbf{F} = m_1\mathbf{a}_1 + m_2\mathbf{a}_2 + m_3\mathbf{a}_3, \tag{8-3}$$

where $\mathbf{F}$ is the sum of all of the *external* forces that act on the system of the three particles. Thus the internal forces have disappeared from the analysis.

Equation (8-3) can be simplified further by rewriting its right-hand side as

$$\frac{d^2}{dt^2}(m_1\mathbf{r}_1 + m_2\mathbf{r}_2 + m_3\mathbf{r}_3),$$

where, as usual, $\mathbf{r}_1$, $\mathbf{r}_2$, and $\mathbf{r}_3$ represent the position vectors of the particles. Now we define a vector **R** such that

$$\mathbf{R} = \frac{m_1\mathbf{r}_1 + m_2\mathbf{r}_2 + m_3\mathbf{r}_3}{m_1 + m_2 + m_3} = \frac{1}{M}(m_1\mathbf{r}_1 + m_2\mathbf{r}_2 + m_3\mathbf{r}_3), \tag{8-4}$$

in which we have abbreviated the total mass of the system $m_1 + m_2 + m_3$ by M. The new vector **R** is called the position vector of the *center of mass* of the system of particles. Since, by Eq. (8–4),

$$M\mathbf{R} = m_1\mathbf{r}_1 + m_2\mathbf{r}_2 + m_3\mathbf{r}_3,$$

and

$$\begin{aligned} M\frac{d^2\mathbf{R}}{dt^2} &= \frac{d^2}{dt^2}(m_1\mathbf{r}_1 + m_2\mathbf{r}_2 + m_3\mathbf{r}_3) \\ &= m_1\frac{d^2\mathbf{r}_1}{dt^2} + m_2\frac{d^2\mathbf{r}_2}{dt^2} + m_3\frac{d^2\mathbf{r}_3}{dt^2} \\ &= m_1\mathbf{a}_1 + m_2\mathbf{a}_2 + m_3\mathbf{a}_3, \end{aligned}$$

we can write Eq. (8–3) in the simple form

$$\mathbf{F} = M\frac{d^2\mathbf{R}}{dt^2} = M\mathbf{A}.$$

Here **A** is the acceleration of the center of mass; in other words, we have found that *the motion of the center of mass of the system of particles is exactly the same as if the entire mass of the system were concentrated at its center of mass, with the net force acting on it restricted to the total external force only.*

If the extended body, instead of being composed of only three particles, consists of n particles, by exactly the same reasoning as we used above we obtain the result that

$$\mathbf{F} = M\frac{d^2\mathbf{R}}{dt^2} = M\mathbf{A}, \tag{8-5}$$

where now **F**, M and **R** are defined more generally as

$$\mathbf{F} = \sum_{i=1}^{n} \mathbf{F}_i^e, \qquad M = \sum_{i=1}^{n} m_i, \qquad \mathbf{R} = \frac{1}{M}\sum_{i=1}^{n} m_i\mathbf{r}_i. \tag{8-6}$$

The symbol $\sum$ represents "the sum of," and the first and last equations of (8–6) represent vector sums.

8–2 Center of mass

Before we can apply Eq. (8–5) to actual objects and benefit from its elegant simplicity, we must be able to calculate the positions of the centers of mass of the objects. The last formula in (8–6) tells us everything we need to know about how to find the center of mass, of course, but a few examples are nevertheless in order.

Example: Where is the center of mass of a system of two particles of masses m_a and m_b a distance d apart?

Solution: First we have to establish the coordinate system to be used in expressing the location of the center of mass. It makes no real difference where we place the origin of coordinates, but certain positions make matters easier. For instance, in the present example, by placing the x-axis along the line joining the two particles (Fig. 8–2), the y- and z-coordinates may be ignored. If we select the position of particle a as the origin, the center of mass has the x-coordinate

$$R_x = \frac{m_a x_a + m_b x_b}{m_a + m_b} = \left(\frac{m_b}{m_a + m_b}\right) d,$$

since $x_a = 0$ and $x_b = d$. The larger m_b is by comparison with m_a, the closer the center of mass lies to b. (A balanced seesaw, with the heavier child nearer the point of support in order for the center of mass to be directly above it, is an illustration of this result.) As a check, we note that in the limit of $m_a = 0$, $R_x = d$, and in the limit of $m_b = 0$, $R_x = 0$. Locating the origin elsewhere would give a different equation for R_x, but the actual location in space of the center of mass would be unchanged.

Example: Three particles of the same mass (Fig. 8–3) are situated in a plane at the positions (0, 0), (1, 2), and (2, 0). Where is their center of mass C?

Solution: Applying Eq. (8–4), we have for each of the components of **R**,

$$R_x = \frac{m_a x_a + m_b x_b + m_c x_c}{m_a + m_b + m_c} = \frac{m(0) + m(1) + m(2)}{3m} = 1,$$

$$R_y = \frac{m_a y_a + m_b y_b + m_c y_c}{m_a + m_b + m_c} = \frac{m(0) + m(2) + m(0)}{3m} = \frac{2}{3},$$

$$R_z = 0.$$

Hence the center of mass C has the coordinates $(1, \frac{2}{3})$.

If there are so many particles that the system may be thought of as a continuous distribution of mass, the finite sum of Eq. (8–6) must be replaced by the integral

$$\mathbf{R} = \frac{1}{M}\int \mathbf{r}\, dM. \tag{8–7}$$

A more convenient form of this equation involves the mass density ρ, the mass per unit volume (or the mass per unit area if a two-dimensional object is being considered, or the mass per unit length for a one-dimensional object). In terms of ρ the mass of an infinitesimal volume element is $dM = \rho\, dV$. (For a two-dimensional object, $dM = \rho\, dA$, and $dM = \rho\, dl$ for a one-dimensional distribution, where ρ is the mass per unit area or the mass per unit length.) In general, ρ is a

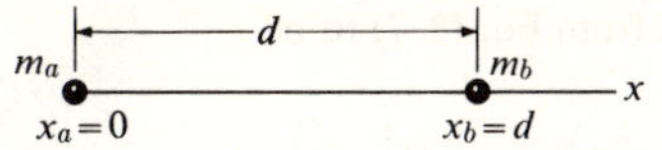

FIG. 8-2. Coordinate system for determining center of mass for two particles.

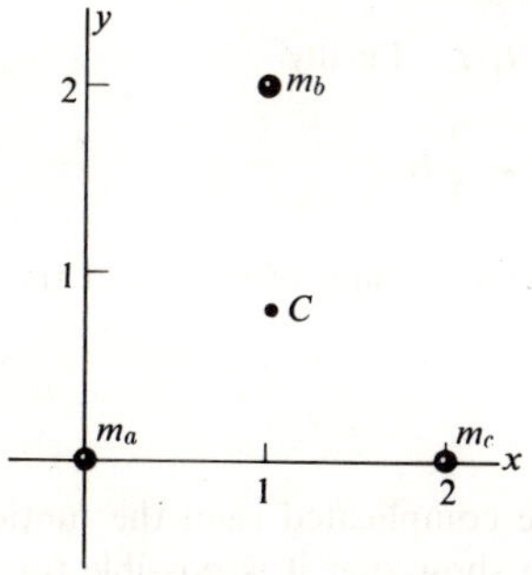

FIG. 8-3. Coordinate system for determining position of center of mass for three particles.

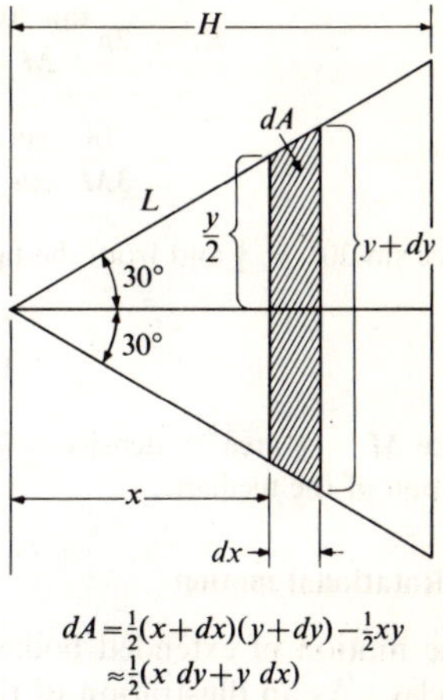

FIG. 8-4. Finding the center of mass of an equilateral triangle.

function of position $\rho(\mathbf{r})$, and

$$\mathbf{R} = \frac{1}{M}\int \mathbf{r}\rho(\mathbf{r})\,dV, \tag{8-8}$$

where

$$M = \int \rho\,dV \quad \text{or} \quad \int \rho\,dA,\ \int \rho\,dl.$$

Example: Where is the center of mass of a plane sheet of material cut in the shape of an equilateral triangle made of material of uniform density?

Solution: It is clear from Fig. 8-4 that by symmetry the center of mass must lie along the altitude. We apply Eq. (8-7) in this case. The area of the strip between x and $x + dx$ is $dA = \frac{1}{2}(x + dx)(y + dy) - \frac{1}{2}xy$. Since the density is uniform, we have for the mass of the strip,

$$dM = \tfrac{1}{2}(x\,dy + y\,dx)\rho,$$

neglecting infinitesimals of higher order. From Fig. 8-4, we have

$$\frac{y/2}{x} = \tan 30°, \qquad dy = 2\,dx \tan 30°,$$

and

$$dM = 2\rho \tan 30° x\,dx.$$

The position of the center of mass R_x is found from Eq. (8-7) to be

$$R_x = 2\rho\frac{\tan 30^\circ}{M}\int_0^H x\,dx = \frac{2\rho}{M}\tan 30^\circ\frac{H^3}{3}$$

$$= \frac{2\rho}{3M}\cdot\frac{\sin 30^\circ}{\cos 30^\circ}\cdot H^3 = \frac{2}{3}\frac{\rho}{M}\cdot\frac{1}{2}\cdot\frac{1}{H/L}\cdot H^3,$$

since $\sin 30^\circ = \frac{1}{2}$ and from the figure $\cos 30^\circ = H/L$. Finally

$$R_x = \frac{2}{3}\cdot\frac{\rho}{M}\cdot\frac{1}{2}LH\cdot H = \frac{2}{3}H,$$

since $M = \frac{1}{2}$ area $\times$ density $= \frac{1}{2}LH \times \rho$. Thus the center of mass is at the intersection of the medians.

8-3 Rotational motion

The motion of extended bodies can be more complicated than the motion of particles. As an illustration of this fact we can show that it is possible for us to have a body on which the total external force is zero, so that its center of mass is not accelerated, and yet be neither stationary nor in motion with a uniform speed. Suppose that we apply two equal and opposite forces to the flywheel shown in Fig. 8-5. The vector sum of these forces is zero, and therefore the center of mass of the flywheel undergoes no acceleration. However, the wheel rotates, and all its constituent particles except the ones on its axis of rotation are accelerated. (By axis of rotation is meant the line of particles in a rotating body which do not move.) In order to describe fully the behavior of rigid bodies we must investigate rotational as well as translational motion.

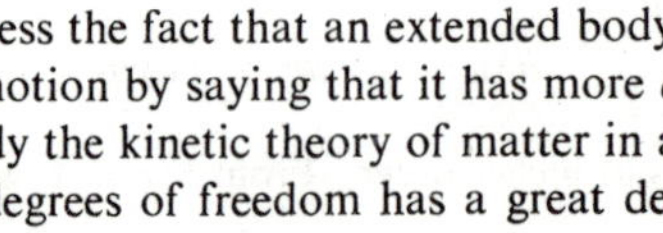

FIG. 8-5. Forces acting on flywheel to produce rotation but no acceleration of center of mass.

It is customary to express the fact that an extended body can execute rotational as well as translational motion by saying that it has more *degrees of freedom* than a particle. When we study the kinetic theory of matter in a later chapter, we shall find that the notion of degrees of freedom has a great deal of significance. Extended bodies may, in general, execute types of motion other than rotational and translational; for example, a deformable body may change its shape without turning or shifting the position of its center of mass.

The unrestricted motion of an extended body is extremely complicated. The treatment of this problem in all its generality is beyond the scope of this book.

What we shall do is to try to be as general as possible at first, and then when the going gets rough we shall specialize, confining ourselves to situations which are within our powers to analyze at this stage, and which apply to many important situations.

The first such specialization is that we consider only a *rigid body,* whose constituent particles retain their relative positions. This means that the body cannot be deformed. But this is still very complicated. To visualize the complexity of rigid-body motion, we might think of the earth, which rotates daily on its axis while moving through space on its annual revolution about the sun. The direction of the axis of the earth in space is not fixed either—it wobbles a bit. Every point on the earth's surface describes a very complicated path through space.

Many complicated motions can be decomposed into a translation of the center of mass plus a rotation about an axis through the center of mass, and we can analyze these two cases separately. The problem of a moving axis of rotation will be beyond our capabilities for the most part, and we shall neglect it.

8-4 Torque

Before we consider the rotational motion of a rigid body, it is helpful to introduce a quantity which plays the same role in this motion as that which force plays in the case of translational motion.

Let us consider the problem of opening a door. Clearly we must apply a force in order to produce motion in the door, but where should we install the doorknob, and in what direction should we exert the force on the doorknob in order that the force be a minimum? Experience tells us the answers to both these questions. The axis of rotation of a door is a line through its hinges, as in Fig. 8-6, and the farther from the axis we attach the doorknob, the less force is needed to cause the door to open with a given acceleration. If we try to open (or close) a door by pushing near the hinges, we must use a good deal more force than is necessary if we push at the outer edge of the door. The vertical position of the knob makes no difference; the important quantity is the distance from the axis of rotation to the point of application of the force. Were we to perform quantitative experiments, we would find that the angular acceleration of the door is directly proportional both to the magnitude **F** of the applied force and to **r**, the radius vector from the axis of rotation to the point where **F** is exerted (Fig. 8-7). These proportionalities assume that the same angle

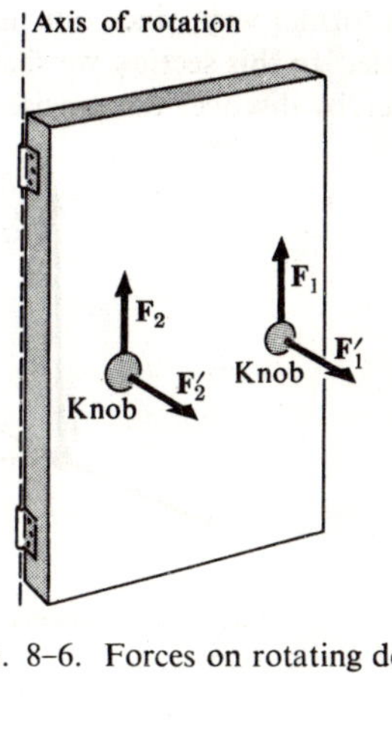

FIG. 8-6. Forces on rotating door.

θ between **F** and **r** is used each time, because this angle also plays a part in determining the angular acceleration of the door. If we push on the doorknob parallel to the plane of the door—up or down, toward or away from the hinges—no motion at all takes place. However, whenever we push perpendicular to the plane of the door we achieve very rapid motion.

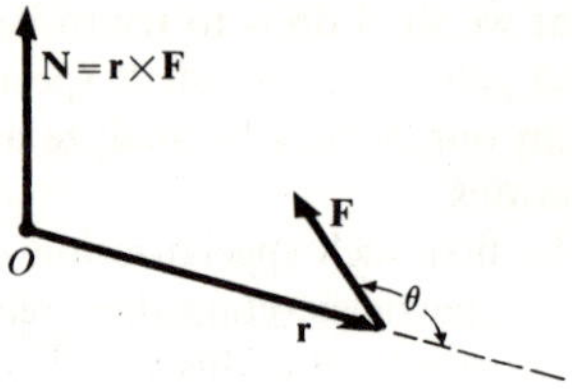

FIG. 8-7. Definition of torque.

From our knowledge of vector analysis, experimental data of the above kind suggest that the three variables (**F**, **r**, and θ) we have observed to influence the rotational motion of a rigid body (here the door) should be related according to the formula

$$\mathbf{N} = \mathbf{r} \times \mathbf{F}. \tag{8-9}$$

This would make the angular acceleration proportional to **r**, **F**, and a function of the angle θ which is zero when $\theta = 0$. The value of the quantity **N**, called *torque* or *moment of a force*, is what determines the angular acceleration of a rigid body about a particular axis of rotation. The individual values of **r**, **F**, and θ are not significant, but only the vector product as defined in Eq. (8-9).

From the definition of torque in Eq. (8-9), it is clear that its units are those of force × distance, which are newton-meters (n-m) in the mks system and pound-feet (lb-ft) in the British system.

8-5 Dynamics of rotation

While we have associated torque with angular acceleration, we have not done so in a formal way, since our intention was to make their relationship intuitively plausible. In this section we shall derive the equations of motion for a rigid body and thereby discover the precise role of **N** in angular motion.

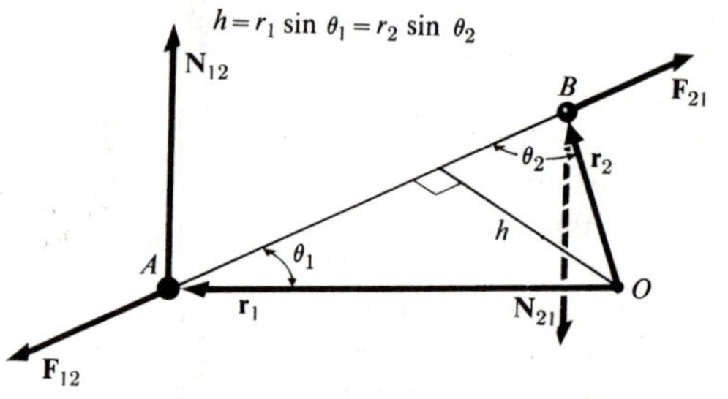

FIG. 8-8. Cancellation of torques due to internal forces.

When we speak of a rigid body, we mean one in which the relative positions of its constituent particles do not change in the course of any motion of the body. Most "hard" objects that we encounter seem to have this characteristic, but if we were to make exceedingly careful measurements we would find that even the hardest of them become distorted when forces are applied. We shall neglect such departures from rigidity here.

Let us return to our previous example of a body composed of three particles, with the additional assumption that these particles are rigidly held together. With the help of Fig. 8–8, which shows two of these particles, we can show that the torques about any axis of rotation due to the interparticle forces cancel one another. Consider $\mathbf{F}_{12}$ and $\mathbf{F}_{21}$; the torque about the point O of the body of $\mathbf{F}_{12}$ is

$$\mathbf{N}_{21} = \mathbf{r}_2 \times \mathbf{F}_{21},$$

and the torque about O of $\mathbf{F}_{21}$ is

$$\mathbf{N}_{12} = \mathbf{r}_1 \times \mathbf{F}_{12}.$$

From Newton's third law, we have $\mathbf{F}_{12} = -\mathbf{F}_{21}$, and, as the diagram shows,

$$r_1 \sin\theta_1 = r_2 \sin\theta_2 = h,$$

where h is the altitude of the triangle $O12$. The two torques therefore have the same magnitude but opposite directions, and

$$\mathbf{N}_{12} + \mathbf{N}_{21} = 0.$$

In the same way we can show that all of the other torques due to internal forces within the body cancel out in pairs, so that their resultant torque vanishes.

Now we write down the second law of motion for each of the three particles, neglecting the internal forces:

$$\mathbf{F}_1 = m_1 \frac{d\mathbf{v}_1}{dt}, \qquad \mathbf{F}_2 = m_2 \frac{d\mathbf{v}_2}{dt}, \qquad \mathbf{F}_3 = m_3 \frac{d\mathbf{v}_3}{dt}.$$

Let us take the vector product of both sides of each of the above equations with the corresponding radius vector $\mathbf{r}_i$ perpendicular to the axis of rotation of the system to the ith particle. The three vector products are

$$\mathbf{N}_1 = \mathbf{r}_1 \times \mathbf{F}_1 = m_1\mathbf{r}_1 \times \frac{d\mathbf{v}_1}{dt},$$

$$\mathbf{N}_2 = \mathbf{r}_2 \times \mathbf{F}_2 = m_2\mathbf{r}_2 \times \frac{d\mathbf{v}_2}{dt},$$

$$\mathbf{N}_3 = \mathbf{r}_3 \times \mathbf{F}_3 = m_3\mathbf{r}_3 \times \frac{d\mathbf{v}_3}{dt}.$$

The left-hand side of each equation is by definition the torque exerted on each particle about the axis of rotation of the entire system of three particles.

If we have a total of n particles composing the rigid body, rather than just three, we can write for the torque $\mathbf{N}_i$ on the ith particle

$$\mathbf{N}_i = \mathbf{r}_i \times \mathbf{F}_i = m_i\mathbf{r}_i \times \frac{d\mathbf{v}_i}{dt}.$$

Adding all the individual torques, we find that the resultant torque $\mathbf{N}$ on the entire system of n particles about its center of mass is

$$\mathbf{N} = \sum_{i=1}^{n} \mathbf{r}_i \times \mathbf{F}_i = \sum_{i=1}^{n} m_i\mathbf{r}_i \times \frac{d\mathbf{v}_i}{dt}. \tag{8-10}$$

We can rewrite the above expression for $\mathbf{N}$, the external torque, in the form

$$\mathbf{N} = \frac{d}{dt}\sum_{i=1}^{n} m_i\mathbf{r}_i \times \mathbf{v}_i = \sum_{i=1}^{n} m_i\mathbf{r}_i \times \frac{d\mathbf{v}_i}{dt} + \sum_{i=1}^{n} m_i\frac{d\mathbf{r}_i}{dt} \times \mathbf{v}_i. \tag{8-11}$$

Equation (8–11) is equivalent to (8–10) because

$$\frac{d\mathbf{r}_i}{dt} \times \mathbf{v}_i = \mathbf{v}_i \times \mathbf{v}_i = 0,$$

from the definition of the vector product.

The right-hand side of Eq. (8–11) may be simplified, since the angular velocity $\boldsymbol{\omega}$ of each particle of the rigid body is the same. That is [Eq. (4–11)],

$$\mathbf{v}_i = \boldsymbol{\omega} \times \mathbf{r}_i,$$

where $\boldsymbol{\omega}$, the angular velocity vector, has the same magnitude and direction for each of the n particles. Hence

$$\mathbf{N} = \frac{d}{dt}\sum_{i=1}^{n} m_i\mathbf{r}_i \times (\boldsymbol{\omega} \times \mathbf{r}_i). \tag{8-12}$$

8–6 Motion about a fixed axis

Equation (8–12) is a very complicated expression indeed. We propose to consider some straightforward mass distributions that show how complicated (8–12) can get, and then consider only situations in which it reduces to a simple formula.

First we rewrite (8–12) in terms of a new quantity,

$$\mathbf{N} = \frac{d}{dt}\mathbf{L}, \tag{8-13}$$

where

$$\mathbf{L} = \sum_{i=1}^{n} m_i\mathbf{r}_i \times (\boldsymbol{\omega} \times \mathbf{r}_i). \tag{8-14}$$

The quantity **L** is called the *angular momentum* of the system. The right-hand side of (8–14) can be simplified. In problem 23 of Chapter 2 we found that the double vector product could be written as

$$\mathbf{r}_i \times (\boldsymbol{\omega} \times \mathbf{r}_i) = r_i^2\boldsymbol{\omega} - (\mathbf{r}_i \cdot \boldsymbol{\omega})\mathbf{r}_i,$$

and so

$$\mathbf{L} = \sum_{i=1}^{n} m_i[r_i^2\boldsymbol{\omega} - (\mathbf{r}_i \cdot \boldsymbol{\omega})\mathbf{r}_i]. \tag{8–15}$$

Let us apply (8–15) to the simplest possible example to see how complex it is. A single particle of mass m is rotating in a plane (parallel to the xy-plane) perpendicular to the z-axis in Fig. 8–9. The angular momentum has components along the x-, y-, and z-axes. Let $\boldsymbol{\omega} = \omega\mathbf{k}$; the angular velocity points in the z-direction. Then from Eq. (8–15), we have

$$\begin{aligned} L_z &= m(\omega r^2 - \omega z^2) = m\omega(x^2 + y^2 + z^2 - z^2) \\ &= m\omega(x^2 + y^2) = m\omega d^2. \end{aligned} \tag{8–16}$$

The result is that L_z is equal to $m\omega$ times the square of the perpendicular distance of the particle from the axis of rotation, d. Using (8–15), and noting that $\boldsymbol{\omega}$ has only a z-component, we obtain

$$L_x = -m\omega zx, \qquad L_y = -m\omega zy. \tag{8–17}$$

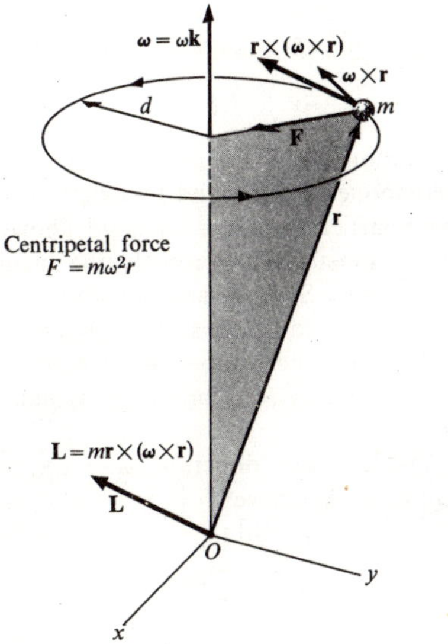

FIG. 8–9. The angular momentum of a particle moving in a circular path.

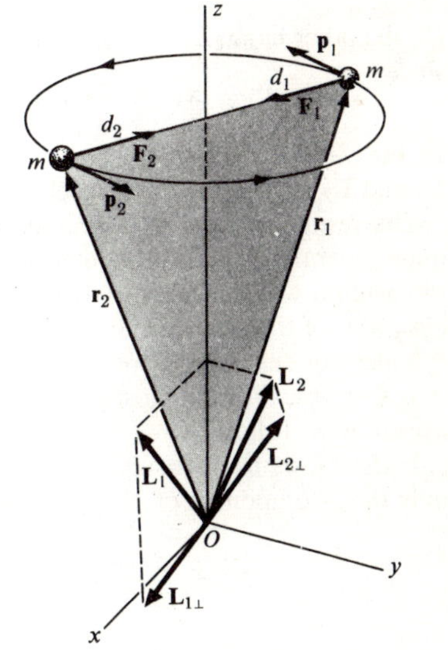

FIG. 8–10. The angular momentum of two identical particles moving in a circular path.

The angular momentum about the z-axis is a constant. The angular momenta about the x- and y-axes are not constant, since x and y vary during the motion.

The centripetal force that keeps the particle moving in a circle provides the x- and y-components of the torque. By referring to (8–13), we see that an external torque in the x- and y-direction implies a changing x- and y-component of angular momentum in time. The centripetal force **F** does not provide any z-component of angular momentum, since it lies in a plane parallel to the xy-plane (Fig. 8–9). Thus L_z is a constant, since the z-component of the torque is zero.

In most applications great simplifications occur in understanding rigid-body motions when the angular momentum is parallel to the angular velocity. We would prefer to avoid terms like (8–17). This can happen in several ways, for instance, if the axis of rotation passes through a symmetry axis of the body. Suppose we have two particles of the same mass at opposite ends of the diameter, moving in a circular path in a plane parallel to the xy-plane (Fig. 8–10). We compute L_z again, and find that

$$\begin{aligned} L_z &= m\omega[(r_1^2 - z_1^2) + (r_2^2 - z_2^2)] \\ &= m\omega[(x_1^2 + y_1^2) + (x_2^2 + y_2^2)] \\ &= m\omega(d_1^2 + d_2^2) \\ &= \sum_{i=1}^{n} m\omega d_i^2 = 2m\omega d^2. \end{aligned}$$

On the other hand, $L_x = L_{1x} + L_{2x}$ and $L_y = L_{1y} + L_{2y}$ are zero, from (8–17), since

$$x_1 = -x_2, \qquad y_1 = -y_2, \qquad z_1 = z_2.$$

In Fig. 8–10 we see that the sum of the components of the angular momentum $\mathbf{L}_1$ and $\mathbf{L}_2$ in the xy-plane cancel. The z-components of $\mathbf{L}_1$ and $\mathbf{L}_2$ add.

The result we have obtained for the two-particle problem analyzed above is quite general. If the axis of rotation is some symmetry axis of a body, the angular momentum will be parallel to the angular velocity. Such an axis is called a *principal axis* of the system. The diameter of a sphere and the axis of a right circular cylinder are examples of principal axes. Here only the component of **L** parallel to $\boldsymbol{\omega}$ survives. For the most part we shall consider only rotations about symmetry axes in what follows.

If the z-axis is a principal axis and we take $\boldsymbol{\omega}$ in that direction, $\boldsymbol{\omega} = \omega\mathbf{k}$, and only the z-component of **N** is different from zero. We have

$$N_z = \frac{d}{dt}\sum_{i=1}^{n} m_i(r_i^2 - z_i^2)\omega\mathbf{k} = \frac{d}{dt}\sum_{i=1}^{n} (m_i d_i^2)\omega,$$

where d_i^2 is the square of the distance of the ith particle from the axis of rotation.

Since $\sum_{i=1}^{n} m_i d_i^2$ is constant in a rigid body, then

$$\mathbf{N} = \sum_{i=1}^{n} m_i d_i^2 \frac{d\omega}{dt}. \tag{8-18}$$

The scalar quantity

$$I = \sum_{i=1}^{n} m_i d_i^2 \tag{8-19}$$

is called the *moment of inertia* of the rigid body about the axis of symmetry. We shall discuss its calculation in the following section. It has the same value regardless of the body's state of motion, and depends only on the distribution of mass and the axis of rotation. In terms of the moment of inertia I of the body and its angular acceleration,

$$\alpha = d\omega/dt.$$

We may finally write

$$\mathbf{N} = I\alpha. \tag{8-20}$$

The product of the moment of inertia and the angular acceleration of a rigid body equals the external torque applied to it. This equation is reminiscent of Newton's second law of motion,

$$\mathbf{F} = M\mathbf{A},$$

as applied to the linear motion of a rigid body, and has a similar significance in the rotational motion of such a body. Evidently the moment of inertia I plays a part in rotational motion analogous to that of mass M in linear motion, just as torque $\mathbf{N}$ and force $\mathbf{F}$, and angular acceleration α and linear acceleration $\mathbf{A}$ are analogous, respectively.

We have confined our derivations of the formulas for rigid-body motion to rotations about a fixed principal axis. The axis of rotation need not be a principal one, of course, and for that matter the body need not be symmetrical: A potato can be made to rotate about a randomly placed skewer. In the general case the rigid axis exerts forces on the rigid body which provide the necessary torques to satisfy the fundamental equations $\mathbf{N} = (d/dt)\mathbf{L}$ for angular momenta other than those along the axis of rotation. We have already seen this occur when a single particle rotates in a plane. Since these forces are perpendicular to the axis of rotation they will not produce torques along the axis. If we consider the axis and the body as one system and we are not concerned with internal forces, the only relevant angular momentum will be along the axis of rotation, and we need consider only the external torques about this axis. The internal forces are very important in technology whenever, for example, high-speed machinery must be balanced, but we shall not concern ourselves with this problem.

We have simplified the general equations of motion for rigid-body motion to consider cases only when the angular momentum $\mathbf{L} = I\omega$, where I is a scalar and

L is parallel to $\boldsymbol{\omega}$. The general relationship between angular momentum and angular velocity can be written as $\mathbf{L} = \mathbf{I}:\boldsymbol{\omega}$. The quantity **I** in this equation is not a scalar, but mixes in various components of angular momenta for each component of **L**. In this case **L** and $\boldsymbol{\omega}$ are not parallel, and **I** is a more complicated kind of quantity called the *moment of inertia tensor*.

8-7 Moment of inertia

We can see that the moment of inertia of a body depends on the distribution of mass in the body with respect to the axis of rotation. For instance, we might compare the two disks of Fig. 8-11; one is a disk of iron surrounded by a wooden ring, while the other, of identical total mass and dimensions, consists of a wooden core surrounded by an iron ring. The latter has the greater moment of inertia because, since each mass element is weighted by the square of its distance from the axis, the heavier iron on the outside has a greater effect on I than when it forms the core of the body. The dependence of I on the position of the axis is illustrated by the examples below. Note that the radius vector $\mathbf{r}_i$ is always taken as the distance of the ith particle of the body from the axis of rotation. In the metric system I is expressed in kilograms·meter^2, and in the British system in slugs·foot^2.

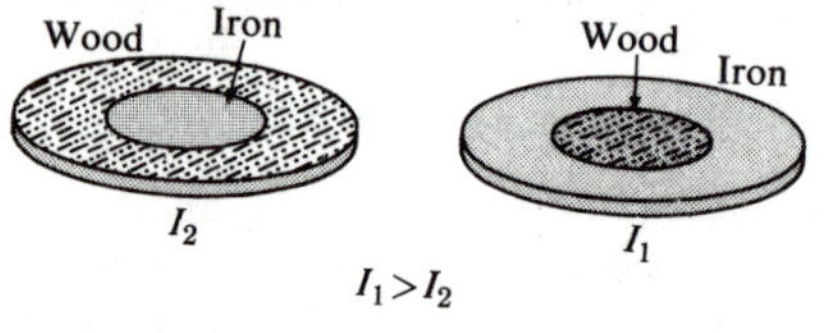

FIG. 8-11. Comparison of moments of inertia of disks of equal sizes and masses with different distribution of mass.

Example: Find the moment of inertia of a thin ring of mass M and average radius R about an axis passing through its center and perpendicular to the plane of the ring.

Solution: As in Fig. 8-12, we proceed by subdividing the ring into n segments, each of which is at a distance R from the axis. Hence

$$\begin{aligned} I &= m_1R^2 + m_2R^2 + m_3R^2 + \cdots + m_nR^2 \\ &= (m_1 + m_2 + m_3 + \cdots + m_n)R^2. \end{aligned}$$

But the sum of the masses of the segments is the same as the total mass M of the ring, and so

$$I = MR^2. \tag{8-21}$$

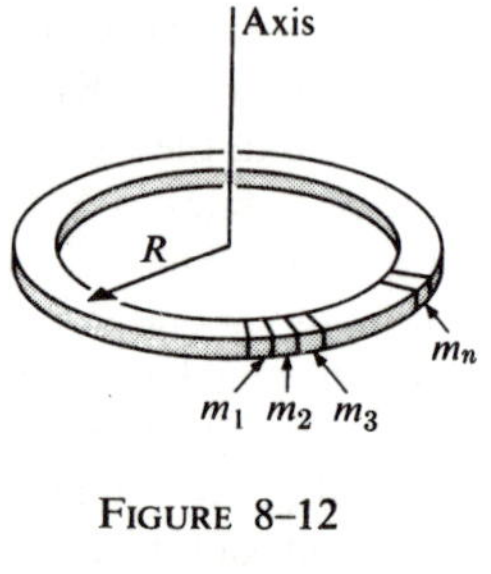

FIGURE 8-12

Example: Find the moment of inertia of a weightless rod 4 ft long that has 10-lb iron balls at each end about an axis perpendicular to the rod and passing through its midpoint.

Solution: From Eq. (8–15), with the help of Fig. 8–13,

$$I = \sum_i m_i r_i^2 = m_1 r_1^2 + m_2 r_2^2$$

$$= 2 \times (\tfrac{10}{32} \times 2^2) \text{ slug-ft}^2$$

$$= 2.5 \text{ slug-ft}^2.$$

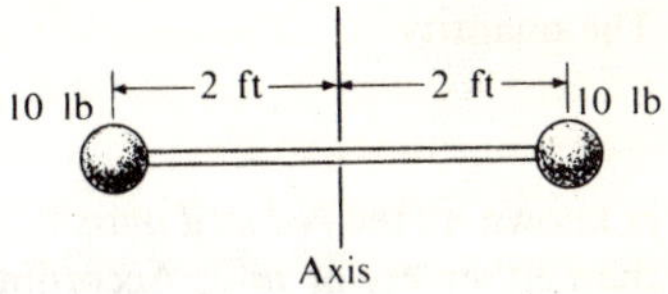

FIGURE 8–13

An object that consists of two particles at either end of a weightless rod—a sort of idealized dumbbell—is worth special attention because it is a reasonably good approximation of both a simplified model of a one-electron atom and of a diatomic molecule. Such an object can rotate about its center of mass, and we shall require a formula for its moment of inertia when we consider atoms and molecules in later chapters. The center of mass of the system is determined by the condition that

$$m_1 r_1 = m_2 r_2, \tag{8–22}$$

where r_1 and r_2 are respectively the distances of the particles of masses m_1 and m_2 from the center of mass (Fig. 8–14). If the particles are a distance l apart,

$$r_1 + r_2 = l.$$

The moment of inertia of the system about an axis through the system's center of mass perpendicular to the line joining the particles is

$$I = m_1 r_1^2 + m_2 r_2^2. \tag{8–23}$$

We can obtain a more convenient expression for I by proceeding as follows. First we multiply Eq. (8–23) by $(m_1 + m_2)/(m_1 + m_2)$, to obtain

$$I = \frac{m_1^2 r_1^2 + m_1 m_2 r_1^2 + m_1 m_2 r_2^2 + m_2^2 r_2^2}{m_1 + m_2}. \tag{8–24}$$

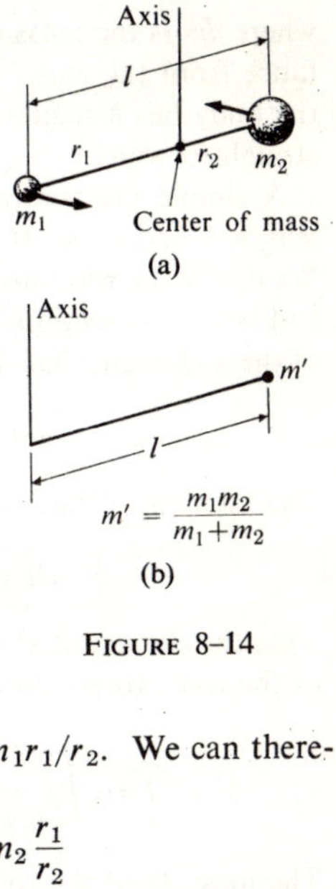

FIGURE 8–14

From Eq. (8–22) we see that $m_1 = m_2 r_2 / r_1$ and $m_2 = m_1 r_1 / r_2$. We can therefore substitute

$$m_1^2 = m_1 m_2 \frac{r_2}{r_1} \quad \text{and} \quad m_2^2 = m_1 m_2 \frac{r_1}{r_2}$$

in the first and last terms of the numerator of Eq. (8–24), with the result that

$$I = \frac{m_1 m_2}{m_1 + m_2}(r_1^2 + 2 r_1 r_2 + r_2^2) = \frac{m_1 m_2}{m_1 + m_2}(r_1 + r_2)^2$$

$$= \frac{m_1 m_2}{m_1 + m_2} l^2 = m' l^2. \tag{8–25}$$

The quantity

$$m' = \frac{m_1 m_2}{m_1 + m_2} \tag{8–26}$$

is known as the *reduced mass* of the system of two particles, because m' is smaller than either m_1 or m_2. According to Eq. (8–25), a pair of rigidly connected particles that are a distance l apart have a moment of inertia about their center of mass equal to that of a single particle of mass m' that rotates the distance l from its axis of rotation.

When a body consists of a continuous distribution of matter, we can calculate the moment of inertia by subdividing the body into many particles. The more particles we imagine it to contain, the more accurate will the computed value of its moment of inertia be. In the limit of the finest possible subdivision, the summation of Eq. (8–19) becomes the integral

$$I = \int r^2\, dm, \tag{8–27}$$

where dm is the mass of each of the infinitesimal elements of mass and r is its distance from the axis. The integral is taken over the volume of the body, and, if the body has a regular shape and is of a uniform density, its evaluation is usually straightforward.

A simple illustration of the use of Eq. (8–27) is the case of a uniform rod of length L and mass M which rotates about an axis through its center (Fig. 8–15). We divide the rod into elements of length dx. If the rod has cross-sectional area A and density ρ, each of these elements has the volume $A\,dx$ and the mass

$$dm = \rho A\, dx.$$

The moment of inertia dI of the element is

$$dI = x^2 \rho A\, dx,$$

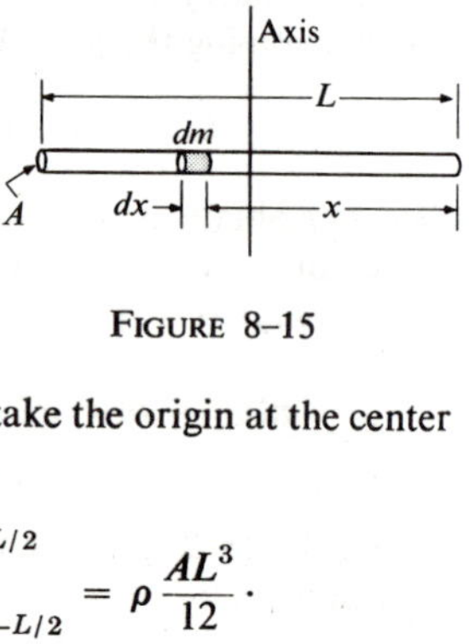

FIGURE 8–15

since the element is at the distance x from the axis. We take the origin at the center of the rod. Integrating dI from $-L/2$ to $L/2$, we have

$$I = \int_{-L/2}^{L/2} dI = \rho A \int_{-L/2}^{L/2} x^2\, dx = \rho A \left.\frac{x^3}{3}\right|_{-L/2}^{L/2} = \rho \frac{AL^3}{12}.$$

The mass M of the rod is the product of its density ρ and its volume AL,

$$M = \rho A L,$$

and so the moment of inertia of a rod about an axis perpendicular to the rod and passing through its center is

$$I = \tfrac{1}{12} M L^2. \tag{8–28}$$

The element of mass dm that is most appropriate to employ depends on the body under consideration. For instance, the simplest way of finding the moment of inertia of a solid disc about its axis of symmetry is to choose a ring dr thick as the element of mass, as in Fig. 8-16. If the disc has the uniform density ρ and is h thick, and the ring is r in radius, the volume of the ring is $(2\pi r\, dr)h$, and its mass is

$$dm = 2\pi rh\rho\, dr.$$

From Eq. (8-20), the moment of inertia dI of the ring is

$$dI = r^2\, dm = 2\pi\rho hr^3\, dr.$$

Integrating dI from $r = 0$ to $r = R$, the radius of the disc, we obtain

$$I = \int_0^R dI = 2\pi\rho h \int_0^R r^3\, dr = 2\pi\rho h \frac{R^4}{4}.$$

Since the disc has area πR^2, its mass is

$$M = \rho\pi R^2 h,$$

and its moment of inertia is simply

$$I = \tfrac{1}{2}MR^2. \tag{8-29}$$

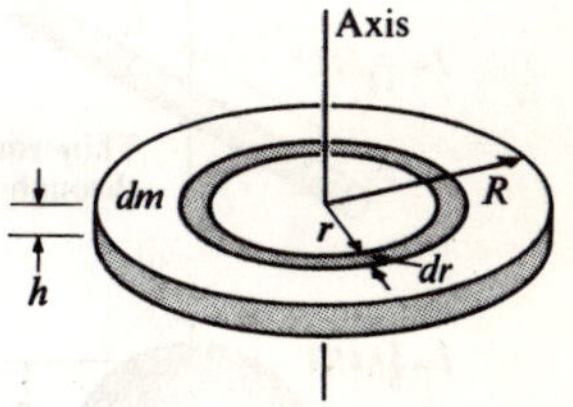

FIGURE 8-16

Figure 8-17 gives the moments of inertia of several regularly shaped bodies in terms of the total mass M, and dimensions of each.

It is occasionally convenient to express the moment of inertia of an object in terms of the distance its mass, contracted to a point, must be placed from the axis of rotation in order to give the same value of I. This distance is called the *radius of gyration k*. In terms of k the moment of inertia is

$$I = Mk^2, \tag{8-30}$$

so that when we know I and M we can always compute the radius of gyration.

Example: What is the radius of gyration of a uniform rod rotating about an axis perpendicular to the rod and passing through its center?

Solution: Rewriting Eq. (8-30), we obtain

$$k = \sqrt{I/M}, \qquad I = \tfrac{1}{12}ML^2.$$

Therefore

$$k = \sqrt{ML^2/12M} = L/2\sqrt{3}.$$

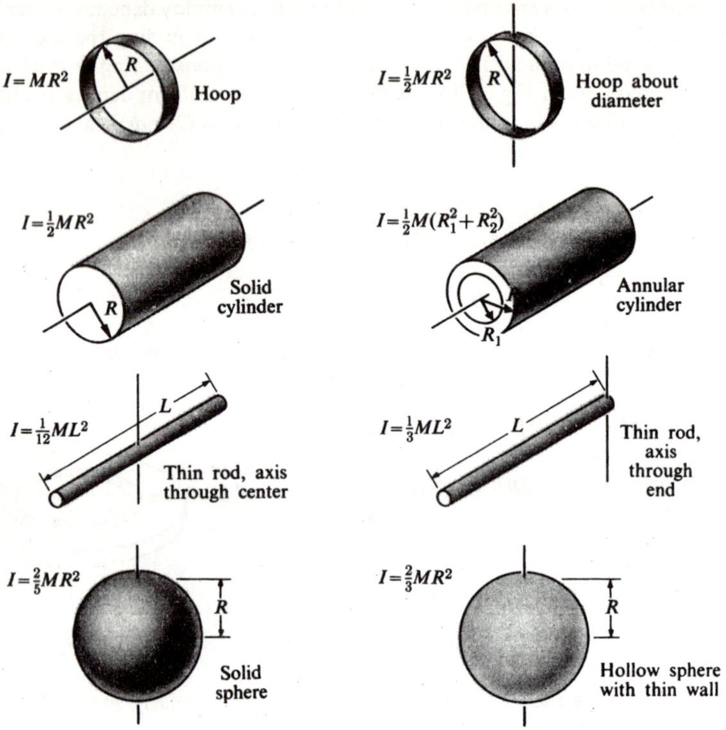

FIG. 8-17. Moments of inertia of various bodies about indicated axes.

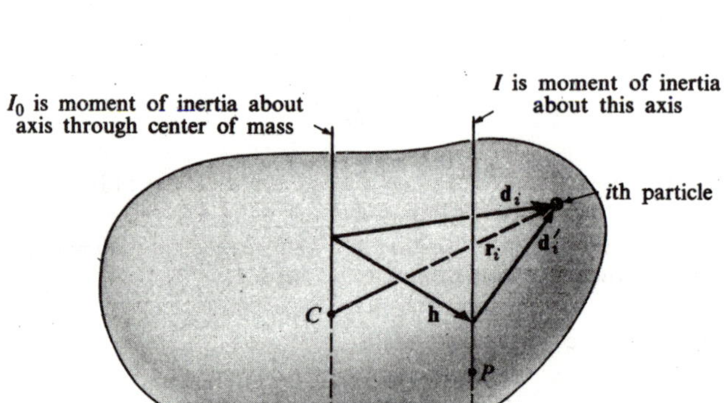

FIG. 8-18. Transfer theorem for moments of inertia.

8–8 Transfer theorem for moments of inertia

There is a useful relationship between the moment of inertia of a body about an axis through its center of mass and its moment of inertia about a parallel axis at a distance h from the first. The relationship is

$$I = I_0 + Mh^2, \tag{8–31}$$

where I_0 is the moment of inertia about the axis passing through the center of mass. The proof is quite simple (Fig. 8–18). We take as the origin the center of mass of the system. The moment of inertia about the parallel axis through P is given by

$$I = \sum_{i=1}^{n} m_i d_i'^2. \tag{8–32}$$

We define the moment of inertia as a sum, although the integral representation for continuously distributed masses give the same result. From Fig. 8–18,

$$\mathbf{d}_i' = \mathbf{d}_i - \mathbf{h}. \tag{8–33}$$

Inserting (8–33) into (8–32), we have

$$\begin{aligned} I &= \sum_{i=1}^{n} m_i(\mathbf{d}_i - \mathbf{h})^2 \\ &= \sum_{i=1}^{n} m_i d_i^2 - 2\mathbf{h}\cdot\sum_{i=1}^{n} m_i \mathbf{d}_i + h^2 \sum_{i=1}^{n} m_i. \end{aligned} \tag{8–34}$$

The middle term is intimately related to the definition of the position of the center of mass

$$\mathbf{R} = \frac{1}{M}\sum_{i=1}^{n} m_i \mathbf{r}_i.$$

But the center of mass is at the origin, so that

$$\mathbf{R} = 0 = \sum m_i \mathbf{r}_i.$$

The expression $\mathbf{h}\cdot\sum_i m_i \mathbf{d}_i$ is the component of $\sum_i m_i \mathbf{r}_i$ in the direction of $\mathbf{h}$; $\mathbf{d}_i$ is the projection of $\mathbf{r}_i$ in the direction of $\mathbf{d}_i$ (see Fig. 8–18). Therefore $\mathbf{h}_i\cdot\mathbf{d}_i$ is the projection of $\mathbf{d}_i$, and therefore of $\mathbf{r}_i$ in the direction of $\mathbf{h}$. Since $\sum_i m_i \mathbf{r}_i = 0$, its projections along any line, in particular along $\mathbf{h}$, is also zero. Thus Eq. (8–34) reduces to

$$I = I_0 + Mh^2,$$

since $\sum_{i=1}^{n} m_i = M$. This was the theorem to be proved.

Example: What is the moment of inertia of a rod of length L about an axis passing through one end?

Solution: We have previously found that the moment of inertia about an axis passing through the center of mass is

$$I_0 = \frac{ML^2}{12}.$$

By the transfer theorem,

$$I = I_0 + M\left(\frac{L}{2}\right)^2 = \tfrac{1}{12}ML^2 + \tfrac{1}{4}ML^2 = \frac{ML^2}{3}.$$

8-9 Equilibrium of a rigid body

In Chapter 5 we saw that in order for a particle to be in equilibrium (i.e., not be accelerated), it was necessary that the sum of the forces acting upon it be zero. Earlier in this chapter, however, we found that the sum of the forces acting on a rigid body—no longer a simple particle—could equal zero and yet parts of the body could be accelerated. Of course, if the resultant $\mathbf{F} = 0$, the center of mass of the body is not accelerated, but the body could be undergoing an angular acceleration about the center of mass if a torque were present. Thus there are more conditions for the equilibrium of a rigid body than for the corresponding particle: we must have both

$$\mathbf{F} = \sum_{i=1}^{n} \mathbf{F}_i = 0, \tag{8-35}$$

$$\mathbf{N} = \sum_{i=1}^{n} \mathbf{N}_i = 0. \tag{8-36}$$

In fact, Eqs. (8-35) and (8-36) represent six conditions that have to be satisfied for an extended rigid body to be in equilibrium. This is hardly surprising, since the body has six degrees of freedom available to it, three for rotation and three for translation. A particle has only three degrees of freedom associated with translation. An extended body in a plane has three degrees of freedom also—two for translation in the plane and one for rotation about an axis perpendicular to the plane.

In applying Eq. (8-36), the question frequently arises as to which axis one should choose for calculating the moments or torques. We can show that if the body is in equilibrium about one axis, it is also in equilibrium about all axes parallel to the first. Suppose that

$$\sum_{i=1}^{n} \mathbf{F}_i = 0$$

and

$$\sum_{i=1}^{m} \mathbf{N}_i = 0$$

about some axis passing through A (Fig. 8-19). Now consider another axis displaced from A by an amount $\mathbf{h}$. The moments about A are given by

$$\mathbf{N}_A = \sum_{i=1}^{n} (\mathbf{r}_i \times \mathbf{F}_i).$$

The moments about B are given by

$$\mathbf{N}_B = \sum_{i=1}^{n} (\mathbf{r}'_i \times \mathbf{F}_i) = \sum_{i=1}^{n} (\mathbf{r}_i + \mathbf{h}) \times \mathbf{F}_i$$

$$= \sum_{i=1}^{n} (\mathbf{r}_i \times \mathbf{F}_i) + \mathbf{h} \times \sum_{i=1}^{n} \mathbf{F}_i$$

$$= \mathbf{N}_A + \mathbf{h} \times \sum_{i=1}^{n} \mathbf{F}_i. \qquad (8\text{–}37)$$

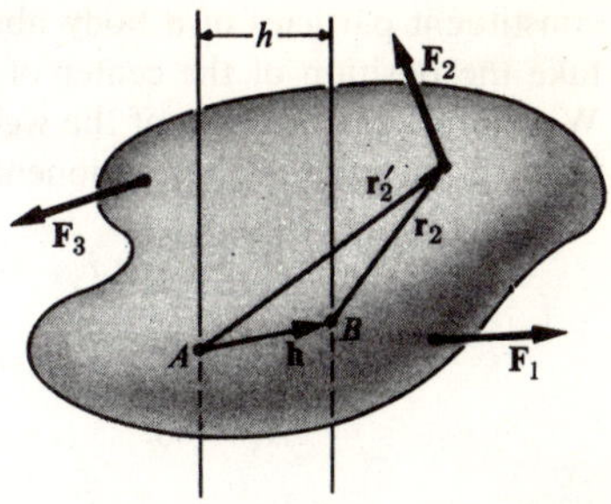

FIG. 8–19. Moments about parallel axes.

But since $\mathbf{F} = \sum_{i=1}^{n} \mathbf{F}_i = 0$, we have that $\mathbf{N}_B = \mathbf{N}_A$. The theorem is thus proved, since $\mathbf{N}_A$ is zero about any axis if the body is in equilibrium.

Equation (8–37) actually is a statement of a more general theorem which is sometimes useful. *The torque due to forces about axis B is the same as the torque of the same forces about a parallel displaced axis A plus the torque due to the sum of forces applied to A about B.*

We can use this theorem to show that the weight of a body can be considered, so far as torques are concerned, as acting at the center of mass. The weight of a body consists of the weight of each of the constituent particles all acting parallel to each other. If g is the same for all particles, the torques of the weights of the

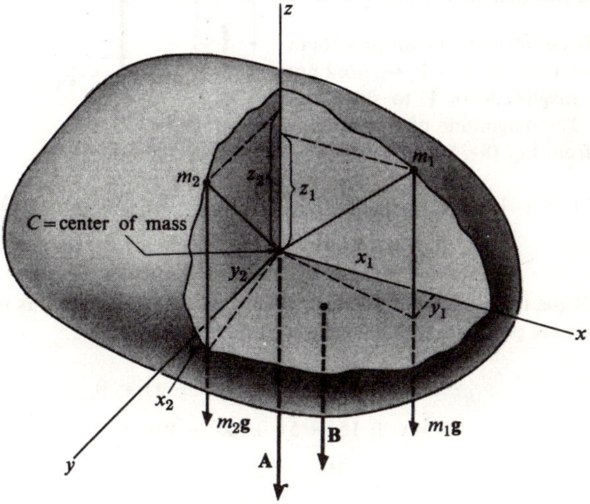

FIG. 8–20. Center of mass is center of gravity.

constituent particles of a body about its center of mass are zero. To prove this, take the position of the center of mass as the origin of coordinates (Fig. 8-20). We choose the direction of the weights to be the z-axis, so that all forces act only in the z-direction. The components of the torque are

$$N_x = \sum y_i F_{iz} = \sum y_i m_i g = g \sum m_i y_i,$$

$$N_y = -\sum x_i F_{iz} = -\sum x_i m_i g = -g \sum m_i x_i,$$

$$N_z = 0. \tag{8-38}$$

[See Eq. (8-10).] Hence the magnitude of $\mathbf{N}$ is proportional to $\mathbf{R}$, where $\mathbf{R}$ is the position vector of the center of mass. Since the center of mass is at the origin, $\mathbf{R} = 0$, and all the components of torque about the center of mass are zero.

Using the theorem of Eq. (8-37) with $\mathbf{N}_A = 0$, we see that the torque due to the weight about any axis is equal to that of the weight concentrated at the center of mass. For this reason the center of mass is called the *center of gravity*. The latter is an ambiguous term, however, and does not emphasize that it applies only when the weight vectors are all parallel and that g is constant for the body.

The actual application of these ideas in practice is simple, as the following examples will show.

Example: A uniform rod 6 ft long (Fig. 8-21) weighs 5 lb. A weight of 10 lb is placed at one end and a weight of 30 lb at the other. At what point should a man lift the rod in order that it be balanced?

Solution: If we calculate the torques about the point of application of $\mathbf{F}$, we need not know the magnitude of $\mathbf{F}$ to solve this problem. The magnitude of $\mathbf{F}$ can easily be found from Eq. (8-35):

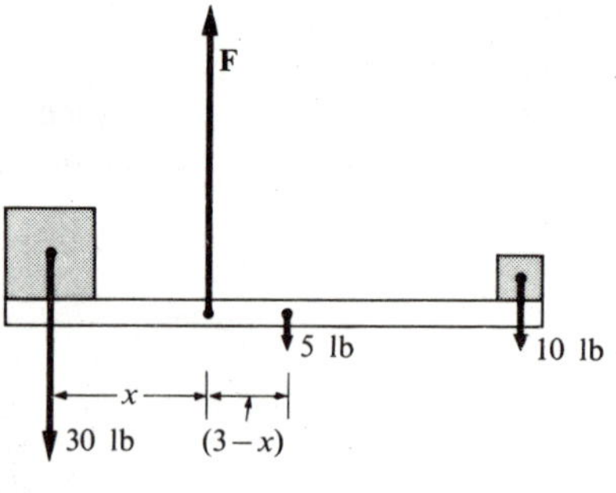

FIG. 8-21. Example of taking moments.

$$30 + 5 + 10 - \mathbf{F} = 0,$$

$$\mathbf{F} = 45 \text{ lb}.$$

The sum of the torques about an axis passing through this point is as follows (clockwise torques are positive, counterclockwise negative):

$$-30x + 5(3 - x) + 10(6 - x) = 0,$$

$$30x + 15 - 5x + 60 - 10x = 0,$$

$$45x = 75$$

$$x = 1\tfrac{2}{3} \text{ ft}.$$

Example: A uniform ladder of length 20 ft (Fig. 8-22) rests against a frictionless wall at an angle of 60° with the horizontal. The ladder weighs 30 lb. A 150-lb man is 4 ft from the top of the ladder. What is the reaction force at each end of the ladder?

Solution: There is only a reaction perpendicular to a frictionless wall. We call the reaction F_R. The components of the reaction at the base are F_H and F_V. Applying the condition (8-35), $\sum_i F_i = 0$. The horizontal components yield

$$F_R + F_H = 0,$$

$$F_R = -F_H.$$

For the vertical components,

$$F_V + 150 + 30 = 0,$$

$$F_V = -180 \text{ lb}.$$

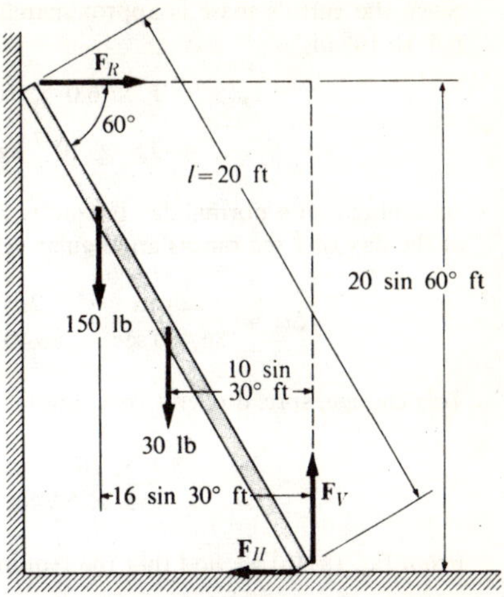

FIG. 8-22. Equilibrium of a ladder leaning against a smooth wall.

If we calculate the torques about the foot of the ladder, then, since the r's for these forces are zero (we call clockwise torques positive and counterclockwise torques negative), and since the torque is $\mathbf{r} \times \mathbf{F}$, we have

$$F_R \cdot 20 \sin 60 = 30 \times 10 \sin 30 + 150 \times 16 \sin 30,$$

$$F_R = \frac{1}{10\sqrt{3}}(150 + 1200) = \frac{135}{\sqrt{3}} = 86.5 \text{ lb}.$$

Knowing F_R, we can find F_H. The resultant of F_H and F_V can be found by vectorial addition.

8-10 Dynamics of rigid body motion

The theory we have developed enables us to solve many interesting dynamical problems. What usually goes into the solution of these problems is a combination of Eq. (8-20) and Newton's second law of motion (8-5) as applied to rigid bodies. Some examples will clarify how these are used.

Example: How much torque must be applied to the earth in order to increase the length of the day by 1 sec per year? Assume that the earth is a perfect sphere of uniform density.

Solution: From Fig. 8-17, we note that for a sphere

$$I = \tfrac{2}{5}Mr^2.$$

Since the earth's mass is approximately 6.0×10^{24} kg and its radius approximately 6.4×10^6 m,

$$I_{\text{earth}} = \tfrac{2}{5} \times 6.0 \times 10^{24} \times (6.4 \times 10^6)^2 \text{ kg-m}^2$$
$$= 9.8 \times 10^{37} \text{ kg-m}^2.$$

Assuming that a normal day is exactly 24 hr or 86,400 sec long, a change in the length of the day of 1 sec means an angular velocity change of

$$\Delta\omega = \frac{2\pi \text{ rad}}{86{,}400 \text{ sec}} - \frac{2\pi \text{ rad}}{86{,}401 \text{ sec}} = 8.4 \times 10^{-10} \text{ rad/sec}.$$

This change, spread over a year, implies an angular acceleration of

$$\alpha = \frac{\Delta\omega}{\Delta t} = \frac{8.4 \times 10^{-10} \text{ rad/sec}}{365 \times 86{,}400 \text{ sec}} = 2.7 \times 10^{-17} \text{ rad/sec}^2.$$

From Eq. (8–16) we find that the required torque is

$$N = I\alpha = 9.8 \times 10^{37} \times 2.7 \times 10^{-17} \text{ kg-m}^2/\text{sec}^2$$
$$= 2.6 \times 10^{21} \text{ n-m},$$

a rather large torque.

Example: A circular cylinder has an inextensible string wrapped around it, as in Fig. 8–23. What is the linear acceleration of the cylinder when it is released?

Solution: Call the tension in the string F. The translation of the center of mass is described by Eq. (8–5):

$$Mg - F = MA, \qquad (8\text{–}39)$$

where A is the acceleration of the center of mass perpendicular to the earth. Applying (8–20), we note that the torque about the axis of the cylinder is supplied only by the tension in the string:

$$FR = I\alpha.$$

The string is inextensible, so the linear acceleration and the angular acceleration are related by

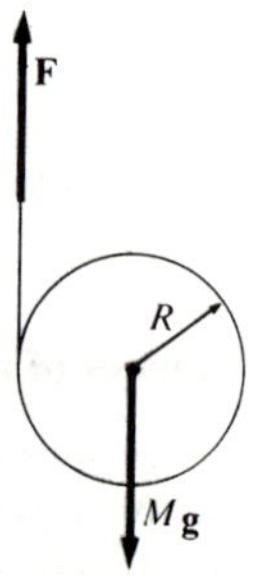

FIGURE 8–23

$$A = R\alpha.$$

Finally, from Table 8–17, $I = \frac{1}{2}MR^2$. Substituting into (8–39), we have

$$Mg - F = MA, \qquad FR = \tfrac{1}{2}MR^2 \cdot \frac{A}{R},$$
$$F = \tfrac{1}{2}MA, \qquad A = \tfrac{2}{3}g.$$

Example: A pulley of radius R and mass M has an inextensible string placed over it (Fig. 8-24). Masses m_1 and m_2 ($m_2 > m_1$) are attached to the string. What is the acceleration of the system? What are the tensions in the string?

Solution: Call the tensions in the string T_1 and T_2. The mass m_2 suffers an acceleration given by

$$m_2 g - T_2 = m_2 a,$$

and m_1 is accelerated according to

$$T_1 - m_1 g = m_1 a.$$

The a's are the same, since the string is inextensible. The torque on the pulley is provided by the tensions in the string. Since

$$\mathbf{N} = (T_2 - T_1)R$$
$$= \tfrac{1}{2}MR^2 \cdot \frac{a}{R},$$
$$T_2 - T_1 = \tfrac{1}{2}Ma.$$

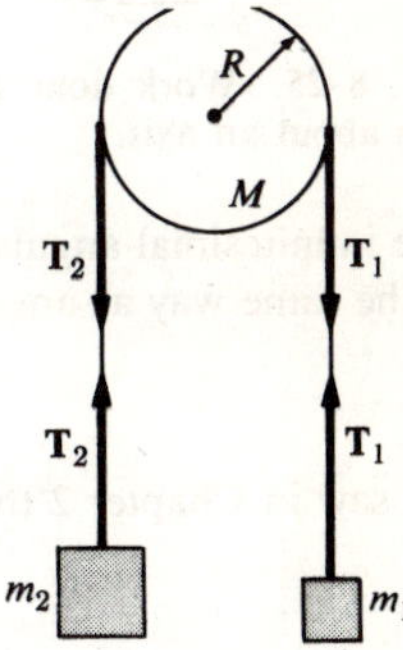

FIG. 8-24. Atwood's machine, with pulley which has non-negligible mass.

Adding the first two equations above therefore yields

$$-(T_2 - T_1) + (m_2 - m_1)g$$
$$= -\tfrac{1}{2}Ma + (m_2 - m_1)g = (m_1 + m_2)a;$$

$$a = \frac{m_2 - m_1}{m_1 + m_2 + \tfrac{1}{2}M}\, g.$$

The problem is similar to the Atwood's machine. The finite mass of the pulley reduces the acceleration of the system, as compared with the Atwood's machine which has a massless pulley.

8-11 Rotational energy and work

In Chapters 6 and 7 we explored the relationship between work and energy for single particles. We can define the work done by external forces on rigid bodies, and we can calculate their kinetic and potential energies also. The easiest way to do this is to add up the work done on the individual particles of which the rigid body is composed, inserting the condition for rigidity while doing so.

The work done by external forces $\mathbf{F}_i$ on a particle if it rotates through an angle $d\theta$ (Fig. 8-25) is

$$dW_i = \mathbf{F}_i \cdot d\mathbf{S}_i. \tag{8-40}$$

For rotations about a fixed axis,

$$d\mathbf{S}_i = d\boldsymbol{\theta} \times \mathbf{r}_i. \tag{8-41}$$

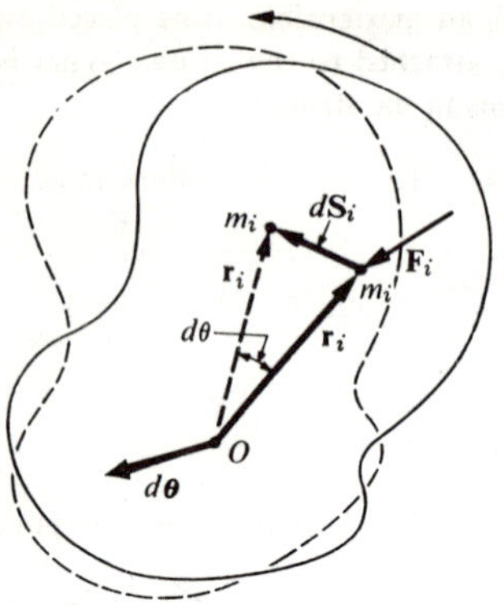

FIG. 8-25. Work done in arbitrary rotation about an axis.

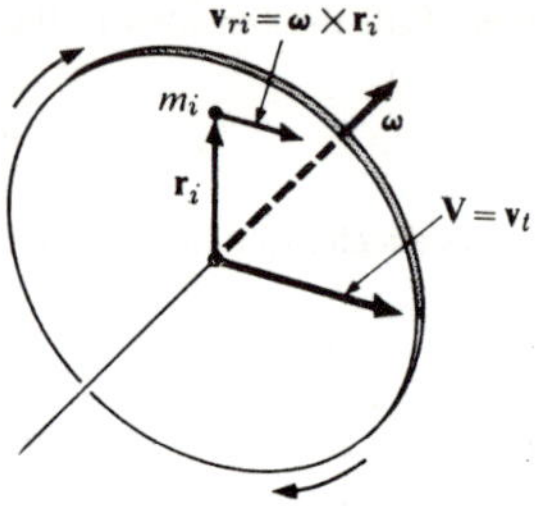

FIG. 8-26. Translational velocity of a portion of a rotating body.

The infinitesimal angular displacement is a vector, and its direction is determined in the same way as angular velocity. Inserting (8-41) into (8-40), we have

$$dW_i = \mathbf{F}_i \cdot d\boldsymbol{\theta} \times \mathbf{r}_i.$$

We saw in Chapter 2 that $\mathbf{A} \cdot \mathbf{B} \times \mathbf{C} = \mathbf{B} \cdot \mathbf{C} \times \mathbf{A}$; hence

$$dW_i = d\boldsymbol{\theta} \cdot (\mathbf{r}_i \times \mathbf{F}_i) = \mathbf{N}_i \cdot d\boldsymbol{\theta},$$

$$W = \sum_{i=1}^{n} \int \mathbf{N}_i \cdot d\boldsymbol{\theta}.$$

The integral is over the entire angular displacement.

In the case of a rigid body composed of many particles, the total energy of the body is the sum of the individual energies of all the particles. Thus the kinetic energy T of a rigid body with n constituent particles is

$$T = \sum_{i=1}^{n} \tfrac{1}{2} m_i v_i^2. \tag{8-42}$$

Now each particle may be thought of as having two distinct velocities superimposed upon one another: a translational velocity $\mathbf{v}_t$ equal to the velocity $\mathbf{V}$ of the center of mass of the body, and a velocity $\mathbf{v}_r$ due to the motion of the particle around an axis through the center of mass (Fig. 8-26). In terms of the angular velocity ω of the particle about the axis of rotation, which has the same value for all the n particles in the body,

$$\mathbf{v}_r = \boldsymbol{\omega} \times \mathbf{r}_i,$$

where, as before, $\mathbf{r}_i$ is the radius vector from the axis of rotation to the particle we are discussing.

The resultant velocity $\mathbf{v}_i$ of the ith particle of the rigid body, then, consists of two parts and is given by

$$\mathbf{v}_i = (\mathbf{v}_i)_t + (\mathbf{v}_i)_r = \mathbf{V} + \boldsymbol{\omega} \times \mathbf{r}_i. \tag{8–43}$$

Substituting the value of $\mathbf{v}_i$ from this equation into the expression for the kinetic energy of the entire body, Eq. (8–40), we find that

$$T = \frac{1}{2}\sum_{i=1}^{n} m_i(\mathbf{V} + \boldsymbol{\omega} \times \mathbf{r}_i)^2 = \frac{1}{2}\sum_{i=1}^{n} m_i[V^2 + 2\mathbf{V}\cdot\boldsymbol{\omega}\times\mathbf{r}_i + (\boldsymbol{\omega}\times\mathbf{r}_i)^2].$$

If the rigid body has the kind of symmetry about the axis of rotation discussed in Section 8–5 and illustrated by Fig. 8–10, the summation

$$\sum_{i=1}^{n} 2\mathbf{V}\cdot\boldsymbol{\omega}\times\mathbf{r}_i$$

over all the particles of the body equals zero. The reason is that, for every particle with a certain value of $\mathbf{r}_i \times \boldsymbol{\omega}$, there is a corresponding one on the opposite side of the axis of rotation whose $\mathbf{r}_i \times \boldsymbol{\omega}$ has the same magnitude but the opposite direction. Thus the $\mathbf{r}_i \times \boldsymbol{\omega}$ cancel out in pairs when summed over the entire body, and

$$T = \frac{1}{2}\sum_{i=1}^{n} m_i[V^2 + (\boldsymbol{\omega}\times\mathbf{r}_i)^2].$$

A further simplification is possible under most circumstances:

$$(\boldsymbol{\omega}\times\mathbf{r}_i)^2 = \omega^2 r_i^2 \sin^2\theta_i,$$

where θ_i is the angle between $\boldsymbol{\omega}$ and $\mathbf{r}_i$. The expression $r_i \sin\theta_i$ (Fig. 8–27) is just the perpendicular distance d_i of the ith particle from the axis of rotation. We have then that

$$(\boldsymbol{\omega}\times\mathbf{r}_i)^2 = \omega^2 d_i^2.$$

We may therefore write for the kinetic energy of a rigid body,

$$T = \frac{1}{2}\sum_{i=1}^{n} m_i V^2 + \frac{1}{2}\sum_{i=1}^{n} m_i d_i^2 \omega^2.$$

But the quantity $\sum_{i=1}^{n} m_i$ is just the total mass M of the body, and $\sum_{i=1}^{n} m_i d_i^2$ is its moment of inertia I. Thus

$$T = \tfrac{1}{2}Mv^2 + \tfrac{1}{2}I\omega^2. \tag{8–44}$$

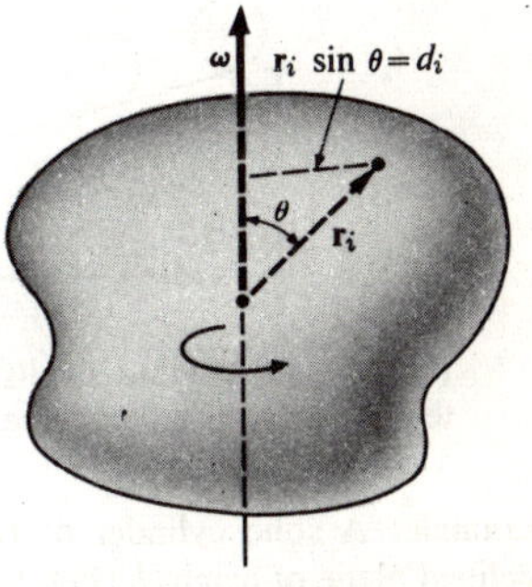

FIG. 8–27. Calculation of $(\omega \times r_i)^2$.

The total kinetic energy of a rigid body is the sum of the kinetic energy of the translational motion of a mass M located at its center of mass and the kinetic energy of its rotation about the center of mass. Even if $V = 0$, so that the center of mass is fixed in space, the body can have kinetic energy if it is rotating.

Example: Compute the rotational kinetic energy of the earth.

Solution: In an earlier example we found the earth's moment of inertia to be

$$I = 9.8 \times 10^{37} \text{ kg-m}^2.$$

The angular velocity of the earth is 7.3×10^{-5} rad/sec, corresponding to one rotation per day. Hence its rotational kinetic energy is

$$\begin{aligned} T &= \tfrac{1}{2} I\omega^2 \\ &= \tfrac{1}{2} \times 9.8 \times 10^{37} \text{ kg-m}^2 \times (7.3 \times 10^{-5} \text{ rad/sec})^2 \\ &= 2.6 \times 10^{29} \text{ joules.} \end{aligned}$$

It is interesting to compare this figure with the translational kinetic energy of the earth due to its annual revolution around the sun. Since the average orbital speed of the earth is 3×10^4 m/sec, its translational kinetic energy is

$$\begin{aligned} T &= \tfrac{1}{2} mv^2 \\ &= \tfrac{1}{2} \times 6 \times 10^{24} \text{ kg} \times (3 \times 10^4 \text{ m/sec})^2 \\ &= 2.7 \times 10^{33} \text{ joules.} \end{aligned}$$

The kinetic energy of the earth's orbital motion is about 10,000 times greater than that of its rotational motion.

It is also frequently possible to consider the *potential* energy of a rigid body as the potential energy of a particle having the mass M of the entire body that is situated at the body's center of mass.

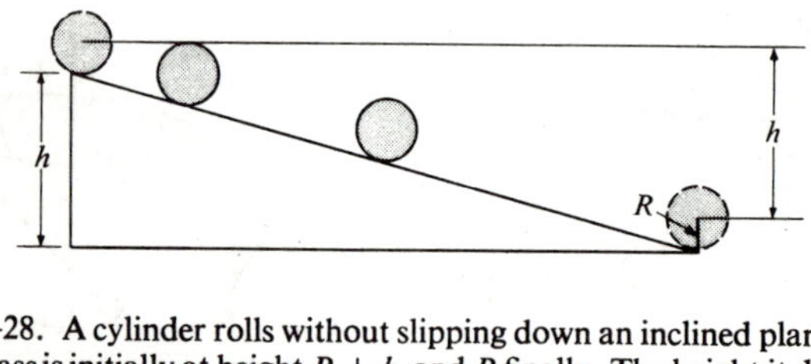

FIG. 8-28. A cylinder rolls without slipping down an inclined plane. Center of mass is initially at height $R + h$, and R finally. The height it drops is h.

Example: A solid cylinder of radius R and mass M rolls without slipping down an inclined plane of height h (Fig. 8-28). If it starts from rest at the top of the plane, what is its velocity at the bottom?

Solution: At the top of the inclined plane the cylinder has potential energy only, while at the bottom it has kinetic energy only. Applying the law of conservation of energy, we have

$$Mgh = \tfrac{1}{2}MV^2 + \tfrac{1}{2}I\omega^2.$$

For a cylinder, $I = \frac{1}{2}MR^2$, and $\omega R = V$ if there is no slippage; therefore

$$Mgh = \tfrac{1}{2}MV^2 + \tfrac{1}{4}MV^2 = \tfrac{3}{4}MV^2,$$

and the velocity at the bottom is $V = \sqrt{\frac{4}{3}gh}$. If the body *slid* down the inclined plane instead of rolling down,

$$Mgh = \tfrac{1}{2}MV^2 \quad \text{and} \quad V = \sqrt{2gh}.$$

Hence the cylinder travels more slowly when it rolls, owing to the fact that some of the available energy appears as rotational energy.

When no frictional or other dissipative forces are acting on a rigid body, mechanical energy is conserved just as in the case of a single particle. There is, however, one situation in which, even though an external dissipative force acts, the mechanical energy of a rigid body is *still* conserved. This occurs when frictional forces cause an object to roll rather than slide along a surface. The reason that energy is conserved under these circumstances is that the frictional forces do no work: they always act parallel to the surface along which the body moves, and (Fig. 8–29) there is no instantaneous displacement parallel to the surface when the body is rolling. Since work is the product of force and displacement, rolling motion conserves energy, provided that no sliding takes place.

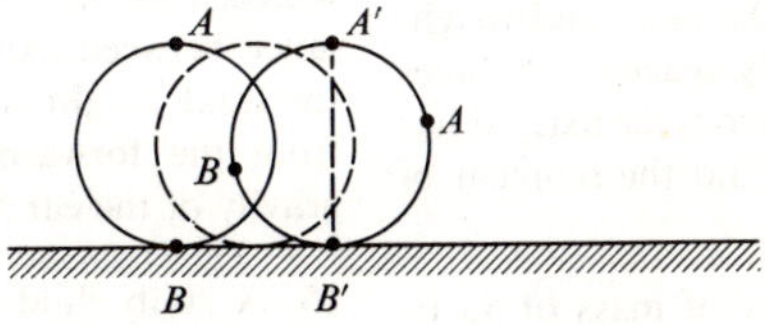

FIG. 8–29. A cylinder rolls without slipping on a plane. The frictional force does no work.

Of course, experience tells us that frictional forces do dissipate energy when bodies are rolling. If the bodies are rigid, they are in contact only along a line. Real bodies are not rigid. The surfaces in contact deform, and the frictional forces do operate through some distance and do work.

Problems

1. Find the center of mass of a system of three masses located at the vertices of an equilateral triangle whose side is 1 m long, if the ratio of the masses is 1:2:3.

2. What is the position of the center of mass of a circular disk of radius R from which has been cut a circular hole of radius $R/2$ tangent to the edge of the disk?

3. Find the center of mass of a uniform hemisphere.

4. A mass m_1 is moving with velocity $\mathbf{v}_1$ toward another mass m_2, which is stationary. Find the velocity of the center of mass of the system.

5. Find the moment of inertia of a hollow right circular cylinder whose inner radius is R_1, and outer radius R_2.

6. What is the moment of inertia for a right circular cone about its axis? The radius of the base is r.

7. Find the moment of inertia of a square about a diagonal.

8. A steel disk 2 ft in diameter and weighing 200 lb has six evenly spaced 2-in. holes whose centers are 9 in. from the axis. It has a central 4-in. hole. Find the moment of inertia about the axis.

9. Show that the center of mass of an irregular object can be found by suspending it from two different points and finding the intersection of the two plumb lines drawn from the point of suspension.

10. The Bohr atom is an electron whose mass is 9×10^{-31} kg rotating in a circle about a proton. According to the model the angular momentum of the electron in its lowest state is approximately 10^{-34} joule-sec and the radius of the orbit is 0.5×10^{-10} m. What is the linear speed of the electron in its orbit?

11. A thin uniform rod of length l is free to rotate about a horizontal axis at one end. If it were released from a horizontal position, what would its angular velocity be when it was vertical?

12. A fishing rod 3 m long is attached to a pivoting holder at its lower end. The rod is 50° above the horizontal. A fisherman grasps the rod 1 m from its lower end. What horizontal force must the fisherman exert to keep the rod at this angle when reeling in a 20-lb fish from directly below the rod's upper end?

13. A horizontal beam 6 m long projects from the wall of a building. A guy wire that makes a 40° angle with the horizontal is attached to the outer end of the beam. When a weight of 100 n is attached to the end of the beam, what is the tension in the guy wire? (Neglect the weight of the beam.)

14. The front wheels of a certain car are found to support 1600 lb and the rear wheels 1400 lb. The car's wheelbase (distance between axles) is 8 ft. (a) What is the total weight of the car? (b) How far from the forward axle is the center of gravity of the car?

15. A 30-lb child and a 50-lb child sit at opposite ends of a 12-ft seesaw pivoted at its center. Where should a third child weighing 35 lb sit in order to balance the seesaw?

16. A ladder 12 ft long rests against a vertical frictionless wall with its lower end 4 ft from the wall. The ladder weighs 40 lb and its center of gravity is at its center. If the ladder is to stay in place, what must be the minimum coefficient of friction between the bottom of the ladder and the ground?

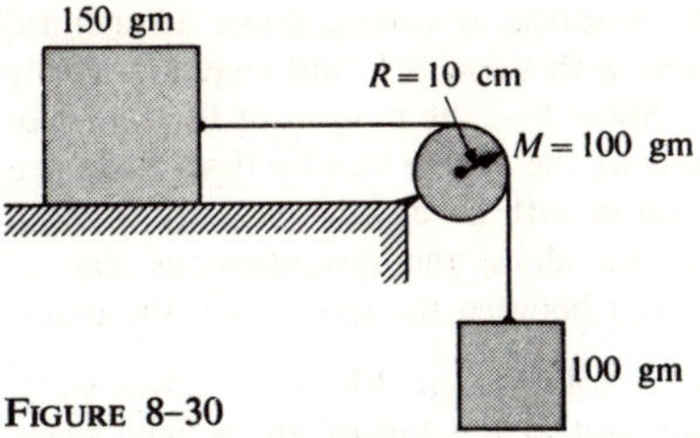

FIGURE 8–30

17. Find the acceleration of the system and the tension in the cord shown in Fig. 8–30. Neglect friction between table and block. The pulley rotates and its moment of inertia is $\frac{1}{2}MR^2$.

18. The mass of the uniform pulley shown in Fig. 8–31 is 10 lb. Find the acceleration of the system and the tension in the ropes. Neglect friction.

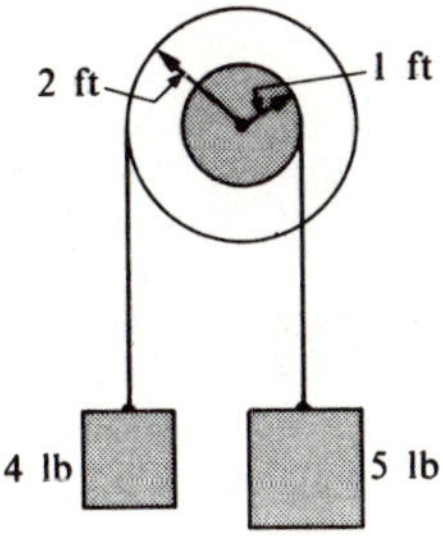

FIGURE 8–31

19. A spool of thread rests on a horizontal table (see Fig. 8–32). Which way will it roll when a force **F** is applied horizontally to the thread?

Moment of inertia $= \frac{1}{4}MR^2$

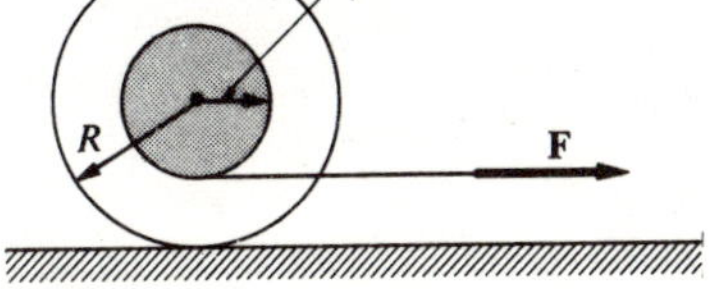

FIGURE 8–32

20. A billiard ball is thrown so that it slides along a table without spinning, with velocity v. Show that it will begin to roll without sliding when its velocity is $\frac{5}{7}v$. The coefficient of friction is constant.

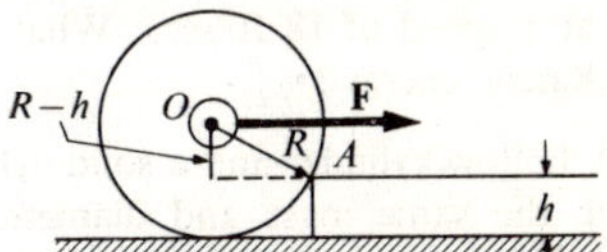

FIGURE 8–33

21. A wheel of radius R and mass m is against a step of height h (see Fig. 8–33). Show that the minimum force applied at the axle parallel to the ground necessary to make the wheel climb the step is

$$mg\sqrt{\frac{h(2R - h)}{(R - h)^2}}.$$

22. A rope 1 m long is wound around the rim of a drum of radius 12 cm and moment of inertia 0.02 kg-m^2. The rope is pulled with a force of 2.5 n. (a) Assuming that the drum is free to rotate without friction, and starts from rest, what is its final angular velocity? (b) What is its final kinetic energy? (c) How much work is done by the force?

23. A 3-kg hoop 1 m in diameter rolls down an inclined plane 10 m long that is at an angle of 20° with the horizontal. (a) What is the angular velocity of the hoop at the bottom of the plane? (b) What is its linear velocity? (c) What is its rotational kinetic energy? (d) What is its total kinetic energy?

24. The baton of a drum majorette is a 0.7-lb uniform rod 2 ft long. What is its kinetic energy when it is twirled at an angular velocity of 10 rad/sec?

25. A 200-kg cylindrical flywheel 0.3 m in radius is acted upon by a torque of

20 n-m. (a) If it starts from rest, how much time is required to accelerate it to an angular velocity of 10 rad/sec? (b) What is its kinetic energy at this angular velocity?

26. A 16-lb bowling ball 1 ft in diameter rolls at a speed of 18 ft/sec. What is its total kinetic energy?

27. A hollow cylinder and a solid cylinder having the same mass and diameter are released from rest simultaneously at the top of an inclined plane. Which reaches the bottom first?

28. A sphere is rolling down an inclined plane with velocity V and angular velocity ω. Show that the motion of the center of mass of the sphere can be described by a rotation with an angular velocity ω about an axis along the instantaneous line of contact between the sphere and the plane.

29. A solid sphere 10 cm in radius starts from rest at the top of an inclined plane 10 m long. It rolls without slipping, and reaches the bottom in 7 sec. What angle does the plane make with the horizontal?

9

Conservation Theorems for Systems of Particles

We learned of the great usefulness of a conservation theorem in Chapter 7, where the conservation of mechanical energy was considered. The significant thing about this theorem is that it often permits us to gain an insight into the motion of a particle going from one place to another under the influence of a force without going into detail about its trajectory. When we deal with a system of particles rather than an individual particle, we can obtain other conservation theorems in addition to that of conservation of energy, which together may be of considerable help in analyzing the motion of the system. These theorems also apply to single particles, of course, but in that case they have only limited value. In this chapter we shall discuss the conservation of two vector quantities, linear momentum and angular momentum; work and energy, as we recall, are both scalar quantities.

9–1 Conservation of linear momentum

The *linear momentum* **P** of a system of *n* particles is defined as

$$\mathbf{P} = \sum_{i=1}^{n} m_i \mathbf{v}_i, \tag{9–1}$$

where $\mathbf{v}_i$ is the velocity of the *i*th particle whose mass is m_i. The linear momentum is simply the vector sum of the products of m and $\mathbf{v}$ for each of the particles making up the system.

In the last chapter we found that the second law of motion when applied to a system of particles can be written

$$\mathbf{F} = \frac{d^2}{dt^2} \sum_{i=1}^{n} m_i \mathbf{r}_i, \tag{9–2}$$

where **F** is the resultant of all the external forces acting on the system. Since

$$\mathbf{v}_i = \frac{d\mathbf{r}_i}{dt},$$

if we assume that the mass of each particle does not change with time we can express (9–2) in terms of the linear momentum **P** of the system simply as

$$\mathbf{F} = \frac{d\mathbf{P}}{dt}. \tag{9–3}$$

According to this equation, the rate of change of the momentum of a system is equal to the applied external force.

When the resultant **F** of the external forces is zero,

$$\frac{d\mathbf{P}}{dt} = 0 \quad \text{and} \quad \mathbf{P} = \text{constant}.$$

When the sum of the external forces acting on a system of particles vanishes, the total linear momentum of the system remains constant. This statement is known as the theorem of the *conservation of linear momentum.* It is important to note that this theorem can be applied only when there is no resultant external force present.

A similar theorem holds for individual particles. If we call the momentum of a particle **p**, where

$$\mathbf{p} = m\mathbf{v}, \tag{9–4}$$

then, since $\mathbf{F} = m\mathbf{a} = m(d\mathbf{v}/dt)$, in terms of momentum, $\mathbf{F} = d\mathbf{p}/dt$. When no force is exerted on the particle, $d\mathbf{p}/dt = 0$, and its momentum remains constant in time. For single particles, conservation of linear momentum is not a very useful theorem, since it is identical with the first law of motion: A particle continues at rest or in uniform motion in a straight line in the absence of an applied force.

Although for single particles conservation of momentum has no consequences of note, it is much more useful when more than one particle is involved. Let us examine a simple situation. Suppose we have a particle of mass m, initially at rest, and it suddenly breaks up into two particles of masses m_1 and m_2. The forces acting on the particle that caused it to break up were internal ones, and no external force was present. Since m had the initial momentum $\mathbf{p} = 0$, the final momentum of m_1 and m_2, taken together, must be zero. Hence

$$m_1\mathbf{v}_1 + m_2\mathbf{v}_2 = 0, \qquad \mathbf{v}_2 = -\frac{m_1}{m_2}\mathbf{v}_1,$$

where $\mathbf{v}_1$ and $\mathbf{v}_2$ are the final velocities of the two particles. We note immediately that these velocities must be in opposite directions along the same line.

Example: A rifle weighing 5 lb fires a bullet weighing 0.03 lb at a muzzle speed of 2000 ft/sec (Fig. 9–1). What is the recoil speed of the rifle?

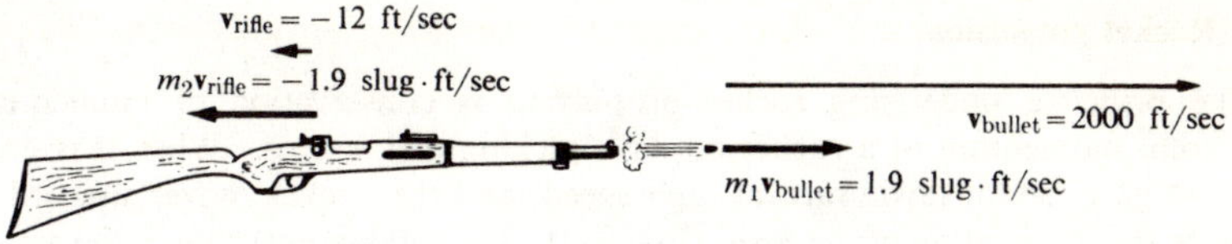

FIG. 9–1. Recoil of a rifle which has been fired.

Solution: From the above equation

$$v_{\text{rifle}} = -\frac{m_{\text{bullet}}}{m_{\text{rifle}}} v_{\text{bullet}} = -\left(\frac{0.03}{5}\right) 2000 \text{ ft/sec} = -12 \text{ ft/sec}.$$

It is essential to keep in mind the vector properties of momentum. In adding up the individual momenta $m\mathbf{v}$ we must consider the directions of the velocities and not merely add them algebraically. Sometimes the problem under consideration involves motion along a straight line, as in the example above, but in general it will be in two or three dimensions, and we must be sure to take this into account by a vector calculation.

Example: Two cars are traveling at right angles to each other (Fig. 9–2). One weighs 2000 lb and is going east at 30 mi/hr; the other, weighing 3000 lb, is traveling north at 20 mi/hr. They collide and stick together. What is the velocity of the wreck after the collision?

Solution: Call the W–E direction the positive x-axis and the S–N axis y. The final mass is 5000 lb. Momentum is conserved, which implies that the x- and y-components of the total momentum are separately conserved. Thus, in units convenient for the problem though not, strictly speaking, proper momentum units,

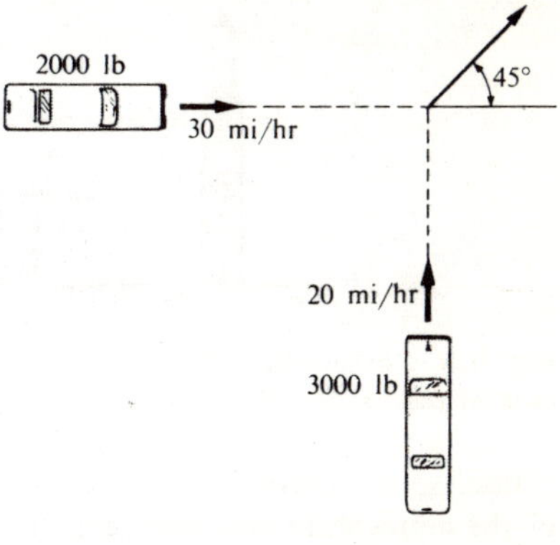

FIG. 9–2. Collision of two cars traveling at right angles to each other.

$$x\text{-component: } 2000 \times 30 \text{ lb-mi/hr} = 5000 \times V_x \text{ lb-mi/hr};$$
$$y\text{-component: } 3000 \times 20 \text{ lb-mi/hr} = 5000 \times V_y \text{ lb-mi/hr}.$$

Therefore

$$V_x = 12 \text{ mi/hr}, \qquad V_y = 12 \text{ mi/hr}.$$

The magnitude of the final velocity is

$$V = \sqrt{V_x^2 + V_y^2} = 12\sqrt{2} \cong 17 \text{ mi/hr}.$$

The direction is along a line making a 45° angle with the x-axis, which is northeast.

9–2 Rocket propulsion

The principle underlying rocket propulsion is conservation of momentum. The total momentum of a rocket on its launching pad is zero. When it is fired, exhaust gases shoot downward at high speed, and the rocket moves upward to balance the momentum of the gases (Fig. 9–3). The total momentum of the system of rocket plus exhaust ideally retains its initial value of zero if we neglect external forces. Rockets do not operate by "pushing" against their launching pads, the air, or anything else; in fact, they perform best in space, where there is no atmosphere to impede their motion. The energy of the rocket and its exhaust comes from chemical energy stored in the rocket fuel.

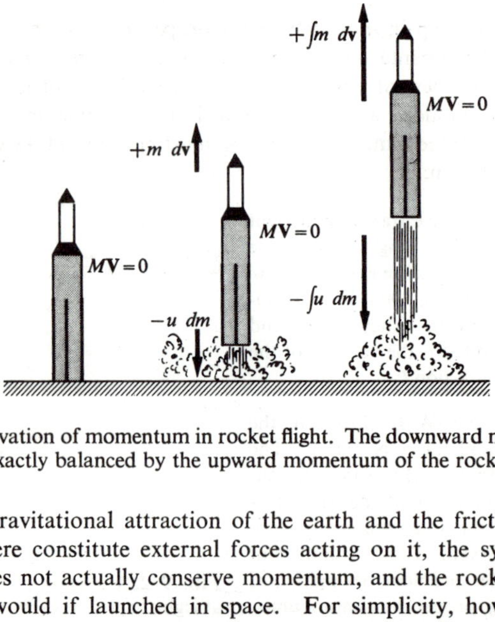

FIG. 9–3. Conservation of momentum in rocket flight. The downward momentum of the exhaust gases is exactly balanced by the upward momentum of the rocket itself.

Because the gravitational attraction of the earth and the frictional resistance of the atmosphere constitute external forces acting on it, the system of rocket plus exhaust does not actually conserve momentum, and the rocket climbs more slowly than it would if launched in space. For simplicity, however, we shall examine the latter situation. In a particular time interval dt the rocket ejects a mass dm of exhaust gas downward at a speed u relative to the rocket (Fig. 9–4). Since dm was initially part of the rocket, the change in momentum it undergoes is $u\,dm$ with respect to its momentum before leaving the rocket. Meanwhile the rocket's upward speed increases by dv for a momentum change of $m\,dv$, where m, the rocket mass at that instant, is large compared with dm. For momentum to be conserved, the momentum changes experienced by rocket and exhaust must be equal in magnitude. Taking into account that the momenta are in opposite directions, we have

$$u\,dm = -m\,dv, \qquad dv = -u\frac{dm}{m}\cdot \tag{9–5}$$

Under the assumption that the exhaust speed u is constant, which is approximately true in practice, we can integrate dv from the initial rocket speed $v = 0$ at launching to the final speed v, and the quantity $u\,dm/m$ from the initial rocket mass m_0 to the final mass m. Hence

$$\int_0^v dv = -u\int_{m_0}^{m} \frac{dm}{m} = -u[\ln m]_{m_0}^{m},$$

$$v = u(\ln m_0 - \ln m) = u \ln \frac{m_0}{m}. \tag{9-6}$$

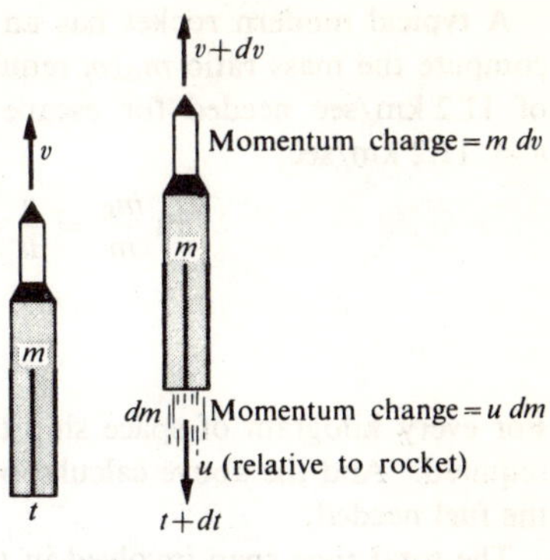

FIG. 9–4. Change of momentum of rocket in time dt.

The ultimate speed of the rocket is directly proportional to the speed u of its exhaust gases, and to the natural logarithm of the ratio of the rocket's initial mass to its final mass after the fuel has been consumed.

Rocket propulsion is a gradual rather than an instantaneous process. If the entire fuel supply of a rocket, whose mass is $m_0 - m$, could be ejected all at once at the speed u, we would have, from conservation of momentum,

$$mv = (m_0 - m)u, \qquad v = u\left(\frac{m_0}{m} - 1\right).$$

This speed is greater than the actual one,

$$v = u \ln \frac{m_0}{m},$$

as we can see from the graphs of these formulas in Fig. 9–5.

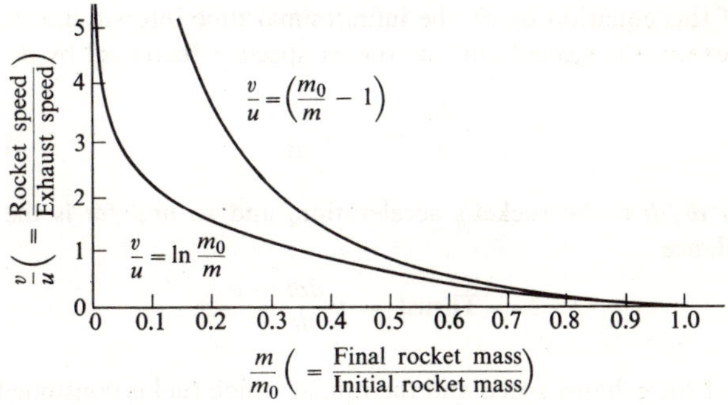

FIG. 9–5. The variation of rocket speed with the ratio between its final and initial masses. The upper curve corresponds to the sudden ejection of all the rocket's fuel, and the lower one to its gradual ejection.

A typical modern rocket has an exhaust velocity of about 2 km/sec. Let us compute the mass ratio m_0/m required of a rocket if it is to achieve the velocity of 11.2 km/sec needed for escape from the earth. With $u = 2$ km/sec and $v = 11.2$ km/sec,

$$\ln \frac{m_0}{m} = \frac{v}{u} = \frac{11.2\text{ km/sec}}{2\text{ km/sec}} = 5.6,$$

$$\frac{m_0}{m} = 270\,!$$

For every kilogram of space ship that leaves the earth, almost 270 kg of fuel is required. And the above calculation omits air resistance, which further increases the fuel needed.

The total time span involved in the acceleration does not appear in Eq. (9–6), and is irrelevant in the absence of gravity; to convince ourselves that this is reasonable, we might imagine that the rocket coasts at constant speed between discrete ejections of exhaust. Near the earth, however, gravity exerts a downward pull on the rocket that slows it down by an amount that depends on the rocket's time of flight t. If we consider a rocket of limited vertical range, so that g does not change significantly, the speed at fuel burnout will be reduced by gt:

$$v = u \ln \frac{m_0}{m} - gt.$$

When the fuel is exhausted the upward acceleration stops, and the downward acceleration of gravity slows the rocket until it comes to rest and then begins to descend. A rocket of greater altitude capabilities, of course, is proportionately less affected by gravity, since g decreases as $1/r^2$ (as we shall see in Chapter 11), where r is the distance from the center of the earth.

The upward force exerted on a rocket by the expulsion of exhaust gases is called *thrust*. We can determine the thrust of a rocket from Eq. (9–5). By dividing both sides of this equation by dt, the infinitesimal time interval during which the mass dm of exhaust is ejected and the rocket speed is increased by dv, we obtain

$$u \frac{dm}{dt} = -m \frac{dv}{dt}.$$

The quantity dv/dt is the rocket's acceleration, and so $m\, dv/dt$ is the force that acts on it. Hence

$$\text{Thrust} = u \frac{dm}{dt}, \tag{9–7}$$

the product of the exhaust speed and the rate at which fuel is consumed. At takeoff the engines of a typical large missile provide a total thrust of about 400,000 lb; if the exhaust speed u is 6000 ft/sec, the kerosene and liquid-oxygen fuel of the rocket are being used up at the rate

$$\frac{dm}{dt} = \frac{400{,}000 \text{ lb}}{6000 \text{ ft/sec}} = 67 \text{ slugs/sec},$$

which is more than one ton per second. When a rocket is launched from the earth's surface, the thrust of its engines must exceed its weight m_0g, in order for it to rise from the ground. If the thrust remains constant, the acceleration of the rocket will increase as its mass diminishes owing to the consumption of fuel. Final accelerations of several times g in magnitude are usual.

9-3 Inelastic collisions

The law of conservation of momentum is indispensable in dealing with collisions between particles. In such cases no external forces act on the participants, and therefore their total momentum before they collide equals their total momentum afterward. The essential effect of the collision is to redistribute the total momentum of the particles.

If we know only the masses and initial velocities of the bodies involved in a collision, however, momentum conservation by itself does not yield a unique result for their subsequent motion. An indeterminate amount of kinetic energy may be lost to heat, sound, or other forms of energy when the bodies interact, and, though the requirement that momentum be conserved does set limits to the result of the collision, it can go no farther. These limiting cases, though, are worth examining. At one extreme are *completely elastic collisions* in which kinetic energy is conserved. During the actual collision, to be sure, some kinetic energy becomes elastic potential energy as the bodies are deformed by the impact, but all this energy is returned as the bodies move apart. At the other extreme are *completely inelastic collisions* in which the bodies stick together permanently upon impact. The kinetic energy loss in a completely inelastic collision is the maximum possible consistent with momentum conservation.

Let us examine a completely inelastic collision between two bodies. Figure 9-6 shows a 5-kg lump of clay moving with a velocity of 10 m/sec to the left that strikes a 6-kg lump of clay moving with a velocity of 12 m/sec to the right. The

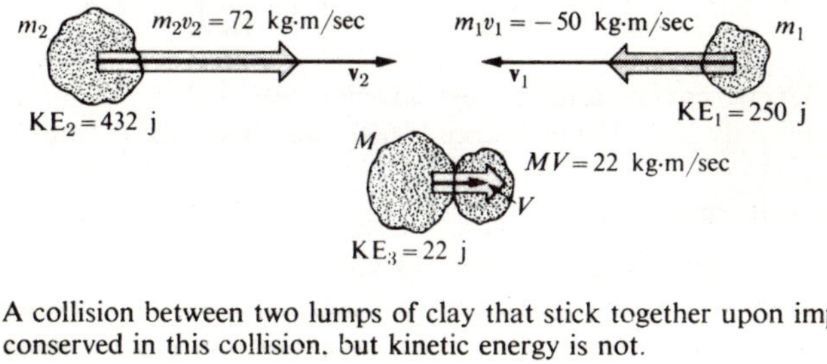

FIG. 9-6. A collision between two lumps of clay that stick together upon impact. Momentum is conserved in this collision, but kinetic energy is not.

two lumps stick together after the collision. If we call the mass of the final body M and its velocity V, conservation of momentum requires that

$$MV = m_1 v_1 + m_2 v_2. \tag{9-8}$$

Here we have

$$m_1 = 5\text{ kg}, \qquad m_2 = 6\text{ kg},$$
$$M = m_1 + m_2 = 11\text{ kg},$$
$$v_1 = -10\,\frac{\text{m}}{\text{sec}}, \qquad v_2 = 12\,\frac{\text{m}}{\text{sec}},$$

where we have adopted the convention that motion to the right is plus and motion to the left is minus. Using these values in Eq. (9-8), we have

$$11V\text{ kg} = -50\,\frac{\text{kg·m}}{\text{sec}} + 72\,\frac{\text{kg·m}}{\text{sec}},$$

$$V = \frac{22\text{ kg·m/sec}}{11\text{ kg}} = 2\,\frac{\text{m}}{\text{sec}};$$

the final body moves to the right with a velocity of 2 m/sec. We have assumed that mass is conserved in this collision, an assumption that is a valid one at the low velocities involved in this example.

Energy and momentum are entirely independent concepts. In the above problem the lumps of clay before the collision had the kinetic energies

$$\text{KE}_1 = \frac{1}{2}m_1v_1^2 = \frac{1}{2} \times 5\text{ kg} \times \left(-10\,\frac{\text{m}}{\text{sec}}\right)^2 = 250\text{ j},$$

$$\text{KE}_2 = \frac{1}{2}m_2v_2^2 = \frac{1}{2} \times 6\text{ kg} \times \left(12\,\frac{m}{\text{sec}}\right)^2 = 432\text{ j}.$$

After the collision the new lump of clay has the kinetic energy

$$\text{KE}_3 = \frac{1}{2}MV^2 = \frac{1}{2} \times 11\text{ kg} \times \left(2\,\frac{\text{m}}{\text{sec}}\right)^2 = 22\text{ j}.$$

The total kinetic energy prior to the collision was 432 + 250 or 682 j, while afterward it is only 22 j. The difference of 660 j was dissipated during the collision.

9-4 Elastic collisions

In contrast to the above situation, no kinetic energy is lost in a completely elastic collision. An interesting example is a head-on collision between two bodies whose masses are m_1 and m_2. This means that their motions are along a straight line before and after the impact. Calling the initial and final speeds of the bodies

v_1, v_2 and v_1', v_2', respectively, conservation of momentum and energy requires that

	Before		After
Momentum:	$m_1v_1 + m_2v_2$	$=$	$m_1v_1' + m_2v_2'$;
Energy:	$\frac{1}{2}m_1v_1^2 + \frac{1}{2}m_2v_2^2$	$=$	$\frac{1}{2}m_1v_1'^2 + \frac{1}{2}m_2v_2'^2$.

The first of these equations can be rewritten in the form

$$m_1(v_1 - v_1') = m_2(v_2' - v_2), \tag{9–9}$$

and the second in the form

$$m_1(v_1^2 - v_1'^2) = m_2(v_2'^2 - v_2^2),$$

$$m_1(v_1 + v_1')(v_1 - v_1') = m_2(v_2' + v_2)(v_2' - v_2). \tag{9–10}$$

Now we divide Eq. (9–10) by Eq. (9–9), and thereby obtain

$$v_1 + v_1' = v_2' + v_2$$

or

$$(v_2' - v_1') = -(v_2 - v_1). \tag{9–11}$$

Since $(v_2 - v_1)$ is the velocity of m_2 relative to m_1 before the collision and $(v_2' - v_1')$ its velocity relative to m_1 after the collision, this result means that the effect of a completely elastic head-on collision between two bodies is to reverse their relative velocity without changing its magnitude. Thus the relative velocity of approach is equal to the relative velocity of recession.

When the mass m_2 is initially at rest, $v_2 = 0$, and conservation of momentum requires that

$$m_1v_1 = m_1v_1' + m_2v_2'. \tag{9–12}$$

The relative velocity of the two bodies is changed only in direction by the collision, and so

$$v_1 = v_2' - v_1'. \tag{9–13}$$

Combining Eqs. (9–12) and (9–13), we find that the final velocity of v_1' of m_1 is

$$v_1' = \frac{m_1 - m_2}{m_1 + m_2}v_1, \tag{9–14}$$

and the final velocity v_2' of m_2, which was initially at rest, is

$$v_2' = \frac{2m_1}{m_1 + m_2}v_1. \tag{9–15}$$

The above results permit us to draw some general conclusions. If m_1 is less than m_2, v_1' is in the opposite direction to v_1: the lighter body rebounds from the larger one (Fig. 9–7a). A handball striking a wall is an extreme example, where,

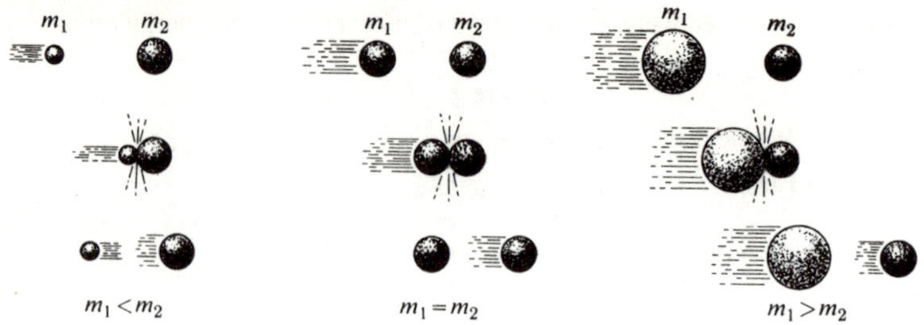

FIG. 9-7. The results of an elastic, head-on collision between a moving body of mass m_1 and a stationary body of mass m_2 depend on the ratio of their masses.

since m_2 is virtually infinite compared with m_1, $v_1' = -v_1$. In the event that $m_1 = m_2$, $v_1' = 0$, and $v_2' = v_1$, the colliding body stops, while the struck one moves off with the same speed as the incident one and in the same direction (Fig. 9-7b). When m_1 is greater than m_2, the colliding body continues on in the same direction after the impact but with reduced speed, while the struck body moves ahead of it at a faster pace (Fig. 9-7c). When m_1 is much greater than m_2, the colliding body loses little speed, while the struck one is imparted a speed nearly twice v_1.

The ratio between the kinetic energy $\frac{1}{2}m_2 v_2'^2$ transferred to the initially stationary body and the kinetic energy $\frac{1}{2}m_1 v_1^2$ of the colliding body can be found with the help of Eq. (9-15). We have

$$\frac{KE_2}{KE_1} = \frac{(1/2)m_2 v_2'^2}{(1/2)m_1 v_1^2} = \frac{m_2}{m_1}\frac{4m_1^2}{(m_1 + m_2)^2}$$

$$= \frac{4(m_2/m_1)}{(1 + m_2/m_1)^2}. \quad (9\text{-}16)$$

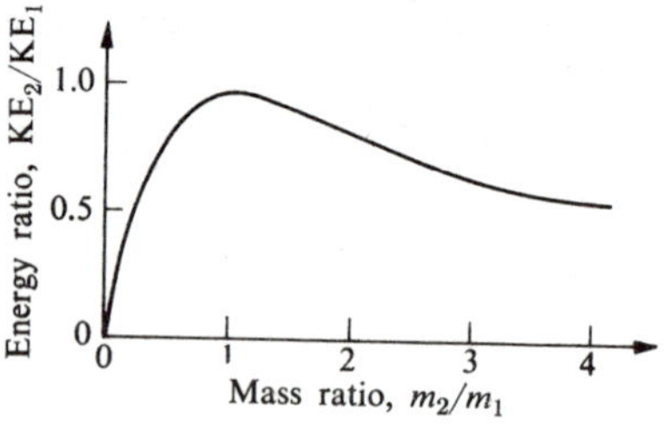

FIG. 9-8. Kinetic energy transferred to a body in a head-on collision, as a function of the ratio of the masses.

This formula is plotted in Fig. 9-8. Evidently the transfer of energy is a maximum for $m_1 = m_2$, although the greatest relative acceleration of the struck body occurs when $m_1 \gg m_2$.

In the construction of nuclear reactors it is frequently necessary to have the neutrons, which are produced at high energies, slowed down very quickly so that they will be more effective in producing further fissions. The considerations of this section show why the moderator, the material used to slow down the neutrons, is a material chosen because it has certain atoms whose masses are as small as possible. Protons, which have the same mass as neutrons, lead to the most pronounced decrease of speed in collisions. But there are other factors operating: for instance, the tendency of protons to capture neutrons to form stable nuclei. Hence deuterium (a form of hydrogen whose nuclei each consist of a proton and a neutron already bound together) and carbon are frequently used. Very heavy elements would be almost completely ineffective.

9–5 Impulse

When one body collides with another, the total momentum of the system remains constant. However, the individual momenta of the participating bodies certainly change during the actual collision, and it is interesting to look into the factors involved in these changes.

For a single particle, as we know, the second law of motion states that

$$\mathbf{F} = \frac{d\mathbf{p}}{dt}.$$

We can rewrite this as

$$d\mathbf{p} = \mathbf{F}\, dt,$$

and integrate both sides to obtain

$$\mathbf{p}_{\text{final}} - \mathbf{p}_{\text{initial}} = \int_{t_i}^{t_f} \mathbf{F}\, dt. \tag{9–17}$$

In other words, the change in the momentum of a particle equals the integral of the resultant force acting on it over the time interval in which the momentum changed. The quantity

$$\int_{t_i}^{t_f} \mathbf{F}\, dt$$

is called *impulse*; when a force acts on a particle, the resulting change in the particle's momentum is equal to the impulse of the force.

During a collision between two bodies, the bodies exert forces on one another, and the impulses of these forces cause each of the particles' momenta to change. But the force $\mathbf{F}_{AB}$ exerted by particle A on particle B is, according to Newton's third law, equal in magnitude but in the opposite direction to the force $\mathbf{F}_{BA}$ exerted by particle B on particle A at any instant of time; that is

$$\mathbf{F}_{AB} = -\mathbf{F}_{BA}.$$

Clearly the impulse imparted to particle A is equal and opposite to that imparted to particle B, so that the changes of momenta are also equal and opposite. The sum of the initial momenta of the two particles is the same as the sum of their momenta after the collision.

The concept of impulse is most useful either when a constant force is acting or when we know exactly how the force varies with time. Then we can measure the momentum change suffered by the body acted upon by the force. In such situations we can determine the time of action of the force, a fact which might not otherwise be obtainable.

9-6 Angular momentum

We have already seen a number of analogies between linear and angular motion. While the parallelism is not always exact—a finite angular displacement, for instance, is not a vector, although a finite linear displacement is—the forms of the various kinematical equations are identical in both linear and angular measure. Thus the linear displacement x from the origin of a particle restricted to move along the x-axis is

$$x = x_0 + v_0 t + \tfrac{1}{2}at^2$$

a time t after it has passed the point x_0 with the velocity v_0 while undergoing the constant acceleration a, and the corresponding equation for angular displacement θ (when θ, ω_0, and α all have a common axis) is

$$\theta = \theta_0 + \omega_0 t + \tfrac{1}{2}\alpha t^2,$$

where the symbols have their usual meanings. Such parallelisms are to be found in dynamics as well as in kinematics.

The rotational analog of linear momentum is called *angular momentum*, a concept we introduced in Chapter 8, and it is also a vector. Like linear momentum, angular momentum is most useful in treating the motion of a system of particles rather than that of a single particle. However, in order to make clear just what is involved in angular momentum, we shall begin by defining it for a single particle.

If we consider a particle moving along an arbitrary path, as in Fig. 9-9, its angular momentum $\mathbf{L}$ about some point O is defined as

$$\mathbf{L} = \mathbf{r} \times \mathbf{p}. \tag{9-18}$$

The radius vector $\mathbf{r}$ is drawn from O to the position of the particle at a particular instant, and $\mathbf{p}$ is the linear momentum of the particle at that instant. The simplest case is that of a particle moving in a circle with a uniform speed v; here $\mathbf{v}$ is always

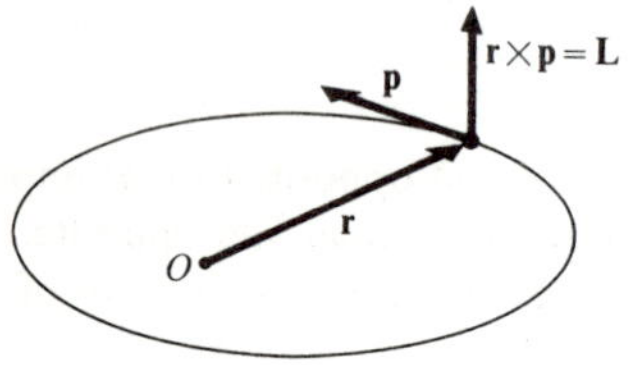

FIG. 9-9. Definition of angular momentum of particle.

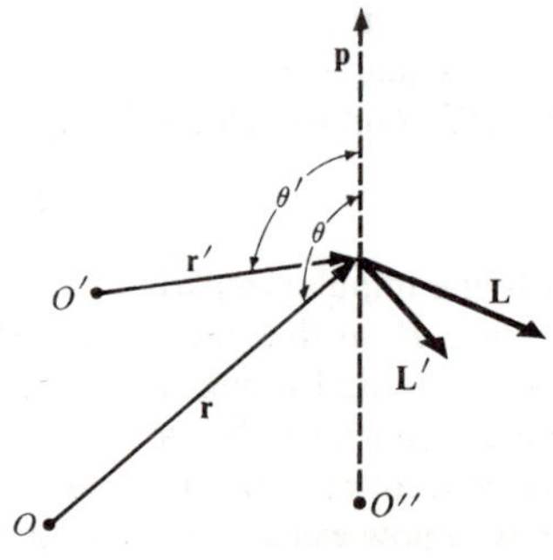

FIG. 9-10. Definition of angular momentum shows that it is different for different axes about which it is taken.

perpendicular to $\mathbf{r}$, and $L = rp$. In general, the magnitude of $\mathbf{L}$ is $rp \sin \theta$, where θ is the angle between r and p. From the definition of the vector product in Chapter 2, the angular momentum vector $\mathbf{L}$ is perpendicular to the plane in which $\mathbf{r}$ and $\mathbf{p}$ lie.

The angular momentum of a particle evidently depends on the choice of the point O about which it is taken. If we select another point O' (Fig. 9–10), both $\mathbf{r}$ and θ are different from before, and a different value of $\mathbf{L}$ is obtained. There are even points about which the angular momentum vanishes; for example, computing $\mathbf{L}$ about the position of the particle itself yields $\mathbf{L} = 0$, since the radius vector $\mathbf{r}$ is then zero. If the point chosen lies along the direction of $\mathbf{p}$, such as O'' in Fig. 9–10, the angle $\theta = 0$ and $\mathbf{L} = 0$. The angular momentum (taken about any fixed point) of a particle moving along a complicated path varies as the particle changes its position.

As we saw in the previous chapter, the angular momentum of a system of n particles is the vector sum of the angular momenta of its constituent particles, so that

$$\mathbf{L} = \sum_{i=1}^{n} m_i \mathbf{r}_i \times \mathbf{v}_i. \tag{9–19}$$

We shall now review some of the things we said in Chapter 8. If the system of particles is a rigid body, so that a single angular velocity $\boldsymbol{\omega}$ is all that is needed to describe its rotational motion, the angular momentum of the body becomes

$$\mathbf{L} = \sum_{i=1}^{n} m_i \mathbf{r}_i \times \mathbf{v}_i = \sum_{i=1}^{n} m_i \mathbf{r}_i \times (\mathbf{r}_i \times \boldsymbol{\omega})$$

$$= \sum_{i=1}^{n} m_i [\mathbf{r}_i^2 \boldsymbol{\omega} - (\mathbf{r}_i \cdot \boldsymbol{\omega}) \mathbf{r}_i]. \tag{9–20}$$

In general the angular momentum $\mathbf{L}$ of a body is not parallel to its angular velocity $\boldsymbol{\omega}$. However, if the axis of rotation is a principal axis,

$$\mathbf{L} = \sum_{i=1}^{n} m_i \, d_i^2 \boldsymbol{\omega}$$

and

$$\mathbf{L} = I\boldsymbol{\omega}. \tag{9–21}$$

(As in Chapter 8, the quantity

$$I = \sum_{i=1}^{n} m_i \, d_i^2$$

is the moment of inertia of the rigid body.) Equation (9–21) states that the angular momentum $\mathbf{L}$ of a rigid body of moment of inertia I is the product of I and the body's angular velocity $\boldsymbol{\omega}$.

9-7 Conservation of angular momentum

In the preceding chapter we obtained an equation relating the rate of change $d\omega/dt$ of the angular velocity of a rigid body with appropriate symmetry about its axis of rotation with the applied torque **N** of the forces acting upon the body. This was

$$\mathbf{N} = I\frac{d\omega}{dt},$$

which we can rewrite as

$$\mathbf{N} = \frac{d}{dt}(I\omega), \qquad \mathbf{N} = \frac{d\mathbf{L}}{dt}. \tag{9-22}$$

The rate of change of the angular momentum of a rigid body is equal to the sum of the external torques on it.

If the external torque **N** is zero, then

$$\frac{d\mathbf{L}}{dt} = 0 \qquad \text{and} \qquad \mathbf{L} = I\omega = \text{constant}.$$

When the sum of the external torques on a rigid body vanishes, the total angular momentum of the system remains constant. This statement is known as the theorem of *conservation of angular momentum.*

I large, ω small

I small, ω large

FIG. 9-11. Example of the conservation of angular momentum.

The conservation of angular momentum is employed by a skater doing a spin. The skater starts turning with her arms extended and then pulls them in close to her sides, whereupon she spins much faster (Fig. 9-11). By drawing in her arms the skater is actually decreasing her moment of inertia, and to keep the angular momentum the same, her angular velocity increases. The same is true of a tumbler who brings his knees against his chest in the course of a somersault.

A spinning football is another example of the use of the conservation of angular momentum. Angular momentum is imparted to a football by throwing it so that it spins about its long axis (Fig. 9–12). External torques are virtually absent and the angular momentum remains constant during the football's flight. Here **L** is a vector and its direction may not vary, and so the ball cannot travel end over end.

FIG. 9–12. Constant angular momentum of spinning football. If the ball were traveling end over end, the angular momentum would vary.

Often forces act on a system but produce no torques. For example, if we have a central force, it produces no torque about its center, since **r** and **F** are parallel. The earth and sun together constitute such a system. We can conclude therefore that the angular momentum of the earth about the sun is a constant. In Chapter 11 we shall see the important consequences of this fact.

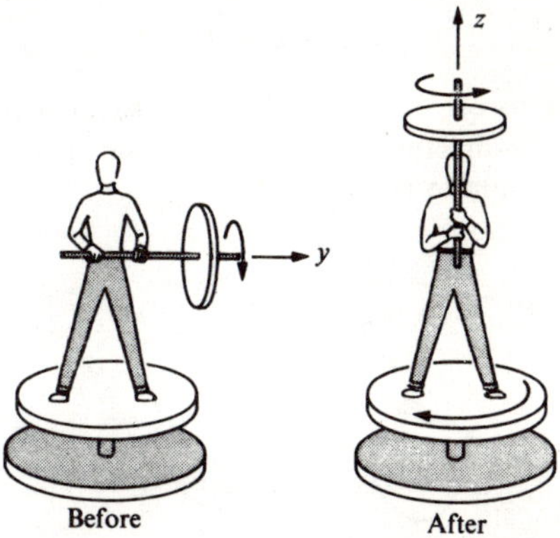

FIG. 9–13. Before: axis is horizontal; momentum about vertical axis = 0. After: axis is vertical; momentum about vertical axis is still zero; man spins in direction opposite to spin of the wheel.

The most startling applications of the conservation of angular momentum are (a) a rotating wheel and (b) a gyroscope. Suppose that we have a man, standing on a platform which can rotate about the z-axis, who holds a rotating wheel in his hand; the axis of the wheel points in the y-direction (Fig. 9–13). The component of angular momentum about the z-axis is zero. If somehow the wheel is turned so that its axis is parallel to the z-axis, the man must rotate in a direction opposite to the wheel's rotation in order for the angular momentum about the z-axis to remain zero.

It is interesting to see how one can turn the wheel in this direction and precisely what force causes the man to start rotating. In Fig. 9–14 we have a diagram which indicates the original position of the angular momentum. We wish to rotate the axle of the wheel through the angle $d\theta$. In the time dt the angular momentum turns through an angle $d\theta$ to the axis. The rate of rotation of $\mathbf{L}_0$ is an angular velocity in the x-direction, and is equal to $\Omega = d\theta/dt$. The change $d\mathbf{L}$ is a vector pointing in the z-direction. The torque must therefore point in the z-direction and be equal to

$$N_z = \frac{dL}{dt} = L_0 \frac{d\theta}{dt} = L_0\Omega,$$

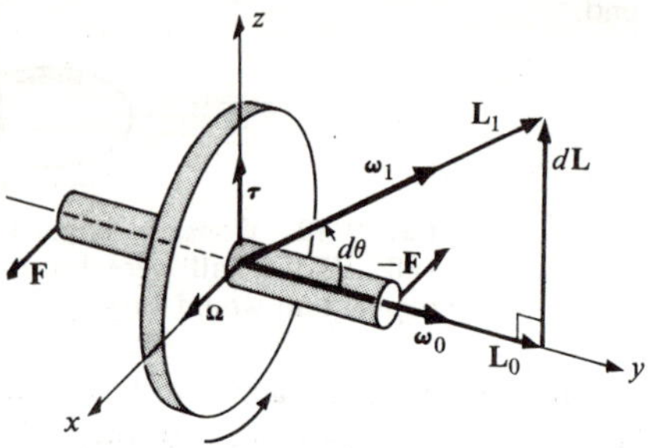

FIG. 9–14. A model of a gyroscope.

since the magnitude of $\mathbf{L}$ does not change. The above equation written in vector form is

$$\mathbf{N} = \mathbf{\Omega} \times \mathbf{L}. \tag{9–23}$$

In order to provide a torque in the z-direction, the man must exert equal and opposite forces in the x-direction. It is the reaction to these forces (Newton's third law) which causes him to rotate.

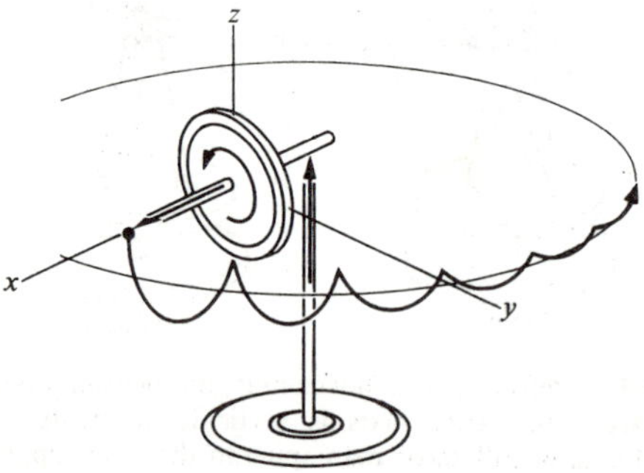

FIG. 9–15. Actual motion of tip of axis of gyroscope under gravity just after releasing axis previously held fixed. The wobbly motion (called a nutation) is not explained in the text and is due to frictional forces.

A gyroscope is essentially a rapidly spinning wheel. When placed on a pivot (Fig. 9–15), it does not fall because of the gravitational force. The gravitational force plus pivot provide a torque in the y-direction. The initial angular momentum is in the x-direction. The gyroscope will start to rotate (precess is the correct word) with an angular velocity in the z-direction according to Eq. (9–23).

Later, in Chapter 24, we shall analyze mathematically the precession of an angular momentum due to a torque at right angles to it, but the law of the conservation of angular momentum enables us to gain some physical insight into this phenomenon at this time.

Problems

1. Two boys, one weighing 60 lb and the other 90 lb, are at rest on frictionless roller skates. The larger boy pushes the other so that the latter rolls away at a speed of 5 mi/hr. What is the effect of this action on the larger boy himself?

2. A driverless car weighing 4000 lb is moving along a road at a speed of 50 mi/hr. In order to stop the car, a tank weighing 16,000 lb collides with it, head-on. (a) What should the tank's speed be in order that both tank and car come to a stop as a result of the collision? (b) How much kinetic energy is dissipated in the collision if the tank has the speed of part (a)?

3. A freight car weighing 12 tons rolls at 5 ft/sec along a horizontal railroad track. It collides with another freight car weighing 16 tons which is standing at rest on the track, and the two cars couple together. What is the speed of the cars after the collision?

4. A hunter has a rifle that can fire 0.06-kg bullets with a muzzle velocity of 900 m/sec. A 40-kg leopard springs at him at a speed of 10 m/sec. How many bullets must the hunter fire into the leopard in order to stop him in his tracks?

5. A 0.5-kg stone moving at 4 m/sec overtakes a 4-kg lump of clay moving at 1 m/sec. The stone becomes embedded in the clay. (a) What is the speed of the composite body after the collision? (b) How much kinetic energy is lost?

6. A 0.5-kg stone moving north at 4 m/sec collides with a 4-kg lump of clay moving west at 1 m/sec. The stone becomes embedded in the clay. (a) What is the velocity (speed and direction) of the composite body after the collision? (b) How much kinetic energy is lost?

7. A neutron of mass 1.67×10^{-27} kg and speed 10^5 km/sec collides with a stationary deuteron of mass 3.34×10^{-27} kg. The two particles stick together. What is the speed of the composite particle (called a *triton*)?

8. A neutron of mass 1.67×10^{-27} kg and speed 10^5 m/sec collides with a stationary deuteron of mass 3.34×10^{-27} kg. The particles do not stick together, and no kinetic energy is lost in the collision. What is the subsequent speed of each particle?

9. A 160-lb man is sliding on the frictionless surface of a frozen pond at a speed of 0.8 ft/sec. He is struck by a snowball weighing 0.5 lb whose velocity is 60 ft/sec in a direction perpendicular to his motion. What is the man's final speed?

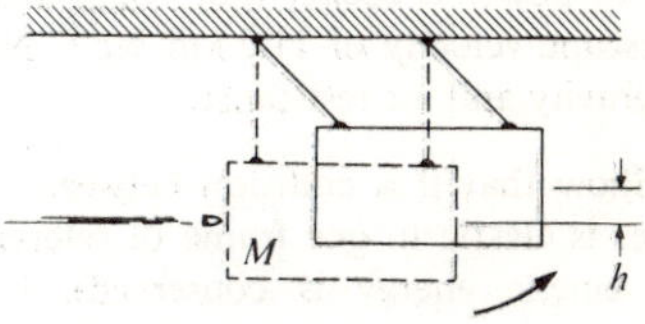

FIG. 9–16. A ballistic pendulum.

10. A block of wood of mass M is suspended from the ceiling. A bullet of mass m traveling with velocity v is shot at it, and is embedded in the wood. Find the maxi-

mum height to which the block will rise. How can this device (called a ballistic pendulum) be used to measure the bullet's velocity on impact? (See Fig. 9–16.)

11. A moving sphere collides, *not* head-on, with an identical one at rest. Show that they go off at right angles to each other's motion (neglect any rotations).

12. A shell is shot into the air (neglect air resistance). At the top of its trajectory it splits into two equal masses, one of which returns to the point of firing. Where will the other fragment land?

13. A boat is at rest in quiet water. A man in the boat walks from one end to the other. The boat is 10 ft long and weighs 300 lb. The man weighs 150 lb. What distance will the boat move?

14. A rocket motor consumes 100 kg of fuel per second, exhausting it with a speed of 5×10^3 m/sec. What force is exerted on the rocket?

15. A rocket fired from the earth's surface ejects 1% of its mass at a speed of 2000 m/sec in the first second. Find the average acceleration of the rocket in the first second.

16. If an ion engine with an exhaust velocity of 10^8 m/sec could be constructed, what percentage of the initial mass would have to be fuel if the engine were to be used to power a rocket that could attain the escape velocity of 11.2 km/sec? Neglect gravity and air resistance.

17. Show that if a collision between two bodies is elastic in one frame of reference (the kinetic energy is conserved), it is elastic in any other frame of reference moving with constant velocity with respect to it.

18. In a nuclear reactor the fast neutrons produced during fission are slowed down by elastic collisions with the atomic nuclei of the "moderator." Two substances often used as moderators are deuterium, whose nuclei have masses approximately double that of the neutron, and carbon, whose nuclei have masses approximately twelve times that of the neutron. (These substances are used because they have little tendency to absorb neutrons.) Find the percentage of the initial energy lost by a neutron colliding head-on with (a) a stationary deuterium nucleus, and (b) a stationary carbon nucleus.

19. The amount of energy transferred to a stationary body of mass m_2 by a collision with a moving body of mass m_1 can be increased by interposing another body of mass m_3 between them, so that m_1 strikes m_3 which in turn strikes m_2. Show that the energy transferred to m_2 is a maximum when $m_3 = \sqrt{m_1 m_2}$.

20. A uniform bar of length L is supported horizontally at rest. It is suddenly released, and at the same instant is struck a sharp blow vertically upward, close to one end. Show that the upward velocity of the center of mass of the bar is $(\omega L/6) - gt$, where ω = angular velocity about the center of mass.

21. A skater has a moment of inertia equal to 400 lb-ft^2 when his arms are extended and 120 lb-ft^2 when his arms are folded. His initial angular velocity is 3 rev/sec with arms extended. (a) What is his final angular velocity? (b) What are his initial and final kinetic energies? (c) Explain why these can be unequal.

22. If the earth's radius were to shrink by $\frac{1}{2}$%, what would be the change in the length of the day?

23. Two 0.4-kg balls are joined by a 1-m string and set whirling through the air at 5 rev/sec about a vertical axis through the center of the string. After a period of time, the string stretches to 1.2 m. (a) What is the new angular velocity? (b) What was the change in kinetic energy?

24. A gyroscope consists of a disc of mass 1 lb and radius 5 in. It spins with an angular velocity of 50 rev/sec. It is pivoted at a distance of 7 in. from the center of the disc and the axis of rotation makes an angle of 30° with the horizontal. What is the precessional velocity?

25. A solid sphere is mounted so that it can rotate about a diameter. It is pivoted at the end of the diameter, with the diameter horizontal. If the angular velocity is ω, what is the precessional velocity?

10

Simple Harmonic Motion

The term *simple harmonic motion* is used to describe the most fundamental kind of oscillatory behavior. It is an idealization of happenings in the real world, and, like other successful idealizations in physics, it permits us to gain insights into actual events, insights which would not be possible if we had to take into account from the start all the complications that are usually present. Virtually all periodic phenomena—that is, phenomena which repeat themselves at regular intervals—are either forms of simple harmonic motion or are the results of several different simple harmonic motions superimposed on one another. Because periodic phenomena are so common in nature, playing important roles within the atom as well as on a larger scale, it is appropriate for us to consider simple harmonic motion in some detail. In fact, even nonperiodic phenomena may sometimes be considered in terms of simple harmonic motions in order to help understand the physical processes involved or to facilitate mathematical manipulation.

10–1 The oscillating spring

A particle oscillates in simple harmonic motion when it is acted on by a force whose magnitude is proportional to the displacement of the particle from an equilibrium position and whose direction is always toward that position. We can specify this *linear restoring force* mathematically by writing it in the form

$$\mathbf{F} = -k\mathbf{r}, \tag{10–1}$$

where $\mathbf{r}$ is the radius vector from the equilibrium position to the particle and k is a constant. Equation (10–1) is obeyed fairly well by elastic solids (e.g., rubber, most metals). Coil springs in particular usually follow Eq. (10–1) quite closely, provided that they are not extended or compressed too much, and by studying the motion of such a spring we shall be able to discover some of the characteristics of simple harmonic motion.

We know from experience that the more a spring is extended or compressed, the harder it is further to compress or extend it. This behavior is contained in (10–1), with a large value of k corresponding to a stiff spring and a small value to a weak spring. Let us start by placing the spring on a frictionless horizontal surface with one end fixed and a particle attached to the other end (Fig. 10–1). When the spring is at $r = 0$ no force acts on the mass, and if it is initially at rest it remains at rest. Now we pull the particle out a distance A, elongating the spring, and then release it. The initial force acting is $\mathbf{F} = -k\mathbf{A}$, which tends to return the particle to the origin. As the particle moves it is accelerated, and even though the force on it decreases as it approaches the origin, its velocity increases, though with a diminishing acceleration. Finally, back at the origin, $\mathbf{F} = 0$, and the moving particle continues past the origin with a momentarily constant velocity. On the other side of the origin it encounters a force due to the compression of the spring which is in the direction opposite to its velocity, and it is slowed down. Ultimately the particle comes to a stop at $-A$ and then, under the influence of the compressed spring, begins to accelerate back toward the origin. In the absence of friction the particle will oscillate back and forth between A and $-A$ indefinitely.

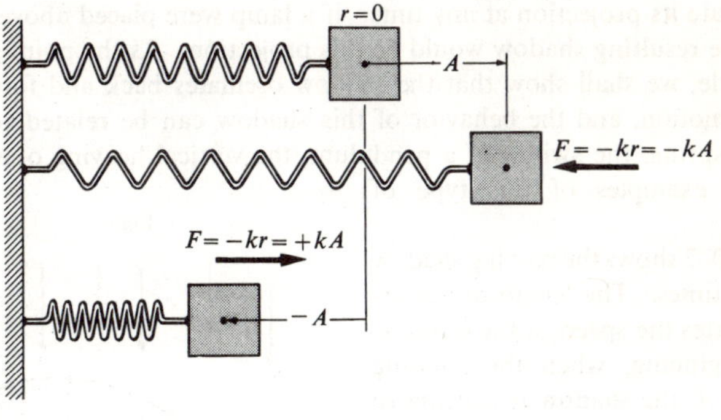

FIG. 10–1. The force exerted when a spring is displaced is opposite in direction to the displacement.

It is evident that, in cases of simple harmonic motion, we must deal with constantly changing force, acceleration, velocity, and position. There are certain regularities in the motion, though: The oscillating particle is confined to a definite region of space; its velocity is a maximum at the origin and zero at the ends of its path; the acceleration is a maximum at each end of the path, is always directed toward the origin, and is zero at the origin; and the motion repeats itself at definite intervals.

Before going any further, let us define three quantities we shall find useful in describing simple harmonic motion. The first is the *amplitude* of the motion, which is the maximum displacement of the particle from the origin. In Fig. 10–1 the amplitude is A (*not* $2A$). The second quantity is the *period* of the motion, the time T required by the particle to go through a complete cycle and return to its starting point. This would be the time involved in going from A to $-A$ and back to A once more, for example. The third quantity is the *frequency* of the motion, which is the number f of complete cycles the particle experiences per unit time. The usual way of expressing frequency is in cycles/second. Frequency f and period T are related by the formula

$$f = \frac{1}{T}. \tag{10-2}$$

The terms amplitude, cycle, period, and frequency may be applied to any phenomenon which recurs at regular intervals, even though it may not be pure simple harmonic motion.

10-2 Relation of simple harmonic to circular motion

The simple harmonic motion of a particle is exactly the same as that of the projection on a line of the motion of a particle moving in a circular path with uniform angular velocity, as in Fig. 10-2. Below the circle in this diagram is a straight line on which, by dropping a perpendicular from the rotating particle, we can locate its projection at any time. If a lamp were placed above the moving particle, the resulting shadow would be this projection. As the point goes around on the circle, we shall show that the shadow oscillates back and forth in simple harmonic motion, and the behavior of this shadow can be related to the vibrations of a spring, the swings of a pendulum, the vertical heaving of a ship at sea, and other examples of this type of motion.

Figure 10-2 shows the moving shadow at various times. The length of the arrows indicates the speed of the shadow. At the beginning, when the rotating point is at A, the shadow is starting to move to the left. It moves faster and faster until it is in the center of its path, whereupon it begins to slow down until it has come to a stop at the left-hand end of the path. Then, as the rotating point continues on, the shadow starts to move to the right, again going faster as it approaches the center. Once past the center it slows down, until back at the starting point it comes to a momentary stop before repeating the cycle. This is precisely the type of motion we found to be exhibited by a particle attached to a horizontal spring.

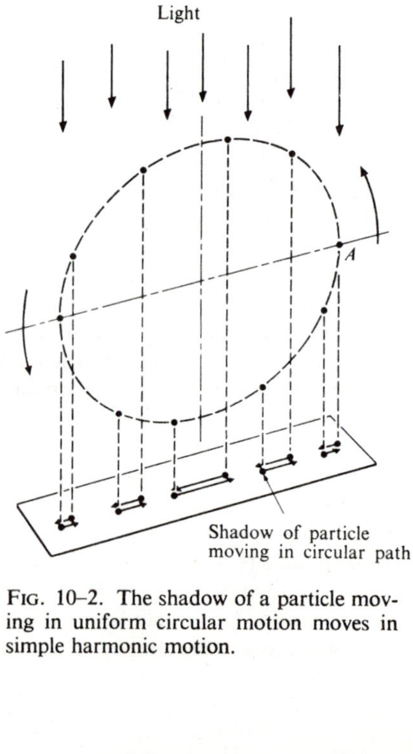

FIG. 10-2. The shadow of a particle moving in uniform circular motion moves in simple harmonic motion.

While the rotating particle has a constant speed along its circular trajectory, its shadow does not. The reason for

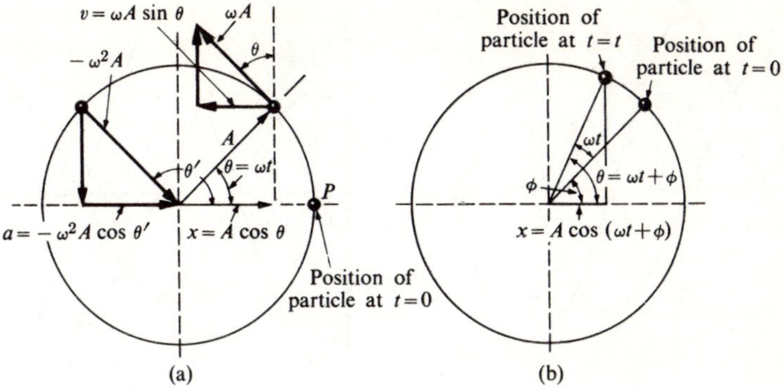

FIG. 10–3. The relationships among the displacement, velocity, and acceleration of a particle moving in uniform circular motion and its projection on a diameter. The projection exhibits simple harmonic motion.

this is that the shadow's velocity represents only the horizontal component of the particle's velocity, which is continually varying as the particle changes its direction of motion in going around the circle.

Let us relate the displacement x of the shadow from the center, its velocity v, and its acceleration a to the angle θ of Fig. 10–3(a). If the radius of the reference circle is A and the angular velocity of the particle is ω, the particle's speed at any time is ωA and its centripetal acceleration $\omega^2 A$. Hence

$$x = A\cos\theta, \qquad v = \omega A \sin\theta, \qquad a = -\omega^2 A\cos\theta.$$

Since the particle's angular velocity is constant, the angle θ is related to the time interval t from the start of its motion at the point P by the equation

$$\theta = \omega t. \tag{10–3}$$

We may therefore express x, v, and a as functions of time t by the formulas

$$x = A\cos\omega t, \tag{10–4}$$

$$v = \omega A \sin\omega t, \tag{10–5}$$

$$a = -\omega^2 A\cos\omega t. \tag{10–6}$$

Equations (10–4) and (10–5) are plotted in Fig. 10–4; clearly the characteristics of simple harmonic motion which we observed in the case of an oscillating spring—such as zero velocity when the acceleration is a maximum, and vice versa—are fulfilled in the motion described by the above equations.

We need not restrict ourselves only to those situations in which the motion begins (or we start making our observations) at the particular point corresponding to the particle being located at $\theta = 0$ in Fig. 10–3(a). If the position of the par-

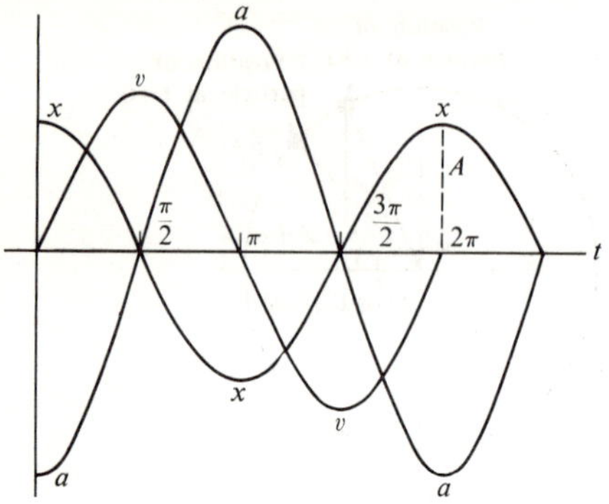

FIG. 10–4. The acceleration, velocity, and displacement of a particle undergoing simple harmonic motion as a function of time (phase angle = 0).

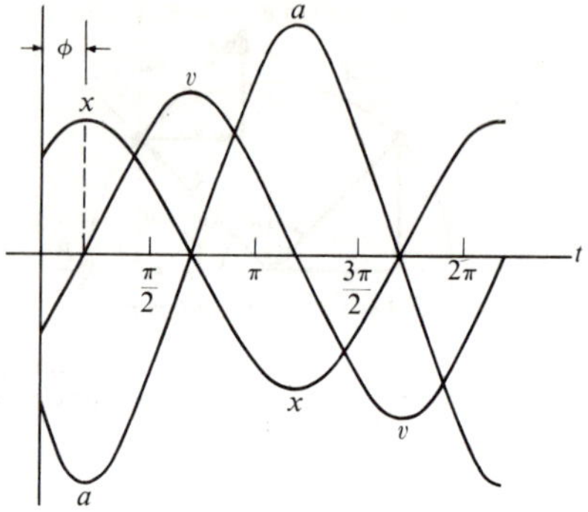

FIG. 10–5. The acceleration, velocity, and displacement of a particle undergoing simple harmonic motion as a function of time (phase angle = ϕ).

ticle corresponds to $\theta = \phi$ at $t = 0$ instead of to $\theta = 0$, at some future time $t = t$

$$\theta = \omega t + \phi. \tag{10–7}$$

The initial angle ϕ is called the *phase angle* of the motion (Fig. 10–3b). Making use of the more general expression for θ given by Eq. (10–7), the equations describing simple harmonic motion become

$$x = A \cos(\omega t + \phi) \tag{10–8}$$

$$v = -\omega A \sin(\omega t + \phi) \tag{10–9}$$

$$a = -\omega^2 A \cos(\omega t + \phi). \tag{10–10}$$

When these equations are plotted, as in Fig. 10–5, it is clear that the resulting curves differ from those of Fig. 10–4 only in that those of the former are shifted horizontally a distance corresponding to the angle ϕ.

10–3 Period and frequency

Because both displacement x and acceleration a are proportional to $\cos(\omega t + \phi)$, there is a constant ratio between x and a at all times during a given simple harmonic motion. We have, from Eqs. (10–8) and (10–10), that

$$a = -\omega^2 x; \tag{10–11}$$

the acceleration of the oscillating particle is proportional to its displacement from the origin and is in the opposite direction.

The rotating particle requires a time $2\pi/\omega$ to go through a complete circle. Now a complete circumnavigation corresponds to a complete cycle in the motion of the shadow, so that the period T of the latter corresponds to the time $2\pi/\omega$. Another way of looking at this is to note that in the course of a period T the particle sweeps through an angle of 2π radians: thus

$$\theta - \theta_0 = \omega(t - t_0), \qquad 2\pi = \omega T,$$

and

$$T = \frac{2\pi}{\omega}. \tag{10–12}$$

Because the frequency f is equal to $1/T$,

$$f = \frac{\omega}{2\pi}. \tag{10–13}$$

The fact that T and f may be so easily determined (given the angular velocity ω) is convenient, since Eq. (10–11) permits us, in any specified physical situation involving simple harmonic motion, to find ω if a and x are known at any particular instant. Because

$$\omega = \sqrt{-a/x}, \tag{10–14}$$

we may express T and f directly in terms of simultaneous values of a and x as

$$T = 2\pi\sqrt{-x/a} \tag{10–15}$$

$$f = \frac{1}{2\pi}\sqrt{-\frac{a}{x}}. \tag{10–16}$$

Example: A particle of mass 1.5 kg is attached to a horizontal massless spring whose restoring force is proportional to its displacement. The spring constant is 6 n/m, and no frictional forces are present. Find the frequency of the oscillations the system undergoes if it is displaced from its equilibrium position.

Solution: The magnitude of the restoring force F exerted by the spring is related to the magnitude of the displacement x and the spring constant k by the equation

$$F = -kx,$$

which is the one-dimensional form of Eq. (10–1). From the second law of motion,

$$F = ma,$$

we have, in view of the above expressions,

$$ma = -kx \qquad \text{or} \qquad \frac{a}{x} = -\frac{k}{m} = -\frac{6 \text{ n/m}}{1.5 \text{ kg}} = -4 \text{ sec}^{-2}.$$

Hence ω and f are, respectively,

$$\omega = \sqrt{\frac{k}{m}} = 2 \text{ rad/sec}, \qquad f = \frac{1}{2\pi}\sqrt{\frac{k}{m}} = \frac{1}{\pi} \text{ cycles/sec}.$$

Example: A pendulum ideally consists of a bob of mass m suspended from a frictionless pivot with a massless, inextensible string of length L (Fig. 10–6). Find the frequency of such a pendulum when the angular displacement is small.

Solution: There are a number of different approaches to this problem. We shall examine the subject here from the same point of view as we used in the preceding example, and determine the linear displacement and acceleration of the bob as a function of the angle θ between the string and the vertical. In Fig. 10–6, when the angle θ is small, the chord and arc between the bob and its equilibrium position are very nearly equal, so that the bob's displacement s is

$$s = L\theta.$$

The acceleration of the bob is, of course, that of gravity, namely g directed vertically downward. The component a of this acceleration which points toward the equilibrium position of the bob is, from the diagram,

$$a = -g\theta.$$

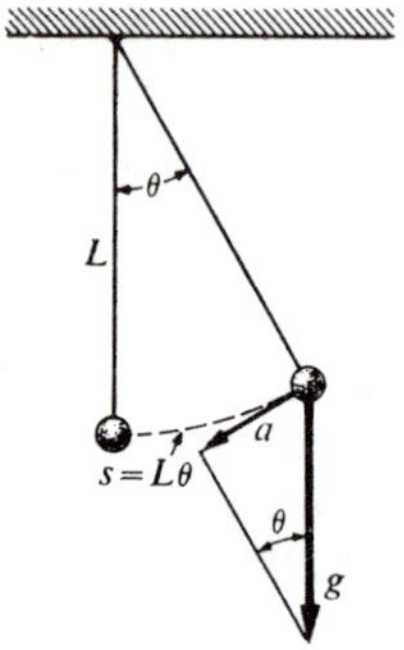

FIG. 10–6. Simple pendulum.

(Here, for simplicity, we let $g = |g| = 9.8\ \text{m/sec}^2 = 32\ \text{ft/sec}^2$, omitting the minus sign which we have associated with its numerical value in the past. This permits us always to have a with the opposite sign of x in treating the pendulum.) Thus

$$\frac{a}{s} = -\frac{g}{L},$$

and the pendulum frequency is

$$f = \frac{1}{2\pi}\sqrt{\frac{g}{L}};$$

the mass of the bob is irrelevant, and the only variable parameter determining the frequency is the length of the string.

10–4 Analytic treatment of simple harmonic motion

It should be possible to derive—directly from the restoring force law

$$\mathbf{F} = -k\mathbf{r}, \tag{10–1}$$

and from Newton's laws of motion—all the details of simple harmonic motion. Let us do this for a one-dimensional harmonic oscillator, in which the only coordinate is x. Applying the second law of motion to Eq. (10–1), we have

$$-kx = m\frac{d^2x}{dt^2}. \tag{10–17}$$

By solving this equation we determine the precise manner in which the position of the oscillating particle varies with time, and, knowing x as a function of time, we may then find expressions for v and a merely by differentiating with respect to t. We begin by rewriting Eq. (10–17) in the form

$$m\frac{d^2x}{dt^2} + kx = 0,$$

and then multiply both sides by dx/dt to obtain

$$m\frac{dx}{dt}\frac{d^2x}{dt^2} + k\frac{dx}{dt}x = 0.$$

This is equivalent to

$$\frac{m}{2}\frac{d}{dt}\left(\frac{dx}{dt}\right)^2 + \frac{k}{2}\frac{d}{dt}(x^2) = 0$$

or

$$\frac{d}{dt}\left(\frac{m}{2}\frac{dx^2}{dt} + \frac{k}{2}x^2\right) = 0.$$

Integrating, we obtain

$$\frac{m}{2}\left(\frac{dx}{dt}\right)^2 + \frac{k}{2}x^2 = C, \tag{10–18}$$

which we recognize as the condition for the conservation of energy. The first term is the kinetic energy of the oscillating particle, and the second is the potential energy of the spring when displaced by the distance x from its equilibrium position (Chapter 7).

Now we solve Eq. (10–18) for dx/dt, and find that

$$\frac{dx}{dt} = \left(\frac{2C - kx^2}{m}\right)^{1/2}.$$

Separating variables and letting $2C/k = A^2$, we have

$$\frac{dx}{\sqrt{k/m}\,(A^2 - x^2)^{1/2}} = dt, \qquad \frac{dx}{(A^2 - x^2)^{1/2}} = \sqrt{\frac{k}{m}}\,dt.$$

To proceed, we refer to a table of integrals for help with the left-hand side. There we find the standard integral

$$\int \frac{dx}{(A^2 - x^2)^{1/2}} = \sin^{-1}\left(\frac{x}{A}\right) + \text{constant},$$

and so

$$\sin^{-1}\frac{x}{A} = \sqrt{\frac{k}{m}}\,t + \phi,$$

where ϕ is a constant of integration which we have already called the phase angle. Solving for x, we obtain

$$x = A \sin\left(\sqrt{\frac{k}{m}}\, t + \phi\right). \tag{10-19}$$

It is not difficult to determine the amplitude and period of the simple harmonic motion described by Eq. (10-19). Since the value of the sine function varies from -1 to $+1$, the particle oscillates between $-A$ and $+A$, and thus the amplitude is A. For the period T, let us consider two times, t_0 and t_1, that differ by exactly T. That is,

$$t_1 - t_0 = T.$$

By the definition of the period, the position of the particle must be the same at t_1 as it was at t_0, which means that the sine must have the same value at both times. This can happen only if the argument $(\sqrt{k/m}\, t + \phi)$ increases by exactly 2π (or some multiple of 2π) in the interval, and therefore we have that

$$\sqrt{\frac{k}{m}}\, t_1 + \phi = \sqrt{\frac{k}{m}}\, t_0 + \phi + 2\pi.$$

Hence

$$\sqrt{\frac{k}{m}}\,(t_1 - t_0) = \sqrt{\frac{k}{m}}\, T = 2\pi,$$

so that the period is

$$T = 2\pi\sqrt{\frac{m}{k}},$$

as before. We note that the period depends only on the mass of the particle and the constant k, and not on the amplitude of the motion. The time required for a complete oscillation is the same, whether the particle barely moves or describes wide excursions on either side of the equilibrium position. This behavior finds practical application in the regulation of a clock by a pendulum or of a watch by a balance wheel attached to a spring.

The frequency of a particle in simple harmonic motion is

$$f = \frac{1}{T} = \frac{1}{2\pi}\sqrt{\frac{k}{m}},$$

and in terms of f we can write, in place of Eq. (10-19), that

$$x = A \sin(2\pi f t + \phi). \tag{10-20}$$

As we explained before, the quantity ϕ, the phase of the motion, tells us the position at which the particle was located at the time $t = 0$ when we began making measurements. For example, if the particle of Fig. 10-7 is at the right-hand end of its trajectory when $t = 0$, we have

$$A = A \sin \phi, \qquad \phi = \pi/2.$$

We have therefore, for the complete description of the particle's motion,

$$x = A \sin (2\pi ft + \pi/2)$$
$$= A \cos 2\pi ft.$$

If at $t = 0$ the particle is at the origin, then

$$0 = A \sin \phi, \qquad \phi = 0,$$

and

$$x = A \sin 2\pi ft.$$

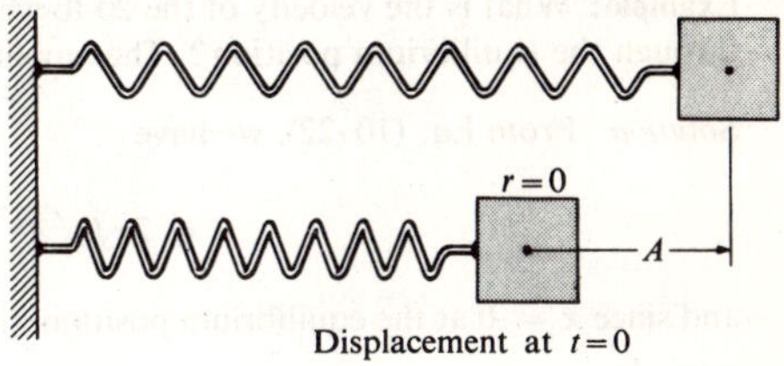

FIG. 10–7. Initial conditions for determining phase of simple harmonic motion.

The velocity of a particle executing simple harmonic motion may be found by differentiating Eq. (10–20);

$$v = \frac{dx}{dt} = 2\pi fA \cos (2\pi ft + \phi). \tag{10–21}$$

Since $\sin^2 \theta + \cos^2 \theta = 1$, this may be rewritten as

$$v = 2\pi f\sqrt{A^2 - A^2 \sin^2 (2\pi ft + \phi)}$$
$$= 2\pi f\sqrt{A^2 - x^2}. \tag{10–22}$$

Differentiating once more yields the acceleration,

$$a = \frac{dv}{dt} = -4\pi^2 f^2 x. \tag{10–23}$$

According to this equation, the acceleration of the particle, and therefore the force on it, is proportional to the displacement and in the opposite direction. This, we remember, is the definition of simple harmonic motion with which we began.

Example: An uncalibrated spring scale is found to have a 1-sec period of oscillation when a 20-lb weight is suspended from it. How much does the spring elongate when a 50-lb weight is suspended from it?

Solution: Since the period is given by

$$T = 2\pi \sqrt{\frac{m}{k}} = 2\pi \sqrt{\frac{w/g}{k}},$$

we find that

$$k = 25 \text{ lb/ft}.$$

The elongation when a 50-lb weight is attached is therefore

$$x = \frac{F}{k} = 2 \text{ ft}.$$

Example: What is the velocity of the 20-lb weight in the above example when it passes through the equilibrium position? The amplitude of the oscillation is 6 in.

Solution: From Eq. (10–22), we have

$$v = 2\pi f\sqrt{A^2 - x^2},$$

and since $x = 0$ at the equilibrium position, the velocity at the equilibrium position is:

$$v = 2\pi fA = \pi \text{ ft/sec}.$$

10–5 Energy considerations

At any instant the total energy of the system of particle plus spring consists of the kinetic energy of the former and the potential energy of the latter. As we saw in Eq. (10–18),

$$E = \tfrac{1}{2}mv^2 + \tfrac{1}{2}kx^2,$$

which we may rewrite as

$$E = \tfrac{1}{2}m4\pi^2 f^2(A^2 - x^2) + \tfrac{1}{2}kx^2 = \tfrac{1}{2}kA^2, \tag{10–24}$$

since $f = \sqrt{k/m}/2\pi$. This equation states that the total energy equals the potential energy at the maximum displacement $x = +A$, when the kinetic energy is zero. This comes as no surprise, of course, but it is encouraging to find that our theoretical picture is self-consistent. As the particle moves toward the origin the potential energy decreases, until at the origin there is only kinetic energy. Past the origin on the other side, the potential energy grows at the expense of the kinetic energy. We may regard simple harmonic motion, then, as a constant interchange of a fixed amount of energy between the moving particle and the medium in which it travels, which is the spring.

Example: Find the frequency of vibration of a pendulum for small displacements of the bob from the vertical, using the principle of conservation of energy.

Solution: We begin by calculating the potential energy V of the bob when the string is displaced from the vertical by the angle θ (Fig. 10–8). This is

$$V = mgh = mgL(1 - \cos\theta).$$

Since

$$\cos\theta = 1 - \frac{\theta^2}{2!} + \frac{\theta^4}{4!} - \cdots,$$

if θ is small we have that

$$\cos\theta = 1 - \frac{\theta^2}{2},$$

approximately. Hence

$$V \cong \tfrac{1}{2} mgL\theta^2,$$

or, since the displacement s of the bob is equal to $L\theta$, in this approximation

$$V = \frac{\frac{1}{2}mg}{L} s^2.$$

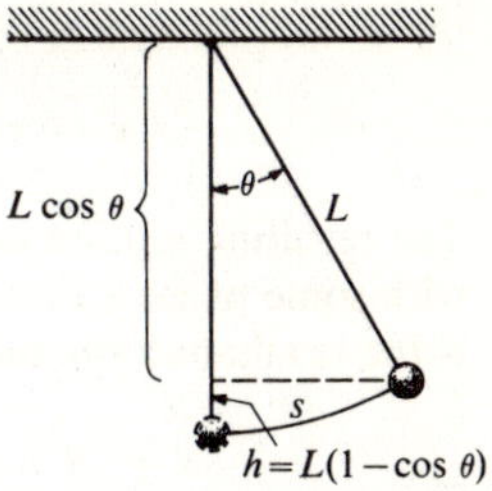

FIG. 10–8. Alternative derivation of pendulum formula.

This is in the same form as the potential-energy part of Eq. (10–18), and we infer from a comparison of the two that the restoring-force constant for the simple pendulum is

$$k = \frac{mg}{L}.$$

Hence

$$f = \frac{1}{2\pi}\sqrt{\frac{k}{m}} = \frac{1}{2\pi}\sqrt{\frac{g}{L}},$$

as before.

10–6 Addition of simple harmonic motions

We have already mentioned that simple harmonic motions are often encountered in combination; a particle, for example, may be attached to two springs (or even more) as shown in Fig. 10–9. The resulting motion of the particle will depend on the orientation of the springs, on their frequencies of oscillation when connected to the particle, on their amplitudes, and on their relative phase. We shall postpone considering cases in which the superposed harmonic motions are of different frequencies, and here treat a simple situation in which two simple harmonic motions of the same frequency are combined along the same line. We shall deal with the situation in which two identical massless springs are attached to a particle on either side so that, at any instant, their elongations are different, which introduces a phase difference ϕ between the motions of the springs.

We can write for the displacement of one of the springs

$$x_1 = A_1 \cos\omega t \qquad (10\text{–}25)$$

and for the other

$$x_2 = A_2 \cos(\omega t + \phi). \qquad (10\text{–}26)$$

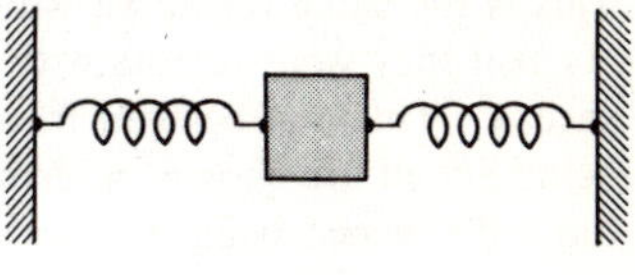

FIG. 10–9. Addition of two simple harmonic motions in the same straight line.

Here A_1 and A_2 are the amplitudes of the motions, ω the common angular frequency, and ϕ

the phase difference. The sum of the two displacements is

$$x = x_1 + x_2 = A_1 \cos \omega t + A_2 \cos (\omega t + \phi).$$

The resulting motion is simple harmonic, with the same angular frequency ω but with some phase δ that is different from the phases of 0 for x_1 and ϕ for x_2. If A is the resultant amplitude,

$$x = A \cos (\omega t + \delta) = A_1 \cos \omega t + A_2 \cos (\omega t + \phi). \tag{10–27}$$

By expanding the cosine functions with the help of elementary trigonometry, we find that

$$A \cos \delta \cos \omega t - A \sin \delta \sin \omega t$$
$$= A_1 \cos \omega t + A_2 \cos \phi \cos \omega t - A_2 \sin \phi \sin \omega t.$$

In order for this equation to hold for all t, the coefficients of $\sin \omega t$ and $\cos \omega t$ on each side of the equation must be equal separately. That is, we break the above equation into two individual ones:

$$A \cos \delta \cos \omega t = A_1 \cos \omega t + A_2 \cos \phi \cos \omega t$$

and

$$-A \sin \delta \sin \omega t = -A_2 \sin \phi \sin \omega t.$$

Dividing the first equation through by $\cos \omega t$ and the second by $\sin \omega t$, we obtain

$$A \cos \delta = A_1 + A_2 \cos \phi, \qquad A \sin \delta = A_2 \sin \phi. \tag{10–28}$$

Squaring these equations and adding them, we have

$$A^2 = A_1^2 + A_2^2 + 2A_1A_2 \cos \phi. \tag{10–29}$$

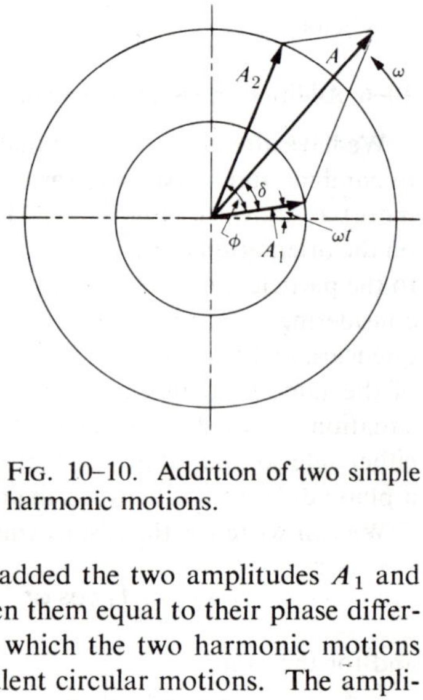

FIG. 10–10. Addition of two simple harmonic motions.

This is the same result we would obtain if we added the two amplitudes A_1 and A_2 as if they were vectors with the angle between them equal to their phase difference ϕ. This can be seen from Fig. 10–10, in which the two harmonic motions being added are shown in terms of their equivalent circular motions. The amplitudes do indeed behave as though they are vectors displaced with respect to one another, provided that they rotate with the same angular velocity ω and thereby keep the same relative position.

Equation (10–28) and its vector diagram illustrate a very interesting property of harmonic motion. Let us suppose that the two amplitudes A_1 and A_2 are equal to, say, A'. If $\phi = 0$, so that they are also in phase, then

$$A^2 = A'^2 + A'^2 + 2A'^2 = 4A'^2,$$

$$A = 2A';$$

the resulting vibration has twice the amplitude of either of the vibrations that are added together. However, if $\phi = \pi$ radians $= 180°$, the two vibrations are completely out of phase, and

$$A^2 = A'^2 + A'^2 - 2A'^2 = 0.$$

The resulting amplitude is 0, and the two vibrations have completely cancelled each other out. Figure 10–11 shows this cancellation graphically.

When the amplitudes A_1 and A_2 are different, Eq. (10–28) must be used in determining the resulting amplitude. In general, though, if the phase angle between the two motions is $\phi = 0$, so that they are in phase,

$$A = A_1 + A_2,$$

while if they are out of phase by $\phi = \pi$ radians, then

$$A = |A_1 - A_2|.$$

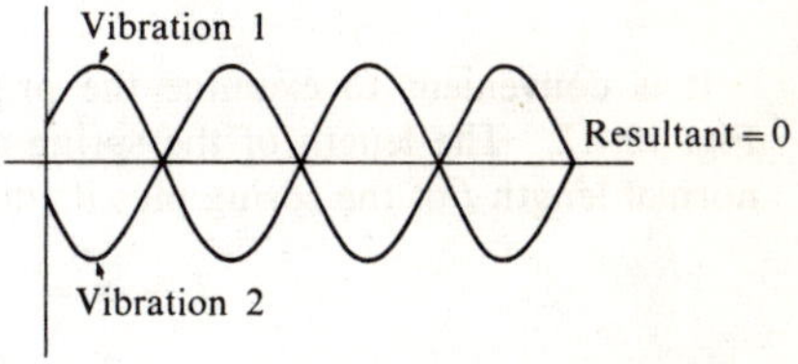

FIG. 10–11. The addition of two simple harmonic motions, of the same frequency and amplitude, which are 180° out of phase.

The diagram of Fig. 10–10 may also be used to determine the phase angle δ of the resultant harmonic motion. From the law of sines, we have

$$\frac{A_1}{\sin(\phi - \delta)} = \frac{A_2}{\sin \delta},$$

from which we obtain

$$\delta = \tan^{-1} \frac{A_2 \sin \phi}{A_1 + A_2 \cos \phi}. \tag{10–30}$$

This same result may be found directly from Eqs. (10–28).

10–7 Two-body oscillator

Figure 10–12 shows two bodies of masses m_1 and m_2 joined by a spring, and we imagine that they are on a horizontal, frictionless surface so that the only forces that affect their motions are those which the spring exerts. In the absence of external forces the linear momentum of the system must remain constant, and the oscillations of the bodies therefore cannot affect the motion of their center of mass. We assume that the center of mass is at rest. Thus m_1 and m_2 vibrate back and forth relative to their center of mass in opposite directions: if m_1 has the velocity $\mathbf{v}_1$ at some instant, m_2 must have a velocity $\mathbf{v}_2$ at that instant such that

$$m_1\mathbf{v}_1 + m_2\mathbf{v}_2 = 0,$$

$$\mathbf{v}_2 = -\frac{m_1}{m_2}\mathbf{v}_1. \tag{10-31}$$

When m_1 moves to the left, m_2 moves to the right, and both reach the respective extremes of their motions at the same time. What is the frequency of these oscillations?

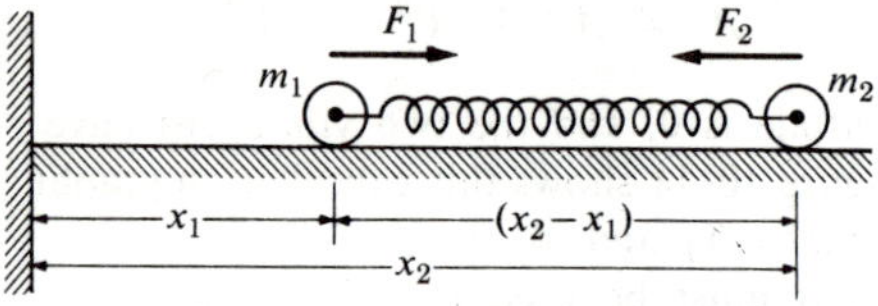

FIGURE 10-12

It is convenient to examine the problem by using the coordinate system of Fig. 10-12. The length of the spring at any time is $x_2 - x_1$, which equals the normal length l of the spring plus its displacement x:

$$l + x = x_2 - x_1,$$

$$x = x_2 - x_1 - l. \tag{10-32}$$

The displacement x is positive when the spring is extended and negative when it is compressed. The forces F_1 and F_2 which the spring exerts on each body are equal in magnitude and opposite in direction, and so

$$F_1 = kx, \qquad F_2 = -kx, \tag{10-33}$$

where k is the force constant of the spring. From the second law of motion, $\mathbf{F} = m\mathbf{a}$, we find that the bodies must obey the equations

$$m_1a_1 = m_1\frac{d^2x_1}{dt^2} = kx, \tag{10-34}$$

$$m_2a_2 = m_2\frac{d^2x_2}{dt^2} = -kx. \tag{10-35}$$

These equations can be combined by multiplying the first by m_2 and the second by m_1, and then subtracting the first from the second. This procedure yields

$$m_1m_2\frac{d^2x_2}{dt^2} - m_1m_2\frac{d^2x_1}{dt^2} = -m_1kx - m_2kx,$$

which can be rewritten

$$\frac{m_1m_2}{m_1+m_2}\left(\frac{d^2x_2}{dt^2} - \frac{d^2x_1}{dt^2}\right) = -kx. \tag{10-36}$$

Now we differentiate Eq. (10–32) twice with respect to time. Since l, the normal spring length, is a constant, we obtain

$$\frac{d^2x}{dt^2} = \frac{d^2x_2}{dt^2} - \frac{d^2x_1}{dt^2}.$$

Equation (10–36) can therefore be stated entirely in terms of x, the spring displacement:

$$\frac{m_1m_2}{m_1 + m_2}\frac{d^2x}{dt^2} = -kx,$$

$$m'a = -kx. \tag{10–37}$$

The quantity

$$m' = \frac{m_1m_2}{m_1 + m_2} \tag{10–38}$$

is called the *reduced mass* of the system, and a is the acceleration a particle of mass m' would have if it were attached to a spring of force constant k which is displaced by x from its equilibrium position. Hence the motion of a two-body oscillator is described by the same equation as that of a single body of mass m' attached to the same spring. We therefore conclude that the frequency of the two-body oscillator is

$$f = \frac{1}{2\pi}\sqrt{\frac{k}{m'}}. \tag{10–39}$$

The notion of the reduced mass of a system of two particles was also introduced in Chapter 8, where we saw that the rotation of two linked particles separated by a distance l about their center of mass is equivalent to that of a single particle of mass m' rotating about an axis at a distance l from it.

The relative energies of the two vibrating bodies are not difficult to determine. If v_{1m} and v_{2m} are the maximum speeds of m_1 and m_2, the ratio of their total energies (kinetic and potential, since PE $= 0$ when the speeds are maxima) is

$$\frac{E_1}{E_2} = \frac{\frac{1}{2}m_1v_{1m}^2}{\frac{1}{2}m_2v_{2m}^2}.$$

From Eq. (10–31), $v_{2m} = -(m_1/m_2)v_{1m}$, and

$$\frac{E_1}{E_2} = \frac{m_2}{m_1}. \tag{10–40}$$

The energies of the oscillating bodies are inversely proportional to their masses: the heavy body has *less* energy than the lighter one.

Problems

1. A force of 0.5 lb is required to push a Jack-in-the-box into its box, an operation in which the spring is compressed 4 in. If the Jack-in-the-box weighs 0.2 lb, what will its maximum speed be when it pops out?

2. A toy rifle employs a spring whose force constant is 30 lb/ft. In use, the spring is compressed 2 in., and when released, it propels a lead ball weighing 0.01 lb. What is the speed with which the ball leaves the rifle?

3. A 5-kg object is dropped on a vertical spring from a height of 2 m. The force constant of the spring is 1500 n/m. What is the maximum compression the spring will suffer?

4. A body moves in simple harmonic motion. The amplitude of the motion is 10 cm and its period is $\frac{1}{10}$ sec. The displacement is 3 cm when $t = 0.05$ sec. Write an expression for the displacement of the body as a function of time.

5. When a 2-lb weight is suspended from a spring, the spring stretches by 3 in. If the weight oscillates up and down, what is its period? What is its frequency?

6. A body whose mass is 1 kg hangs from a spring. When the body is pulled down 5 cm from its equilibrium position and released, it oscillates once per second. (a) What is the force constant of the spring? (b) What is the body's speed when it passes through its equilibrium position? (c) What is the maximum acceleration of the body?

7. A body whose mass is 0.2 slug is suspended from a spring and oscillates with a frequency of 5 cycles/sec and an amplitude of 0.5 ft. (a) What is the total energy of the motion? (b) What is the maximum acceleration of the body?

8. A body whose mass is 0.4 kg is suspended from a spring and oscillates with a period of 2 sec. By how much will the spring contract when the body is removed?

9. A body whose mass is 0.005 kg is in simple harmonic motion with a period of 0.04 sec and an amplitude of 0.01 m. (a) What is its maximum acceleration? (b) What is the maximum force on the body? (c) What is its acceleration when it is 0.005 m from its equilibrium position? (d) What is the force on it at that point?

10. Show that a spring whose constant is k and which has a mass m connected to it will oscillate with the same frequency whether the mass is displaced on a horizontal frictionless surface or is hung near the surface of the earth. Compare the energies of the system in both cases.

11. A chandelier is suspended from a high ceiling with a cable 20 ft long. What is its period of oscillation?

12. A pendulum whose length is 1.53 m oscillates 24 times per minute in a particular location. What is the acceleration of gravity there?

13. A pendulum has a length of 1 ft. Find its period when it is suspended in (a) a stationary elevator; (b) an elevator falling at the constant speed of 20 ft/sec; (c) an elevator falling at the constant acceleration of 7 ft/sec^2.

14. What is the frequency of a pendulum whose normal period is T when it is in an elevator in free fall?

15. A pendulum 0.82 ft long has a period of 1 sec in a certain place. What is the value of g there?

16. A harmonic oscillator has a period of 0.2 sec and an amplitude of 10 cm. Find

the velocity of the body when it passes through the equilibrium position.

17. Prove that the average speed of a particle undergoing simple harmonic motion is $2v_{max}/\pi$.

18. Make a plot of velocity versus displacement for a particle moving in a simple harmonic motion.

19. A body of mass m is connected to two springs whose constants are k_1 and k_2, each of length l, one on each side of the body. The springs are stretched so that their total length is $3l$. What is the equilibrium position of the mass, and what is its frequency?

20. Repeat Problem 19 for a case in which the springs are connected end to end on one side of the body.

21. A harmonic oscillator consists of a particle of mass m attached to two springs whose force constants are k_1 and k_2. The springs are side by side, and both join the particle to the fixed support. Find the period of the oscillations.

22. A hammer strikes the particle m_1 of Fig. 10–12 from the left and suddenly imparts to it an impulse of J. The system of two particles plus weightless spring moves off to the right while the particles simultaneously oscillate relative to their center of mass. Show that the ratio between the energy of oscillation and the kinetic energy of translational motion of the center of mass is m_2/m_1.

23. A mass oscillates according to the equation $x = 0.01 \cos (\pi t + \frac{1}{4}\pi)$, where x is in meters and t in seconds. Find (a) the frequency and amplitude of the motion, and (b) the displacement of the mass at $t = 0$ and at $t = 2.75$ sec.

24. A particle of mass 0.2 kg is attached to a horizontal spring of force constant 80 n/m. The spring is pulled out a distance of 0.05 m and released at $t = 1.0$ sec. Find an equation that gives the position of the particle at any time.

25. A particle is subjected to two simple harmonic motions of the same frequency ω in the same straight line. The amplitudes of the motions are the same and equal to A, and the phase difference between the two is 45°. Write down a formula for the displacement of the particle as a function of time, and plot it.

26. In the text we discussed the addition of simple harmonic motions in the same line. Simple harmonic motions can also be added in a plane. In that case the x- and y-coordinates are given by

$$x = A \sin (\omega_1 t + \phi),$$
$$y = B \sin (\omega_2 t + \Upsilon).$$

Find the curve the particle traverses under these circumstances when

(a) $A = B, \quad \omega_1 = \omega_2, \quad \phi = \Upsilon$
(b) $A = B, \quad \omega_1 = \omega_2, \quad \phi = \Upsilon + \pi/2$
(c) $A \neq B, \quad \omega_1 = \omega_2, \quad \phi = \Upsilon + \pi/2$
(d) $A \neq B, \quad \omega_1 = \omega_2, \quad \phi = \Upsilon + \pi/4$

Plot curves for x and y as functions of t in each case.

27. Prove Eq. (10–30).

28. What is the displacement relative to the amplitude of a body in simple harmonic motion at the moment its kinetic energy is three times its potential energy?

11

Gravitational Forces

The laws of motion which Newton postulated describe the relationships between the kinematical behavior of bodies and the forces exerted on them. One of Newton's chief motivations in developing these laws was his desire to prove the existence of a universal force which acts between all masses, and to discover the properties of this force. Such a *gravitational force* had been suspected even before Newton's time, by Galileo for instance, and Newton's contemporaries were aware of the problem, but it remained for him to show that the force which causes apples to fall to the ground is of the same kind and obeys the same laws as the force which holds the planets in their courses around the sun.

Gravitation was the first of the relatively few basic forces in nature to be understood. A major occupation of physics, continuing in the present day, has been the discovery of these forces, their properties, the effects they are responsible for, and their interrelationships.

11-1 Kepler's laws

For thousands of years before Newton, the sky had been the subject of intense scrutiny, and by the seventeenth century astronomical data of surprising precision was available. From the planetary data then in existence Johannes Kepler was

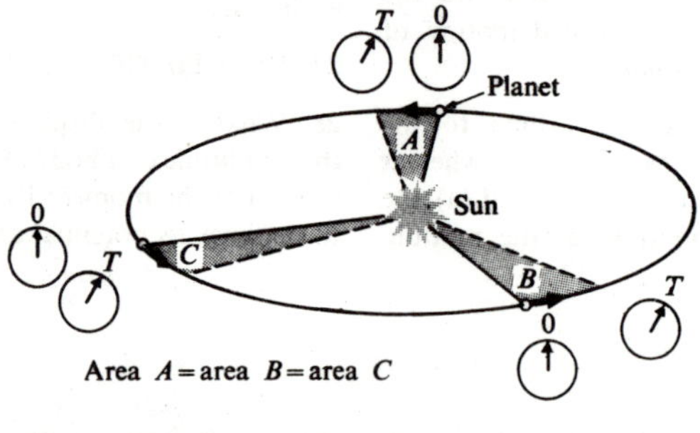

FIG. 11-1. Kepler's second law: Equal areas are swept out by the radius vector of a planet in equal times.

able to arrive at three famous laws concerning the motions of the planets:

(1) The planets all travel around the sun in elliptical orbits having the sun at one focus. (The eccentricities of the ellipses are quite small, so that the orbits are very nearly circular.) Implicit in this law is the fact that the paths of the planets lie in planes containing the sun.

(2) The radius vector from the sun to any planet sweeps out equal areas in equal times (Fig. 11-1).

(3) The ratio between the cube of the major axis (or radius if we approximate the orbits as circles) of a planet's orbit and the square of its period of revolution has the same value for all the planets.

Let us see how Kepler's three empirical laws, together with the laws of motion, enable us to formulate a theory of universal gravitation.

11-2 Conservation of angular momentum for particles

In Chapter 9 we discussed the relationship between torque and angular momentum in the case of rigid body rotation. There is a similar relationship which is applicable to single-particle motion, with the same consequence of conservation of angular momentum in the absence of external torques. Let us consider an arbitrary fixed point O in space and a particle of mass m moving with velocity v relative to O (Fig. 11-2). The angular momentum $\mathbf{L}$ of the particle about the point O is, by definition,

$$\mathbf{L} = \mathbf{r} \times \mathbf{p}. \tag{11-1}$$

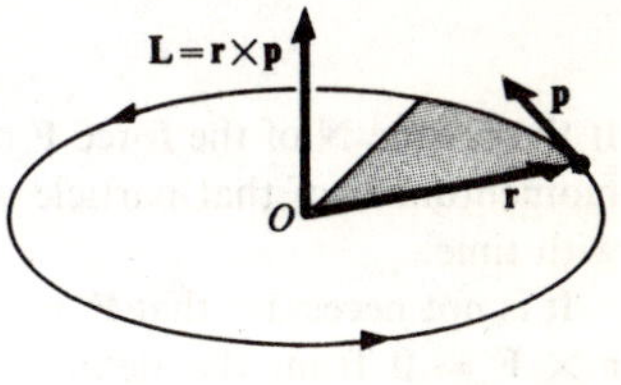

FIG. 11-2. Angular momentum of $\mathbf{p}$ about O of particle moving in a plane.

The value of $\mathbf{L}$ depends on our selection of the point O. Here $\mathbf{r}$ is interpreted as being the radius vector from an arbitrary point in space O rather than the radius vector from the axis of rotation of a system of particles, and $\mathbf{p}$ is the linear momentum $m\mathbf{v}$.

It is a simple matter to obtain a conservation theorem for angular momentum for the motion of a single particle. According to the second law of motion,

$$\mathbf{F} = \frac{d\mathbf{p}}{dt}. \tag{11-2}$$

If we take the vector product of each side of Eq. (11-2) with the radius vector $\mathbf{r}$, we have

$$\mathbf{r} \times \mathbf{F} = \mathbf{r} \times \frac{d\mathbf{p}}{dt}$$

$$= \mathbf{r} \times m\frac{d\mathbf{v}}{dt}.$$

Now

$$\frac{d}{dt}(\mathbf{r} \times \mathbf{v}) = \frac{d\mathbf{r}}{dt} \times \mathbf{v} + \mathbf{r} \times \frac{d\mathbf{v}}{dt}$$

$$= \mathbf{v} \times \mathbf{v} + \mathbf{r} \times \frac{d\mathbf{v}}{dt}$$

$$= \mathbf{r} \times \frac{d\mathbf{v}}{dt},$$

since $d\mathbf{r}/dt = \mathbf{v}$ and $\mathbf{v} \times \mathbf{v} = 0$. Hence

$$\mathbf{r} \times \mathbf{F} = m\frac{d}{dt}(\mathbf{r} \times \mathbf{v})$$

$$= \frac{d}{dt}(\mathbf{r} \times \mathbf{p}). \tag{11-3}$$

Since

$$\mathbf{L} = \mathbf{r} \times \mathbf{p} = \text{angular momentum},$$

$$\mathbf{N} = \mathbf{r} \times \mathbf{F} = \text{torque},$$

Eq. (11–3) may be written as

$$\mathbf{N} = \frac{d\mathbf{L}}{dt}. \tag{11-4}$$

If the torque $\mathbf{N}$ of the force $\mathbf{F}$ on a particle about the point O is zero, the angular momentum $\mathbf{L}$ of that particle about the point O is a constant and does not vary with time.

It is not necessary that $\mathbf{F} = 0$ in order for $\mathbf{r} \times \mathbf{F}$ to be 0; if $\mathbf{r}$ and $\mathbf{F}$ are parallel, $\mathbf{r} \times \mathbf{F} = 0$ from the definition of the vector product. Whenever we have a central force acting on a particle, the angular momentum of the particle about the origin of the force remains constant in the motion of the particle. And, as an important corollary, whenever we find a situation in which a particle maintains a constant angular momentum about a particular point, we may conclude that the only forces that might be acting on the particle must originate at that point and be directed along the radius vector from the point to the particle.

11–3 Kepler's first and second laws

Since the planets do not move along straight lines, forces must be present which act on them. Newton postulated that the force acting on a planet which deflects it from rectilinear motion originates in the sun. Such a force is a central force, and therefore the torque on a planet revolving about the sun is always zero. Angular momentum is conserved in this event, so that, since $\mathbf{L}$ is a vector, it must always point in the same direction. Consequently each planetary orbit should lie in a plane, as actually occurs. Besides confirming Newton's postulate in a qualitative way, Kepler's first law permits the deduction of the precise manner in which the attractive solar force on a planet varies with distance, a result we shall obtain more readily from Kepler's third law.

Kepler's second law, the law of areas, follows from conservation of angular momentum. We have

$$\mathbf{L} = \text{constant} = \mathbf{r} \times \mathbf{p}$$
$$= m\mathbf{r} \times \mathbf{v}$$
$$= m\mathbf{r} \times \frac{d\mathbf{s}}{dt}, \qquad (11\text{–}5)$$

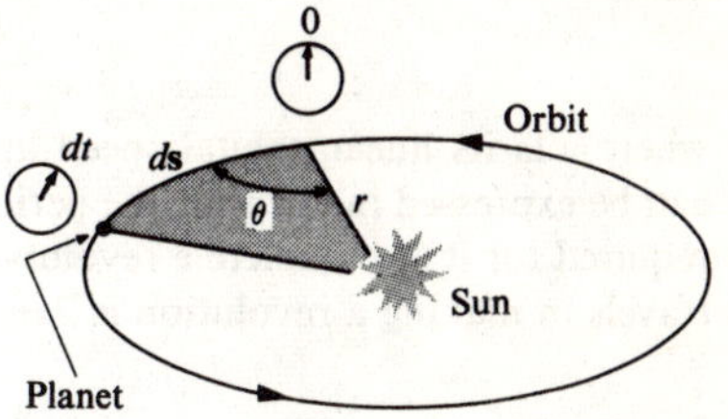

FIGURE 11-3

where $d\mathbf{s}$ is the displacement vector measured along the elliptical path of the planet. Now the area A swept out by a radius vector of length r in a time dt (Fig. 11–3) is given by

$$A = \tfrac{1}{2} r\, ds \sin\theta.$$

The area may be considered as a vector perpendicular to the plane of $\mathbf{r}$ and $d\mathbf{s}$, and thus

$$\mathbf{A} = \tfrac{1}{2}\mathbf{r} \times d\mathbf{s},$$

where $d\mathbf{s}$ is the displacement of the tip of the radius vector in dt. Hence the rate at which the radius vector sweeps out the area is

$$\frac{d\mathbf{A}}{dt} = \tfrac{1}{2}\mathbf{r} \times \frac{d\mathbf{s}}{dt} = \frac{1}{2m}(\mathbf{r} \times m\mathbf{v})$$
$$= \frac{1}{2m}(\mathbf{r} \times \mathbf{p}) = \frac{\mathbf{L}}{2m} = \text{constant}, \qquad (11\text{–}6)$$

since $\mathbf{L}$ and m, the angular momentum and the mass of the planet, respectively, are both constant. Hence the radius vector to a body acted on by a central force from the center of force sweeps out equal areas in equal time.

11–4 Variation with distance

From Kepler's third law we can determine the way in which gravitational forces vary with the distance of the interacting bodies. Kepler's first two laws, as we have seen, are consistent with the simple assumption that it is the sun that provides the attractive force which keeps the planets in elliptical orbits. We might make various guesses as to the form of the law of gravitation and then check them against the third law, but a more direct method is to begin with the third law itself.

For simplicity we shall assume that the sun's mass is infinite compared with that of the planets, and that the planetary orbits are circles centered at the sun.

The centripetal acceleration each planet experiences is

$$a = \frac{v^2}{r}, \tag{11-7}$$

where v is its linear orbital speed and r its distance from the sun. The speed v can be expressed in terms of the period T of the planet's motion, which is the time required for it to complete a revolution about the sun. Since the distance a planet travels in making a revolution is $2\pi r$,

$$v = \frac{2\pi r}{T}. \tag{11-8}$$

Hence the acceleration may be expressed as

$$a = \frac{4\pi^2 r}{T^2}. \tag{11-9}$$

Now Kepler's third law states that the ratio between the cube of the orbital radius r and the square of the period T is a constant for all the planets; that is,

$$\frac{r^3}{T^2} = C. \tag{11-10}$$

We can express the acceleration of a planet in terms of the constant C by substituting for the period T in Eq. (11-9) its value from Eq. (11-10), so that

$$a = \frac{4\pi^2 C}{r^2}. \tag{11-11}$$

The acceleration of each planet, and therefore the force exerted on it, varies inversely with the square of its distance from the sun.

11-5 The law of universal gravitation

While we know that gravitational forces are central forces and vary with the inverse square of the distance between interacting bodies, we do not know on what factors the constant C of Kepler's third law depends. Since the constant is the same for all planets, it will depend on some property of the attracting body (the sun), presumably on its mass. Here Newton took the considerable step of hypothesizing that the force keeping the planets in their course is a manifestation of a universal law, and has the same nature as the force which keeps the moon in its orbit about the earth and which is responsible for the attractive force pulling objects to the earth's surface.

Let us consider the earth as a center of gravitational force. The gravitational force which produces the acceleration of Eq. (11-11) is an inverse-square one, and it should apply to the moon's motion about the earth (with the constant C assuming a different value). That is, we assume that these bodies act as though they are particles in their interaction, in the same way that we assume a similar behavior in the case of the planets and the sun. What we want is to explain the attraction between the earth and the objects at its surface in terms of the same force that holds the moon in its orbit about the earth. Here we seem to run into trouble, because it is not obvious that a relatively small object near the earth and the earth itself can be thought of in their effects on one another as being simply point masses. This was one problem faced by Newton. A possible method of attacking it is to consider the inverse-square force required by Eq. (11–11) as applying between each particle of the earth and a single particle at the position of the object whose acceleration is to be calculated, and then to integrate the force over the volumes of the earth and moon. We shall indicate an alternative procedure for dealing with this problem in Section 11–7, and here only state the result for the present situation: a spherically symmetric system of particles exerting a gravitational force on an external body acts as though a single center of force were present whose location is the center of mass of the system.

With this information in hand, Eq. (11–11) indicates that the acceleration g due to gravity experienced by an object at the earth's surface should be

$$g = \frac{4\pi^2 C'}{R_e^2}, \tag{11–12}$$

where R_e is the radius of the earth. The constant C' may be found by applying Kepler's third law to the moon's motion, since the acceleration of the moon and particles near the earth involve the same constant; we have therefore that

$$C' = \frac{R_m^3}{T^2},$$

where R_m is the radius of the moon's orbit and T its period. Thus

$$g = \frac{4\pi^2 R_m^3}{T^2 R_e^2}.$$

All the quantities on the right-hand side of this equation can be measured; the values that have been obtained are

$$R_e = 6.37 \times 10^6 m, \qquad R_m = 3.83 \times 10^8 m,$$

and

$$T = 2.36 \times 10^6 \text{ sec}.$$

Hence the theoretical figure is

$$g = 9.81 \text{ m/sec}^2,$$

in complete agreement with the experimental value.

Having shown that the phenomenon of gravitation obeys the same law everywhere, we are in a position to derive the equation governing that law. From

Eq. (11-11), we obtain the force the earth exerts on the moon:

$$F = m_m a_m = \frac{4\pi^2 m_m C'}{R_m^2}, \qquad (11\text{-}13)$$

where m_m is the moon's mass. According to the third law of motion the moon exerts a force of the same magnitude on the earth, in this case given by

$$F = m_e a_e = \frac{4\pi^2 m_e C''}{R_m^2},$$

where m_e is the earth's mass. The constant C'' depends on the relevant properties of the moon. Equating the two expressions for F,

$$\frac{C'}{m_e} = \frac{C''}{m_m},$$

which entitles us to suppose that C/m is a universal constant. If we let

$$\frac{C}{m} = \frac{G}{4\pi^2}, \qquad (11\text{-}14)$$

where G is some constant, the gravitational force law between two bodies of masses m_1 and m_2 that are separated by the distance r can be inferred from (11-13) and (11-14) to be

$$F = \frac{Gm_1m_2}{r^2}. \qquad (11\text{-}15)$$

11-6 Constant of gravitation

The constant of gravitation G cannot be found from astronomical data alone. The only way of determining G is to perform an experiment to measure directly the gravitational force between two objects of known mass. Such a measurement was first performed by Cavendish in 1798. He used an especially sensitive type of balance, called a torsion balance, to find the force of attraction between two small spheres attached to the ends of a suspended rod, and two larger stationary spheres. By first measuring the spring constant of the suspension wire (Fig. 11-4) so that the force required to twist it to a given extent is known, the deflection that results when the large spheres are brought near can be converted into the force involved. A variant of this method was used in 1930 in the most recent determination of G. The accepted value of G is

$$G = 6.670 \times 10^{-11}\ \mathrm{n \cdot m^2/kg^2}.$$

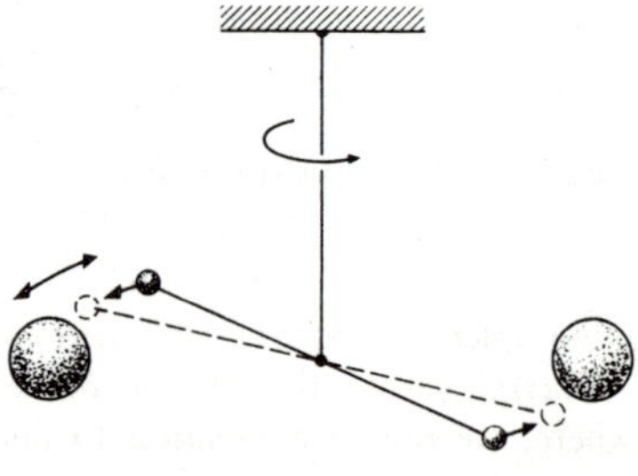

FIG. 11-4. The Cavendish experiment.

11-7 Gauss' theorem

We now turn to the question of whether it is correct to consider astronomical bodies such as the earth and sun as point particles in discussing their gravitational interactions. What we shall do is to assume that the law of universal gravitation as expressed in Eq. (11-15) is correct when applied to particles, and then see if this law still holds when we are treating spherical objects as though they were point particles with all their mass concentrated at their centers. Let us begin with a particle of mass m, and determine the force on a unit mass a distance R away. This force is

$$F = -Gm/R^2, \tag{11-16}$$

and it is the same anywhere on the surface of an imaginary sphere of radius R centered at the particle (Fig. 11-5). The minus sign in Eq. (11-16) describes the attractive nature of the gravitational force. Now if we integrate this force over this imaginary sphere we have

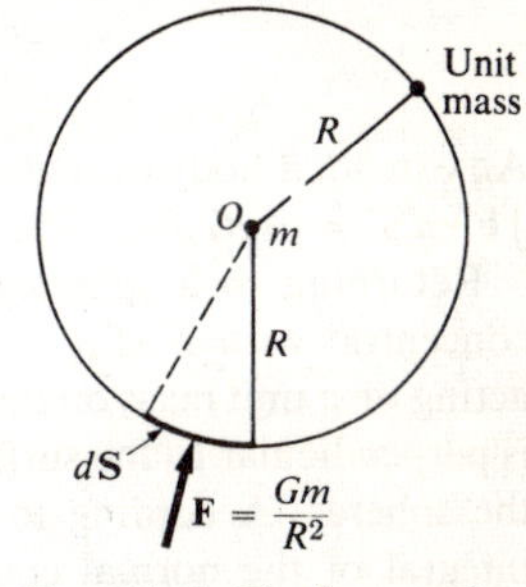

FIG. 11-5. The force acting on a particle of unit mass on the surface of a sphere surrounding a mass m.

$$\oint \mathbf{F} \cdot d\mathbf{S} = \frac{Gm}{R^2} \times 4\pi R^2 = 4\pi Gm. \tag{11-17}$$

(Since **F** acts radially, it is always antiparallel to an element of surface $d\mathbf{S}$ on a sphere whose direction, by definition, is radially outward.) The important result embodied in Eq. (11-17) is that this integral is independent of the radius of the sphere over which the integral in (11-17) is carried out. This arises, of course, from the fact that the force varies inversely with the square of the distance, while the area varies directly with the square of the distance. Equation (11-17) is known as *Gauss' theorem*, and states that the integral of $\mathbf{F} \cdot d\mathbf{S}$ taken around any closed surface—not just a sphere—is equal to $4\pi Gm$, where m is the mass enclosed by the surface. To see this, consider the mass m surrounded by a surface S of arbitrary shape, and construct an infinitesimal cone with its vertex at the position of the mass. The cone intercepts the surface in an area $d\mathbf{S}$ (Fig. 11-6). The solid

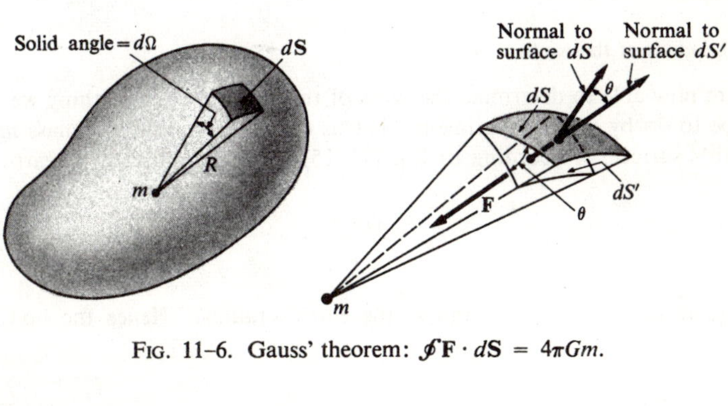

FIG. 11-6. Gauss' theorem: $\oint \mathbf{F} \cdot d\mathbf{S} = 4\pi Gm$.

angle subtended at m by dA is $d\Omega$. The relationship between **F**, the normal to the surface dS, and the projection dS' of dS on a plane perpendicular to **F** is shown in Fig. 11-6. We have

$$R^2\, d\Omega = dS' = dS \cos\theta.$$

Since $\mathbf{F} \cdot d\mathbf{S}' = F\, dS \cos\theta$, we have

$$\int \mathbf{F} \cdot d\mathbf{S}' = \int \frac{Gm_1}{R^2} \cos\theta \times \frac{R^2}{\cos\theta}\, d\Omega$$
$$= 4\pi G m_1.$$

An extended body of mass m consists of many infinitesimal masses dm. For each $\int \mathbf{F} \cdot d\mathbf{S}' = 4\pi G\, dm$. The total integral is therefore equal to $4\pi G \sum dm = 4\pi G m$.

Returning to a spherical body of mass m, let us draw a sphere of radius R concentric with it (Fig. 11-7). If the body is uniform, the gravitational force acting on a unit mass on the surface of the sphere is perpendicular to the surface and constant over the sphere. According to Gauss' theorem, the integral of the normal component of the force acting on a unit mass is $4\pi m G$, where m is the total enclosed mass. Thus we have for the force on a unit mass on the sphere

$$\oint \mathbf{F} \cdot d\mathbf{S} = -F 4\pi R^2 = 4\pi G m$$

or

$$F = -\frac{Gm}{R^2}. \qquad (11\text{–}18)$$

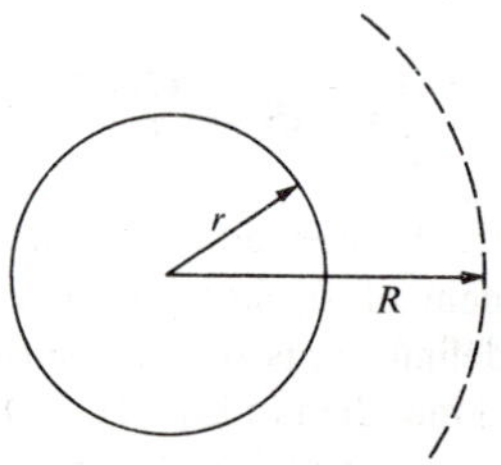

FIG. 11-7. Determining the force exerted by an extended spherical body using Gauss' theorem.

Equation (11-18) is identical with Eq. (11-16), which gave the force per unit mass a distance R from a mass m. We are therefore entitled to consider that spherical bodies behave gravitationally precisely as if they are point particles located at their centers.

11-8 The mass of the earth

We are now able to determine the mass of the earth itself, something we could not hope to do by direct experiment. Let us consider an object of mass m near the earth's surface. According to Eq. (11-15), it is attracted to the earth with a force

$$F_g = \frac{Gmm_e}{r_e^2},$$

where m_e is the earth's mass and r_e the earth's radius. Hence the body will experience an acceleration a given by

$$F_g = ma,$$

or, substituting for the gravitational force F_g,

$$\frac{Gmm_e}{r_e^2} = ma.$$

Canceling the mass m on both sides of this equation and solving for a, we have

$$a = \frac{Gm_e}{r_e^2}.$$

But this acceleration is the known acceleration of gravity g! If we substitute g for a, we find that

$$g = \frac{Gm_e}{r_e^2}$$

and

$$m_e = \frac{gr_e^2}{G}.$$

Inserting the numerical values

$$g = 9.8\,\frac{\text{m}}{\text{sec}^2}, \qquad r_e = 6.38 \times 10^6\ \text{m}, \qquad G = 6.67 \times 10^{-11}\,\frac{\text{n·m}^2}{\text{kg}^2},$$

we find that

$$m_e = \frac{9.8\ \text{m/sec}^2 \times (6.38 \times 10^6\ \text{m})^2}{6.67 \times 10^{-11}\ \text{n·m}^2/\text{kg}^2}$$

$$= 5.98 \times 10^{24}\ \text{kg}.$$

The earth's mass is 5.98×10^{24} kg, or 6.6×10^{21} tons.

11-9 The gravitational field

In everyday life, forces seem to be transmitted by "direct contact": something pushes or pulls something else. Gravitational forces, however, act in the absence of direct contact and are able to produce their effects through millions of miles of empty space. There are two ways of regarding this situation. The first is merely to think of gravitational phenomena as representing the mutual interaction of separated centers of force, as "action at a distance." In the other view the interaction is considered as involving the region in which the centers of force are located: the presence of a single mass alters the properties of the region around it, setting up a *gravitational field* which stands ready to interact with any other masses brought into it. Thus the sun is surrounded by a gravitational field, and it is the

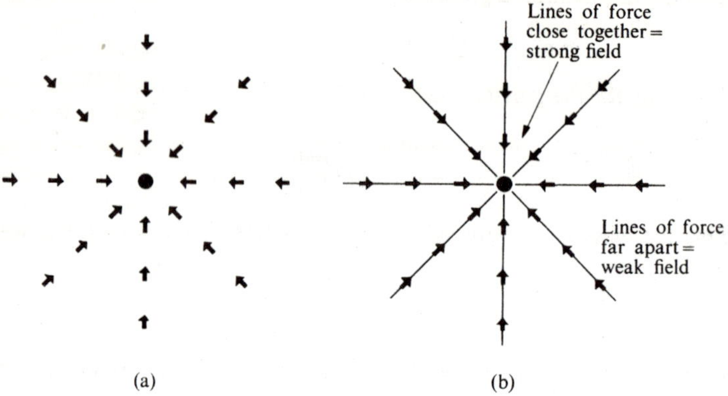

FIG. 11-8. Lines of force of an isolated mass.

force exerted *by this field* on the planets which provides the centripetal force that holds them in their orbits. A number of different factors have led to the adoption of the field concept to replace the older one of action at a distance, and we shall consider several of them in later chapters.

To help visualize what is meant by a gravitational field we may use *lines of force*. These are constructed as shown in Fig. 11-8: at several points in space we draw arrows in the direction in which a small test mass would go if allowed to move freely there. Then we connect these arrows so as to form smooth curves, called lines of force, arranging their concentration in the vicinity of any point to be proportional to the magnitude of the force on the test mass at that point. The resulting picture may be thought of as a portrait of the gravitational field. If we were to place a mass in the field, it would experience a force along whatever line of force it was on, with the direction of the arrow and with a magnitude depending on how close together the lines of force are in its vicinity. Figure 11-9 shows the lines of force near two spheres that are close together. A body in the center will never reach either, since the net force on it is zero. At distances large relative to the diameters of the spheres and their separation, the field reduces to that of a single point mass.

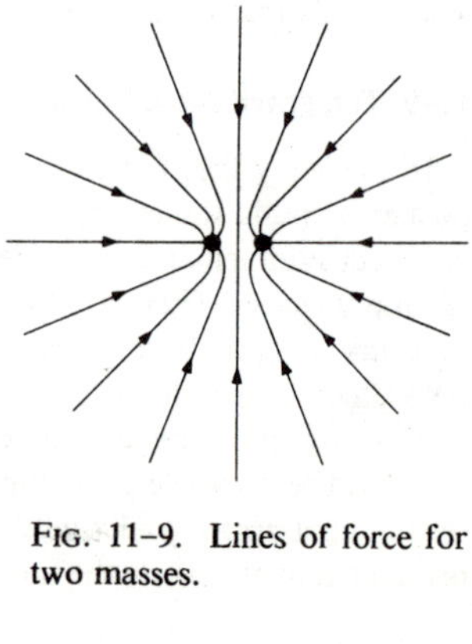

FIG. 11-9. Lines of force for two masses.

It is important, of course, to keep in mind that lines of force do not actually exist threading space; they are no more than a device for giving intuitive form to our thinking about gravitational fields. We shall apply the notion of lines of force to other fields of force in addition to gravitational fields. Electric and magnetic fields, for instance, are often described in this manner.

The gravitational force is one of the relatively few *fundamental forces* which exist in nature. Fundamental forces are those which cannot be explained in terms of others. For example, frictional forces and the forces which hold molecules and atoms together are electromagnetic in origin. Electromagnetic forces, nuclear forces, and weak interactions take their place with gravitational forces as the only four fundamental forces that are known; it is remarkable that so much of our complex universe can be explained in terms of so few forces. The fact that gravitational force is one of those few points up how important it is. A large part of the remainder of this book will be devoted to the discussion of fundamental forces.

Problems

1. A man weighs 160 lb at the earth's surface. (a) What would he weigh at an altitude of one earth radius? (b) What would his mass be at that altitude?

2. The radius of the earth is 6.4×10^6 m. What is the acceleration of a meteor when it is 8×10^6 m from the center of the earth?

3. Even if the earth were a perfect sphere, the experimental value of g, the acceleration of gravity, would be least at the equator. Why?

4. Is the gravitational force between the earth and the sun the same at all times of the year? Explain.

5. The moon's mass is 1.2% of the earth's mass. Is the gravitational force exerted by the moon on the earth less than, the same as, or more than the gravitational force exerted by the earth on the moon? Explain.

6. The earth's equatorial radius is about 21 km (1 km = 10^3 m) greater than its polar radius, a phenomenon known as the "equatorial bulge." (a) How is this fact related to the daily rotation of the earth on its axis? (b) If the earth were to rotate twice as fast as it now does, would you expect the equatorial bulge to be more than, the same as, or less than it is now? Explain.

7. The mass of the planet Jupiter is 1.9×10^{27} kg and that of the sun is 2.0×10^{30} kg. The average distance between them is 7.8×10^{11} m. (a) What is the gravitational force the sun exerts on Jupiter? (b) Assuming that Jupiter has a circular orbit, what must its speed be for the orbit to be a stable one?

8. The moon's radius is 27% of the earth's radius and its mass is 1.2% of the earth's mass. (a) What is the acceleration of gravity (in ft/sec^2) on the surface of the moon? (b) How much would a boy weigh there whose weight on the earth is 100 lb?

9. The mass of the planet Jupiter is 1.9×10^{27} kg and its radius is 7.0×10^7 m. What is the acceleration of gravity (in m/sec^2) on the surface of Jupiter?

10. A 2-kg mass is 1 m away from a 5-kg mass. What is the gravitational force (a) that the 5-kg mass exerts on the 2-kg mass; (b) that the 2-kg mass exerts on the 5-kg mass? (c) If both masses are free to move, what are their respective accelerations in the absence of other forces?

11. (a) Calculate the force the sun exerts on 1 kg of water at a point on the earth nearest the sun and also at the earth's center. (b) Calculate the force the moon exerts on 1 kg of water at the same locations. (c) On the basis of these results, explain why the moon is more effective than the sun at causing tides.

12. Sketch the pattern of gravitational lines of force near (a) an egg-shaped body of uniform composition; (b) three identical point masses at the corners of an equilateral triangle; (c) a flat, square mass (like a tabletop).

13. Can lines of force ever intersect in space? Explain.

14. Use Gauss' theorem to show that a particle in a shaft along the radius of the earth below the surface is attracted to the center by a force proportional to its distance from the center, assuming that the earth is of uniform density. If a shaft were dug through the earth, show that a particle dropped into the shaft would oscillate with simple harmonic motion. Calculate its period.

15. With what force would a mass M be attracted to a hollow sphere of radius R and density ρ at a distance d from the center of the original sphere, as in Fig. 11–10?

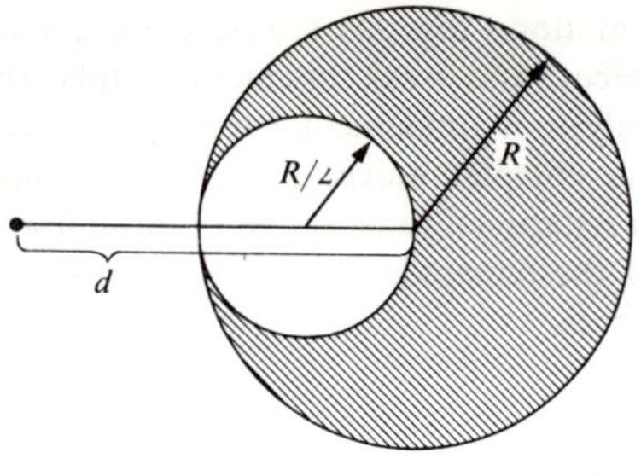

FIGURE 11–10

16. An artificial satellite of mass 100 kg is in a circular orbit at 500 km above the earth's surface. (The radius of the earth is 6500 km.) What is its speed and period of revolution?

17. With what minimum speed must a particle be ejected from the earth's surface in order for it to escape the earth's gravitational field?

12

The Properties of Matter

Newton's laws of motion are entirely valid within the limits of our everyday experience. As we saw in the previous chapter, these laws are also successful in interpreting the larger-scale phenomena of the solar system. But will we still find agreement with experiment when the laws of motion are applied to microscopic phenomena below the threshold of our direct perception? Specifically, can we show that the properties of matter in bulk with which we are familiar follow from the behavior of its individual molecules (which chemistry tells us are its essential ingredients) according to Newtonian physics? A great deal of thought and experiment has gone into answering this question, and the resulting *kinetic theory of matter* has been, as we shall learn, very successful in correlating the microscopic and macroscopic aspects of matter. Beginning with this chapter we shall study some of the basic properties of matter which the kinetic theory must explain; later we shall see how some of these basic properties are explained by kinetic theory.

12–1 The states of matter

The most primitive observation we can make about matter is that it exists in three states: solid, liquid, and gas. The particular state that a sample of matter assumes depends on its physical environment as well as on its chemical composition. Thus water, which we usually think of as a liquid, under some circumstances solidifies into ice and under others may vaporize into steam; iron, ordinarily a solid, may be melted into a liquid without too much difficulty; and carbon dioxide, normally a gas, is also familiar as solid "dry ice."

It is not hard to characterize the various states of matter qualitatively. We may conveniently regard samples of solid matter as being rigid, in that they tend to maintain their shapes, and as having fairly well-defined volumes. Liquid samples, while they share the approximately constant volumes of solids, assume the shape of whatever containers they are placed in. Gas samples have neither fixed volumes nor fixed shapes; a gas always fills its container, and so its volume and shape are those of the container. Liquids and gases have many properties in common, and together are categorized as *fluids.*

Recently a "fourth state of matter" called plasma has come under close scrutiny. A plasma is a gas an appreciable number of whose constituent particles have been ionized. The consequent presence of electric and magnetic forces in plasmas make them behave in ways not characteristic of ordinary gases. Much of the matter in the universe is in the plasma state, and the study of plasmas is an active branch of current physics research.

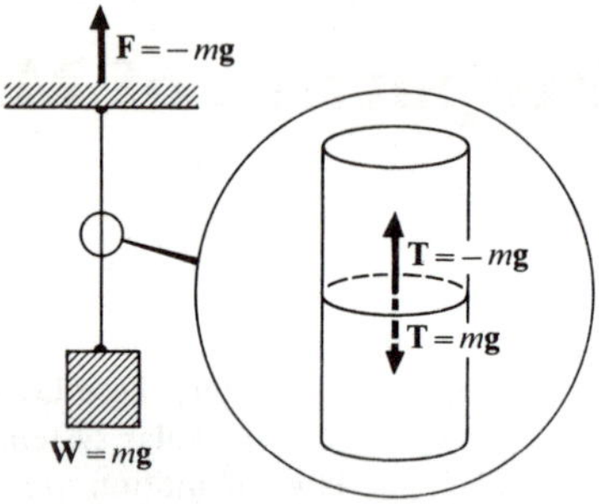

FIG. 12-1. Forces in a wire supporting a weight. Note that the effect is to transmit a force equal to the weight from the bottom of the wire to its point of support.

12-2 Elastic properties of solids

We shall now consider the distinction between solids, liquids, and gases in a more detailed way. We saw in an earlier chapter that, if we suspend a weight from a wire attached to the ceiling, the force of the weight can be transmitted through the wire. Any cross section (Fig. 12-1) of the wire perpendicular to the direction of the applied force has equal and opposite forces acting across it, and is thus in equilibrium. When we consider the cross section at the ceiling, the internal force must be balanced by a force exerted by the ceiling. What causes the body to exert across any cross section a force which can balance the external force? There are internal forces which keep a solid together, and the external force deforms the body—in this case stretches it. The elongation of the body brings additional forces into play (just as in the case of a spring) which balance the external force. These forces are known as *stresses*; when the body is elongated, the stresses are called *tensile stresses.* A similar analysis of what happens when we squeeze the two ends of a bar together would show that the bar would be diminished in size, and as a result would cause other stresses, known as *compressive stresses*, to be set up inside it. If the internal structure of a solid is such that it cannot produce the necessary compressive or tensile stresses to balance the external force, it crumples, breaks, or fails in some other way.

The response of a body to each of the above stresses depends on its composition, shape, temperature, and so on, but as a general rule the amount of deformation is directly proportional to the applied force provided that the force does not exceed a certain limit. In the case of tension, for example, we might find that supporting a 20-lb weight with a certain thin wire causes the wire to stretch 0.1 in. (Fig. 12-2). Doubling the weight to 40 lb therefore will produce a total elongation of 0.2 in., tripling the weight to 60 lb will produce a total elongation of 0.3 in., and so on. When the force is removed, the wire returns to its original length. This proportionality is called *Hooke's law*, and may be written

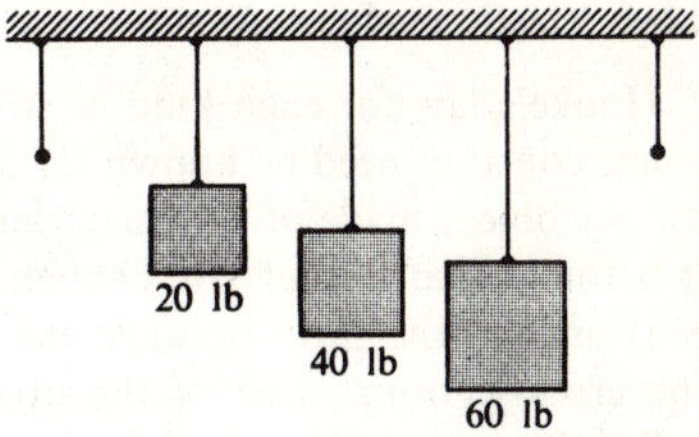

FIG. 12-2. The elongation of a wire is proportional to the stress applied to it, provided that the elastic limit is not exceeded. When the stress is removed, the wire returns to its original length. Elongation has been exaggerated here.

$$F = ks, \qquad (12\text{–}1)$$

where F is the applied force, which is equal and opposite to the tensile force, s the resulting elongation, and k a constant whose value depends on the nature and dimensions of the object under stress. The force constant k is higher for materials such as steel than it is for materials such as rubber; it is directly proportional to the cross-sectional area of the object, so that a thick wire of a given material has a higher value of k than a thin wire of the same material. Relationships similar to Eq. (12-1) apply in compression also.

The term *elastic limit* refers to the maximum deformation an object can undergo as the result of stress forces without being permanently altered. When its elastic limit is exceeded, the object may or may not be far from rupture. Brittle substances like glass or cast iron break at or near their elastic limits. For example, a glass rod whose cross-sectional area is 1 in^2 obeys Hooke's law so long as the tension or compression forces on it are less than about 10,000 lb, but it will break if this figure is exceeded. Most metals may be deformed considerably beyond their elastic limits, a property known as *ductility*. Copper is a very ductile metal, and while a copper bar whose cross-sectional area is 1 in^2 reaches its elastic limit when a force of about 6000 lb is applied, it will not rupture until the force has reached perhaps 33,000 lb. A graph of the elongation of an iron rod as a function of the tension applied to it is shown in Fig. 12-3. At first the graph is a straight line, corresponding to Hooke's law, but at the elastic limit it flattens out, corresponding to the shift from a reversible to an irreversible change in length. A moderate increase in tension beyond the elastic limit causes the rod to break.

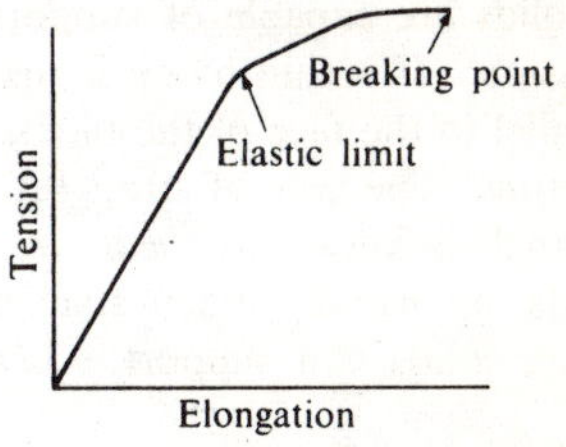

FIG. 12-3. Graph of the elongation of an iron rod as a function of the tension applied to it.

12-3 Elastic moduli

Hooke's law for each kind of stress can be expressed in such a way that only a single constant need be known for a particular material in order to relate the stress on *any* object made of this material to the resulting deformation. Experimentally it is found that the relative change in size or shape of a body under stress is proportional to the ratio between the applied force and the cross-sectional area of the object. This ratio is called the *stress.* Thus with a given force a thick rod will stretch less than a thin rod. In the case of a rod of initial length L_0 and cross-sectional area A, a tension or compression force F (Fig. 12-4) produces a change ΔL in the length of the rod according to the relationship

$$Y\frac{\Delta L}{L_0} = \frac{F}{A}. \tag{12-2}$$

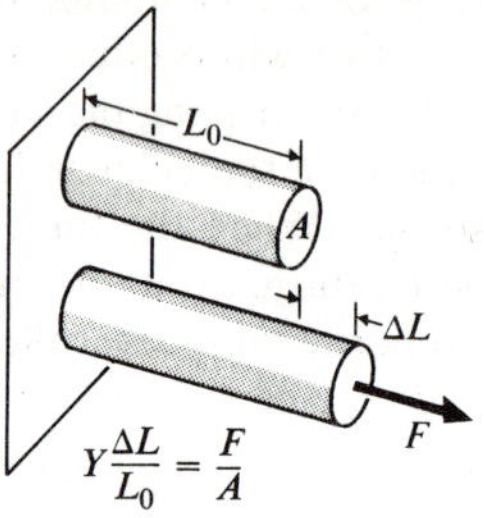

FIG. 12-4. Tension.

The rod increases in length by ΔL if it is in tension and decreases by that amount if it is in compression. The ratio $\Delta L/L_0$ is called the *strain.* The value of the quantity Y, which is called *Young's modulus,* depends on the composition of the rod. Young's moduli for a number of common substances are given in Table 12-1; it is customary to express Y in units of lb/in^2 in the British system. Equation (12-2) is valid only when the elastic limit is not exceeded, of course.

Solids are capable of supporting internal stresses other than tension or compression. If a thin block is inserted in a vise (Fig. 12-5) and a force is applied parallel to the face of the block, it does not elongate or compress, but distorts by twisting. The type of stress set up by the twisting or turning of the face through an angle is known as *shear stress.* The shear stresses which can be supported by solids are usually much smaller than either the compressive or tensile stresses which solids can support. *Liquids and gases ideally cannot support any shear.*

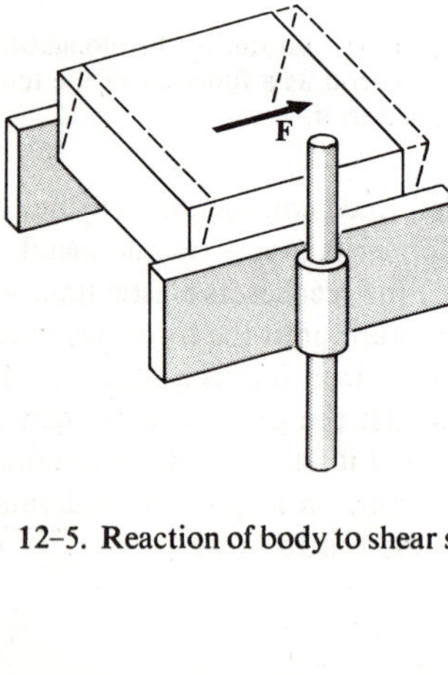

FIG. 12-5. Reaction of body to shear stress.

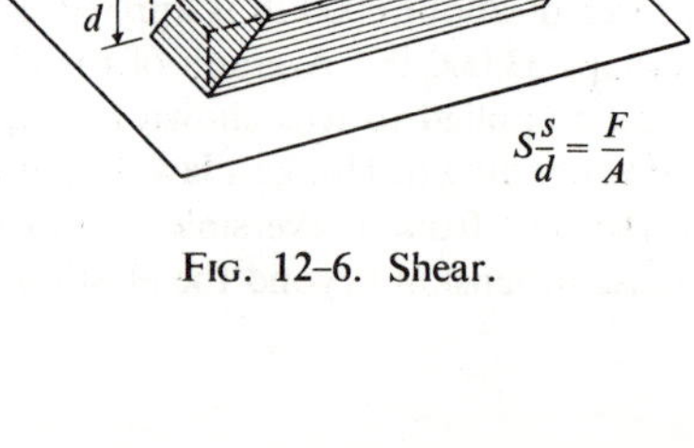

FIG. 12-6. Shear.

That is, if an external couple is exerted on a fluid, it will move under this external couple and cannot set up an internal force equal and opposite to it.

Shear stresses change the shape of an object on which they act. Let us consider a block of thickness d whose lower face is fixed in place and on whose upper face the force F acts (Fig. 12–6). A measure of the relative distortion of the block caused by the shear stress is the ratio s/d between the displacement s of the block's faces and the distance d between them. The greater the area A of these faces, the less they will be displaced by the shear force F. The analog of Eq. (12–2) for shear stresses is therefore

$$S\frac{s}{d} = \frac{F}{A}. \tag{12–3}$$

Here the applied forces are *parallel* to the faces on which they act, and not perpendicular as in the case of tension and compression. The quantity S is called the *shear modulus*, and values of S for various substances are given in Table 12–1. It is interesting to note that the shear modulus of any material is usually a good deal less than its Young's modulus. This means that it is easier to slide the atoms of a solid past one another than it is to pull them apart or squeeze them together.

An elastic modulus which is related to Young's modulus and which measures volume deformations is called the *bulk modulus*, and it is especially useful in describing the elastic properties of liquids.

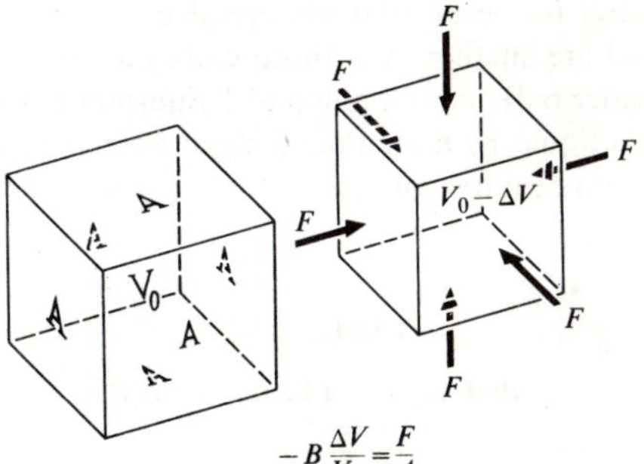

$$-B\frac{\Delta V}{V_0} = \frac{F}{A}$$

FIG. 12–7. Bulk compression.

Let us consider a volume upon whose entire surface a compressive force acts. Its volume decreases by some amount ΔV from its original volume of V_0. Only those force components that are perpendicular to the body's surface where they act are effective in compressing it, since the parallel components lead only to shear stresses. If the compression force per unit area F/A is the same over the entire surface of the body, as in Fig. 12–7, the following relationship holds:

$$-B\frac{\Delta V}{V_0} = \frac{F}{A}. \tag{12–4}$$

TABLE 12-1

Elastic moduli

Material	Young's modulus, Y		Shear modulus, S		Bulk modulus, B	
	$10^{10}\ \frac{n}{m^2}$	$10^6\ \frac{lb}{in^2}$	$10^{10}\ \frac{n}{m^2}$	$10^6\ \frac{lb}{in^2}$	$10^{10}\ \frac{n}{m^2}$	$10^6\ \frac{lb}{in^2}$
Aluminum	7.0	10	2.4	3.4	7.0	10
Brass	9.1	13	3.6	5.1	6.1	8.5
Copper	11	16	4.2	6.0	14	20
Glass	5.5	7.8	2.3	3.3	3.7	5.2
Iron	9.1	13	7.0	10	10	14
Lead	1.6	2.3	0.56	0.8	0.77	1.1
Nickel	21	30	7.7	11	26	34
Steel	20	29	8.4	12	16	23

The minus sign corresponds to the fact that an increase in force leads to a decrease in volume. The quantity B is the aforementioned bulk modulus; typical values of B are given in Table 12–1.

Liquids can support neither tensions nor shears, but they do tend to resist compression. Bulk moduli for several liquids are given in Table 12–2. Interatomic forces within a liquid are smaller than those within a solid, which is reflected in the considerably smaller bulk moduli of liquids. Substantial forces are nevertheless needed to compress a liquid by more than a slight amount; to compress a volume of water by one percent requires an inward force per unit area of 3300 lb/in^2.

TABLE 12-2

Bulk moduli of liquids at 20°C

Liquid	Bulk modulus, B	
	$10^9\ \frac{n}{m^2}$	$10^5\ \frac{lb}{in^2}$
Alcohol, ethyl	0.90	1.3
Benzene	1.05	1.5
Kerosene	1.3	1.9
Mercury	26	38
Oil, lubricating	1.7	2.5
Sulfuric acid	3.0	4.3
Water	2.3	3.3

12-4 Stresses and pressure

In the previous section, we have seen that the deformations a body undergoes are a function of the ratio F/A, the stresses set up in the body. We have tacitly neglected the vector properties of the force, and also the vector properties of the area (it has a direction along the outwardly directed normal). But the vector properties are important, since the strain when **F** is parallel to **A** is different from the case when **F** is perpendicular to **A**, when there is a shearing stress. In order really to predict an arbitrary strain, we must know the effect of three components of force applied to three mutually perpendicular areas in a body. Thus the stress has nine different values corresponding to F_x, F_y, F_z, each applied across A_x, A_y, A_z. Quantities of this sort are called *tensors*, but we shall not discuss them further.

In a solid, whenever a force is exerted in the body we have a combination of tension (or compression) and shear. If we consider an arbitrary area (Fig. 12–8), the component of force perpendicular to the area (parallel to its normal) produces tension (or compression). The component of force which is parallel to the area produces shear. The solid deforms to accommodate these stresses. In the event that we have an ideal fluid, the shear stress must always disappear, since the liquid cannot support a shear. Thus, no matter what the force distribution on a liquid, there are only forces perpendicular to areas, no matter how they are oriented; and at every point the force per unit area must be the same. Hence at every point in an ideal fluid the stress F/A is the same in all directions, and the directional properties of the stress have disappeared. When this happens we call this stress a *pressure*. The use of this term is confined almost always to the discussion of fluids because of their property of not being able to support shear stresses.

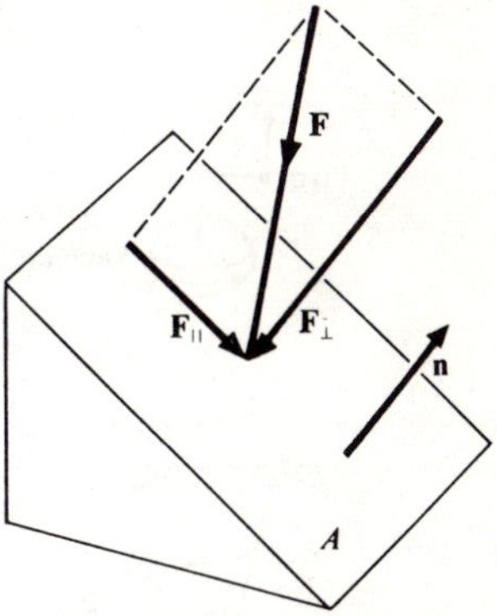

$\mathbf{F}_{\perp}$ produces compression
$\mathbf{F}_{\parallel}$ produces shear

FIG. 12–8. An arbitrary force acting in an arbitrary direction produces compression (or tension) and shear simultaneously.

The lack of rigidity exhibited by fluids has three significant consequences:

(1) The forces a fluid exerts on the walls of its container, and vice versa, always act perpendicular to the walls. If this were not so, any sideways force by a fluid on a wall would be met, according to the third law of motion, by a sideways force back on the fluid, which would cause the fluid to move constantly parallel to the wall. But fluids may be at rest in containers of any shape, and so the sole forces they can exert on their containers must be perpendicular to the walls of the latter.

(2) An external pressure exerted on a fluid is transmitted uniformly throughout the volume of the fluid. If this were not so, the fluid would flow from a region of

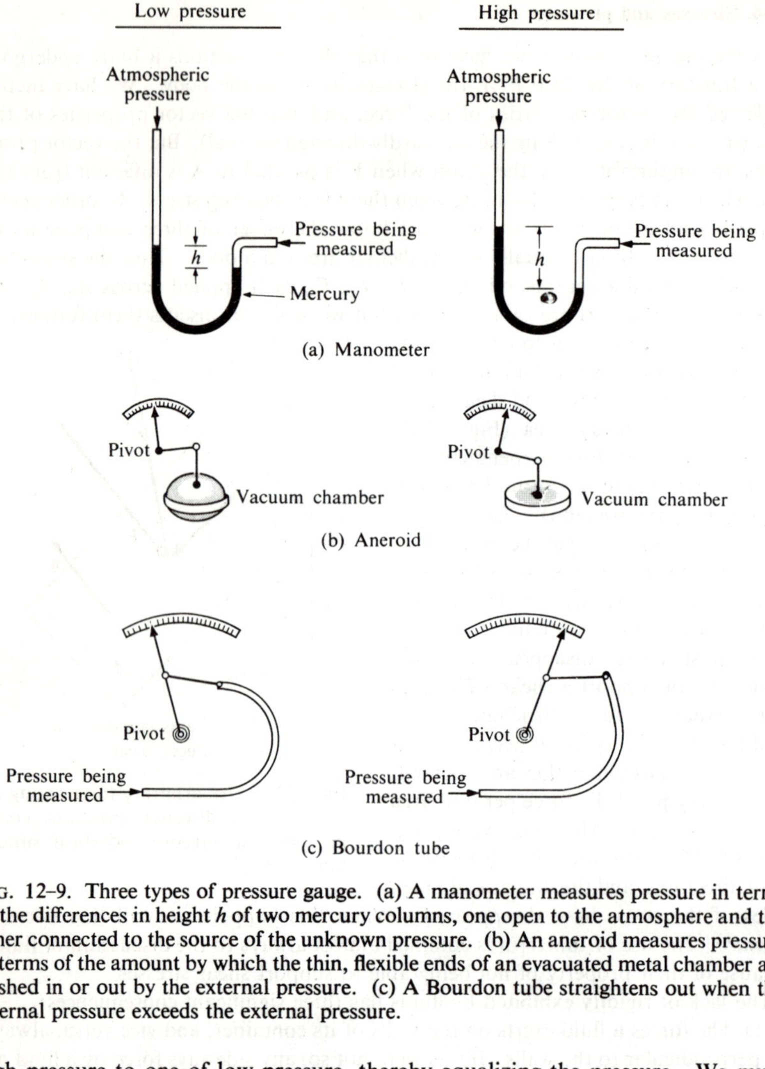

FIG. 12-9. Three types of pressure gauge. (a) A manometer measures pressure in terms of the differences in height h of two mercury columns, one open to the atmosphere and the other connected to the source of the unknown pressure. (b) An aneroid measures pressure in terms of the amount by which the thin, flexible ends of an evacuated metal chamber are pushed in or out by the external pressure. (c) A Bourdon tube straightens out when the internal pressure exceeds the external pressure.

high pressure to one of low pressure, thereby equalizing the pressure. We must keep in mind, however, that the above refers to a pressure imposed from outside the fluid. The fluid at the bottom of a container is always under greater pressure than that at the top owing to the weight of the overlying fluid. Such pressure differences are ordinarily significant only for liquids.

(3) At any depth in a fluid the pressure is the same in all directions. If this were not so, again, the fluid would flow in such a way as to equalize the pressure.

Pressure may be measured in a number of ways, three of which are illustrated in Fig. 12–9. Usually what is directly determined is the difference between the unknown pressure and atmospheric pressure. This difference is called the *gauge pressure*, while the true pressure is called the *absolute pressure*. That is,

$$\text{Absolute pressure} = \text{gauge pressure} + \text{atmospheric pressure.}$$

Thus a tire inflated to a gauge pressure of 24 $\mathrm{lb/in^2}$ contains air at an absolute pressure of 39 $\mathrm{lb/in^2}$, since sea-level atmospheric pressure is 14.7 $\mathrm{lb/in^2}$.

The metric unit of pressure is the $\mathrm{n/m^2}$, sometimes called the *pascal*, and the British unit of pressure is the $\mathrm{lb/ft^2}$. Unfortunately a number of other units are in more common use: the $\mathrm{lb/in^2}$, equal to 144 $\mathrm{lb/ft^2}$; the *millibar* (mb), equal to 100 $\mathrm{n/m^2}$; the *atmosphere* (atm), representing the average pressure exerted by the atmosphere at sea level, equal to 1.013×10^5 $\mathrm{n/m^2}$ or 14.7 $\mathrm{lb/in^2}$; and the inch and the millimeter of mercury (mm Hg), representing the pressures exerted by columns of mercury of these heights, equal respectively to 3.39×10^3 and 133 $\mathrm{n/m^2}$.

12–5 Pressure and depth

The pressure inside a volume of fluid depends on the depth below the surface, since the deeper we descend, the greater the weight of the overlying fluid. Suppose that we have a tank of height h and of cross-sectional area A which is filled with a fluid of density ρ (Fig. 12–10). The *density* of a substance is its mass per unit volume. Densities of various substances are given in Table 12–3. The volume of the tank is

$$V = Ah,$$

so that the mass of fluid contained in it is

$$m = \rho V = \rho A h.$$

The weight of the fluid in the tank is therefore

$$w = mg = \rho g A h.$$

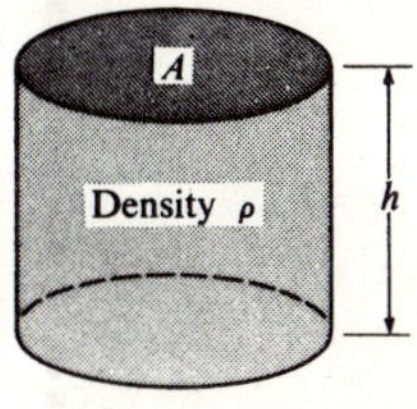

FIGURE 12–10

The pressure Δp which the fluid exerts on the bottom of the tank is its weight divided by the cross-sectional area of the tank, with the result that

$$\Delta p = \frac{F}{A} = \frac{w}{A} = \rho g h. \tag{12–5}$$

The pressure difference between the top and the bottom of the tank is directly proportional to the height of the fluid column and to the fluid density. This result also applies to *any* depth h in a fluid, whether at the bottom or not, since the fluid beneath that depth does not contribute to the weight pressing down there.

TABLE 12–3

Densities of various substances at atmospheric pressure and indicated temperature

Solids, 20°C	10^3 kg/m^3	slugs/ft^3	Liquids, 20°C	10^3 kg/m^3	slugs/ft^3	Gases, 0°C	kg/m^3	10^{-3} slugs/ft^3
Aluminum	2.70	5.25	Alcohol, ethyl	0.79	1.53	Air	1.29	2.50
Balsa wood	0.13	0.25	Benzene	0.88	1.71	Ammonia	0.76	1.47
Brass	8.7	17	Bromine	3.19	6.18	Argon	1.78	3.45
Concrete	2.3	4.5	Gasoline	0.68	1.32	Carbon dioxide	1.96	3.80
Copper	8.89	17.3	Kerosene	0.80	1.55	Chlorine	3.16	6.13
Glass	2.6	5.0	Mercury	13.6	26.4	Helium	0.18	0.35
Gold	19.3	37.5	Oil, lubricating	0.90	1.75	Hydrogen	0.090	0.17
Ice	0.92	1.8	Sulfuric acid	1.83	3.55	Nitrogen	1.25	2.42
Iron	7.8	15	Water, pure	1.00	1.94	Oxygen	1.43	2.78
Lead	11.3	22.0	Water, sea	1.03	2.00	Propane	2.02	3.92
Oak	0.72	1.4						
Steel	7.8	15						

The *total* pressure within a fluid, of course, also depends on the pressure p_{atm} exerted on its surface by the atmosphere or, perhaps, by a piston. For example, the pressure at a depth of 8 ft in a swimming pool filled with water (density 1.94 slugs/ft^3) is

$$p = p_{\text{atm}} + \rho g h. \tag{12–6}$$

Here

$$\rho g h = 1.94 \frac{\text{slugs}}{\text{ft}^3} \times 32 \frac{\text{ft}}{\text{sec}^2} \times 8 \text{ ft} \times \frac{1}{(12 \text{ in/ft})^2} = 3.4 \frac{\text{lb}}{\text{in}^2},$$

and so

$$p = (14.7 + 3.4) \frac{\text{lb}}{\text{in}^2} = 18.1 \frac{\text{lb}}{\text{in}^2}.$$

In the event a *weight density* (such as 62.5 lb/ft^3 in the case of water) is known, it is substituted for ρg in Eq. (12–5).

The famous and useful *Archimedes' principle* may be derived from Eq. (12–5), together with the fact that the pressure at a point in a fluid is the same in all directions. Let us consider a block of height H and of cross-sectional area A that is completely submerged in a fluid of density ρ so that its top is h below the surface, as in Fig. 12–11. The downward pressure p_1 on the top of the block is

$$p_1 = p_{\text{atm}} + \rho g h,$$

and the upward pressure p_2 on the bottom of the block is

$$p_2 = p_{\text{atm}} + \rho g(h + H).$$

Since the downward force the fluid exerts on the block is

$$F_1 = A p_1,$$

while the upward force is

$$F_2 = A p_2,$$

the *net* force on the block is

$$F = F_2 - F_1 = A(p_2 - p_1) = A \rho g H.$$

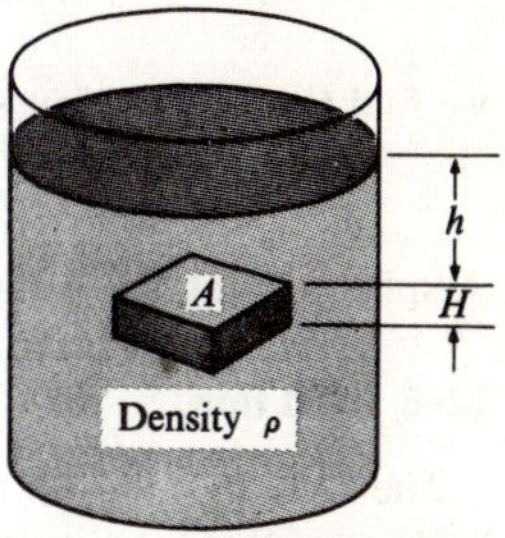

FIG. 12–11. Archimedes' principle states that the buoyant force on a submerged object is equal to the weight of liquid displaced.

Since the pressure on the bottom of the block is greater than that on the top, the net force is upward, and is known as the *buoyant force* on the block. Because the volume of the block is

$$V = AH,$$

we have the simple result that

$$F = V \rho g \tag{12–7}$$

buoyant force = weight of displaced fluid.

The buoyant force on a submerged object is equal to the weight of fluid displaced by the object. This is Archimedes' principle.

Archimedes' principle enables us to determine only the buoyant force on a submerged object, not the resultant force on it. If the weight of the object is greater than the buoyant force on it, it will sink; if its weight is less than the buoyant force, it will rise. As an exercise, let us compute the volume V a helium-filled balloon must have if it is to support a mass of 1000 kg. This mass includes the mass of the balloon itself, but *not* that of the helium inside. The total weight to be supported is therefore

$$w = mg = (1000\text{ kg} + V\rho_{\text{He}})g,$$

where ρ_{He} is the density of helium at atmospheric pressure. The buoyant force on the balloon is

$$F = V\rho_{\text{air}}g.$$

Hence the condition for equilibrium is

$$F = w, \qquad V\rho_{\text{air}}g = (1000\text{ kg} + V\rho_{\text{He}})g, \qquad V = \frac{1000\text{ kg}}{\rho_{\text{air}} - \rho_{\text{He}}}.$$

Since

$$\rho_{\text{air}} = 1.29\ \frac{\text{kg}}{\text{m}^3}, \qquad \rho_{\text{He}} = 0.18\ \frac{\text{kg}}{\text{m}^3},$$

we find that the volume of the balloon must be

$$V = 9.1 \times 10^3\text{m}^3.$$

A spherical balloon of this volume has a radius of about 13 meters.

12-6 Fluid flow

The study of fluids in motion is one of the more difficult branches of mechanics because of the diversity of phenomena that may occur. However, the fundamental aspects of fluid flow can be understood on the basis of a simple model that is reasonably realistic in many cases. This model involves liquids that are incompressible and exhibit no internal friction or viscosity. In the absence of viscosity, layers of fluid slide freely past one another and past other surfaces, so that our model applies to such liquids as water, but not to such liquids as molasses. Another approximation we shall make is that the fluid undergoes *streamline flow*

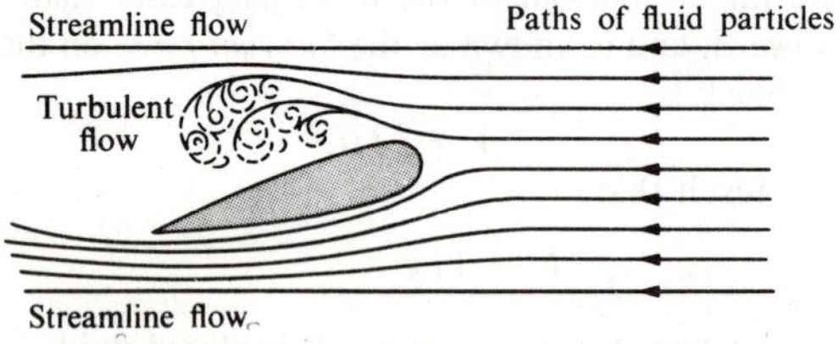

FIG. 12-12. Streamline and turbulent flows around an obstacle.

exclusively. In streamline flow, which is illustrated in Fig. 12–12, every particle of liquid passing a particular point follows the same path (called a *streamline*) as the particles that passed that point previously. Furthermore, the direction in which the individual fluid particles move is always the same as the direction in which the fluid as a whole moves. At the other extreme is *turbulent flow*, which is characterized by the presence of whirls and eddies, such as those in a cloud of cigarette smoke or at the foot of a waterfall. Turbulence generally occurs at high speeds and when there are obstructions or sharp bends in the path of the fluid.

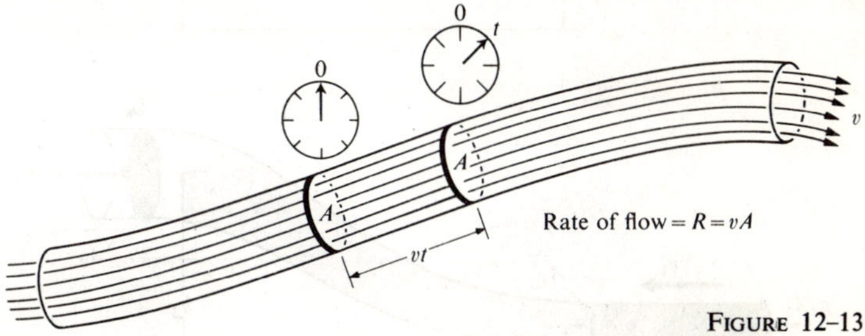

FIGURE 12–13

The volume of liquid that flows through a pipe per unit time is easy to compute and is often of interest. If the average speed of the liquid in the pipe of Fig. 12–13 is v, each part of the stream travels the distance vt in the time interval t. Let us call the cross-sectional area of the pipe A. The volume of liquid transported the distance vt in the time t is vt multiplied by the area A, or vtA. Therefore the rate of flow R of liquid through the pipe is

$$R = \frac{dV}{dt} = \frac{vtA}{t} = vA, \tag{12-8}$$

the product of the liquid speed and the cross-sectional area of the pipe. The rate R is the volume of liquid passing through the pipe per unit time, and, for convenience in many applications, is often expressed in units such as gal/min instead of ft^3/sec or m^3/sec. If the pipe size varies, the speed of the liquid also varies so as to keep R constant. Hence a liquid flows faster through a constriction in a pipe and slower through a dilation. As in Fig. 12–14, streamlines drawn close together signify rapid motion, while streamlines far apart signify slow motion.

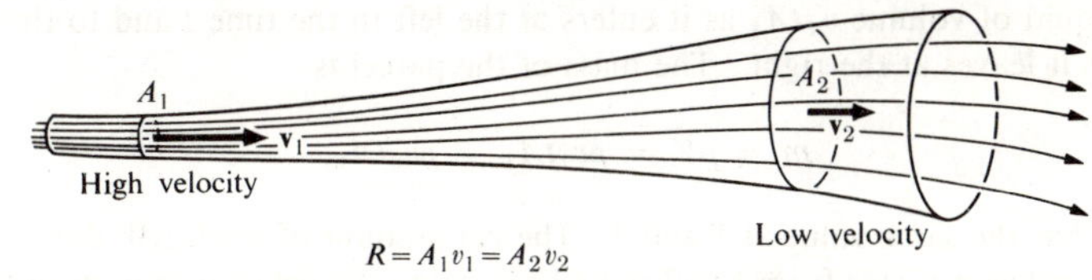

FIG. 12–14. In streamline flow, liquid speed is inversely proportional to the cross-sectional area of the pipe.

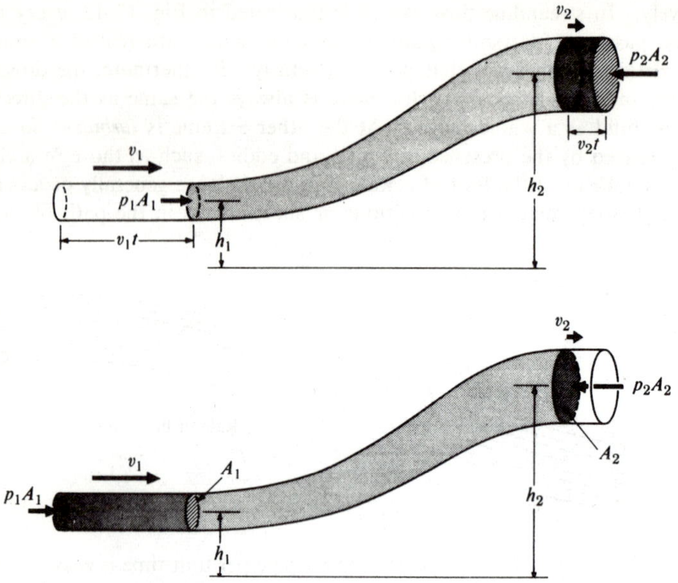

$$p_1 + \rho g h_1 + \tfrac{1}{2}\rho v_1^2 = p_2 + \rho g h_2 + \tfrac{1}{2}\rho v_2^2$$

FIG. 12–15. Bernoulli's equation.

12–7 Bernoulli's equation

When a liquid flowing through a pipe enters a region where the pipe diameter is reduced, its speed increases. A change in speed involves an acceleration; this means that a net force must be acting on the liquid. This force can only arise from a difference in pressure between the different parts of the pipe. Evidently the pressure in the part of the pipe having a large diameter is the greater, since the liquid increases in speed on its way to the constriction. Thus we expect a relationship between the pressure in a moving liquid and its speed.

Figure 12–15 shows a curved pipe of nonuniform cross section through which a liquid flows. Let us apply the principle of conservation of energy to a parcel of the liquid of volume v_1tA_1 as it enters at the left in the time t and to the same parcel as it leaves at the right. The mass of the parcel is

$$m = \rho V = \rho v_1 t A_1 = \rho v_2 t A_2, \tag{12–9}$$

since vA has the same value at 1 and 2. The net amount of work ΔW done on the liquid parcel as it passes from 1 to 2 must be equal to the net change in its potential energy ΔPE as its height goes from h_1 to h_2, plus the net change in its kinetic energy

ΔKE as its speed goes from v_1 to v_2. That is,

$$\Delta W = \Delta \text{PE} + \Delta \text{KE}. \tag{12-10}$$

The work done *on* the parcel at 1 is the force p_1A_1 on it multiplied by the distance v_1t through which the force acts. The work done *by* the parcel at 2 is the force p_2A_2 multiplied by the distance v_2t through which the force acts. The *net* work done on the parcel is therefore

$$\Delta W = p_1A_1v_1t - p_2A_2v_2t,$$

which we can write as

$$\Delta W = \frac{p_1m}{\rho} - \frac{p_2m}{\rho}$$

by virtue of Eq. (12–9). The change in the potential energy of the parcel in going from 1 to 2 is

$$\Delta \text{PE} = mgh_2 - mgh_1,$$

and the change in its kinetic energy is

$$\Delta \text{KE} = \tfrac{1}{2}mv_2^2 - \tfrac{1}{2}mv_1^2.$$

Inserting these expressions for ΔW, ΔPE, and ΔKE into Eq. (12–10), we obtain

$$\frac{p_1m}{\rho} - \frac{p_2m}{\rho} = mgh_2 - mgh_1 + \tfrac{1}{2}mv_2^2 - \tfrac{1}{2}mv_1^2.$$

When we divide through by the common factor m, multiply by ρ, and rearrange terms, the result is

$$p_1 + \rho gh_1 + \tfrac{1}{2}\rho v_1^2 = p_2 + \rho gh_2 + \tfrac{1}{2}\rho v_2^2. \tag{12-11}$$

Equation (12–11) is known as *Bernoulli's equation*, after Daniel Bernoulli (1700–1782), who first derived it. According to Bernoulli's equation the quantity

$$p + \rho gh + \tfrac{1}{2}\rho v^2$$

has the same value at all points in a liquid undergoing streamline flow.

In many situations the speed, pressure, or height of a liquid is constant, and simplified forms of Bernoulli's equation hold. Thus when a liquid column is stationary, we see that the pressure difference between two depths in it is

$$p_2 - p_1 = \rho g(h_1 - h_2), \tag{12-12}$$

which is just what Eq. (12–5) states. Evidently the latter formula is included in Bernoulli's equation.

Another straightforward result occurs in the event $p_1 = p_2$. As an example, Fig. 12–16 illustrates a liquid emerging from an orifice at the bottom of a tank. The liquid pressure equals atmospheric pressure at the top of the tank and at the orifice. If the orifice is small compared with the cross section of the tank, the liquid level in the tank will fall slowly enough for the liquid speed at the top of the tank to be assumed zero. If the speed of the liquid as it leaves the orifice is v and the difference in height between the top of the liquid and the orifice is h, Bernoulli's equation reduces to

$$\tfrac{1}{2}\rho v^2 = \rho g h,$$

$$v = \sqrt{2gh}. \qquad (12\text{-}13)$$

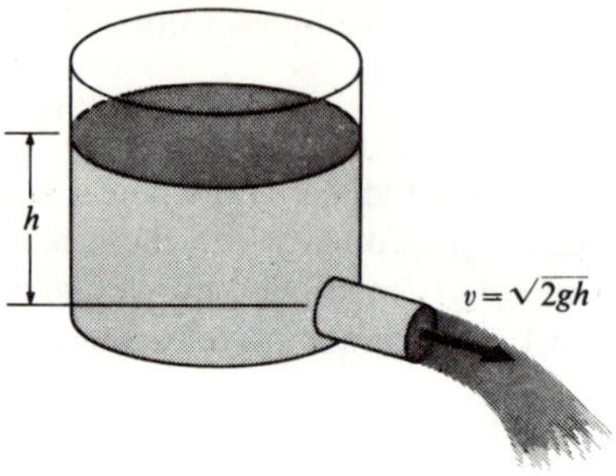

FIG. 12-16. Torricelli's theorem.

The speed with which the liquid is discharged is the same as the speed of a body falling from rest from the height h. This result is called *Torricelli's theorem*, and, like the relationship between pressure and depth, it is a special case of Bernoulli's equation. The rate at which liquid flows through the orifice may be found from Eq. (12-13) if the orifice area A is known. The volume of liquid being discharged per unit time is

$$R = vA = A\sqrt{2gh}.$$

As an illustration of this formula, we note that water will flow through a hole 1 cm^2 in area at the bottom of a tank in which the water level is 3 meters high at the rate of

$$R = 1\ \text{cm}^2 \times \frac{1}{10^4\ \text{cm}^2/\text{m}^2} \times \sqrt{2 \times 9.8\ \text{m/sec}^2 \times 3\ \text{m}}$$

$$= 7.7 \times 10^{-4}\ \text{m}^3/\text{sec},$$

which is a little less than a quart per second, an appreciable amount.

The most interesting special case of Bernoulli's equation occurs when there is no change in height during the motion of the liquid (Fig. 12-17). Here

$$p_1 + \tfrac{1}{2}\rho v_1^2 = p_2 + \tfrac{1}{2}\rho v_2^2, \qquad (12\text{-}14)$$

which means that the pressure in the liquid is least where the speed is greatest, and vice versa. There are many applications of this principle. A familiar one is the lifting force produced by the flow of air past the wing of an airplane, as in Fig. 12-18. (Air is, of course, a compressible fluid and so does not fit our model exactly, but the behavior predicted by Bernoulli's equation is not a bad approximation for gases at moderate speeds.) Air moving past the upper surface of the wing must travel faster than air moving past the lower surface; this is indicated by the closeness of the streamlines near the former. The difference in speed leads to a decreased pressure over the top of the wing, a pressure which is equivalent to a suction force that lifts the wing. The greater the difference in air

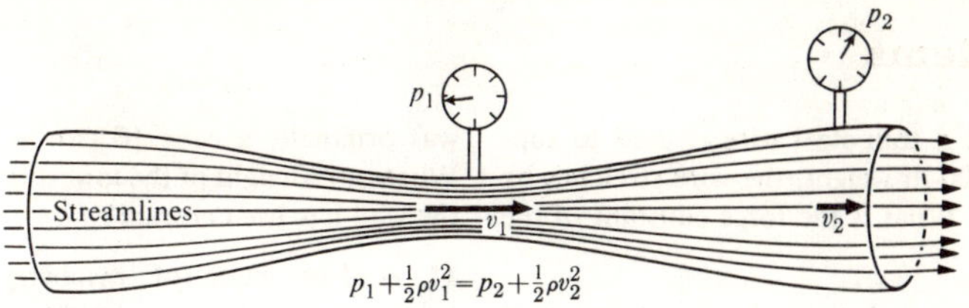

FIG. 12-17. In a horizontal pipe, the pressure is greatest when the velocity is least, and vice versa.

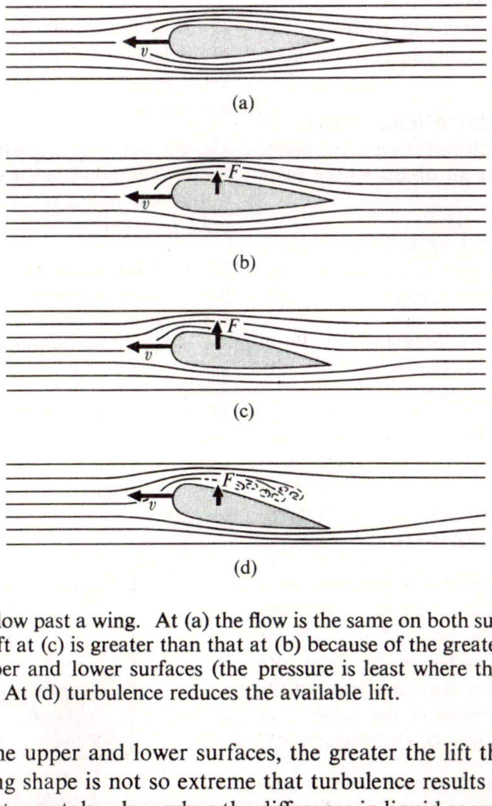

FIG. 12-18. Air flow past a wing. At (a) the flow is the same on both surfaces, so that no lift results. The lift at (c) is greater than that at (b) because of the greater pressure difference between upper and lower surfaces (the pressure is least where the streamlines are closest together). At (d) turbulence reduces the available lift.

speeds around the upper and lower surfaces, the greater the lift that is produced, provided the wing shape is not so extreme that turbulence results (Fig. 12-18).

A curious effect may take place when the difference in liquid speeds at two nearby places is so great that the absolute pressure where the liquid flows most rapidly becomes negative. This can happen when the propeller of a boat rotates at a high speed. The result is a phenomenon called *cavitation* in which cavities ("holes") appear in the water. Cavitation causes the efficiency of the propeller to drop sharply.

Problems

1. When a thin steel wire is used to support a 100-lb weight, the wire stretches by 0.02 in. What is the force constant of the wire?

2. When a coil spring is used to support a 12-kg object, the spring stretches by 4 cm. What is the force constant of the spring?

3. A coil spring has a force constant of 1000 n/m. How much will it stretch when it is used to support an object whose mass is 8 kg?

4. A coil spring has a force constant of 2 lb/ft. How much will it stretch when it is used to support an object whose mass is 0.5 slug?

5. A steel wire 1 m long and 1 mm square in cross section supports a mass of 6 kg. By how much does it stretch?

6. A copper wire 5 ft long and 0.05 in. in diameter supports a weight of 17 lb. By how much does it stretch?

7. A nylon rope $\frac{1}{2}$ in. in diameter breaks when a force of 5000 lb is applied to it. What would you estimate for the breaking strength of a nylon rope (a) $\frac{1}{4}$ in. in diameter? (b) $\frac{3}{4}$ in. in diameter?

8. A steel cable whose cross-sectional area is 0.4 in^2 supports an elevator weighing 1 ton. The elastic limit of the cable is 4×10^4 lb/in^2. What is the maximum upward acceleration that can be given the elevator if the tension in the cable is to be no more than 20% of the elastic limit?

9. An iron pipe 10 ft long is used to support a sagging floor. The inside diameter of the pipe is 4 in. and its outside diameter is 5 in. When the force on it is 3000 lb, by how much is it compressed?

10. A wall of lead bricks 1 m high is used to shield a sample of radium. Each brick was originally a cube 10 cm on an edge. What is the height of the lowest brick when the wall has been erected?

11. A 1-in. cube of raspberry-flavored gelatin on a table is subjected to a shearing force of 0.1 lb. The upper surface is displaced by 0.2 in. What is the shear modulus of the gelatin?

12. A 50-kg woman balances on the heel of her left shoe, which is 1 cm in diameter. What pressure (in atm) does she exert on the ground?

13. The force on a phonograph needle is 1.2 n. The point has a circular cross section whose radius is 0.1 mm. What is the pressure it exerts on the record (in atm)?

14. A piston weighing 12 n rests on a sample of gas in a cylinder 5 cm in diameter. (a) What is the gauge pressure in the gas? (b) What is the absolute pressure in the gas?

15. An irregular nugget of gold is dropped into a filled glass of water, and 2 in^3 of water overflow. Gold is worth \$35 per ounce of weight. How much is the nugget worth?

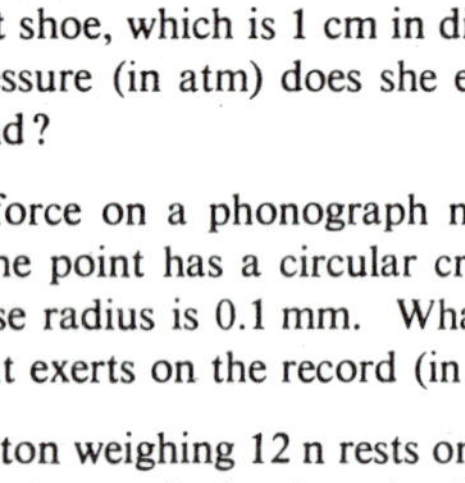

FIGURE 12–19

16. Deduce Archimedes' principle from the principle of the conservation of energy as follows: When a submerged ball of mass m is lifted from position 1 to position 2 (Fig. 12–19), its potential energy is in-

creased by mgh. The energy of the "ball-water" system is decreased by ρVgh, due to the fact that the water occupying position 2 must descend to position 1.

17. A wooden cube is floating in water. Show that when it is depressed and released it will oscillate with simple harmonic motion. Calculate its frequency in terms of its density, the density of water and the length of its side.

18. Calculate the density of sea water at a depth of 3 mi.

19. Calculate the density of mercury at the bottom of a mercury column 1.3 m high.

20. What is the minimum area of an ice floe 3 in. thick that can support a 120-lb girl without getting her feet wet? The floe is in a fresh-water lake.

21. A barge 120 ft long and 20 ft wide weighs 20 tons. What is the depth of sea water required to float it?

22. An ice cube floats in a glass of water filled to the brim. What will happen when the ice melts?

23. What percentage of the volume of an iceberg is below the surface of the sea?

24. A sailboat has a ton of lead ballast attached to the bottom of its keel, which is 6 ft below the water surface when the boat is level. What is the torque exerted by this ballast when the boat is heeled by 20° from the vertical?

25. A 30-kg balloon is filled with 100 m^3 of hydrogen. How much force is needed to hold it down?

26. Two spheres of the same diameter but of different mass are dropped from a tower. If air resistance is the same for both, which will reach the ground first?

27. The densities of people are slightly less than that of water. Assuming that these densities are the same, compute the buoyant force of the atmosphere on a 160-lb man.

28. Water flows through a hose whose internal diameter is 1 cm at a speed of 1 m/sec. What should the diameter of the nozzle be if the water is to emerge at 5 m/sec?

29. Why is it impossible to remove filter paper from a funnel by blowing into the narrow end of the funnel?

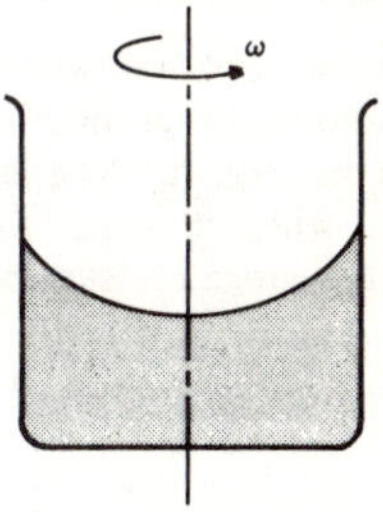

FIGURE 12–20

30. Find the formula of the cross section of the surface of water which is rotating with constant angular velocity ω about an axis perpendicular to the surface (Fig. 12–20).

31. A small opening of cross section A is at a height h below the surface of a liquid of density ρ. What is the force acting on the vessel? (See Fig. 12–21.)

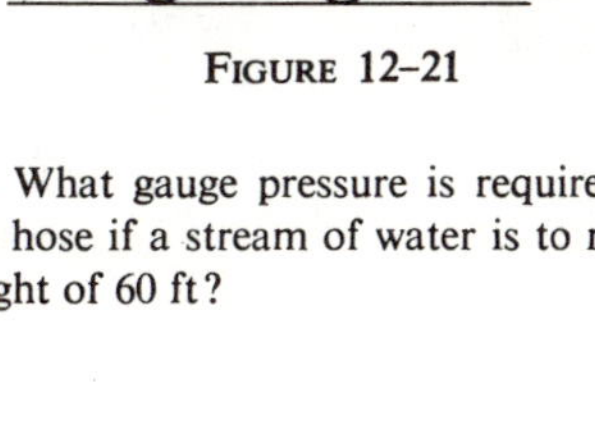

FIGURE 12–21

32. What gauge pressure is required in a fire hose if a stream of water is to reach a height of 60 ft?

33. A tank of height H is filled with water. A hole is made a distance y from the top. How far from the tank does the water strike the ground?

34. A horizontal pipe 1 in. in diameter carries water at the rate of 10 ft^3/min. What is the speed of the water in the pipe? It passes through a constriction 0.3 in. in diameter. What is the speed of the water in the constriction, and what is the pressure difference in the constriction and in the main body of the pipe?

35. A horizontal stream of water leaves an orifice 1 m above the ground and strikes the ground 2 m away. (a) What is the speed of the water when it leaves the orifice? (b) What is the gauge pressure behind it?

36. A horizontal tube whose cross section is 10 cm^2 has a constriction equal to 3 cm^2 in area. The difference of pressure between the main part of the tube and the constriction is 7 cm of Hg. What is the velocity of water flowing in the tube?

37. A barrel filled with kerosene is 4 ft high. A crack 1 in. long and 0.07 in. wide appears at its base. How many pounds of kerosene flow out per minute?

38. The brain of a man is approximately $\frac{1}{3}$ m above his heart, while this distance is approximately 2 m in a giraffe. What is the pressure required to circulate blood between the heart and brain of (a) a man? (b) a giraffe? The density of blood is 1.1×10^3 kg/m^3.

13

Heat

In Chapter 7 we saw that under certain circumstances mechanical energy was not conserved. Thus if we pull a block along a plane and there is friction between the two, the kinetic energy of the block will not be equal to the work done but will be less.

We can either abandon the concept of conservation of energy, or we can assume that the conservation theorem is correct and investigate further to see what happened to the excess energy. The route taken by physics has been the latter one, and thus far no experiments have been performed in which the energies (including mass energy) at the beginning and end of any process could not be balanced.

What happens to the "missing" energy in a process depends on the nature of the matter involved. Either we can trace the disposition of the excess energy by looking into the constituents of macroscopic bodies and seeing whether they have gained energy, or we can construct a theory for the energy imbalance without specifying exactly how the building blocks of macroscopic bodies, namely the molecules, interact with one another. The former procedure of tracing the energy to the molecules is called the *kinetic theory of matter*, and will be discussed in Chapter 14. This is a somewhat more difficult method, since it requires us to have some detailed knowledge about interatomic and intermolecular forces. In this chapter we shall follow the latter course to see how we can account for the energy imbalance *without* any specific model of the microscopic nature of matter. In order to do this we shall first discuss what we mean by temperature. This is not surprising, since one of the effects we note experimentally when two bodies slide over one another in the presence of friction is that they become hot.

13–1 Temperature

The object of a temperature measurement is to specify in a precise way something which we recognize physiologically as a sensation of hot or cold. Our senses are able to determine which of two things is hotter than the other, but such a qualitative distinction is hardly adequate. What we would like to do is exploit some definite physical property which varies with temperature. Such a characteristic is called a *thermometric property*. The dimensions of a solid constitute a useful thermometric property, as do the electrical resistance of a metal, the volume of a liquid, the electromotive force at a junction of dissimilar metals, the color of a body when it is hot enough to glow, the pressure of a gas kept at constant volume, the volume of a gas kept at constant pressure, and so on.

Once we select a particular thermometric property, we have reduced the measurement of temperature to the measurement of that property. Now we need a unit of temperature for expressing temperature differences, and a reference temperature with which to compare our readings. From experience we know that there are certain natural systems which remain at constant temperatures. The most important of these from a practical point of view are a mixture of ice and water and a mixture of steam and water. How can we be sure that their temperatures do not change? The answer is that none of the thermometric properties we know of vary in any way in the mixtures of these systems, and for this reason (and also because they are so readily duplicated) ice-and-water mixtures and steam-and-water mixtures are convenient to use as fixed points in constructing a scale of temperature.

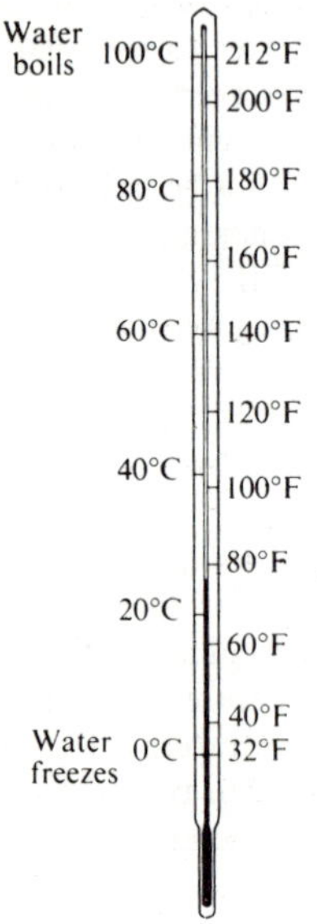

FIG. 13–1. A comparison of the fahrenheit and celsius temperature scales.

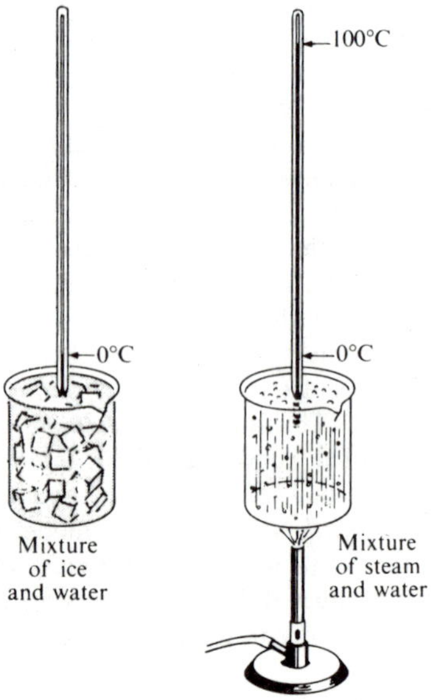

FIG. 13–2. Calibrating a thermometer on the celsius scale. A mixture of ice and water is, by definition, at 0°C and a mixture of steam and water is, again by definition, at 100°C.

On the celsius or centigrade scale the temperature of melting ice is, by definition, the zero point. The unit of temperature, the celsius degree, is specified as $\frac{1}{100}$ of the difference in a linear thermometric property of matter between the temperatures of melting ice and boiling water. (By a linear property we mean one that is directly proportional to the temperature.) Hence the steam point of water is 100°C (Fig. 13-1).

On the fahrenheit scale the temperature of melting ice is taken as 32°. The unit of temperature, the fahrenheit degree, is $\frac{1}{180}$ of the difference in a linear thermometric property between the ice and steam points, which puts the steam point of water at 212°F.

To calibrate a thermometer, say an ordinary mercury thermometer, we first plunge it into a mixture of ice and water. When the mercury column has come to rest, we mark the position of its top 0°C on the glass (Fig. 13-2). Then we plunge it into a mixture of steam and water, and when the mercury column has again come to rest, we mark the new position of the top of the mercury column 100°C. Finally we divide the interval between the 0°C and 100°C markings into 100 equal parts, each representing a change in temperature of 1°C, and extend the scale with divisions of the same length beyond 0°C and 100°C as far as is convenient. In doing this we have, of course, assumed that changes in the length of the mercury column are always directly proportional to the changes in temperature that brought them about.

Fahrenheit degrees are five-ninths as large as celsius degrees, for reasons enumerated above and illustrated in Fig. 13-1. With the formulas

$$°F = \tfrac{9}{5}°C + 32° \qquad \text{and} \qquad °C = \tfrac{5}{9}(°F - 32°), \tag{13-1}$$

we can convert temperatures from one scale to the other. For instance, a fahrenheit temperature of 70°F, which is normal room temperature, on the celsius scale is

$$\tfrac{5}{9}(70° - 32°) = 21°C.$$

If we were to define the centigrade (or fahrenheit) scale as we have indicated, we would find that the results are somewhat different depending on the particular thermometric property that has been used. If we look at a quality mercury thermometer we note that the degree marks are crowded together more at the lower than at the higher temperatures. The graduations are so arranged because the mercury column does not expand uniformly with the temperature. Mercury thermometers are calibrated in terms of the pressure of a gas at constant volume, and in fact the centigrade and fahrenheit scales are defined in terms of a constant-volume hydrogen thermometer. There are theoretical and experimental reasons for believing that, at low pressures, the pressure of a gas does increase uniformly with temperature. In order to investigate this more thoroughly we shall discuss the expansion of solids and gases when their temperatures increase.

13-2 Thermal expansion

When the temperature of a sample of matter changes, the dimensions of the sample also change. This is a familiar effect: Sidewalks buckle on a hot summer day, the mercury column rises in a heated thermometer, warm air rises because, owing to its expansion, it is lighter than surrounding cool air. It is a matter of experience that, when the temperature of a solid or liquid is changed by an amount Δt, its fractional volume change turns out to be proportional to Δt, if Δt is small. We are therefore able to express the change in volume ΔV of a sample whose original volume was V_0 as

$$\Delta V = \beta V_0 \, \Delta t, \tag{13-2}$$

where the constant of proportionality β is called the *coefficient of volume expansion.* Table 13-1 contains values of β for some common materials.

TABLE 13-1

Coefficients of volume expansion

Substance	Coefficient, $\times 10^{-4}/°C$
Ethyl alcohol	11
Glass (average)	0.2
Glycerin	5.1
Ice	0.5
Mercury	1.8
Pyrex glass	0.09
Water	2.1

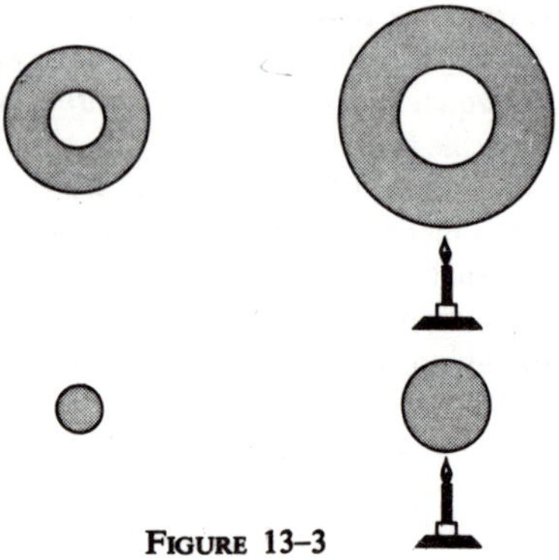

FIGURE 13-3

Example: Calculate the volume of water that overflows when a Pyrex beaker filled to the brim with 250 cm^3 of water at 20°C is heated to 60°C.

Solution: First we note that a cavity in a body expands or contracts by precisely as much as a solid object having the same volume as the cavity and having the composition of the body (Fig. 13-3). This means that we can write for the increase in capacity of the beaker,

$$V_b = \beta_P V_b \, \Delta t = 0.09 \times \frac{10^{-4}}{\text{C}} \times 250 \text{ cm}^3 \times 40°\text{C} = 0.09 \text{ cm}^3.$$

The increase in the volume of the water is

$$\Delta V_w = \beta_w V_w \, \Delta t = 2.1 \times \frac{10^{-4}}{°\text{C}} \times 250 \text{ cm}^3 \times 40°\text{C} = 2.1 \text{ cm}^3,$$

and so the volume of water that overflows is

$$\Delta V_w - \Delta V_b = 2.0 \text{ cm}^3.$$

If all we are interested in is the linear change in one of the dimensions of a solid, say the length of a wire or rod, we can obtain the corresponding *coefficient of linear expansion* from Eq. (13-2) in the following way. We consider a cube of some uniform material initially L_0 long on a side and increase its temperature an amount Δt. Now each side is longer by ΔL. Since the initial volume of the cube was $V_0 = L_0^3$, its change in volume is

$$\Delta V = V - V_0 = (L_0 + \Delta L)^3 - L_0^3$$
$$= 3L_0^2 \Delta L + 3L_0 \Delta L^2 + \Delta L^3.$$

Since ΔL is always much smaller than L_0, the second and third terms of the above equation for ΔV are negligible compared with the first term. Substituting $\Delta V = 3L_0^2 \Delta L$ into Eq. (13-2), we find that, since $V_0 = L_0^3$, then

$$3L_0^2 \Delta L = \beta L_0^3 \Delta t, \qquad \Delta L = \frac{\beta}{3} L_0 \Delta t,$$

or, letting $\beta/3 = \alpha$,

$$\Delta L = \alpha L_0 \Delta t. \tag{13-3}$$

The coefficient of linear expansion α is $\frac{1}{3}$ of the coefficient of volume expansion β. While we have considered a cubical solid body in obtaining Eq. (13-3), it holds for a body of any shape. The same is true of Eq. (13-2), which is correct not only for bodies of any shape but also for hollow bodies. As we said in the example, a glass bottle expands at precisely the same rate as a solid glass object of the same size and outside shape.

It is not difficult to calculate the stresses exerted by an expanding or contracting body. The procedure is to determine the stress that would have to be applied in order to produce the strain caused by a change in temperature. For example, the reduction in the volume of a solid when it is subjected to a pressure p is, from Eq. (12-3),

$$B\Delta V = -V_0 p,$$

while the volume reduction after a temperature decrease $-\Delta t$ is

$$\Delta V = -\beta V_0 \Delta t.$$

Hence

$$p = \beta B \Delta t$$

is the pressure necessary to compress the solid to the same extent as a temperature decrease Δt. The inverse is also true: A solid expanding due to a rise in temperature exerts the above pressure p in all directions.

Let us consider a steel girder 50 ft long whose temperature changes from 5°C to 25°C, an increase of 20°C (Fig. 13-4). The coefficient of linear expansion of

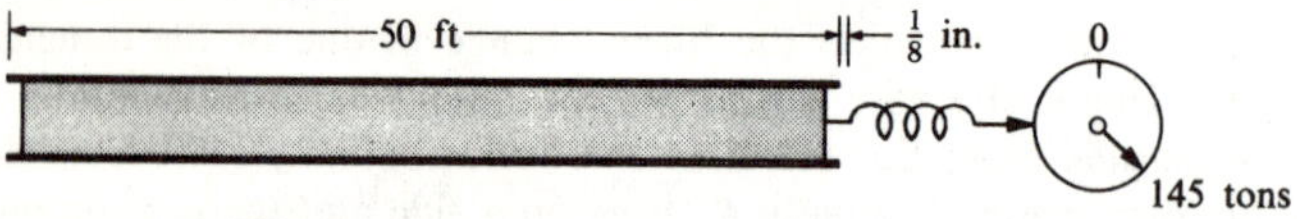

FIG. 13-4. A girder 50 ft long expands only $\frac{1}{8}$ in. when heated from 5° to 25°C, but it exerts a force of 145 tons.

steel is 1.2×10^{-5}/°C, and so

$$\Delta L = \alpha L_0 \, \Delta t = 1.2 \times \frac{10^{-5}}{°\text{C}} \times 50 \text{ ft} \times 20°\text{C} = 0.012 \text{ ft},$$

which is a little over $\frac{1}{8}$ inch. This does not seem a very significant change until we compute the force exerted by the expanding girder, which is equal to the force required to stretch it by 0.012 ft. To do this we refer to Eq. (12-2), which gives the change in length ΔL of a rod that is subjected to a tension or compression force F as

$$\Delta L = \frac{L_0}{Y} \frac{F}{A},$$

where A is the cross-sectional area of the rod and Y is Young's modulus for the material of the rod. If the girder has a cross-sectional area of 40 in^2, and a Young's modulus of 3×10^7 lb/in^2,

$$F = \frac{YA \, \Delta L}{L_0} = \frac{3 \times 10^7 \text{ lb/in}^2 \times 40 \text{ in}^2 \times 0.012 \text{ ft}}{50 \text{ ft}}$$

$$= 2.9 \times 10^5 \text{ lb}.$$

Thus a force of 290,000 lb, which is 145 tons, is associated with the expansion of the girder!

13-3 Boyle's law

A peculiar difficulty arises when we attempt to measure the coefficient of volume expansion of a gas. Unlike solids and liquids, gases do not have specific volumes at a particular temperature, but expand to fill their containers. The only way to change the volume of a gas is to change the size of its container. However, even though its volume may remain the same, another property of a confined gas varies with its temperature, namely the pressure it exerts on the container walls. An illustration of this effect is that the air pressure in an automobile tire drops in cold weather and increases in warm.

When the temperature of a sample of gas is held constant, the absolute pressure it exerts on its container is inversely proportional to the volume of the container. Expanding the container lowers the pressure; shrinking the container raises the pressure. This property of gases is called *Boyle's law* (Fig. 13-5) after its discoverer, Robert Boyle (1627–1691). We can write Boyle's law as

$$pV = \text{constant (for constant } t) \quad (13\text{–}4)$$

or, alternatively, as

$$p_1V_1 = p_2V_2. \quad (13\text{–}5)$$

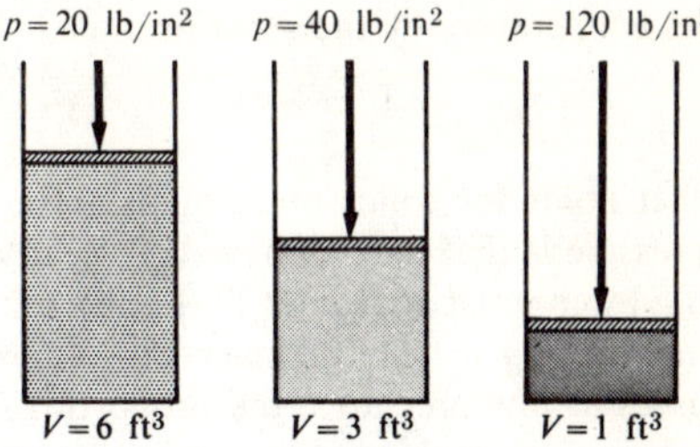

FIG. 13–5. Boyle's law.

In the latter case p_1 is the absolute gas pressure when its volume is V_1, and p_2 is the gas pressure when its volume is V_2; the temperature must be the same in states 1 and 2 for Eqs. (13–4) and (13–5) to hold.

We can use Boyle's law to find how much air at the sea-level atmospheric pressure of 15 lb/in^2 can be stored in the 12-ft^3 tank of an air compressor which can withstand an absolute pressure of 100 lb/in^2. Letting state 1 represent the air in the compressor tank and state 2 the air at atmospheric pressure,

$$V_2 = \frac{p_1}{p_2} V_1 = \frac{100 \text{ lb/in}^2 \times 12 \text{ ft}^3}{15 \text{ lb/in}^2} = 80 \text{ ft}^3.$$

A total of 80 ft^3 of air at atmospheric pressure can be stored in the 12-ft^3 compressor tank at the high pressure.

13–4 Charles' law

Now let us see what happens to a gas when its temperature is changed. As we said earlier, if the volume of a gas is held constant, the pressure it exerts on its container depends on its temperature. If we hold the gas pressure constant, its volume should vary with temperature. When this prediction is experimentally tested, which was first done over 150 years ago by Charles and Gay-Lussac, it is found that the change in volume ΔV of a gas sample is related to a change Δt in

TABLE 13–2

Coefficients of linear expansion

Substance	Coefficient, $\times 10^{-5}/°C$	Substance	Coefficient, $\times 10^{-5}/°C$
Aluminum	2.4	Lead	3.0
Brass	2.4	Quartz	0.05
Concrete	0.7–1.2	Silver	2.0
Copper	1.7	Steel	1.2
Iron	1.2		

its temperature by the same formula,

$$\Delta V = \beta V_0 \, \Delta t, \tag{13-6}$$

that holds for solids and liquids. The significant thing about gases at constant pressure is that they *all* have very nearly the same coefficient of volume expansion β_0; by contrast, as Tables 13-1 and 13-2 indicate, the thermal coefficients for solids and liquids may have markedly different values for different substances. For gases at low pressures, the proportionality between ΔV and Δt holds for a large range of Δt, making it an ideal substance for defining a temperature scale.

At 0°C the coefficient of volume expansion β_0 of all gases is

$$\beta_0 = \frac{1/273.15}{°\text{C}}.$$

If we vary the temperature of a gas sample while holding its pressure constant, its volume changes by 1/273.15 of its volume at 0°C for each 1°C temperature change. A child's balloon filled with air whose volume at 0°C is 1.000 m^3 has a volume of 1.037 m^3 at 10°C and 0.963 m^3 at −10°C (Fig. 13-6).

A natural question to ask here is: What happens when the balloon is cooled to −273.15°C? At that temperature the air in the balloon should have lost 273.15/273.15 of its volume at 0°C, and therefore would vanish entirely! Actually, all gases condense into the liquid state at temperatures above −273.15°C, so the question has no physical meaning. But −273.15°C is still a significant temperature. If we set up a new temperature scale, the *absolute temperature scale*, and designate −273.15°C as the zero point on this scale, the volume of a gas is always proportional to its absolute temperature. To help in understanding this statement we refer to Fig. 13-7, which is a plot of the volume of a particular sample of a gas at constant pressure versus its temperature in both the celsius and absolute scales.

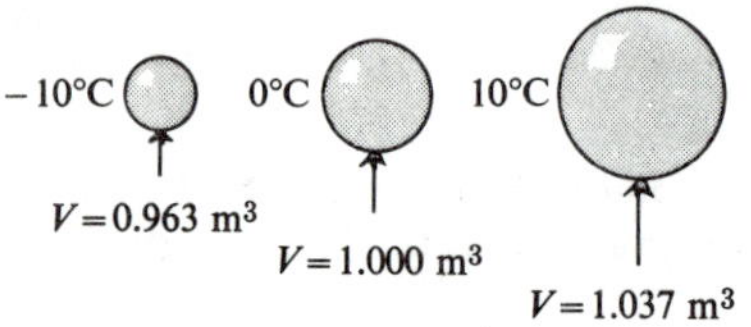

FIG. 13-6. Charles' law.

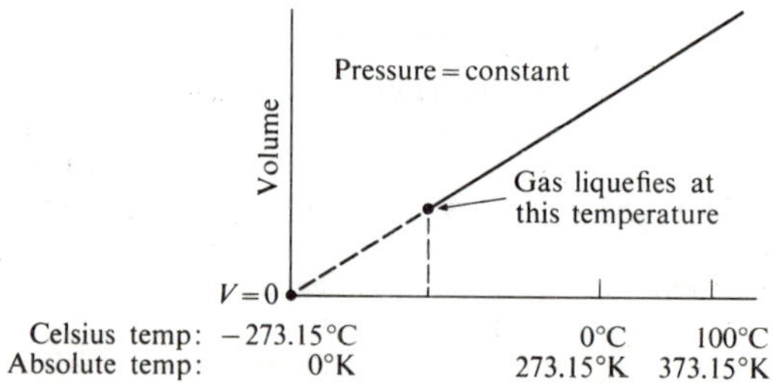

FIG. 13-7. The volume of a gas at constant pressure is directly proportional to its absolute temperature.

The proportionality between volume and absolute temperature is evident. At constant pressure, then, *Charles' law* states that

$$V/T = \text{constant}, \tag{13-7}$$

or, alternatively,

$$\frac{V_1}{T_1} = \frac{V_2}{T_2}, \tag{13-8}$$

where the subscripts 1 and 2 refer to different states of the gas.

13-5 The equation of state of an ideal gas

We are now in a position to relate the pressure, volume, and temperature of an ideal gas with a single equation. Let us consider a gas sample whose volume is V_0 at atmospheric pressure p_0 and at temperature 0°C. We heat the gas, keeping its pressure constant at p_0, and in this heating process the volume of the gas increases to V' and its temperature to t. (Figure 13-8 is a pressure-versus-volume graph of this whole process.) Then the gas is compressed at the constant temperature t until its volume drops to V while its pressure increases to p.

Our procedure is to apply Boyle's and Charles' laws to the entire process of first heating the gas at constant pressure and then compressing it at constant temperature. By Boyle's law,

$$p_0V' = pV, \tag{13-9}$$

since the measurements of p_0, V', p, and V were all made at the same temperature t. By Charles' law [Eq. (13-6)],

$$\Delta V = \beta_0 V_0 t,$$

$$V' = V_0(1 + \beta_0 t), \tag{13-10}$$

since the pressure was held constant at p_0 during the initial thermal expansion of the gas. Multiplying Eq. (13-10) by p_0 yields

$$p_0V' = p_0V_0(1 + \beta_0 t),$$

or, substituting pV for p_0V' according to Eq. (13-9),

$$pV = p_0V_0(1 + \beta_0 t)$$
$$= p_0V_0\beta\left(t + \frac{1}{\beta_0}\right). \tag{13-11}$$

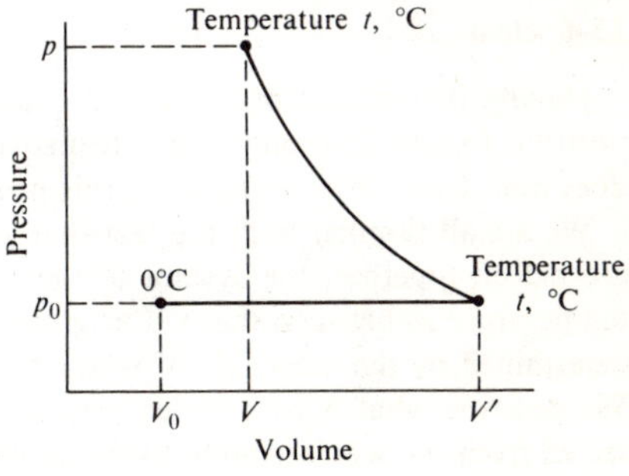

FIG. 13-8. Cycle of heating and compression used for calculating the equation of state of an ideal gas.

In Eq. (13-11) the quantities p_0 (atmospheric pressure) and β_0 are both constants. The value of V_0, the gas volume at p_0 and $t = 0$°C, depends on the mass of the sample and on its composition. For a given mass, V_0 is different for different gases, but one kilomole (abbreviated kmole) of *any* gas exhibits the volume 22.4 m^3 at atmospheric pressure and 0°C. A *kilomole* of a substance is one kilogram-molecular weight of it; i.e., that quantity of the substance whose mass in kilograms is equal to its molecular weight (see Chapter 14). Hence we find that, if the gas sample we are describing contains n kmoles,

$$p_0V_0\beta_0 = n \times 1.013 \times 10^5 \text{ n/m}^2 \times 22.4 \text{ m}^3 \times \frac{1}{273.15°\text{C}} = nR,$$

where

$$R = 8.31 \times 10^3 \text{ j/kmole °C}. \tag{13–12}$$

The quantity R has the same value for all gases, and is known as the *universal gas constant.*

When we substitute nR for $p_0V_0\beta_0$ in Eq. (13–11) and use the numerical value of β_0, we find that

$$pV = nR(t + 273.15°\text{C}). \tag{13–13}$$

This expression relates the pressure, volume, and temperature of n kmoles of an ideal gas. To simplify it, we can use the *absolute temperature scale*, defined earlier by the relationship

$$T = t + 273.15°\text{C}, \tag{13–14}$$

which merely shifts the zero point of the absolute scale to -273.15°C. Then

$$pV = nRT, \tag{13–15}$$

which is called the *equation of state of an ideal gas.* It is remarkable that so simple a formula is able to specify completely the behavior of an ideal gas, and still more remarkable that we may consider all gases as ideal gases without encountering too much error, except at high pressure values.

13–6 Heat

Having defined temperature and found ways to measure it, we are now in a position to continue our efforts to prove that the conservation of energy indeed does hold true. The next step in this program is to define what we mean by *heat.*

We are all familiar with the fact that when two bodies at different temperatures are placed together, the system of the two bodies comes to equilibrium at some temperature in between the two original temperatures, the final temperature being determined by the materials of which the bodies are composed, and their masses. We describe what occurs under these circumstances by saying that heat is transferred from the warmer body to the cooler one. This is what is always done when we heat a body: Heating a pot of water consists of bringing it in contact with a flame which is at a higher temperature. We can define heat or heat energy as *energy that is transferred from one body to another in thermal contact with it by virtue of a temperature difference between the two.* We have assumed in this definition that there is a transfer of energy from the warmer to the cooler body. The experimental reason for this assumption will become clearer in Section 13–9.

We now require a definition of a heat unit and some method of measuring it. The unit of heat in the metric system is the *kilocalorie* (kcal), which is the amount of heat necessary to change the temperature of one kilogram of water from 14.5°C to 15.5°C. Actually, the amount of heat required to change the temperature of 1 kg of water by 1°C is virtually independent of the location of the temperature interval (provided there is no change of phase), and we shall neglect the difference.

Although we shall not employ them here, two other heat units are in common use: the *calorie* and the *British thermal unit*, abbreviated cal and Btu, respectively. The calorie is 10^{-3} kcal (0.001 kcal), and the Btu is the amount of heat required to raise the temperature of 1 lb of water by 1°F; it is equal to 0.252 kcal.

13–7 Heat transfer and heat conductivity

It is appropriate to interrupt the principal argument at this point to discuss the mechanics of transferring heat from a hot to a cool body. There are three methods by which this can be accomplished: *convection*, *conduction*, and *radiation*. In the first two methods, the hot and the cold body are connected by matter. In convective heat transfer, the agent transferring the heat literally moves from the hot region to the cool one: a radiator in a room circulates heat in this way. The air in contact with the radiator is heated and expands as a result. The hot air then rises and flows through the room, while cold air replaces it near the radiator. Convective heat flow is difficult to analyze and not very important for our purposes, although it has great engineering applicability.

In conductive transfer of heat, one end of the object doing the conduction (Fig. 13–9) is hot and the other cold. The object does not undergo any visible

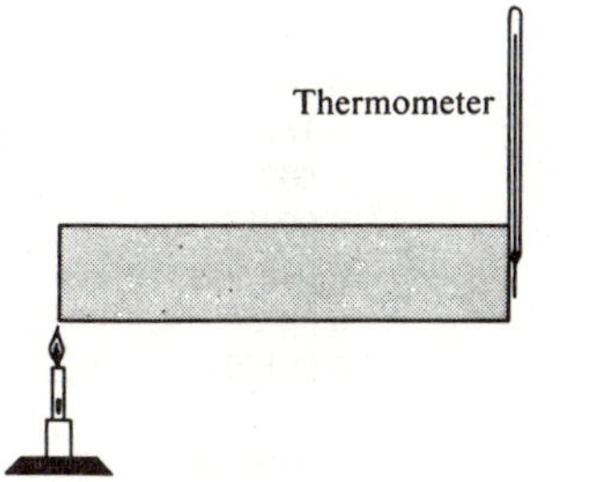

FIG. 13–9. In the conduction of heat, the body doing the conducting undergoes no change once a steady state is reached.

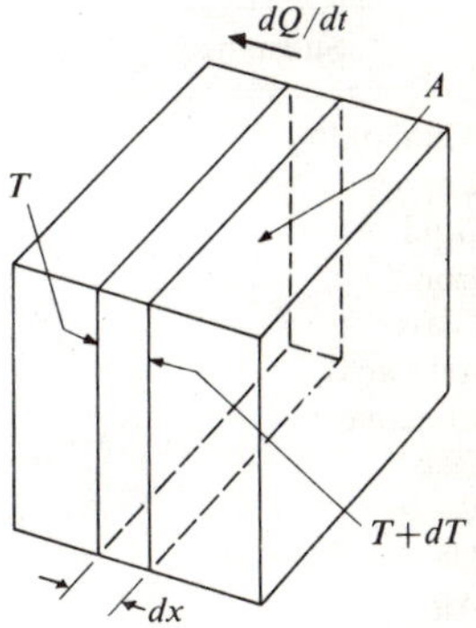

FIG. 13–10. Conduction of heat across a slab of thickness dx with temperature difference dT across the slab.

change during the conduction once a steady state is reached. The mechanism responsible for the transfer of heat by conduction involves the microscopic constituents of the conductor—the atoms, molecules, and electrons of which it is composed. The experimental facts about conduction of heat can be summarized as follows: If we limit ourselves to conduction in one direction only, we may express the rate of heat flow across a slab of matter of area A (Fig. 13–10) and thickness dx, where the temperature difference between the two faces is dT, as dQ/dt. Experimentally it is found that dQ/dt is proportional to A and to the temperature gradient dT/dx, so that

$$\frac{dQ}{dt} = -KA\frac{dT}{dx}, \qquad (13\text{–}16)$$

where K, the *conductivity* of the material, is characteristic of the particular material involved, and varies somewhat with temperature. (The minus sign expresses the fact that the direction of the heat flow is toward the lower temperature.) The conductivity K may be expressed in various units, the most common being kcal/cm-sec-°C. Table 13–3 is a list of the conductivities of several different substances, expressed in these units. The larger the value of K, the better the conductor.

The transfer of heat by radiation does not require any material connection between the hot body and the cold one. Essentially radiation involves a transfer of heat by means of electromagnetic waves traveling in space. The sun, for example, transfers its energy to the earth by this method, through the "vacuum" of interplanetary space. Radiation is an important process about which we shall have more to say later, after we have studied electromagnetism.

TABLE 13–3

Thermal conductivities

Substance	Temperature	Conductivity, K
Aluminum	30°C	0.479×10^{-3} kcal/cm-sec-°C
Copper	18°C	0.918
Gold	17°C	0.705
Iron	18°C	0.161
Glass	0°C	0.002
Asbestos	0°C	0.00215
Concrete	0°C	0.002
Oak	20°C	0.00058
Water	0°C	0.00150
Ice	0°C	0.0053
Air	0°C	0.0000566
Argon	0°C	0.0000388
Carbon dioxide	0°C	0.0000332
Hydrogen	0°C	0.000416

13–8 Specific heat

Other substances respond to the addition or removal of heat with temperature changes greater than that of an equal mass of water. One kcal of heat increases the temperature of 1 kg of helium (its volume held constant) by 1.3°C, of 1 kg of ice by 2°C, and of 1 kg of gold by 33°C; 1 kg of water by definition increases in temperature by 1°C when 1 kcal of heat is added to it. The *specific heat* (symbol c) of a substance refers to what we might think of as its thermal inertia; c is the amount of heat required to change the temperature of 1 kg of a particular substance by 1°C. Table 13–4 is a list of specific heats for various substances. A high specific heat implies a relatively small change in temperature for a given change in heat, just as a large inertial mass implies a relatively small acceleration when a given force is applied.

Use of specific heats enables us to measure heat more easily, since we can calibrate the heat transferred to substances other than water in terms of the kilocalorie. With the help of specific heat we can obtain a formula for the quantity of heat Q involved when a mass m of a substance undergoes a change in temperature of ΔT. The formula is simply

$$Q = mc\,\Delta T. \tag{13–17}$$

To cool 14 kg of aluminum from 80°C to 15°C, for instance, involves the quantity of heat

$$Q = mc\,\Delta T = 14\text{ kg} \times 0.22\,\frac{\text{kcal}}{\text{kg·°C}} \times (-65°\text{C}) = -200\text{ kcal};$$

TABLE 13–4

Specific heats of various substances*

Substance	Specific heat, kcal/kg·°C	Substance	Specific heat, kcal/kg·°C
Alcohol (ethyl)	0.58	Marble	0.21
Aluminum	0.22	Mercury	0.033
Copper	0.093	Silver	0.056
Glass	0.20	Steam	0.48
Gold	0.030	Sulfuric acid	0.27
Granite	0.19	Turpentine	0.42
Ice	0.50	Water	1.00
Iron	0.11	Wood	0.42
Lead	0.030	Zinc	0.092

* The actual values vary somewhat with temperature; the ones given in the table represent averages.

the minus sign means that this quantity of heat is to be removed from the aluminum to achieve the temperature change of $-65°C$.

Example: We pour 0.2 kg of coffee at 90° into a 0.15-kg cup at 20°C. Assuming that no heat is transferred to or from the outside, what is the final temperature of the coffee?

Solution: We assume that the specific heat of coffee is that of water, and that the specific heat of the cup is that of glass. To solve this problem, we begin by noting that, since heat is energy, and this is the only energy involved in this problem, the conservation of energy gives

$$\text{Heat gained by cup} = \text{heat lost by coffee.}$$

If the final temperature of both coffee and cup is t,

$$\begin{aligned}\text{Heat gained by cup} &= m_{\text{cup}}c_{\text{cup}}\,(t - 20°\text{C})\\ &= 0.15\ \text{kg} \times 0.20\,\frac{\text{kcal}}{\text{kg}\cdot°\text{C}} \times (t - 20°\text{C})\\ &= (0.03t - 0.6)\ \text{kcal}\end{aligned}$$

and

$$\begin{aligned}\text{Heat lost by coffee} &= m_{\text{coffee}}c_{\text{coffee}}\,(90°\text{C} - t)\\ &= 0.2\ \text{kg} \times 1.0\,\frac{\text{kcal}}{\text{kg}\cdot°\text{C}} \times (90°\text{C} - t)\\ &= (18 - 0.2t)\ \text{kcal.}\end{aligned}$$

Hence, equating the heat gained by the cup with the heat lost by the coffee,

$$\begin{aligned}0.03t - 0.6 &= 18 - 0.2t,\\ 0.23t &= 18.6,\\ t &= 79°\text{C}.\end{aligned}$$

The temperature of the coffee will drop 9°C as it warms the cup.

13-9 Mechanical equivalent of heat

We are now in a position to reevaluate the conservation of energy. The rubbing together of two blocks causes a temperature rise in the blocks. From a knowledge of the specific heats of the blocks and a measurement of the temperature rise, we could conclude that this is equivalent to a certain amount of heat energy, and by further experiment we could then see whether the combination of mechanical energy and heat energy is conserved.

The importance of performing such a quantitative experiment was recognized by James Prescott Joule (1818–1889), who demonstrated convincingly the equivalence of mechanical and heat energy. Joule used an apparatus similar to that sketched in Fig. 13–11: a descending weight causes paddle wheels inside an insulated container to rotate, thereby stirring the liquid within the container.

The stirring must be done against the frictional resistance of the liquid (the viscosity of a liquid is a measure of this resistance), and the heat equivalence can be determined by measuring the increase in the liquid's temperature and by knowing its specific heat. The results are always the same, no matter what the nature of the liquid, its volume, or the total amount of work done on it by the falling weight: 1 kcal of heat appears for every 4185 joules of work done. The quantity

$$J = 4185 \frac{\text{j}}{\text{kcal}}$$

is accordingly known as the *mechanical equivalent of heat.* The same result was obtained by other experimenters using other methods of doing work on systems, including electrical work. The equivalence of heat and mechanical energy was thus firmly established.

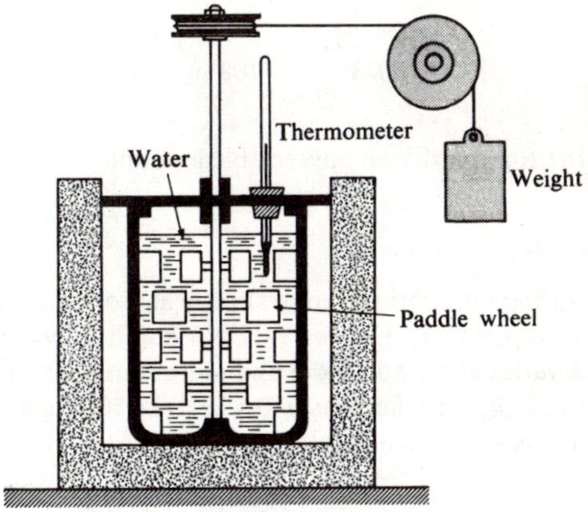

FIG. 13–11. Joule's method for measuring the mechanical equivalent of heat. As the weight descends it turns paddle wheels inside an insulated container of water, so that its potential energy is converted first into the kinetic energy of the paddle wheels and then into heat energy of the water. When the energy lost by the descending weight is divided by the increase in the heat content of the water, the mechanical equivalent of heat is found to be 4185 j/kcal.

As a simple application of the mechanical equivalent of heat, we can compute the difference in temperature of water at the top and at the bottom of a waterfall. In the course of dropping from a height h, a mass m of water loses the potential energy

$$\text{PE} = mgh.$$

This potential energy appears as kinetic energy in the falling water, and we shall assume that it is all dissipated as the heat

$$Q = mc\,\Delta t$$

at the bottom of the fall. To convert Q, which is in kilocalories, to joules, we multiply it by J, since the units of J are j/kcal. Hence

$$\mathrm{PE} = JQ, \qquad mgh = Jmc\,\Delta t,$$

and the rise in temperature of the water is

$$\Delta t = \frac{gh}{Jc} = 0.0023h\,\frac{^\circ\mathrm{C}}{\mathrm{m}}\cdot$$

We note that the mass m of water that we have been considering turns out to be irrelevant, as it should. We can apply the general formula we have obtained to any waterfall whose height h we know. For example, the Victoria Falls on the Zambesi River are 108 meters high, so that the water at the foot of the falls is

$$\Delta t = 0.0023\,\frac{^\circ\mathrm{C}}{\mathrm{m}} \times 108\ \mathrm{m} = 0.25^\circ\mathrm{C}$$

warmer than at the top, hardly an appreciable amount.

13-10 Change of state

Not always does the addition or removal of heat from a sample of matter lead to a change in its temperature. Figure 13-12 is a graph showing how the temperature of 1 kg of ice varies as we add more and more heat to it. The ice was initially at -50°C, and, since its specific heat is 0.5 kcal/kg·°C, each kilocalorie of heat that is supplied increases its temperature by 2°C. At 0°C, however, the steady

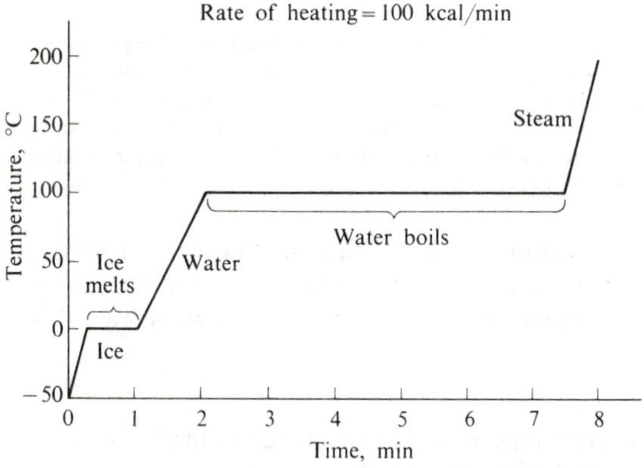

FIG. 13-12. A graph of temperature versus time for 1 kg of water, initially ice at -50°C, to which heat is being added at the constant rate of 100 kcal/min.

temperature rise ceases, and the temperature remains at 0°C while we add 80 kcal more. This is the amount of heat needed to melt 1 kg of ice at 0°C to water at 0°C. Then, when all the ice has turned to water, the temperature goes up once more as further heat is supplied. Since the specific heat of water is 1 kcal/kg·°C, there is now a rise of 1°C per kilocalorie of heat. This rate of change is less than that of ice, since the specific heat of water is greater than that of ice, and so the slope of the graph is smaller. When 100°C is reached, the temperature again stays constant despite further heating until a total of 540 kcal is added. This is the amount of heat needed to vaporize 1 kg of water at 100°C to steam at 100°C. After all the water has become steam, its temperature increases again. The specific heat of steam is 0.48 kcal/kg·°C, so the temperature increase is 2.1°C per kilocalorie of heat, a steeper slope than the ones corresponding to ice or water.

The *heat of fusion* of a substance is the amount of heat that must be supplied to change 1 kg of it at its melting point from the solid to the liquid state; the same amount of heat must be removed from 1 kg of the substance in the liquid state at its melting point to change it to a solid. The usual symbol for heat of fusion is L_f.

TABLE 13-5

Heats of fusion and vaporization and melting and boiling points of various substances

Substance	Melting point, °C	L_f, kcal/kg	Boiling point, °C	L_v, kcal/kg
Alcohol (ethyl)	−114	25	78	204
Bismuth	271	12.5	920	190
Bromine	−7	16	60	43
Lead	330	5.9	1170	175
Lithium	186	160	1336	511
Mercury	−39	2.8	358	71
Nitrogen	−210	6.1	−196	48
Oxygen	−219	3.3	−183	51
Sulfuric acid	8.6	39	326	122
Water	0	80	100	540
Zinc	420	24	918	475

The *heat of vaporization* of a substance is the amount of heat that must be supplied to change 1 kg of it at its boiling point from the liquid to the gaseous (or vapor) state; the same amount of heat must be removed from 1 kg of the substance in the gaseous state at its boiling point to change it into a liquid. The usual symbol for heat of vaporization is L_v. The heats of fusion and vaporization for a number of substances are listed in Table 13-5, together with their melting and boiling points.

Under certain circumstances most substances can change directly from the solid to the vapor state, or vice versa, processes both called *sublimation.* For example, "dry ice" (solid carbon dioxide) evaporates directly to gaseous carbon dioxide at temperatures above $-78.5°C$, and does not pass through the liquid state. With the exception of carbon dioxide and a few other substances, however, sublimation does not occur at ordinary temperatures except at pressures well below that of the atmosphere.

Example: What is the minimum amount of ice at $-10°C$ that must be added to 0.5 kg of water at 20°C in order to bring the water down to 0°C?

Solution: We begin, as before, with the statement of heat conservation,

$$\text{Heat gained by ice} = \text{heat lost by water.}$$

The heat Q_1 gained by the unknown mass of ice in going from $-10°C$ to its melting point at 0°C is

$$Q_1 = m_{\text{ice}}c_{\text{ice}}\,\Delta t_{\text{ice}} = m_{\text{ice}} \times 0.50\,\frac{\text{kcal}}{\text{kg}\cdot°\text{C}} \times 10°\text{C},$$

and the heat Q_2 gained by the ice in melting at 0°C is

$$Q_2 = m_{\text{ice}}Lr_{\text{ice}} = m_{\text{ice}} \times 80\,\frac{\text{kcal}}{\text{kg}}\cdot$$

Hence

$$\text{Heat gained by ice} = Q_1 + Q_2 = m_{\text{ice}}(5 + 80)\,\frac{\text{kcal}}{\text{kg}}\cdot$$

The heat lost by the water in cooling to 0°C is

$$\begin{aligned}\text{Heat lost by water} &= m_{\text{water}}c_{\text{water}}\,\Delta t_{\text{water}}\\ &= 0.5\text{ kg} \times 1\,\frac{\text{kcal}}{\text{kg}\cdot°\text{C}} \times 20°\text{C}\\ &= 10\text{ kcal.}\end{aligned}$$

Equating the heat gained with the heat lost,

$$85m_{\text{ice}}\,\frac{\text{kcal}}{\text{kg}} = 10\text{ kcal}, \qquad m_{\text{ice}} = 0.118\text{ kg.}$$

13-11 The triple point

Let us examine the effect of pressure on changes of state. Figure 13-13 shows how the boiling point of water varies with pressure, behavior typical of other liquids as well. The upper limit of this *vaporization curve*, which occurs at a temperature of 374°C and a pressure of 218 atm, is known as the *critical point.* A substance cannot exist in the liquid state at a temperature above that of its

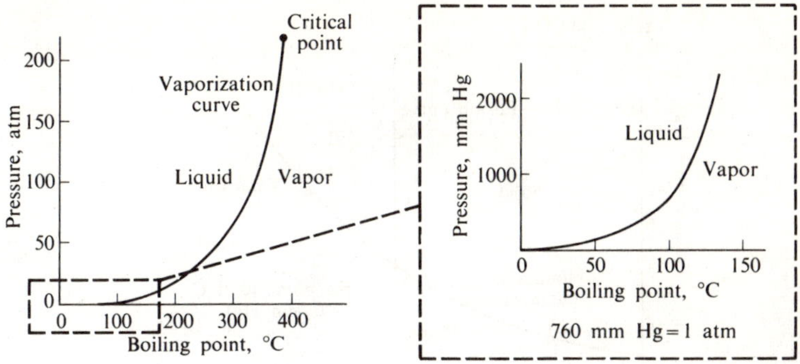

FIG. 13–13. The vaporization curve of water, showing how the boiling point of water varies with pressure.

critical point, regardless of how great the pressure may be. Helium has the lowest critical temperature, −268°C, and is therefore a gas at all temperatures above that.

The melting points of solids also depend on pressure, although to a smaller extent than the boiling points. The variation of the melting point of ice with pressure is shown in the *fusion curve* of Fig. 13–14. Ice, together with gallium and bismuth, is unique in that its melting point *decreases* with increasing pressure; the melting points of all other substances increase with increasing pressure. Hence it is possible to melt ice by applying pressure to it as well as by heating it. An ice skater makes use of this fact in an interesting way. His entire weight is supported by skate blades of very small area, and the resulting pressure on the ice may exceed 1000 atm. The ice under the blades melts because of the great pressure, creating a thin film of water that acts as an efficient lubricant. On unusually cold days even such pressures may not be sufficient to melt the ice, and skating then becomes impossible.

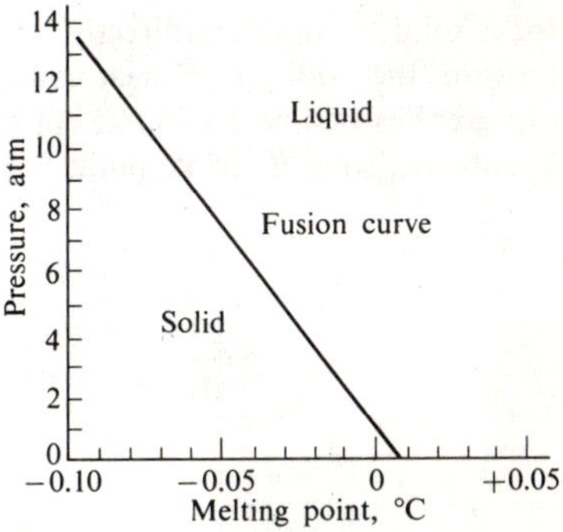

FIG. 13–14. The fusion curve of water, showing how the melting point of ice varies with pressure.

The fusion and vaporization curves of water intersect at a temperature of 0.01°C and a pressure of 4.6 mm Hg, as shown on the combined plot of Fig. 13–15. Along the fusion curve both ice and water can simultaneously exist, and along the vaporization curve both water and water vapor can simultaneously exist; hence under conditions corresponding to those of the intersection of the two

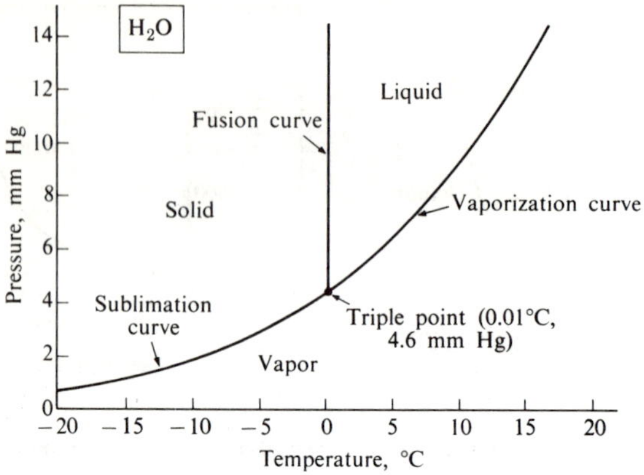

FIG. 13-15. Triple-point diagram of water.

curves, the solid, liquid, and vapor states of water can all exist together. This intersection is accordingly called the *triple point* of water.

At pressures below that of its triple point, no substance can exist as a liquid. The dividing line on a pressure-temperature graph between the solid and vapor states is called the *sublimation curve*, since it represents the conditions required for a solid to vaporize directly or a vapor to solidify directly. At atmospheric pressure the addition of heat causes ordinary ice to melt, since the triple point of water lies well below 1 atm, but the addition of heat causes solid carbon dioxide to sublime, since its triple point lies above 1 atm (Fig. 13-16).

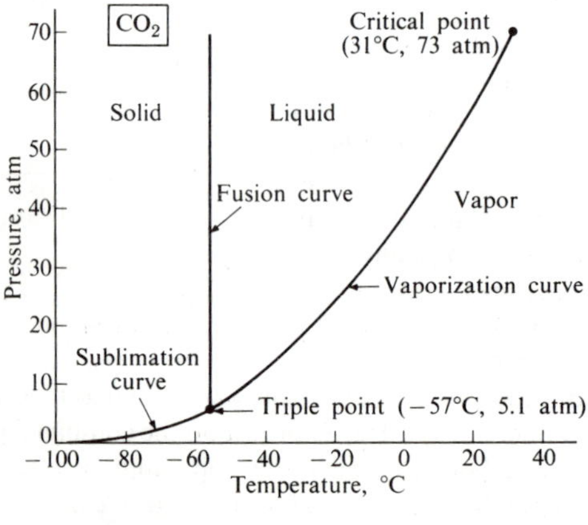

FIG. 13-16. Triple-point diagram of carbon dioxide.

Problems

1. The melting point of lead is 330°C and its boiling point is 1170°C. Express these temperatures on the fahrenheit scale.

2. The normal temperature of the human body is 98.6°F. What is this temperature on the celsius scale?

3. At what temperature would celsius and fahrenheit thermometers give the same reading?

4. Mercury freezes at −40°C. What is this temperature on the fahrenheit scale?

5. Dry ice (solid carbon dioxide) vaporizes at −112°F. What is this temperature on the celsius scale?

6. The absolute temperature scale described in the text is based on the celsius scale. An absolute temperature scale, called the rankine scale, may also be constructed, based on the fahrenheit scale. Temperatures in this scale are designated °R. (a) What is the fahrenheit temperature of absolute zero? (b) Devise equations for converting temperatures in °F to °R and vice versa, similar to Eq. (13–1). (c) What is the rankine temperature of the freezing and boiling points of water?

7. A solid steel rail is 1 mi long. Estimate how high the center of the rail will be in the desert, if the ends are fixed and the temperature rises 100°F between midnight and noon.

8. How large a gap should be left between steel rails that are 10 m long when laid at 20°C if they are to just barely touch at 30°C?

9. A rod 2 m long expands by 1 mm when heated from 8°C to 72°C. What is the coefficient of linear expansion of the material from which the rod is made?

10. The outside diameter of a wheel is 1.000 m. An iron tire for this wheel has an inside diameter of 0.992 m at 20°C. To what temperature must the tire be heated in order for it to fit over the wheel?

11. A Pyrex beaker is filled to the brim with 250 cm^3 of glycerin at 15°C. How much glycerin overflows at 25°C?

12. A Pyrex flask holds 500 cm^3 of mercury at 0°C. How much mercury will run out when it is heated at 80°C?

13. Vodka that is "100 proof" is a mixture of half ethyl alcohol and half water. How much profit per quart will a merchant make if he buys it at $5.00 per quart at −5°C and sells it at $5.00 per quart at 25°C?

14. In the construction of a light bulb, wires are led through the glass at the base by means of airtight seals. If the wires are made of copper, what will happen when the light is turned on and the bulb heats up? What must be true for a wire to be successfully used for this purpose?

15. A steel tape measure is calibrated at 20°C. A reading of 27.1 m is obtained when it is used to determine the width of a building at 5°C. What is the true width of the building?

16. A sample of gas occupies 2 m^3 at 300°K and an absolute pressure of 2×10^5 n/m^2. (a) What is its pressure at the same temperature when it has been compressed to a volume of 1 m^3? (b) What is its volume at the same temperature when its pressure has been decreased to 1.5×10^5 n/m^2? (c) What is its volume when it is at a temperature of 400°K and a pressure of 2×10^5 n/m^2?

17. A helium balloon has a volume of 100 cm^3 at the surface of the earth when the temperature is 10°C. What is its volume at an altitude of 10 mi, where the pressure is 95 mm of Hg and the temperature is −50°C?

18. A sample of gas occupies 100 cm^3 at 0°C and 1 atm pressure. What is its volume (a) at 50°C and 1 atm pressure; (b) at 0°C and 2.2 atm pressure; (c) at 50°C and 2.2 atm pressure?

19. An automobile tire contains air at a gauge pressure of 24 lb/in^2 at 0°C. If the volume of the tire is unchanged, what will the pressure be when the temperature has increased to 27°C?

20. A sample of gas occupies a volume of 8 ft^3 at a temperature of 400°R and 1 atm pressure. What is its volume at 500°R and 1 atm pressure? (See Problem 6.)

21. A small amount of water is placed in a can and brought to a boil. The can is then tightly covered and removed from the stove. Explain why the can collapses after it has cooled for a few minutes.

22. One end of a copper bar 0.001 m^2 in cross section and 0.5 m long is in contact with steam at atmospheric pressure. The other end is in contact with ice at 0°C. How many grams of ice will be melted in $\frac{1}{4}$ hr?

23. An iron steam pipe is 1 m long. It has an internal diameter of 3 cm and is $\frac{1}{2}$ cm thick. The outside surface is at 75°C. How much heat is lost through it per hour?

24. How much heat per day is lost by conduction through the walls of a concrete building whose area is 100 m^2 and thickness $\frac{1}{4}$ m? The inside temperature is 25°C and the outside 0°C.

25. A glass pane 5 mm thick has an area of 3 m^2. The inside face of the pane is at a temperature of 20°C and its outside face is −5°C. How much heat is conducted through the pane per day?

26. How much heat must be added to 1 kg of copper to raise its temperature from 20°C to 100°C?

27. Why will the engine of a car whose cooling system is filled with an alcohol antifreeze be more likely to overheat in the summer than one whose cooling system is filled with water?

28. A 60-kg woman is on a diet that provides her with 2500 kcal daily. If a corresponding amount of heat were added to 60 kg of water at 37°C, what would its final temperature be?

29. A 0.6-kg copper container holds 1.5 kg of water at 20°C. A 0.1-kg iron ball at 120°C is dropped into the water. What is the final temperature of the water?

30. A 0.1-kg piece of silver is taken from a bath of hot oil and placed in a 0.08-kg glass jar containing 0.2 kg of water at 15°C. The temperature of the water increases by 8°C. What was the temperature of the oil?

31. What will be the temperature rise of water which falls 200 m and in which all the kinetic energy is converted into heat?

32. A wooden beam is pushed across a stone floor by a force just sufficient to overcome friction. The coefficient of friction is 0.4. Assuming that all the work done against friction goes into heating the beam, what is its rise in temperature for each meter that it is pushed?

33. The rate at which work is being done is often specified in *watts*, one watt equaling one joule per second. (a) How many kcal/sec are equivalent to 1 watt? (b) How many kcal are evolved by a 60-watt light bulb per hour? (c) Radiant energy from the sun arrives at the earth's surface at the approximate rate of 20 $kcal/m^2$ per minute. Express this rate in $watts/m^2$.

34. A 1-kg block of ice at 0°C falls into a lake whose water is also at 0°C, and 0.01 kg of ice melts. What was the altitude from which the ice fell?

35. A lead bullet at 100°C strikes a steel plate and melts. What was its minimum speed?

36. A 50-kg block of ice at 0°C is pushed across a wooden floor also at 0°C for a distance of 20 m. A total of 25 gm of ice melts as a result of the friction of the block on the floor. What is the coefficient of friction in this case?

37. How much more heat must be added to 1 kg of ice at 0°C to convert it to steam at 100°C than is required to raise the temperature of 1 kg of water from 0°C to 100°C?

38. How much steam at 120°C is required to melt 0.5 kg of ice at 0°C?

39. How much ice at −10°C is required to cool a mixture of 0.1 kg ethyl alcohol and 0.1 kg water from 20°C to 5°C?

40. A 5-kg iron bar is taken from a forge at a temperature of 1000°C and plunged into a pail containing 10 kg of water at 60°C. How much steam is produced?

41. By mistake, 0.2 kg of water at 0°C is poured into a vessel containing liquid nitrogen at −196°C. How much nitrogen vaporizes?

42. The pressure and temperature of the atmosphere at 115,000 ft are 4 mm Hg and −23°C, respectively. What is the state of water under those conditions? of carbon dioxide?

43. Why does food cook faster in a pressure cooker?

44. Water expands slightly when it freezes. What connection might you suspect between this fact and the fact that the melting point of ice decreases with increasing pressure?

45. Carbon dioxide is usually shipped in tanks under a pressure of approximately 1000 lb/in^2. At 20°C, is the carbon dioxide a solid, liquid, or gas?

14
Kinetic-Molecular Theory of Matter

In the previous chapter we have seen that by a combination of theory and experiment it was possible to demonstrate the equivalence of mechanical work and heat energy without making any detailed assumptions about the nature of matter. In this chapter we shall show that equivalent results can be obtained by making a plausible hypothesis about the nature of the constituents of matter and the forces between them and using Newton's laws in the microscopic realm. Apart from the intrinsic beauty of this approach, it has the virtue of demonstrating that the mechanical laws which hold for astronomical phenomena also are applicable to some extent to the microcosm.

14–1 Molecular theory of matter

The idea that matter is not infinitely divisible, that all substances are composed of characteristic individual particles, is an ancient one. The ultimate particles of any substance are called *molecules.* Although molecules may be further broken down, when this happens the resulting atoms no longer are representative of the original substance. The molecules of a compound consist of the *atoms* of its constituent elements joined together in a definite ratio. Thus each molecule of water contains two hydrogen atoms and one oxygen atom. While the ultimate particles of elements are atoms, many elemental gases consist of molecules rather than atoms. Oxygen molecules, for instance, contain two oxygen atoms each. The molecules of other gases, such as helium and argon, are single atoms. Figure 14–1 shows schematically the composition of some common molecules.

The masses of atoms and molecules are usually expressed in units such that the most abundant type of carbon atom has a mass of precisely 12. (The existence of different kinds of atoms, known as *isotopes*, of the same chemical species is discussed in Chapter 36.) It is customary to refer to masses in these units as *atomic masses* and *molecular masses.* Appendix IV contains a list of the atomic masses of the elements; we note that ordinary carbon, some of whose atoms are heavier than the most abundant kind, has an atomic mass of 12.01. If we know the composition of a compound, we can determine the corresponding molecular mass.

Name of molecule	Chemical formula
Helium	He
Oxygen	O_2
Water	H_2O
Salt	NaCl
Carbon tetrachloride	CCl_4
Ethyl alcohol	C_2H_5OH

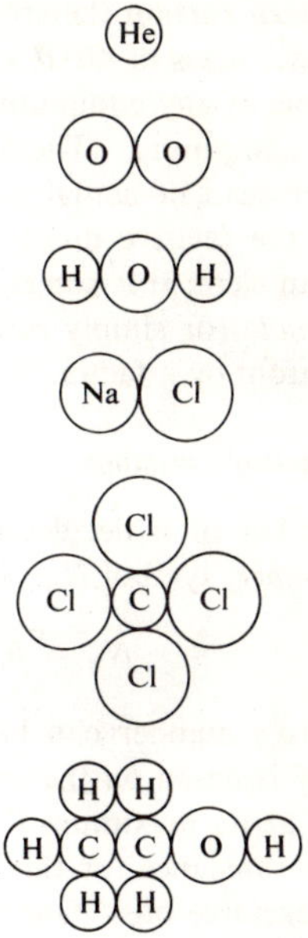

FIG. 14–1. The molecular compositions of several common substances.

The carbon atom has an actual mass of 1.992×10^{-26} kg, and so

$$\begin{aligned}1 \text{ atomic mass unit} &= \frac{1.992 \times 10^{-26}\text{ kg}}{12}\\ &= 1.660 \times 10^{-27}\text{ kg}.\end{aligned}$$

The molecular mass of a compound is the total of the atomic masses of the elements it contains multiplied respectively by the numbers of their atoms in each molecule of the compound. Thus the molecular mass of gaseous oxygen is 32, since its formula is O_2 and the atomic mass of oxygen is 16. To find the molecular mass of the more complicated compound, ethyl alcohol (C_2H_5OH), we proceed as follows:

$$\begin{aligned}2\,\text{C} &= 2 \times 12.01 &&= 24.02\\ 6\,\text{H} &= 6 \times 1.008 &&= 6.05\\ 1\,\text{O} &= 1 \times 16.00 &&= \underline{16.00}\\ & &\text{total} &= 46.07\end{aligned}$$

The molecular mass of ethyl alcohol is 46.07.

A *kilomole* (abbreviated *kmole*) of any element or compound is that quantity of it whose mass is equal to its atomic or molecular mass expressed in kilograms. One kmole of carbon therefore has a mass of 12.01 kg, and one kmole of ethyl alcohol has a mass of 46.07 kg. The usefulness of the kmole arises from the fact that a kmole of any compound contains exactly as many molecules as a kmole of any other compound. The kmole is widely used in chemistry as a measure of quantity, instead of actual mass, because the primary interest of the chemist is usually in the relative numbers of atoms and molecules that react together. A kmole of an element is sometimes referred to as a *kilogram-atom* of that element; the *gram-mole* (or simply *mole*) and *gram-atom* are smaller than the kmole and kilogram-atom by a factor of 10^3, since there are 10^3 grams in a kilogram.

14-2 Avogadro's number

The number of molecules in a kmole is a universal constant known as *Avogadro's number*, symbol N_0. Its value is

$$N_0 = 6.025 \times 10^{26} \text{ molecules/kmole.}$$

Avogadro's number can be determined in a variety of ways. A particularly simple way is based on the spreading of an oil film on calm water, where we assume that at its maximum thickness the film is exactly one molecule thick. A convenient substance for this purpose is oleic acid, a constituent of most animal fats and vegetable oils. One milligram (10^{-6} kg) of oleic acid placed on a water surface covers an area of about 1.4 m^2 under the most favorable conditions. The density of oleic acid is 895 kg/m^3, so that 1 mg occupies a volume of $1.12 \times 10^{-9}\ \text{m}^3$. Hence the thickness of the film is

$$a = \frac{1.12 \times 10^{-9}\ \text{m}^3}{1.4\ \text{m}^2} = 8.0 \times 10^{-10}\ \text{m.}$$

Under the assumption that this length is characteristic of all the dimensions of the oleic acid molecule, the volume occupied by each molecule is

$$a^3 = 5.1 \times 10^{-28}\ \text{m}^2.$$

From the formula for oleic acid, $CH_3(CH_2)_7CH{=}CH(CH_2)_7COOH$, we find that its molecular mass is 282.5. The volume per kmole is therefore

$$\frac{\text{volume}}{\text{kmole}} = \frac{\text{molecular mass}}{\text{density}} = \frac{282.5\ \text{kg/kmole}}{895\ \text{kg/m}^3} = 0.316\ \text{m}^3\text{/kmole.}$$

Since we suppose that each molecule has a volume of $5.1 \times 10^{-28}\ \text{m}^3$, the number of molecules per kmole of oleic acid is

$$N_0 = \frac{0.316\ \text{m}^3\text{/kmole}}{5.1 \times 10^{-28}\ \text{m}^3\text{/molecule}} = 6.2 \times 10^{26} \text{ molecules/kmole.}$$

Thus a quite primitive experiment yields a value for Avogadro's number not far from its actual value of 6.025×10^{26} molecules/kmole. More accurate methods of determining N_0 involve x-ray diffraction and electrolysis, as we shall learn in Chapters 21 and 35.

The specimens of matter we normally deal with contain immense numbers of atoms and molecules, and actually counting them is out of the question. However, if we know the mass of a sample of a particular element or compound, we can find the number N of atoms or molecules it contains by using the definition of the kmole together with Avogadro's number. Since

$$N = nN_0, \tag{14-1}$$

where n is the number of kmoles in the sample and N_0 is Avogadro's number, we have

$$N = \frac{\text{mass of sample}}{\text{molecular weight}} \times N_0. \tag{14-2}$$

A kilogram of ethyl alcohol, for instance, contains

$$N = \frac{1 \text{ kg}}{46.07 \text{ kg/kmole}} \times 6.025 \times 10^{26} \text{ molecules/kmole}$$

$$= 1.34 \times 10^{25} \text{ molecules.}$$

It is an experimental fact that under identical conditions of pressure and temperature, equal volumes of gases contain equal numbers of molecules. The reason for this is that gas molecules themselves have negligible volumes compared with the sizes of their containers; gas molecules fill their containers and exert pressure on their walls because of their random motions, an idea we shall explore in detail in the remainder of this chapter. Under the standard conditions of atmospheric pressure and at 0°C—conditions often referred to by the abbreviation STP—one kmole of any gas occupies a volume of 22.4 m^3. We have used this fact in Chapter 13 to evaluate the universal gas constant R.

14-3 Kinetic theory of an ideal gas

We shall now apply the basic notion of kinetic theory to the case of an ideal gas. Three assumptions must be made:

(1) An ideal gas consists of molecules which can undergo *elastic* (i.e., energy-conserving) collisions with one another and with the walls of their container.

(2) The gas molecules are sufficiently far apart most of the time so that they do not exert any forces on one another except during collisions, and therefore have kinetic but not potential energy.

(3) The distribution of speeds among the gas molecules at equilibrium does not change with time. The same number of molecules always has a particular speed, although the specific molecules having this speed may be different at different times. To be precise about what we mean, we would say that, on the average, as

many molecules of some initial velocity v_1 emerge from collisions with the higher velocity v_2 as there are molecules whose initial velocity is v_2 and which are slowed down to v_1 in other collisions. While in dealing with large numbers of identical particles there is the disadvantage that we cannot follow the motion of each particle individually, the fact that the numbers are so great permits us to use a statistical treatment to discuss the *average* behavior of all the particles.

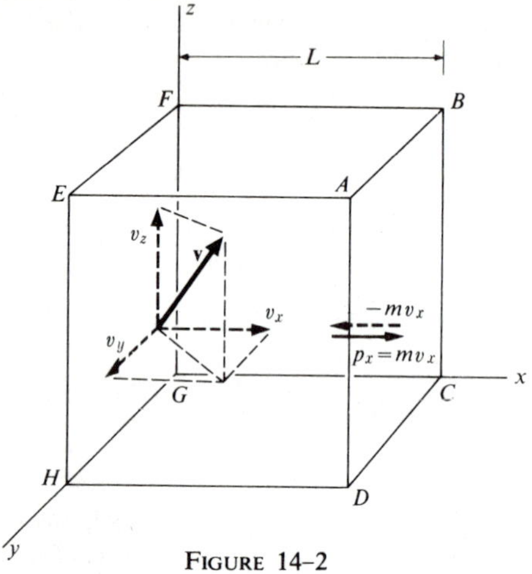

FIGURE 14-2

Let us consider a total of N ideal gas molecules, each of mass m, contained in a cubical box of length L on each edge (Fig. 14-2). When a molecule collides with one of the faces of the box and rebounds, a force is exerted on that face of the box; the result of many such forces continually acting on all the walls of the box is a uniform pressure on the walls. In order to compute this pressure in terms of the behavior of the individual molecules, we begin by resolving the velocity $\mathbf{v}$ of any molecule at any instant into the components v_x, v_y, v_z, as shown in the diagram. The molecule has the momentum in the x-direction mv_x when it strikes the face $ABCD$ and $-mv_x$ when it bounces off after the collision, a momentum change of

$$\Delta mv = 2mv_x.$$

Since, from Newton's second law of motion, force is equal to the rate of change of momentum,

$$\mathbf{F} = \frac{d}{dt}(m\mathbf{v}) = \frac{\Delta m\mathbf{v}}{\Delta t},$$

we can find the average force on $ABCD$ by dividing the momentum change Δmv by the time Δt required for the molecule to go from $ABCD$ to $EFGH$ and back to $ABCD$. This time interval is twice the distance between $ABCD$ and $EFGH$, which is $2L$, divided by the molecule's x-component of velocity v_x, so that

$$t = 2L/v_x.$$

Hence the average force f on $ABCD$ due to the single molecule whose x-component of velocity is v_x turns out to be

$$f = \frac{\Delta mv}{\Delta t} = \frac{2mv_x}{2L/v_x} = \frac{mv_x^2}{L}. \tag{14–3}$$

Not all the gas molecules have the same x-component of velocity. A number N_1 of them have the velocity component v_{x1}, a different number N_2 the different velocity component v_{x2}, and so on. Equation (14–3) involves the square of the x-component of velocity, so let us form an average of this square for all

$$N = N_1 + N_2 + N_3 + \cdots$$

of the molecules in the box:

$$\overline{v_x^2} = \frac{N_1v_{x1}^2 + N_2v_{x2}^2 + N_3v_{x3}^2 + \cdots}{N}. \tag{14–4}$$

The magnitude of the *total* force on $ABCD$ and $EFGH$ as well is accordingly

$$F = \frac{Nm\overline{v_x^2}}{L}. \tag{14–5}$$

What is the pressure on $ABCD$? Since p is, by definition, force per unit area acting perpendicular to a surface, the pressure on $ABCD$ is F divided by the area L^2, so that

$$p = \frac{F}{A} = \frac{Nm\overline{v_x^2}}{L^3} \tag{14–6}$$

or, since the volume V of the box is L^3, we may write

$$p = \frac{Nm\overline{v_x^2}}{V}, \qquad pV = Nm\overline{v_x^2} = 2N \cdot \tfrac{1}{2}m\overline{v_x^2}. \tag{14–7}$$

Identical reasoning gives

$$pV = 2N \cdot \tfrac{1}{2}m\overline{v_y^2}$$

for the sides $BCGF$ and $ADHE$ and

$$pV = 2N \cdot \tfrac{1}{2}m\overline{v_z^2}$$

for the sides $ABFE$ and $DCGH$.

Thus far we have taken into account only collisions between the gas molecules and the container walls. However, the above analysis does not change when we include the possibility of intermolecular encounters which alter the molecular momenta without affecting the walls. This is a consequence of our third assumption, that the *distribution* of velocities does not change with time, despite changes in the velocities of individual molecules. Even though many intermolecular collisions occur, the net effect of all of them is to leave the system as a whole unaffected.

The molecular directions of motion are purely random, and so

$$\overline{v_x^2} = \overline{v_y^2} = \overline{v_z^2}.$$

Furthermore, by definition,

$$\overline{v^2} = \overline{v_x^2} + \overline{v_y^2} + \overline{v_z^2}.$$

Combining the above two equations, we find that

$$\overline{v_x^2} = \overline{v_y^2} = \overline{v_z^2} = \tfrac{1}{3}\overline{v^2}. \tag{14-8}$$

The mean square of each average velocity component is precisely one-third of the mean square average speed of the molecules. Equation (14-7) now takes the form

$$pV = \tfrac{2}{3}N \cdot \tfrac{1}{2}m\overline{v^2}. \tag{14-9}$$

Since $\frac{1}{2}m\overline{v^2}$ is the average kinetic energy of each molecule in the gas, and the molecules, in this model, have no potential energy, we may write

$$\overline{KE} = \tfrac{1}{2}m\overline{v^2} = \overline{\tfrac{1}{2}mv^2}$$

for the average total kinetic energy of each molecule. Equation (14-9) thus reads

$$pV = \tfrac{2}{3}N\,\overline{(\tfrac{1}{2}mv^2)} = \tfrac{2}{3}N\,\overline{KE}, \tag{14-10}$$

where we recognize $N\,\overline{KE}$ as being the total internal energy in our ideal gas sample containing N molecules.

Equation (14-10) was derived for a cubical box, but it is possible, though more complicated, to obtain the same equation for a container of any shape whatever. It is often helpful to assume a specific geometrical configuration in investigating a problem, and if the result contains no geometrical reference, in many cases it is a perfectly general one restricted only by the other initial conditions.

14-4 Molecular kinetic energy

Equation (14-10) is the closest that the kinetic theory of gases can come to the ideal gas law,

$$pV = nRT. \tag{13-15}$$

The two formulas agree, however, if we assume that

The average kinetic energy of the molecules of a gas is proportional to the absolute temperature of the gas.

This assumption means that we set Eqs. (14–10) and (13–15) equal to each other. Then we have

$$\tfrac{2}{3}N\,\overline{\text{KE}} = nRT \qquad \text{and} \qquad \overline{\text{KE}} = \frac{3}{2}\frac{R}{N/n}T. \tag{14–11}$$

Now N is the number of molecules and n is the number of kmoles in the sample, so their ratio, N/n, is equal to the number of molecules in a kmole, which is Avogadro's number N_0. The quantity

$$\begin{aligned} k &= \frac{R}{N/n} = \frac{R}{N_0} \\ &= \frac{8.32 \times 10^3 \text{ j/kmole·°K}}{6.025 \times 10^{26} \text{ molecules/kmole}} \\ &= 1.38 \times 10^{-23} \text{ j/molecule·°K} \end{aligned}$$

is therefore the universal gas constant expressed in terms of molecules rather than in terms of kmoles. It is known as *Boltzmann's constant*, after the nineteenth-century Austrian physicist Ludwig Boltzmann. The symbol for Boltzmann's constant is k, and its value is

$$k = 1.38 \times 10^{-23}\text{ j/°K}.$$

In terms of Boltzmann's constant,

$$\overline{\text{KE}} = \tfrac{3}{2}kT; \tag{14–12}$$

$\tfrac{3}{2}kT$ is the average kinetic energy of the molecules of all gases at the absolute temperature T.

The interpretation of absolute zero in terms of the elementary kinetic theory of gases is a simple one: it is that temperature at which all molecular movement ceases.

The kinetic theory of gases leads us directly to the ideal gas law, but, as we said earlier, the ideal gas law is only a good approximation to reality. If we examine the initial assumptions we made in order to obtain the kinetic theory, it is easy to see why we should expect discrepancies between theory and experiment. For example, we assumed explicitly that gas molecules have volumes so small as to be negligible, that they exert no forces on one another except in actual collisions, that these collisions conserve kinetic energy, and that the molecules have no kinetic energy of rotation or internal vibration. When the kinetic theory is worked out starting from more realistic assumptions, the results are in excellent agreement with observational data.

14–5 Molecular speeds

We can use Eq. (14–12) to compute an average speed of gas molecules whose mass is m:

$$\overline{\tfrac{1}{2}mv^2} = \tfrac{3}{2}kT,$$

$$v_{\text{rms}} = \sqrt{\overline{v^2}} = \sqrt{\frac{3kT}{m}}. \tag{14–13}$$

The above speed is denoted v_{rms} because it is the square root of the mean of the squared molecular speeds—the "root-mean-square" speed—and therefore different from the simple arithmetical average speed $\bar{v}$. To emphasize their difference with a simple example, we can evaluate both kinds of average for an assembly of two molecules, one with a speed of 1 m/sec and the other with a speed of 3 m/sec. We find that

$$\bar{v} = \frac{v_1 + v_2}{2} = \frac{(1 + 3)\text{ m/sec}}{2} = 2\text{ m/sec},$$

whereas

$$v_{\text{rms}} = \sqrt{\frac{v_1^2 + v_2^2}{2}} = \sqrt{\frac{1^2 + 3^2}{2}}\text{ m/sec} = \sqrt{5}\text{ m/sec} = 2.24\text{ m/sec},$$

so that v_{rms} and $\bar{v}$ are not the same. The relationship between $\bar{v}$ and v_{rms} depends on the distribution in molecular speeds under consideration. For the distribution of molecular speeds found in a gas,

$$v_{\text{rms}} \approx 1.09\bar{v},$$

so that the root-mean-square speed of Eq. (14–13) is about 9% greater than the arithmetical average speed $\bar{v}$. Let us apply Eq. (14–13) to the molecules of oxygen, one of the major constituents of air. As we saw earlier, the molecular mass of oxygen is 2×16 or 32, one atomic mass unit is equivalent to a mass of 1.660×10^{-27} kg, and so each oxygen molecule has a mass of

$$m_{\text{oxygen}} = 32 \times 1.660 \times 10^{-27}\text{ kg} = 5.31 \times 10^{-26}\text{ kg}.$$

At an absolute temperature of 273°K (corresponding to 0°C), the root-mean-square speed of oxygen molecules is therefore

$$v_{\text{rms}} = \sqrt{\frac{3 \times 1.38 \times 10^{-23}\text{ j/°K} \times 273\text{°K}}{5.31 \times 10^{-26}\text{ kg}}} = 4.61 \times 10^2\text{ m/sec},$$

which is a little over 1000 mi/hr! Evidently molecular speeds are very large compared with those of the macroscopic bodies familiar to us.

It is important to keep in mind that actual molecular speeds vary considerably on either side of v_{rms}. Figure 14–3 is a graph showing the actual distribution of molecular speeds in oxygen at 73°K and at 273°K, and in hydrogen at 273°K.

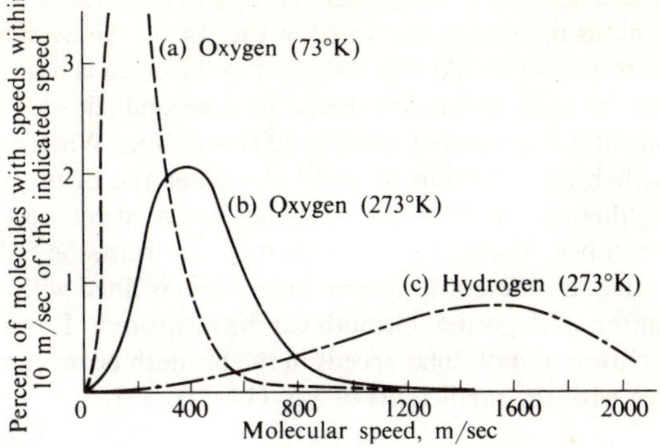

FIG. 14-3. The distribution of molecular speeds in oxygen at 73°K (−200°C), in oxygen at 273°K (0°C), and in hydrogen at 273°K. The average molecular speed increases with temperature and decreases with molecular weight; hence molecular speeds in oxygen at 73°K (curve *a*) are on the average less than at 273°K (curve *b*), and at the same temperature molecular speeds in hydrogen (curve *c*) are on the average greater than in oxygen (curve *b*) since oxygen molecules are heavier. At the same temperature the average molecular *energy* is the same for all gases.

The basic shape of these curves is the same, and is given by the formula

$$N_v\,dv = 4\pi N\left(\frac{m}{2\pi kT}\right)^{3/2} v^2 e^{-mv^2/2kT}\,dv. \tag{14–14}$$

Here $N_v\,dv$ is the number of molecules whose speeds are between v and $v + dv$, N is the total number of molecules present, m is the molecular mass, T is the absolute temperature, and k is Boltzmann's constant. Equation (14–14) was obtained a century ago by James Clerk Maxwell on the assumption of wholly random molecular motions with frequent collisions. The curves of Fig. 14–3 are not symmetrical because the lower limit to v is fixed at $v = 0$ while there is, in principle, no upper limit; actually, as the curves show, the likelihood of speeds many times greater than v_{rms} is small.

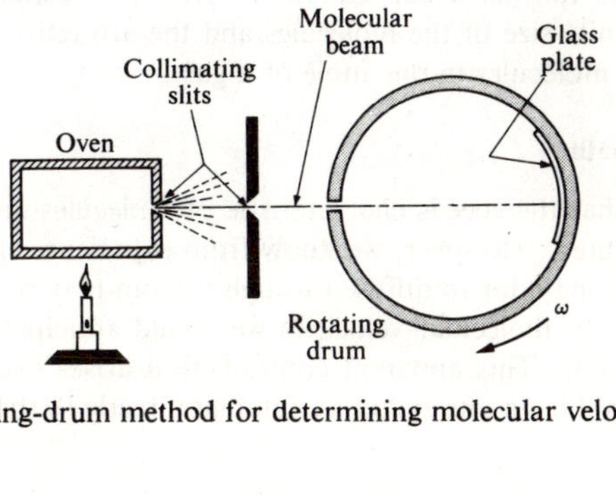

FIG. 14-4. Rotating-drum method for determining molecular velocity distribution.

The Maxwell distribution of molecular speeds can be verified directly in several ways. One ingenious method is illustrated in Fig. 14–4. An oven is used to heat a sample of bismuth to 800°C. At this temperature bismuth is a gas, and it escapes through a slit in the oven and is collimated by a second slit some distance away. Near the second slit is a drum that spins at 6000 rev/min. When a slit in the drum faces the bismuth beam, a group of molecules enter the drum together. These molecules have different speeds, and accordingly arrive at the opposite side of the drum at different times. Owing to the rotation of the drum, the fast and slow molecules are deposited at different locations. The drum is lined with a glass plate, so that the distribution of deposited bismuth can be measured. From these measurements the distribution of molecular speeds in the bismuth beam can be determined. The results agree with the predictions of Eq. (14–14).

14–6 The van der Waals equation

The empirical equation of state,

$$pV = nRT = NkT,$$

is in such good agreement with the kinetic theory of gases that we might hope that the kinetic theory could suggest improvements in it that would make it a still better representation of experimental results. This is actually the case. Two aspects of gases which we have neglected to consider in our discussion of kinetic theory are the finite (though small) volumes occupied by the molecules themselves, and intermolecular forces. Actual gas molecules have weak attractive forces acting among them which become strongly repulsive at very small distances. In liquids and solids the attractive forces are relatively strong, and dominate the behavior of the molecules, but we cannot assume that these forces completely vanish when matter is in the gaseous state. A modified equation of state, for one mole of a gas, which takes into account the finite sizes of molecules and attractive forces among them is

$$\left(P + \frac{a}{V^2}\right)(V - b) = kT, \tag{14–15}$$

which is called the van der Waals equation. Here a is a numerical factor which depends on the finite size of the molecules and the attractive forces, and b is the volume of all the molecules in one mole of a gas.

14–7 Mean free path

We have seen that the speeds characteristic of molecules are considerable even at room temperatures. However, we know from experience that the time needed by a gas with a strong odor to diffuse through a room may well be a few minutes, while on the basis of molecular velocities we would anticipate a time of a small fraction of a second. This apparent contradiction arises because gas molecules are not able to travel from one end of a container directly to the other, but undergo numerous deflecting collisions. The paths of individual molecules resemble random patterns like that of Fig. 14-5. While molecular speeds are quite high, the effective velocity of a molecule in a given direction may be quite low. The mean distance a molecule travels between collisions—its *mean free path*—is an important parameter in the description of a gas by means of kinetic theory.

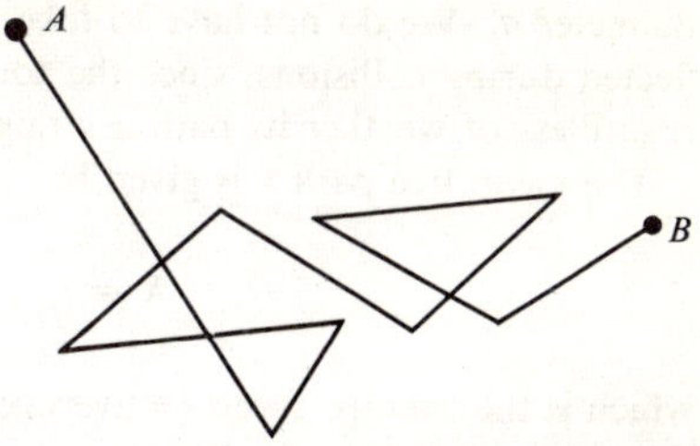

FIG. 14-5. In drifting from A to B, the molecule makes many collisions.

The mean free path in a particular gas sample depends on the number density n in the gas; that is, on the number of molecules per unit volume. The higher the number density, the more probable collisions become, and the shorter is the mean free path. The mean free path depends also on molecular sizes, with large molecules implying short mean free paths. In order to compute the mean free path in a gas, we require a somewhat more elaborate model of a gas than we employed in determining the equation of state. We shall assume that all the molecules of a gas move with constant speeds relative to one another, the value of the speed being the average molecular speed $\bar{v}$. We shall also assume that all the gas molecules are identical hard spheres with diameters a. Our object is to calculate the number of collisions per second made by each molecule; knowing the molecular speeds, we can then obtain the average distance the molecule travels between collisions. From Fig. 14-6, it is clear that the number of encounters which a given molecule will make with other molecules in one second is equal to the number of molecules within a cylinder of length $\bar{v}$ (the distance the molecule goes in one second) and diameter $2a$. This number of molecules is

$$N = n\bar{v}\pi a^2,$$

where n is the number of molecules per unit volume in the gas and $\bar{v}\pi a^2$ is four times the volume of the cylinder of space swept out each second by a molecule of

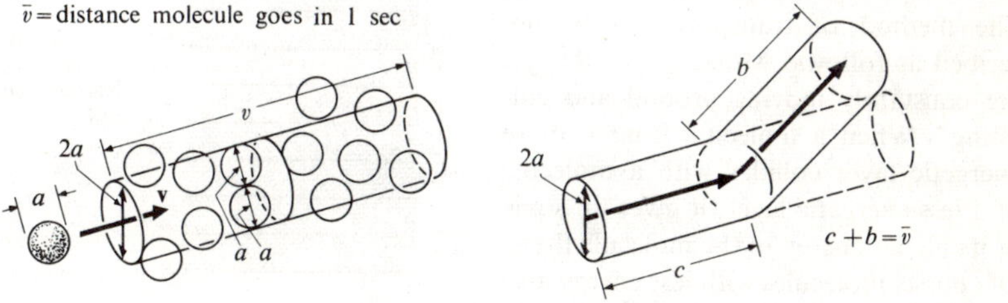

Number of collisions a molecule makes in 1 sec = density times volume of cylinder = $n\bar{v}\pi a^2$

Cylinder during deflection after collision

FIG. 14-6. Calculation of mean free path.

diameter a. We do not have to take into account the fact that molecules are deflected during collisions, since the total volume each one sweeps out is the same regardless of whether its path is straight or not.

The mean free path λ is given by

$$\lambda = \frac{\bar{v}}{N}, \qquad \lambda = \frac{1}{n\pi a^2}, \tag{14-16}$$

which is the average speed (=average distance traveled in one second) divided by the average number of collisions per second. If we take explicitly into account the distribution of velocities present in a gas, instead of merely assuming that all the molecules have the same relative velocity, we obtain the more exact result

$$\lambda = \frac{1}{\sqrt{2}\, n\pi a^2}. \tag{14-17}$$

14-8 Thermal conductivity

As an example of the usefulness of the idea of mean free path, let us consider the kinetic theory interpretation of the thermal conductivity of a gas. In Chapter 13 we learned that the rate at which heat per unit area was conducted was proportional to the temperature gradient; the coefficient of proportionality is the coefficient of thermal conductivity. The defining equation of the coefficient of thermal conductivity is

$$\frac{1}{A}\frac{dQ}{dt} = -K\frac{dT}{dz}, \tag{13-16}$$

where K is the coefficient.

In order to calculate K for an ideal gas, we assume that the gas sample consists of layers along the z-axis, each with different values of the internal energy (and therefore with different values of the temperature). That is, the internal energy is a function of z (Fig. 14-7). We attribute the thermal conductivity to the transport of energy along the z-axis. The method of transport can be described as follows: Molecules of the gas are constantly moving around and colliding. When a molecule from a more energetic layer collides with a molecule of a less energetic layer, it gives up some of its higher energy to the molecule there. Of course, molecules with less energy also collide with molecules of greater energy and degrade the energy of the higher layers, but the net result of all these collisions

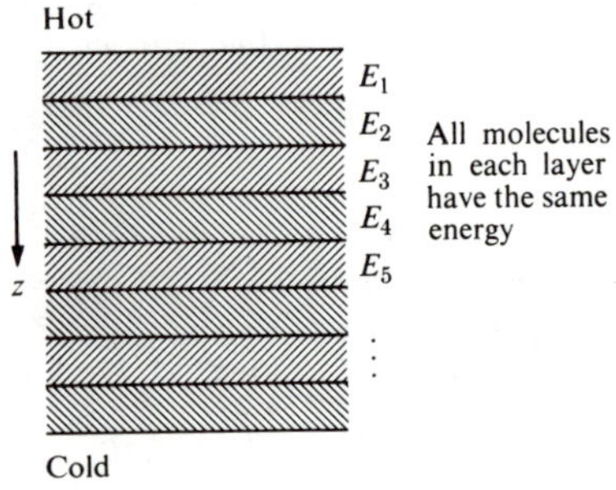

FIG. 14-7. A distribution of energy which depends only on z.

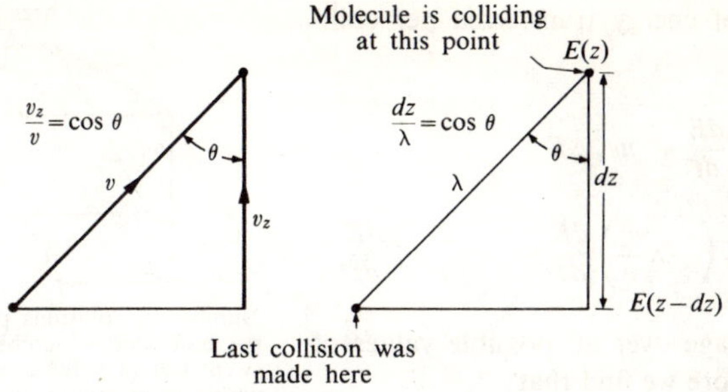

FIG. 14-8. Calculation of λ in terms of v and v_z.

is to have more higher-energy molecules transport energy to the cooler regions than vice versa.

Next we fasten on some point whose coordinates are x, y, z, a point where a molecule is making a collision. This molecule has traveled one mean free path since its last collision (Fig. 14-8). The energy transferred in this collision is the difference between the energy at $z - dz$ and z, assuming that the mean free path is small enough to consider distances between collisions as infinitesimal. If this is the case, we have

$$E(z - dz) = E(z) - dz\frac{dE}{dz}\cdot \tag{14-18}$$

We now must find an expression for dz. From Fig. 14-8 we see that

$$dz = \lambda \cos \theta. \tag{14-19}$$

We also have another expression for $\cos \theta$, since it is equal to the ratio of the z-component of velocity to its speed,

$$\cos \theta = \frac{v_z}{v},$$

so from Eq. (14-19) we have

$$dz = \lambda\frac{v_z}{v}\cdot \tag{14-20}$$

The net transport of energy per collision is therefore

$$\Delta E = E(z - dz) - E(z) = -dz\frac{dE}{dz} = -\lambda\frac{v_z}{v}\frac{dE}{dz}\cdot \tag{14-21}$$

To calculate the coefficient of thermal conductivity, we must know the energy transported across a unit area per unit time. This energy is equal to the number of collisions made in a volume of unit cross section and of height v_z (Fig. 14-9) times the amount of energy transferred per collision, namely

$$\frac{1}{A}\frac{dQ}{dt} = \frac{1}{A}\frac{dE}{dt} = nv_z\,\Delta E$$

$$= nv_z\left(-\lambda\frac{v_z}{v}\right)\frac{dE}{dz} = -\frac{n\lambda}{v}v_z^2\frac{dE}{dz}.$$

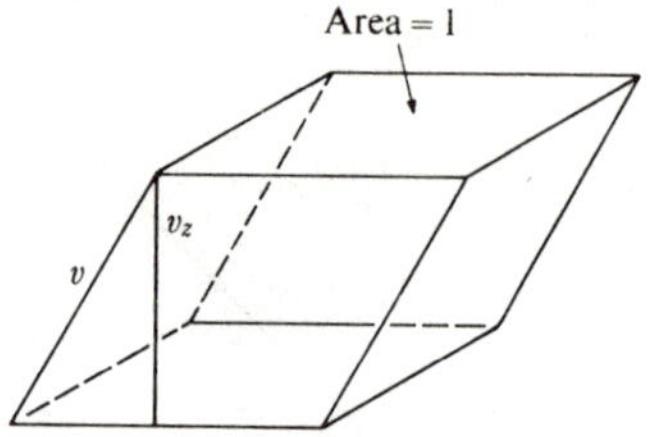

Number of collisions per unit area per unit time = number of collisions swept out in volume of slant height v whose cross section is unity = nv_z

FIGURE 14-9

Now we average over all possible values of v_z^2, and as before we find that

$$\overline{v_z^2} = \tfrac{1}{3}\overline{v^2}.$$

We also assume that $(\bar{v})^2 = (\overline{v^2})$. Hence the energy transported per unit area per unit time is

$$\frac{1}{A}\frac{dQ}{dt} = -\tfrac{1}{3}\lambda n\bar{v}\frac{dE}{dz}. \tag{14-22}$$

Since the energy per molecule of an ideal, monatomic gas is, by Eq. (14-12),

$$E = \tfrac{3}{2}kT, \qquad \frac{dE}{dz} = \tfrac{3}{2}k\frac{dT}{dz} \qquad (k = \text{Boltzmann's constant}),$$

we have

$$\frac{1}{A}\frac{dQ}{dt} = -\tfrac{1}{2}\lambda n\bar{v}k\frac{dT}{dz}.$$

Comparison of this expression with Eq. (13-16) shows that the coefficient of thermal conductivity is given by

$$K = \tfrac{1}{2}n\bar{v}k\lambda. \tag{14-23}$$

A more detailed calculation taking into account the true distribution of velocities would give the same dependence on n, $\bar{v}$, k, and λ as Eq. (14-23), but the coefficient would be 0.749 instead of $\frac{1}{2}$, in agreement with experiment.

Several features of Eq. (14-23) are of interest. For instance, because

$$\lambda = \frac{1}{\sqrt{2}\,n\pi a^2},$$

we find that

$$K = \frac{0.749}{\sqrt{2}}\frac{\bar{v}k}{\pi a^2}. \tag{14-24}$$

There is no dependence of the coefficient of conductivity on gas density! Intuitively we would anticipate that dense gases would exhibit greater conductivity than more diffuse ones, but this is evidently not correct. Experimentally K is independent of gas density down to about $\frac{1}{60}$ of an atmosphere of pressure; below this low pressure, mean free paths become so long that most molecular collisions occur with the container walls rather than with other gas molecules. Another observation that we can make is that conductivity varies with $\sqrt{T}$, since the factor $\bar{v}$ is proportional to $\sqrt{T}$. Thus the conductivity of a gas *increases* with temperature.

14–9 Kinetic theory of matter

The kinetic theory does not meet with as much quantitative success when its notions are applied to the solid and liquid states. However, the concept that the internal energy of a body resides in the kinetic energy of its molecules helps in understanding a variety of phenomena in these states.

The random motion of water molecules led to an important event in the history of science. In 1827 the British botanist Robert Brown noticed that pollen grains in water are in continual, agitated movement. Similar *Brownian motion* is apparent whenever very small particles are suspended in a fluid medium, for example smoke particles in air. According to kinetic theory, Brownian motion originates in the bombardment of the particles by molecules of the fluid. This bombardment is completely random, with successive molecular impacts coming from different directions and contributing different impulses to the particles. Albert Einstein, in 1905, found that he could account for Brownian motion quantitatively by assuming that, as a result of continual collisions with fluid molecules, the particles themselves have an average kinetic energy of $\frac{3}{2}kT$. Surprising as it may seem, this was the first direct verification of the reality of molecules, and it convinced many distinguished scientists who had previously been reluctant to believe that such things actually exist.

Another kinetic-molecular phenomenon characteristic of the liquid state is evaporation. A dish of water well below its boiling point of 100°C will nevertheless gradually turn into vapor, growing colder as it does so. The faster the evaporation, the more pronounced the cooling effect; alcohol and ether chill the skin upon contact because of their extreme volatility. This behavior follows from the distribution of molecular speeds in a liquid. While not identical with that found in a gas, this distribution resembles those shown in Fig. 14–3 in that certain fractions of the molecules in any sample have much greater and much smaller velocities than the average. The fastest molecules have enough energy to escape through the liquid surface despite the attractive forces of the other molecules. The molecules left behind redistribute the available energy in collisions among themselves, but, because the most energetic ones escape, the average energy that remains is less than before and the liquid is now at a lower temperature.

Thermal expansion in a solid also has a straightforward explanation in terms of kinetic theory. Most solids are crystalline in nature, which means that the various atoms that compose them form a regular arrangement in space (Chapter 35). (In some crystalline solids the basic constituents are whole molecules, rather than individual atoms, but for convenience we shall refer to them here as atoms in general.) The atoms behave as though they are joined together by tiny springs, thereby accounting for Hooke's law, and constantly oscillate about their equilibrium positions. Figure 14–10 shows how the atomic potential energy of a solid varies with interatomic spacing. The normal spacing a corresponds to the lower portion of the curve, where the energy per atom is least. The amplitude of the vibrations is determined by the width of the curve: When the atomic separation is a minimum or a maximum, the energy of a pair of adjacent atoms is wholly potential, as in the case of a harmonic oscillator at each end of its path, while in the middle their energy is wholly kinetic. The average interatomic spacing is what determines a, and hence the dimensions of the solid. When additional energy of thermal origin is added, the atomic spacing alternates through a wider range than before. If the potential energy curve were symmetrical about the distance a, no change in the dimensions of the solid would occur, since a would always be halfway between the two parts of the curve. However, the attractive and repulsive forces between atoms vary with distance in different ways: The repulsive force increases more rapidly as the atoms move closer together than the attractive force increases as the atoms move farther apart. Consequently, as shown in Fig. 14–10, the average interatomic spacing becomes larger as the thermal energy of the solid increases and the amplitudes of the atomic vibrations increase. As it happens, changes in the average interatomic spacing are very nearly proportional to changes in temperature, leading to the linear thermal expansion formula of Eq. (13–3).

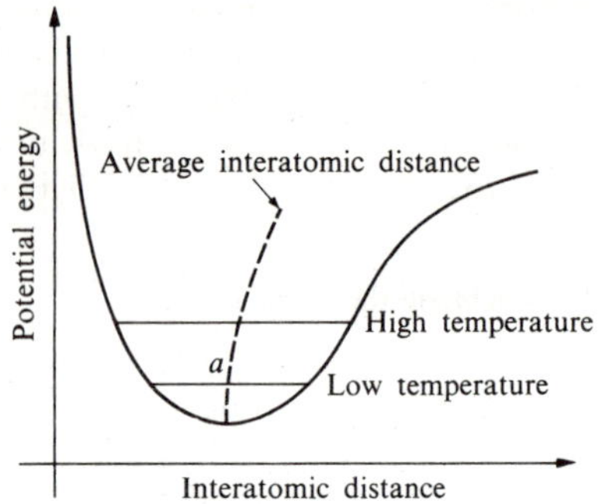

FIG. 14–10. The atomic potential energy of a crystal as a function of the spacing of its constituent atoms. The increase in the average interatomic distance with energy is the cause of thermal expansion in a solid.

Generally speaking, the stronger the atomic binding forces, the steeper and narrower the potential energy curve is, and hence the less pronounced the thermal expansion. As we shall see in Chapter 35, covalent and ionic bonds are very strong, metallic bonds intermediate, and molecular bonds (in which whole molecules constitute the crystal) weakest. Accordingly we are not surprised to find that the coefficient of thermal expansion of diamond, a covalent solid, is $0.11 \times 10^{-5}/°C$. That of iron, a metal, is $1.2 \times 10^{-5}/°C$, and that of ice, a molecular solid, is $5.1 \times 10^{-5}/°C$. Some solids do not have their atoms arranged symmetrically, and their potential energy curves are different in different directions. Thus the expansion coefficient of a quartz crystal parallel to a certain direction is $1.4 \times 10^{-5}/°C$, while it is $0.9 \times 10^{-5}/°C$ perpendicular to this direction. ("Fused quartz" is a glasslike noncrystalline form of quartz which has the very low thermal expansion coefficient of $0.05 \times 10^{-5}/°C$ in all directions.)

Problems

In the following, neglect difference between average speed and root-mean-square speed where necessary.

1. (a) Find the mass of a 0.1-kmole sample of gaseous oxygen. (b) How many molecules are present in the sample? (c) What volume does the sample occupy at STP?

2. (a) How many molecules are present in 0.0042 kg of gaseous nitrogen? (b) What volume does the sample occupy at STP?

3. How many atoms are there in 1 lb of copper?

4. (a) Find the molecular mass of sodium sulfate, Na_2SO_4. (b) How many pounds of sodium are present in a ton of sodium sulfate?

5. When one kmole of potassium chlorate, $KClO_3$, is heated, 1.5 kmoles of O_2 gas are evolved. How many kilograms of potassium chlorate must be decomposed to produce 1 m^3 of oxygen at STP?

6. Two vessels of the same size are at the same temperature. One of them holds 1 kg of H_2 gas and the other holds 1 kg of N_2 gas. (a) Which vessel contains more molecules? How many times more? (b) Which vessel is under the greater pressure? How many times greater? (c) In which vessel are the average molecular speeds greater? How many times greater?

7. (a) What is the average kinetic energy of the molecules of a gas at 0°C? (b) at 100°C?

8. Actual molecules attract one another slightly. Would the gas pressure be greater or less than the value computed for an ideal gas of the same internal energy? Why?

9. In terms of the kinetic theory, explain why the pressure of a gas in a closed container increases when the gas is heated.

10. Which of the following gases has (a) the highest average molecular speed at a given temperature? (b) the lowest average molecular speed? CO_2, UF_6, H_2, He, Xe, NH_3.

11. The average speed of a hydrogen molecule at room temperature and atmospheric pressure is 1 mi/sec. (a) What happens to the average speed of such a molecule if the temperature remains constant but the pressure is doubled? (b) What is the average speed of a nitrogen molecule (N_2) at room temperature and atmospheric pressure? Compute this from the above figure for hydrogen.

12. (a) Find the average speed of carbon dioxide (CO_2) molecules at 100°C. (b) At what temperature would this speed be doubled?

13. The minimum speed which a body must have if it is to leave the earth permanently is 11,200 m/sec (36,800 ft/sec). This speed is called the escape velocity. At the top of the earth's atmosphere the temperature is approximately 0°C. With the help of Fig. 14–3, explain why hydrogen escapes from the atmosphere more readily than oxygen.

14. One of the assumptions of the kinetic theory is that the average distance between molecules is much greater than the dimensions of the molecules themselves. Oxygen and nitrogen molecules are roughly 2×10^{-10} m in diameter, and there are 2.7×10^{25} molecules in a cubic meter of air at room temperature and atmospheric pressure. (a) On the average, how far apart are the molecules in air? (b) How many molecular diameters is their average separation?

15. The average speed of air molecules is roughly 4×10^2 m/sec, and the average

distance an air molecule goes between collisions with other molecules is about 10^{-7} m. What is the average number of collisions an air molecule makes per second?

16. A vessel contains 1 kg of chlorine at 0°C and a pressure of 1 atm. When another kg of chlorine is added to the vessel at the same temperature, the pressure rises to 2 atm. Explain the change in pressure on the basis of the kinetic theory of gases.

17. Silver is a vapor at 1500°K. What is the average speed of silver atoms in a vapor at this temperature?

18. A molecule of methane (molecular mass 16) is a sphere whose volume is five times that of an argon atom (atomic mass 40). What is the ratio of the thermal conductivity of methane and argon at a given temperature?

19. The thermal conductivity of argon at 0°C is

$$394 \times 10^{-8}\ \text{kcal}\cdot\text{m}^{-1}\cdot\text{sec}^{-1}\cdot{}^{\circ}\text{K}^{-1}.$$

What is the mean free path if the pressure is 1 atm? At what pressure would the mean free path be 1 mm?

20. One pan of a balance is bombarded continuously with shot, each of mass $\frac{1}{4}$ gm, velocity 80 cm/sec, at the rate of 100/sec. Assuming that the collisions are elastic and that the shot is incident normally, what mass must be placed in the other pan if it is to remain in balance?

21. Demonstrate that one can obtain Eq. (14–10) for the product pV by analyzing the collisions of a particle in a spherical box whose average speed is $\bar{v}$. [*Hint:* Calculate the average time between collisions and the component of momentum transferred to the walls per collision.]

22. What is the ratio of the average translational velocities of the molecules in gaseous hydrogen and oxygen at the same temperature? What is the ratio of the average translational kinetic energies?

23. Explain qualitatively why the coefficient of heat conduction is independent of the pressure.

24. At what temperature would the root-mean-square velocity of hydrogen gas be equal to the escape velocity on the moon?

25. Explain qualitatively in what limits the van der Waals equation goes over into the ideal gas law.

26. There is a critical isotherm in the van der Waals gas which is defined as that pressure, volume, and temperature for which

$$\left(\frac{\partial p}{\partial V}\right)_{T=\text{const}} = 0,$$

$$\left(\frac{\partial^2 p}{\partial V^2}\right)_{T=\text{const}} = 0.$$

The critical pressure, volume, and temperature are denoted by p_c, v_c, T_c. (a) Show that

$$p_c = a/27b^2,$$

$$v_c = 3b,$$

$$T_c = 8a/27Rb.$$

(b) Express the van der Waals equation in terms of p_c, v_c, and T_c, instead of the constants a and b. (c) Plot the critical isotherm. Plot an isotherm in which

$$p \text{ and } v \gg p_c \text{ and } v_c,$$

and one in which

$$p,\ v < p_c,\ v_c.$$

27. Derive a formula for the bulk modulus of an ideal gas.

28. Derive a formula for the bulk modulus of a van der Waals gas.

29. The fraction of molecules in a gas at temperature T whose speed is between v and $v + dv$ is given by

$$f(v) = Av^2 e^{-mv^2/2kT}.$$

(a) Calculate A by integrating $f(v)$ from 0 to ∞ and setting it equal to 1. (b) Calculate the average speed, the most probable speed, the root-mean-square velocity, and compare them. (c) Calculate the average kinetic energy. Note that

$$\int_0^\infty e^{-x^2}\,dx = \tfrac{1}{2}\sqrt{\pi};$$

$$\int_0^\infty x^n e^{-x^2} = \tfrac{1}{2}(n-1)\int_0^\infty x^{n-2} e^{-x^2}\,dx.$$

This distribution is known as the Maxwell-Boltzmann distribution.

30. What is the mean free path of a molecule whose radius is 5×10^{-8} cm at atmospheric pressure and 0°C?

15

The First Law of Thermodynamics

In the last two chapters we have seen that one can consistently define a form of energy, called heat, which flows by itself from a hot to a cold body. From a kinetic theory point of view this energy is distributed among the constituents of the cooler body and appears as the energy (kinetic or potential) of these constituents. This energy is the internal energy of the body, and part of it is recognizable as the temperature of the body.

We can *decrease* the internal energy by allowing the body to do work. For example, when we let a gas expand suddenly, it cools and does work while expanding. The internal energy can also be decreased by extracting heat from the body. We can *increase* the internal energy by doing work on a body or by transferring heat to it. Thus work and heat can be transformed into internal energy, and vice versa.

There are limitations in nature on processes that connect heat, work, and internal energy. The study of these processes and limitations is called *thermodynamics*, and we shall explore it in this and the next chapters. Although thermodynamics is developed as a discipline without regard to any model of the structure of matter, we shall find it helpful during the discussion to refer to the kinetic-molecular picture.

15–1 The first law of thermodynamics

If we wish to write an equation expressing conservation of energy in processes involving heat, it is convenient if we think in terms of a "system" whose interactions with its environment may be measured. The system may be a container of a gas, an automobile, the solar system, and so forth, the important thing being its existence as an entity separate from its surroundings. A system may give out heat or absorb it; it may do external work or have work done on it. At the same time its internal energy may change. The usual way of expressing the relationship between these quantities in any process is called the *first law of thermodynamics*, and is written

$$U_0 + Q = U + W$$

or

$$Q = U - U_0 + W. \tag{15-1}$$

In this equation Q represents the heat *added* to the system, U_0 is its initial internal energy and U its final internal energy, and W is the work done *by* the system in the process. Should the system *release* heat, Q in Eq. (15-1) is taken as negative, and similarly work done *on* the system is taken as negative. All the quantities in Eq. (15-1) must, of course, be expressed in the same units, whether calories, Btu, joules, or ft-lb.

In the case of an infinitesimal process, the first law of thermodynamics becomes

$$dQ = dU + dW. \tag{15-2}$$

In this form the first law is of great importance, in the sense that it applies to *any* system undergoing *any* infinitesimal process; but it is too general to be really useful. Let us investigate its application to processes involving a gas.

We shall begin by considering a gas sample confined to a cylinder with a movable piston (Fig. 15-1). The gas exerts a certain force F on the piston, and the work done by the gas on the piston when the latter moves out a distance dx is

$$dW = F\,dx.$$

If the gas pressure on the piston is p and the piston area is A, the force on the piston is

$$F = pA$$

and

$$dW = pA\,dx.$$

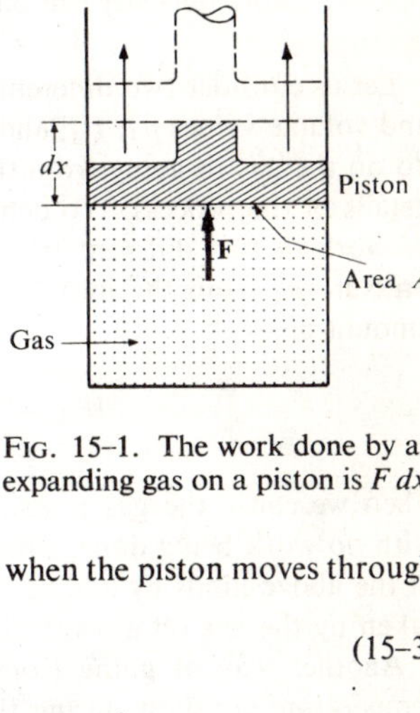

FIG. 15-1. The work done by an expanding gas on a piston is $F\,dx$.

But $A\,dx$ is the increase dV in the volume of the gas when the piston moves through the distance dx, and so

$$dW = p\,dV. \tag{15-3}$$

Hence the first law of thermodynamics as applied to a gas is

$$dQ = dU + p\,dV. \tag{15-4}$$

The state of any system is determined in general by three thermodynamic parameters, which, for a gas, are its pressure, volume, and temperature. A relationship among these three parameters is called an *equation of state*. Thus

$$pV = nRT \tag{15-5}$$

is the equation of state of an ideal gas. If we know the equation of state of whatever system we are dealing with, only two thermodynamic parameters need be specified, since the third can be found from the equation of state.

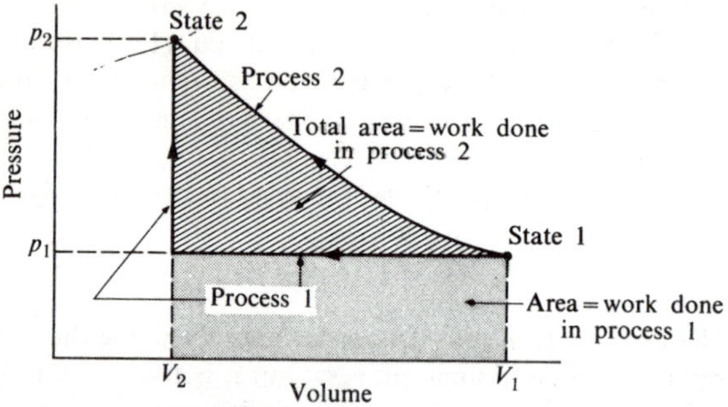

FIG. 15-2. Work is done in compressing a gas. The amount of work done depends on the details of the process.

Let us consider two different states of an ideal gas characterized by the pressure and volume values p_1, V_1, and p_2, V_2 (Fig. 15-2). The amount of work we must do on the gas in going from the state p_1, V_1 to p_2, V_2 is *not* independent of the details of the process. To demonstrate this, we shall compute the above amount of work in two different processes. First we compress the gas at the constant pressure p_1 from the initial volume V_1 to the final volume V_2, performing the amount of work

$$W_A = \int_{V_1}^{V_2} p\,dV = p_1(V_2 - V_1). \tag{15-6}$$

Then we cause the gas pressure to increase from p_1 to p_2 at constant volume, with no work being done, since there is no motion. (We can perform operations of the above kinds by adding or removing heat during the operations.) The path taken by the gas on a p-versus-V diagram is shown in Fig. 15-2.

Another way of going from p_1, V_1 to p_2, V_2 involves maintaining the gas temperature constant during the process. Then, making use of the equation of state to eliminate p, we have

$$W_B = \int_{V_1}^{V_2} p\,dV = nRT\int_{V_1}^{V_2} \frac{dV}{V}$$

$$= nRT \ln \frac{V_2}{V_1} = p_1V_1\left(\ln \frac{V_2}{V_1}\right), \tag{15-7}$$

which is quite different from W_A. Thus the work done on (or by) a system during a thermodynamic process depends on the path taken between the end points of the process as well as on the values of the end points. Mathematically, therefore, the differential of work is not an exact differential, since the definite integral of an exact differential is the difference of its value at the two end points. To emphasize this fact we shall write it as $đW$ instead of dW.

We know from our previous discussion of mechanics that, in the absence of dissipative forces, the work done on a particle or system of particles depends solely on the initial and final states of the particle or system, and not on the details of how the state was changed. It takes 6 ft-lb of work to raise a 1-lb weight 6 ft above the ground, regardless of whether the weight is raised vertically by this distance or carried around the world (without friction) in the course of raising it. Since dU is the change in the internal energy of a system during an infinitesimal process, and on the molecular level there are no dissipative forces, dU is an exact differential. Because the differential of heat is the sum of dU, an exact differential, and $đW$, an inexact differential, the differential of heat is itself inexact and we must write it as $đQ$. Hence the proper way of writing the first law of thermodynamics is

$$đQ = dU + đW \tag{15–8}$$

in general, and

$$đQ = dU + p\,dV \tag{15–9}$$

for a gas.

15–2 Specific heats of gases

In Chapter 13 the specific heat c of a substance was defined as the amount of heat that must be added to or removed from one kilogram of it to cause a change in temperature of 1°C. This simple statement cannot be used without qualification in the case of a gas, whose pressure or volume or both must change when the temperature changes. For example, the balloon of Fig. 13–6 expands when heat is added to raise its temperature from -10°C to $+10$°C, and the gas within has performed work in moving the balloon wall outward against the elastic forces of the rubber. Some of the heat goes into this work as well as into raising the gas temperature. On the other hand, a gas sample confined in a rigid container cannot change in volume, and the pressure changes that occur when heat is added to or removed from it do not involve work. Thus the specific heat of a gas whose volume remains constant is always less than its specific heat under constant pressure.

When we are considering specific heat in terms of kinetic theory it is convenient to use the kilomole as the unit of quantity instead of the kilogram, since kilomoles of all substances contain the same number of molecules. We shall denote molar specific heat by the symbol C to distinguish it from specific heat per unit mass, c, and we shall use a subscript to indicate when a particular variable is held constant. Accordingly the specific heat of a gas per kmole at constant volume is written C_V. At constant volume the amount of heat that must be added to 1 kmole of a gas to increase its temperature by dT is $C_V\,dT$. Since no work is done by the gas, all of this heat goes into molecular energy. Thus the average energy of each of the gas molecules increases from $\frac{3}{2}kT$ to $\frac{3}{2}k(T + dT)$, a change of $\frac{3}{2}k\,dT$. There are N_0 (Avogadro's number) molecules in the kilomole of gas under consideration, and so

$$\text{Heat added} = \text{increase in total molecular energy;}$$

$$C_V\,dT = \tfrac{3}{2}N_0k\,dT.$$

The quantity N_0k is, as we saw in Section 14–4, equal to the universal gas constant R, and so, dividing the above equation through by dT, we obtain

$$C_V = \tfrac{3}{2}R. \tag{15–10}$$

To change the temperature of 1 kmole of an ideal gas by 1°K requires the addition or removal of $\frac{3}{2}R$ of heat. Since $R = 8.32 \times 10^3$ j/kmole·°K or 1.99 kcal/kmole·°K, we expect that, in heat units,

$$C_V = \tfrac{3}{2} \times 1.99 \text{ kcal/kmole·°K} = 2.98 \text{ kcal/kmole·°K}.$$

How well is this prediction fulfilled by real gases? Table 15–1 is a list of the molar specific heats at constant volume for a number of common gases and vapors. Evidently theory and experiment agree reasonably well for gases composed of monatomic molecules, but gases with more complicated molecules all have higher heat capacities than $\frac{3}{2}R$. The agreement for monatomic gases indicates that we are certainly on the right track, but some relevant characteristic of polyatomic molecules has been neglected in deriving Eq. (15–10).

When we measure the temperature of a gas, what we actually determine is the average kinetic energy of translational motion of its constituent molecules. Since

TABLE 15–1

Approximate molar specific heats at constant volume and at room temperature

Gas	C_V
Helium (He)	3.02 kcal/kmole·°K
Argon (Ar)	3.0
Mercury vapor (Hg)	3.0
Hydrogen (H_2)	4.88
Oxygen (O_2)	4.99
Carbon monoxide (CO)	5.02
Chlorine (Cl_2)	6.01
Sulfur dioxide (SO_2)	7.5
Ethane (C_2H_6)	10.3
Ether [$(C_2H_5)_2O$]	30.8

we assumed that a molecule is a minute structureless body whose mass can be regarded as being concentrated at a single point, this is the only kind of energy it can have. Adding $\frac{3}{2}R$ to 1 kmole of a monatomic gas causes its temperature to rise by almost exactly 1°K, but to increase the temperature of 1 kmole of a polyatomic gas by 1°K requires *more* than $\frac{3}{2}R$ of heat. Somewhere in the latter kind of gas is a repository of energy not found in a monatomic gas. It is not hard to imagine what such a repository might be. Even though we apparently are entitled to think of an *atom* as a point mass, we may not be justified in regarding a *combination of atoms* in so simple a manner. If a combination of atoms has extension in space, it can possess rotational kinetic energy. A diatomic molecule, for instance, is like a tiny dumbbell (Fig. 15-3), and, because it has moments of inertia about two perpendicular axes, can have kinetic energy associated with rotation about those axes. (Because we are considering atoms as point particles from the point of view of kinetic theory, a diatomic molecule has no moment of inertia about a line joining its two atoms. We shall see in Chapter 34 that this approximation is justified.) In the course of collision with the container walls and with other molecules, a diatomic molecule is set into rotation, and molecular rotation therefore absorbs some of the total energy content of the gas.

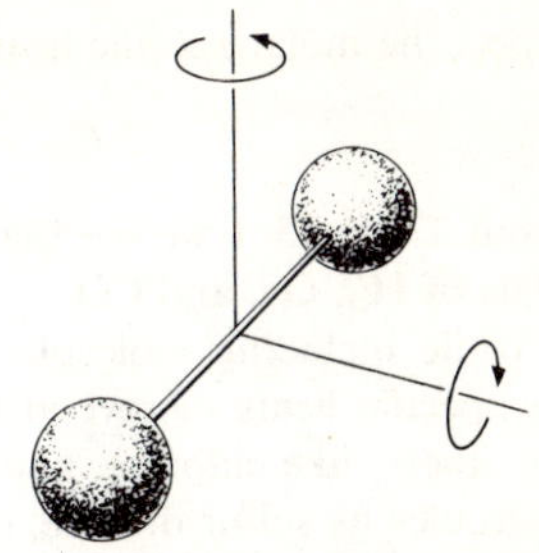

FIG. 15-3. A diatomic molecule.

A monatomic gas molecule may be said to have three *degrees of freedom*, corresponding to its ability to have independent motion in the x-, y-, and z-directions. By comparison, a particle restricted to the x-axis has only a single degree of freedom, while a particle fixed in place has no degrees of freedom. Because the pressure in a gas is the same in all directions, the molecules must be moving about in a random manner, and the three degrees of freedom share the available average molecular energy of $\frac{3}{2}kT$ equally. (The sharing is on an average basis too, of course.) Thus we can associate $\frac{1}{2}kT$ of energy with each one of the degrees of freedom of translational motion of a monatomic gas molecule:

$$\overline{\text{KE}}_x = \tfrac{1}{2}kT, \qquad \overline{\text{KE}}_y = \tfrac{1}{2}kT, \qquad \overline{\text{KE}}_z = \tfrac{1}{2}kT.$$

The principle of *equipartition of energy* is a generalization of the above result. According to this principle, in an assembly of molecules the same amount of energy is associated with each degree of freedom. In this sense we may regard a degree of freedom as a mode of energy possession. A diatomic molecule therefore has a total of five degrees of freedom, three of translation and two of rotation; the latter correspond to the two independent rotational motions it can undergo. Each of these degrees of freedom represents an average energy of $\frac{1}{2}kT$ per molecule, or a total of $\frac{5}{2}kT$. In an assembly of N_0 diatomic molecules, then, the total molecular energy content is $\frac{5}{2}N_0kT$ or $\frac{5}{2}RT$. By analogy with Eq. (15-10) we expect the molar specific heat for a diatomic gas to be

$$C_V = \tfrac{5}{2}R = 4.97 \text{ kcal/kmole}\cdot{}^\circ\text{K}. \tag{15-11}$$

From Table 15-1 we see that this figure is indeed in agreement with the specific heats of H_2, O_2, and CO.

While including molecular rotation does yield an adequate understanding of the specific heats of several diatomic molecules, it still predicts too low a value for others, like chlorine, another diatomic molecule, and for such more complex molecules as sulfur dioxide, ethane, and ether, even when an additional rotational degree of freedom is included to take into account their three-dimensional structures.

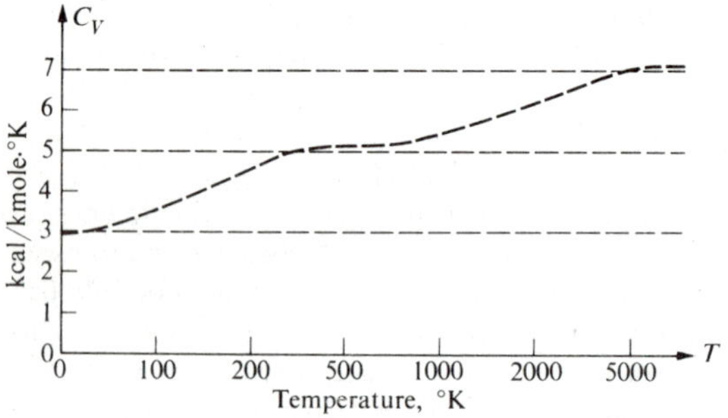

FIG. 15-4. Temperature versus specific heat at constant volume of hydrogen gas. (A logarithmic scale is used for temperature.)

A hint as to the nature of the additional degrees of freedom possessed by complicated molecules can be obtained by studying the variation of C_V with temperature for a diatomic gas. Figure 15-4 shows this variation for hydrogen. The heat capacity of H_2 is approximately 5 kcal/kmole·°K only between about 250°K and 750°K, dropping to 3 kcal/kmole·°K below 250°K and rising to 7 kcal/kmole·°K above 750°K. We can understand the reduced value of C_V at the lower temperatures by assuming that, for some reason, only translational motion is possible in a very cold gas. At the higher temperatures we can assume that the H_2 molecule also vibrates like two balls joined by a spring (Fig. 15-5). The vibration involves both kinetic and potential energies, adding two more degrees of freedom to the five of translation plus rotation. The heat capacity at high temperatures accordingly becomes $\tfrac{7}{2}R$ or 6.96 kcal/kmole·°K, almost exactly what is observed. Every link between the atoms in a molecule is therefore capable of adding two degrees of freedom and increasing the molar heat capacity by R; so the high specific heats of polyatomic gases have a plausible explanation.

We apparently have been able to understand the diverse specific heats of real gases by making no more than a reasonable extension of the simple kinetic-theory model of an ideal gas. However, while doing so, we have actually found something wholly contradictory to the fundamental assumption of the kinetic theory, which is that molecules can be treated as though they obey the ordinary laws of mechanics. If a diatomic molecule rotates and vibrates at high temperatures, why does it not do so at *all* temperatures? The behavior exhibited in Fig. 15-4 is really quite remarkable. Kinetic theory can explain any *one* of the 3, 5, and 7 kcal/kmole·°K specific heat values of H_2, but not the occurrence of all three in different temperature ranges. We shall discover the reason for the nonclassical aspects of specific heats in Chapter 34, where molecular physics is considered in the light of quantum theory. Adequate as the kinetic theory of gases is for many purposes, there is nevertheless a limit to its range of application.

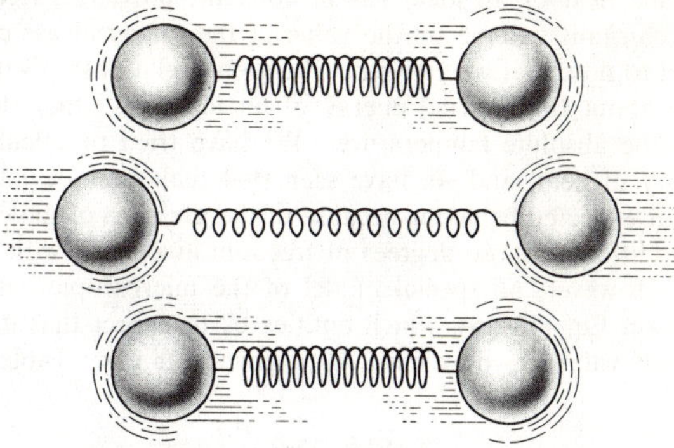

FIGURE 15-5

It is easy to obtain a relationship between the specific heat of an ideal gas at constant volume, C_V, and its specific heat at constant pressure, C_p, by using the first law of thermodynamics. When n kmoles of an ideal gas are maintained at constant pressure, the heat $đQ$ that is involved in a change of dT in the gas temperature is

$$đQ = nC_p\,dT. \tag{15-12}$$

Inserting $đQ = nC_p\,dT$ and $dU = nC_V\,dT$ in Eq. (15-4), we have

$$đQ = dU + p\,dV,$$

$$nC_p\,dT = nC_V\,dT + p\,dV,$$

and

$$n(C_p - C_V)\,dT = p\,dV. \tag{15-13}$$

Evidently C_p is greater than C_V, which reflects the fact that $p\,dV$ of work must be done at the expense of the incoming heat to expand the gas volume by dV; or, if the gas is cooled and its volume contracts, the heat given off is more than the internal energy decrease because work is done *on* the gas when it is compressed. We can evaluate $p\,dV$ by differentiating the ideal gas law, $pV = nRT$, which yields

$$p\,dV + V\,dp = nR\,dT. \tag{15-14}$$

At constant pressure, $dp = 0$, and so

$$p\,dV = nR\,dT, \qquad \text{at constant pressure.}$$

Substituting $nR\,dT$ for $p\,dV$ in Eq. (15–13), we have

$$n(C_p - C_V)\,dT = nR\,dT,$$

$$C_p - C_V = R. \tag{15–15}$$

The molar specific heat of an ideal gas at constant pressure exceeds its molar specific heat at constant volume by the value of the universal gas constant R.

It is important to note that we arrived at the above relationship without making any assumption about the internal energy of an ideal gas other than that it is proportional to the absolute temperature. We have tried to calculate C_V from the kinetic theory of heat, and we have seen that real gases deviate by varying extents from their expected behavior because of the presence of molecular degrees of freedom in addition to those degrees of freedom associated with pure translational motion. However, no special model of the microscopic nature of a gas was used to obtain Eq. (15–15), which entitles us to expect that it should hold regardless of what value C_V may have for a particular gas. Table 15–2, which

TABLE 15–2

Approximate molar specific heats of gases at room temperatures

Gas	C_p	C_V	$C_p - C_V$	$\gamma = C_p/C_V$
Helium (He)	5.00	3.02	1.98	1.66
Argon (Ar)	5.0	3.0	2.0	1.67
Mercury vapor (Hg)	5.0	3.0	2.0	1.67
Hydrogen (H_2)	6.8	4.88	2.0	1.41
Oxygen (O_2)	7.0	5.0	2.0	1.40
Carbon monoxide (CO)	7.01	5.02	1.99	1.40
Chlorine (Cl_2)	8.15	6.01	2.14	1.36
Sulfur dioxide (SO_2)	9.7	7.5	2.2	1.29
Ethane (C_2H_6)	12.4	10.3	2.1	1.20
Ether [$(C_2H_5)_2O$]	33.3	30.8	2.5	1.08

lists C_p and C_V for a number of real gases, shows that $C_p - C_V$ is indeed very close to $R = 1.99$ kcal/kmole·°K regardless of molecular structure. This table also shows the ratio between C_p and C_V for these gases; this ratio is often abbreviated γ (Greek letter *gamma*), so that

$$\gamma = \frac{C_p}{C_V}. \tag{15–16}$$

15-3 Isothermal processes

A thermodynamic process in which the temperature of the substance that undergoes a change of some kind remains constant is called an *isothermal* process. If the substance is a gas confined to a container, we can imagine that the container is a good conductor of heat and is immersed in a constant-temperature reservoir of heat. Such a reservoir might be a large quantity of water whose total heat capacity is so great that any heat flowing from the gas into the reservoir, or from the reservoir into the gas, does not change the reservoir temperature. No process that occurs in the real world is ever wholly isothermal, just as no actual gas exactly resembles an ideal gas, but many processes are quite close to being isothermal. We shall require a knowledge of the isothermal behavior of ideal gases in order to analyze the operation of the Carnot engine, a hypothetical engine of considerable importance because its efficiency is the highest possible in any given set of circumstances. (See Chapter 16.)

Let us determine the amount of work done by an ideal gas that expands isothermally from the initial volume of V_1 to the final volume V_2. The pressure, temperature, and volume of n moles of an ideal gas are related by the ideal gas law

$$pV = nRT.$$

If T is constant, we have Boyle's law, $pV =$ constant. When the gas volume V increases during the expansion, p must drop in order that their product remain unchanged. Figure 15-6 is a graph of p versus V for an expanding gas at a certain temperature. From Eq. (15-3), we know that

$$đW = p\,dV,$$

so that, integrating both sides, we obtain

$$W_{12} = \int_{V_1}^{V_2} p\,dV. \qquad (15\text{-}16)$$

This integral is the area under the curve of Fig. 15-2, and has been evaluated in Eq. (15-7). The result is

$$W_{12} = nRT \ln \frac{V_2}{V_1}. \qquad (15\text{-}17)$$

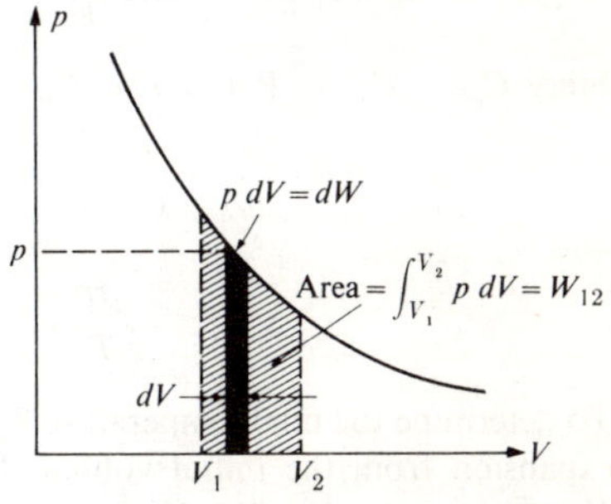

FIG. 15-6. Pressure versus volume for an ideal gas sample at constant temperature.

While we have obtained this formula for the work done by an ideal gas in an isothermal expansion, the same formula holds for an isothermal compression. In the latter case, V_2 is less than V_1, and, since the logarithm of a number less than 1 is negative, W_{12} is negative. A negative value of W means, according to our sign convention, that the work is done *on* the gas, which of course is necessary in order to compress it.

15–4 Adiabatic processes

A system which is at the other extreme from one in such intimate contact with a heat reservoir that its temperature never changes is a system so completely isolated from its surroundings that heat can neither enter nor leave it. Any process undergone by a system of this kind is called an *adiabatic* process. Most rapid thermodynamic processes are approximately adiabatic, since heat transfer takes time and the process may be completed before an appreciable amount of heat has passed through the walls of the system.

In an adiabatic process, $đQ = 0$ by definition, so that the first law of thermodynamics becomes

$$dU = -p\,dV.$$

Any work done *by* the system reduces its internal energy, and any work done *on* the system increases its internal energy. When we do work on the air in a tire pump by pushing down on the piston, the pump becomes hot because an increase in internal energy means a rise in temperature. The temperature change is related to dU by $dU = nC_V\,dT$, where n is the number of moles of gas present and C_V is its molar specific heat at constant volume. By combining the above expressions for dU, we obtain

$$nC_V\,dT = -p\,dV.$$

Substituting for the gas pressure p from the ideal gas law yields

$$nC_V\,dT = nRT\frac{dV}{V} \quad \text{or} \quad \frac{dT}{T} = -\frac{R}{C_V}\frac{dV}{V}.$$

Since $C_p - C_V = R$ and $\gamma = C_p/C_V$, then

$$\frac{R}{C_V} = \frac{C_p - C_V}{C_V} = \gamma - 1,$$

and

$$\frac{dT}{T} = -(\gamma - 1)\frac{dV}{V}. \tag{15–18}$$

To determine the final temperature T_2 of a gas sample that undergoes an adiabatic expansion from the initial volume V_1 and temperature T_1 to the final volume V_2, we integrate Eq. (15–18) between these states:

$$\int_{T_1}^{T_2}\frac{dT}{T} = -(\gamma - 1)\int_{V_1}^{V_2}\frac{dV}{V},$$

$$\ln T_2 - \ln T_1 = (\gamma - 1)(\ln V_1 - \ln V_2),$$

$$\ln\frac{T_2}{T_1} = (\gamma - 1)\ln\frac{V_1}{V_2}, \qquad \frac{T_2}{T_1} = \left(\frac{V_1}{V_2}\right)^{\gamma-1},$$

$$T_1V_1^{\gamma-1} = T_2V_2^{\gamma-1}. \tag{15–19}$$

In an adiabatic process the quantity $TV^{\gamma-1}$ remains constant.

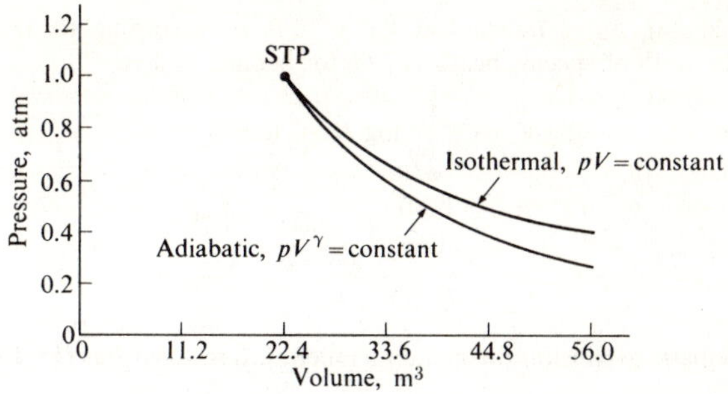

FIG. 15-7. Isothermal and adiabatic processes in 1 kmole of an ideal gas. The isothermal curve corresponds to a constant temperature of 273°K (0°C).

We can use the ideal gas law to replace T in Eq. (15-19). Since at the beginning and at the end of the process, $p_1V_1 = nRT_1$ and $p_2V_2 = nRT_2$, respectively, it follows that

$$\frac{p_1V_1}{p_2V_2} = \frac{T_1}{T_2},$$

and so

$$p_1V_1^\gamma = p_2V_2^\gamma. \tag{15-20}$$

This result may be compared with the isothermal case, where $p_1V_1 = p_2V_2$. Figure 15-7 is a comparison of the isothermal and adiabatic expansions of 1 kmole of an ideal gas starting from STP. The adiabatic expansion results in a lower final pressure, which reflects the loss of the internal energy that has been converted into work by the expanding gas. This lost internal energy is not replenished from an outside reservoir, as it is in the isothermal expansion. Thus the gas temperature drops and, since V_2 is the same in both cases, the final pressure is correspondingly lower.

Example: The rapid filling of a balloon from a tank of compressed gas is approximately adiabatic; however, after the expanded gas has warmed up to its original temperature we can regard the entire process as isothermal. Consider a 10-ft^3 tank that contains helium at a pressure of 20 atm. Find the volume of the balloon immediately after filling, and at some time later. [*Note:* This balloon is one used to make measurements in the upper atmosphere.]

Solution: When the helium has filled the balloon, its pressure will have dropped to 1 atm. From Eq. (15-20),

$$p_1V_1^\gamma = p_2V_2^\gamma,$$

$$V_2 = \left(\frac{p_1}{p_2}\right)^{1/\gamma} V_1,$$

$$\log V_2 = \frac{1}{\gamma}\log\left(\frac{p_1}{p_2}\right) + \log V_1.$$

Here $p_1 = 20$ atm, $p_2 = 1$ atm, and $V_1 = 10\text{ ft}^3$. According to Table 15–2, the value of γ, the ratio of specific heats, is 1.66 for helium, and so

$$\log V_2 = \frac{1}{1.66}\log 20 + \log 10$$

$$= \frac{1.301}{1.66} + 1.000 = 1.784;$$

$$V_2 = 61\text{ ft}^3.$$

After the adiabatic expansion the helium temperature is, from Eq. (15–19),

$$T_2 = T_1\left(\frac{V_1}{V_2}\right)^{\gamma-1},$$

$$\log T_2 = \log T_1 + (\gamma - 1)\log\frac{V_1}{V_2}\cdot$$

If the atmospheric temperature is 300°K (27°C or 81°F), T_1 has this value, and substituting for T_1, V_1 and V_2 in the above equation yields $T_2 = 92$°K (−181°C or −295°F). The adiabatic expansion is very effective in cooling the gas. After a while the helium will absorb heat from the atmosphere and return to its original temperature of 300°K, so that the isothermal formula $p_1V_1 = p_2V_2$ applies to the whole process. Hence

$$V_2 = \left(\frac{p_1}{p_2}\right)V_1 = \frac{20\text{ atm}}{1\text{ atm}} \times 10\text{ ft}^3 = 200\text{ ft}^3.$$

The greater part of the total expansion occurs during the warming of the helium after it has expanded adiabatically. The volume of the balloon is 51 ft³ and 190 ft³ after each process since it is still attached to the 10 ft³ helium tank.

From the point of view of the kinetic theory it is clear that the gas temperature must increase during an adiabatic compression and decrease during an adiabatic expansion. (In terms of conservation of energy, of course, this occurs because during a compression work is done *on* the gas, while in an expansion work is done by the molecules of the gas.) If the piston moves inward with velocity $-\mathbf{v}$, and the velocity of a molecule with respect to the cylinder walls is $\mathbf{c}$ (Fig. 15–8), the relative velocity of the molecule with respect to the piston is $\mathbf{c} + \mathbf{v}$. The molecule

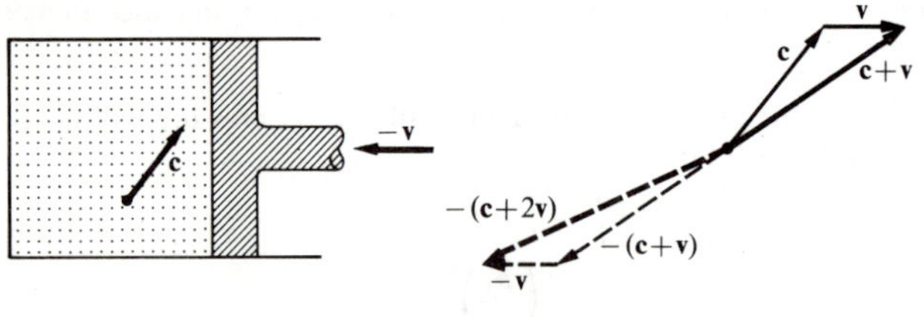

FIG. 15–8. A molecule gains a velocity 2**v** when colliding with a piston compressing the gas that moves with the velocity **v**.

will rebound from the piston after the two collide with the relative velocity $-(\mathbf{c} + \mathbf{v})$, so that its new velocity with respect to the container is $-(\mathbf{c} + 2\mathbf{v})$. Thus each gas molecule increases in energy as a result of collisions with the piston as it moves into the cylinder, and this energy increase manifests itself as an increase in temperature. Similarly, molecules colliding with an outward-moving piston during an expansion lose kinetic energy, thus reducing the gas temperature.

Problems

1. One kmole of an ideal gas expands adiabatically. Express the work done during the expansion in terms of the specific heat at constant volume of the gas, assumed constant, and in terms of the initial and final temperatures.

2. One kmole of a gas at STP for which $\gamma = 1.3$ is adiabatically compressed to half its initial volume. Determine the ratio of the final pressure and temperature to the initial values.

3. (a) Does a gas perform the most work when it expands isobarically (that is, at constant pressure), isothermally, or adiabatically? (b) In which process does it perform the least work? (c) Which process requires the greatest heat input?

4. Find the work done by n moles of an ideal gas ($C_V = 3R/2$, $C_p = 5R/2$) when it expands to twice its initial volume (a) isobarically, (b) isothermally, and (c) adiabatically. Express the results in terms of the initial temperature T of the gas.

5. A gas-filled flask of volume V has a narrow neck of cross-sectional area A (Fig. 15–9). A ball of mass m is placed in the neck, and the ball comes to equilibrium with the gas. Assume that the compression of the gas is adiabatic for any displacement, and (a) show that the ball would oscillate in simple harmonic motion. [*Hint:* Calculate dp in terms of dV for an adiabatic process; relate dp to the force applied to the ball and relate dV to its displacement.) (b) Calculate its period. (c) If $V = 10$ liters, the cross-sectional area $2\ \text{cm}^2$, the initial pressure 10^6 dynes/cm^2, and if the ball has a mass of 0.30 gm, what is the frequency of oscillation for a monatomic gas?

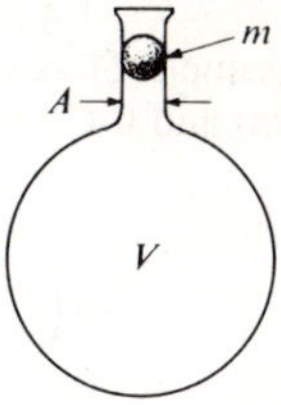

FIGURE 15–9

6. Show that, for a van der Waals gas, where $a/v^2 \ll p$ and c_V is independent of temperature, $T(v - b)^{R/c_V}$ is a constant in an adiabatic expansion.

7. Two identical flasks are filled with hydrogen, the left-hand one at 0°C and the right-hand one at 20°C (Fig. 15–10). A drop of mercury separates them. If both flasks are heated to increase their temperature 10°C, which way will the mercury move? If the cross section of the tube is 1 cm in diameter, and the volume is 0.1 liter, how much will it move?

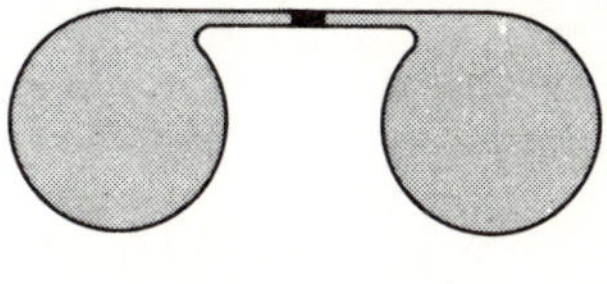

FIGURE 15–10

8. The volume of a gas at a pressure of 10^3 n/m^2 expands from 1 to 2 liters. Compare the pressure change in an adiabatic and an isothermal process.

9. What is the temperature change of an ideal gas when it is compressed adiabatically to one-third of its original volume, starting at 20°C?

10. A volume of oxygen of 10 gm is compressed adiabatically from 0°C at atmospheric pressure to a pressure of 5 atm. How much work is done on the gas? What is the new volume and new temperature?

11. How much heat is required to raise the temperature of a liter of ether from 50 to 60°C at atmospheric pressure? Repeat for chlorine and helium.

12. A 100-gm sample of oxygen is at a pressure of 3 atm and 0°C. Its final state is 1 atm and 300°C. Find the work done, the heat absorbed, and the change in internal energy for the following processes which bring the gas from its initial to its final state: (a) Constant pressure followed by constant volume, (b) constant volume followed by constant pressure. Assume that the sample behaves like an ideal gas.

13. If a chlorine molecule at 300°K were to rotate at an angular frequency predicted by the equipartition theorem, what would be the centripetal force? (Take the separation to be $= 2 \times 10^{-8}$ cm and the mass of the chlorine atom $= 6 \times 10^{-28}$ kg.)

14. If a chlorine molecule at 1000°K were to vibrate along its line of centers according to the equipartition theorem, what would the amplitude of the vibration be? Assume the effective spring constant of chlorine to be 2.5 n/m.

16

The Second Law of Thermodynamics

Heat can be transferred from one place to another in a variety of ways, but, as we know, it flows from regions of high temperature to regions of low temperature. To reverse this natural flow of heat—for instance, by extracting cold air from a box and exhausting it to warmer air outside—requires that work be done, in the above case by a machine called a refrigerator. The first law of thermodynamics gives us no indication that heat should exhibit this preferred direction of flow; like its counterpart in dynamics, the law of conservation of energy, the first law is invaluable for understanding certain aspects of the behavior of various systems, but by itself it does not go far enough.

There are many other processes that are entirely possible according to the first law, yet which never occur. For example, a piece of ice that is dropped to the ground from a sufficiently great height melts, the required heat coming from the conversion of its initial potential energy to kinetic energy and then into internal energy. Reversing this process would have a puddle of water spontaneously rising from the ground while turning into ice, acquiring kinetic energy from its latent heat of fusion as it freezes. Though an event of this kind is hardly likely, what we know thus far about thermodynamics does not prohibit it. The general theorem which has been found to govern the possibility of occurrence of such hypothetical processes is called the *second law of thermodynamics*, and we shall consider this law in some detail in this chapter.

16–1 The second law of thermodynamics

A statement of the second law of thermodynamics which specifically forbids the type of process which we have described above is due to Lord Kelvin. It states that *it is impossible by means of any inanimate agency to derive mechanical work from any portion of matter by cooling it below the lowest temperatures of its surroundings*. This statement takes care of the example given above. The puddle of water cannot freeze (be cooled below the temperature of the air) and do work (rise up). It also says that a boat cannot travel across the ocean by extracting energy from the internal energy of the ocean and thus cooling it. There are no prohibitions of these processes according to the first law of thermodynamics. The ocean at a finite temperature contains an immense amount of internal energy, but it is not available for conversion into work in this way.

An *engine* is a device for converting heat energy into work. The way a typical engine operates is to absorb heat from some substance or reservoir at high temperature, to permit the working substance to do work and thereby be cooled, and then to reject some heat to a reservoir at a lower temperature. For example, in a steam engine a boiler heats up the steam (absorbs heat from high-temperature reservoir); the steam then expands against a piston (does work) and finally rejects the cooler steam into the air (which is the reservoir at low temperature).

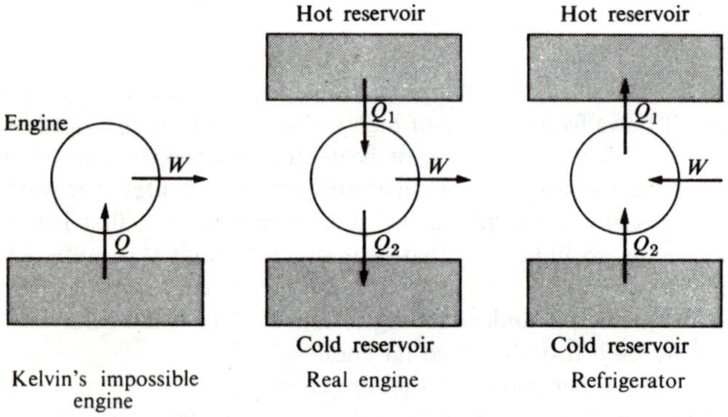

FIG. 16–1. Examples of some engines, real and impossible. A refrigerator is an engine running backward.

Figure 16–1 depicts both Kelvin's impossible engine and a real engine. An engine which is a continuous device is represented by a circle since it must operate in a cycle and its internal energy cannot undergo a net change. By the first law of thermodynamics, $Q_1 - Q_2 = W$ for a real engine. A refrigerator, also depicted in Fig. 16–1, is an engine which runs backward. Work is done on the refrigerator, and the net effect is to cool a cooler reservoir (a room, for instance, if the refrigerator is an air conditioner) and heat a hotter reservoir (the outdoors).

Another statement of the second law of thermodynamics, equivalent to the Kelvin statement, is one due to Clausius: *It is impossible for a self-acting machine to convey heat continuously from one body to another which is at a higher temperature.* A refrigerator which does convey heat from a lower to a higher temperature does not violate this statement of the law, since it is not self-acting but has work done on it. The Clausius statement is merely an expression of the result of our experience that heat energy goes from hot to cold bodies, but not the other way.

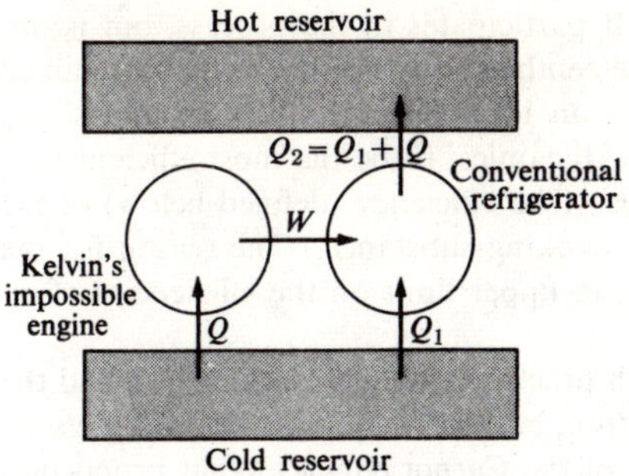

FIG. 16-2. Kelvin's statement of the second law implies the Clausius statement.

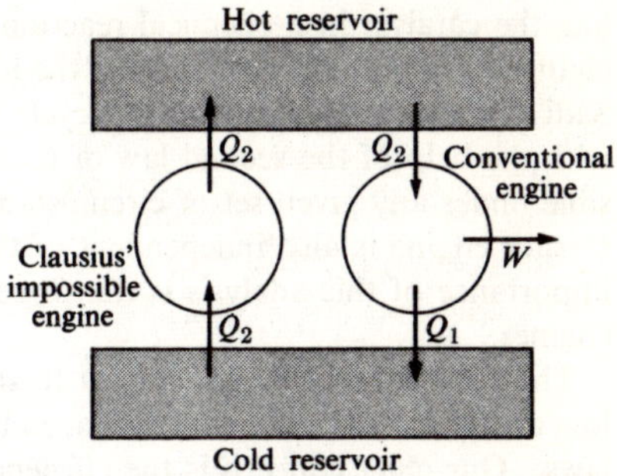

FIG. 16-3. The Clausius statement of the second law implies the Kelvin statement.

The first law of thermodynamics tells us nothing about the direction of heat flow. The second law does.

One can show that the Kelvin and Clausius statements are equivalent to one another by proving that if the Kelvin statement is not true, neither is the Clausius statement. Then we show that if the Clausius statement is not true, neither is the Kelvin statement.

Suppose that the Kelvin statement is false and that we have an engine which removes heat from a reservoir and does work. We now permit the impossible Kelvin engine to run a conventional engine backward as a refrigerator (Fig. 16-2), removing heat Q_1 from the cold reservoir and delivering Q_2 to the hot reservoir. Since the internal energy of these engines remains the same, $W = Q$ and $W = Q_2 - Q_1$. Therefore $Q_2 = Q_1 + Q$. The combination of the two devices is a self-acting device which removes a quantity of heat Q_2 from a cold reservoir and delivers it to the hot one in violation of the Clausius statement.

Assuming that the Clausius statement is false, we construct an engine which takes Q_2 (Fig. 16-3) from a cold reservoir and delivers it to a hot one, and we operate it simultaneously with a conventional engine which removes Q_2 from a hot reservoir, does work W, and delivers Q_1 to the cold reservoir. The combination of these devices removes heat $Q_2 - Q_1$ from a cold reservoir and does work W. The two devices working together therefore violate the Kelvin statement.

16-2 The Carnot engine

A heat engine is a device which converts heat energy from a reservoir to mechanical energy; by the second law of thermodynamics, it must also exhaust some of its heat input to another reservoir at a lower temperature than the first. A heat engine usually operates in a cycle which takes some working substance through successive changes in pressure, volume, and temperature that periodically bring it back to its initial state. Such a working substance functions, in a sense, like the catalyst in a chemical reaction: It participates in the process, but is unchanged by it. In the early part of the nineteenth century a French engineer named Sadi Carnot devised a working cycle for an ideal engine which we can show, with the help of the second law of thermodynamics, to be the most efficient possible under any given set of circumstances. The efficiency (defined below) of the Carnot engine is also independent of the working substance. The reason for the importance of this analysis is that it sets an upper limit on the efficiencies of all engines.

The *Carnot engine* is not subject to such practical difficulties as friction and the loss of stored heat by conduction or radiation, but naturally must obey all physical laws. Our chief concern is the efficiency of the Carnot engine: What proportion of the heat supplied to it can it transform into mechanical energy? Of course, all the effects we are neglecting reduce the performance of actual engines below that of the Carnot engine, but it is of great interest to find out the ultimate limits on engine efficiency.

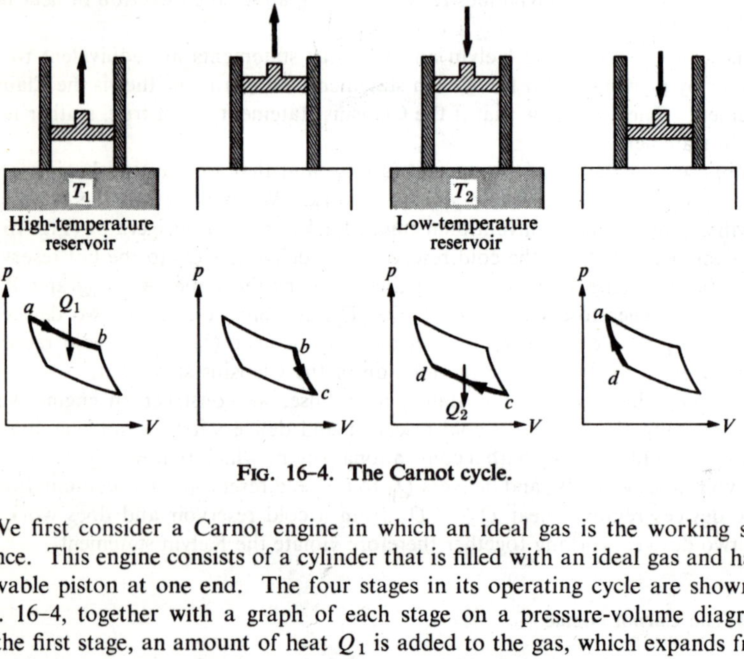

FIG. 16–4. The Carnot cycle.

We first consider a Carnot engine in which an ideal gas is the working substance. This engine consists of a cylinder that is filled with an ideal gas and has a movable piston at one end. The four stages in its operating cycle are shown in Fig. 16–4, together with a graph of each stage on a pressure-volume diagram. In the first stage, an amount of heat Q_1 is added to the gas, which expands from its initial volume V_a to the final volume V_b while remaining at its original temperature T_1. During this isothermal expansion, the pressure drops from p_a to p_b. Then the heat source is removed, and the gas is allowed to expand further adiabatically to V_c, which involves a further pressure drop to p_c. This second expansion takes place at the expense of the energy stored in the gas as heat, and so its temperature falls from T_1 to T_2. During these expansions the piston exerts a force on whatever it is attached to, and thereby performs work.

Having done work in pushing the piston outward, the engine must now be returned to its initial state in order for it to be able to do further work. The third stage in the operation of a Carnot engine involves an isothermal compression of the gas at the constant temperature T_2, during which an amount of heat Q_2 is given off. The volume of the gas is reduced to V_d while its pressure increases to p_d. In the final stage, the gas is returned to its original temperature, pressure, and volume by an adiabatic compression during which heat is neither added to it nor removed from it. In compressing the gas a force had to be exerted on the piston, which means that work had to be done on the engine by some external agency.

In its cycle the Carnot engine performs some net amount of work W, which is the difference between the work it does during the two expansions and the work done on it during the two compressions. The work done is proportional to the area enclosed in a p-V plot of the cycles. It has taken in the heat Q_1 and ejected the heat Q_2; we observe that the heat Q_2 *must* be ejected if the engine is to return to its initial state to await another cycle. The efficiency of any engine is

$$\text{Eff} = \frac{\text{work output}}{\text{energy input}}, \tag{16-1}$$

so that for a Carnot engine,

$$\text{Eff} = \frac{W}{Q_1}. \tag{16-2}$$

According to the first law of thermodynamics,

$$W = Q_1 - Q_2,$$

since at the end of the cycle the gas has the same properties it began with. Hence

$$\text{Eff} = \frac{Q_1 - Q_2}{Q_1} = 1 - \frac{Q_2}{Q_1}. \tag{16-3}$$

The smaller the ratio of the ejected heat Q_2 to the absorbed heat Q_1, the more efficient the engine.

We can evaluate the ratio Q_2/Q_1 in terms of the temperatures T_1 and T_2 of the high- and low-temperature reservoirs between which the engine operates. During the isothermal expansion from a to b there is no change in the internal energy of the gas, and so all of the heat Q_1 goes into work done in the expansion. Hence, from Eq. (15-17),

$$Q_1 = \int_a^b p\,dV = nRT_1 \ln \frac{V_b}{V_a}.$$

Similarly, the heat Q_2 given off by the engine during the isothermal compression from c to d is

$$Q_2 = -\int_c^d p\,dV = nRT_2 \ln \frac{V_c}{V_d}.$$

The ratio Q_2/Q_1 is therefore

$$\frac{Q_2}{Q_1} = \frac{T_2}{T_1}\left[\frac{\ln(V_c/V_d)}{\ln(V_b/V_a)}\right].$$

Now we draw on our knowledge of the adiabatic portions of the Carnot cycle. For the adiabatic expansion from b to c we have, from Eq. (15–19),

$$T_1 V_b^{\gamma-1} = T_2 V_c^{\gamma-1},$$

and for the adiabatic compression from d to a,

$$T_2 V_d^{\gamma-1} = T_1 V_a^{\gamma-1}.$$

Hence

$$V_b = V_c \left(\frac{T_2}{T_1}\right)^{1-\gamma},$$

$$V_d = V_a \left(\frac{T_1}{T_2}\right)^{1-\gamma},$$

and

$$\left[\frac{\ln(V_c/V_d)}{\ln(V_b/V_a)}\right] = \left[\frac{\ln(V_c/V_a)(T_2/T_1)^{1-\gamma}}{\ln(V_c/V_a)(T_2/T_1)^{1-\gamma}}\right] = 1.$$

Thus

$$\frac{Q_2}{Q_1} = \frac{T_2}{T_1}; \tag{16–4}$$

the heat Q transferred to or from a Carnot engine is directly proportional to the absolute temperature T of the reservoir with which it is in contact. The efficiency of a Carnot engine may therefore be written

$$\text{Eff} = 1 - \frac{T_2}{T_1}. \tag{16–5a}$$

The smaller the ratio between T_2 and T_1, the more efficient the engine. Another way of expressing the efficiency is

$$\text{Eff} = \frac{T_1 - T_2}{T_1} = \frac{\Delta T}{T_1}. \tag{16–5b}$$

For an engine, the greater the temperature differences between the reservoirs and the colder the reservoir at the lower temperature, the greater the efficiency of the engine.

16-3 Efficiency of a Carnot engine

We can now show that all Carnot engines, independent of their working substance, must have the same efficiency if they operate between the same two reservoirs. We show this by proving that if it were not so, the Clausius statement of the second law of thermodynamics would be violated.

Suppose that we have two such engines and that the efficiency E_1 of the first is greater than the second E_2. Since the Carnot engine is reversible, we reverse the second and let the first drive it as a refrigerator (Fig. 16-5). The efficiencies of the engines are, by Eq. (16-3),

$$E_1 = \frac{Q_1 - Q_2}{Q_1}; \qquad E_2 = \frac{Q_1' - Q_2'}{Q_1'}.$$

Since W is the same for both engines,

$$W = Q_1 - Q_2 = Q_1' - Q_2'. \quad (16\text{-}6)$$

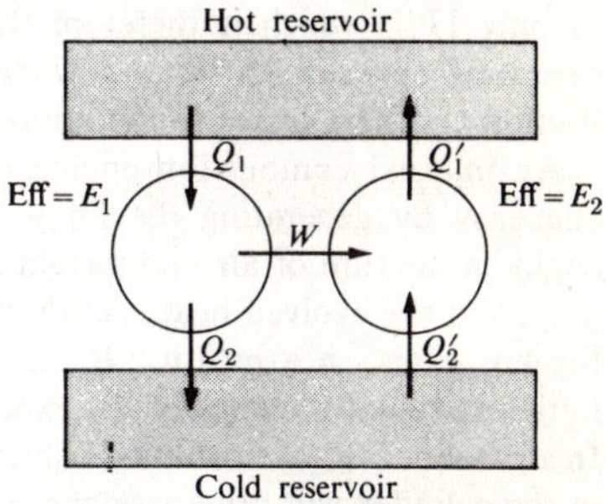

FIG. 16-5. All Carnot engines operating between the same two reservoirs have the same efficiencies.

By hypothesis, $E_1 > E_2$, and

$$\frac{Q_1 - Q_2}{Q_1} = \frac{W}{Q_1} > \frac{W}{Q_1'}, \qquad \frac{1}{Q_1} > \frac{1}{Q_1'}, \qquad Q_1' > Q_1. \quad (16\text{-}7)$$

From Eqs. (16-6) and (16-7), $Q_2' > Q_2$. Therefore

$$Q_2' - Q_2 > 0$$

and

$$Q_1' - Q_1 > 0.$$

But $Q_2' - Q_2$ is the amount of heat removed from the cold reservoir and $Q_1' - Q_1$ is the amount of heat delivered to the hot reservoir. Thus the two engines together remove heat from a cold reservoir and deliver it to a hot one without work of external origin being done—a violation of the Clausius statement of the second law. Thus our hypothesis is incorrect and $E_2 \geq E_1$. Similarly, if engine 2 were to run engine 1 backward, one could prove $E_2 \leq E_1$. Thus the two efficiencies must be equal, $E_1 = E_2$.

If we had an irreversible engine, by permitting it to run a Carnot engine backward we could prove $E_{\text{Carnot}} \geq E_{\text{irreversible}}$. Since we cannot reverse an irreversible engine, we see that the Carnot engine is the most efficient that can be run between two reservoirs. This is a surprising result since the maximum theoretical engine efficiencies are so low.

16–4 Practical engines

Practical heat engines have low efficiencies. One reason for their inability to make more use of their energy inputs is the unavailability of heat reservoirs at sufficiently high and low absolute temperatures for the ratio T_2/T_1 to be small. Steam enters a typical steam engine at about 450°K and is discharged at about 373°K. A Carnot engine operating between these temperatures has an efficiency of only 17%, which is therefore the upper limit to the efficiency of *any* engine operating between 450°K and 373°K. And, of course, the inevitable presence of friction and heat losses to the atmosphere reduces the efficiency even further.

An internal combustion engine is able to achieve a relatively high operating efficiency by generating the input heat within the engine itself. In a gasoline engine a mixture of air and gasoline vapor is ignited in each cylinder by a spark plug, and the evolved heat is converted into mechanical energy by the pressure of the hot gases on a piston. It can be shown that the greater the ratio between initial and final volumes of the expanding gases, the greater the engine efficiency. In a gasoline engine this ratio is limited to 7 or 8 to 1, since the gasoline-air mixture in the cylinder will otherwise spontaneously ignite during its compression before the end of the stroke is reached. The more efficient Diesel engine circumvents this difficulty by compressing only air and injecting fuel oil into the hot, compressed air at the instant the piston has reached the end of its travel. No spark plug is required. The compression ratio in a Diesel engine might be 16 to 1. The effective values of T_1 and T_2 in a modern Diesel engine are perhaps 1800°K and 800°K, respectively, for a Carnot efficiency of 56%.

16–5 Absolute thermodynamic temperature scale

According to Eq. (16–4),

$$\frac{Q_2}{Q_1} = \frac{T_2}{T_1}.$$

Thus the ratio between the heat input and heat output of a Carnot engine is equal to the ratio between the temperatures of the input and output reservoirs, a result independent of the working substance. This feature of a Carnot engine permits us to define in principle a temperature scale which, unlike the other ones we have considered, is independent of any specific properties of matter, such as coefficients of thermal expansion and the conditions under which changes of state occur. This scale is called the *absolute thermodynamic* or *Kelvin* temperature scale. Since the temperature 273.15°K is assigned to the freezing point of water, the absolute thermodynamic and the absolute celsius scales are identical if we take the size of the Kelvin degree as $\frac{1}{100}$ of the difference between the temperature of boiling water and ice at STP. The absolute scale has the important distinction, as we said, of being independent of the properties of matter.

According to Eqs. (16–5), a reservoir at absolute zero would permit us to operate a Carnot engine with perfect efficiency, exhausting no energy to the reservoir. However, the second law of thermodynamics prohibits the existence of an engine which completely converts its heat input into work; hence a reservoir at absolute zero would violate the second law. Methods have been developed for reaching temperatures of less than 1°K, but it is impossible for a temperature of 0°K to be achieved.

16–6 Entropy

The second law of thermodynamics is rather awkward to describe, and we should like to be able to express it mathematically. In order to do this, we introduce a new concept, called *entropy*. We can describe the second law in terms of changes in entropy. The quantity

$$dS = \frac{dQ}{T} \tag{16–8}$$

is called the change in entropy of a working substance during a reversible process involving the transfer of an amount of heat dQ at a temperature T. If the process is irreversible, we must calculate dS by making the same changes in the system by an equivalent reversible process.

We first calculate the change of entropy of a substance undergoing a Carnot cycle. There are no changes in the entropy in the adiabatic portions of the cycle, since for these $dQ = 0$. During the isothermal expansion of Fig. 16–4, then, the entropy change is

$$\Delta S_1 = \frac{Q_1}{T_1}. \tag{16–9}$$

During the isothermal compression, the entropy change is

$$\Delta S_2 = -\frac{Q_2}{T_2}. \tag{16–10}$$

The entropy change is negative during the compression because heat leaves the system. From Eq. (16–4), we have that

$$\frac{Q_1}{T_1} = \frac{Q_2}{T_2}. \tag{16–11}$$

The total entropy change, by virtue of Eqs. (16–9), (16–10), and (16–11) is therefore

$$\Delta S = \Delta S_1 + \Delta S_2 = 0. \tag{16–12}$$

A complete Carnot cycle involves no change in entropy. Now a Carnot engine is the most efficient possible engine operating between T_1 and T_2. An actual engine which operated between reservoirs at these temperatures and had the same heat input Q_1 would have a larger heat exhaust Q_2, and therefore

$$|dS_2| > |dS_1|$$

or

$$dS > 0. \tag{16–13}$$

During the operation of any engine less efficient than a Carnot engine, there is an increase in entropy. Equations (16–12) and (16–13) can be combined in the form

$$dS \geq 0. \tag{16–14}$$

The statement that the change in entropy for any process [Eq. (16–14)] is greater than or equal to zero is a consequence of the second law of thermodynamics, and is equivalent to it. We have demonstrated it only for a Carnot cycle, but the result is of much greater generality. The second law of thermodynamics is equivalent to the statement that for any process in the universe the entropy cannot decrease. Of course we can decrease the entropy of a portion of the universe, as, for example, during the isothermal compression of the gas in a Carnot cycle. Such a process is accompanied, however, by a corresponding increase in other parts of the universe (for example, the heat reservoirs), and the net result is that the entropy of the entire system—gas plus reservoirs—increases, and Eq. (16–14) holds.

16–7 The second law and kinetic theory

It is interesting to consider the meaning of the second law of thermodynamics on the molecular level. Let us consider a heat reservoir at a high temperature and a heat reservoir at a low temperature. The molecules of each are in equilibrium with the molecular energy distributions shown in Fig. 16–6(a) and (b). If we consider the two reservoirs as a single system, the molecular energy distribution in the system is like that of Fig. 16–6(c). This distribution is, in a statistical sense, very improbable; if they are mixed together, the molecules of the two reservoirs would soon blend their energies in collisions to attain the equilibrium distribution of Fig. 16–6(d), which corresponds to a temperature intermediate between the initial ones of the two reservoirs. We note the important fact that the

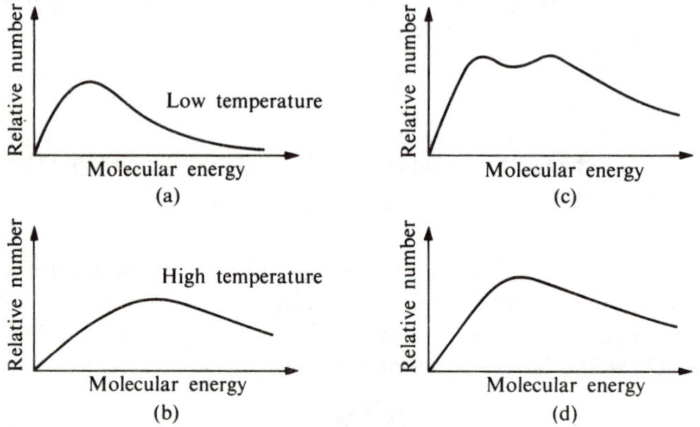

FIGURE 16–6

total energy contents of the distributions of Fig. 16–6(c) and (d) are identical; the only distinction between them is the manner in which the energy is allotted to the molecules on the average. However, it is just this distinction with which the second law of thermodynamics is concerned, because this law states that a system of two heat reservoirs at different temperatures can be made to yield a net work output by an engine operating in a cycle, while a single heat reservoir, no matter how much energy it contains, cannot be made to perform any net work using such an engine. A system of molecules whose energies are distributed in the most probable way is "dead" thermodynamically, while a system having a different distribution of molecular energies is capable of doing mechanical work.

The distribution of molecular velocities represented by Fig. 16–6(c) represents a more "ordered" one than the distribution of Fig. 16–6(d). The concept of entropy on a molecular level is closely related to the concept of order. The more disordered the state, the greater its entropy. We shall not go into this question in any more detail here. It is, however, possible to define the concept of order mathematically and to define entropy precisely in terms of this concept.

The increasing entropy of the universe considered as a single system of molecules would seem to predict that its lifetime is finite. We may consider that the evolution of the universe is a result of the flow of heat from high-temperature reservoirs (the stars) to low-temperature reservoirs (everything else). The creation of new high-temperature stars is always at the expense of other sources of entropy in the universe if the second law of thermodynamics is to be believed. Ultimately the entire universe will be at the same temperature, and all its constituent particles will have the same average energy, a condition sometimes called the "heat death" of the universe.

Problems

1. What is the maximum possible efficiency of an engine that obtains heat at 200°C and exhausts heat at 80°C?

2. An engine operating between 300°C and 60°C is 15% efficient. What would its efficiency be if it were a Carnot engine?

3. A Carnot engine takes in 10^3 kcal of heat from a reservoir at 327°C and exhausts heat to a reservoir at 127°C. How much work does it do?

4. One of the most efficient engines ever developed operates between about 2000°K and 700°K. Its actual efficiency is 40%. What percentage of its maximum possible efficiency is this?

5. Three designs for a heat engine to operate between 450°K and 300°K are proposed. Design *A* is claimed to require a heat input of 0.2 kcal for each 1000 j of work output, design *B* a heat input of 0.6 kcal, and design *C* a heat input of 0.8 kcal. Which design would you choose and why?

6. The sun's corona is a very dilute gas at a temperature of about 10^6°K that is believed to extend into interplanetary space at least as far as the earth's orbit. Why

can we not use the corona as the high-temperature reservoir of a heat engine in an earth satellite?

7. An attempt is made to cool a kitchen during the summertime by leaving the refrigerator door open and closing the kitchen door and windows. What will happen and why?

8. In an attempt to cool a room in the summer, a man turns on an electric fan and leaves the room. Will the room be cooler when he returns?

9. A Carnot refrigerator (which is a Carnot engine operating in reverse) extracts heat from a freezer at $-5°C$ and exhausts it at 25°C. How much work per kcal of heat extracted is required?

10. A Carnot refrigerator is used to make 1 kg of ice at $-10°C$ from 1 kg of water at 20°C, which is also the temperature of the kitchen. How many joules of work must be done?

11. Plot a Carnot cycle with temperature as ordinate and volume as abscissa.

12. We have shown that, for a Carnot cycle,

$$\frac{Q_1}{T_1} + \frac{Q_2}{T_2} = 0.$$

It is true that

$$\oint \frac{dQ}{T} = 0$$

for any closed reversible cycle. Indicate how this might be demonstrated.

13. An ideal gas, $\gamma = 1.50$, is used as the working substance in a cylinder undergoing a Carnot cycle. The temperature of the hot reservoir is 240°C. At this temperature it is expanded isothermally from a pressure of 10 atm and a volume of one liter to a pressure of 4 atm and a volume of 2.5 liters. (a) Between what limits of pressure and volume does the gas operate with a cold reservoir at a temperature of 50°C? (b) Calculate the heat taken from the hot reservoir and rejected to the cold one (at 50°C) in each cycle. (c) Calculate the efficiency of the reversible engine. (d) The piston area is 200 cm^2 and a force of 1 n is needed to overcome friction between the piston and the cylinder. Calculate the reduction in efficiency.

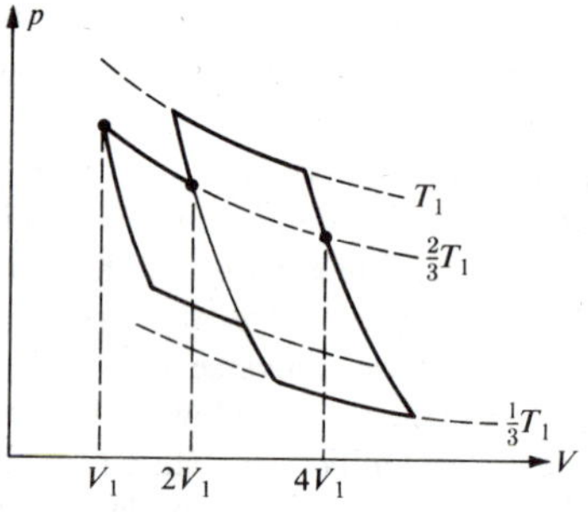

FIGURE 16–7

14. Determine the efficiency of the cycle (Fig. 16–7) which consists of adiabatics followed by isothermals. Compare the efficiency of an engine operating in this cycle to a single Carnot cycle operating between the highest and lowest temperatures.

15. An electric refrigerator has a heat leakage of 3 kcal/min. The cooling element is at $-10°C$ and the radiator at 40°C. Its efficiency is 40% of that of a Carnot engine between these reservoirs. What is the power consumption?

16. An engine runs at 650 rev/min and takes in 10 kcal at each revolution. The temperature of the boiler is 180°C and that of the exhaust 100°C. Given that its efficiency is 48% of that of an ideal Carnot engine, compute the power in kilowatts.

17

Electricity and Magnetism

In our study of the properties of matter we found it necessary to conclude that molecules, instead of behaving as do the ideal billiard balls of the simple kinetic theory of gases, actually do exert forces on one another. The force between two molecules is attractive when they are some distance apart and repulsive when they are close together. The attractive force is too large to be gravitational in nature, and of course the repulsive force cannot be accounted for in this way either. We now know that these are examples of *electrical forces*, whose importance follows from the facts that particles having electrical properties are the basic constituents of all matter, and that electrical forces, on the whole, are much stronger than gravitational ones. While gravity significantly affects the behavior of only relatively large aggregates of matter, electrical forces predominate in the world of microscopic phenomena. Magnetism, which, as we shall see, is closely related to electricity, manifests itself everywhere—from the interior of the atom to the vast reaches of galactic space. A knowledge of electricity and magnetism is essential to the physical scientist if he is to understand the workings of the universe.

17–1 Conservation of charge

Despite the fact that electrified particles compose all matter, electrical forces seldom spontaneously display themselves. Something must be done to matter before its electrical properties become evident. Since the time of Thales it has been known that, if a piece of amber is rubbed with some fur, the amber can be made to attract light objects such as bits of paper or pith balls. If we rub a glass rod with a silk cloth, we can produce a similar effect. The amber and glass in their new states are said to be electrified, and they can communicate some of their electrification to other bodies by direct contact. Thus we can perform experiments to determine the properties of electrification. For example, we might touch an amber rod that has been rubbed with fur to each of two pith balls, and then bring the balls close together. What we observe is that the two balls *repel* each other strongly, and if they are attached to strings, as in Fig. 17–1, they will stay apart despite the gravitational force tending to pull them down and hence together. If we electrify two similar balls with a glass rod that has been rubbed with silk, the same thing happens. However, if one pith ball is electrified by the amber and the other one by the glass, the two are found to *attract* each other.

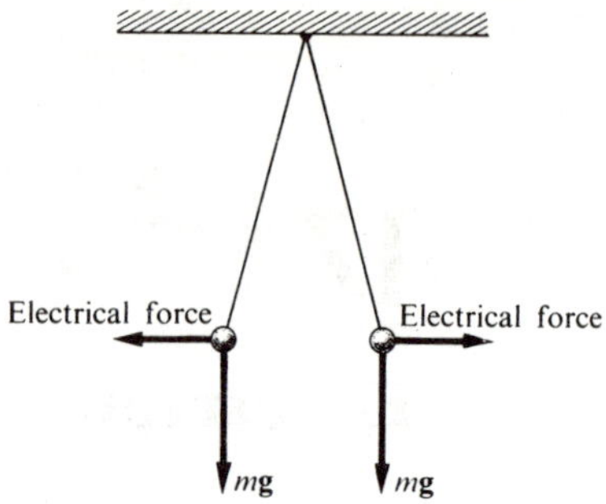

FIG. 17-1. Similarly charged pith balls repel each other.

Experiments such as the ones described above were carried out in the eighteenth century, and led to the conclusion that there are two different kinds of electricity. No matter how electrification was produced, it always behaved as if it had originated either in a glass or in an amber rod, and these types were called vitreous and resinous electricity, respectively. *There is a repulsive force acting between bodies similarly electrified, while if they are oppositely electrified, the force is an attractive one.* In contrast, gravitational forces are always attractive.

Benjamin Franklin arrived at a highly significant conclusion on the basis of some experiments he performed. He noted that when amber is rubbed with fur, the fur as well as the amber becomes electrified. Furthermore, the fur is electrified vitreously, while the amber becomes electrified resinously. The same effect, although reversed, occurs in the case of glass and silk cloth. Franklin surmised that equal amounts of each kind of electricity are always produced, and that the effect of the rubbing was to separate them. Actually, Franklin's own experiments were hardly quantitative enough to permit so sweeping a statement, but more precise measurements have proven his hypothesis correct. We may summarize this discovery by saying that electricity can be neither created nor destroyed, but equal amounts of the two different types can be separated.

Today we speak of an electrified body as being *charged*, with *positive charge* corresponding to vitreous electrification and *negative* charge to resinous electrification. This usage is very convenient, since we can determine the exact amount and kind of charge present somewhere merely by adding the various charges together algebraically. When the net charge on a body is zero, it is called neutral. A neutral body remains that way when equal amounts of positive and negative charge are either added or removed. We may restate the conclusion of the previous paragraph by saying that *charge is conserved.* The conservation of charge, together with the other conservation laws we have already learned and are about to learn, is one of the foundation stones underlying all physical theory.

17-2 Coulomb's law

By means of a direct measurement, Charles Coulomb (1736–1806) discovered that the force one charge exerts upon another in air is directly proportional to the product of their magnitudes and inversely proportional to the square of their

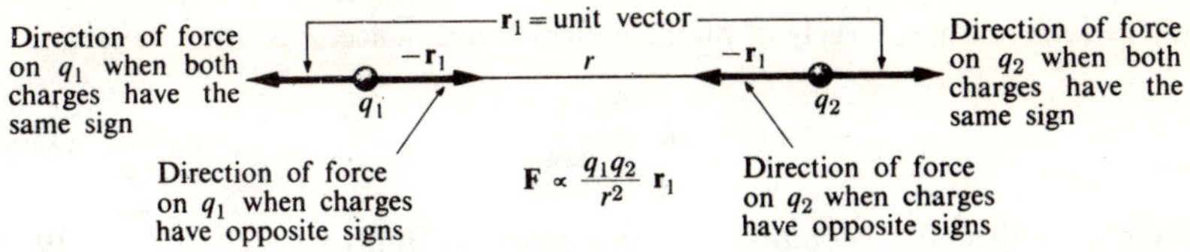

FIG. 17–2. Coulomb's law.

separation (Fig. 17–2). That is,

$$\mathbf{F} = k\frac{q_1 q_2}{r^2}\mathbf{r}_1, \tag{17–1}$$

where k is a constant of proportionality and r is the distance between the two charges q_1 and q_2; $\mathbf{r}_1$ is a unit vector in an outward sense along the line joining the charges. Equation (17–1) is called Coulomb's law, and it expresses the fact that the force between q_1 and q_2 is in the $+\mathbf{r}_1$ direction (which implies repulsion) when q_1 and q_2 have the same sign, while it is in the $-\mathbf{r}_1$ direction (which implies attraction) when the two charges have opposite signs.

The law of force between charges is similar to the law of gravitation, in that both electrical and gravitational forces are inversely proportional to the square of the distance between the centers of force, and in that in both cases the forces are exerted along the radius vector joining the centers of force. They differ, of course, in that gravitational forces are always attractive. Because electrical forces may be repulsive as well as attractive, the effects of a particular charge may be canceled out by other charges. Hence we can shield against electrical phenomena, while we cannot against gravitational ones.

We must now provide units for the various quantities appearing in Eq. (17–1), and establish the value of k. In the cgs system, $\mathbf{F}$ is in dynes and r in centimeters, and the unit of charge is defined by taking $k = 1$. The resulting *electrostatic unit of charge*, or esu, is that identical charge on each of two bodies that results in a force between them of 1 dyne when their separation is 1 cm. The esu, although buttressed by tradition and still often used, is far too small for practical applications. In addition, electricity and magnetism are intimately related, and it is awkward to use the cgs system of units in applications involving both. A much better scheme is to introduce an arbitrary unit of charge, the *coulomb* (abbreviated *coul*), and to use it with mks mechanical units. The coulomb has the virtue of being part of the "practical" system of electrical units used in all engineering work. The relationship between the coulomb and the esu of charge is

$$1 \text{ coul} = 2.9979 \times 10^9 \text{ esu}.$$

Expressing $\mathbf{F}$ in newtons, r in meters, and q in coulombs, the constant k of Eq. (17–1) equals 8.9874×10^9 n·m^2/coul^2, but for most purposes it is adequate to let

$$k = 9 \times 10^9 \text{ n·m}^2/\text{coul}^2. \tag{17–2}$$

For various reasons, largely of mathematical convenience, k is often written

$$k = \frac{1}{4\pi\epsilon_0}, \tag{17–3}$$

where ϵ_0, called the *permittivity of free space*, is approximately 8.85×10^{-12} $\text{coul}^2/\text{n}\cdot\text{m}^2$. We shall adopt this usage, rewriting Coulomb's law as

$$\mathbf{F} = \frac{1}{4\pi\epsilon_0}\frac{q_1 q_2}{r^2}\mathbf{r}_1. \tag{17–4}$$

It is often puzzling for the student to understand how Coulomb was able to discover the law which bears his name, and then have this law define the unit of charge. How did he know that the force was proportional to the charge when he could not measure the unit of charge? The first thing Coulomb determined was that the force using the *same* charges varied inversely as the square of the distance. Next he determined that if he put together two equal charges, the force of attraction to a third charge which remained the same doubled, thus establishing that the force was proportional to the first and not to some other power of the charge. Equal charges could be determined as those charges that at a fixed distance from a third charge gave the same force.

The coulomb is an extremely large unit. The force between two charges of 1 coul each a distance of 1 m apart is

$$F = \frac{9 \times 10^9\,\text{n}\cdot\text{m}^2/\text{coul}^2 \times 1\,\text{coul} \times 1\,\text{coul}}{1\,\text{m}^2}$$

$$= 9 \times 10^9\,\text{n},$$

which is about 2 billion lb! Because electrostatic forces are so powerful, even the most highly charged bodies that can be produced seldom contain more than a minute fraction of a coulomb.

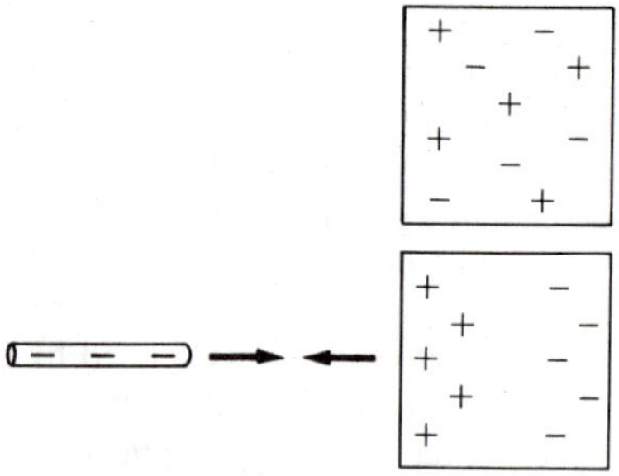

FIG. 17–3. How a charged body attracts an uncharged one.

Let us return for a moment to the first experiment we mentioned, in which a charged amber rod attracted bits of paper. Since the paper bits were originally uncharged, how could the rod exert a force on them? The explanation is shown in Fig. 17–3. When the negatively charged rod is brought near the paper, some of the negative charges in the paper which are not tightly bound in place move as far away as they can from the rod, while some of the positive charges which are not tightly bound move toward the rod. Because electrical forces vary inversely with distance, the attraction between the rod and the closer positive charges is greater than the repulsion between the rod and the farther negative charges, and so the paper moves toward the rod. Only a small amount of charge separation actually occurs; and so, with little force available, only very light objects can be attracted in this way.

Example: The Bohr model of the hydrogen atom consists of an electron rotating about a proton (Fig. 17-4). The charge on the electron is -1.60×10^{-19} coul and that on the proton is $+1.60 \times 10^{-19}$ coul. If the radius of the electron's orbit is 0.53×10^{-10} m, find the angular velocity the electron must have for the electrostatic force on it to equal the centripetal force required for its motion in such an orbit. The electron mass is 9.11×10^{-31} kg.

Solution: Equating the electrostatic and centripetal forces on the electron, we have

$$-\frac{1}{4\pi\epsilon_0}\frac{e^2}{r^2} = -m\omega^2 r,$$

where we have designated the charge on the electron as e. The minus signs on both sides signify an inward force, i.e., toward the proton. Solving for the angular velocity ω, we have

$$\omega = \left(\frac{1}{4\pi\epsilon_0}\frac{e^2}{r^3 m}\right)^{1/2}$$

$$= \left(9 \times 10^9 \frac{\text{n}\cdot\text{m}^2}{\text{coul}^2} \times \frac{(1.60 \times 10^{-19}\ \text{coul})^2}{(0.53 \times 10^{-10}\ \text{m})^3 \times 9.11 \times 10^{-31}\ \text{kg}}\right)^{1/2}$$

$$= 4.12 \times 10^{16}\ \text{rad/sec}.$$

When more than two charges are present, the force on any one of them may be calculated by vectorially adding the forces exerted by each of the others. This principle of superposition is valid whenever electric or magnetic forces act.

Example: Three charges, $q_1 = +7 \times 10^{-5}$ coul, $q_2 = -4 \times 10^{-5}$ coul, and $q_3 = -5 \times 10^{-5}$ coul, are placed as shown in Fig. 17-5 at the vertices of a triangle whose sides measure 3, 4, and 5 cm. Determine the magnitude and direction of the force on q_3.

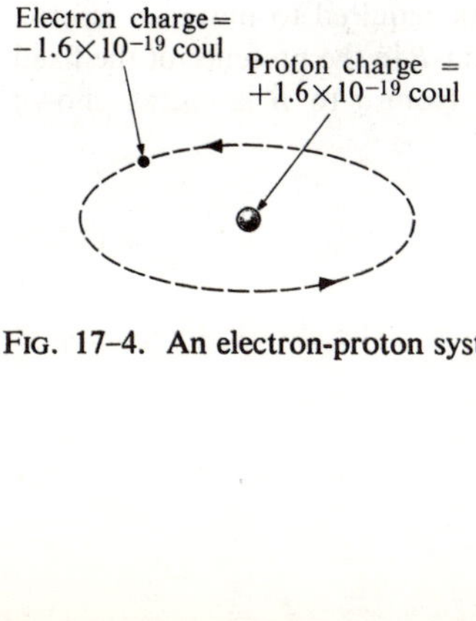

FIG. 17-4. An electron-proton system.

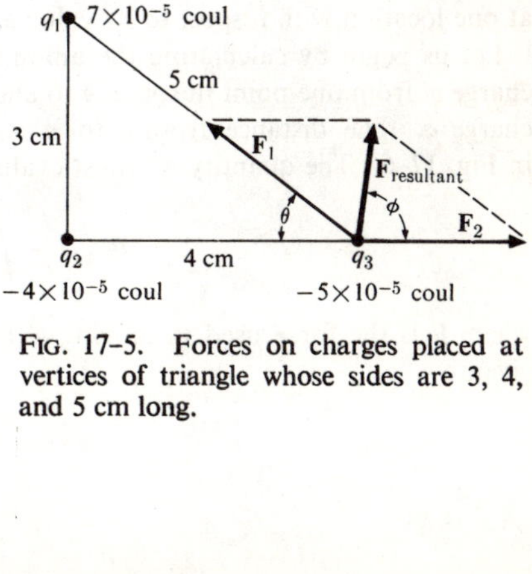

FIG. 17-5. Forces on charges placed at vertices of triangle whose sides are 3, 4, and 5 cm long.

Solution: Let us call $\mathbf{F}_1$ the force exerted on q_3 by q_1, and $\mathbf{F}_2$ the force exerted on q_3 by q_2. Then, by Eq. (17-4), the magnitudes of the forces are

$$F_1 = \frac{q_1 q_3}{4\pi\epsilon_0 r_1^2} = 12.6 \times 10^3 \text{ n},$$

$$F_2 = \frac{q_2 q_3}{4\pi\epsilon_0 r_2^2} = 11.3 \times 10^3 \text{ n}.$$

The directions of $\mathbf{F}_1$ and $\mathbf{F}_2$ are parallel to the 5- and 4-cm sides of the triangle, as shown. In order to evaluate the net force $\mathbf{F}$ on q_3, we shall resolve $\mathbf{F}_1$ and $\mathbf{F}_2$ into x- and y-components. The results are

$$F_{1y} = F_1 \sin\theta = \tfrac{3}{5}F_1 = 7.56 \times 10^3 \text{ n},$$

$$F_{1x} = F_1 \cos\theta = \tfrac{4}{5}F_1 = -10.1 \times 10^3 \text{ n},$$

$$F_{2y} = 0,$$

$$F_{2x} = 11.3 \times 10^3 \text{ n}.$$

Hence

$$F = \sqrt{F_x^2 + F_y^2} = 7.65 \times 10^3 \text{ n},$$

and the angle between $\mathbf{F}$ and the x-axis is

$$\phi = \tan^{-1}\frac{F_y}{F_x} = \tan^{-1} 6.30 = 81°.$$

17-3 Electric potential energy

Our initial discussion of mechanics was based on force and motion, but we soon found that the concept of energy provided a more powerful approach. The concept of energy is also advantageous in understanding electrical phenomena. Coulomb's law, like the law of gravitation, describes central forces whose magnitudes depend on the distances between interacting bodies. In both cases the work done in going from one place to another is independent of the actual path taken, being determined only by the endpoints, and so the potential energy of a mass or charge at one location with respect to its value at another location can be specified.

Let us begin by calculating the amount of work required to move an electric charge q' from one point in space A to another point B in the presence of the fixed charge q. The distance from q to A is r_A, and from q to B is r_B, as shown in Fig. 17-6. The quantity we must evaluate is

$$W_{AB} = \int_A^B \mathbf{F} \cdot d\mathbf{s}, \tag{17-5}$$

where $\mathbf{F}$ is the force used in moving q' from A to B. The electric force *on* q' is given by

$$\frac{1}{4\pi\epsilon_0}\frac{qq'}{r^2}\mathbf{r}_1,$$

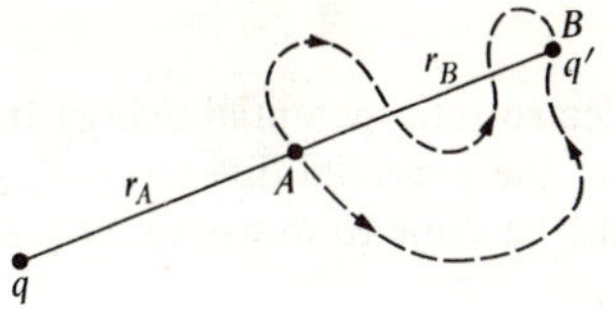

FIG. 17-6. Work done in moving charge q' from A to B is independent of path taken.

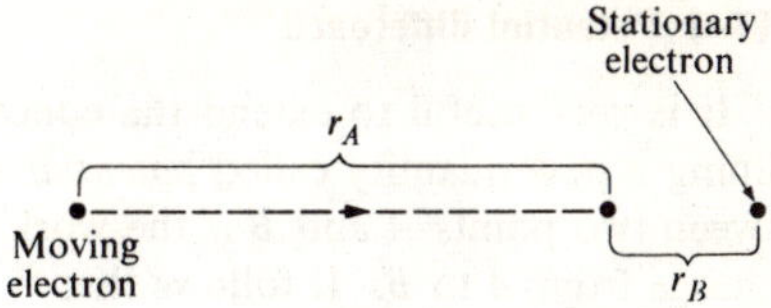

FIG. 17-7. A moving electron, as it approaches a stationary one, transforms its kinetic energy into electrical potential energy.

and hence the force we must use in finding W_{AB} is the *negative* of this, namely the force *we* must exert to produce the change in position of q'. Therefore

$$\mathbf{F} = -\frac{1}{4\pi\epsilon_0}\frac{qq'}{r^2}\mathbf{r}_1$$

and

$$W_{AB} = -\int_{r_A}^{r_B}\frac{1}{4\pi\epsilon_0}\frac{qq'}{r^2}\,dr = \frac{qq'}{4\pi\epsilon_0}\left(\frac{1}{r_B} - \frac{1}{r_A}\right). \tag{17-6}$$

Because W_{AB} depends only on r_A and r_B we could have used *any* other path from A to B, even the ones represented by dashed lines in Fig. 17-6, which is characteristic of conservative force fields (see Chapter 7).

Example: An electron having the initial velocity 10^3 cm/sec is directed at another electron, whose position is fixed, from a distance of 100 cm. How close to the stationary electron will the other one approach before it stops and reverses its path?

Solution: As the moving electron travels toward the stationary one, its kinetic energy is transformed into potential energy. In effect, it does work on the electric field of the fixed electron. Finally, when all the initial kinetic energy is gone, the electron stops and, like a stone thrown upward, begins moving in the direction it came from (Fig. 17-7). Letting r_A be the initial separation of the electrons and r_B the final one, and being careful to have all quantities in mks units, we obtain

$$W_{AB} = \tfrac{1}{2}mv^2,$$

$$\frac{e^2}{4\pi\epsilon_0}\left(\frac{1}{r_B} - \frac{1}{r_A}\right) = \tfrac{1}{2}mv^2,$$

$$r_B = \frac{e^2 r_A}{2\pi\epsilon_0 m v^2 r_A + e^2} = 0.33\text{ m}.$$

The existence of two kinds of electrical force, attractive and repulsive, leads to situations with no parallel in the case of gravitation. If q is positive and q' is also positive in Fig. 17-6, the potential energy of q' at B is less than it is at A; if q' is negative, however, its potential energy at B is greater than it is at A. Since A is closer to q than to B, in a comparable gravitational problem the potential energy at B would always exceed that at A.

17-4 Potential difference

It is very useful to extend the concept of electrostatic potential energy by defining a new quantity called *potential difference:* the potential difference V_{AB} between two points A and B is the work that must be done to move a unit positive charge from A to B. It follows that

$$V_{AB} = W_{AB}/q'. \tag{17-7}$$

If the potential difference is positive, we have done work on the charge and its potential energy at B is greater than it was at A; if the potential difference is negative, the electric field has done work on the charge and its potential energy at B is lower than it was at A. The potential difference V is expressed in joules/coulomb from its definition, and is so often used that it has been given the special unit of the *volt:*

$$1 \text{ volt} = 1 \text{ joule/coulomb}.$$

In the vicinity of an isolated charge q, we can find the potential difference between two points, located r_A and r_B, respectively, from q, with the help of Eq. (17-6):

$$V_{AB} = \frac{W_{AB}}{q'} = \frac{q}{4\pi\epsilon_0}\left(\frac{1}{r_B} - \frac{1}{r_A}\right). \tag{17-8}$$

17-5 Potential

It is sometimes more convenient to think in terms of the potential at a point than of the potential difference between two points. We recall that when we were dealing with gravitation near the earth, it often proved helpful to define the potential energy of any object at the earth's surface as zero. This enabled us to express the potential energy of an elevated object as simply mgh, instead of the more general expression $mg(h - h_0)$ in which we would always have to specify a reference height h_0. We can do the same thing in the case of electric forces, where it is most convenient to choose as the zero of potential its value at infinity, that is, its value an infinite distance from all charges. Adopting this convention, we may now speak of the potential at a point, with the understanding that what we really mean is the difference between the potential there and at infinity. For the potential of a point charge, letting $r_A = \infty$ and $r_B = r$, we have

$$V = \frac{q}{4\pi\epsilon_0 r}. \tag{17-9}$$

The potential at a point due to a number of charges is just the algebraic sum of the individual potentials there, since V is a scalar, so that

$$V = \frac{1}{4\pi\epsilon_0}\sum_{i=1}^{n}\frac{q_i}{r_i}, \tag{17-10}$$

where r_i is the distance from the ith charge to the point in question.

Example: One of the most important charge configurations in physics is the dipole, which consists of two equal and opposite charges a certain distance apart. Find the potential at a point whose distance from the dipole is large compared with the separation of the charges.

Solution: In Fig. 17–8 the charges have been placed at $x = -d$ and $x = +d$, and the point P is in the xy-plane. (The results will be no different if P has a z-coordinate as well, but the calculation is more complicated in that case.) The potential at P due to the two charges is

$$V = \frac{1}{4\pi\epsilon_0}\left(\frac{q}{r_1} - \frac{q}{r_2}\right),$$

which may be rewritten

$$V = \frac{q}{4\pi\epsilon_0}\,\frac{r_2 - r_1}{r_1 r_2}.$$

When $2d$ is much smaller than r_1 and r_2,

$$\frac{r_2 - r_1}{r_1 r_2} \approx \frac{r_2 - r_1}{r^2},$$

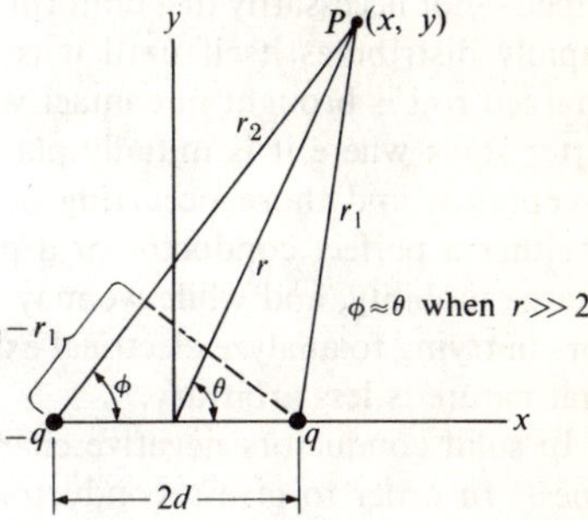

FIG. 17–8. Potential of a dipole.

where the symbol $\approx$ means "is approximately equal to," and r is the distance from the center of the dipole to P. From Fig. 17–8, since $\phi \approx \theta$,

$$r_2 - r_1 \approx 2d \cos\theta,$$

and

$$V = \frac{2qd\cos\theta}{4\pi\epsilon_0 r^2}.$$

The quantity $2qd$, which is the product of the magnitude of one of the charges and the distance between them, is called the *dipole moment p*. In terms of p,

$$V = \frac{p\cos\theta}{4\pi\epsilon_0 r^2} = \frac{pr\cos\theta}{4\pi\epsilon_0 r^3} \tag{17–11}$$

$$= \frac{px}{4\pi\epsilon_0 (x^2 + y^2)^{3/2}}. \tag{17–12}$$

The region near a positive charge is one of high potential, and that near a negative charge a region of low potential (in fact, of negative potential). A positive charge will move spontaneously from a position at high potential to one of lower potential, while a negative charge does the opposite. This is simply another way of stating that like charges repel and unlike charges attract, but with the advantage of facilitating calculations.

17–6 Conductors and insulators

The molecules which constitute matter are, as we have said, composed of positive and negative electric charges, normally in equal quantities, so that ordinary matter is neutral. It is customary to classify all substances as either conductors or insulators. The distinction between them is that charges are free to move about in a conductor, while they cannot do so in an insulator. When a charged rod is touched to a conductor, the entire surface of the conductor becomes electrified—not necessarily in a uniform manner, but the essential fact is that the charge rapidly distributes itself until it is in equilibrium. On the other hand, when a charged rod is brought in contact with an insulator, any charge transferred to the latter stays where it is initially placed for a long period of time. With very few exceptions, and those occurring only under unusual circumstances, no material is either a perfect conductor or a perfect insulator. Instead there are degrees of charge mobility, and while we may think in terms of ideal insulators and conductors in trying to analyze electrical experiments in general terms, we must remember that nature is less arbitrary.

In solid conductors negative charges have much greater mobility than positive ones. In order to give a conductor a negative charge, negative charges must be supplied to it; to give it a positive charge, negative charges must be removed.

An imaginary surface in space all of whose points are at the same potential is called an *equipotential surface*. The surface of a conductor is always an equipotential surface because the charges on it have distributed themselves in such a way that the force on each charge is zero. The work required to displace a charge infinitesimally is therefore also zero, and every point on the surface is at the same potential as neighboring points. Hence the entire surface of the conductor is at the same potential.

17–7 Electric current

Moving charges constitute an electric current. Such currents are of great importance from a practical point of view, as well as from the point of view of understanding physical phenomena, and a substantial part of the remainder of this book is devoted, in one way or another, to discussing electric currents and their effects. In the remainder of this chapter we shall restrict ourselves to steady currents in conductors.

The vector *current density* $\mathbf{j}$ is useful in describing electric currents. It is defined as the rate at which positive charge passes through a unit area which is perpendicular to the direction of flow. In most applications in this book we shall be interested in the total charge passing through a surface of area A perpendicular to the direction of $\mathbf{j}$. In such cases the concept of the total *current* i is important and

$$i = \frac{dq}{dt} = jA. \tag{17–13}$$

Here dq/dt is the total rate of flow of charge.

If the area is not perpendicular to the direction of **j**, but **j** is constant over the area, the current is given by Fig. 17–9:

$$i = A\mathbf{j} \cdot \mathbf{n}. \qquad (17\text{–}14)$$

If the flow is not uniform,

$$i = \int_A \mathbf{j} \cdot \mathbf{n}\, dA, \qquad (17\text{–}15)$$

where **n** is a unit vector perpendicular to the area A.

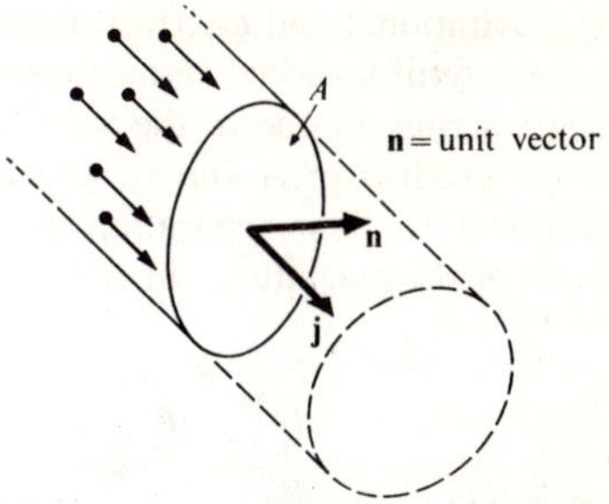

FIG. 17–9. Current density due to motion of charges.

In the case of a conducting wire, where the charges are restricted to the wire, we may speak of current as the rate at which charge travels past a point in the wire. From Eq. (17–13), the unit of current is the coulomb/second, which is called the *ampere:*

$$1 \text{ ampere} = 1 \frac{\text{coulomb}}{\text{second}}.$$

The direction of the current i is conventionally taken as the direction of *positive* current flow, that is, the direction in which positive charges would have to move to produce the observed current.

Positive charges move spontaneously from a region of high potential to a region of low potential. In doing so they tend to lower the high value of the potential and to increase the low value. In order for a steady current to flow, some external agency is required to maintain the potential difference. We shall have more to say about steady currents, and how they are maintained, in Chapter 20.

Currents produce chemical effects, heating effects, and magnetic effects. All these will be discussed in subsequent chapters. In the next section we shall have an initial look into magnetic forces.

17–8 Magnetic forces

A current flowing in a wire reveals itself in several ways; for example, by producing an increase in the temperature of the wire. Another effect is found when two current-carrying wires are placed parallel to one another (Fig. 17–10): When both currents flow in the same direction, there is a force of attraction between the wires, while if the currents flow in opposite directions, the force between the wires is repulsive. Until now the only forces we have encountered that were able to act at a distance from the seat of force were those that were gravitational or electrical in origin. Neither of these can be responsible for the force between current-carrying wires, since the masses of the wires do not change when current flows through them nor is there any net charge on the wires under such circumstances. What we are dealing with here is called *magnetic force*, and while similar in some respects to gravitational and electrical forces, it has a number of unique features that require careful scrutiny. Experimental measurements show that the force *per unit length* acting on one of the wires due to the current in the other is proportional to the product of i_1 and i_2, and inversely proportional to their separation R. That is, in terms of the magnitudes involved,

$$\frac{F}{l} = \frac{k' i_1 i_2}{R}, \qquad (17\text{–}16)$$

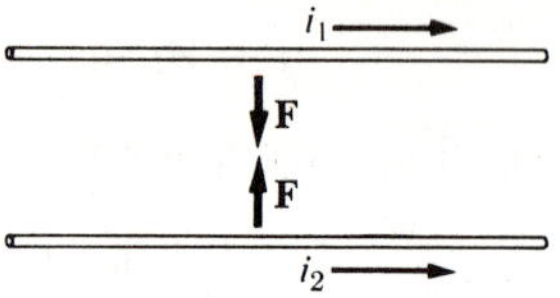

FIG. 17–10. Two parallel currents have a force of attraction between them.

where k' is a constant appropriate to the system of units employed. In the mks system, with F/l in newtons/meter, R in meters, and i in amperes, k' has the assigned value 2×10^{-7} n/amp^2, and Eq. (17–16) is used to *define* the ampere in terms of the force per unit length between two current-carrying wires. Thus the ampere is the mks electrical standard, and the coulomb, the quantity of charge transferred in one second by a current of one ampere, is a derived unit.

In the rationalized mks system of units, which is the system we are using, k' is written

$$k' = \frac{\mu_0}{2\pi}. \qquad (17\text{–}17)$$

Hence μ_0, called the *permeability of free space*, is equal to

$$\mu_0 = 4\pi \times 10^{-7} \frac{\text{newton}}{\text{amp}^2} = 12.57 \times 10^{-7} \frac{\text{newton}}{\text{amp}^2}. \qquad (17\text{–}18)$$

All magnetic effects can be ultimately traced to currents. Even in the familiar case of magnetized iron bars, the magnetic effects arise from currents. Here we are dealing with currents on the atomic level rather than macroscopic currents. We shall consider electric and magnetic phenomena and their peculiar entwinement in much of the remainder of this book, since the properties and behavior of matter are in large measure consequences of these phenomena.

Problems

1. An inflated balloon is charged by rubbing it with fur. Will it stick more readily to a conducting or to an insulating wall, and why?

2. Two charges of unknown magnitude and sign are observed to repel one another with a force of 0.1 n when they are 5 cm apart. What will the force be when they are (a) 10 cm apart? (b) 50 cm apart? (c) 1 cm apart?

3. A particle carrying a charge of $+6 \times 10^{-5}$ coul is located halfway between two other charges, one of $+1 \times 10^{-4}$ coul and the other of -1×10^{-4} coul, that are

0.40 m apart. All three charges lie on the same straight line. What is the magnitude and direction of the force on the $+6 \times 10^{-5}$ coul charge?

4. A small sphere carrying a charge of $+2 \times 10^{-4}$ coul is 0.1 m from another small sphere carrying a charge of -5×10^{-4} coul. What is the magnitude and direction of the force exerted by the -5×10^{-4} coul charge on the $+2 \times 10^{-4}$ coul charge?

5. How far apart should two electrons be if the force each exerts on the other is to equal the weight of an electron?

6. A test charge of -5×10^{-5} coul is placed between two other charges so that it is 5 cm from a charge of -3×10^{-5} coul and 10 cm from a charge of -6×10^{-5} coul. The three charges lie along a straight line. What is the magnitude and direction of the force on the test charge?

7. Three charges, $+q$, $+q$, and $-q$, are at the vertices of an equilateral triangle a long on each side. Find the magnitude and direction of the force on one of the positive charges.

8. Two identical pith balls are at the ends of strings 10 cm long. The mass of each of the balls is 2×10^{-3} kg. The system comes to equilibrium when each string makes an angle of 30° with the vertical. What is the charge on each ball, assuming that they are equal?

9. Assume that the charges of the earth and the sun are not quite neutralized, and that the net charges are of the same sign (negative) and equal in magnitude. What must the charge on each be in order that the coulomb force will just cancel the gravitational force? How many free electrons does this charge correspond to?

10. Calculate the potential energy of an electron in an atom from the following data: Positive charge of nucleus is 1.6×10^{-19} coul and is equal to the negative charge on the electron. Radius of the orbit is 0.5×10^{-10} m. The coulomb force supplies the centripetal force and the mass of the electron is 9×10^{-31} kg. Calculate the kinetic energy and compare to the potential energy.

11. Two thin insulated rods are 0.8 m long. Both rods have small balls at either end with a positive charge of 2.5×10^{-7} coul at one end and an equal negative charge at the other. They are placed parallel to each other with the plus ends pointed the same way, 0.6 m apart. What force does one rod exert on the other?

12. Calculate the work required to arrive at the configuration of a charge Q_A, Q_B and Q_C placed at the vertices of a triangle A, B, and C, respectively. The sides of the triangle are of length $AB = c$, $AC = b$, $CB = a$.

13. Calculate the force on an electric dipole of strength p at a distance R from a charge $+Q$ perpendicular to the axis of the dipole. [*Hint:* Take a positive charge q separated a distance l from a negative charge $-q$ such that $ql = p$. Calculate the force, and then let $l \to 0$ and $q \to \infty$ in such a manner that ql remains equal to p.]

14. Repeat the previous problem, with $+Q$ along the axis of the dipole.

15. Calculate the potential at a point 120 cm from the positive charge along a direction perpendicular to the line connecting two charges of the same magnitude (3×10^{-7} coul) but opposite sign, 0.5 m apart. Compare the result with the one obtained from the approximate formula derived in the text.

16. Repeat the calculation in the previous problem for a point along the line connecting the charges and 120 cm from the midpoint.

17. A charge of 5×10^{-6} coul is separated from an equal and opposite charge by 1 m. Draw a graph of the potential as a function of the distance from the positive charge along the line joining the two. Start at a point 0.1 m from the positive charge and end 0.1 m from the negative charge. What is the surface in space that is at zero potential?

18. Two charges of magnitude 6×10^{-8} coul and of the same sign are placed 1 m apart. (a) Calculate the potential at a point midway between the two charges. (b) Calculate the potential at a point 1 cm to one side of the midpoint. Repeat for a point $\frac{1}{2}$ cm to one side of the midpoint. (c) Show that the rate of change of the potential approaches the negative of the force acting on a unit positive charge placed at the center.

19. Repeat the previous problem, with charges of opposite sign.

20. A current of 3×10^{-10} amp can be detected by a sensitive current-measuring meter. The charge on an electron is 1.6×10^{-19} coul. How many electrons/second does this correspond to? If the electrons are in a uniform beam and the cross section of the beam is 10^{-4} m^2, what is the density of electrons in the beam, expressed in electrons/cm^2?

21. Protons in cosmic rays strike the earth's upper atmosphere at a rate averaged over the earth's surface of 0.15 protons per cm^2/sec. If the radius of the earth is 6.4×10^6 m, what is the total cosmic-ray current received by the earth? What is the average current density?

22. A cathode is a right circular cylinder whose radius is 0.1 cm and which is 2 cm high. It emits 10^{12} electrons per second, which are all made to pass through the surface of a concentric cylinder of radius 2 cm and height 2 cm. What is the current through the outer cylinder and what is the current density there?

23. Calculate the force per unit length between two current-carrying wires 0.1 m apart, each having a cross-sectional area of 10^{-2} cm^2. The current density in one is 3 amp/cm^2 and the current density in the other is 5 amp/cm^2.

18

The Electric Field

Gravitational, electrostatic, and magnetic forces have in common the property of acting between bodies not in direct contact with each other. While it may seem peculiar at first, this property is actually very familiar: the force on Newton's proverbial apple was exerted by the earth without any direct contact between apple and earth. How can these forces act through empty space? To scientists of previous centuries the answer was an actual mechanical system—not directly observed but nevertheless present even in a vacuum—a system that transmitted forces from one body to another as a chain of pushes or pulls from one element of the mechanical system to the next. Today we discard notions that cannot be empirically verified, and instead speak of distortions in space called *force fields.* We have no idea what a force field "looks like," but take the position that if the behavior of such a field can be completely specified by describing some ideal experiments for the determination of such a field, and if this specification agrees with experiment, we "know" all that there is to know about how it acts in empty space. Implicit in the concept of a force field is the presumption that when a mass, a charge, or a current element exists somewhere, the space in its neighborhood is changed in such a way that another mass, charge, or current element, respectively, will experience a force when it is in this neighborhood. Of course, a gravitational field will act only on the mass of a particle, the electric field only on the charge, and the magnetic field only on the current element. The field exists even though nothing may be there for it to act upon. The older concept of action at a distance held to the contrary view that the force one body exerts on another represents a direct interaction between the bodies, with the intervening space somehow transmitting the force between them but not being affected by the presence of a single body. Action at a distance is not so useful a concept as the field concept, partly because the reality of force fields has been demonstrated in a number of convincing ways.

We shall see later that the fields which we define have energy and momentum associated with them, and that this energy and momentum obey the usual conservation theorems associated with these quantities. Also there is a finite velocity of propagation of these fields. According to Newton's law of gravitation or

Coulomb's law, whenever two masses or two charges exist in space there is at all times a force between them. If we were to envisage a situation in which charges are created—a situation, by the way, which can exist—according to Coulomb's law as soon as the charges are created there is a force between them and the other charges in space. We shall see, however, when we discuss relativity theory that it takes a finite time for the force to make itself felt. This is described by saying that there is a finite time required for the propagation of the electric field. Anything which has a finite velocity, energy, and momentum must be accorded some reality.

Another reason for using the concept of an electric field is that charges are not the only sources for this field, and the concept of the electric field serves to unify all phenomena which result in forces on electric charges. The concept of a force field is fundamental to our understanding of the natural world, and it is worth our close study in this and subsequent chapters.

18-1 The electric field

We learned in the previous chapter that a charge q_1 is acted on by the force

$$\mathbf{F} = \frac{1}{4\pi\epsilon_0}\frac{q_1 q_2}{r^2}\mathbf{r}_1 \tag{18-1}$$

when it is a distance r from another charge q_2. The unit vector $\mathbf{r}_1$ points from q_1 opposite to a line joining it with q_2 (Fig. 18-1). What we want to do now is specify the electric field around q_2 in such a way that, if we place q_1 the distance r from q_2, its interaction with the electric field results in the force $\mathbf{F}$ of Eq. (18-1). Our specification is therefore as follows: the electric field intensity $\mathbf{E}$ at any point in space is the force that a unit positive charge would experience if placed there. Electric field intensity is a vector quantity whose direction is that of the force on the hypothetical unit positive charge. In equation form,

$$\mathbf{E} = \mathbf{F}/q, \tag{18-2}$$

where $\mathbf{F}$ is the force on any arbitrary charge q. The units of $\mathbf{E}$ are newtons/coulomb (n/coul); since

$$1 \text{ volt} = 1 \text{ joule/coulomb} \quad \text{and} \quad 1 \text{ joule} = 1 \text{ newton-meter},$$

it is clear that $\mathbf{E}$ can also be expressed in volts/meter, which is conventional usage.

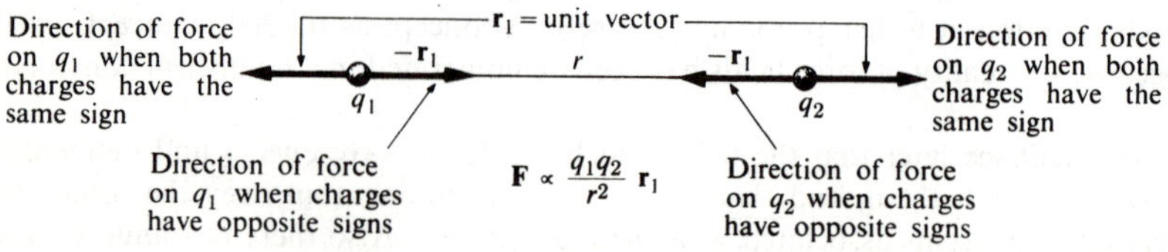

FIG. 18-1. Coulomb's law.

Equation (18–2) is not entirely satisfactory as a definition of electric field intensity, because the actual introduction of a finite charge q in a region containing matter may induce a separation of charge in the matter that would modify $\mathbf{F}$ by an amount depending on irrelevant factors, such as the magnitude of q and the nature of the matter. The reader will recall that the influence of a charge on a neutral paper caused an attraction between the two. In other words, Eq. (18–2) is not a good *operational definition*, since it does not tell us how to measure $\mathbf{E}$ unambiguously. A better definition is

$$\mathbf{E} = \lim_{q \to 0} \frac{\mathbf{F}}{q}. \tag{18–3}$$

The electric field intensity $\mathbf{E}$ at a point in space is the limiting value of the ratio between the force $\mathbf{F}$ on a charge q at that point and the value of q as q approaches 0. [Strictly speaking, we can never find the limit of Eq. (18–3) because there is a certain minimum electric charge in nature, called the charge on the electron, and q can never be less than this finite quantity. Practically, of course, the charge on the electron is so minute that Eq. (18–3) is entirely adequate for most situations.] So long as we restrict ourselves to electric charges in vacuum, or, to a good approximation, in air or other dilute gases, the less precise definition of Eq. (18–2) may legitimately be used.

The electric field at a point in the vicinity of a charge or distribution of charges may be calculated from Eqs. (18–1) and (18–2). We imagine that we bring a positive test charge q' to the point in question, and compute the resultant force $\mathbf{F}$ on it due to the other charges; then we divide $\mathbf{F}$ by q'. At a distance r from a single charge q the electric field intensity is therefore

$$\mathbf{E} = \frac{\mathbf{F}}{q'} = \frac{1}{q'}\frac{1}{4\pi\epsilon_0}\frac{qq'}{r^2}\mathbf{r}_1 = \frac{1}{4\pi\epsilon_0}\frac{q}{r^2}\mathbf{r}_1. \tag{18–4}$$

When there is more than one charge producing the force on the test charge q', we must calculate the *vector sum* of the separate forces $\mathbf{F}_i$; if each charge q_i is at a distance r_i from the point where $\mathbf{E}$ is to be determined, then

$$\mathbf{F}_i = \frac{1}{4\pi\epsilon_0}\frac{q_i q'}{r_i^2}\mathbf{r}_{1i}.$$

The vector $\mathbf{r}_{1i}$ is a unit vector drawn from q_i in the direction of the point where E is determined, and

$$\mathbf{E} = \frac{\sum_{i=1}^{n} \mathbf{F}_i}{q'} \quad \text{(vector sum)}$$

$$= \frac{1}{4\pi\epsilon_0}\sum_{i=1}^{n}\frac{q_i}{r_i^2}\mathbf{r}_{1i} \quad \text{(vector sum)}, \tag{18–5}$$

where the total number of charges contributing to $\mathbf{E}$ is n.

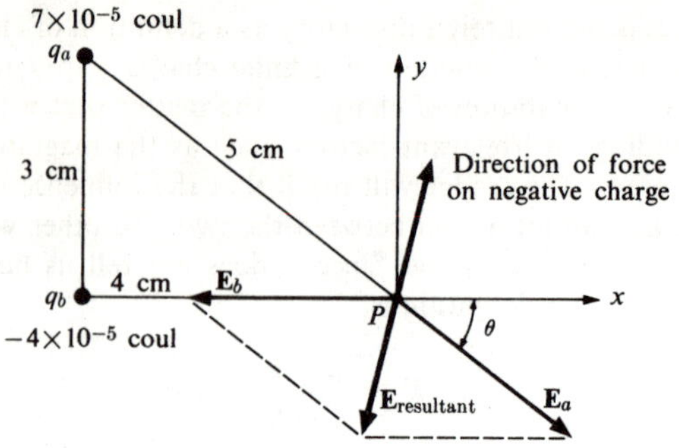

FIG. 18–2. A vector calculation of the electric field due to a charge distribution.

Example: Two charges, $q_a = +7 \times 10^{-5}$ coul and $q_b = -4 \times 10^{-5}$ coul, are placed as shown in Fig. 18–2. Determine the electric field at the point P, and use it to find the force that a charge of -5×10^{-5} coul would experience if placed at P.

Solution: The contribution to the electric field at P by the charge q_a is

$$\mathbf{E}_a = \frac{q_a}{4\pi\epsilon_0 r_a^2}\mathbf{r}_{1a} = +2.52\mathbf{r}_{1a} \text{ v/m},$$

and the contribution by the charge q_b is

$$\mathbf{E}_b = \frac{q_a}{4\pi\epsilon_0 r_b^2}\mathbf{r}_{1b} = 2.26\mathbf{r}_{1b} \text{ v/m}.$$

What we require is the *vector sum* of $\mathbf{E}_a$ and $\mathbf{E}_b$, so we first resolve each into x- and y-components:

$$E_{ax} = E_a \cos\theta = +2.02 \text{ v/m},$$

$$E_{ay} = E_a \sin\theta = -1.51 \text{ v/m},$$

$$E_{bx} = -2.26 \text{ v/m},$$

$$E_{by} = 0.$$

Hence

$$E_x = E_{ax} + E_{bx} = -0.24 \text{ v/m},$$

$$E_y = E_{ay} + E_{by} = -1.51 \text{ v/m},$$

and

$$E = \sqrt{E_x^2 + E_y^2} = 1.53 \text{ v/m}.$$

The force on a charge $q = -5 \times 10^{-5}$ coul at the point P is, in magnitude,

$$F = qE = 7.65 \times 10^3 \text{ n}.$$

Note that this is the same as the force on q_3 in the second illustrative problem in Section 17–2. Here, however, by finding **E** at the point P, we can at once compute the force on *any* charge at P.

18–2 Continuous distributions of charge

It is often necessary for us to calculate the electric field intensity due to a continuous distribution of charge instead of a set of discrete charges. In this case we first state the definition of the electric field in differential form,

$$d\mathbf{E} = (1/4\pi\epsilon_0)(dq/r^2)\mathbf{r}_1. \tag{18–6}$$

This equation states that an infinitesimal charge dq produces an infinitesimal field $d\mathbf{E}$ according to the same prescription that holds for a finite charge. To find the effect of a finite continuous distribution of charge, we add the effects of the infinitesimal contributions, vectorially as we have done earlier, but instead of a finite sum we have an integral to consider. The magnitude of each field contribution and the direction of the unit vector both depend on the position vector $\mathbf{r}'$ of the infinitesimal charge. The distance between the position of the infinitesimal charge and the position at which we are evaluating the field is $|\mathbf{r} - \mathbf{r}'|$. The procedure for accomplishing this calculation is, in principle, the same as that of Eq. (18–5), except that, instead of adding vectorially the contributions to $\mathbf{E}$ of the n charges q_i, we integrate the contributions of each infinitesimal charge element dq, so that

$$d\mathbf{E} = \frac{1}{4\pi\epsilon_0}\frac{dq(\mathbf{r}')}{|\mathbf{r} - \mathbf{r}'|^2}\mathbf{r}_1',$$

$$\mathbf{E} = \frac{1}{4\pi\epsilon_0}\int\frac{dq(\mathbf{r}')}{|\mathbf{r} - \mathbf{r}'|^2}\mathbf{r}_1', \tag{18–7}$$

where $\mathbf{r}_1$ is a unit vector in the direction of the electric field produced by $dq(\mathbf{r}')$. The integration is to be taken over the entire charged region. The above integral is a rather complex affair, since we must take the directions of each $d\mathbf{E}$ into account. The only reasonable method of performing the operation required in Eq. (18–7) is to split $\mathbf{E}$ into three components, E_x, E_y, E_z, for example, perform an ordinary scalar integration for each, and then combine them by

$$E = \sqrt{E_x^2 + E_y^2 + E_z^2}.$$

Example: Determine the electric field of a thin, uniformly charged ring lying in the xy-plane with its center at the origin at any point on the z-axis, as shown in Fig. 18–3.

Solution: We start by computing the contribution $d\mathbf{E}$ to the electric field at the point P due to the element of charge dq on the portion of the ring dL. If the linear charge density in the ring is λ, then

$$dq = \lambda\, dL \qquad \text{and} \qquad d\mathbf{E}_L = \frac{\lambda\, dL}{4\pi\epsilon_0 r^2}.$$

We need take into account only the component of $d\mathbf{E}$ along the z-axis, since the components perpendicular to z cancel out when we integrate around the ring. Hence,

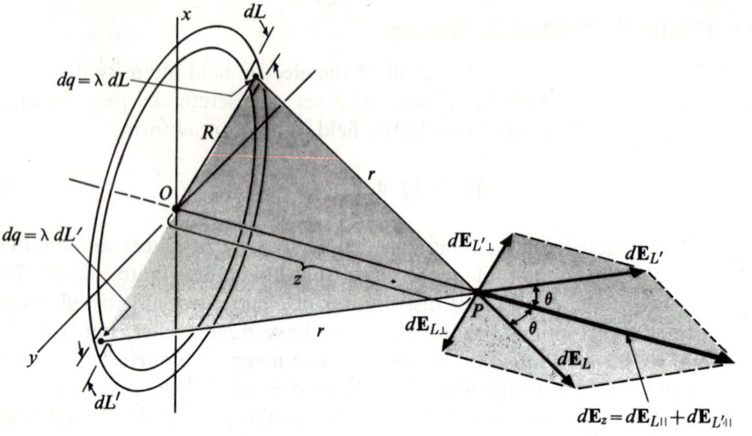

FIG. 18–3. The transverse components of the electric field of a uniformly charged ring cancel out at a point on the axis of the ring.

with R the radius of the ring,

$$E = \int_0^{2\pi R} \lambda \frac{dL \cos\theta}{4\pi\epsilon_0 r^2} = \frac{\lambda RZ}{2\epsilon_0(R^2 + Z^2)^{3/2}}\,\mathbf{Z}_1 = \frac{qZ}{4\pi\epsilon_0(R^2 + Z^2)^{3/2}}\,\mathbf{Z}_1,$$

which is the electric field intensity at the point $z = Z$; q is the total charge in the ring, and $\mathbf{Z}_1$ is a unit vector in the direction of positive Z. In carrying out the integration we note that θ and r both remain constant.

18–3 Gauss' theorem

An electric field is a vector field, which means that at every point within it the field strength has a magnitude and a direction. Generally we have represented individual force by an arrow pointing in the direction of the force, an arrow whose length is proportional to the force's magnitude. In trying to represent a vector field, however, it is more appropriate to use continuous lines in place of arrows, with the direction of the lines indicating the direction of the field and the density

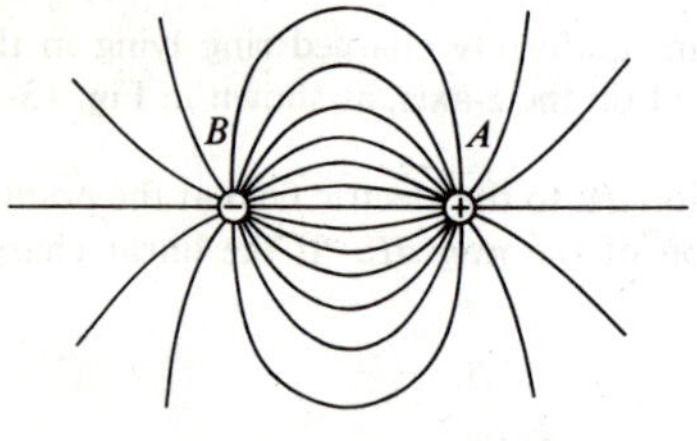

FIG. 18–4. Lines of force for two oppositely charged particles.

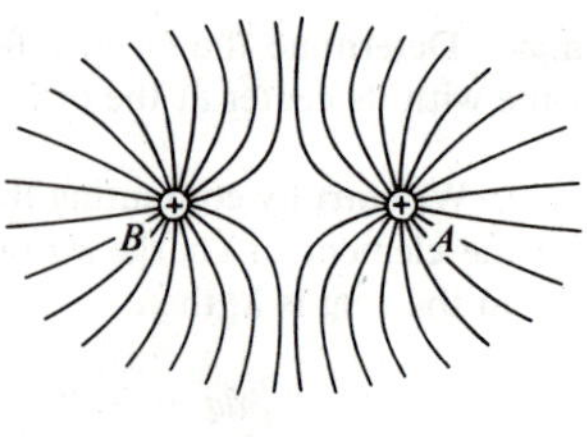

FIG. 18–5. Lines of force for two charges of the same sign.

of the lines indicating the field intensity. In this way we can readily illustrate the essential features of a specific vector field. The lines themselves, as we recall from Chapter 11, are called *lines of force*. In Fig. 18–4 are pictured the lines of force surrounding two charges of equal magnitude and opposite sign separated by a small distance, and Fig. 18–5 shows the same situation, except that the charges here have the same sign.

The concept of lines of force, which was introduced by Faraday, may be related quantitatively to the electric field strength at a certain point. Let us consider an infinitesimal area dA perpendicular to the lines of force dN there (Fig. 18–6). Since we have drawn the lines of force so that their density is proportional to E,

$$E = \alpha \frac{dN}{dA}, \tag{18–8}$$

where α is an arbitrary constant.

The number of lines of force that emerge from a charge q can be calculated by first integrating Eq. (18–8),

$$\alpha N = \int \mathbf{E} \cdot \mathbf{n} \, dA,$$

where $\mathbf{n}$ is a unit vector perpendicular to the surface dA. Since the area used in Eq. (18–8) must be perpendicular to the lines of force, only the component of $\mathbf{n}$ in the direction of $\mathbf{E}$ contributes to αN. This is the origin of $\mathbf{E} \cdot \mathbf{n}$ in the integrand. The component of $\mathbf{E}$ in the direction of the normal $\mathbf{n}$ is called E_n, and we have

$$\alpha N = \int E_n \, dA, \tag{18–9}$$

where E_n is the component of $\mathbf{E}$ normal to dA. In order to intercept all the lines of force leaving q, we shall perform the integration over the surface of a sphere of radius R having q at its center (Fig. 18–6). In this case the angle between $\mathbf{E}$ and the normal to the surface is always 90°, and according to Eq. (18–7),

$$E_n = E = \frac{1}{4\pi\epsilon_0} \frac{q}{R^2}.$$

Hence Eq. (18–9) becomes

$$\alpha N = \frac{1}{4\pi\epsilon_0} \frac{q}{R^2} \cdot 4\pi R^2 = \frac{q}{\epsilon_0}.$$

If we now assign the constant α the value $\alpha = 1/\epsilon_0$, we have that the number N of lines of force leaving the charge q is

$$N = q. \tag{18–10}$$

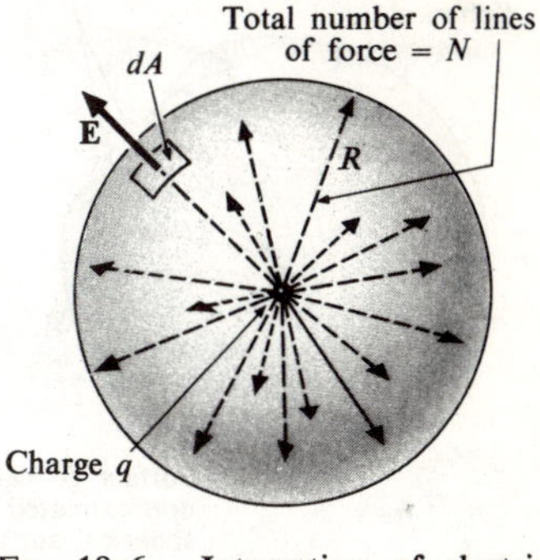

FIG. 18–6. Integration of electric field over a sphere surrounding an isolated charge. Here $E_n = \mathbf{E}$, since the field is radial.

Equation (18–9) becomes, with the new definition of α,

$$N = \epsilon_0 \int E_n \, dA. \tag{18–11}$$

The idea of lines of force, while it is helpful in thinking about electric fields, is nevertheless a completely artificial way of describing the situation. However, because we have specified what we mean by N in a precise mathematical form, we are entitled to make use of any conclusions we have obtained with the help of lines of force. Let us combine Eqs. (18–10) and (18–11), and thus we obtain

$$q = \epsilon_0 \int_{\substack{\text{closed}\\ \text{surface}}} E_n \, dA. \tag{18–12}$$

This equation states that when the electric field intensity normal to some closed surface is integrated over that surface and multiplied by ϵ_0, the result is exactly equal to the charge enclosed within the surface. While we have not actually proved that this is correct except for the situation in which the surface we integrate E_n over is a sphere with q at its center, it is possible to show that the statement of Eq. (18–12) is indeed true, no matter what the shape of the surface A or the location within A of the various charges whose sum is q.

This statement can be made plausible by noting that the integration over combinations of spherical surfaces (Fig. 18–7) also yields the result that $\epsilon_0 \int E_n \, dA = q$. This is true because the integration over the radial lines is equal to zero, since $\mathbf{n}$ is perpendicular to the field. Finally, complicated surfaces can be approximated by combinations of spherical and radial portions.

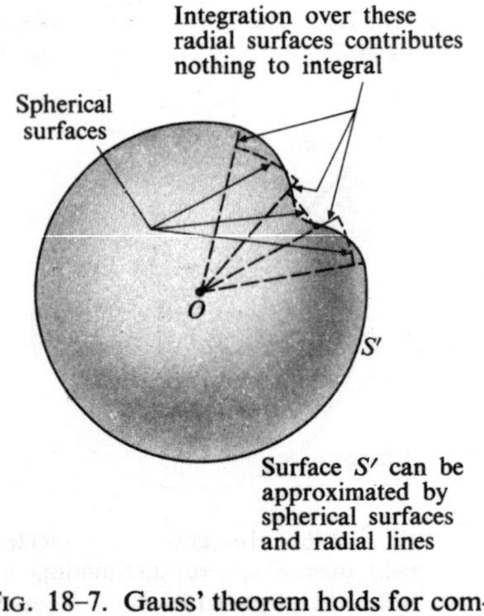

FIG. 18–7. Gauss' theorem holds for combinations of spherical surfaces, as shown.

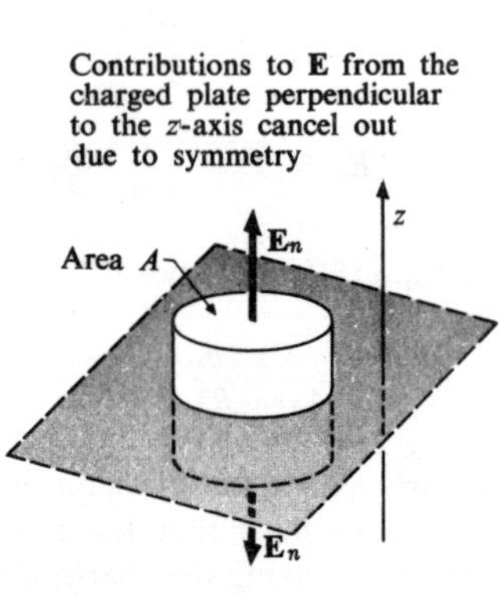

FIG. 18–8. Electric field in the vicinity of an infinite charged plane.

Equation (18-12) is called *Gauss' theorem;* it is the electrical analog of the Gauss theorem for gravitation which we derived in Section 11-7.

Gauss' theorem is useful chiefly in determining the electric fields surrounding symmetrical charge distributions. Consider, for example, the electric field of an infinite uniform plane sheet of charge covering the entire xy-plane (Fig. 18-8). The charge density is σ coul/m^2. Let us choose a cylinder as the closed Gaussian surface over which to integrate E_n. The electric field intensity is the same on both sides of the sheet of charge, and is everywhere directed parallel to the z-axis because the contributions of components of **E** perpendicular to this axis cancel out. Since **E** has no component perpendicular to the sides of the cylinder, we need integrate only over its ends. The result is that

$$q = 2\epsilon_0 \int_0^A E_n\, dA = 2\epsilon_0 EA,$$

where A is the area on one of the ends. Since the charge q enclosed by the cylinder is σA,

$$E = \frac{\sigma}{2\epsilon_0} \cdot \qquad (18\text{-}13)$$

The electric field of an infinite plane sheet of charge is proportional to the charge density only, and is constant everywhere regardless of distance from the sheet. Equation (18-13) is approximately valid near a finite charge sheet, provided that the perpendicular distance from the point in question to the sheet is small compared with the distance from that point to the edges of the sheet.

18-4 Charged conductors

We can use Gauss' theorem to investigate the final location of charges placed on a conducting body. Ultimately the charges must come to rest somewhere in or on the body. This conclusion follows from the experimental facts that charge is always associated with mass (the moving charges in a conductor are electrons; a deficiency of electrons acts as a positive charge and a surplus as a negative charge), and that, as was mentioned earlier, even the best natural conductors exhibit some resistance to the motion of charges in them. Under these circumstances the principle of conservation of energy would be violated if the charges were to continue to move about indefinitely. When the charges slow down and stop, their final arrangement must be one in which there is no force on any of them. But if **F** is zero, since

$$\mathbf{E} = \mathbf{F}/q,$$

there can be no electric field within the conducting body. Now we draw a Gaussian surface just underneath the surface of the conductor (Fig. 18-9). The electric field $\mathbf{E} = 0$ at this surface, and therefore by Gauss' theorem there can be no charge inside it. It can only be inferred from this that when charge is present in a conductor it must reside on the *outside surface* of the conductor.

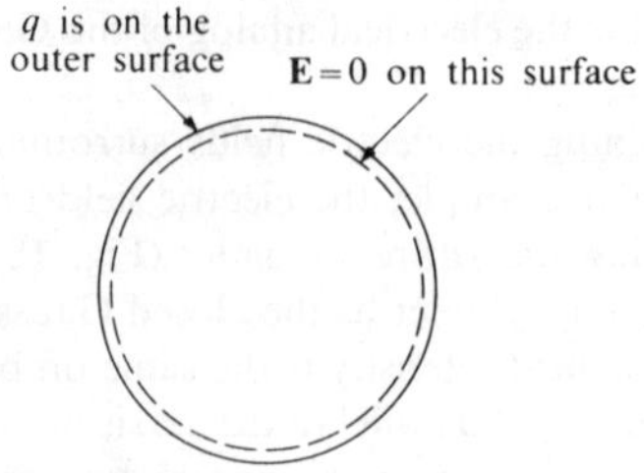

FIG. 18–9. The charge on a conductor is always on its outer surface.

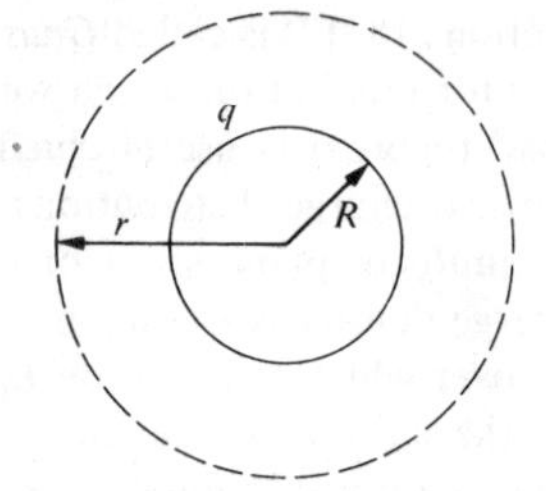

FIG. 18–10. Use of Gauss' theorem to calculate field at a distance r from uniformly charged sphere.

Example: Find the electric field intensity around the charged conducting sphere shown in Fig. 18–10.

Solution: Here the charge distributes itself evenly over the sphere's surface and the surrounding electric field—from symmetry, together with the vector nature of **E**—is everywhere directed radially. (It is outward when the charge is positive and inward when it is negative.) We draw our Gaussian surface as a sphere of radius r outside the charged sphere. Here **E** is perpendicular to this surface (since it is radial), and so

$$q = \epsilon_0 \int_0^{4\pi r^2} E_n \, dA = 4\pi\epsilon_0 r^2 E$$

or

$$E = \frac{1}{4\pi\epsilon_0}\frac{q}{r^2}, \qquad r > \text{radius of sphere},$$

where q is the total charge on the sphere. This result means that a charged conducting sphere sets up an external electric field identical to that set up by a single charge equal to the total charge on the sphere located at the center of the sphere. The radius of the sphere is not relevant except that, of course, inside the sphere $\mathbf{E} = 0$.

18–5 Potential

We now have two different ways of describing the situation that exists at a point in space when a charge placed there experiences a force. One employs the scalar quantity *electric potential*, the other the vector quantity *electric field intensity*. These quantities are related in a manner analogous to that in which potential energy is related to force in mechanics: **E** is always directly proportional to the force a test charge experiences at a point, and V to the potential energy it has with respect to some reference point. In the last chapter, in which we introduced electric potential, we confined ourselves to describing it in terms of the individual charges responsible for establishing differences in potential. We shall now discuss potential in terms of the electric field itself, independently of the sources of the field.

When a charge q is placed in an electric field $\mathbf{E}$, it experiences a force

$$\mathbf{F} = q\mathbf{E}.$$

To displace the charge by $d\mathbf{s}$ therefore requires the work

$$dW = \mathbf{F} \cdot d\mathbf{s} = q\mathbf{E} \cdot d\mathbf{s}$$

to be done *by the charge on the field.* (If dW turns out negative, this means that the field $\mathbf{E}$ has done work *on the charge.*) The work dW is the negative of the change in the potential energy qV of the charge, according to the principle of conservation of energy. Hence

$$qV = -q\mathbf{E} \cdot d\mathbf{s}.$$

If we write E_s for the component of $\mathbf{E}$ in the direction of $d\mathbf{s}$,

$$\mathbf{E} \cdot d\mathbf{s} = E_s\, ds$$

and

$$dV = -E_s\, ds.$$

We can rewrite this last equation as

$$E_s = -\frac{\partial V}{\partial s}. \tag{18–14}$$

The component of electric field intensity in any direction at a point is the negative of the partial derivative of the electric potential at that point taken with respect to the coordinate in that direction.

(In calculating a partial derivative with respect to a particular variable, the rules for ordinary differentiation are followed, except that the other variables are treated as constants.)

In a cartesian coordinate system the entire field $\mathbf{E}$ is given by its components

$$E_x = -\frac{\partial V}{\partial x}, \qquad E_y = -\frac{\partial V}{\partial y}, \qquad E_z = -\frac{\partial V}{\partial z}. \tag{18–15}$$

In order to determine $\mathbf{E}$ at a point from a knowledge of V, it is evidently not sufficient merely to know the value of V at that point—we must also know how V varies in the region surrounding the point in order to compute the derivatives.

Example: Determine the electric field of an electric dipole at a distance large relative to the separation of the charges in the dipole.

Solution: In Section 17–5 we found that the potential of a dipole of moment p located at the origin and directed along the x-axis is given by

$$V = \frac{px}{4\pi\epsilon_0(x^2 + y^2)^{3/2}}.$$

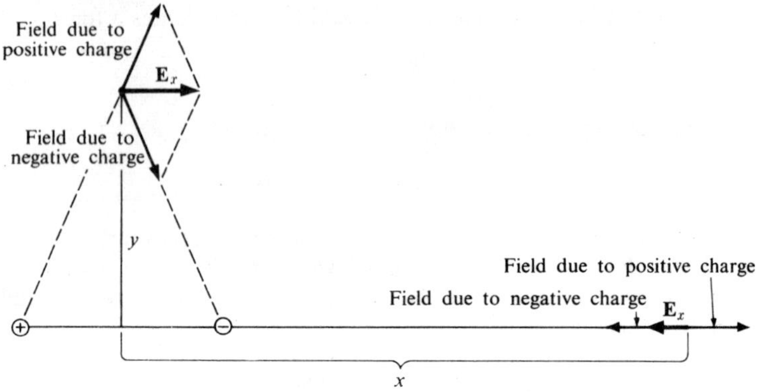

FIG. 18–11. Field of an electric dipole.

The components of **E** at the point (x, y, z) are therefore

$$E_x = -\frac{\partial V}{\partial x} = \frac{p}{4\pi\epsilon_0}\frac{2x^2 - y^2}{(x^2 + y^2)^{5/2}},$$

$$E_y = -\frac{\partial V}{\partial y} = \frac{p}{4\pi\epsilon_0}\frac{3xy}{(x^2 + y^2)^{5/2}},$$

$$E_z = -\frac{\partial V}{\partial z} = 0. \tag{18-16}$$

The electric field of a dipole aligned along the x-axis can be evaluated simply for two points, one on the x-axis and one on the y-axis. In each case there is only one component of the field different from zero, namely E_x. The magnitude of the electric field in each case is

$$E_x = \frac{p}{4\pi\epsilon_0}\frac{2}{x^3} \qquad \text{(on the } x\text{-axis where } y = 0)$$

$$E_x = -\frac{p}{4\pi\epsilon_0}\frac{1}{y^3} \qquad \text{(on the } y\text{-axis where } x = 0). \tag{18-17}$$

The field of a dipole is proportional to the dipole moment, and varies as the inverse-cube power of the distance from the dipole. The direction of the field is shown in Fig. 18–11.

It is interesting to investigate the different aspects of an electric field revealed by the field intensity **E** and the potential V. Consider first a point midway between two identical positive charges q a distance d apart (Fig. 18–12). Here

$$E = 0, \qquad V = \frac{q}{\pi\epsilon_0 d}.$$

FIG. 18-12. Electric potential and electric field midway between two charges.

If we replace one of the charges by a negative charge of the same magnitude,

$$E = \frac{2q}{\pi\epsilon_0 d^2}, \qquad V = 0.$$

In the first case a test charge placed at the specified point would feel no force, but work must have been done in bringing it to that position. Because electric fields are conservative, the same amount of work is recoverable. In the second case a test charge would certainly feel a force, but owing to its location no net work had to be done to bring it there from infinity.

18-6 Electrical and gravitational fields

By this time it should be obvious that there are many similarities between static electrical phenomena and gravitational phenomena. In the latter case we chose to analyze the situation in terms of direct interaction, in the former we chose to use the field concept. In principle the field concept could have been developed in similar detail for gravitational phenomena also.

The entire discussion of potential energy in mechanics could be rephrased now in terms of the concepts of the potential of the gravitational field. We shall not do this here, but it might be instructive to refer to the discussion of gravitation to see how it would go. The principal differences arise from the fact that there are two different types of electrical charges and there are possibilities of charges repelling as well as attracting. In this sense the gravitational case is simpler. However, only in their static aspects is there a resemblance between electrical and gravitational fields. The dynamical aspects of electric and gravitational fields are quite different and few analogies are possible.

Problems

1. Two charges, one of 1.5×10^{-6} coul and the other of 3×10^{-6} coul, are 0.2 m apart. Where is the electric field in their vicinity equal to zero?

2. What is the electric field intensity 0.4 m from a charge of $+7 \times 10^{-5}$ coul?

3. Two charges of $+4 \times 10^{-6}$ coul and $+8 \times 10^{-6}$ coul are 2 m apart. What is the electric field intensity halfway between them?

4. A body whose mass is 10^{-6} kg carries a charge of $+10^{-6}$ coul. What is the magnitude of an electric field that can hold the body suspended in equilibrium near the earth?

5. A particle carrying a charge of 10^{-5} coul starts moving from rest in a uniform electric field whose intensity is 50 volts/m. (a) What is the force on the particle? (b) How much kinetic energy will the particle have after it has moved 5 m?

6. Four charges of $+1 \times 10^{-8}$ coul are at the corners of a square 1 m long on each side. (a) Find the potential at the center of the square. (b) Find the electric field intensity at the center of the square.

7. Four charges of $+5, -8, -10, +8 \times 10^{-8}$ coul are at the corners of a square 0.2 m on a side. Find the electric field intensity at the position of the charge -8×10^{-8} coul. What is the force on this charge?

8. A straight wire of infinite length carries a charge of λ coul/m. Show that the magnitude and direction of the electric field a distance r from the wire are, respectively, $E = \lambda/2\pi\epsilon_0 r$ and perpendicular to the wire.

9. Show that the potential difference between two points r_A and r_B from the wire of the preceding problem is given by $V_{AB} = (\lambda/2\pi\epsilon_0) \ln (r_B/r_A)$.

10. An infinite cylinder of radius a is uniformly filled with charge whose density is ρ coul/m^3. Use Gauss' theorem to find the electric field inside the cylinder.

11. A charge Q is uniformly distributed throughout a spherical volume of radius R. Calculate the electric field inside the sphere as a function of r, the distance from the center. Plot E as a function of r for $0 \leq r \leq \infty$.

12. Two charges, one of 4×10^{-8} coul and the other of -1×10^{-8} coul, are placed 0.1 m apart. The former charge is placed to the left of the latter, which is at the origin. (a) What is the potential as a function of x for $x > 0$? (b) What is the electric field as a function of x for $x > 0$? (c) What is the potential as a function of y for positive y along a line perpendicular to the x-axis at $x = 0.1$ m? (d) Derive an expression for the components of the electric field along this line.

13. A thin wire is bent in the form of a circle of radius r and subtending an angle θ at the center. The wire carries a charge ρ per unit length. Derive an expression for the electric field intensity at the origin.

14. A straight wire of length 10 cm is charged uniformly with a charge density of 10^{-6} coul/m. (a) What is the potential and what is the electric field as functions of the distance R along the perpendicular bisector of the wire? (b) Repeat (a) for distances R along the axis, with the center as the origin, for $R > 5$ cm.

15. The radial component of an electric field (E_r) is given by $-\partial V/\partial r$; the tangential component (E_θ) is given by $-(1/r)(\partial V/\partial \theta)$. Express the potential of a dipole in polar coordinates and calculate E_r and E_θ. Check your results against those given in the text at $\theta = 0$ and $\theta = \pi/2$.

16. The potential is given by the expression

$$V = \frac{ax}{x^2 + y^2},$$

with a constant. Find the x- and y-components of the force on a unit positive charge.

17. The potential in a region $a < x^2 + y^2 < b$ is given by

$$V = -a \ln (x^2 + y^2).$$

(a) Draw the curves in the xy-plane along which the potential is constant. (b) What are the x- and y-components of the electric field? (c) What are the r- and θ-components of the electric field?

18. The potential along the x-axis is given by

$$V = 10\left(\frac{3}{6x + 4} - \frac{1}{10x + 4}\right) \text{ volts.}$$

An electron starts from $x = \infty$ with zero kinetic energy. What is its speed at $x = 0$? What is the electric field at $x = 0$?

19

The Magnetic Field

In the previous chapter we learned that all electrical effects have their ultimate origin in static electric charges. When charges are in motion other effects are observed; the forces that bring these effects about are called *magnetic forces.* We were introduced to the magnetic forces that electric currents (which are charges in motion) exert on one another in Chapter 17, and we shall now consider these forces in some detail.

Let us first review our treatment of electrostatics, so that the similarities and differences between electric and magnetic phenomena will appear in sharp relief. Coulomb's experiments showed that the force $d\mathbf{F}$ between two point charges dq_1 and dq_2 is given by the equation

$$d\mathbf{F} = \frac{dq_1\, dq_2}{4\pi\epsilon_0 r^3}\mathbf{r}, \tag{19–1}$$

where $\mathbf{r}$ is the radius vector joining the charges and ϵ_0 is a universal constant equal to $8.85 \times 10^{-9}\,\text{coul}^2/\text{n·m}^2$. [Note that $\mathbf{r}/r^3 = \mathbf{r}_1/r^2$, so Eq. (19–1) is identical with Eq. (17–4), though expressed in slightly different form. We shall find the notation of Eq. (19–1) more convenient in developing electromagnetism.] Electric forces are better described in terms of the action of a field than as the result of direct action at a distance between centers of force. We define the electric field $\mathbf{E}$ at a point in space as the derivative of the electric force $\mathbf{F}$ there on a charge with respect to charge q taken in the direction of the force, so that

$$\mathbf{E} = \frac{d\mathbf{F}}{dq}. \tag{19–2}$$

Hence the electric field $d\mathbf{E}$ at a distance r from a charge dq is

$$d\mathbf{E} = \frac{1}{4\pi\epsilon_0}\frac{dq}{r^3}\mathbf{r}.$$

We can now, by Gauss' theorem, relate the integral of the electric field over a closed surface to the total charge contained within the surface:

$$\int_{\substack{\text{closed}\\ \text{surface}}} E_n\, dA = \frac{q}{\epsilon_0}. \tag{19–3}$$

Finally, because the electric field is conservative, the concept of electric potential turns out to be a useful one. From a knowledge of the electric field in a region of space we can deduce the electric potential at any point within it, and from a knowledge of the potential we can infer the field intensity; we are therefore able to adopt either approach in a particular calculation.

19–1 Ampere's law

If we are to follow, in the magnetic case, the program outlined above, we must begin by writing down a law of force between infinitesimal current elements. This law cannot be obtained by direct experiment, since it is impossible to isolate a current element in the way a small charge can be isolated. What we seek is that elementary law of force between current elements which yields the experimental results for the interaction between finite currents. Such a law, analogous to Coulomb's law in electrostatics, is: The force on the infinitesimal current element $i_1\, d\mathbf{l}_1$ due to the current element $i_2\, d\mathbf{l}_2$ the distance $\mathbf{r}$ away is given by

$$d\mathbf{F} = \frac{\mu_0}{4\pi} \frac{i_1\, d\mathbf{l}_1 \times (i_2\, d\mathbf{l}_2 \times \mathbf{r})}{r^3}. \tag{19–4}$$

This is Ampere's law. In Eq. (19–4), the force is to be expressed in newtons, the length in meters, and the currents in amperes. Since all the fundamental units in Eq. (19–4) are specified, μ_0 is a universal constant which can be deduced from measurements, just as ϵ_0 was in electrostatics. It is equal to

$$\mu_0 = 4\pi \times 10^{-7}\, \frac{\text{newton}}{\text{ampere}^2}. \tag{19–5}$$

Ampere's law is a very complicated one to visualize. In order to gain some familiarity with it, we shall use it to determine the force between two currents. Consider two parallel infinite currents a distance R apart, which flow in the same direction, as in Fig. 19–1. Before calculating the force per unit length between the two current elements, we first calculate the value of the quantity

$$d\mathbf{B} = \frac{\mu_0}{4\pi} \frac{i_2\, d\mathbf{l}_2 \times \mathbf{r}}{r^3} \tag{19–6}$$

at the position of the current element dl_1; $d\mathbf{B}$ will turn out to be the magnetic analog of the electric field element $d\mathbf{E}$. The magnitude dB is

$$dB = \frac{\mu_0}{4\pi} \frac{i_2}{r^2} \sin\theta\, dl_2. \tag{19–7}$$

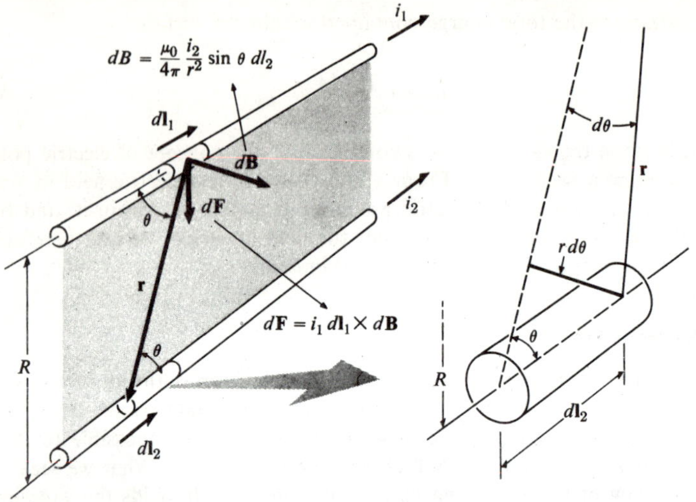

FIG. 19-1. Calculation of the force between parallel currents.

The vector $d\mathbf{B}$ is perpendicular to the plane in which the two wires lie, and points as shown at the position of dl_1, no matter where the current element dl_2 is. We can therefore calculate a value of $d\mathbf{B}$ at the position of dl_1 by adding the effects of all current elements dl_2 along the length of the second wire as if they were scalar quantities. This is a simple problem in integration, and can be carried out by expressing the variables r and dl_2 in terms of a single angle θ and the fixed distance R. From the geometry of Fig. 19-1, we see that

$$dl_2 = (r\,d\theta)\csc\theta = r\csc\theta\,d\theta$$

$$= \frac{r(R\csc\theta)\,d\theta}{R}$$

$$= \frac{r^2\,d\theta}{R}. \tag{19-8}$$

From Eq. (19-7),

$$dB = \frac{\mu_0}{4\pi}\frac{i_2}{r^2}\sin\theta\,dl_2$$

$$= \frac{\mu_0}{4\pi}\frac{i_2}{r^2}\sin\theta\frac{r^2\,d\theta}{R} \tag{19-9}$$

$$= \frac{\mu_0}{4\pi}\frac{i_2}{R}\sin\theta\,d\theta. \tag{19-10}$$

We must integrate from $\theta = 0$ to $\theta = \pi$. Carrying out this integration, we obtain

$$B = \frac{i_2\mu_0}{4\pi R}\int_0^\pi \sin\theta\, d\theta = \frac{\mu_0}{2\pi}\frac{i_2}{R}. \tag{19–11}$$

The direction of the force acting on the current element dl_1 is found from the law of multiplication of vectors to be toward the other wire. The direction of this force is independent of the location of dl_1, and we may add all the force contributions of the current elements dl_2 in scalar fashion. Since $\mathbf{B}$ is perpendicular to $d\mathbf{l}_1$ and $\sin(\pi/2) = 1$, we have for the force

$$dF = \frac{\mu_0}{2\pi}\frac{i_1 i_2}{R}\, dl_1. \tag{19–12}$$

The force per unit length, dF/dl_1, is therefore

$$\frac{dF}{dl_1} = \frac{\mu_0}{2\pi}\frac{i_1 i_2}{R}. \tag{19–13}$$

This is precisely the force per unit length which we previously gave [Eq. (17–16)] for the case of two parallel infinite currents. The force is an attractive one for currents which are flowing in the same direction and a repulsive one for currents which are flowing in opposite directions.

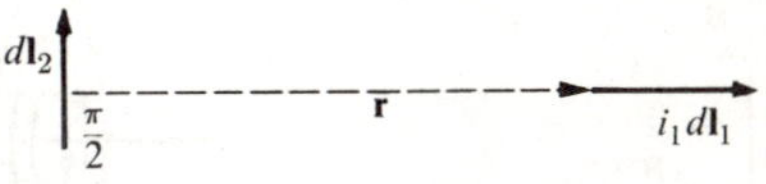

FIG. 19–2. For infinitesimal current elements, Newton's third law is not necessarily satisfied.

An unusual aspect of Ampere's law is that, for infinitesimal current elements, Newton's third law of motion is not satisfied. Let us consider two infinitesimal currents placed perpendicular to each other, as in Fig. 19–2. The force on the element $i_2\, dl_2$ is zero, since the sine of the angle between the vector $\mathbf{r}$ and the vector $i_1\, d\mathbf{l}_1$ is zero. On the other hand, the force on $i_1\, dl_1$ due to $i_2\, dl_2$ is *not* zero, since the angle between $i_2\, d\mathbf{l}_2$ and $\mathbf{r}$ is 90°, and the sine of 90° is 1. The action and reaction forces are quite different. The mathematical reason for the failure of this law is that the current elements $i_1\, d\mathbf{l}_1$ and $i_2\, d\mathbf{l}_2$ do not appear in Ampere's law in a symmetrical way. In other words, we *cannot* exchange $i_1\, d\mathbf{l}_1$ and $i_2\, d\mathbf{l}_2$ in Eq. (19–4) and get the same result, while we *can* exchange dq_1 and dq_2 in Coulomb's law, Eq. (19–1), without affecting $d\mathbf{F}$. However, if we integrate $d\mathbf{F}$ around an entire circuit of which $i_1\, d\mathbf{l}_1$ and $i_2\, d\mathbf{l}_2$ are simply parts, the total resultant force equals zero. We shall return to this interesting feature of magnetic forces later in this chapter.

19–2 Magnetic induction

Continuing with our approach, we now attempt to describe the force between two current elements as an interaction between a field and a current element. We are evidently able to express the force on the current element $i_1\, d\mathbf{l}_1$ as

$$d\mathbf{F} = i_1\, d\mathbf{l}_1 \times d\mathbf{B}. \tag{19–14}$$

The quantity **B** is called *magnetic induction*, or alternatively, *magnetic flux density*. The unit of magnetic induction is the weber/meter^2, which is the same as the newton/amp-meter. By comparing Eq. (19–14) with Ampere's law, we see that the magnetic induction $d\mathbf{B}$ due to a current element $i_2\, d\mathbf{l}_2$ at $\mathbf{r}$ from the current element is given by

$$d\mathbf{B} = \frac{\mu_0}{4\pi}\,\frac{i_2\, d\mathbf{l}_2 \times \mathbf{r}}{r^3}. \tag{19–15}$$

The magnetic induction corresponds in magnetic phenomena to the electric field in static electrical phenomena.

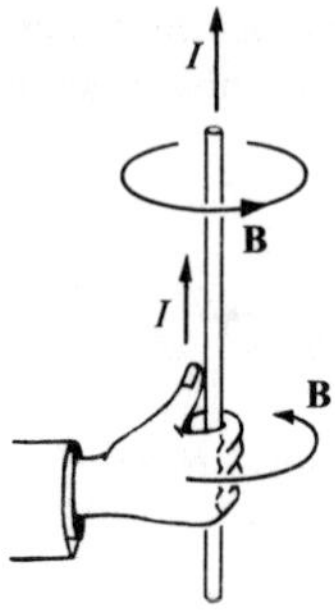

FIG. 19–3. The right-hand rule.

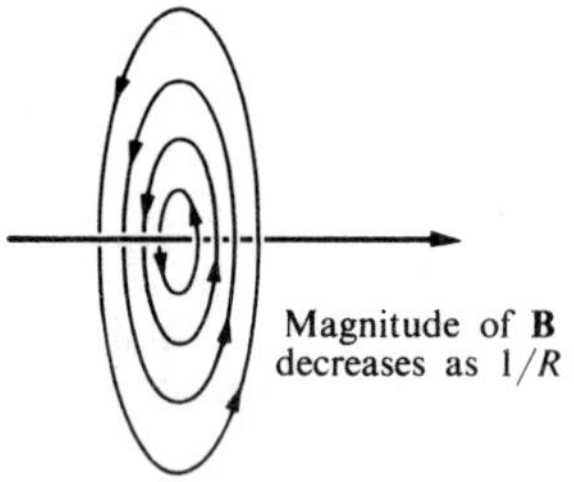

FIG. 19–4. Magnetic inductions surrounding wire carrying current.

A convenient method for obtaining the direction of the magnetic field around a long straight current is the so-called right-hand rule: The thumb of the right hand is placed in the direction of the current and the fingers are in the direction of **B**. This procedure is shown in Fig. 19–3. The magnitude of **B** decreases inversely as the distance from the current. In Fig. 19–4 we have drawn the lines of magnetic induction surrounding a current-carrying wire. The lines of magnetic induction are closed; this is in contrast to the electric lines of force which originate on positive charges and terminate on negative ones. There are no sources for magnetic induction in the sense that lines of force originate on them. The property that lines of magnetic induction be closed is not peculiar to the geometrical configuration we have considered here, but holds for all currents.

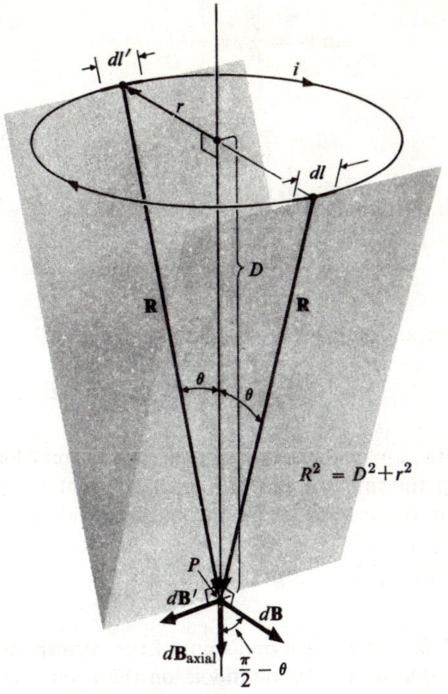

FIG. 19–5. The calculation of magnetic induction due to a circular current.

In order to become more familiar with the formula defining magnetic induction, we shall now calculate the magnetic induction at a point P on the axis of a circular current loop (Fig. 19–5).

The loop carries the current i, its radius is r, and P is a distance D away from the center of the loop. The contribution of the current flowing through the element of wire dl to the magnetic field at P is $d\mathbf{B}$, which has the direction shown in the diagram. Since the contribution to the magnetic field of the wire element dl' diametrically opposite to dl has a component perpendicular to the axis of exactly the same magnitude but of opposite direction to $d\mathbf{B}$, the sum of the perpendicular magnetic components cancels out. This leaves only an axial component,

$$dB_{\text{axial}} = dB \cos\left(\frac{\pi}{2} - \theta\right) = dB \sin\theta$$

to be summed over the loop. Thus

$$dB_{\text{axial}} = dB \sin\theta = \frac{\mu_0}{4\pi} \frac{i\,dl \sin\theta}{D^2 + r^2}. \tag{19–16}$$

Since

$$\sin\theta = \frac{r}{R} = \frac{r}{\sqrt{D^2 + r^2}},$$

we have

$$dB = \frac{\mu_0}{4\pi}\frac{i\,dl\,r}{(D^2 + r^2)^{3/2}}. \tag{19-17}$$

Integrating dl from 0 to $2\pi r$ (the circumference of the loop), we find that

$$B = \frac{\mu_0}{2}\frac{ir^2}{(D^2 + r^2)^{3/2}}. \tag{19-18}$$

Two special cases are of interest. In the first, we set $D = 0$, and find that

$$B = \frac{\mu_0}{2}\frac{i}{r} \tag{19-19}$$

is the magnetic induction at the center of a circular current loop.

Next we consider the situation in which the field point P is very far away from the current loop. In this case ($r \ll D$), the magnetic induction is given by

$$B = \frac{\mu_0}{2}\frac{ir^2}{D^3}. \tag{19-20}$$

The dependence of B on $1/D^3$ is reminiscent of the corresponding dependence of the electric field E due to an electric dipole on the inverse cube of the distance from the dipole. By rewriting Eq. (19-20) in the form

$$B = \frac{\mu_0}{2\pi}\frac{i\pi r^2}{D^3},$$

we see that, if we define the *magnetic dipole moment* of the current loop as

$$p = i\pi r^2, \tag{19-21}$$

we can express the magnetic induction on the loop axis at the distance D from the loop as

$$B = \frac{\mu_0}{2\pi}\frac{p}{D^3}. \tag{19-22}$$

The quantity πr^2 in Eq. (19-21) is the area A enclosed by the current loop. It is possible to show that, in general, the magnetic moment of a current loop of any shape is

$$p = iA, \tag{19-23}$$

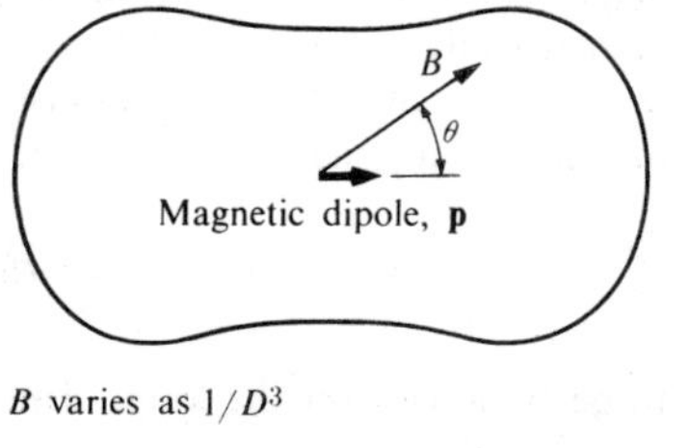

FIG. 19-6. B as a function of θ for fixed distance D near a dipole.

where A is the area enclosed by the loop and i the current. The direction of the vector $\mathbf{p}$ is that of the vector $\mathbf{A}$; that is, perpendicular to the plane of the loop and in the sense given by the right-hand rule. The magnetic field in the vicinity of a magnetic dipole has the same spatial dependence as the electric field of an electric dipole: Eq. (18–7) may be used in the magnetic case simply by substituting $\mathbf{B}$ for $\mathbf{E}$ and $\mu_0 p/2\pi$ for $p/4\pi\epsilon_0$. Figure 19–6 is a plot of B versus θ for a fixed value of the distance D from a dipole. The strength of the field decreases as $1/D^3$.

19–3 Ampere's theorem

In order to establish a theorem analogous to Gauss' theorem for magnetic phenomena, we first calculate the integral of the magnetic induction around a circular path surrounding an infinitely long wire carrying the current i. The result is

$$\oint \mathbf{B} \cdot d\mathbf{s} = \frac{\mu_0}{2\pi}\frac{i}{r}(2\pi r) = \mu_0 i,$$

where $d\mathbf{s}$ is an infinitesimal element of the path. Since the final result does not depend on the radius of the circle, we surmise that the result is independent of the exact shape of the path used in going around the wire. The expression

$$\oint \mathbf{B} \cdot d\mathbf{s} = \mu_0 i \tag{19–24}$$

is, in fact, a general theorem that holds true regardless of the exact path of integration. This theorem is called *Ampere's theorem.* The quantity i can represent the algebraic sum of the currents threading the closed curve, as well as a single current. Equation 19–24 finds its chief utility in computing the magnetic field

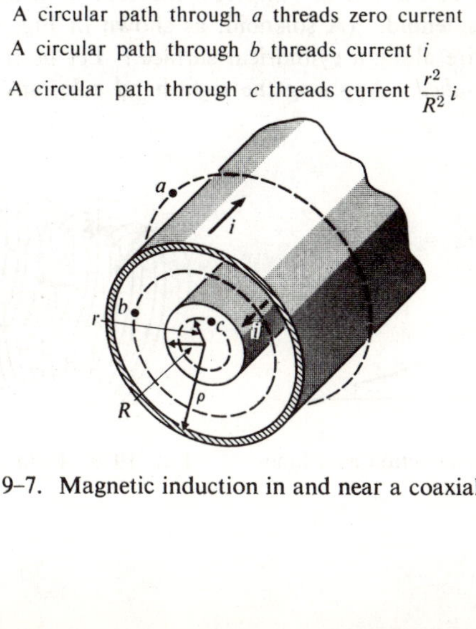

FIG. 19–7. Magnetic induction in and near a coaxial cable.

of a current distribution, just as its electrostatic analog, Gauss' theorem, was useful in computing the electric field of a charge distribution. We shall show, with two examples, how Eq. (19–24) is applied.

A coaxial cable, as in Fig. 19–7, consists of a hollow conductor concentric with a solid one, with equal but opposite currents i flowing in each. What is the magnetic induction at the points a, b, and c? Since the total current enclosed by a curve through a is zero, $B_a = 0$, from Ampere's theorem. A closed curve through b, however, is threaded by the current i in the inner wire, so that

$$B_b = \frac{\mu_0 i}{2\pi r}, \tag{19–25}$$

where r is the distance from b to the axis of the wire. In order to determine B_c, we must first determine the amount of current flowing within a circular cross section of the wire of radius r. Letting R be the outer radius of the wire, and assuming that the current density within the wire is uniform, we have

$$\frac{i_e}{i_t} = \frac{\text{enclosed area}}{\text{total area}} = \frac{\pi r^2}{\pi R^2} \quad \text{or} \quad i_e = \frac{r^2 i}{R^2},$$

and hence

$$B_c = \frac{\mu_0 i r}{2\pi R^2}. \tag{19–26}$$

Figure 19–8 is a graph of B as a function of distance r from the axis of the coaxial conductor. We find the direction of B by the right-hand rule, with thumb of the right hand placed in the direction of i. The lines of flux consist of concentric circles about the wire axis in the direction of the curled fingers of that hand. The same result is obtained by noting the direction of $d\mathbf{l} \times \mathbf{r}$.

Another example of the use of Ampere's theorem is in finding the magnetic induction inside a solenoid. (A solenoid, as shown in Fig. 19–9, is a series of uniform turns of wire along a cylindrical surface.) Let us choose as our closed curve the rectangle $abcd$. Applying the right-hand rule, and assuming that the

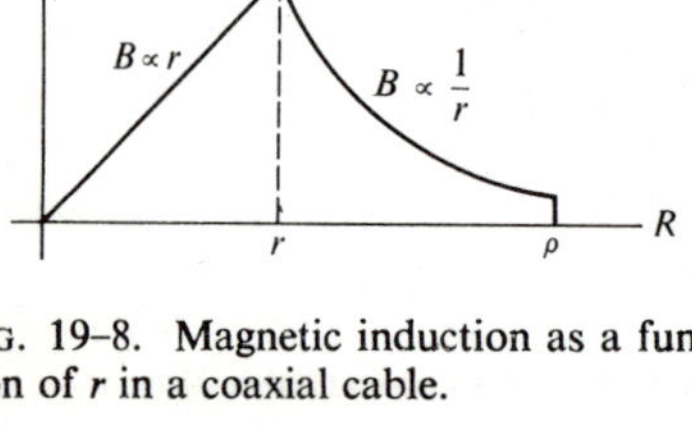

FIG. 19–8. Magnetic induction as a function of r in a coaxial cable.

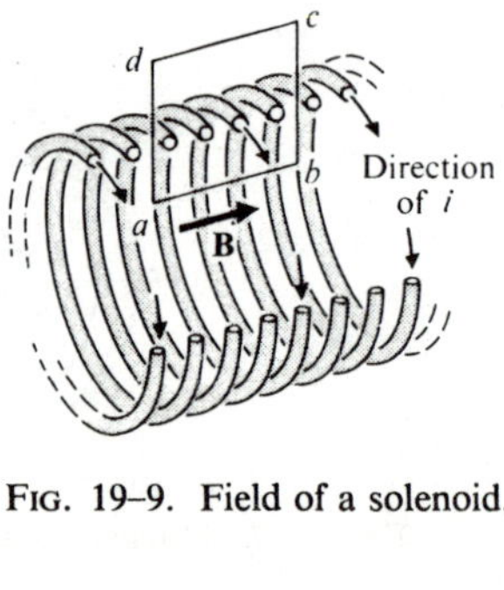

FIG. 19–9. Field of a solenoid.

solenoid is in effect infinite in length (which means, in practice, that the length is many times greater than its diameter), we see that the lines of flux within the solenoid are straight and parallel to the axis. Hence $\mathbf{B} \cdot d\mathbf{s} = 0$ along *ad* and *bc*, where **B** is perpendicular to these sides of the rectangular loop. If we made *ad* and *bc* long enough, **B** would be as close as we chose to zero along *dc*. Hence only *ab* contributes to $\oint \mathbf{B} \cdot d\mathbf{s}$. If there are N turns of wire between a and b, the current enclosed by *abcd* is Ni, and

$$\oint \mathbf{B} \cdot d\mathbf{s} = \int_{ab} B\, dl = Bl,$$

where l is the length of *ab*. By Ampere's theorem,

$$\oint \mathbf{B} \cdot d\mathbf{s} = \mu_0 Ni.$$

Hence

$$Bl = \mu_0 Ni$$

and

$$B = \mu_0 \frac{Ni}{l} = \mu_0 ni, \tag{19–27}$$

(where n is the number of turns per unit length) is the induction within the solenoid.

19–4 Comparison of electrostatics and electromagnetism

If we were to continue our parallel development of electromagnetism and electrostatics, we should now proceed to introduce a potential for magnetic phenomena. Unfortunately the existence of Ampere's theorem makes this more difficult in the case of magnetism than in the case of electrostatics. The integral of the magnetic induction around a closed path is zero or $\mu_0 i$, depending on whether the path has encircled a current or not. Thus one cannot define a potential so easily here. It turns out that one *can* define a potential from which one can compute magnetic induction, but it is a much more complex affair than in the electrostatic situation.

The second difference between electrostatics and electromagnetism arises from the fact that the field lines of the magnetic induction are always closed. This means that there are no magnetic sources in the same sense as there are static electric sources. The sources of static electric fields are charges, and the lines of force originate on them. The sources of magnetic induction are moving charges. No one has yet isolated sources of magnetism on which magnetic lines of force originate. The sources of magnetism we know are currents, the magnetic lines of force are closed, and because of this the nature of the magnetic field is quite different from that of the electric field.

Let us return to the failure of Newton's third law of motion in magnetic effects. This is another feature of electromagnetism with no electric analog. Considerable thought has been given to the status of this law in relation to Ampere's law, the conclusion being that momentum must be a property of magnetic induction.

On the basis of this assumption, the principle of the equality of action and reaction forces holds, because the magnetic field itself is now involved as well as the current elements themselves. The field concept was originally introduced as a convenience, but we now see that the magnetic field must have some degree of physical reality since momentum is associated with its presence. We shall see later that a magnetic field (like all fields) also contains energy. We tend to think that only particles can have energy and momentum; the fact that magnetic fields are in this respect similar to particles has noteworthy implications which we shall explore subsequently.

19-5 The interaction of a current and a magnetic field

We began our discussion of magnetic phenomena in this chapter with a discussion of the force between two current elements, which we saw could be expressed as an interaction between a current and a magnetic induction. In this section we shall discuss some further interactions of a current configuration and a magnetic induction. For simplicity we consider, as in Fig. 19-10, a rectangular loop of wire *abcd* which is carrying a current i in a clockwise direction such that $\mathbf{B}$ lies in the plane of the loop. As indicated in the figure, the length of the loop is s and its width is t. We now wish to calculate the net force on the loop carrying the current due to the presence of magnetic induction. The forces on sides *ab* and *cd* are zero, since the force is equal to $\mathbf{i} \times \mathbf{B}$ and the current and the magnetic induction are directed parallel to one another. The force on the side *bc* is also given by $\mathbf{i} \times \mathbf{B}$, and its direction is found, by the rule for cross products, to be out of the paper. The magnitude of the force on *ad* is the same as the magnitude on the side *bc* and is into the paper. Consequently the net force on the loop is equal to zero. However, there is a resultant *torque* N on it equal in magnitude

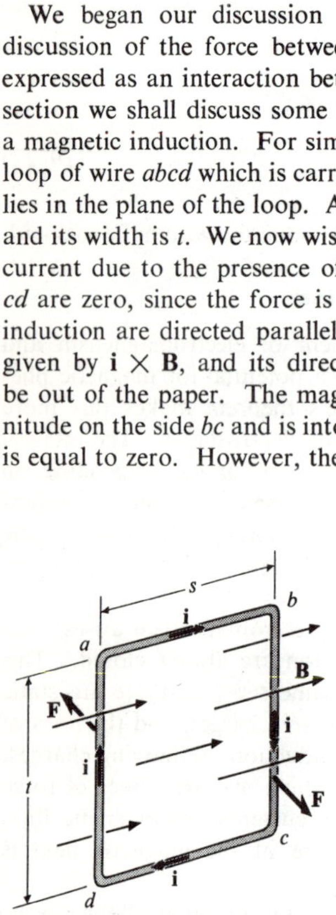

FIG. 19-10. Forces on a current-carrying coil perpendicular to a magnetic induction **B**.

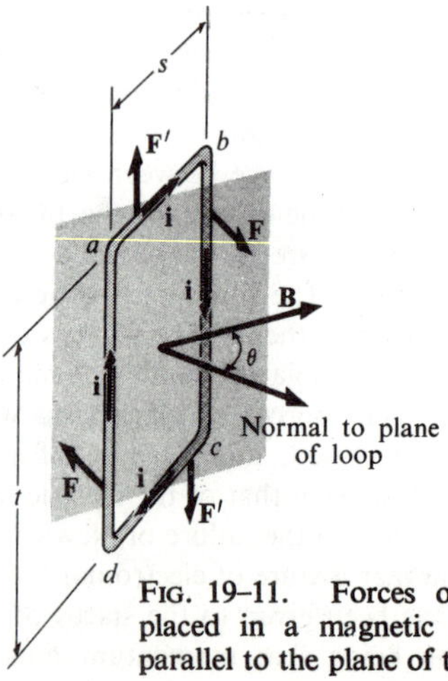

FIG. 19-11. Forces on a coil placed in a magnetic field not parallel to the plane of the coil.

to $stiB$. That is, since the area A of the loop is st,

$$N = iBA. \tag{19-28}$$

If the normal to the plane of the loop makes an angle θ with the direction of the magnetic field, as in Fig. 19-11, the forces on the sides ab and cd are no longer equal to zero, but are in the directions indicated in the figure. The force on ab is equal and opposite to the force on side cd, so that the net forces on sides ab and cd are equal and opposite, and contribute neither to the net force nor the net torque on the loop. The forces on the sides bc and ad, which are in the same directions as they were in the previous example, contribute nothing to the net force on the loop, but *do* contribute a torque, this time equal to

$$N_A = iBt\frac{s}{2}\sin\theta = \text{torque due to force on } bc,$$

$$N_B = iBt\frac{s}{2}\sin\theta = \text{torque due to force on } ad.$$

The lever arm is no longer $s/2$, but $(s/2)\sin\theta$. Since N_A and N_B are in the same direction, the total torque $N = N_A + N_B$ equals

$$N = iBts\sin\theta.$$

This equation can be rewritten

$$N = iBA\sin\theta. \tag{19-29}$$

We see that a loop of wire placed in a uniform magnetic field with the axis of the loop perpendicular to the magnetic field will rotate until the axis points in the direction of **B**. If we were to again define a magnetic moment for this loop of magnitude equal to iA whose direction is perpendicular to the plane of the loop and related to the direction of the current by the right-hand rule, we could further simplify the expression for the torque and write it as just

$$\mathbf{N} = \mathbf{p} \times \mathbf{B}. \tag{19-30}$$

This result is independent of the particular geometrical configuration of the loop, and, as one would suspect, can be derived for a plane loop of any shape. If a loop of N turns is involved instead of a single loop, its magnetic moment is NiA. To summarize, if a loop of magnetic moment **p** is placed in a uniform magnetic field, it will rotate until the magnetic moment of the loop is parallel to the direction of the magnetic induction **B**, and

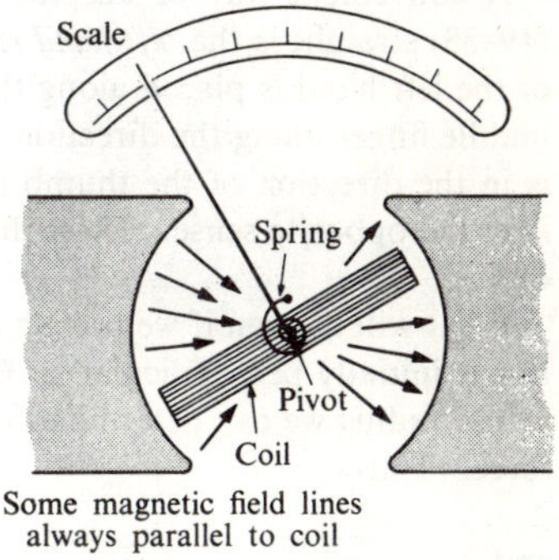

FIG. 19-12. Galvanometer.

then will remain in equilibrium. The magnitude of the torque is directly proportional to the magnitude of the current in the loop.

The principle of operation of current-measuring devices (called *galvanometers* or *ammeters*) is just the one we have been describing. A coil of wire is mounted as in Fig. 19–12 so that it is free to rotate in a magnetic field of the configuration illustrated. The torque on the coil is proportional to the current passing through it (and independent of θ, since the field is always perpendicular to the coil). When we provide a restoring torque, proportional to the angular displacement, with a helical spring, this displacement will be a measure of the current.

19–6 The Lorentz force

Until this point we have considered only the interaction between a magnetic field and an electric current. Since currents actually consist of moving charges, we should be able to obtain an expression for the force on an individual moving charge in a magnetic field. The force is known as the *Lorentz force*, and it can be deduced from the force on a current element in the following way. Since

$$i = dq/dt, \qquad (19\text{–}31)$$

and

$$d\mathbf{l}/dt = \mathbf{v}, \qquad (19\text{–}32)$$

Eq. (19–14) may be rewritten

$$d\mathbf{F} = dq(\mathbf{v} \times \mathbf{B}),$$

or simply

$$\mathbf{F} = q(\mathbf{v} \times \mathbf{B}). \qquad (19\text{–}33)$$

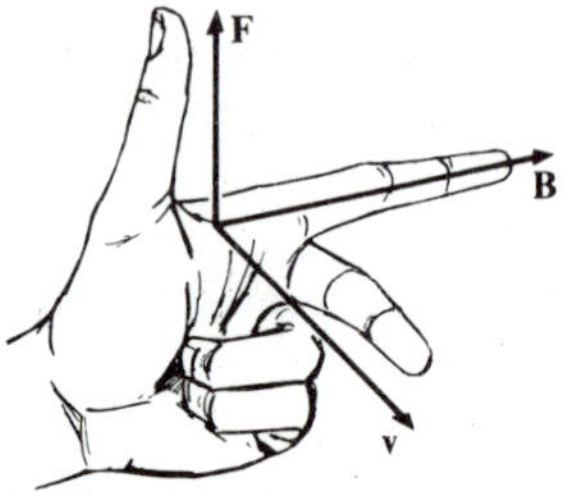

FIG. 19–13. Left-hand rule.

In Eq. (19–33), **F**, the Lorentz force, is the force experienced by a particle of charge q which is moving with the velocity $\mathbf{v}$ in a magnetic field of induction **B**.

A convenient way of keeping the directions of the various quantities in Eq. (19–33) straight is the *left-hand rule* for the Lorentz force: When the index finger of the left hand is placed along the direction of the magnetic induction **B** and the middle finger along the direction of $\mathbf{v}$, the force **F** on a positively charged particle is in the direction of the thumb (Fig. 19–13). With a negative particle the force is in the opposite sense. These directions can also be deduced from the definition of **F**.

What will happen if we project a charged particle into a uniform magnetic field? If $\mathbf{v}$ is initially perpendicular to **B**, the particle's trajectory will consist of a circle whose radius we can determine by equating the Lorentz force with the centripetal force. Thus

$$qvB = mv^2/R$$

and

$$R = mv/qB. \qquad (19\text{–}34)$$

If **v** makes the angle θ with **B**, the particle moves in a helical path of radius

$$R = \frac{mv}{qB}\sin\theta$$

as in Fig. 19–14.

No matter what the initial relative orientations of **v** and **B**, the effect of the Lorentz force is to change the direction of **v** but not its magnitude. We should like to believe that the total momentum of field and particle is conserved whenever **v** changes because of the Lorentz force. The only way this can occur is for the magnetic field itself to possess momentum which can be transferred to and from the moving charge. The *system of moving charge plus magnetic field* obeys all the conservation laws of mechanics.

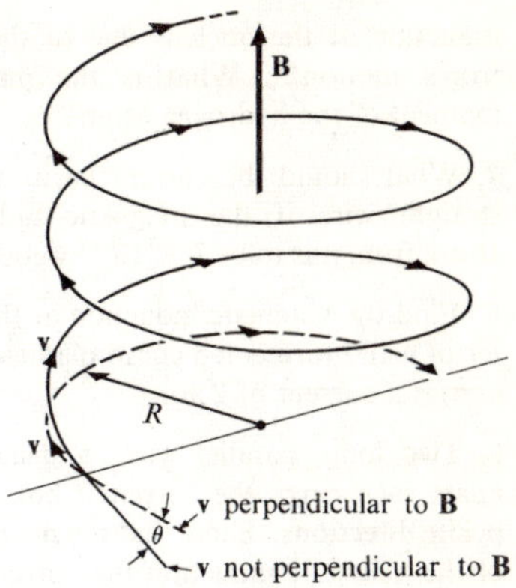

FIG. 19–14. Motion of positively charged particle projected perpendicular to **B** and not perpendicular to **B**.

Historically the Lorentz force on moving charges provided the means whereby the electron was recognized as a distinct entity. The experiments on *cathode rays* performed by J. J. Thomson which established this result will be discussed in Chapter 21.

Problems

1. A current-carrying wire is in a magnetic field. (a) What angle should the wire make with **B** for the force on it to be zero? (b) What should the angle be for the force to be a maximum?

2. A wire 1 m long is perpendicular to a magnetic field of induction 5×10^{-2} weber/m^2. What is the force on the wire when it carries a current of 2 amp?

3. A certain size of copper wire has a mass of 5 kg per 100 m. A length of this wire is placed horizontally in a 3×10^{-2} weber/m^2 magnetic field whose lines of force are horizontal and perpendicular to the direction of the wire. What should the current in the wire be in order for it to be supported against gravity by the magnetic force on it?

4. What should the current be in a wire loop 1 m in diameter if the magnetic induction at the center of the loop is to be 1 weber/m^2?

5. A 50-turn circular coil has a radius of 10 cm. What is the magnetic induction at the center of the coil when the current in it is 0.7 amp?

6. According to a particular model of the hydrogen atom, its electron whose charge is 1.6×10^{-19} coul circles the nucleus 7×10^{15} times per second in an orbit 5×10^{-11} m in radius. What is the magnetic

induction at the nucleus due to the electron's motion? What is the magnetic moment of the hydrogen atom?

7. What should the current be in a long, straight wire if the magnetic induction 10 cm from it is to be 2×10^{-2} weber/m^2?

8. Find the magnetic induction at the center of a 100-turn coil 5 cm in diameter that carries a current of 2 amp.

9. Two long, parallel wires a distance d apart each carry the current i but in opposite directions. Show that the magnitude of the magnetic induction they produce at a point a distance x from the plane of the wires equidistant from each is

$$B = 2\mu_0 i/\pi(4x^2 + d^2)^{1/2}.$$

10. A particle of charge q is moving with a speed v perpendicular to an electric field of intensity E. What should be the magnitude and direction of a magnetic field in the same region that will exactly cancel out the effects of the electric field on the moving charge?

11. An electron moving through an electric field of intensity 500 volts/m and a magnetic field of induction 0.1 weber/m^2 experiences no force. The two fields and the electron's direction of motion are all mutually perpendicular. What is the speed of the electron?

12. What is the radius of curvature of the path of an electron whose velocity is 3×10^6 m/sec in a magnetic field of 0.02 weber/m^2?

13. Compute the radius of curvature in the earth's magnetic field at sea level, assuming $B = 3 \times 10^{-5}$ weber/m^2, of (a) a proton whose speed is 2×10^7 m/sec, and (b) an electron of the same speed. (Proton mass is 1.67×10^{-27} kg; proton charge is $+1.60 \times 10^{-19}$ coul. The electron mass is 9.11×10^{-31} kg; the electron charge is 1.60×10^{-19} coul.)

14. An electron in a television picture tube travels at a speed of 3×10^7 m/sec and is acted upon both by gravity and by the earth's magnetic field. Which of these exerts the greater force on the electron? (See Problem 13.)

15. A charge of $+10^{-6}$ coul is moving with a speed of 5×10^2 m/sec along a path parallel to a long, straight wire and 0.1 m from it. The wire carries a current of 2 amp in the same direction as the velocity of the charge. What is the magnitude and direction of the force on the charge?

16. A long, straight wire carries a current of 100 amp. (a) What is the force on an electron traveling parallel to the wire, in the opposite direction to the current, at a speed of 10^7 m/sec when it is 10 cm from the wire? (b) Find the force on the electron under the above circumstances when it is traveling perpendicularly toward the wire.

17. What is the magnetic induction at the center of a square loop of length l carrying a current i?

18. Two infinitely long wires a distance of 0.01 m apart carry currents of 20 amp each in opposite directions. (a) What is the force per unit length on each wire? (b) Derive an expression for the magnetic induction in a plane perpendicular to the plane of the wires. Express the answer as a function of y, the distance from the plane of the wires, and x, the coordinate perpendicular to y in the plane. (c) Repeat (b) for the case in which the currents are in the same direction.

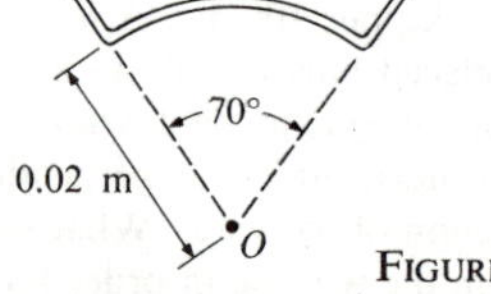

FIGURE 19–15

19. Find the magnetic induction at position O in Fig. 19–15.

20. A uniformly charged loop of radius R carrying a charge q is rotated with an angular velocity $\boldsymbol{\omega}$. The axis of rotation is perpendicular to the plane of the loop and passes through its center. What is the magnetic induction at the center of the loop? What is the magnetic moment of the loop?

21. Compute the magnetic induction at a point P on the axis of a closely wound solenoid of finite length carrying a current i by considering the magnetic effect of a length dx of the solenoid when the number of turns per unit length is n. Express the answer in terms of n, i, θ_1, and θ_2 (see Fig. 19–16).

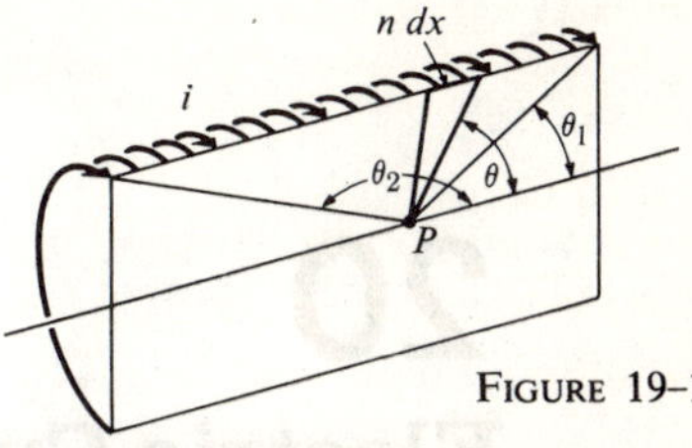

FIGURE 19–16

22. A galvanometer coil has a circular cross section and radius 0.01 m. What is the maximum torque on the coil in an external field of 0.005 webers/m^2 when the coil carries a current of 5 amp?

23. Use Ampere's theorem to calculate the magnetic intensity on either side of an infinite plane sheet of current whose linear current density is j and which is directed parallel to the plane.

20
Electric Current

We have seen that electric currents exhibit magnetic behavior. Hardly less significant are the thermal and chemical effects associated with electric currents, which, apart from their practical importance, are valuable tools for investigating the microscopic nature of matter. In this chapter we shall consider various aspects of steady current flow in conducting materials.

20–1 Introduction

If we establish a difference of potential between the ends of a conductor, charge will flow through the conductor. The rate dq/dt at which charge moves past a point in the conductor has been defined as the current i:

$$i = dq/dt. \tag{20–1}$$

The unit of current is the coulomb/second, which is called the *ampere*. The direction of the current i is conventionally taken as the direction of *positive* current flow, that is, the direction in which positive charges would have to move to produce the observed current. Actual currents in conductors on a microscopic scale may consist of the flow of negative electrons or positive charge. All problems can be analyzed, however, in terms of the conventional current we have defined, which we shall call simply current from now on. It is unnecessary to distinguish between "conventional current" and "electron current."

Electrons are able to move about freely in a metal for reasons we shall consider in a later chapter. We may note in passing that the mobility of electrons in a metal is responsible for the metal's ability to conduct heat as well as electric current. It is not necessary to have a solid conductor in order to have current flow. Currents can also flow through liquids, gases, and even in a vacuum under the proper circumstances. All one requires is the presence of charges that are free to move, and a difference of potential. A mobile negative charge flows spontaneously from a region of low potential to a region of high potential, since establishing a potential difference is equivalent to setting up an electric field which exerts a force on a negative charge opposite to the field direction (Fig. 20–1). In a vacuum tube

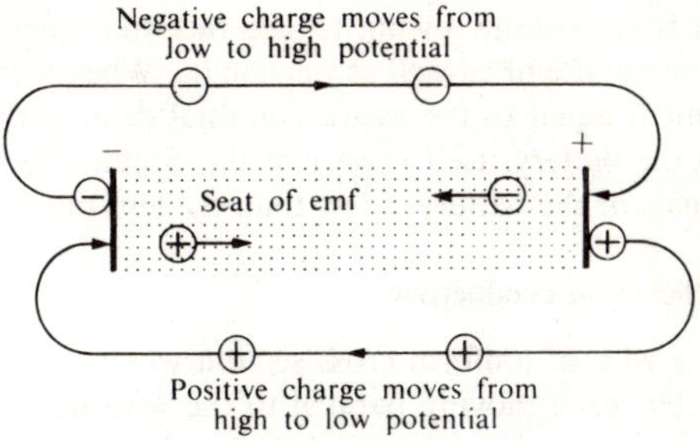

FIG. 20–1. Within a seat of electromotive force, the potential energy of mobile charges is increased.

such as that employed in electronics, the source of mobile charge is the cathode from which electrons are literally boiled off.

Unless the difference in potential between the ends of the conductor is maintained by some external means, the entire conductor ultimately reaches the same potential and the current ceases. However, if we remove the positive charges from the low-potential end as fast as they arrive and do work on them by bringing them back to the high-potential end to replenish the supply there (Fig. 20–1), a steady-state situation results in which the current is constant and does not change with time.

Every time a charge is taken around a circuit, work is done on it. An electrostatic field cannot provide this work, since it is conservative; the net work done by a conservative force field over a closed path is zero. To maintain a steady current, then, we require nonelectrostatic sources of electric fields, which are called sources of electromotive force. Such sources are *batteries*, which convert chemical energy into electrical energy, and *generators*, which convert mechanical energy into electrical energy. (Other sources, for example *thermocouples*, which convert heat into electrical energy, also exist, but are of less practical importance.) Electromotive force is usually abbreviated emf. The electromotive force in a circuit is the amount of work done per unit positive charge by the sources of emf on the charges in a circuit, in carrying them around the circuit.

Even though work is done on charges in a circuit by the sources of emf, we know from experience that energy does not accumulate there. The energy supplied by a battery to an external circuit is dissipated in the circuit, partly as heat and partly as electromagnetic radiation (both of which we shall discuss subsequently). Of course, if there are motors in a circuit, part of the energy appears as mechanical energy. As in electrostatics, the amount of energy per unit positive charge dissipated between two points in the external circuit is called the *potential difference* between those points. The distinction between emf and potential difference is that in the former case work is done on the charge, whereas in the latter case the charge does work on the circuit. Both these quantities are measured in *volts*, since they represent work per unit charge.

This distinction is important, owing to the fact that energy dissipation may occur within the source of emf as well as outside it. When a battery causes a current to flow, its emf is equal to the entire potential drop in the complete circuit. However, because the battery itself is part of the circuit, the potential difference between the terminals of the battery is less than the emf of the battery.

20–2 Elementary theory of conductivity

Let us consider a wire of uniform cross section which contains n charged particles per cubic meter, each moving parallel to the wire with a velocity v. If the charge on each particle is e, in one second the total charge that passes any cross section of the wire per unit time is the amount of charge contained in a cylinder of cross section A and velocity v, since all the charge in the cylinder must have passed through the cross section in one second. The current in the wire at this point is therefore

$$i = nevA. \tag{20–2}$$

If an electron is in a uniform electric field, it is accelerated by the field and its decrease in potential energy is accompanied by an increase in its kinetic energy. During the process of metallic conduction, however, this is not the case. The kinetic energy of an electron on the average remains the same, and the additional energy is given to the conductor itself as heat. The origin of this behavior lies in the fact that, after an electron has been accelerated in the field for a short time t, it undergoes a collision with an atom of the conductor and gives up some of its energy to the atom. This results in a rise in the temperature of the conductor. The motion of the charge as it proceeds along the wire is shown schematically in Fig. 20–2. The velocity of an individual electron is accordingly greater than the average velocity with which the charges drift from one end of the conductor to the other. The velocity of drift is the average increase in velocity an electron acquires between collisions. This is the velocity v which should be put into Eq. (20–2) when calculating the current.

The acceleration of a particle of charge e and mass m in an electric field E is, from $F = ma$,

$$a = eE/m.$$

Hence the velocity it acquires after a time t is

$$v = at = eEt/m.$$

The time between collisions is characteristic of the particular metal. The current in the wire is therefore, from Eq. (20–2),

$$i = \frac{ne^2EtA}{m}. \tag{20–3}$$

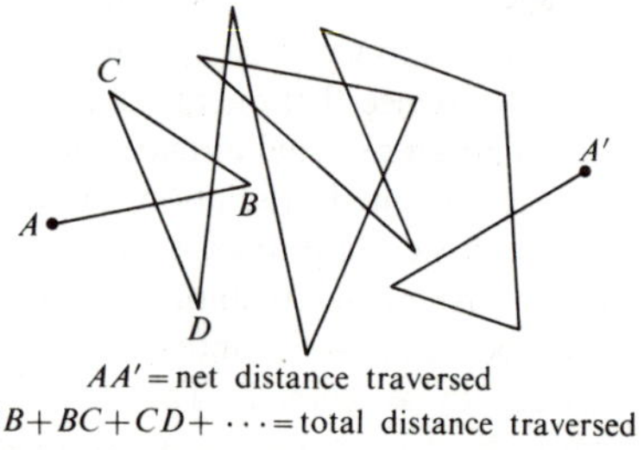

FIG. 20–2. Motion of an electron in a conductor.

If we have a uniform electric field, the potential difference V between the ends of the wire is equal to LE, where L is the length of the wire. Thus we may write Eq. (20–3) as

$$i = \frac{ne^2tAV}{mL}, \tag{20–4}$$

or, equivalently, as

$$V = \frac{L}{A}\left(\frac{m}{ne^2t}\right)i, \tag{20–5}$$

$$V = \frac{L}{A}\rho i. \tag{20–6}$$

Equation (20–6) tells us that the potential difference V between the ends of the wire is proportional to the current i in the wire. The proportionality constant has two parts: one, L/A, which depends on the geometry of the conductor, and another, ρ, which depends on the physical properties of the conducting material through the parameters n, the density of charge, and t, the time between collisions. The quantity ρ is called the *resistivity* of the material. Resistivity must be determined experimentally, since n and t cannot be measured with any accuracy.

The proportionality between V and i is a result of the collisions of the electrons with atoms of the metal. If no collisions occurred, then the smallest potential difference would accelerate the electrons and their velocity would increase without limit. By Eq. (20–2), we would then have an infinite current for any potential difference.

20–3 Ohm's law

It is convenient to write Eq. (20–6) as

$$V = iR, \tag{20–7}$$

where R, given by

$$R = (L/A)\rho, \tag{20–8}$$

is called the *resistance* of a given conductor. Since for a given wire R is observed to be constant over a wide range of current values for metallic conductors, the value of R in a particular case determines the ratio V/i. The unit of R is the *ohm*, where

$$1 \text{ ohm} = 1 \frac{\text{volt}}{\text{amp}}.$$

Equation (20–7) is known as *Ohm's law* in honor of Georg Ohm, who first formulated it.

Typical resistivities at 20°C are given in Table 20–1. The majority of substances fall into the extreme categories of conductors and nonconductors (or *insulators*), but a few, such as carbon, silicon, and germanium, are intermediate in their resistivity. These substances are called *semiconductors*.

TABLE 20-1

Approximate resistivities at 20°C

Substance	Resistivity $\times 10^{-8}$ ohm-m	Substance	Resistivity $\times 10^{-8}$ ohm-m
Aluminum	2.83	Mercury	94
Carbon	3.5×10^3	Phosphorus	1×10^{12}
Copper	1.62	Platinum	11.1
Germanium	2×10^8	Silver	1.60
Gold	2.44	Sulfur	1×10^{17}
Iron	9–13	Nichrome	100
Lead	20.4		

In metals, resistivity increases with temperature. The reason for this behavior is that the higher the temperature, the larger the amplitude of the motion of the atoms in the metal, and the more frequent the collisions with the moving electrons. The reduction in t causes ρ to increase. Semiconductors, on the other hand, are characterized by a supply of mobile electrons much less than that of metals. The principal effect of a rise in the temperature of a semiconductor is an increase in the number of electrons available for carrying current, and the consequent increase in n more than counterbalances the simultaneous reduction in t. Hence the resistivity of a semiconductor decreases with temperature.

Equations (20–6) and (20–7) are useful when the current, resistivity, and potential differences are constant in the conductor. For some applications when this does not hold it is convenient to express Ohm's law in terms of quantities which can be determined as a function of position within the conductor.

We start with Eq. (20–6) and consider an infinitesimal volume element, $d\tau$, around x, y, z whose cross section is dA and whose thickness is dx; dx is in the direction of the electric field at the point x, y, z (Fig. 20–3). In this case, we express the potential difference, according to (20–6), as

$$-dV = V(x + dx) - V(x) = \rho \frac{dx}{dA} di_x.$$

We have written $-dV$ since, for positive current flow toward positive x, $V(x + dx)$ must be less than $V(x)$ (positive current flows from higher to lower potential). We have therefore

$$\frac{di_x}{dA} = -\frac{1}{\rho}\frac{dV}{dx}.$$

But the electric field E_x in the direction of the flow of current is

$$E_x = -\frac{dV}{dx}.$$

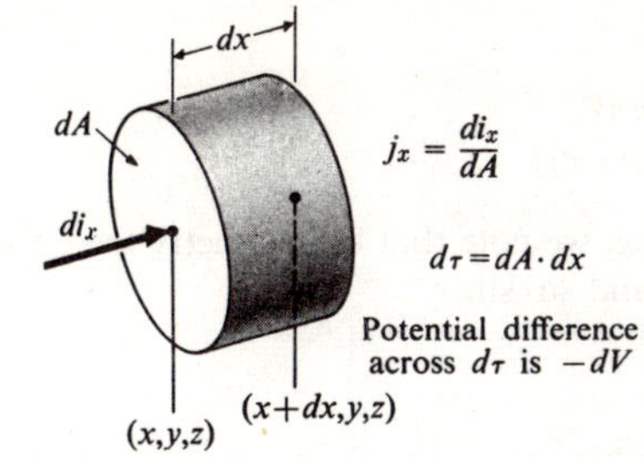

FIG. 20-3. The definition of current density.

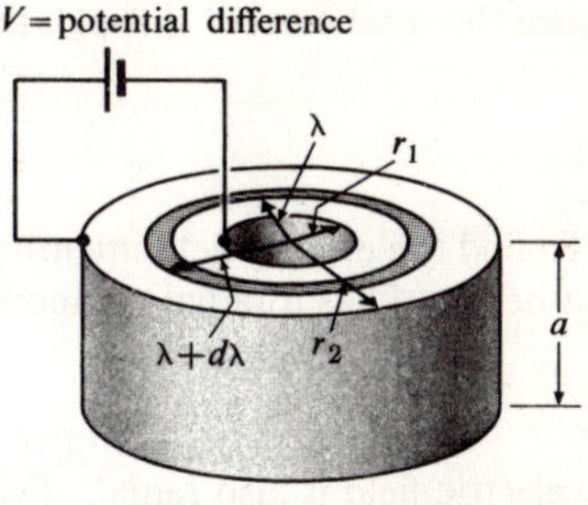

FIG. 20-4. Radial resistance of a cylinder.

The quantity di_x/dA is the current flow per unit area, which we have called the *current density* j_x. This j_x is the component of a vector quantity in the direction of the current, and its magnitude is the current per unit area perpendicular to this direction. We have, therefore, that

$$j_x = \frac{1}{\rho} E_x. \tag{20–9a}$$

When current flows in an arbitrary direction, not along the x-axis, Eq. (20–9a) can be generalized and written in the vector form

$$\mathbf{j} = \frac{1}{\rho} \mathbf{E}. \tag{20–9b}$$

By defining the conductivity of the material σ as $\sigma = 1/\rho$, we can write Eq. (20–9b) as

$$\mathbf{j} = \sigma \mathbf{E}. \tag{20–10}$$

Equation (20–10) is the vector form of Ohm's law. The current density is in the direction of the electric field. Equation (20–10) holds everywhere in a body that is conducting electricity.

As an illustrative example let us consider a metal ring of resistivity ρ whose inner radius is r_1, whose outer radius is r_2, and whose length is a (Fig. 20–4). If we establish a potential difference V between the inner and outer surfaces of the ring, what is the current that flows through the ring? What is the electric field intensity midway between the inner and outer boundaries of the ring?

Our first step is to calculate the radial resistance R of the ring. The resistance dR of a strip between λ and $\lambda + d\lambda$ is, according to Eq. (20–8),

$$dR = \rho \frac{d\lambda}{A} = \rho \frac{d\lambda}{2\pi\lambda a}.$$

Integrating, we have

$$R = \frac{\rho}{2\pi a} \int_{r_1}^{r_2} \frac{d\lambda}{\lambda} = \frac{\rho}{2\pi a} \ln \frac{r_2}{r_1}.$$

Hence the total current through the ring is

$$i = \frac{V}{R} = \frac{2\pi a V}{\rho \ln (r_2/r_1)}.$$

To find the electric field intensity in the ring, we note that by symmetry the current density **j** has a radial component only, and so, since

$$\mathbf{j} = \sigma \mathbf{E},$$

the electric field is also radial. From their definitions we have, for the magnitude of **j** at the midpoint of the ring, where $r = (r_1 + r_2)/2$,

$$j = \frac{i}{2\pi[(r_1 + r_2)/2]a}$$

while

$$\sigma = 1/\rho,$$

and we therefore find, for the electric field intensity in the center of the ring,

$$E = \frac{2V}{(r_1 + r_2) \ln (r_2/r_1)}.$$

20–4 Resistors in combination

For most purposes it is convenient to consider the various resistive elements in an electrical circuit in the form of individual resistors connected by resistanceless wires. There are many ways in which resistors can be put together, but almost all of them consist of a combination of *series* (Fig. 20–5) and *parallel* (Fig. 20–6) arrangements. It is often helpful to know the *equivalent resistance* of a group of resistors in some particular arrangement, that is, the value of the single resistor that can be substituted for the group without changing the characteristics of the circuit.

Let us first examine three resistors, R_1, R_2, and R_3, connected in series as in Fig. 20–5. The potential differences across the resistors are V_1, V_2, and V_3 respectively, and V, the potential difference across the combination, is

$$V = V_1 + V_2 + V_3.$$

The same current i flows through all of the resistors. Therefore, by Ohm's law, $V_1 = iR_1$, $V_2 = iR_2$, and $V_3 = iR_3$. If the equivalent resistance of the combination is R_s, the total potential difference across the three resistors in series is $V = iR_s$. Hence

$$iR_s = iR_1 + iR_2 + iR_3,$$

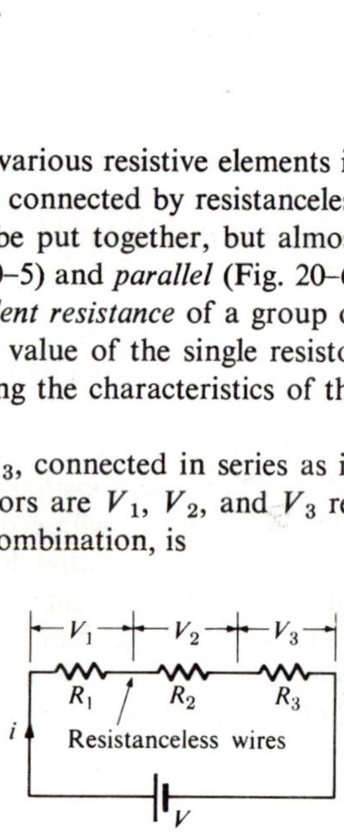

FIG. 20–5. Series arrangement of resistors.

and, dividing by the common factor i,

$$R_s = R_1 + R_2 + R_3. \tag{20-11}$$

Hence in a series arrangement the equivalent resistance is the sum of the individual resistances.

In Fig. 20-6 these resistors are connected in parallel. Here they all have the same potential difference V across their terminals, but each carries a different current. The total current i is the sum of the individual currents, that is,

$$i = i_1 + i_2 + i_3.$$

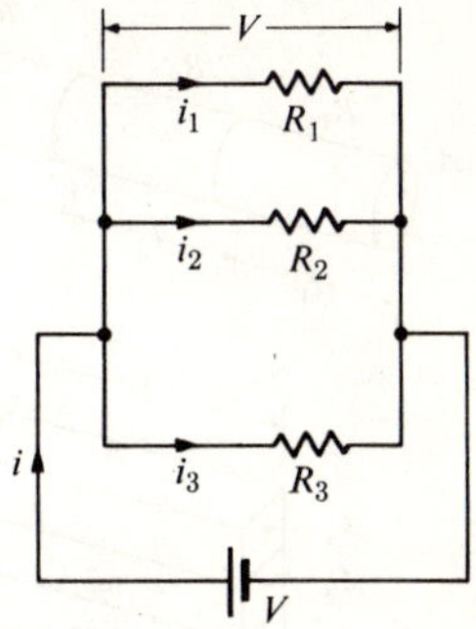

FIG. 20-6. Parallel arrangement of resistors.

Again applying Ohm's law to each resistor, $i_1 = V/R_1$, $i_2 = V/R_2$, and $i_3 = V/R_3$. In terms of the equivalent resistance R_p, the total current i is $i = V/R_p$. Combining these equations,

$$\frac{V}{R_p} = \frac{V}{R_1} + \frac{V}{R_2} + \frac{V}{R_3},$$

so that

$$\frac{1}{R_p} = \frac{1}{R_1} + \frac{1}{R_2} + \frac{1}{R_3}. \tag{20-12}$$

In a parallel arrangement the reciprocal of the equivalent resistance is the sum of the reciprocals of the individual resistances. Here the equivalent resistance is always less than the lowest resistance present.

The results we have obtained for series and parallel arrangements of resistors may be derived directly from the basic formula for the resistance of a single conductor,

$$R = \rho \frac{L}{A}.$$

Connecting resistors in series has the same effect as increasing L, while connecting them in parallel has the same effect as increasing A (Fig. 20-7).

As an example, we shall consider an 11-ohm resistor connected to a battery having an emf of 6 v and an internal resistance of 1 ohm. How much current flows in the circuit, and what is the potential difference across the terminals of the battery?

The total resistance in this circuit consists of the resistor R and the internal resistance r in series. Hence

$$i = \frac{E}{R + r} = 0.5 \text{ amp.}$$

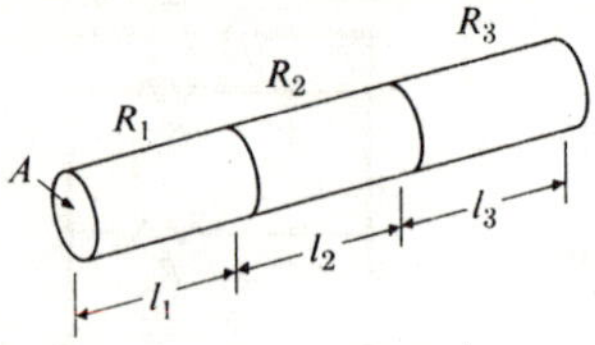

$$l_1+l_2+l_3=L$$

$$R = \rho\frac{L}{A} = \rho\left(\frac{l_1}{A}+\frac{l_2}{A}+\frac{l_3}{A}\right) = \rho(R_1+R_2+R_3)$$

(a)

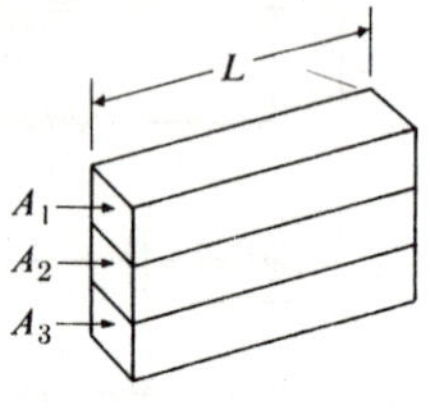

$$A_1+A_2+A_3=A$$

$$R = \rho\frac{L}{A} = \rho\frac{L}{A_1+A_2+A_3}$$

$$\frac{1}{R} = \frac{1}{\rho}\left(\frac{A_1}{L}+\frac{A_2}{L}+\frac{A_3}{L}\right) = \frac{1}{\rho L/A_1}+\frac{1}{\rho L/A_2}+\frac{1}{\rho L/A_3}$$

$$= \frac{1}{\rho}\left(\frac{1}{R_1}+\frac{1}{R_2}+\frac{1}{R_3}\right)$$

(b)

FIG. 20–7. Series and parallel resistors.

The potential difference V across the terminals of the battery may be found in two ways. We can determine V from the iR drop in the external resistor, so that

$$V = iR = 5.5 \text{ v},$$

or we can subtract from the battery emf of 6 v the ir drop in its internal resistance, so that

$$V = E - ir = 5.5 \text{ v}.$$

We note that the potential difference is less than the emf, which, of course, must always be true.

For further practice, let us find the equivalent resistance of the circuit in Fig. 20–8. The method of attacking this problem is shown schematically in Fig. 20–9. First we compute the equivalent resistance R' of R_2 and R_3 in parallel:

$$R' = \frac{R_2R_3}{R_2+R_3} = \frac{6}{7} = 0.86 \text{ ohm}.$$

The R' and R_1 are in series, and their equivalent resistance R'' is just

$$R'' = R' + R_1 = 3.86 \text{ ohms}.$$

Now, in step (3), R'' and R_4 are in parallel, so the equivalent resistance R of the entire circuit is

$$R = \frac{R''R_4}{R''+R_4} = 3.06 \text{ ohms}.$$

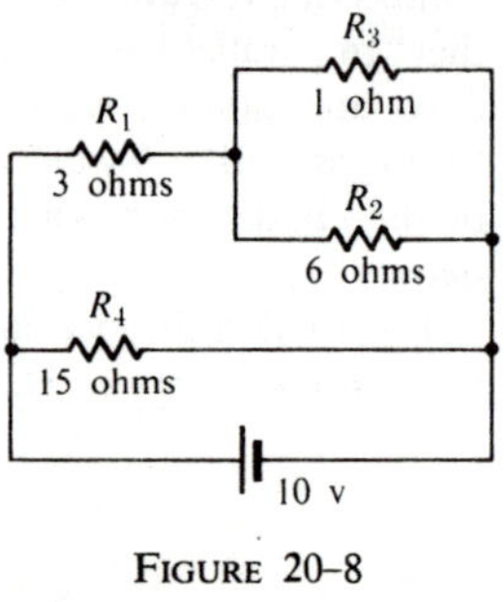

FIGURE 20–8

Let us find the current through R_3 in this circuit if the entire circuit has a potential difference of 10 v placed across it. The current through R'' is, by Ohm's law,

$$i'' = \frac{V}{R''} = \frac{10}{3.86} = 2.59 \text{ amp.}$$

This current goes through both R_1 and R', and so the voltage drop V' across the equivalent resistance R' is

$$V' = i''R' = 2.59 \times 0.86 = 2.23 \text{ v.}$$

Hence the current through R_3 is given by

$$i_3 = \frac{V'}{R_3} = \frac{2.23}{1} = 2.23 \text{ amp.}$$

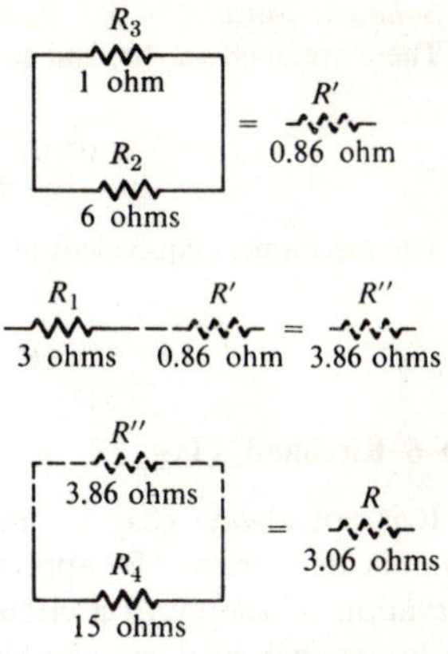

FIG. 20-9. Steps in finding equivalent resistance of circuit of Fig. 20-8.

In solving this problem we have, in effect, gone backward from the single equivalent resistance to the combination of R_1 and R', and then to the original circuit. This procedure is characteristic when currents through individual resistors in a circuit are to be found.

20-5 Power

Energy is dissipated by the moving charges that constitute a current in an electrical circuit as rapidly as the sources of emf in the circuit do work on them. In fact, we found this to be the reason underlying Ohm's law in Section 20-2. The energy dW lost by a charge dq in moving through a potential difference V is

$$dW = V\,dq.$$

Hence the rate of energy loss dW/dt, which equals the power input P, is

$$P = \frac{dW}{dt} = V\frac{dq}{dt} = Vi. \qquad (20\text{–}13)$$

When V is in volts and i in amperes, P is in watts, the mks unit of power. Applying Ohm's law, we may express Eq. (20-13) in the three equivalent forms

$$P = Vi = i^2R = \frac{V^2}{R},$$

in which R is the resistance of the part of the circuit under consideration.

Example: An incandescent lamp of resistance 144 ohms is placed across a potential difference of 120 v. How much heat, in kilocalories, does it produce in 1 hr?

Solution: Since $P = V^2/R$, the rate at which the lamp develops heat is 100 watts. There are 3600 sec/hr, and so

$$W = \int_0^{3600} P\,dt = 3.6 \times 10^5\ \text{j}.$$

The mechanical equivalent of heat is 4186 j/kcal, and therefore

$$W = \frac{3.6 \times 10^5\ \text{j}}{4186\ \text{j/kcal}} = 86\ \text{kcal}.$$

20–6 Kirchhoff's laws

It is not always easy to determine the currents that flow in various parts of a complicated circuit. By applying the principles of conservation of charge and conservation of energy to a circuit in a systematic manner, however, even the most obdurate such problem can be made tractable.

Let us begin with the fact that the total charge in the universe is conserved. A consequence of this is that if we take a volume element somewhere, the difference between the current that flows into it in a given time and the current that flows out of it in the same time equals the net charge that has accumulated there. From experiment it is known that charge never accumulates at any point in a circuit. Thus *the total current flowing into a junction of two or more wires is equal to the total current flowing out of the junction.* This statement is called Kirchhoff's first law.

We have learned that when a current passes through a resistor a certain amount of energy is dissipated. In order to maintain the current at a constant value in a closed loop, then, the total energy lost in the resistances must be equal to the work done by the sources of emf in the loop. From this we conclude that *in going around a closed loop the sum of the potential differences is equal to the sum of the emf's in the loop*, which is Kirchhoff's second law.

In tracing a circuit it is essential to keep in mind these sign conventions: (1) a potential difference is taken as positive in the direction the current is assumed to flow, and negative in the direction opposite to the current; and (2) an emf is taken as positive if we meet the negative terminal in going around the loop, and negative if we meet the positive terminal.

In actually applying Kirchhoff's laws, the first step is to *assume* a current direction in each loop. (If we have chosen the correct one, when we solve for its value a positive figure will result. If we have made an incorrect choice, we will find a negative figure of the correct magnitude in which the minus sign reveals our mistake.) Having assumed currents in each loop, we now apply Kirchhoff's first law to each junction, which provides a number of equations relating the various currents. Then, applying Kirchhoff's second law to each loop, we have additional equations, this time relating the currents, emf's, and resistances in the circuit. In all there will always be as many equations as there are unknown currents in the typical problem in which the emf's and resistances are given, and solving the problem is merely a matter of algebra.

Example: In the circuit shown in Fig. 20–10, let us find the current through each resistor. (External resistances are labeled R_i and internal ones r_i.)

Solution: Applying Kirchhoff's first law to junction A or B, we have

$$i_1 + i_2 = i_3.$$

In the left-hand loop, Kirchhoff's second law gives

$$i_1R_1 + i_1r_1 + i_3r_2 = E_1 + E_2,$$

and in the right-hand loop

$$i_3r_2 + i_2r_3 + i_2R_2 = E_2 - E_3.$$

These equations may be solved for i_1 and i_2 with little difficulty.

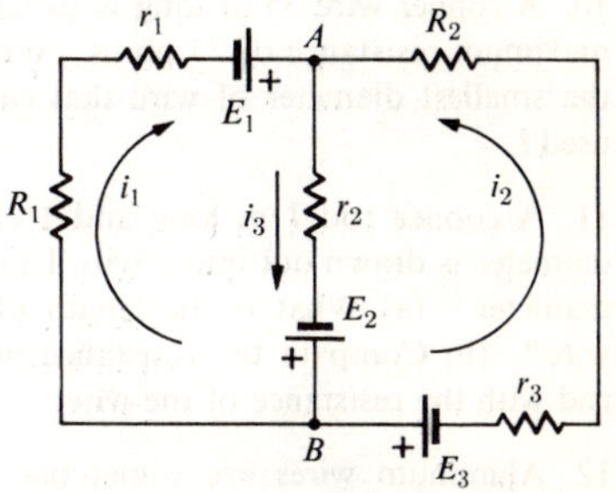

FIGURE 20–10

Problems

1. A certain 12-v storage battery is rated at 80 amp·hr, which means that when it is fully charged it can deliver a current of 1 amp for 80 hr, 2 amp for 40 hr, 80 amp for 1 hr, etc. (a) How many coulombs of charge can this battery deliver? (b) How much energy is stored in it?

2. In a Van de Graaff generator an insulating belt is used to carry charges to a large metal sphere. In a typical generator of this kind, the potential difference between the sphere and the source of the charges is 5×10^6 v. (a) If the belt carries charge to the sphere at a rate of 10^{-3} amp, how much power is required? (b) How much energy will an electron have if it is accelerated by such a potential difference?

3. A current of 3 amp flows through a wire whose ends are at a potential difference of 12 v. How much charge flows through the wire per minute?

4. Approximately 10^{20} electrons/cm participate in conducting electric current in a certain wire. (That is, 10^{20} electrons in each centimeter of the wire are in motion when a current is being carried by the wire.) What is the average drift speed of the electrons when there is a current of 1 amp in the wire?

5. Do bends in a wire affect its electrical resistance? Explain your answer.

6. How many electrons flow through the filament of a 120-v, 60-watt electric light bulb per second?

7. It is sometimes said that an electrical appliance "uses up" electricity. What does such an appliance actually use up in its operation?

8. Design a Nichrome heater to dissipate 200 watts operating at a line of voltage of 110 v.

9. An electric water heater has a resistance of 12 ohms and is operated from a 120-v power line. If no heat escapes from it, how much time is required for it to raise the temperature of 40 kg of water from 15°C to 80°C?

10. A copper wire 35 m long is to have a maximum resistance of 25 ohms. What is the smallest diameter of wire that can be used?

11. A copper rod 1 m long and 1 cm in diameter is drawn out into a wire 1 mm in diameter. (a) What is the length of the wire? (b) Compare the resistance of the rod with the resistance of the wire.

12. Aluminum wires are sometimes used instead of copper wires to transmit electric power. What is the ratio between the masses of an aluminum and a copper wire of the same length whose cross-sectional areas are such that they have the same resistance? (The densities of aluminum and of copper are, respectively, 2.70×10^3 and 8.89×10^3 kg/m^3.)

13. A steady current i is carried radially between two concentric spheres of radius r_1 and r_2. The medium between them has a conductivity σ. What is the current density and electric field midway between them? What is the resistance between the electrodes?

14. Two square plates of area A are immersed in a medium of conductivity σ at a distance d apart. What is the resistance of the combination?

15. Ten cells, each with an emf of 1.7 v, have internal resistance of 2 ohms. What is the current (a) when the cells are connected in series to an external resistance of 100 ohms? (b) when the cells are connected in parallel to the same external resistance? (c) when 5 cells are connected in parallel, and this combination connected in series with the other five cells and the 100-ohm external resistance? (d) What is the power dissipated in the circuit in each case?

16. What is the resistance of a circuit of 12 identical wires, each of resistance 1 ohm, which are arranged as edges of a cube (Fig. 20–11), when the external circuit is connected to the opposite corners a, b? [*Hint:* Imagine that a current i enters the terminal a, and determine how the currents divide up in the various arms. Finally calculate the potential difference along any path from a to b.]

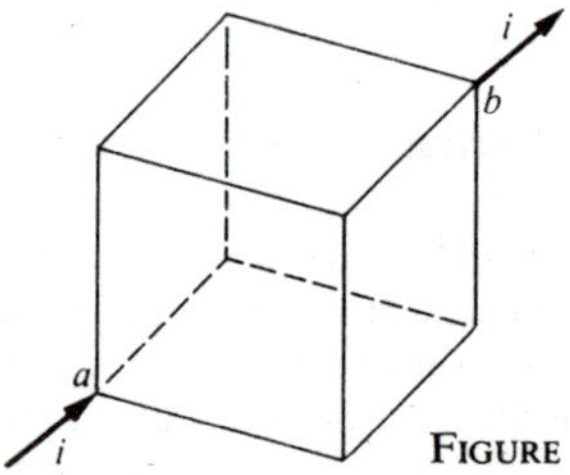

FIGURE 20–11

17. Given two batteries, one with emf 9 v, internal resistance 0.8 ohm, the other 3 v, internal resistance 0.4 ohm. What is the current in the external circuit when these batteries are connected in series to an external resistance of 0.4 ohm? What is the current when the batteries are connected in parallel?

18. A current of 0.5 amp flows to the right in the portion of the circuit shown in Fig. 20–12. Point a is at a potential of

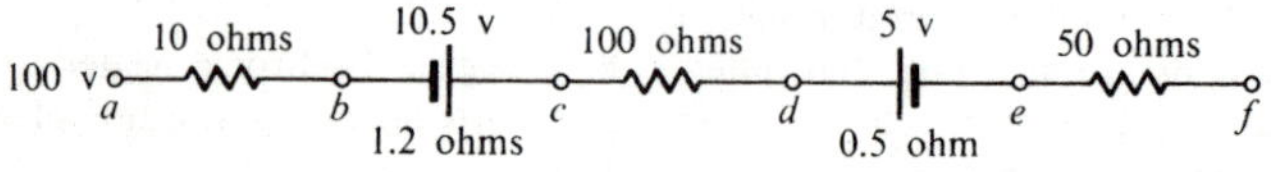

FIGURE 20–12

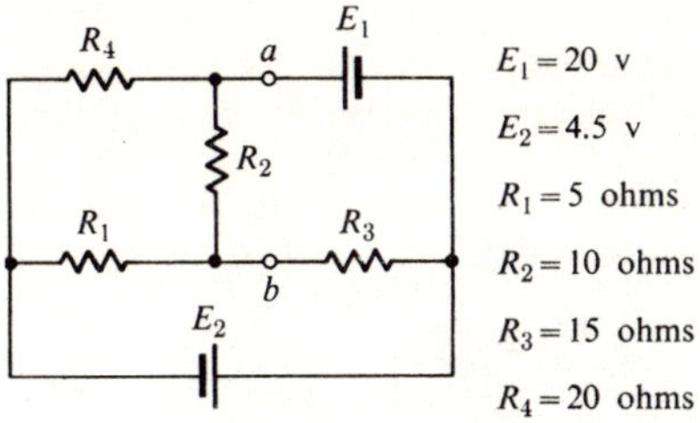

FIGURE 20–13

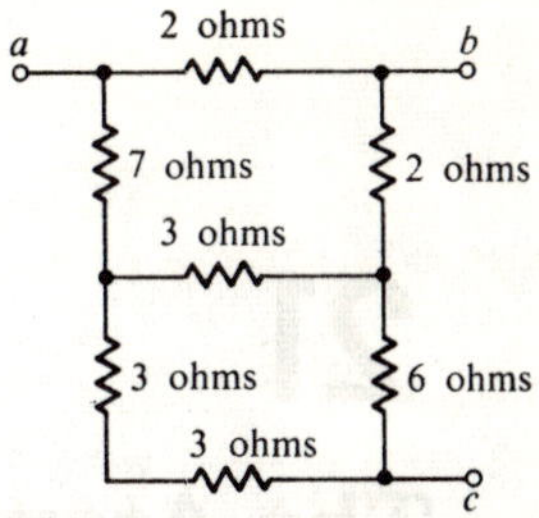

FIGURE 20–15

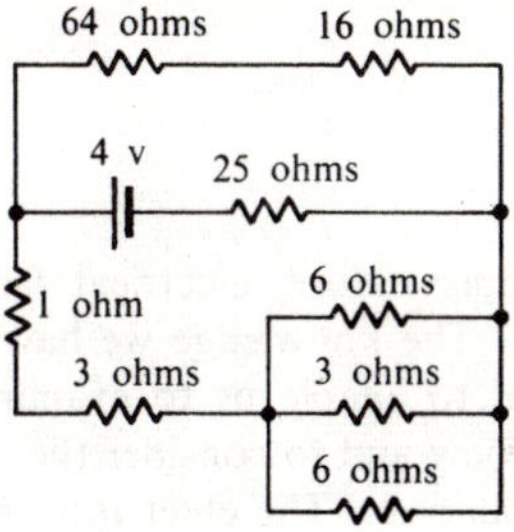

FIGURE 20–14

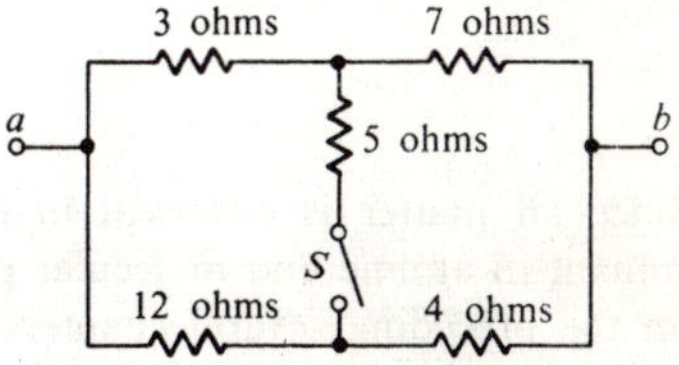

FIGURE 20–16

+100 v. (a) What is the potential at the points b, c, d, e, and f? (b) Repeat for a current flowing in the opposite direction.

19. A galvanometer has an internal resistance of 2000 ohms and gives a full-scale deflection for a current of 2.5×10^{-3} amp. What resistance placed in series will convert it to a voltmeter capable of nearing 100 v at maximum deflection? How could one convert it to an ammeter with a 0–1 amp range?

20. Find the current through E_2 and the potential difference between a and b in Fig. 20–13.

21. Find the current in each branch of the circuit shown in Fig. 20–14.

22. (a) Find the equivalent resistance between the points a and b in Fig. 20–15. (b) Find the equivalent resistance between the points a and c in Fig. 20–15.

23. In Fig. 20–16: (a) Find the equivalent resistance when the switch S is open. (b) Find the equivalent resistance when the switch S is closed.

21

The Atomicity of Charge

Since all matter is electrical in its microscopic nature, electrical forces are dominant in atomic and molecular phenomena. The knowledge we have gained from the preceding several chapters is sufficient to enable us to examine some of these phenomena from an electrical point of view and to consider the methods by which our information on them has been obtained. The chief topic we shall consider now is the quantization of electric charge, the fact that charge is not continuous and capable of being infinitely subdivided but instead always appears in multiples of a certain fundamental value. That charge is quantized is not at all obvious from macroscopic measurements, since the elementary quantum of charge is very small—1.602×10^{-19} coul, in fact. The first indication that charge is quantized came from electrochemical experiments performed by Faraday in the early part of the nineteenth century. Rather than considering these experiments in the way in which they were originally regarded, we shall discuss them in terms of charge quantization.

21–1 Electrolysis

When a pair of electrodes is connected to a source of emf, such as a battery, and immersed in chemically pure water, as in Fig. 21–1(a), no current flows. Chemically pure water (or any other nonconducting liquid) evidently contains no free charges that can migrate from one electrode to the other under the influence of the electric field in the water. However, if a salt such as silver chloride is dissolved in the water, electric current does flow through the external circuit, and can be detected by an ammeter (Fig. 21–1b). The chemical result of the current flow is the liberation of metallic silver at the negative electrode (which is called the *cathode*) and the liberation of gaseous chlorine at the positive electrode (which is called the *anode*). The net result of passing a current through a silver chloride solution, then, is the decomposition of silver chloride into its constituent silver and chlorine atoms; the chlorine atoms combine to form chlorine molecules which are liberated at the anode.

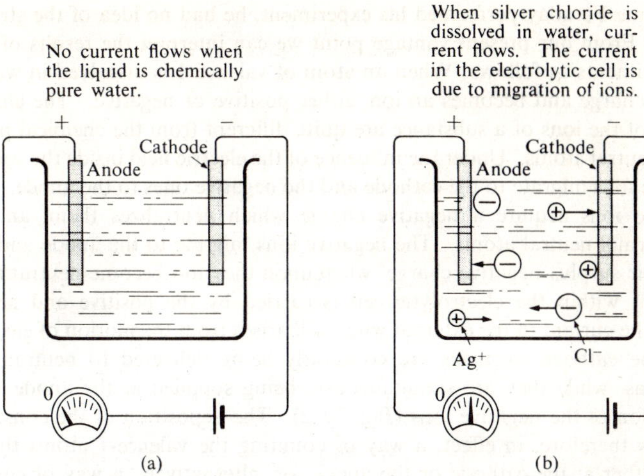

FIG. 21-1. Electrolytic cells.

Faraday, the pioneer investigator of electrochemical effects such as that described above, was curious about whether there is any regularity in the amount of material deposited at the electrodes. The result of his studies of a wide variety of dissolved substances was a simple formula for the amount of material M liberated at either electrode:

$$M = \frac{1}{F}\frac{A}{v}Q. \tag{21-1}$$

In this equation M is the mass deposited, A is the atomic weight of the liberated material, v its valence, Q the total charge that has passed through the solution, and F a universal constant (today called the *faraday*) that is independent of the material involved. The charge Q is the product it of the current i in the circuit and the time t during which it was allowed to flow. From Eq. (21-1) we can see that F is the amount of charge required to deposit a weight (in kilograms) of the substance equal to its atomic weight divided by its valence. This is clear since, when F equals Q, M is just A/v in kilograms. The remarkable constancy of F, which is independent of the electrolyte used and of the mode by which the experiment is performed, indicates that the deposition of chemicals by electrical methods is essentially a counting process. For the moment let us consider only those materials which have the valence 1. It requires F coulombs to deposit 1 kmole weight of the substance, independent of its chemical composition. If we now think about what it is that is constant about a kilomole, we recall that it is *the number of atoms or molecules present*, a vital clue to what is actually taking place in the experiment.

At the time Faraday performed his experiment, he had no idea of the structure of matter. From our present vantage point we can interpret the results of Faraday's experiments as follows: When an atom of valence 1 is dissolved in water, it acquires a charge and becomes an ion, either positive or negative. The chemical properties of the ions of a substance are quite different from the chemical properties of its neutral atoms. Under the influence of the electric field inside the solution, the positive ions migrate to the cathode and the negative ones to the anode. There the positive ions acquire a negative charge which neutralizes them, and they become normal neutral atoms. The negative ions migrate to the anode and there deposit their surplus negative charge, whereupon they too become neutral atoms. The current within the electrolytic cell is carried by the positive and negative charges. The current in the external wires still arises from the motion of electrons, since at the cathode electrons are constantly being delivered to neutralize the positive ions, while they are simultaneously being supplied at the anode by the neutralization of the negative ions (Fig. 21–2). The deposition of the constituent materials is therefore, in effect, a way of counting the valence-1 atoms that are liberated either at the cathode or the anode, or, alternatively, a way of counting the electrons that are either picked up from the circuit or given off to the circuit. In order to deposit 1 kmole, F coulombs of charge must be transferred by the constituents of the electrolytic cell. This picture indicates the significant fact that the charge on all valence-1 atoms in solution is the same.

In the case of valence-2 elements, the transfer of the charge F through a kilomole sample results in the neutralization of half as many ions of either kind when Eq. (21–1) is used with $v = 1$. This implies that valence-2 atoms in solution have double the charge of valence-1 atoms, valence-3 atoms three times the charge, and so forth. This also indicates that, so far as atoms in solution are concerned, there is an atomicity of electric charge: All ionic charges appear as multiples of the charge on the valence-1 ion.

In order to be able to compute the smallest charge that appears on ions in solution, we must know the magnitude of the faraday, F. From the defining equation for the faraday, we see that we must measure the mass M of a particular element that is deposited (for example, the silver which is deposited from a silver chloride solution), the charge Q that has passed through the circuit, and the atomic weight of silver. All these measurements can be made with great accuracy. The best value for F is 9.6522×10^7 coul/kmole. (We shall use the value 9.65×10^7 coul/kmole.) This is the amount of charge, according to our in-

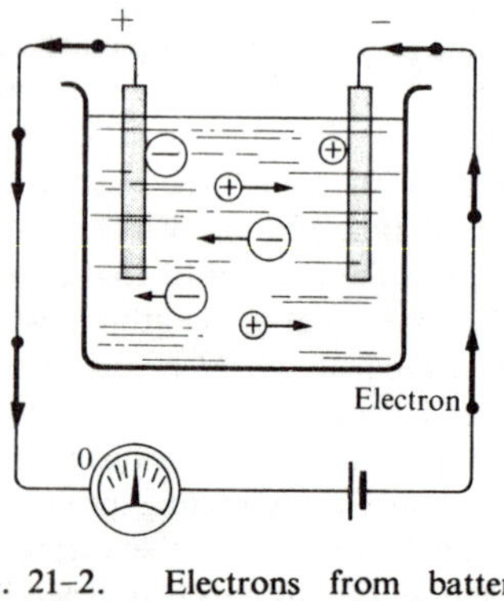

FIG. 21–2. Electrons from battery neutralize charge of positive ions at cathode. Negative ions give up their charges at anode and electrons go to battery.

terpretation, in 6.025×10^{23} (i.e., Avogadro's number) singly charged ions. The charge per ion is then

$$q = \frac{9.65 \times 10^7 \text{ coul/kmole}}{6.025 \times 10^{26} \text{ atoms/kmole}} = 1.602 \times 10^{-19} \text{ coul/ion.} \qquad (21\text{–}2)$$

As we shall see in the next section, this is precisely the value that has been directly obtained for the value of the fundamental negative charge in nature, namely that of the electron. The measurement of the electronic charge is also an extraordinarily accurate one and today the value of Avogadro's number is arrived at by assuming that the electronic and ionic charges are identical, taking the value of the electronic charge as determined in the next section, and using the measured value of the faraday.

Example: How much sodium will be produced if a current of 50 amp flows through molten sodium chloride for 10 min?

Solution: Since $Q = it$, the total charge transferred is

$$Q = 50 \frac{\text{coul}}{\text{sec}} \times 10 \text{ min} \times 60 \frac{\text{sec}}{\text{min}} = 3 \times 10^4 \text{ coul.}$$

A kilomole of sodium contains 22.91 kg, and the valence of sodium is +1. Substituting these figures in Eq. (21–1), we find that

$$m = \frac{Q}{F}\frac{A}{v} = \frac{3 \times 10^4 \text{ coul}}{9.65 \times 10^7 \text{ coul}} \times \frac{22.91 \text{ kg/kmole}}{1} = 7.13 \text{ kg}$$

of metallic sodium will appear at the negative electrode.

21–2 The charge on the electron

As we have seen, Faraday's law of electrolysis suggests the quantization of electric charge. In the latter part of the nineteenth century, much effort was devoted to the direct measurement of the charge on the electron, which experimenters already suspected to be one of the fundamental constituents of all matter. It was observed that when small droplets of oil or water, such as those obtained with the help of an atomizer, are irradiated by ultraviolet light or x-rays, they acquire a small electric charge. If we watch these droplets with the help of a microscope we can follow their motion and measure their speeds with considerable accuracy. In a series of experiments that began in 1906, Robert A. Millikan measured the charge on the electron by using such oil drops (water drops turned out to be too unstable) in a chamber within which he could impose an electric field by means of two parallel plates connected to a source of electromotive force (Fig. 21–3). By adjusting the potential difference V he could make a particular drop remain stationary. This means that the electric force on the droplet was just

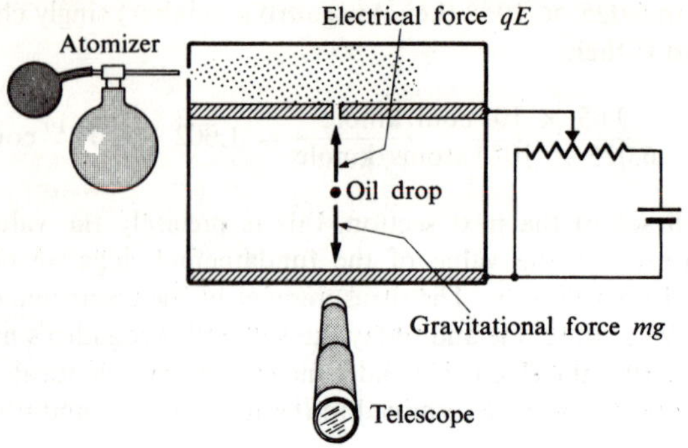

FIG. 21-3. Millikan's oil-drop experiment.

balanced by the gravitational force on the drop, with

$$qE = q\frac{V}{d} = mg, \tag{21-3}$$

q being the charge on the droplet, E the electric field intensity, m the mass of the droplet, and d the separation of the plates. We see that the charge is given in terms of measurable quantities and the mass of the droplet.

Finding the masses of the droplets presents a serious problem, since they are too small (10^{-4} to 10^{-5} cm in diameter) to be weighed by ordinary means. This small size, however, permits their masses to be determined by other means. When the droplets are falling in the chamber, there is a retarding force acting on them, owing to air viscosity, that is proportional to their velocity. A terminal velocity will be attained which is determined by the size of the droplet and by the viscosity of the air. This same property of viscosity is what makes raindrops fall at a constant terminal velocity. The experiment is carried out in the following way. First the electric force on the droplet is increased until the droplet is motionless, then the field is turned off. With the field turned off, the experimenter measures the droplet's terminal velocity which, together with the known viscous properties of the air, enables him to calculate its radius. From a knowledge of the radius of the droplet he can compute its mass, and Eq. (21-3) then permits him to determine the charge on the droplet. Sometimes during the course of the experiment the droplet being observed acquires additional charge, and the gain or loss of charge on the droplet can likewise be measured by readjusting the electric field.

In Millikan's experiments, all the charges that he measured turned out to be small (usually between 2 and 10) multiples of a single charge, which he took as the fundamental irreducible quantum of charge. He took the least common factor of all the measured q values to be the elementary charge, symbolized e; it is now known to be 1.602×10^{-19} coul. Therefore

$$e = 1.602 \times 10^{-19} \text{ coul.}$$

The original experiments of Millikan gave a somewhat smaller number because the relationship between the radius of the droplet and the viscosity of the air was not sufficiently well known. This difficulty has since been rectified and the above value is the accepted one today.

It is difficult to overestimate the importance of this experiment. Not only is the value of the constant important, entering as it does into all theories of matter, but the demonstration of the atomicity of electricity is an immensely important fundamental fact of nature.

21-3 The charge/mass ratio of the electron

The identification of the electron as a particle possessing mass and charge antedates the actual measurement of its charge. The British physicist J. J. Thomson in 1897 measured the ratio of the charge of the electron to its mass, a ratio that is characteristic of this particle.

Thomson's study of the electron came in the course of his experiments on electrical discharges through gases. If two electrodes in a gas-filled chamber are established at a potential difference of several thousand volts, no current flows, since gases are ordinarily nonconductors. If the pressure is reduced, the gas becomes conducting and it begins to glow. When the pressure in the tube is further reduced so that it is 10^{-7} to 10^{-8} of atmospheric pressure, the tube itself begins to glow, as if there were some emanation from the cathode. These emanations were called cathode rays by Thomson, and we now know that they are electrons. One of the reasons for identifying them as electrons is that they can be influenced by a magnetic field in the vicinity of the tube, and are deflected in the same manner that a current of electrons would be deflected. The tube which Thomson used to measure the charge/mass (e/m) ratio of the particles that constitute cathode rays is shown schematically in Fig. 21-4. At the end of the tube is a zinc sulphide screen. The cathode rays cause the zinc sulphide to fluoresce when they strike it, which permits the direction of the beam to be determined, since the rays can be well collimated in the tube by a series of slits.

The experiment is carried out in two parts. First an electric field is established across the plates, and a magnetic field is directed at right angles to the direction

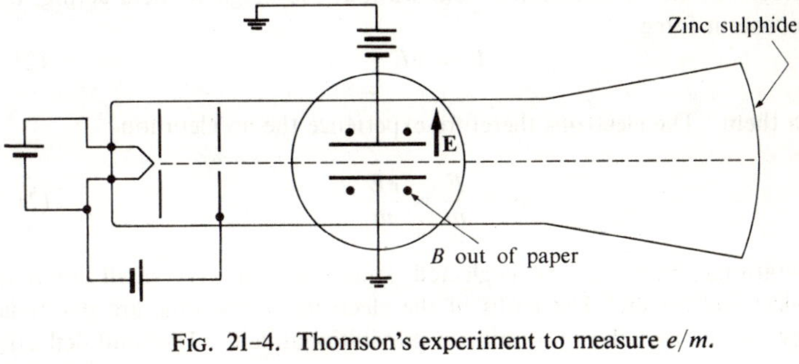

FIG. 21-4. Thomson's experiment to measure e/m.

of the electric field and to the direction of motion of the cathode rays. The important innovation introduced by Thomson in this experiment was to insert the plates producing the electric field inside the cathode tube. Previous experimenters had attempted to perform the experiment with the plates on the outside, and did not succeed. The reason for their failure is the polarization of the glass of the tube, as will be discussed in a subsequent chapter. This polarization seriously affects the resulting field within the tube.

When the magnetic field is out of the paper, as shown, we see that the effect of the magnetic field is to exert a force on the moving electrons which is opposite to the force of the electric field. The fields are now so adjusted that the electric force is just balanced by the magnetic force. When this balance is achieved, the magnetic field is turned off and the deflection of the beam is measured on the zinc sulphide screen. A knowledge of the magnitudes of the electric and magnetic fields whose forces on the cathode rays cancel out, together with the ability to measure the deflection in the absence of the magnetic field, provides us with sufficient information to deduce the ratio of the charge to the mass of the electron.

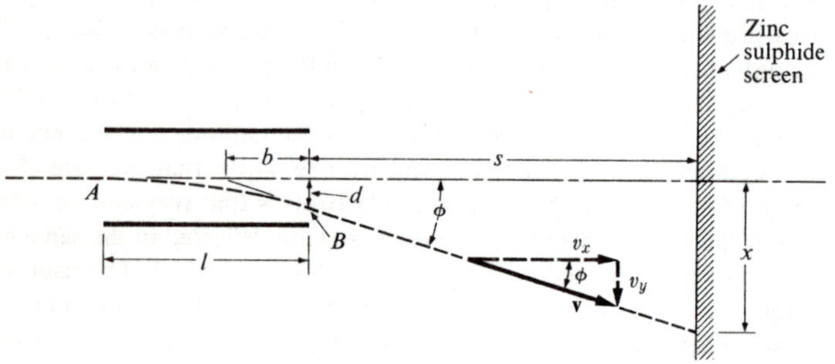

Fig. 21–5. Enlargement of circled region in Fig. 21–4 to show relationship of quantities used to calculate e/m. AB is parabolic path. The quantities b, d, and ϕ are not measurable; l, x, and s are.

While electrons are in the electric field without the magnetic field acting, they have the constant force

$$F = eE \tag{21–4}$$

exerted on them. The electrons therefore experience the acceleration

$$a = \frac{F}{m} = \frac{eE}{m}. \tag{21–5}$$

Here gravitational forces can be neglected, since they are very small compared with the electrical forces. The paths of the electrons in the tube are parabolas, just as they are for particles in a uniform gravitational field. The total deflection

d of the particles in the field E is given by

$$d = \frac{1}{2}at^2 = \frac{1}{2}\frac{eE}{m}\left(\frac{l}{v_x}\right)^2. \tag{21-6}$$

In Eq. (21–6), v_x is the horizontal component of the velocity of an electron and l is the distance the electrons traverse in the electric field. When the electrons leave the field between the plates and enter the field-free region, they move in a straight line along a direction making an angle ϕ with the horizontal. As can be seen from Fig. 21–5,

$$\tan\phi = \frac{v_y}{v_x}. \tag{21-7}$$

If we were to extend the straight line back and call the distance from the ends of the plates to where the line intersects the axis b, we have, since the triangles shown are similar,

$$\frac{x}{d} = \frac{s+b}{b}. \tag{21-8}$$

Here x is the deflection we measure with a constant electric field and no magnetic field, and s is the distance from the ends of the plate to the screen. The distance s is, of course, readily measurable. We note that

$$\frac{b}{d} = \cot\phi. \tag{21-9}$$

The v-component of velocity may be expressed in terms of measurable quantities as

$$v_y = at = \frac{eE}{m}\left(\frac{l}{v_x}\right). \tag{21-10}$$

We may now combine Eqs. (21–6) through (21–10) to obtain

$$\frac{e}{m} = \frac{x}{lE/v_x^2}\,\frac{1}{s+(l/2)}, \tag{21-11}$$

which gives e/m as a function of measurable quantities except for v_x, the x-component of velocity of the electrons. This number can be determined, however, from the first part of the experiment in the following way. Under the conditions that the electric and magnetic forces balance one another, we have

$$Bev_x = eE, \qquad v_x = \frac{E}{B}. \tag{21-12}$$

Thus v_x is equal to the ratio of the magnitudes of these electric and magnetic fields. In practice the electrons do not all have the same velocity. However, their velocities are all so large (of the order of 10^7 m/sec) relative to the deviations from the average that this factor does not seriously affect the accuracy of the experiment. The presently accepted value for the electronic charge/mass ratio is

$$e/m = 1.7589 \times 10^{11}\,\text{coul/kg}.$$

Of course, the fact that electrons are all found to have the same e/m ratio does not by itself establish that they all have the same mass and charge. Millikan's discovery that e, the charge on the electron, has a unique value means that m, the mass of the electron, also has a unique value. The mass of the electron is

$$m_e = \frac{e}{e/m} = 9.1085 \times 10^{-31}\,\text{kg}.$$

The charge/mass ratio of cathode rays is found to be independent of the gas in the tube as well as of the composition of the electrodes, which helps establish the unique character of electrons. The value of this ratio is characteristic of only one particle, the electron, and can be used to identify it whenever q/m is known, even though the particle charge q and mass m may not themselves be known explicitly. The charge/mass ratio of other charged particles can be measured in similar experiments, as we shall see in the next section.

21-4 The mass spectrometer

The least massive atomic constituent which has a positive charge is the *proton*, a particle with a charge of $+e$ and a mass m_p of 1836.13 m_e. The hydrogen atom is a neutral combination of a proton and an electron. In a hydrogen atom the proton is referred to as the *nucleus*, since the electron in such an atom can be crudely visualized as circulating around the proton. Other atoms consist of nuclei whose charges are multiples of the charge of the proton and whose masses are approximately integral multiples of the mass of the proton. Thus the ordinary atom which is next higher in weight after hydrogen is ordinary helium, which consists of a nucleus whose charge is $+2e$ and whose mass is approximately $4m_p$. The He nucleus is surrounded by two electrons which render the entire atom neutral. The atom which is next higher in weight is that of lithium, whose nuclear charge is exactly $+3e$ and whose mass is approximately $6m_p$. The Li nucleus is surrounded by three electrons to achieve neutrality. This pattern persists throughout the entire periodic table.

There are rare types of hydrogen called deuterium and tritium whose charges are $+e$ and whose masses are approximately $2m_p$ and $3m_p$. There is also an unusual kind of helium whose nucleus has a charge of exactly $+2e$, but whose mass is approximately $3m_p$, not $4m_p$; and a rare type of lithium whose nucleus has a charge of $+3e$ and whose mass is approximately $7m_p$, not $6m_p$. The chemical properties of hydrogen, helium, and lithium are determined almost entirely by the electronic structures of their atoms, which are in turn determined by the charges on their nuclei. The different versions of hydrogen, helium, and lithium atoms whose nuclear masses are not the same are called *isotopes* of helium and

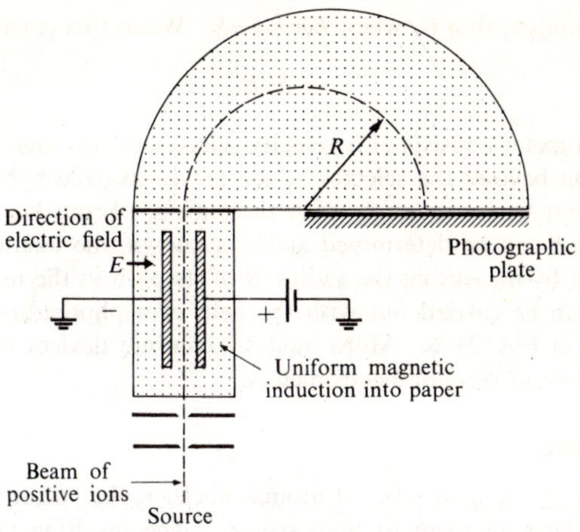

FIG. 21-6. A mass spectrometer.

lithium. In general, isotopes are atoms whose electronic structures are identical but whose nuclear masses differ. The chemical elements as normally found consist of mixtures of various isotopes. When isotopes were discovered and the masses of their nuclei were measured, an important advance was made in the understanding of nuclear structure. In this section we shall discuss one form of *mass spectrometer*, which is a device used for measuring or comparing the masses of isotopes.

When a charged particle enters a uniform magnetic field perpendicular to its direction of motion, it moves in a circular path. The radius of curvature of this path can be found from the fact that the magnetic force on the particle is radial and provides the centripetal force that maintains it in the circular path. Thus if R is the orbital radius (Fig. 21-6),

$$F_{\text{mag}} = F_{\text{cent}}, \qquad Bev = \frac{mv^2}{R},$$

and

$$R = \frac{mv}{eB} = \frac{m}{e}\frac{v}{B}. \tag{21-13}$$

For a given velocity, charge, and magnetic induction the radius of curvature of the path is proportional to the mass of the nucleus. The charge can be restricted to $+e$ by the method used to ionize the original atoms, so that if we wish either to compare or measure masses the chief problem is to have all the ions moving with the same velocity. One method of achieving this result is shown in Fig. 21-6, where we depict a long tube with slits at each end through which each ion passes. In the tube the ion is subjected to crossed electric and magnetic fields. An analysis similar to that of Thomson's experiment indicates that this arrangement will permit particles to pass through all the slits only if their velocities are such that the net forces on them vanish; that is, when $Bev = eE$. When this is true,

$$v = E/B. \tag{21-14}$$

A mass spectrometer consists of a velocity selector of this kind plus a magnetic field in the region beyond the selector which serves to deflect the emerging ions by an amount that varies inversely with their mass. From Eq. (21-13) we see that, since v and B can be determined and e is known, the unknown mass of an ion can be found by measuring the radius R of its path in the magnetic field, an operation that can be carried out with the help of a photographic plate placed where indicated in Fig. 21-6. More modern accurate devices for detecting the ions are electron multipliers or electrometers.

21-5 The cyclotron

The constancy of the q/m ratio of atomic nuclei makes possible an ingenious method for accelerating them to high speeds. The resulting fast particles are essential for the investigation of nuclear properties, as we shall see in a later chapter. The fundamental problem in accelerating particles to high speeds is that a high potential difference is required if they are to receive energy in a single act, and potential differences exceeding several million volts cannot be created in the laboratory for practical reasons. An alternative procedure is to use a smaller potential difference, but to arrange matters so that the particles are exposed to this potential difference many times in succession. The *cyclotron*, developed by E. O. Lawrence in 1931, is a device that accomplishes the latter in a simple way.

Let us return to the case of a particle moving perpendicular to a uniform magnetic field. Equating magnetic and centripetal forces,

$$Bqv = \frac{mv^2}{R},$$

$$v = \frac{q}{m} BR. \tag{21-15}$$

The time t required by the particle to traverse half a complete circle is

$$t = \frac{\pi R}{v} = \frac{\pi m}{Bq}, \tag{21-16}$$

which is independent both of the velocity v and the radius R! The time required for the particle to go from one gap in the split copper cylinder of Fig. 21-7 to the next depends solely on the flux density B of the magnetic field and its charge/mass ratio q/m. Thus if we apply an alternating potential difference across the halves ("dees") of the cylinder whose frequency is just right for the particles to experience an acceleration each time they reach a gap, they will be repeatedly accelerated until their orbits reach the diameter of the dees. This is the principle of the cyclotron.

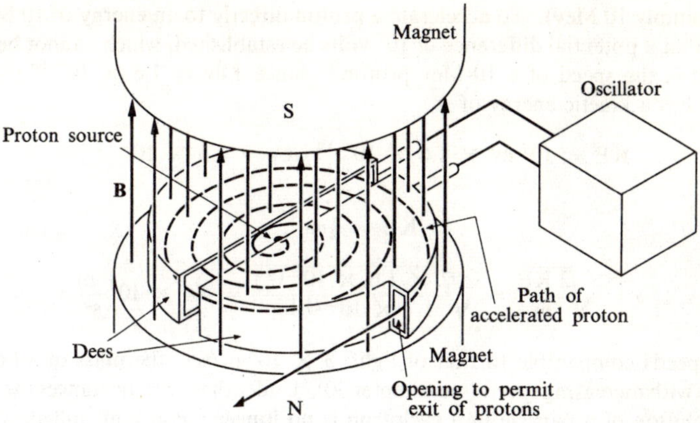

FIG. 21-7. A cyclotron.

A typical cyclotron might have a magnetic induction of 1.5 weber/m^2. If it is accelerating protons, their period is

$$T = \frac{2\pi m}{qB}$$

$$= \frac{2\pi \times 1.7 \times 10^{-27}\,\text{kg}}{1.6 \times 10^{-19}\,\text{coul} \times 1.5\,\text{weber/m}^2}$$

$$= 4.5 \times 10^{-8}\,\text{sec},$$

since the charge of the proton is 1.6×10^{-19} coul and its mass is 1.7×10^{-27} kg. This means that the oscillator should have a frequency of

$$f = \frac{1}{T} = 2.2 \times 10^7\,\text{sec}^{-1},$$

which is 22,000,000 cycles/sec, a perfectly feasible operating frequency. If the potential difference between the dees is 100,000 volts, which is about as high as it can conveniently be made, each proton receives 10^5 ev of energy each time it passes through the gap. The symbol ev stands for *electron volt*. It is convenient to express the energy of the proton in terms of this widely used unit in atomic and nuclear physics. The electron volt is the energy acquired by an electron when it has been accelerated by a potential difference of 1 v, which is

$$1\,\text{ev} = 1.6 \times 10^{-19}\,\text{coul} \times 1\,\text{v} = 1.6 \times 10^{-19}\,\text{j}.$$

In the course of 100 passages, corresponding to 50 complete revolutions, the proton therefore receives a total of 10^7 ev (10,000,000 electron volts, often abbreviated simply 10 Mev). To accelerate a proton directly to an energy of 10 Mev requires that a potential difference of 10^7 volts be established, which cannot be done.

What is the speed of a 10-Mev proton? Since 1 ev = 1.6×10^{-19} j, such a proton has a kinetic energy of

$$\text{KE} = 10^7\ \text{ev} \times 1.6 \times 10^{-19}\ \text{j/ev} = 1.6 \times 10^{-12}\ \text{j}.$$

Hence

$$\text{KE} = \tfrac{1}{2}mv^2,$$

$$v = \sqrt{\frac{2\ \text{KE}}{m}} = \sqrt{\frac{2 \times 1.6 \times 10^{-12}\ \text{j}}{1.7 \times 10^{-27}\ \text{kg}}} = 4.4 \times 10^7\ \frac{\text{m}}{\text{sec}}\cdot$$

At speeds comparable to that of light, 3×10^8 m/sec, the mass of a body increases with increasing speed (see Chapter 30). Under these circumstances the period of revolution of a particle in a cyclotron is no longer a constant, independent of its speed, and it may reach the gap between the dees when the electric field is in such a direction as to decelerate it. The energy limit that this relativistic effect imposes on cyclotrons is several tens of Mev. More sophisticated accelerators have been designed which are able to cope with the mass changes attendant on high particle speeds, and machines are being constructed or are under design that will generate particles with energies of as much as 100,000 Mev (100 Bev).

Problems

1. A uniform current flows for 15 min and deposits 3 gm of silver. What is the current?

2. How many cubic centimeters of hydrogen at 0°C and 760 mm Hg pressure (STP) will a current of 20 amp flowing through a circuit liberate in 2 hr?

3. Three electrochemical cells are placed in series. Aluminum is deposited in the first, copper in the second, and silver in the third. How much aluminum and copper will be deposited when 5 gm of silver are deposited? Suppose that the cells have a resistance ratio of Al:Cu:Ag = 1:2:3, and that they are connected in parallel; what is the answer to the same question? The valence of aluminum is 3, copper 2, and silver 1.

4. What is the speed of an electron whose energy is 50 ev?

5. What is the energy in electron volts of an electron whose speed is 10^6 m/sec?

6. What is the energy, in electron volts, of a potassium atom whose speed is 10^6 m/sec?

7. It is necessary to do 13.6 ev of work to remove the electron from a hydrogen atom. (a) How much energy in joules would be required to remove the electrons from the 6×10^{26} hydrogen atoms in a kilogram of hydrogen? (b) How many kilocalories is this?

8. Electrons with kinetic energies of 500 ev are projected between two parallel charged metal plates in a direction parallel to the surfaces. The plates are at a poten-

tial difference of 5 v and are 5.00 mm apart and 4 cm long. What is the angle the emergent beam makes with the original direction? What value of the magnetic field at right angles to **E** and to the direction of the beam will cause the electrons not to be deflected?

9. A velocity selector uses a magnet to produce a flux density of 0.05 weber/m^2 and a pair of parallel metal plates 1 cm apart to produce a perpendicular electric field. What potential difference should be applied to the plates to permit singly charged ions of speed 5×10^6 m/sec to pass through the selector?

10. A mass spectrometer employs a velocity selector consisting of a magnetic field of flux density 0.04000 weber/m^2 perpendicular to an electric field of 50,000 v/m. The same magnetic field is then used to deflect the ions. Find the radius of curvature of singly charged $_3Li^7$ ions of mass 7.0182 amu in this spectrometer. (1 amu $= 1.66 \times 10^{-27}$ kg.)

11. A beam of singly charged boron atoms is accelerated by a potential difference of 3000 v and enters a magnetic field of flux density 0.2000 weber/m^2 perpendicular to the field. The ions are deflected through 180° and fall on a photographic plate. How far apart will the $_5B^{10}$ (10.016 amu) and $_5B^{11}$ (11.012) isotopes strike the photographic plate?

12. Alpha particles (helium nuclei) are accelerated by an electric field of 450 v in the mass spectrometer described in the text. (a) What should be the value of B in the region after the second slit in order to have a radius of curvature of 0.15 m? (b) Repeat the same calculation for electrons without relativistic correction (see Problem 15). (c) Repeat the calculation with a relativistic mass correction.

13. Nuclei of singly ionized lithium (1500-v Li^6 and Li^7 nuclei of masses 6.0173 and 7.0182 amu, respectively) are projected with energies of 1500 ev into a region with magnetic induction of 0.1 weber/m^2. What will be the difference of the radii of curvature of the paths of the nuclei?

14. At what frequency should a proton cyclotron operate if the magnetic field is 2 webers/m^2?

15. The mass of an electron is given by $m = m_0/\sqrt{1 - v^2/c^2}$, where m_0 is its rest mass, v its velocity, and c the velocity of light $= 3 \times 10^8$ m/sec. The potential difference across the dees in a cyclotron is 1000 ev and the magnetic induction is 1.6 webers/m^2. How should one vary the frequency of oscillation per 180° traversal in order that the potential difference across the dees always be in phase with the electron arrival at the gap?

22

Electromagnetic Induction

We have seen that the flow of an electric current leads to the existence of a magnetic field. Now we may ask whether there is a reciprocal effect, namely the creation of a current by a magnetic field. In order to have a current we must have an emf in a closed circuit, so the question becomes: can a magnetic field produce an emf? This question was investigated for ten years following Oersted's discovery of the magnetic effects of currents in 1820, with apparent negative results. In 1830 Joseph Henry in the United States and in 1831 Michael Faraday in England then discovered that a *changing* magnetic field produces an emf; earlier studies had been confined to static magnetic fields. Faraday is usually given credit for the discovery, since he published his results first. A current may be regarded as consisting of moving charges and thus acts, in effect, as a changing electric field. Hence a changing electric field produces a magnetic field, and as Faraday and Henry found, a changing magnetic field produces an electric field, so that in this sense reciprocity does indeed exist. We shall discuss the phenomenon of electromagnetic induction, as it is called, in this chapter. From our present knowledge of the microscopic properties of matter, we can easily understand one of the experimental arrangements for demonstrating electromagnetic induction, and we shall start with this one. Other methods of exhibiting electromagnetic induction are more difficult to understand microscopically, and we shall not consider them. When we consider how scant was Faraday's and Henry's knowledge of the constituents of matter, it is all the more remarkable that they were able to discover electromagnetic induction.

22–1 Electromagnetic induction

Let us begin by considering a particle of charge q moving with velocity $\mathbf{v}$ perpendicular to the uniform magnetic field $\mathbf{B}$, as shown in Fig. 22–1. According to Eq. (19–33), there is a force

$$\mathbf{F} = q(\mathbf{v} \times \mathbf{B})$$

acting on the particle in the direction indicated in the diagram. But electric field intensity **E** is defined in general as

$$\mathbf{E} = \mathbf{F}/q,$$

so that the effect of a magnetic field **B** on a charged particle of velocity **v** is entirely equivalent to that of the electric field given by

$$\mathbf{E} = \mathbf{v} \times \mathbf{B}. \qquad (22\text{–}1)$$

FIG. 22–1. Relationship between the directions of the velocity and force on a charged particle in a magnetic field.

So far as the motion of a charge is concerned, then, there is no way of distinguishing between the presence of the magnetic field **B** and the presence of the electric field **E** specified by Eq. (22–1).

If instead of an isolated charge we have a metallic wire ab moving with velocity **v** perpendicular to the uniform field **B** (Fig. 22–2), the mobile charges in the wire all experience the effect of the electric field of Eq. (22–1). When we arrange for the wire ab to slide along the sides ad and bc of the U-shaped conductor shown, a current flows through the circuit $abcd$ because of this electric field. The effective emf $\mathcal{E}$ between a and b owing to the motion is

$$\mathcal{E}_{ab} = \mathcal{E} = EL = LvB, \qquad (22\text{–}2)$$

where L is the length of the wire ab. The magnitude of the current that flows depends, of course, on the resistance of the circuit, and may be calculated by Ohm's law with Eq. (22–2) providing the value $\mathcal{E}$. The direction of the current is

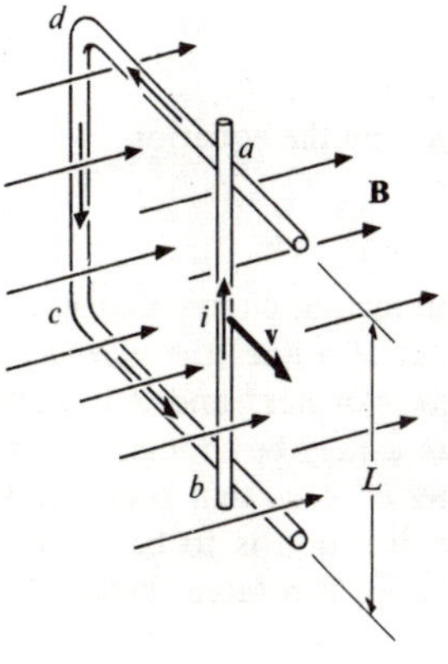

FIG. 22–2. A wire moving in a uniform magnetic field.

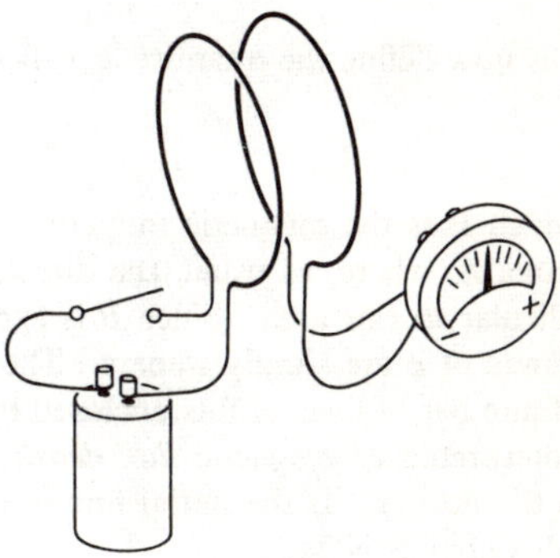

FIG. 22–3. Inducing an emf in a secondary circuit.

as shown in Fig. 22–2. If the velocity of the wire were reversed, the direction of the current would be reversed too.

The above method for obtaining an emf in a circuit in the presence of a magnetic field is not the only possible one. For example, when two wire loops, one connected to a battery and the other to a galvanometer, are placed close together (Fig. 22–3) and the battery is quickly disconnected, the galvanometer needle is momentarily deflected. This indicates that a current and therefore an emf has been induced in the galvanometer circuit. When we reconnect the battery, there is another momentary deflection in the opposite direction, which shows that the induced emf is in the reverse sense. Here, evidently, it is the changing magnetic field during the rise or fall of the current of the interrupted battery circuit that is causing the observed emfs.

It is now necessary to bring together in a single general law both the induction due to motion of a wire in a constant magnetic field and the induction due to a changing magnetic field acting on a stationary conductor. To obtain such a law, we begin by rewriting Eq. (22–2) in the form

$$\mathcal{E} = LvB = L\frac{dx}{dt}B,$$

where x is measured in the direction of $\mathbf{v}$. But the rate at which the area A enclosed by *abcd* is increasing is

$$\frac{dA}{dt} = L\frac{dx}{dt},$$

and so the induced emf in the moving wire is

$$\mathcal{E} = B\frac{dA}{dt}.$$

Let us now define the quantity ϕ, called *magnetic flux*, by the equation

$$\phi = \mathbf{B} \cdot \mathbf{A}, \tag{22–3}$$

in which $\mathbf{B}$ is the magnetic induction through a conducting circuit that encloses the area $\mathbf{A}$, where, as usual, the direction of $\mathbf{A}$ is that of a line (the *normal*) perpendicular to this area. Since B is expressed in webers/meter^2 and A in meter^2, the units of ϕ are simply webers. The magnetic flux ϕ may be thought of as the total number of lines of flux enclosed by the perimeter of A, so that B may in turn be interpreted as *magnetic flux density*. A definite direction is to be associated with the normal. If the initial flux is ϕ and the coil *abcd* is rotated through 180°, the flux change is 2ϕ.

In terms of ϕ, the induced emf $\mathcal{E}$ is

$$\mathcal{E} = \frac{d\phi}{dt}. \tag{22–4}$$

If there are N turns of wire around A,

$$\mathcal{E} = N\frac{d\phi}{dt} = N\frac{d}{dt}(\mathbf{B}\cdot\mathbf{A}). \tag{22–5}$$

In general it is most convenient to use Lenz's law in determining the direction of $\mathcal{E}$, as discussed in the next section. Equation (22–5), formulated by Faraday, covers both the case in which **A** changes with time while **B** is constant, when

$$\mathcal{E} = N\mathbf{B}\cdot\frac{d\mathbf{A}}{dt},$$

and the case in which **A** is fixed and **B** varies, when

$$\mathcal{E} = N\frac{d\mathbf{B}}{dt}\cdot\mathbf{A}.$$

It is also able to treat the simultaneous variation of both quantities.

The concept of magnetic lines of flux, like that of electric lines of force, is a purely artificial one that Faraday introduced as an aid in visualizing magnetic phenomena. Faraday's law of induction may be interpreted as meaning that an emf appears in a conductor whenever lines of flux cut across it, either by virtue of the motion of the conductor or because the flux density is changing. However, since we have precisely defined what we mean by the flux ϕ as the scalar product $\mathbf{B}\cdot\mathbf{A}$, we do not have to rely in any quantitative way on imaginary lines of flux in our work, although they are undoubtedly helpful in picturing certain physical situations.

22–2 Lenz's law

While we have shown that the emf induced by electromagnetic induction is proportional to the rate of change of flux in the circuit, we did not discuss the *direction* of the emf. Often Eq. (22–4) is written $\mathcal{E} = -d\phi/dt$. The minus sign, however, does not really have any precise meaning. What it is supposed to indicate is that *the induced emf produces effects which oppose the action producing the emf.* This statement is known as *Lenz's law.* In our first example of electromagnetic induction, the agency producing the emf is the motion of the conductor. The emf and the current must then be in a direction such that there is a force acting on the wire opposite to its direction of motion. In Fig. 22–2, motion in the direction indicated causes a force to act on the conductor opposite in direction to its velocity. As another example, suppose that we have an emf induced in coil 2 of Fig. 22–4 due to a current in coil 1 in the indicated direction. Since the emf is produced in coil 2 due to the increasing flux through it, the emf in this coil must be in such a direction as to decrease the flux through the coil.

Lenz's law is a consequence of the conservation of energy. If it were not true, then by first starting a current in one coil we should produce an effect which

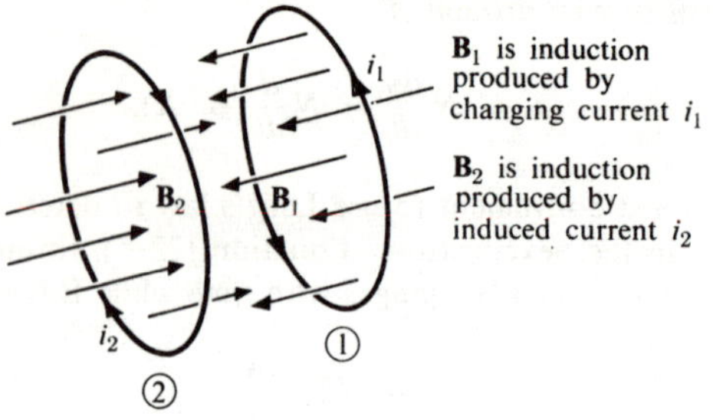

FIG. 22-4. Relation of currents produced by induction in neighboring coils according to Lenz's law.

would enhance the current, which in turn would produce a still larger emf. Thus we should be able to produce an ever-increasing current with a small, fixed expenditure of energy. Since this cannot happen, it follows that Lenz's law must be true. Lenz's law as stated above is very convenient for determining the direction of the induced emf.

22-3 Some applications of electromagnetic induction

With the help of Faraday's law it is possible to measure magnetic flux density **B** in a relatively simple manner. We require a small coil attached to what is known as a ballistic galvanometer. The latter is like an ordinary galvanometer except that its moving coil is constructed with a very large moment of inertia; as a result, when a momentary current is being measured, the deflection is proportional to the total charge that has passed through it. The procedure is to place the search coil in the magnetic field and then to withdraw it rapidly (Fig. 22-5), noting the reading Q of the ballistic galvanometer. The emf induced in the coil is

$$\mathcal{E} = N\frac{d\phi}{dt},$$

where N is the number of turns in the coil. The instantaneous current that flows is

$$i = \frac{\mathcal{E}}{R} = \frac{N}{R}\frac{d\phi}{dt},$$

where R is the total resistance in the circuit, and so, since $i = dq/dt$,

$$N\frac{d\phi}{dt} = R\frac{dq}{dt}.$$

Integrating, we obtain

$$RQ = N\phi = N\mathbf{B}\cdot\mathbf{A} = NBA\cos\theta,$$

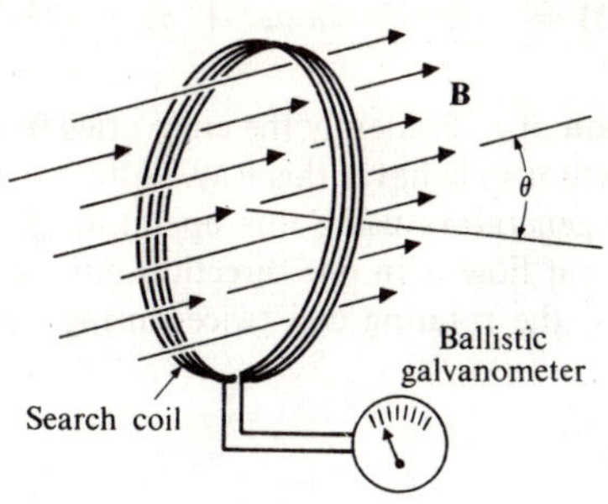

FIG. 22–5. Measuring B, using the law of induction.

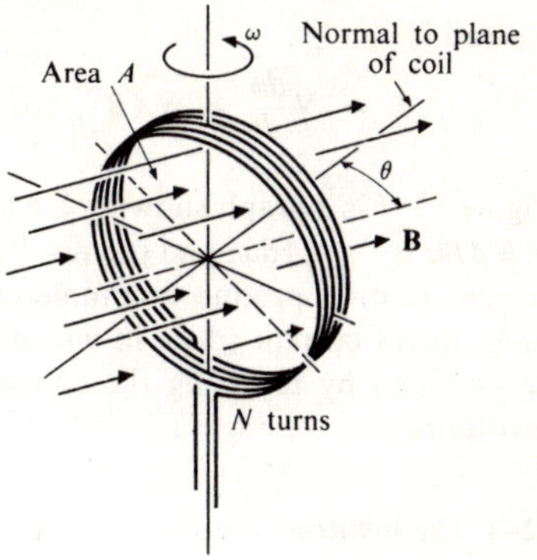

FIG. 22–6. The emf in a rotating coil.

with Q being the total charge that passes through the circuit, A the cross-sectional area of the coil, and θ the angle between the axis of the coil and the direction of **B**. In practice, it is simplest to repeat the experiment with different orientations of the coil until Q is a maximum ($\cos\theta = 1$). Then

$$B = RQ/NA,$$

in which all the quantities on the right-hand side can be readily determined.

Another example of electromagnetic induction is that of a rotating coil in a magnetic field (Fig. 22–6). At any instant the flux through the coil is

$$\phi = AB\cos\theta,$$

where θ is the angle between **B** and the normal to the area A. If the coil is rotating with the constant angular velocity ω,

$$\phi = AB\cos(\omega t + \delta), \tag{22–6}$$

where δ, the phase angle, is the value of θ at $t = 0$. Hence the emf generated in

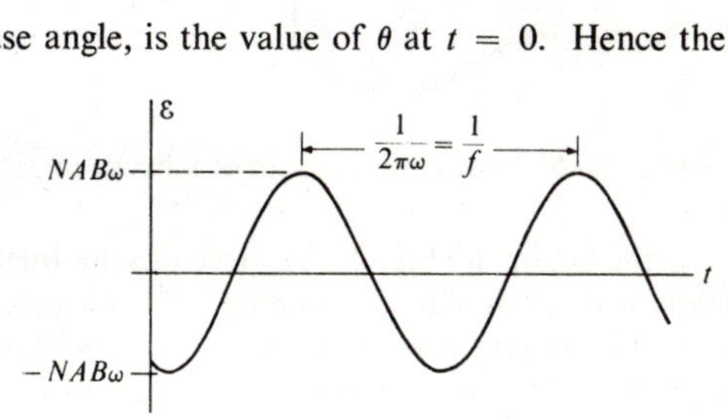

FIG. 22–7. The emf in a rotating coil as a function of time.

the coil is

$$\mathcal{E} = N\frac{d\phi}{dt} = NAB\frac{d}{dt}\cos(\omega t + \delta) = -NAB\omega \sin(\omega t + \delta). \quad (22\text{-}7)$$

Figure 22-7 is a graph showing $\mathcal{E}$ as a function of t. Evidently the emf varies from $+NAB\omega$ to $-NAB\omega$, and the resulting current also behaves this way. Alternating current (a-c) is produced commercially by generators using this operating principle; direct current (d-c), in which the current flow is in one direction only, can be obtained by reversing the connections to the rotating coil twice during each revolution.

22-4 The betatron

An especially interesting application of electromagnetic induction is the betatron, a device for accelerating electrons. Electrons cannot be easily accelerated in a cyclotron, since the operating principle of this latter device is based on the constancy of the ratio of e/m of the electron. The relativistic variation of the mass of an electron with velocity becomes important at such low energies as to render the cyclotron useless for accelerating these particles. The betatron's operation is independent of relativistic considerations.

The operation of the betatron involves simultaneously using a magnetic field to hold moving electrons in a stable orbit while the field itself changes rapidly. The changing flux results in a tangential force that acts on electrons in the orbit.

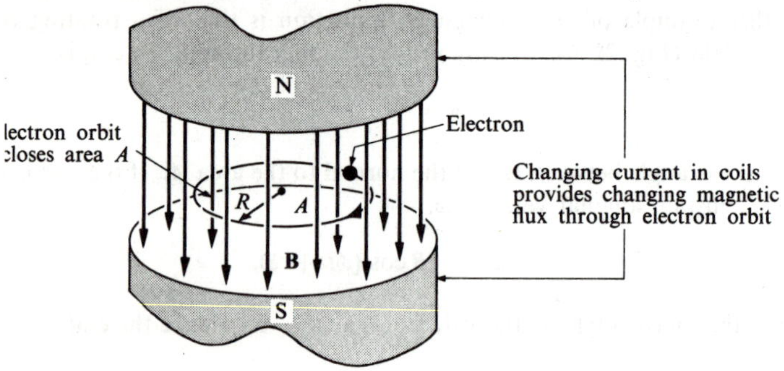

FIG. 22-8. Schematic diagram of a betatron.

In the operation of the betatron (Fig. 22-8), electrons are injected tangent to a circular path of radius R in an evacuated chamber. The magnetic field in this region exerts a centripetal force on the electrons to keep them in orbit. At the time of injection the magnetic induction is varied. According to Faraday's law there is a tangential electric field E_t acting on the electrons during the variation of the magnetic field that causes them to accelerate. The electric field E_t is related to the emf $\mathcal{E}$ due to the changing magnetic flux by

$$\mathcal{E} = 2\pi R E_t,$$

while $\mathcal{E}$ is itself given by

$$\mathcal{E} = \frac{d}{dt}\int d\phi.$$

(We write $\int d\phi$ rather than ϕ, since the field is not constant within the electron orbit.) Hence the force on an electron in a betatron is

$$F = eE_t = \frac{d(mv)}{dt} = \frac{dp}{dt} = \frac{e}{2\pi R}\frac{d}{dt}\int d\phi, \tag{22-8}$$

where $\mathbf{p}$ is the momentum of the electron.

We shall now derive the condition that will ensure that the electrons pursue a circular path of fixed radius R while they are being accelerated. If B is the magnitude of the magnetic induction at the circular path,

$$\frac{mv^2}{R} = evB,$$

from which

$$mv = p = eRB;$$

then

$$\frac{dp}{dt} = eR\frac{dB}{dt}, \tag{22-9}$$

since the radius is constant. From Eqs. (22-8) and (22-9), we have

$$eR\frac{dB}{dt} = \frac{e}{2\pi R}\frac{d}{dt}\int d\phi. \tag{22-10}$$

Integrating with respect to t, we obtain

$$B = \frac{1}{2\pi R^2}\int d\phi = \frac{1}{2}\cdot\frac{1}{A}\int d\phi. \tag{22-11}$$

The condition for a constant orbit of radius R, then, is that the magnetic induction at the position of the orbit should be half the average flux per unit area through the orbit. This result is achieved by shaping the pole pieces of the magnet so that the flux decreases as one goes away from the center, as shown in Fig. 22-9. More extensive analysis shows that this orbit is stable; that is, if electrons are displaced away from it, forces will arise to move them back to the orbit. To extract the electrons after their acceleration, the condition of Eq. (22-11) may be violated by momentarily adding or subtracting some auxiliary flux that will cause the electrons to spiral in or out.

A typical betatron that accelerates electrons to 100 Mev might have an orbit radius of 1 m and a magnetic field changing at the rate of 100 webers/m^2·sec. Hence the orbit area is

$$A = \pi R^2 = 3.1 \text{ m}^2,$$

and the emf, $\mathcal{E}$, per revolution is

$$\mathcal{E} = -A\frac{dB}{dt} = -3.1 \text{ m}^2 \times 100 \frac{\text{webers}}{\text{m}^2\cdot\text{sec}} = 310 \text{ v}.$$

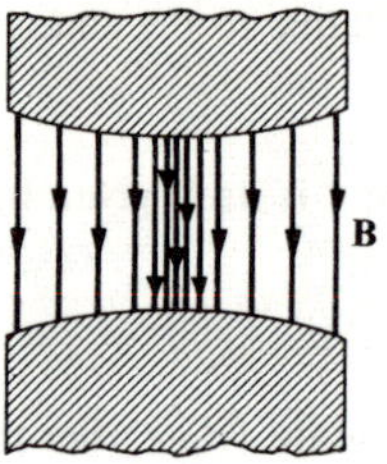

FIG. 22-9. Pole pieces of magnet of betatron. Flux decreases as one proceeds from center.

An electron in this betatron acquires 310 ev each time it makes a complete circle; it must make over 323,000 revolutions before it has an energy of 100 Mev! The relativistic limitation on particle speed that affects the cyclotron does not affect the betatron, since the electrons in the latter are accelerated continuously and do not have to make each turn in precisely the same period of time.

22-5 Mutual and self inductance

Figure 22-10 shows two coils, one connected to a galvanometer and the other to a battery. When the switch is closed a magnetic field is established in space. As this occurs, lines of flux move past the galvanometer coil, inducing the emf of magnitude

$$\mathcal{E} = N\frac{d\phi}{dt}\cdot$$

Since at any instant the total flux is proportional to the current in the battery coil, their rates of charge are also proportional, and

$$\frac{d\phi}{dt} = C\frac{di}{dt},$$

where C is a constant. Hence the emf in the galvanometer circuit is proportional to the time derivative of the current in the battery circuit, so that

$$\mathcal{E}_{\text{gal}} = M\frac{di}{dt}\cdot \qquad (22\text{-}12)$$

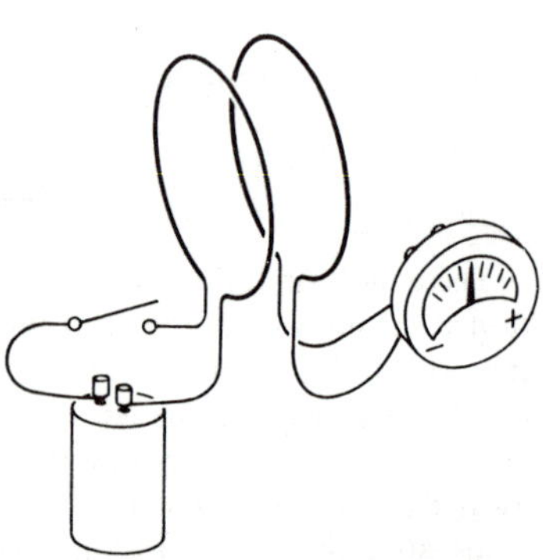

FIG. 22-10. Schematic for defining mutual inductance.

The proportionality constant M is called the *coefficient of mutual inductance*, and may be thought of as a measure of the extent to which a magnetic field set up by one of the coils affects the other. The units of M are evidently volt-sec/amp, which has been given the name of *henry* (in honor of Joseph Henry, the American physicist who discovered electromagnetic induction independently of Faraday). When the coefficient of mutual inductance M between two coils (or other configurations of conductors) is 1 henry, a rate of current change in one of them of 1 amp/sec induces an emf of 1 v in the other. M depends solely on the geometry of the coils, their number of turns, and their relative orientation; it is independent of the current through the coil. However, M does depend on the material around which the coil is wound. Introduction of iron, especially, causes an appreciable increase in M, but we shall assume here that the coil is wound only in air.

Even when there is only a single circuit, a changing current in it causes an emf to be induced owing to the changing flux threading its own circuit. This is also proportional to the rate of change of current,

$$\mathcal{E} = L\frac{di}{dt}, \tag{22–13}$$

where L, the *coefficient of self-inductance*, is once more a function of the geometry of the circuit only, and is also measured in henries. What Eq. (22–13) states is that, in a circuit whose coefficient of self-inductance is 1 henry, when the current through it is varying at the rate of 1 amp/sec, an emf is generated whose magnitude is 1 v. What is the direction of this emf relative to the varying current? By Lenz's law the emf must oppose the change, so that if the current is decreasing, $\mathcal{E}$ tends to reestablish it. Conversely, if the current is increasing, $\mathcal{E}$ tends to keep it down. Of course, these attempts cannot succeed, so that the net effect of self-inductance is to slow down any changes of external origin in the current.

We have indicated that the self-inductance of a coil is due to its geometrical configuration only. It is of interest to calculate this coefficient of self-inductance in some cases. When a current i flows in a circuit of N turns, the induced emf is

$$\mathcal{E} = N\frac{d\phi}{dt} = L\frac{di}{dt}\cdot$$

Integrating this equation, we have

$$L = \frac{N\phi}{i}\cdot \tag{22–14}$$

The constant of integration vanishes, since ϕ is zero when i is zero. Hence in order to calculate the coefficient of self-induction L (often simply called the "inductance"), we must know the flux through a particular circuit as a function of the current in it. In several cases in Chapter 19 we calculated the magnetic induction for a given circuit shape. There we found that the uniform magnetic flux density within a solenoid is equal to

$$B = \mu_0\frac{N}{l}i.$$

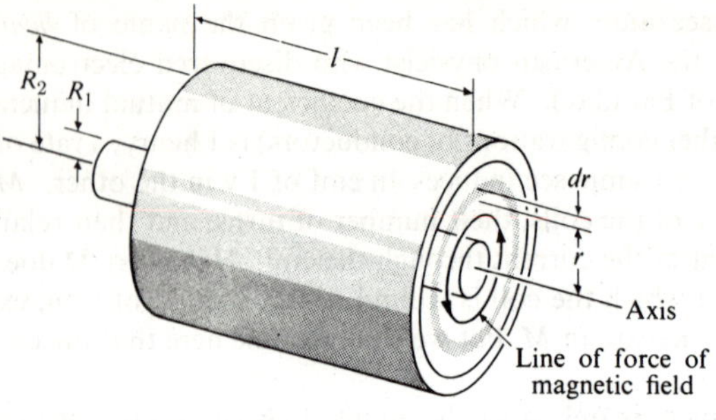

FIG. 22-11. Self-inductance of a coaxial cable.

Therefore the inductance of a solenoid, since $\phi = BA$, is

$$L = \frac{N\phi}{i} = \mu_0 \frac{N^2 A}{l}, \tag{22-15}$$

where A is the cross-sectional area of the solenoid.

Next we consider a coaxial cable. As we determined in Chapter 19, the magnetic induction between the inner and outer conductors is $B = \mu_0 i/2\pi r$, where r is the distance from the axis of the cable (Fig. 22-11). This example is more complicated than the previous one, since the magnetic induction is not uniform in the region we are considering, and we must calculate ϕ by integration. As can be seen from Fig. 22-11, the lines of force of B are circles concentric with the wire. At a distance r from the center of the wire, B is constant. In the element of cable length l the flux through a strip r from the axis that is dr wide is

$$d\phi = Bl\,dr,$$

and

$$\phi = \int_{R_1}^{R_2} Bl\,dr = \int_{R_1}^{R_2} \frac{\mu_0 li}{2\pi}\frac{dr}{r} = \frac{\mu_0 li}{2\pi}\ln\frac{R_2}{R_1},$$

where R_1 is the outer radius of the inner conductor and R_2 the inner radius of the outer conductor. A coaxial cable therefore has the inductance per unit length

$$\frac{L}{l} = \frac{\phi}{li} = \frac{\mu_0}{2\pi}\ln\frac{R_2}{R_1}. \tag{22-16}$$

22-6 Effect of inductance on d-c circuits

An inductance in a direct-current circuit acts to resist changes in the flow of current in a way reminiscent of the effect of a flywheel on a rotating shaft. When the switch in the circuit of Fig. 22-12 is closed, the full emf of the battery does not immediately appear across the resistance R because of the emf in the reverse direction produced in L. This induced emf is always, by Lenz's law, such as to oppose any change in the current; hence if the initial current is zero, the induced emf $\mathcal{E}_i$ tries to maintain it there by opposing the battery emf $\mathcal{E}$. Since at any instant the effective potential difference across R is $\mathcal{E} - \mathcal{E}_i$, we can write

$$\mathcal{E} - \mathcal{E}_i = Ri$$

or

$$\mathcal{E} - L\frac{di}{dt} = Ri. \qquad (22\text{–}17)$$

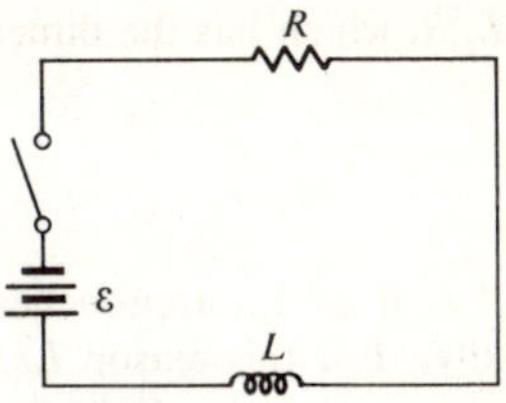

FIG. 22–12. An L/R d-c circuit.

In order to solve Eq. (22–17) for i as a function of time t, we put it in the form

$$L\frac{di}{dt} = \mathcal{E} - Ri, \qquad \frac{L\,di}{\mathcal{E} - Ri} = dt.$$

Integrating both sides, we have

$$\mathcal{E} - \frac{L}{R}[\ln(\mathcal{E} - Ri) - \ln C] = t,$$

where ln C is the constant of integration. This result may be written

$$\mathcal{E} - Ri = Ce^{-Rt/L} \qquad \text{or} \qquad i = \frac{1}{R}(\mathcal{E} - Ce^{-Rt/L}).$$

This constant C may be evaluated by noting that at $t = 0$, $i = 0$, and so

$$0 = \frac{1}{R}(\mathcal{E} - C), \qquad C = \frac{\mathcal{E}}{R}.$$

Thus the final expression for $i(t)$ in a d-c circuit containing an inductance is

$$i = \frac{\mathcal{E}}{R}(1 - e^{-Rt/L}). \qquad (22\text{–}18)$$

When the switch is closed, the current approaches $\mathcal{E}/R$ only asymptotically (Fig. 22–13). The rate at which it does this depends on the value of the ratio

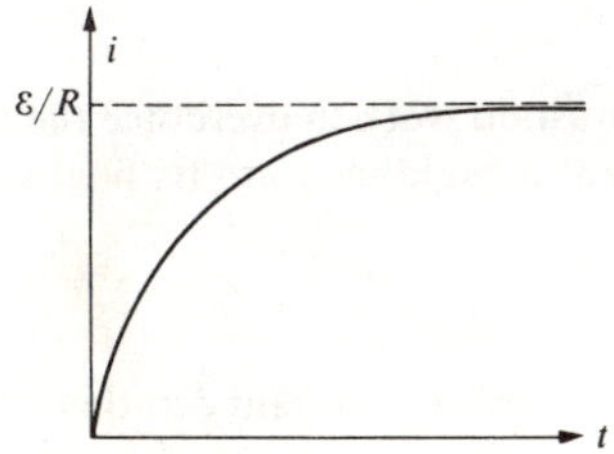

FIG. 22–13. Growth of current in d-c circuit containing an inductance.

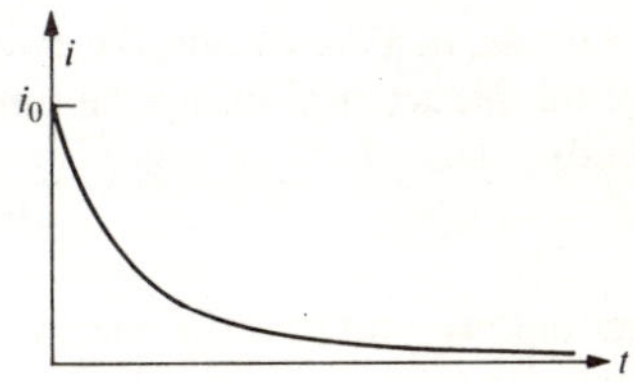

FIG. 22–14. Decay of current in a d-c circuit containing an inductance.

L/R, which has the dimensions

$$\frac{\text{henry}}{\text{ohm}} = \frac{\text{volt-sec/amp}}{\text{volt/amp}} = \text{sec.}$$

If $L/R \gg 1$, i approaches its final value slowly, while if $L/R \ll 1$, it does so rapidly. For this reason L/R is called the *time constant* of a circuit containing both resistance and inductance.

When the switch is opened, the situation is similar. Here $\mathcal{E} = 0$ in Eq. (22–18), and integrating

$$Ri = -L\frac{di}{dt},$$

yields

$$i = i_0 e^{-Rt/L}. \tag{22–19}$$

The current at the instant of opening the switch is i_0, and of course need not be equal to $\mathcal{E}/R$ unless the switch had been closed for a time long compared with L/R. Again the rate at which the current changes is governed by the time constant L/R (Fig. 22–14).

22–7 Energy in a magnetic field

Previously we encountered the puzzling fact that a magnetic field must be thought of as possessing momentum. In this section we shall demonstrate that magnetic fields contain energy as well, and in the next chapter we will find that this is also true for electric fields.

The time rate at which the induced emf

$$\mathcal{E}_i = L\frac{di}{dt},$$

in a circuit of inductance L does work is

$$\frac{dW}{dt} = \mathcal{E}_i i = Li\frac{di}{dt}\cdot$$

For example, in a d-c circuit, the external emf must do work to overcome the back emf of the inductance during the time the current is building up to its final value. Evidently

$$dW = Li\,di.$$

Integrating, we find that the energy required to establish a current i in this circuit is

$$W = \int_0^i Li\,di = \tfrac{1}{2}Li^2. \tag{22–20}$$

This energy is not dissipated, but resides in the form of potential energy in the magnetic field surrounding the circuit. When the applied emf is turned off, the energy $\frac{1}{2}Li^2$ reappears to keep the current of Eq. (22–19) flowing. The slow growth of current in a circuit containing inductance is due to the necessity of simultaneously building up the magnetic field around it; the slow decay of the current is due to the emf induced in the circuit by the collapsing magnetic field.

It is possible to determine the energy density in a magnetic field directly from the magnitude of its flux density B, without knowing what L and i values are involved. Let us do this for a solenoid. The induced emf in a solenoid of cross section A, length l, and number of turns N is

$$\mathcal{E}_i = N\frac{d\phi}{dt} = NA\frac{dB}{dt},$$

and the rate at which it does work is

$$\frac{dW}{dt} = \mathcal{E}_i i = ANi\frac{dB}{dt}.$$

But, from Eq. (19–27),

$$i = \frac{lB}{\mu_0 N} \quad \text{and} \quad dW = \frac{lAB}{\mu_0}\,dB.$$

Hence the total magnetic energy within the solenoid is

$$W = lA\frac{\frac{1}{2}B^2}{\mu_0}. \tag{22–21}$$

The *energy density* w in the solenoid is this energy divided by the volume of the solenoid lA, which is

$$w = \frac{\frac{1}{2}B^2}{\mu_0}. \tag{22–22}$$

This expression contains no reference to the configuration of currents that led to the field B, from which we might suspect that Eq. (22–22) is a general law for the energy density in a magnetic field in free space; a suspicion that can be verified by more advanced methods than we at present command.

Problems

1. A rectangular wire loop 5 cm × 10 cm in size is perpendicular to a magnetic field of induction 10^{-3} weber/m^2. (a) What is the flux through the loop? (b) If the magnetic field drops to zero in 3 sec, what is the potential difference induced between the ends of the loop during that period?

2. A wire of length 10 cm, mass 10 gm, and resistance 0.10 ohm is constrained to fall in a vertical direction, touching a wire frame of negligible resistance (Fig. 22–15). There is a uniform field of 0.1 weber/m^2 directed horizontally. Find the steady-state velocity v of the wire.

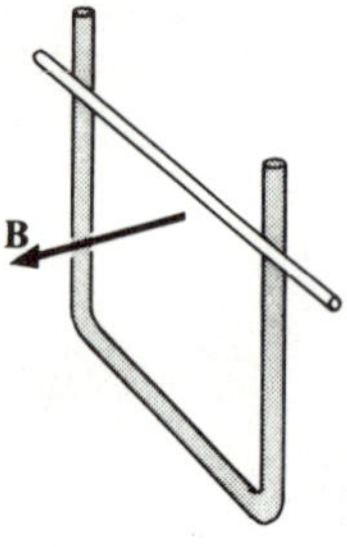

FIGURE 22–15

3. A square wire loop 10 cm on a side is situated so that its plane is perpendicular to a magnetic field. The resistance of the loop is 5 ohms. How rapidly should the magnetic induction change if a current of 2 amps is to flow in the loop?

4. A wire loop is 5 cm in diameter and is situated so that its plane is perpendicular to a magnetic field. How rapidly should the magnetic induction change if a potential difference of 1 v is to appear across the ends of the loop?

5. Explain why, when a wire loop is rotated in a magnetic field about a diameter perpendicular to the direction of the field, the current induced in the loop reverses itself twice in each full rotation.

6. Why is it easy to turn the shaft of a dynamo when it is not connected to an outside circuit, but much harder when such a connection is made?

7. *Magnetic storms* occur when swarms of charged particles from the sun enter the earth's magnetic field. Explain why such storms always produce a decrease in the magnetic field at sea level.

8. A coil whose radius is 10 cm and which has 1000 turns of wire is rotated in a homogeneous magnetic field with an angular velocity of 600 rad/sec. What is the maximum emf in the coil if the field is 0.1 weber/m^2?

9. A circular coil has 40 turns, radius 10 cm, and rotates 200 times per second in a uniform field of 0.10 weber/m^2 about an axis through the diameter perpendicular to the magnetic field. What is the maximum emf and at what position of the coil is the instantaneous emf at a maximum?

10. A wire of infinite length carries a current of 20 amp. A coil whose radius is 1 cm and which has 1000 turns is so placed that its axis is perpendicular to the wire, and its center is 1 cm from the wire. It is moved a distance 2 cm from the wire in 0.01 sec. What is the average emf induced in the coil?

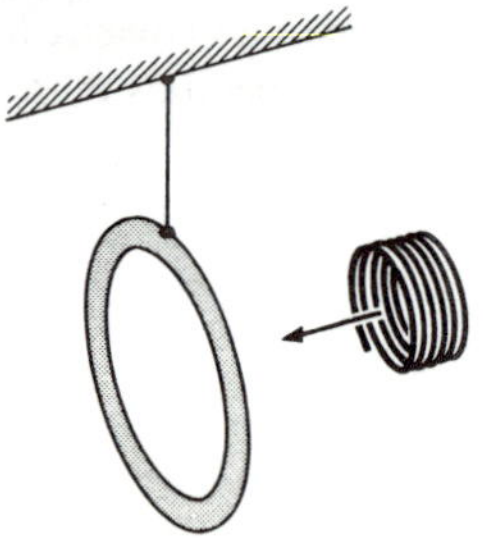

FIGURE 22–16

11. A steel ring is suspended from a ceiling (Fig. 22–16). A current-carrying coil whose magnetic moment is directed down the center of the ring is moved toward it and passed through it. Describe the motion of the ring.

12. The diameter of a ring is 0.2 m, it is wound with 100 turns of wire, and its resistance is 0.1 ohm. It is turned from a position where it is perpendicular to a magnetic field of 2×10^{-4} weber/m^2 to a position in which it is parallel to the magnetic field. What is the total charge which flows through the coil? What charge flows when it is rotated through 180°?

13. The coil in the previous problem is so moved that the rate of change of flux is constant through it. If the motion takes place in 0.01 sec, what is the work required to move it?

14. In what ways does a betatron resemble a cyclotron? In what ways does it differ from a cyclotron?

15. In a particular betatron the direction in which the electrons rotate is clockwise as seen from above. What is the direction of the magnetic field?

16. What must the rate of change of magnetic induction be for a betatron designed to produce 40-Mev electrons if the radius of the orbit is 1 m and it makes 150,000 revolutions before it acquires its ultimate energy?

17. What is the induced emf in a 15-h coil when the current through it is changing at the rate of 4 amp/sec?

18. The magnetic flux through a 20-turn coil drops from 3×10^{-3} weber to 0 in $\frac{1}{4}$ sec. What is the induced potential difference across the ends of the coil?

19. A 2-h coil carries a current of 0.5 amp. (a) How much energy is stored in it? (b) In how much time should the current drop to 0 if an emf of 100 v is to be induced in it?

20. An automobile has a speed of 30 m/sec on a road where the vertical component of the earth's magnetic field is 8×10^{-5} weber/m^2. What is the potential difference between the ends of its axles, which are 2 m long?

21. What is the inductance of a coil whose resistance is 14 ohms and whose time constant is 0.1 sec?

22. Why does an electric motor require more current when it is turned on than when it is running continuously?

23. A 50-mh coil with a resistance of 20 ohms is connected to a 90-v battery. (a) What is the time constant of the circuit? (b) How much energy is stored in the magnetic field of the coil when the current has reached its final value?

24. What is the inductance of a coil 40 cm long and 4 cm in diameter that has 1000 turns of wire?

25. Calculate the mutual inductance per unit length of two circuits as shown in Fig. 22–17.

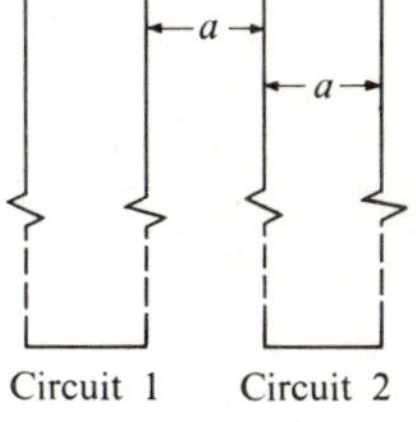

FIGURE 22–17

26. A coaxial cable has an inner radius of $\frac{1}{4}$ cm and an outer radius of $\frac{1}{2}$ cm. What is its inductance per unit length?

27. Two concentric coplanar coils with N_1 and N_2 turns, respectively, have radii r_1 and r_2, with $r_1 \gg r_2$. (The magnetic field of the smaller is uniform over its area.) Find an expression for the mutual inductance of this arrangement.

28. A solenoid 3 cm in radius and 100 cm long is wound uniformly, with 20 turns/cm. What is its self-inductance?

29. A coil 20 cm long and 3 cm in diameter is tightly wound with one layer of copper wire 1 mm in diameter. What is its time constant?

30. A transmission line consists of two long parallel wires, each of radius a, at a distance d apart, $d \gg a$. They carry equal currents in opposite directions. What is its self-inductance/unit length, if one neglects the variation of flux inside the conducting wires?

31. A battery whose emf is $\mathcal{E}$ is connected to a resistanceless coil whose self-inductance is L. Calculate the current in the coil as a function of time.

32. A coil whose self-inductance is 1 mh and whose resistance is 5 ohms is connected to a battery whose emf is 3 v. How long will it take to reach $\frac{7}{8}$ of its ultimate current? After a long time the switch is opened. At what rate does the potential across the coil fall? How long will it take to fall to half its maximum value?

33. Show that the energy of two coils, carrying current i_1 and i_2, is

$$W = \tfrac{1}{2}L_1 i_1^2 + \tfrac{1}{2}L_2 i_2^2 \pm M i_1 i_2,$$

where L_1, L_2, and M are the coefficients of self-inductance and of mutual inductance, respectively.

34. A solenoid has a self-inductance of 10 h. What is its energy when it carries a current of 2 amp?

35. The strongest magnetic fields that have been produced in the laboratory have had flux densities of about 10^2 webers/m^2. (a) How much energy is contained in a 10^3-cm^3 volume of such a field?

36. What is the energy density of the earth's magnetic field at a location where its flux density is 6×10^{-5} weber/m^2?

23

Capacitance and the Electrical Properties of Matter

A device for storing electric charge consists of two conductors separated by a nonconductor, and is called a *capacitor* (or a *condenser*) (Fig. 23–1). In the simplest form of capacitor the conductors are infinite parallel planes separated by a nonconductor or vacuum. The characteristics of a capacitor depend both on its geometrical configuration and on the nature of the nonconductor between its plates. When we vary the nonconductor in a capacitor of known geometry and study the effects, we obtain important information about the electrical properties of matter. Another useful aspect of the capacitor is that it exhibits certain characteristic behavior when used as an element in an electric circuit. We shall explore these topics in this chapter.

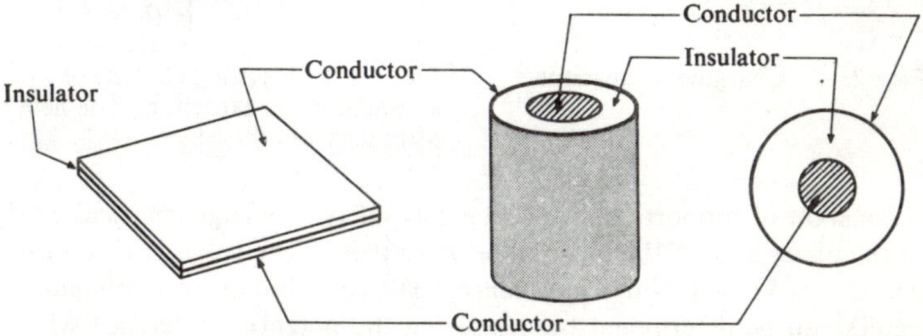

FIG. 23–1. Various forms of capacitors.

23–1 Capacitance

We can calculate the energy of a charged capacitor in much the same way as we calculate the potential energy in a stretched or compressed spring. Let us start by considering an uncharged capacitor whose plates are separated by a vacuum. The net result of charging the capacitor is to transfer a charge of $+Q$ from one plate to the other, so that one has a charge of $-Q$ and the other a charge of $+Q$

(Fig. 23–2). The energy of the charged capacitor is equal to the work done in transferring the charge. If an infinitesimal portion dq of the total charge Q is moved at any time, the work done in moving it is equal to dq multiplied by the potential difference V between the capacitor plates due to the equal and opposite charges Q already upon them. The total work is then equal to the integral of $V\,dq$ taken over the entire process.

Before actually integrating, we must know the relationship between the charge Q present on the plates of a capacitor and the potential difference V between them. For two isolated conductors, if one of the conductors has a charge $+Q$ the other must have a charge $-Q$, since the lines of force originating on positive charges must terminate on negative ones. The isolation of the conductors means that there are no other possible places for the lines of force to go. In this case one observes experimentally that there is a linear relationship between the charge on either of the plates and the potential difference between them, so that

$$Q = CV. \tag{23–1}$$

(If many conductors a, b, c, etc., are present, then the charge on any conductor Q_a is equal to a sum of terms each of which is a constant times the potential difference between conductor a and b, a and c, etc.)

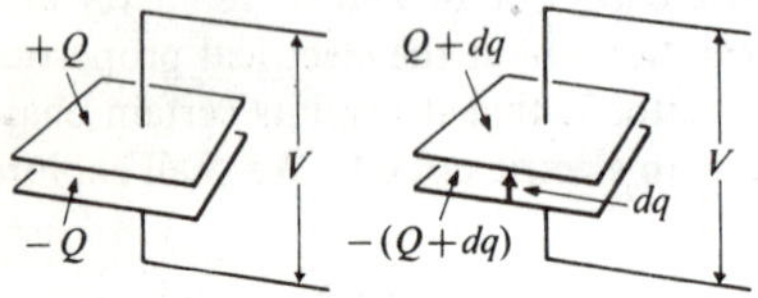

FIG. 23–2. Charging a capacitor.

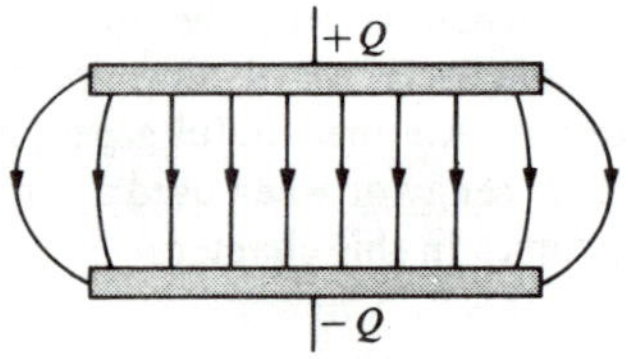

FIG. 23–3. Electric field between plates of a parallel-plate capacitor. The field is uniform far from the edges.

The constant of proportionality C depends solely on the geometrical configuration of the plates and on the medium between them. The value of C for a particular capacitor is known as its *capacitance*. The capacitance of a configuration of conductors can be determined by measuring the potential difference when there is a known charge on the plates. For certain configurations which are especially simple, we can calculate the potential difference as a function of the charge, and hence infer their capacitances. The unit of capacitance is the unit of charge divided by the unit of potential difference, since

$$C = Q/V.$$

In the mks system of units, charge is measured in coulombs and potential difference in volts, so that the unit of capacitance is the coulomb/volt. This unit is commonly called the *farad*;

$$1 \text{ farad} = 1 \text{ coulomb/volt}.$$

Let us calculate the capacitance of some simple configurations of electrodes. Perhaps the simplest is a parallel-plate capacitor in which the separation of the plates is small compared with their size. We must assume relatively small separation, since this calculation is readily performed only when the field between the plates is uniform. For any finite plates there are distortions of the field at the edges (Fig. 23–3), and the assumption of uniform electric field is therefore approximately valid only for small separations. In the latter situation the field can be taken as uniform and the fringing at the edges can be neglected. In Chapter 18 we found, with the aid of Gauss' theorem, that the field intensity near a charged plane conductor has the constant value

$$E = \frac{\sigma}{\epsilon_0}, \tag{23–2}$$

where σ is the charge per unit area and ϵ_0 is the permittivity of free space. For a plate of area A and total charge Q, σ is just

$$\sigma = Q/A, \tag{23–3}$$

and the electric field is therefore equal to

$$E = \frac{Q}{\epsilon_0 A}. \tag{23–4}$$

If the field is uniform, as we assume, the electric field is related to the potential difference between the plates by

$$E = V/d, \tag{23–5}$$

where d is their separation. Thus we have

$$\frac{V}{d} = \frac{Q}{\epsilon_0 A}$$

and

$$Q = \frac{\epsilon_0 A}{d} V. \tag{23–6}$$

In this case we have demonstrated the linear relationship between Q and V. Comparing the last equation with the defining equation for capacitance, we have that

$$C = \frac{\epsilon_0 A}{d} \tag{23–7}$$

for a parallel-plate capacitor. Only the characteristic dimensions of the capacitor enter into this formula.

Typical values for A and d are 10^{-3} m^2 and 10^{-3} m, respectively. Substituting these values in Eq. (23–7) and using the fact that $\epsilon_0 = 8.85 \times 10^{-12}$ coul^2/n·m^2, we find that

$$C = \frac{8.85 \times 10^{-12} \times 10^{-3}}{10^{-3}} \text{ farad} = 8.85 \times 10^{-12} \text{ farad.}$$

It is clear that the farad is an extremely large unit for practical purposes. Usually the *microfarad*, abbreviated μf and equal to one millionth of a farad, or the *micro-microfarad*, abbreviated $\mu\mu$f (also called the picofarad, pf), and equal to one millionth of a microfarad, are used to specify capacitance. That is,

$$1\ \mu\text{f} = 10^{-6} \text{ farad}, \qquad 1\ \mu\mu\text{f (1 pf)} = 10^{-12} \text{ farad.}$$

We note that, as the separation of the parallel plates approaches zero, the capacitance becomes infinite. In a certain sense, then, the capacitance of a single, isolated conductor might be regarded as infinite.

Another configuration whose capacitance is easily calculable is one consisting of two concentric conducting spheres. If the radii of the spheres are a and b, as in Fig. 23–4, and the charge on the inner sphere is $+Q$ and on the outer sphere $-Q$, the electric field in the space between them at a distance r from the common center of the two spheres can be calculated by Gauss' theorem to be

$$E = \frac{Q}{4\pi\epsilon_0 r^2}. \tag{23–8}$$

The potential difference between the spheres may be found by integrating $E\,dr$ from b to a, which yields

$$V = \int_b^a E\,dr = \frac{Q}{4\pi\epsilon_0}\left(\frac{1}{a} - \frac{1}{b}\right). \tag{23–9}$$

Thus, from the definition of capacitance,

$$C = \frac{Q}{V} = \frac{4\pi\epsilon_0 ab}{b - a}. \tag{23–10}$$

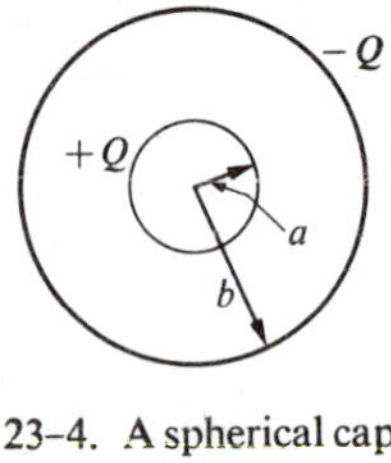

FIG. 23–4. A spherical capacitor.

Again we may note that, as the distance between the spheres approaches zero, the capacitance becomes infinite.

An isolated sphere also has a capacitance. In this case we may regard the second sphere (whose radius is b in our problem) to be at infinity. Taking the limit as b goes to infinity, we see that the capacitance of an isolated sphere is simply proportional to its radius:

$$C = 4\pi\epsilon_0 a. \tag{23–11}$$

The capacitance of the earth, since its radius is equal to 6×10^6 m, is thus approximately 6.7×10^{-4} farad.

23–2 The energy of a charged capacitor

We are now in a position to calculate the work that must be done to add the charge Q to the plates of a capacitor. We begin with the capacitor uncharged, with $V = 0$, and transfer infinitesimal charges dq one at a time from one plate to the other (Fig. 23–5). The more the charge already transferred, the more work we must expend in order to bring additional charges across. This is similar to the process of compressing a spring: the more it has been compressed the more force must be applied to compress it further, and hence the greater the amount of work involved. Suppose we have already carried the total charge q from one plate to the other, so that the potential difference $V = q/C$ exists across them. The work required to transfer the additional charge dq is therefore

$$dW = V\,dq = \frac{q\,dq}{C}.$$

Integrating from $q = 0$ to $q = Q$, we have

$$W = \int_0^Q \frac{q\,dq}{C},$$

$$W = \frac{1}{2}\frac{Q^2}{C}. \tag{23–12}$$

(This resembles the formula for the potential energy of a compressed spring $\frac{1}{2}kx^2$, in which x is the displacement of the spring from equilibrium and k is the spring constant.) By making use of the definition of C, the potential energy of a charged capacitor may also be written

$$W = \tfrac{1}{2}QV, \qquad W = \tfrac{1}{2}CV^2, \tag{23–13}$$

in which V is the potential difference across the fully charged capacitor.

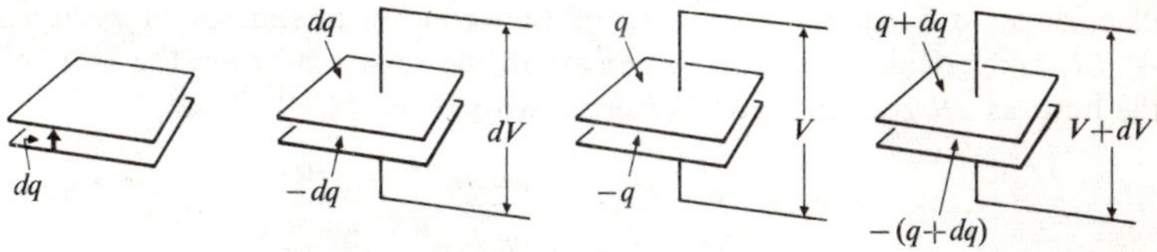

FIG. 23–5. Successive steps in charging a capacitor.

23–3 The energy density of an electric field

In order to cause a separation of charge and hence an electric field, work must be performed. Since all electric fields therefore have energy associated with them, we can regard this energy as residing in the region of space occupied by the field. The density of the energy at any point can be readily expressed in terms of the electric field intensity there. In Chapter 22 we made a similar determination of the energy density in a magnetic field, and found that

$$w = \frac{1}{2}\frac{B^2}{\mu_0}.$$

In the case of an electric field, it is convenient to treat the case of a parallel-plate capacitor, since we are assuming that the field of such a capacitor is uniform and confined to the region between its plates. The total energy of this capacitor when the potential difference V exists across it is

$$W = \tfrac{1}{2}CV^2.$$

Substituting

$$C = \frac{\epsilon_0 A}{d} \qquad \text{and} \qquad V = Ed,$$

where E is the electric field intensity between the plates, we find that

$$W = \tfrac{1}{2}\epsilon_0 AdE^2.$$

The quantity Ad is the volume enclosed by the plates, which is the volume occupied by the field E. This enables us to express the *potential energy per unit volume*, which is the energy density w, of the electric field between the plates of the capacitor as

$$w = \tfrac{1}{2}\epsilon_0 E^2. \tag{23–14}$$

Although we have derived Eq. (23–14) for a specific field configuration, it is a perfectly general result for the energy density of an electric field in a region of space in which the field magnitude is E. While we shall not present a general proof of this statement, as an exercise we shall show that it holds for the more complicated geometry of a spherical capacitor (Fig. 23–6). Here the difficulty is that the electric field is not constant between the plates. To overcome this difficulty, we shall consider a capacitor consisting of two concentric spheres of radii R and $R + dR$, and calculate the energy density in the region between the two spheres in the limit as dR goes to zero. We begin with $W = \frac{1}{2}CV^2$. Here

$$V = E\,dR, \qquad C = \frac{4\pi\epsilon_0 R(R + dR)}{dR},$$

where we obtain the latter formula from Eq. (23–10) by letting $b = R + dR$ and $a = R$. The volume between the plates is, to the first order in dR, $4\pi R^2\,dR$. Therefore the energy density is

$$w = \tfrac{1}{2}4\pi\epsilon_0 \frac{(R^2 + R\,dR)E^2\,dR^2}{dR 4\pi R^2\,dR} = \tfrac{1}{2}\epsilon_0 E^2,$$

in the limit as dR goes to zero; this result is identical to the one we found in the case of a parallel-plate capacitor.

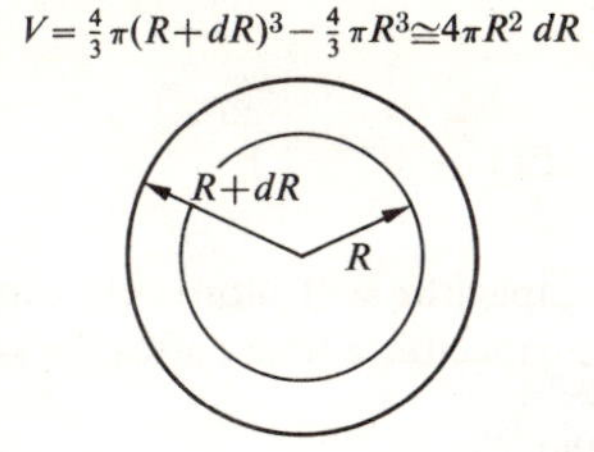

FIG. 23-6. A spherical capacitor of radii R and $R + dR$.

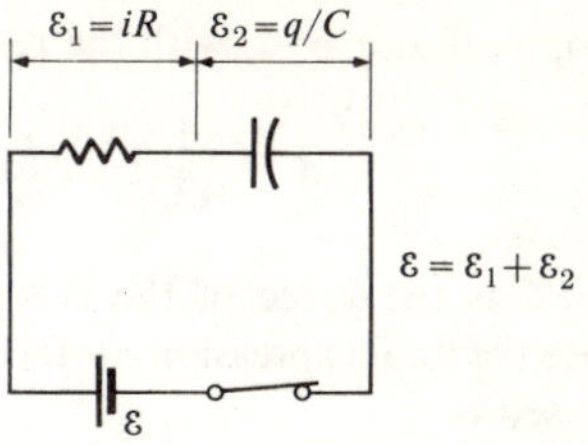

FIG. 23-7. Charging a capacitor in a circuit.

23-4 The capacitor in a d-c circuit

Although electric currents cannot flow *through* a capacitor, transient currents do occur in d-c circuits containing capacitances that enable charges to be built up on the plates of the capacitors present when a source of emf is connected, and to enable the charges to be dissipated when the source of emf is removed. Current can flow, therefore, in such circuits even though no charge passes *through* the capacitors. Let us consider the current i in the circuit of Fig. 23-7 as the capacitor is being charged after the switch is closed. At the instant in which the charge q is already present on the capacitor,

$$\mathcal{E} = iR + \frac{q}{C}, \tag{23-15}$$

which simply states that the emf of the battery equals the potential drop across the resistance R plus the potential difference across C. Since

$$i = dq/dt,$$

we have the differential equation

$$R\frac{dq}{dt} + \frac{q}{C} = \mathcal{E},$$

which may be rewritten

$$\frac{dq}{dt} = \frac{\mathcal{E}C - q}{RC}.$$

Separating variables, we obtain

$$\frac{dq}{\mathcal{E}C - q} = \frac{dt}{RC}.$$

This can be integrated to yield

$$\ln A(\mathcal{E}C - q) = \frac{-t}{RC} \quad \text{or} \quad A(\mathcal{E}C - q) = e^{-t/RC}.$$

The quantity A is a constant of integration which can be evaluated from the fact that $q = 0$ at $t = 0$, with the result that

$$A = \frac{1}{\mathcal{E}C}, \qquad \text{and so} \qquad q = \mathcal{E}C(1 - e^{-t/RC}).$$

But $\mathcal{E}C$ is the value of the charge Q that the capacitor will ultimately acquire. Hence the final expression for the charge q on the capacitor a time t after the switch is closed is

$$q = Q(1 - e^{-t/RC}). \tag{23-16}$$

As t approaches infinity, q approaches Q, as it should. The quantity RC has the dimension of time and determines the rate at which the capacitor charges. It is known as the *time constant* of the circuit. The time constant RC plays the same role in a capacitive circuit as L/R does in an inductive circuit.

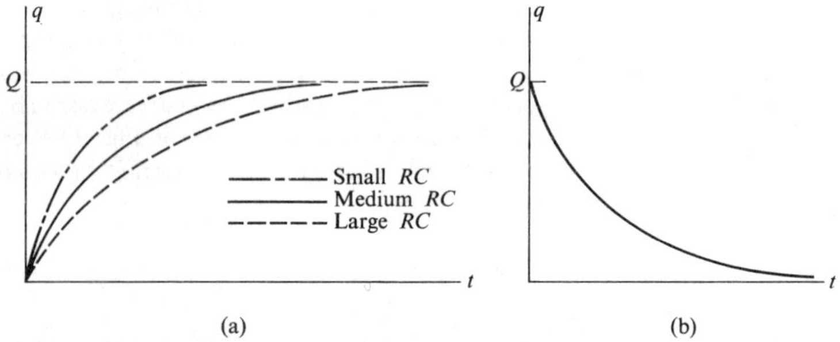

FIG. 23-8. (a) Charging a capacitor, and (b) discharging it.

When the switch is opened, we again employ Eq. (23-15) to find the charge q at time t later. Here, however, $\mathcal{E} = 0$, and

$$R\frac{dq}{dt} + \frac{q}{C} = 0.$$

Solving, we find that

$$\frac{dq}{q} = \frac{-dt}{RC},$$

$$\ln Aq = \frac{-t}{RC},$$

$$q = \frac{1}{A}e^{-t/RC}.$$

The initial conditions in this situation are that $q = Q$ when $t = 0$, and so $A = 1/Q$ and

$$q = Qe^{-t/RC}. \tag{23-17}$$

The charge on C leaks off exponentially. Equations (23–16) and (23–17) are illustrated in Fig. 23–8. During the charging process the current flowing in the circuit is

$$i = \frac{dq}{dt} = \frac{\mathcal{E}}{R} e^{-t/RC}. \tag{23–18}$$

Initially i is $\mathcal{E}/R$, the value given by Ohm's law, and as the capacitor becomes charged the current drops to zero. During the discharge of the capacitor after the switch has been opened,

$$i = -\frac{\mathcal{E}}{R} e^{-t/RC}, \tag{23–19}$$

which is in the direction opposite to the charging current of Eq. (23–18). This is as it should be, since Eq. (23–19) represents the return from the capacitor of the charge Q that had been deposited there according to (23–16) after the switch was closed.

23–5 Displacement current

Our analysis of the transient currents in a circuit containing a capacitance showed that the battery effectively causes charge to be transferred from one of the plates of the capacitor to the other. The current that flows during the charging process is

$$i = \frac{\mathcal{E}}{R} e^{-t/RC}, \tag{23–18}$$

and it flows in the part of the circuit directly connected to the battery. But there is no transfer of charge through the vacuum of the capacitor, and current as we have defined it does not flow there. One of the properties that currents have in conventional d-c circuits is that they are continuous. James C. Maxwell hypothesized that this property of continuity was a basic property of *all* currents, and hence that some kind of current must flow in the region between the plates. The latter is known as a *displacement current*. In order for the current to be continuous, the current flowing in the region between the plates must be the same as in Eq. (23–18). For a parallel-plate capacitor we can rewrite this expression in another form, and the result turns out to be completely general. The electric field between the plates is given by

$$E = q/\epsilon_0 A.$$

But according to Eq. (23–16),

$$q = Q(1 - e^{-t/RC}),$$

and $Q = C\mathcal{E}$. Substituting, we obtain

$$E = \frac{C\mathcal{E}(1 - e^{-t/RC})}{\epsilon_0 A}.$$

Taking the derivative of this with respect to time yields

$$\frac{dE}{dt} = \frac{\mathcal{E}e^{-t/RC}}{R}\,\frac{1}{\epsilon_0 A},$$

so that, by comparison with Eq. (23–18), the displacement current i_{disp} is

$$i_{\text{disp}} = \epsilon_0 A \frac{dE}{dt}. \tag{23–20}$$

Usually it is more useful to express this result in terms of the displacement current density j_{disp}, given by

$$j_{\text{disp}} = \frac{i_{\text{disp}}}{A} = \epsilon_0 \frac{dE}{dt}. \tag{23–21}$$

While we derived Eq. (23–21) for the special case of a parallel-plate capacitor, it is a universally applicable result which implies that a changing electric field in a region of space is equivalent to a current density

$$\mathbf{j} = \epsilon_0 \frac{\partial \mathbf{E}}{\partial t}$$

there. In the most general case **E** and **j** are vectors. Also, since **E** can be a function of both space and time, the partial derivative with respect to time must be taken.

The reality of displacement currents has been demonstrated by measurements of their magnetic effects. The hypothesis of displacement currents has had far-reaching consequences, as we shall see in Chapter 28.

23–6 Capacitors in combination

Capacitors may be connected in a variety of ways in order to alter the effective capacitance of a circuit element, but nearly all combinations can be reduced to series (Fig. 23–9) and parallel (Fig. 23–10) arrangements. As we did in the case of resistances, we shall evaluate the equivalent capacitance of these basic combinations.

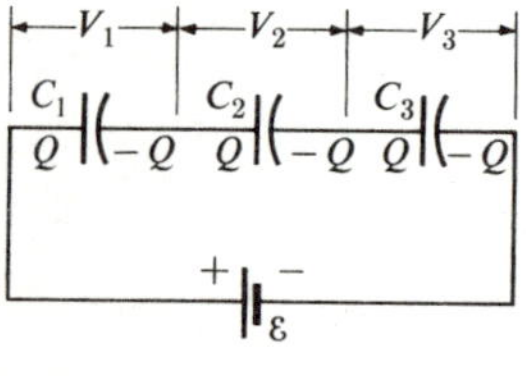

FIG. 23–9. Series arrangement of capacitors.

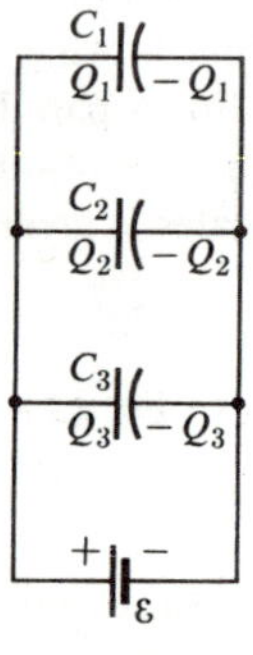

FIG. 23–10. Parallel arrangement of capacitors.

When the capacitors C_1, C_2, and C_3 of Fig. 23-9 are placed across the battery, a current flows until the charge $+Q$ appears on C_1. The lines of force originating on the plate of C_1 that is attached to the battery terminate (ideally) on its opposite plate, inducing the charge $-Q$ there. Since the wire joining C_1 and C_2 was initially neutral, this means that the charge $+Q$ must appear on the plate of C_2 adjacent to C_1. The outer plate of C_2 therefore must possess the charge $-Q$, which is consistent with its being connected to the negative terminal of the battery. This analysis may be extended to treat any number of capacitors in series, with the general result that *the charge on each capacitor in a series circuit is the same.*

The sum of the potential differences V_1, V_2, and V_3 across the capacitors is equal to the emf $\mathcal{E}$ of the battery, that is,

$$\mathcal{E} = V_1 + V_2 + V_3.$$

But

$$V_1 = \frac{Q}{C_1}, \qquad V_2 = \frac{Q}{C_2}, \qquad \text{and} \qquad V_3 = \frac{Q}{C_3},$$

and so

$$\mathcal{E} = \frac{Q}{C_1} + \frac{Q}{C_2} + \frac{Q}{C_3}.$$

If we were to replace C_1, C_2, and C_3 by a single capacitor C,

$$\mathcal{E} = Q/C.$$

Hence the value of the equivalent capacitance C is given by

$$\frac{1}{C} = \frac{1}{C_1} + \frac{1}{C_2} + \frac{1}{C_3},$$

which for the case of n capacitors in series may be written

$$\frac{1}{C} = \sum_{i=1}^{n} \frac{1}{C_i}. \tag{23-22}$$

Capacitors in series combine as resistors in parallel do.

When the capacitors are in parallel, as in Fig. 23-10, *the potential difference V is the same across each of them* and equals the emf of the battery:

$$\mathcal{E} = V_1 = V_2 = V_3, \qquad \mathcal{E} = \frac{Q_1}{C_1} = \frac{Q_2}{C_2} = \frac{Q_3}{C_3}.$$

Here Q_1, Q_2, and Q_3 are the charges which appear on the capacitors, which are different if their capacitances are different. If C_1, C_2, and C_3 are replaced by the single capacitor C, the charge Q induced on it is

$$Q = Q_1 = Q_2 = Q_3.$$

However,

$$Q_1 = C_1\mathcal{E}, \qquad Q_2 = C_2\mathcal{E}, \qquad \text{and} \qquad Q_3 = C_3\mathcal{E},$$

and so

$$C = C_1 + C_2 + C_3.$$

For n capacitors in parallel,

$$C = \sum_{i=1}^{n} C_i. \tag{23-23}$$

Capacitors in parallel combine as resistors in series do.

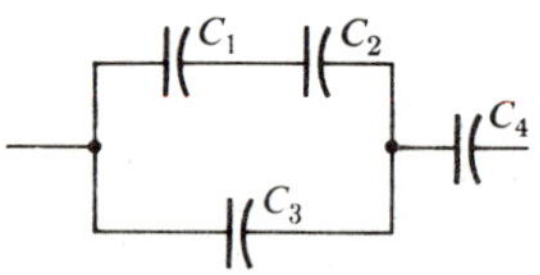

FIG. 23-11. Series-parallel arrangement of capacitors.

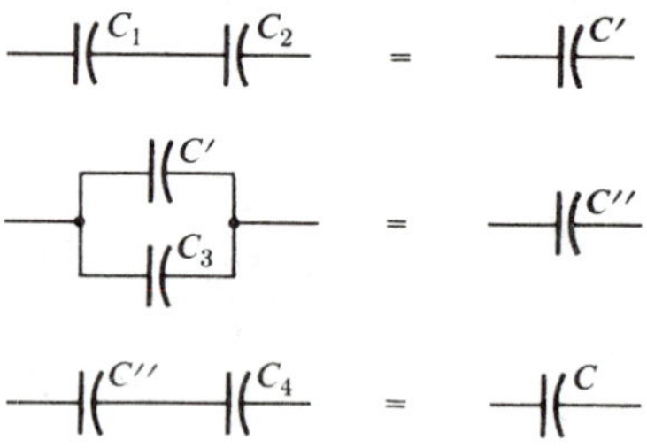

FIG. 23-12. Computation of equivalent capacitance of arrangement in Fig. 23-11.

As an example, let us find the equivalent capacitance of the capacitors shown in Fig. 23-11. The method of attacking this problem is shown schematically in Fig. 23-12. First we compute the equivalent capacitance C' of C_1 and C_2 in series:

$$C' = \frac{C_1C_2}{C_1 + C_2}.$$

C' and C_3 are in parallel, and their equivalent capacitance C'' is therefore

$$C'' = C' + C_3.$$

Since C'' and C_4 are in series, the equivalent capacitance C of the entire circuit is

$$C = \frac{C''C_4}{C'' + C_4}.$$

23-7 Dielectrics

When a nonconducting material such as paraffin or mica is placed between the plates of a capacitor, its capacitance increases. A nonconductor used for this purpose is called a *dielectric*. If, instead, a conductor were placed between the plates of a capacitor, the electrons in the conductor would redistribute themselves so that, for a given charge on the plates, the potential difference between them would be zero. The capacitance of the capacitor would therefore be infinite. In the case of a dielectric, the redistribution of charge that occurs only partially shields the interior of the dielectric. The potential difference between the plates decreases and the capacitance therefore increases with the insertion of the dielectric. The ratio of the capacitance C', after a material is inserted, to the capacitance C when there is a vacuum between the plates is called the *dielectric constant* (or coefficient) of the material. The symbol for dielectric constant is K, so that

$$K = C'/C. \tag{23-24}$$

Another way of describing the effect of introducing a dielectric between the plates of a capacitor makes use of the fact that capacitances can be expressed in general as

$$C = \epsilon L, \tag{23-25}$$

where L is a length characteristic of the geometry of the particular capacitor in question. For a parallel plate capacitor $L = A/d$, and for a spherical capacitor $L = ab/4\pi(b - a)$. The quantity ϵ is called the *permittivity* of the medium between the plates of the capacitor. Since the permittivity of free space is ϵ_0, for any capacitor

$$C = \epsilon_0 L, \qquad C' = \epsilon L,$$

and so

$$C'/C = \epsilon/\epsilon_0 = K.$$

Table 23-1 is a list of the dielectric constants of some common substances.

The observation that its capacitance increases when a dielectric material is placed between the plates of a capacitor permits us to infer something about the change in its energy when this is done. If we insert a dielectric between the plates of an isolated capacitor, the charge on them remains constant while the capacitance increases. Since the energy of a capacitor is $\frac{1}{2}Q^2/C$, the energy of the con-

TABLE 23-1

Dielectric constants (at 20°C except where noted)

Substance	K
Air	1.0006
Air, liquid (−191°C)	1.4
Alcohol, ethyl	26
Amber	2.9
Benzene	2.3
Glass	5–8
Granite	7–9
Ice (−2°C)	94
Mica	2.5–7
Sulfur	3.9
Water	80

figuration must decrease. If, on the other hand, the capacitor is connected to a battery and the voltage across it remains constant when a dielectric is inserted, the charge on it will change. Since the energy of the capacitor is given by $\frac{1}{2}CV^2$, the energy of this configuration will *increase* when the dielectric is inserted. The additional energy in the latter case is supplied by the battery. The reason for the decrease in one case and the increase in the other is due to the properties of the matter which is inserted between the plates. In order to understand these effects it is necessary to examine first some of the basic electrical properties of matter.

23-8 Electrical properties of gases and solids

We have continually alluded to the fact that matter is essentially electrical in nature. To demonstrate this experimentally we need only apply a potential difference between two electrodes in a gas at low pressure in a tube (Fig. 23-13). For small potential differences, no current flows, since the gas is a nonconductor of electricity and has an infinite resistance. As we increase the potential difference, current begins to flow in the tube, indicating that free electrons have appeared which can carry the current. These electrons have their source in the neutral gas atoms in the tube. As we increase the potential difference we can accelerate the few free electrons which are always present until they have enough energy to dislodge other electrons from the neutral gas atoms. The electrons which have been dislodged are accelerated by the electric field and produce still more electrons. In a short time, if we continue to increase the potential difference, an avalanche of electrons is produced and the gas becomes highly conducting. The whole discharge is accompanied by the production of light, whose mechanism for emission we shall take up later. The characteristics of the discharge are a property of the particular gas and of the pressure. This experiment clearly shows that neutral gas atoms consist of combinations of positive and negative charge which are normally bound securely together. If enough energy is supplied to dissociate this combination, the electrons become mobile and the gas becomes conducting.

One of the first experiments to verify directly that solid materials also have electrical properties was performed by Geiger and Marsden in Rutherford's laboratory in 1911. In this experiment, alpha particles were directed against thin

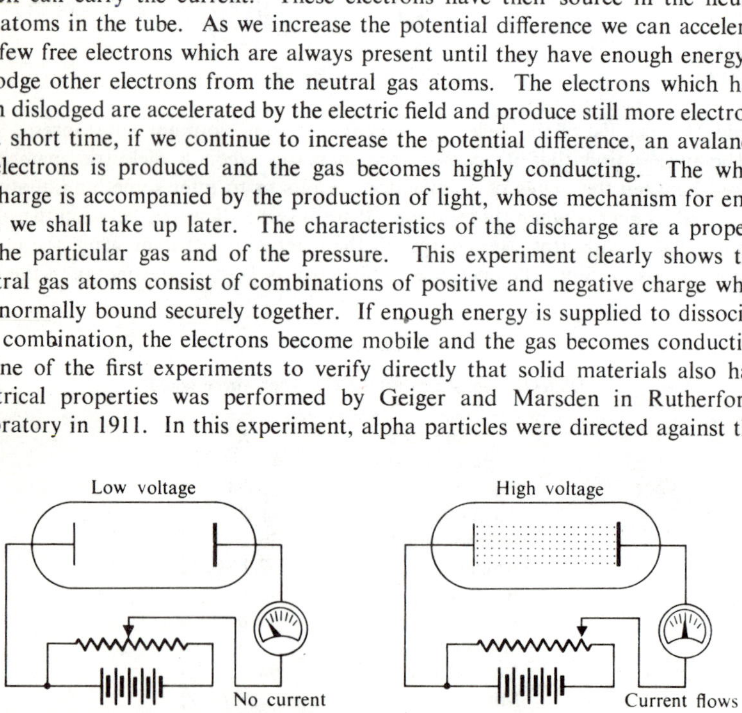

FIG. 23-13. Breakdown of a gas.

metallic foils and their deflections noted. [Alpha particles are helium nuclei which are emanations from radioactive materials (see Chapter 37). Each has a mass approximately 7000 times that of an electron and a positive charge equal in magnitude to twice the electronic charge.] The principal result of the experiment was that the probabilities of deflection of the alpha particles were proportional to $1/\sin^4(\theta/2)$, where θ is the angle of deflection. In Fig. 23–14 we have plotted $1/\sin^4(\theta/2)$ versus θ. Much is to be learned from this distribution law in a qualitative way. First, from our knowledge of the electrical properties of gases, it would not surprise us if the deflections were due to electrical interactions between the alpha particles and the constituents of the foil. The fact that most of the alpha particles are undeflected or very slightly deflected indicates that these particles are either undisturbed or very slightly disturbed in passing through the foil. This implies that most of the space in the foil is empty, so that the particle goes right through.

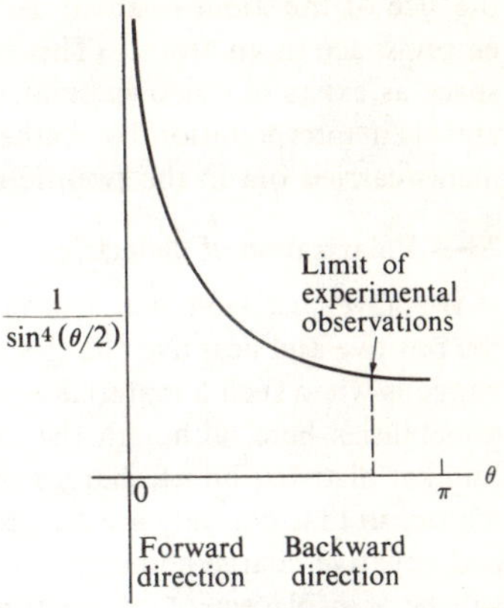

FIG. 23–14. The distribution of particles in Rutherford scattering as a function of angle of deflection θ.

The additional fact that there were some collisions in which the angle of deflection was more than 90° also has some interesting conclusions. Rutherford sometimes referred to alpha particles as "cannonballs" because of their relatively large mass. He reasoned that, in order for these to be deflected through an angle greater than 90° in passing through matter, there must be some equally massive charged particles (charged so that they could exert a force on the charged alpha particles). Electrons, for example, could not deflect the alpha particles very much. The relative infrequency of these large-angle deflections reflects the fact that these massive particles occupy a small region of space, and it is only the occasional encounter when the alpha particle hits this small nucleus of positive charge directly that the large-angle deflection takes place. Rutherford thus arrived at his model of an atom, at whose center was a small concentrated core of positive charge which he called the *nucleus*; surrounding the nucleus at a relatively large distance was the negative charge, in the form of an electron distribution. When Rutherford calculated the scattering of an alpha particle by a positive charge localized at a point, he found that the formula he obtained agreed almost exactly with the experimental results of Geiger and Marsden. The area of disagreement occurred at the very large-angle scatterings. Intuitively we can see that the less the distance of closest approach to the nucleus, the greater will be the angle of scattering. Those deflections which corresponded to a distance of closest approach less than 10^{-15} m were the ones in disagreement with the theory. Rutherford concluded therefore that the size of the nucleus was of this order of magnitude. It is worth noting that the size of the atom is about 10^{-10} m, so that there is a linear factor of 10^5 of empty space in an atom. (This is approximately the same linear factor of empty space as exists in our own solar system.) The Geiger and Marsden experiments, and their interpretation by Rutherford, were some of the most significant experiments carried out in the twentieth century, and indeed in all time.

23–9 Polarization of dielectrics

We have just seen that matter consists essentially of neutral atoms in which the positive and negative charges are held tightly together. We now return to what happens when such a material is placed in an electric field. There are two different possibilities here, although the net result is the same in each. In one case the original distribution of charges in the atoms of the dielectric is spherically symmetric, so that not only are the atoms neutral but centers of gravity of the positive and negative charge coincide. If this material is placed in an electric field, there will be a displacement of positive and negative charge, with the positive charge moving in the direction of the field and the negative charge in the opposite direction. These charges will move only a small distance because internal electric forces are set up when there is a displacement to balance the effects of the external field. A positive and negative charge separated by a small distance is an electrical dipole, and when this material is placed in an external electric field there will thus be an induced dipole moment throughout the volume. The interior of the dielectric will remain neutral; whatever charge is removed from a given volume will be replaced by charge coming from a neighboring volume (Fig. 23–15). There will be free charges at the surface, however.

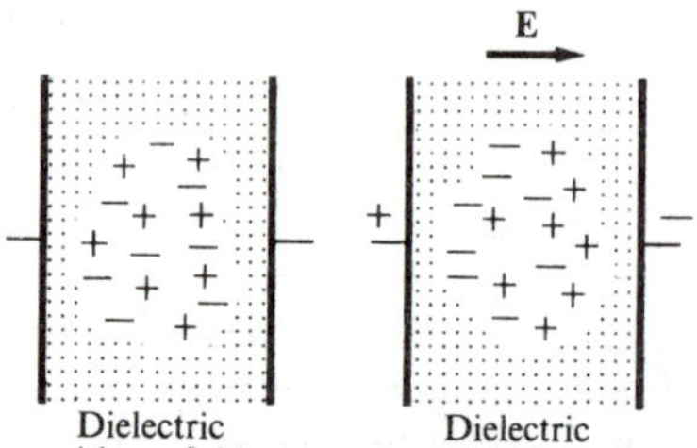

FIG. 23–15. Polarization of dielectric. Molecules have no permanent dipole moment.

The other case concerns charge distributions in the dielectric atoms when the centers of gravity of the positive and negative charge do not coincide. Even though this is the case, the net dipole moment per unit volume is still zero because these dipoles are randomly oriented, and the effect of these various dipoles is to cancel one another out. If such a material is placed in an external field, the electric dipoles behave in a uniform electric field in the same way as magnetic dipoles do in the presence of a uniform magnetic field (see Section 19–5), and they turn so that they are preferentially oriented in the direction of the field; again there is induced a net dipole moment per unit volume. Again there are some charges induced at the surface of the dielectric due to the orientation of the dipoles near the surface (Fig. 23–16).

From these considerations it is possible to see why the energy of capacitors changes as it does when dielectrics are inserted. When we start to insert a dielectric

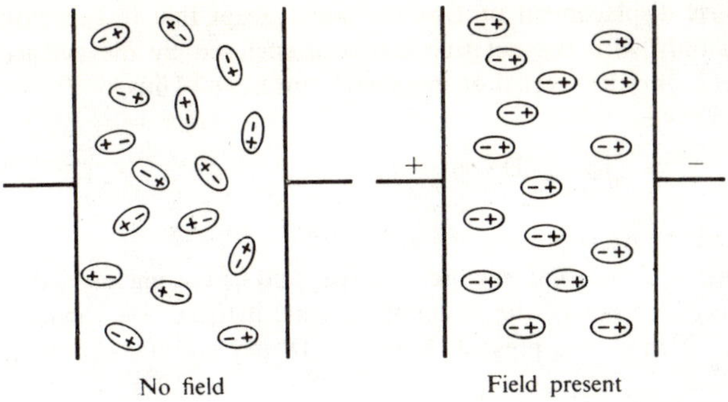

FIG. 23–16. Polarization of dielectric. Molecules have permanent dipole moment.

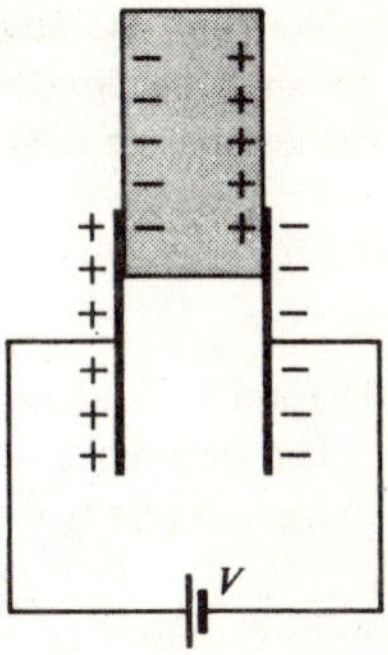

FIG. 23–17. Inserting a dielectric into a capacitor.

between the plates of an isolated capacitor there is an induced charge density on its surface (Fig. 23–17). This leads to a force of attraction between the plates and the dielectric, and causes work to be done on the dielectric as it is pulled in. The work done represents a loss of electrical energy by the system, energy which can be recovered if the dielectric is removed. If the capacitor is connected to a battery, the electric field is maintained between the plates and the dielectric is polarized. This represents an increase in the electrical energy of the system, an increase which resides in the polarized medium. The work done here, of course, is done by the battery.

23–10 Displacement vector

When Maxwell developed the theory of electricity in the nineteenth century, very little was known about the electrical nature of matter. Maxwell did realize that the **E**-vector was not really a suitable one for discussing forces on charged particles in the presence of matter. From our point of view the sources of **E** are *all* the charges which exist, including the charges over which we have no control which exist inside atoms. Maxwell was forced to introduce a *displacement vector* **D** whose sources were the charges which the experimenter does have control of; these are called *free charges* since they are not attached to atoms. The difference between this **D**-vector and the electric field vector **E** represents the distortion of the medium. Previously we found that the integral of the electric field **E** taken over a closed surface is proportional to the total charge Q enclosed by the surface, both free (i.e., macroscopic) charge and atomic (i.e., microscopic) charge. This relationship,

$$\epsilon_0 \int_{\substack{\text{closed}\\ \text{surface}}} \mathbf{E} \cdot d\mathbf{s} = Q, \tag{23–26}$$

was called Gauss' theorem.

We shall define electric displacement in a similar way, except that in this case we shall be concerned only with the net free charge q enclosed by the surface (as before, we may speak of either actual or imaginary surfaces). Thus

$$\int_{\substack{\text{closed}\\\text{surface}}} \mathbf{D}\cdot d\mathbf{s} = q. \tag{23–27}$$

The units of displacement **D** are coulombs/meter^2.

The difference between Q, the total enclosed charge, and q, the enclosed free charge, depends on the properties of the material present; if there is a vacuum, $q = Q$ and $\mathbf{D} = \epsilon_0\mathbf{E}$. If matter is present, however, **D** and $\epsilon_0\mathbf{E}$ are different, and the quantity

$$\mathbf{P} = \mathbf{D} - \epsilon_0\mathbf{E}, \tag{23–28}$$

called the *polarization of the medium*, may be related in a useful way to the physical characteristics of the medium. Here **P**, a vector, is also expressed in coulombs/meter^2.

It is possible to discuss the consequences of polarization of the medium without going into the microscopic processes that occur within the molecules involved. This phenomenological approach is facilitated by the fact that, in most materials, the induced dipole moment per unit volume **P** is directly proportional to the electric field. Hence we can write

$$\mathbf{P} = \chi\mathbf{E}, \tag{23–29}$$

where χ is called the *electric susceptibility* of the substance under consideration. The displacement vector now becomes, from Eq. (23–28),

$$\mathbf{D} = (\epsilon_0 + \chi)\mathbf{E} = \epsilon\mathbf{E}, \tag{23–30}$$

where the quantity ϵ is the permittivity of the substance. Since $\epsilon = \epsilon_0 + \chi$, in a vacuum, when the susceptibility χ is 0, the permittivity becomes that of free space as it should.

What is the electric field produced by a free charge q located inside a material medium? To answer this question, we return to Eq. (23–27),

$$\int_{\substack{\text{closed}\\\text{surface}}} \mathbf{D}\cdot d\mathbf{s} = q,$$

where q is the enclosed free charge. If we have a medium in which $\mathbf{D} = \epsilon\mathbf{E}$, this relationship becomes

$$\epsilon\int_{\substack{\text{closed}\\\text{surface}}} \mathbf{E}\cdot d\mathbf{s} = q,$$

which is identical to Gauss' theorem, with ϵ_0 replaced by ϵ. Thus the electric field **E** surrounding such a charge is

$$\mathbf{E} = \frac{q}{4\pi\epsilon r^3}\mathbf{r} \tag{23–31}$$

a distance r away from the charge. If two charges, q_1 and q_2, are in the medium a distance r apart, the force between them is

$$\mathbf{F} = \frac{1}{4\pi\epsilon}\frac{q_1 q_2}{r^3}\mathbf{r}. \tag{23-32}$$

As we saw earlier, the energy density of an electric field in a vacuum is

$$w = \tfrac{1}{2}\epsilon_0 E^2. \tag{23-33}$$

When the field is present in a dielectric of permittivity ϵ, the energy density is

$$w = \tfrac{1}{2}\epsilon E^2. \tag{23-34}$$

Since $D = \epsilon E$, another way of writing this equation is

$$w = \tfrac{1}{2}DE, \tag{23-35}$$

an equation to which we shall return.

Problems

1. A 25-μf capacitor is connected to a source of potential difference of 1000 v. What is the resulting charge on the capacitor?

2. A 10-μf capacitor can withstand a maximum potential difference of 4000 v. What is the maximum energy it can store?

3. A parallel-plate capacitor of capacitance C is given the charge Q and then disconnected from the circuit. How much work is required to pull the plates of this capacitor to twice their original separation?

4. The plates of a parallel-plate capacitor are 50 cm^2 in area and 1 mm apart. (a) What is its capacitance? (b) When the capacitor is connected to a 45-v battery, what is the charge on either plate? (c) What is the energy of the charged capacitor?

5. The space between the plates of the capacitor of the previous problem is filled with sulfur. Answer questions (a), (b), and (c) for this case.

6. The capacitance of a parallel-plate capacitor is increased from 8 μf to 50 μf when a sheet of glass is inserted between its plates. What is the dielectric constant of the glass?

7. A sheet of mica whose dielectric constant is 5 is placed between the plates of a charged, isolated parallel-plate capacitor. How is the potential difference across the capacitor affected? How is the charge on the capacitor affected?

8. A capacitor with air between its plates is connected to a battery and each of its plates receives a charge of 10^{-4} coul. While still connected to the battery the capacitor is immersed in oil, and a further charge of 10^{-4} coul is added to each plate. What is the dielectric constant of the oil?

9. Calculate the force between the plates of an isolated parallel-plate capacitor by equating the work done in an infinitesimal separation of the plates to the increase of the energy of the capacitor.

10. A 1-μf capacitor is charged to 100 v and then discharged through a 500-ohm resistor. Plot the potential across the capacitor and the resistance as a function of time during the discharge. Plot the current as a function of the time during the discharge. Repeat for a 250-ohm resistor and a 1000-ohm resistor.

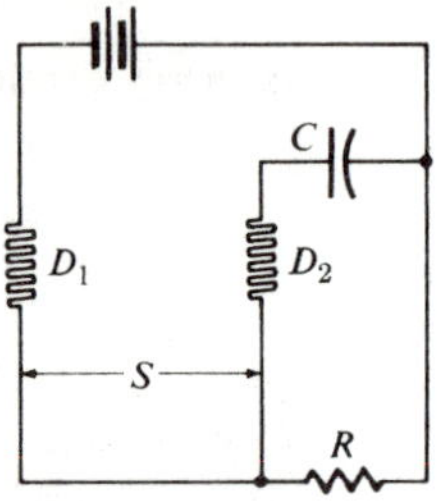

FIGURE 23–18

11. The speed of a rifle bullet is measured by having it break two diaphragms D_1 and D_2 in succession (Fig. 23–18). The breaking of the first one disconnects the fully charged capacitor from the battery and it begins to discharge. The breaking of the second stops the discharge. (a) Derive a formula for the final voltage V_f across the capacitor in terms of R, C, the distance between diaphragms, the bullet's speed v and the initial voltage V_i. (b) If the voltage drops to one-third of its original value, $R = 10^4$ ohms, $C = 1\ \mu$f, and $S = 200$ cm, what is the bullet's speed?

12. A reusable flash bulb requires an energy of 100 j for its discharge. A 450-v battery is used to charge a capacitor for this purpose. The resistance of the charging circuit is 15 ohms. (a) What is the required capacitance? (b) What is the time constant of the circuit?

13. A 5-μf capacitor is connected across a 1000-v battery with wires whose resistance is a total of 5000 ohms. (a) What is the time constant of this circuit? (b) What is the initial current that flows when the battery is connected? (c) How long would it take to charge the capacitor if this current remained constant?

14. A capacitor is charged by connecting it through a resistance to a battery. Show that half the work done by the battery is dissipated as heat.

15. Show that the capacitance per unit length of a coaxial wire is given by

$$C = \frac{2\pi\epsilon_0}{\ln (R_2/R_1)},$$

where R_2 is the radius of the outer conductor and R_1 the radius of the inner conductor.

16. The plates of a parallel-plate capacitor are 1 mm apart, each has an area of 1000 cm^2, and they are kept connected across a 1000-v battery. The medium between the plates is air (or vacuum). (a) What is the force between the plates? (b) If the plates are moved apart until the separation is 3 mm, how much mechanical work is done and how much energy returned to the battery? (c) Check the conservation of energy by also computing the change of energy stored in the field between the plates.

17. In Problem 16, the capacitor is charged and the plates disconnected from the battery. How much work is required to move the plates so they are separated by 3 mm?

18. Find the capacity of a parallel-plate capacitor whose dielectric constant varies uniformly from one plate to the other; the value of the dielectric constant is ϵ_1 at one of the plates and ϵ_2 at the other.

19. In a spherical capacitor whose inner radius is a and outer radius b, the variation of the permittivity is given by $\epsilon = 1/\alpha r + \beta$, where α and β are constants determined by the conditions that $\epsilon = \epsilon_1$ when

$r = a$ and $\epsilon = \epsilon_2$ when $r = b$. (a) What is the capacitance of the system? (b) How does the susceptibility vary with r?

20. A capacitor of 10 μf and one of 20 μf are connected across batteries of 600 v and 1000 v, respectively, and then disconnected. They are then connected with the positive plates together and the negative plates together. (a) What is the initial charge on each capacitor? (b) What is the final charge on each capacitor? (c) How much energy is lost when the connection is made?

21. Repeat Problem 23, with the difference that the positive plates are connected to the negative plates.

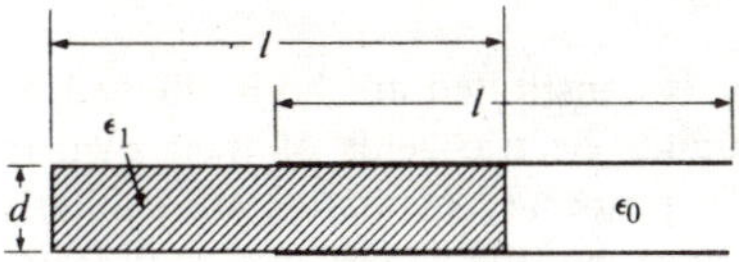

FIGURE 23–19

22. The capacitor shown in Fig. 23–19 consists of square plates whose length on each side is l. The dielectric is also square and of length l and height d. Neglect edge effects. What is the capacitance of the combination? If the potential difference between the plates is V, what is the energy stored? The battery is disconnected and the dielectric is moved by an amount dx. Calculate the force on the dielectric by equating the work to the change of energy due to moving it a distance dx.

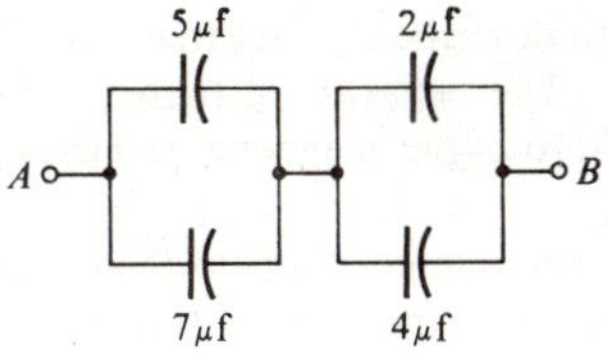

FIGURE 23–20

23. What is the equivalent capacitance between the points A and B in Fig. 23–20?

24. Calculate the potential difference across each capacitor and each resistance in the circuit of Fig. 23–21 if $\mathcal{E} = 110$ v.

Capacitors are in μf; resistances are in ohms.

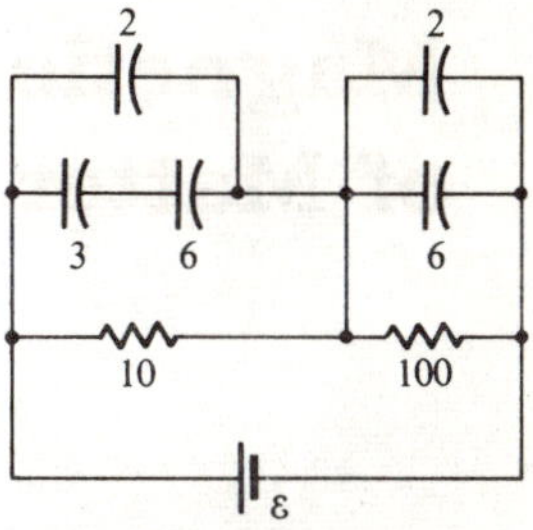

FIGURE 23–21

25. If the radius of a silver nucleus is 7×10^{-15} m, what is the minimum energy that an alpha particle should have to just reach it? The nuclear charge is $+47e$.

26. The dielectric coefficient of hydrogen gas at 0°C and atmospheric pressure is 1.000264. Compute the susceptibility of hydrogen. What is the dipole moment per unit volume when $E = 100$ v/cm?

27. The electric field intensity near the earth's surface is about 100 v/m. How much electrical energy is stored in the lowest kilometer of the atmosphere?

28. Dry air is an insulator provided the electric field intensity in it does not exceed about 3×10^6 v/m. What energy density does this correspond to?

29. An isolated metal sphere of radius R has a charge q. Calculate the radius of a concentric sphere which contains one-half of the total energy of the field.

24

Magnetic Properties of Matter

The atoms and molecules of which matter is constituted are both affected by external electric fields, and in turn may produce electric fields of their own in response to the external ones. The magnetic properties of matter are similarly traceable to the behavior of atoms and molecules, but the variety and complexity of magnetic effects are greater than in the electrical case.

24–1 Sources of atomic magnetism

Matter is essentially electrical in nature, and magnetic effects of matter arise from distributions of currents of the electrical charges in matter. The exact nature of the various electrical currents requires a sophisticated analysis of the properties of matter, but much can be inferred about their magnetic properties without a detailed investigation of the current distributions themselves. Ampère was responsible for the suggestion that atomic currents were responsible for magnetic effects. He proposed this in the nineteenth century, when relatively little was known about the microscopic constituents of matter.

Another source of magnetic properties of matter is a property of the electron known as its spin. In a sense the word is a misnomer because the electron does not really spin on an axis, but the term is used to describe the fact that associated with an electron is a magnetic dipole moment. The behavior of these moments in an external magnetic field contributes to macroscopic magnetic properties of matter.

As a consequence of the fact that magnetic phenomena originate from atomic currents, we should expect that when an external magnetic field is turned on, an emf will result on the atomic level which, by Lenz's law, will be such that the magnetic field of the induced current will tend to cancel the external field. This is known as a *diamagnetic effect*, and always occurs when an external magnetic field is applied to matter.

In some substances there are no net atomic currents and the effects of electron spin cancel out, so that normally there is no magnetic moment per atom. In such cases the only magnetic effect is the diamagnetic one mentioned above. In other substances, however, either due to currents or to uncanceled spins, there is a net magnetic moment per atom. In an external field these will tend to line up in the direction of the field, and the external magnetic field is supplemented by the field of the aligned moments. This is known as a *paramagnetic effect.* It usually exceeds the diamagnetic response which is also present; these substances are said to be *paramagnetic.* Paramagnetic substances increase **B**, diamagnetic ones decrease **B**.

In some substances, notably iron, paramagnetic effects are enhanced greatly (by factors of 10^3 to 10^6) due to the fact that thousands of adjacent atoms are strongly coupled together and act cooperatively. These substances exhibit *ferromagnetism.* Since so many atoms are coupled together, the structure of ferromagnetic materials plays an important role in this effect. Although it is most difficult to analyze, it is also of great technological importance.

24–2 Magnetic energy

In Chapter 19 we learned that a current i flowing in a plane circuit of area A has a magnetic dipole moment of magnitude

$$\mathbf{p} = i\mathbf{A}, \tag{24–1}$$

where the direction of the moment is perpendicular to the plane of the current and related to the direction of the current by the right-hand rule. We also learned that, when a magnetic dipole is placed in a magnetic field, there is a torque

$$\mathbf{N} = -\mathbf{p} \times \mathbf{B} \tag{24–2}$$

that acts. The minus sign means that the torque is such as to try to *restore* the orientation of the dipole to alignment with the field **B**. If a dipole in a uniform magnetic field **B** is rotated through an angle $d\theta$, the work done is

$$dW = N\,d\theta = -pB\sin\theta\,d\theta, \tag{24–3}$$

where θ is the angle between the direction of **p** and **B**. The change in the dipole's potential energy is therefore

$$dV = -dW = pB\sin\theta\,d\theta. \tag{24–4}$$

If the magnitudes of **p** and **B** are independent of θ, we can integrate Eq. (24–4) to obtain for the potential energy V of a magnetic dipole in a magnetic field the expression

$$V = -pB\cos\theta = -\mathbf{p}\cdot\mathbf{B}. \tag{24–5}$$

(We have taken the constant of integration to be 0 when **p** is perpendicular to **B**.)

One can conclude two important things from Eq. (24–5). First, the position of minimum energy of a dipole in a magnetic field is parallel to the field; this is, of course, implicit in Eq. (24–2). Second, we can obtain the force on a magnetic dipole from V by taking its partial derivatives, as mentioned in Chapter 7. Thus the components of **F** in the x-, y-, and z-directions are, respectively,

$$F_x = -\frac{\partial V}{\partial x} = \mathbf{p}\cdot\frac{\partial \mathbf{B}}{\partial x},$$
$$F_y = -\frac{\partial V}{\partial y} = \mathbf{p}\cdot\frac{\partial \mathbf{B}}{\partial y}, \qquad (24\text{–}6)$$
$$F_z = -\frac{\partial V}{\partial z} = \mathbf{p}\cdot\frac{\partial \mathbf{B}}{\partial z}.$$

In a uniform external magnetic field, a magnetic dipole experiences only a torque, not a net force; only if the field is inhomogeneous will there be a net force acting on it.

24–3 Magnetic intensity

The magnetic induction **B** of a magnetic field corresponds to the electric field intensity **E** of an electric field. Both refer to the effective force fields that exist in a region of space, and both are the quantities that must be used in calculating the interactions between the fields and other phenomena. In order to separate out the effect of matter on the electric field, the electric displacement **D** was introduced to represent the field due to macroscopic free charges. If no matter is present,

$$\mathbf{E} = \frac{1}{\epsilon_0}\mathbf{D}.$$

Carrying over this type of analysis into magnetic fields in matter, we must introduce a new quantity referring to the magnetic field of macroscopic currents. This quantity is a vector called **H**, the *magnetic intensity*. In free space,

$$\mathbf{B} = \mu_0\mathbf{H}. \qquad (24\text{–}7)$$

The difference between **B** and $\mu_0\mathbf{H}$ in the presence of matter is a measure of the response of the material to the applied field intensity **H**, and is expressed as

$$\mathbf{B} - \mu_0\mathbf{H} = \mu_0\mathbf{M}, \qquad (24\text{–}8)$$

where **M** is the *magnetic moment per unit volume* induced in the material. Rewriting this expression, we find for the magnetic induction within the material

$$\mathbf{B} = \mu_0(\mathbf{H} + \mathbf{M}). \qquad (24\text{–}9)$$

Both **H** and **M** are expressed in amperes/meter. **H**, like **B**, may be represented diagrammatically by lines of force whose direction at any point is that of **H**.

TABLE 24–1

Magnetic susceptibilities of some common substances

Substance	Susceptibility, mks units
Air	0.037×10^{-5}
Aluminum	2.2
Bismuth	−6.6
Diamond	−2.2
Copper	−0.96
Hydrogen	−0.00021
Lead	−1.7
Mercury	−3.2
Oxygen	0.18
Liquid oxygen	490

In many materials it is found that the magnetic moment per unit volume is proportional to the applied magnetic intensity, so that

$$\mathbf{M} = \chi\mathbf{H}. \tag{24–10}$$

The constant of proportionality χ is called the *magnetic susceptibility* of the material. We may therefore rewrite Eq. (24–7) in the form

$$\mathbf{B} = \mu_0(\mathbf{H} + \chi\mathbf{H}) = \mu_0(1 + \chi)\mathbf{H} = \mu\mathbf{H}. \tag{24–11}$$

The constant μ is called the *permeability* of the material. In free space $\chi = 0$ and $\mu = \mu_0$.

Table 24–1 is a list of the susceptibilities of various substances. A negative susceptibility indicates that the substance is diamagnetic; evidently most common substances are diamagnetic.

Susceptibilities are experimentally determined by measuring the force on a sample placed in an inhomogeneous magnetic field. From the known manner in which the field varies in magnitude as a function of position, together with the force on the sample as determined with a sensitive detector, it is possible to obtain the induced magnetic moment per unit volume.

24–4 Stern-Gerlach experiment

An important experiment performed by Stern and Gerlach in 1921 demonstrated the fact that individual atoms possess magnetic moments. It also demonstrated some other characteristics of the possible directions these moments could assume, which were of fundamental importance in the development of physics of the twentieth century. The method employed involved a low-density beam of neutral atoms. The low density meant that the atoms hardly interacted with one another

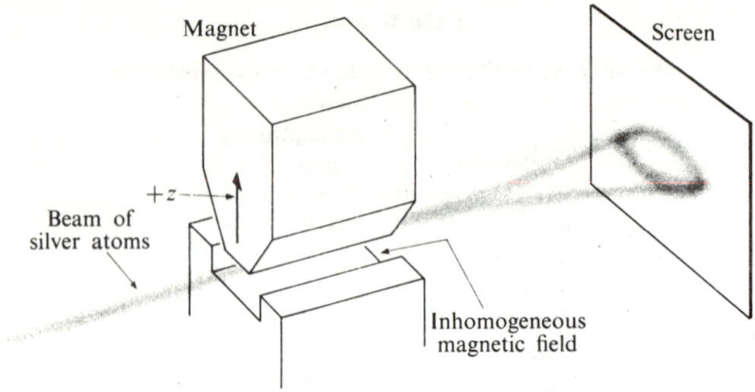

FIG. 24-1. The Stern-Gerlach experiment.

at all, so that the behavior of the atoms themselves could be observed. In their experiment, Stern and Gerlach heated silver in an oven until it vaporized. The silver atoms then emerged in all directions from an orifice in the oven and were collimated by a system of slits. As shown in Fig. 24-1, the beam of silver atoms then passed through an inhomogeneous magnetic field and were ultimately detected by a photographic plate. In the absence of a magnetic field a straight line appears on the plate, indicating that the atomic beam has undergone no deviation. The magnet system used is one that produces a high value of $\partial B_z/\partial z$, where $+z$ is vertically upward in the figure. If an atom of magnetic moment $\mathbf{p}$ is present in this field, the force

$$F_z = \mathbf{p} \cdot \frac{\partial \mathbf{B}}{\partial z} = p_z \frac{\partial B_z}{\partial z} = p \frac{\partial B_z}{\partial z} \cos \theta$$

acts on it, where θ is the angle between $\mathbf{p}$ and the z-axis. It seemed reasonable to suppose that the magnetic moments of the silver atoms emerging from the oven were randomly oriented. If this were so, then the forces on the various atoms would have varied from $+p\,\partial B_z/\partial z$ through 0 (when $\theta = 90°$) to $-p\,\partial B_z/\partial z$, and the effect of the inhomogeneous field would have been to spread the beam out. If $\mathbf{p}$ were 0, of course, no deflections would occur.

What Stern and Gerlach found was that the beam of silver atoms was indeed deviated by the magnetic field, but that instead of a broadened trace on the plane, the beam had split up into two parts (Fig. 24-2). One of the parts corresponded to the alignment of $\mathbf{p}$ in the z-direction, the other to its antialignment; that is, $\theta = 0$ in one case, $\theta = 180°$

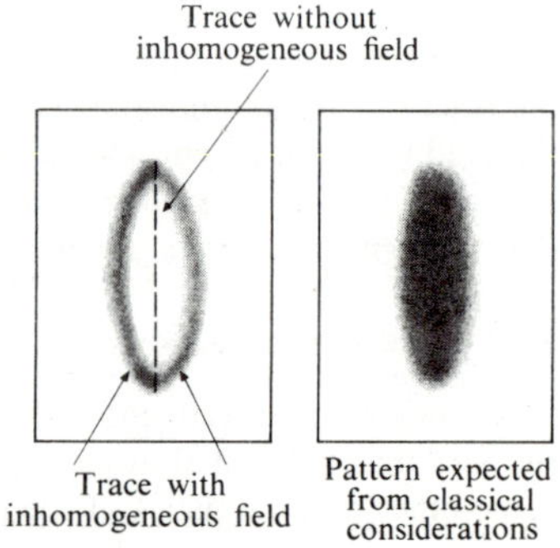

FIG. 24-2. Trace of atomic beam in Stern-Gerlach experiment.

in the other. Thus Stern and Gerlach demonstrated not only that silver atoms possess magnetic moments, but also that these moments can have only two possible orientations in a magnetic field. This latter phenomenon, called *space quantization,* is one of the most significant in atomic physics, and we shall consider it in more detail in Chapter 29.

24–5 Dipole motion in a magnetic field

Magnetic moments on the atomic level are always associated with a mechanical angular momentum. If the moment is due to a circulating electron producing a current, this is obviously true, but there is also an angular momentum associated with an electronic dipole moment due to its spin.

We shall first discuss the motion of a dipole in a constant magnetic field. What happens when a fixed dipole is placed in a uniform magnetic field? This corresponds to a small magnet placed in this field.

We have already seen that a magnetic dipole of moment **p** placed in a field of flux density **B** will experience the torque

$$\mathbf{N} = -\mathbf{p} \times \mathbf{B}, \tag{24–12}$$

whose magnitude is $-pB \sin \theta$, where θ is the angle between **p** and **B**. If the dipole is free to rotate, its response to the torque **N** will be a change in its angular momentum at the rate $d\mathbf{L}/dt$, where

$$\mathbf{N} = d\mathbf{L}/dt. \tag{24–13}$$

If I is the moment of inertia of the dipole about an axis perpendicular to **p**, its angular momentum when its angular velocity is $\boldsymbol{\omega}$ is

$$\mathbf{L} = I\boldsymbol{\omega}, \tag{24–14}$$

and its magnitude is $I\,d\theta/dt$. From Eqs. (24–12) through (24–14), we obtain

$$I\frac{d^2\theta}{dt^2} = -pB \sin \theta.$$

When θ is small—that is, when the dipole is nearly parallel to the magnetic field direction—$\sin \theta$ is very close to θ, and

$$I\frac{d^2\theta}{dt^2} + pB\theta = 0. \tag{24–15}$$

This equation has the same form as that describing the motion of a harmonic oscillator (Chapter 10), provided p is a constant, and its solution is accordingly

$$\theta = A \cos (\sqrt{pB/I}\, t + \phi), \tag{24–16}$$

where A is the amplitude of the oscillations (the maximum angle between **p** and **B**) and ϕ the phase angle (which depends on the value of θ when $t = 0$). The frequency of the angular oscillations is

$$f = \frac{1}{2\pi}\sqrt{\frac{pB}{I}}, \tag{24–17}$$

so that f, which is easily measured in the case of magnetized needles, can be used to compute the magnitude B of the external field. Equation (24–17) forms the basis for one of the standard methods for measuring or comparing magnetic flux densities.

The above results cannot be applied to atomic dipoles whose origin lies in the microscopic currents within them. In the latter case the magnetic moment is proportional to the angular momentum of the charged particle whose motion constitutes the current, so that **p** is not a constant that is independent of time, as assumed above. For a current i which encloses an area **A**, we know that

$$\mathbf{p} = i\mathbf{A}.$$

The current is equal to the charge q that is in motion, multiplied by the number of times per second that it traverses the loop. If ω is the angular velocity of the charge,

$$i = \frac{q}{t} = \frac{q}{2\pi/\omega} = \frac{\omega}{2\pi}\,q.$$

In the case of a circular path of radius r,

$$p = \frac{\omega}{2\pi}\,q\pi r^2 = \frac{q}{2}\,r^2\omega.$$

The direction of **p** is the direction of $\boldsymbol{\omega}$. The particle's angular momentum is

$$\mathbf{L} = I\boldsymbol{\omega} = mr^2\boldsymbol{\omega},$$

and so we have

$$\mathbf{p} = \frac{q}{2m}\,\mathbf{L}.$$

This is indeed the correct result for an orbital electron (for which $q = -e$ and $m = m_e$), but in the case of magnetic moment associated with electron spin, the relationship between magnetic moment and spin angular momentum is given by experiment as

$$\mathbf{p}_{\text{spin}} = -\frac{e}{m}\,\mathbf{L}.$$

In general, we may say that

$$\mathbf{p} = \gamma\mathbf{L}, \tag{24–18}$$

where γ is called the *gyromagnetic ratio*.

We shall now investigate the motion of a magnetic dipole of moment **p** in a field **B**. The torque **N** acting on it is

$$\mathbf{N} = -\mathbf{p} \times \mathbf{B},$$

and, since torque equals the rate of change of angular momentum,

$$\frac{d\mathbf{L}}{dt} = -\mathbf{p} \times \mathbf{B} = -\gamma\mathbf{L} \times \mathbf{B}. \quad (24\text{–}19)$$

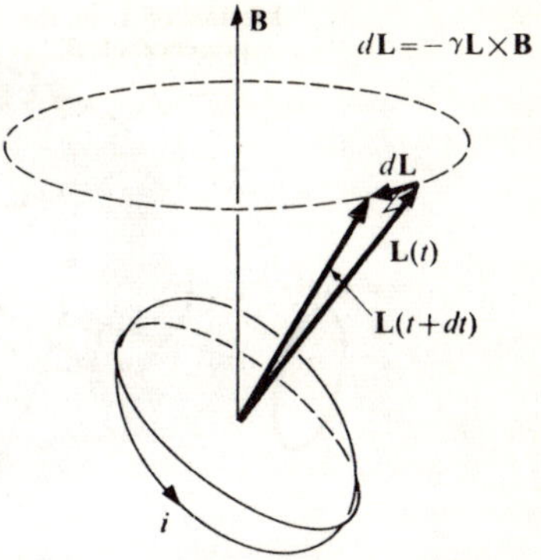

FIG. 24–3. A magnetic moment in a uniform magnetic field **B**. Note that the effect of **B** is to make the end of vector **L** precess around the direction of **B** without changing its magnitude.

What kind of solution does this vector differential equation have? The problem we are concerned with here is the analog of the motion of a gyroscope, which we discussed qualitatively in Chapter 9. As shown in Fig. 24–3, in a time dt the change $d\mathbf{L}$ in **L** must be in a direction perpendicular to both the vector **B** and the vector **L**. In order to satisfy these conditions the end of the vector **L** must move in a circular path about **B** as an axis. Thus **L** does not change its magnitude, but only its direction, as it travels around **B**. This type of motion is called *precession;* it is familiarly found in a spinning top, or a gyroscope, whose axis of rotation describes a cone in space.

24–6 Larmor precession

It is not too difficult to solve Eq. (24–19) and thereby obtain the rate at which the precession takes place. Suppose (Fig. 24–4) that the direction of **B** is along the z-coordinate axis. Splitting Eq. (24–19) into its components, we have for the first one

$$\begin{aligned}\frac{dL_x}{dt} = \frac{d}{dt}(L \sin\theta \cos\phi) &= -\gamma(L_yB_z - L_zB_y)\\ &= -\gamma L_yB\\ &= -\gamma LB \sin\theta \sin\phi. \qquad (24\text{–}20)\end{aligned}$$

In a similar way we find for dL_y/dt,

$$\frac{d}{dt}(L \sin\theta \sin\phi) = -\gamma LB \sin\theta \cos\phi,$$

and for dL_z/dt,

$$\frac{d}{dt}(L \cos\theta) = 0. \qquad (24\text{–}21)$$

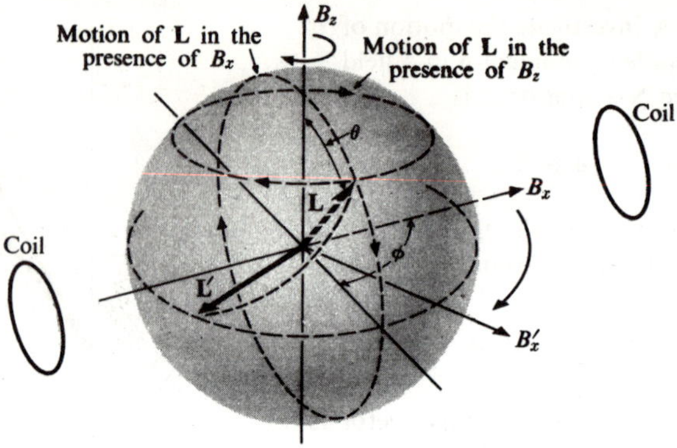

FIG. 24-4. The coils are used to produce an alternating magnetic field perpendicular to the constant field B_z.

From Eq. (24-21) we obtain, since the magnitude of **L** does not vary,

$$\frac{d}{dt}(\cos\theta) = 0, \qquad \theta = \text{const.}$$

This means, as we have indicated, that **L** precesses about **B**. To find the frequency, we refer to Eq. (24-20) and cancel out the $L \sin\theta$ term on both sides, since L and θ are constant; that is,

$$\frac{d}{dt}(L\sin\theta\cos\phi) = -\gamma LB\sin\theta\sin\phi,$$

$$L\sin\theta\frac{d}{dt}(\cos\phi) = -\gamma LB\sin\theta\sin\phi,$$

$$-\sin\phi\frac{d\phi}{dt} = -\gamma B\sin\phi,$$

$$\omega_L = \frac{d\phi}{dt} = \gamma B. \tag{24-22}$$

The angular frequency corresponding to the angular velocity ω_L is called the *Larmor frequency*. The measurement of the Larmor frequency furnishes us with knowledge of the local magnetic field B if γ is known, or the gyromagnetic ratio γ if the local field is known. It can be measured in the following way (see Fig. 24-4): Let us consider a moment precessing about the z-axis when we impose a magnetic field in the x-direction. The moment will now start to precess about the new field in the x-direction. If now the x-directed field were to move around with the same angular frequency with which the moment is precessing about the z-axis, the net result would be that the moment would tip over. If the frequency is different from the Larmor frequency about the z-direction, it would merely wobble a bit and then straighten out. To a coil whose axis is in the x-direction, the flopping moment would be a changing flux and the induced emf could be registered as a current.

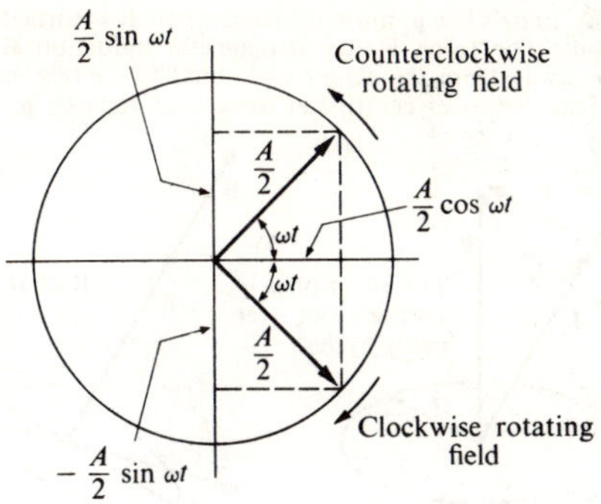

FIG. 24-5. Superposition of clockwise and counterclockwise rotating fields. Sum of two oppositely rotating fields is equivalent to oscillating field in x-direction $= A \sin \omega t$. The y-components cancel.

One can arrange for a rotating magnetic field by noting that two fields

$$B_x = A/2 \cos \omega t, \qquad B_y = A/2 \sin \omega t,$$

and

$$B_x = A/2 \cos \omega t, \qquad B_y = -A/2 \sin \omega t,$$

when superposed, give a field

$$B_x = A \cos \omega t, \qquad B_y = 0.$$

The first of these is a field rotating in a counterclockwise direction, the other in a clockwise direction. Thus a field $B_x = A \cos \omega t$ imposed in the x-direction is the superposition of a clockwise and counterclockwise rotating field (see Fig. 24-5). Only one of these can follow the precessing moment; the other is out of phase and can cause only an occasional wobble whose average effect will vanish.

The experiment is done by establishing an alternating field $B_x = A \cos \omega t$ in the x-direction and then varying the field in the z-direction until an emf is detected in the coil placed in the x-direction. For this value of the external z-field, $\omega = \gamma B_z$. The experiment is done this way because it is easier to vary the magnitude of B_z than the frequency of the oscillating field, although this too can be done in principle. From the known values of ω and B_z, it is possible to infer the moments of the atoms of the substance under study. This technique is known as *paramagnetic resonance*, and is used to investigate the structures of molecules and solids.

24-7 Diamagnetism

Let us examine the origin of diamagnetism quantitatively. We shall consider an electron of charge $-e$ rotating in a circular orbit of radius r [Fig. 24-6(a)]. When an external magnetic field is impressed, the component of the orbiting elec-

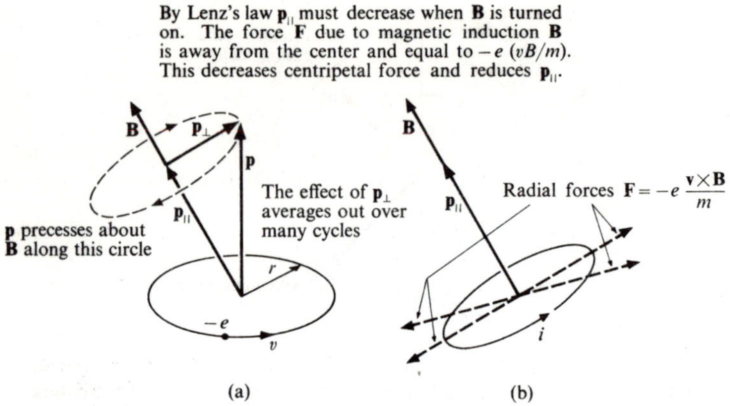

FIG. 24–6. Origin of diamagnetism.

tron's moment perpendicular to **B** will average out to zero, since the moment will precess about **B**; the only component we need consider is that parallel or antiparallel to **B**. When the field is turned on, a current is in effect induced in the electron's orbit. It persists so long as the field is there, since there is no electrical resistance on the atomic level. To compute the induced current, we recall that the emf due to a changing magnetic field is

$$\mathcal{E} = -\frac{d\Phi}{dt},$$

so that here

$$\mathcal{E} = -\pi r^2 \frac{dB}{dt}.$$

The emf is the result of a tangential electric field. Since, by definition, emf is work done per unit charge, the magnitude of this tangential electric field is

$$2\pi r E = \mathcal{E},$$

$$E = -\tfrac{1}{2} r \frac{dB}{dt}.$$

The force on the charge is

$$F = -eE = +\tfrac{1}{2} e r \frac{dB}{dt},$$

and, since $F = m\,dv/dt$,

$$\frac{dv}{dt} = +\frac{er}{2m}\frac{dB}{dt}.$$

If r is constant while the field is being established, the change in the electron's orbital speed during this period is

$$\Delta v = +\frac{er}{2m}\int_0^B \frac{dB}{dt}\,dt = +\frac{er}{2m}B. \tag{24-23}$$

The radius *does* remain the same because the electron's centripetal acceleration is v^2/r. If the speed of the electron changes by Δv, the change in its acceleration is given by the equation

$$\frac{(v + \Delta v)^2}{r} - \frac{v^2}{r} \approx \frac{2v\,\Delta v}{r} = +\frac{ev}{m}B.$$

The acceleration due to the Lorentz force is

$$-\frac{e}{m}(\mathbf{v} \times \mathbf{B}) \approx -\frac{e}{m}vB,$$

so that the field supplies at all times just the correct force to accommodate the new electron speed at the original orbital radius. The direction of the force is also correct. In Fig. 24-6(b), **p** and **B** are parallel, and the electron is decelerated as **B** increases, which reduces its contribution to the magnetic field. If **p** and **B** were antiparallel, the electron would increase in speed and so produce a larger moment that would tend to maintain **B** at its original value. This effect, then, is independent of the orientation of the orbit and is always present when atoms are placed in magnetic fields. As we said earlier, diamagnetism is present even in those substances which behave paramagnetically or ferromagnetically, but it is swamped by the latter effects.

The change in electron speed involves a change in angular momentum of

$$\Delta L = mr^2\,\Delta\omega = mr\,\Delta v = +\tfrac{1}{2}er^2B.$$

The corresponding change in the magnetic moment of the orbiting electron, according to (24-18), is

$$\Delta p = \gamma\,\Delta L = -\frac{e}{m}\cdot\tfrac{1}{2}er^2B = -\frac{e^2}{2m}r^2B.$$

If there are n electrons per unit volume, we know, from the results of Section 24-3, that the magnetic moment per unit volume is related to χ by

$$H\chi = M = n\,\Delta p, \qquad \chi = \frac{n\,\Delta p}{H} = -\frac{ne^2}{2m}r^2\frac{B}{H}.$$

Since χ is small compared with 1, we may replace B/H on the right-hand side by

$$B/H = \mu = \mu_0(1 + \chi) \cong \mu_0.$$

The result for χ is therefore

$$\chi = -\mu_0 \frac{ne^2}{2m}\overline{r^2}. \tag{24-24}$$

The quantity $\overline{r^2}$ is the average value of r^2 in the various orbits. Typical values for the quantities in Eq. (24–24) might be $\overline{r^2} = 10^{-20}$ m, $m = 9 \times 10^{-31}$ kg, $\mu_0 = 12.5 \times 10^{-7}$ weber/amp-m, $e = 1.6 \times 10^{-19}$ coul, and $n = 10^{28}/\text{m}^3$, so that χ has an estimated order of magnitude of 10^{-5}, which is not far from the observed values.

24–8 Paramagnetism

Paramagnetism is more complicated to treat. For an electron moving in a circle of radius r,

$$p = \gamma L = \frac{e}{m} \cdot mr^2\omega = evr. \tag{24-25}$$

Using the typical values for v and r of 10^6 m/sec and 10^{-10} m, respectively, we find p to be roughly 10^{-23} amp-m^2

The effect of an external magnetic field, as we have seen, is to cause the moments to precess. In a solid, all the current elements are interacting with the lattice, so that the precession is interrupted by collisions. The effect of the collision is to cause a certain fraction of the moments to assume a position of minimum energy, namely, to align themselves with the field [see Eq. (24–5)]. At high temperatures there are many disorienting collisions after the alignment has been made; at lower temperatures, once the moment is aligned, it remains aligned, due to the rarity of collisions.

It never happens that an appreciable fraction of the moments are aligned with the field; this would make the paramagnetic effect much larger than observed. The fraction that does line up depends on the ratio of the magnetic energy to the thermal energy. It was found by Langevin to be

$$f = pB/3kT, \tag{24-26}$$

where k is Boltzmann's constant and T is the absolute temperature. Thus

$$M = npf = \frac{np^2B}{3kT} = \chi H.$$

Again, because of the smallness of χ, we replace B/H by μ_0, and

$$\chi = \mu_0 \frac{np^2}{3kT}. \tag{24-27}$$

At room temperature this formula yields $\chi \approx 10^{-3}$ to 10^{-4} for $n \approx 10^{28}$ to 10^{29} The dependence of χ on $1/T$ is known as *Curie's law.*

24-9 Ferromagnetism

As we mentioned in the introduction to this chapter, there is still a third magnetic effect, where the permeability is very much larger than the permeabilities encountered in ordinary paramagnetism. The substances which exhibit these effects are called *ferromagnets.* Ferromagnetism involves the cooperative action of very many dipoles, and because so many atoms are involved, the structure of the ferromagnet plays an important role in its properties.

One important difference between ferromagnets and paramagnets is that one cannot write the proportionality $\mathbf{B} = \mu\mathbf{H}$ for the former; or, to put it another way, μ is not a constant but depends, not only on the values of **B** and **H**, but on their previous history. In particular, it is possible for $\mathbf{H} = 0$ and $\mathbf{B} \neq 0$, which is the reason that one can find permanently magnetized samples without any inducing field. These magnetized samples, by the way, oscillate in a small external field as described by Eq. (24–17).

The behavior of ferromagnets has very important technological consequences in the design of electrical machinery, transformers, and indeed in the design of magnets themselves. From an engineering point of view these substances are of immense importance, but we cannot consider them in detail because too much space would be needed. However, one remarkable aspect of ferromagnetism should be discussed, even in passing, and that is *hysteresis*, the tendency of ferromagnetic materials to retain an initial magnetization. Figure 24–7 shows how B varies with H for an initially unmagnetized sample of annealed iron placed in a coil. At first B rises rapidly with increasing H, indicating a high permeability. When $H = 100$ amp/m, for instance, $B = 0.7$ weber/m^2 and

$$\mu = \frac{B}{H} = \frac{0.7 \text{ weber/m}^2}{100 \text{ amp/m}} = 7 \times 10^{-3} \frac{\text{weber}}{\text{amp/m}}.$$

Since the permeability of free space is $\mu_0 = 12.6 \times 10^{-7}$ weber/amp·m, the presence of the iron in the coil has increased the magnetic induction to over 5500 times what it would be if the coil were in vacuum. However, μ is not constant. When $H = 1000$ amp/m, ten times greater, $B = 1.6$ weber/m^2, little more than double its magnitude at $H = 100$ amp/m. Now

$$\mu = \frac{B}{H} = \frac{1.6 \text{ weber/m}^2}{1000 \text{ amp/m}}$$

$$= 1.6 \times 10^{-3} \frac{\text{weber}}{\text{amp/m}},$$

which is smaller than before though still substantially greater than μ_0.

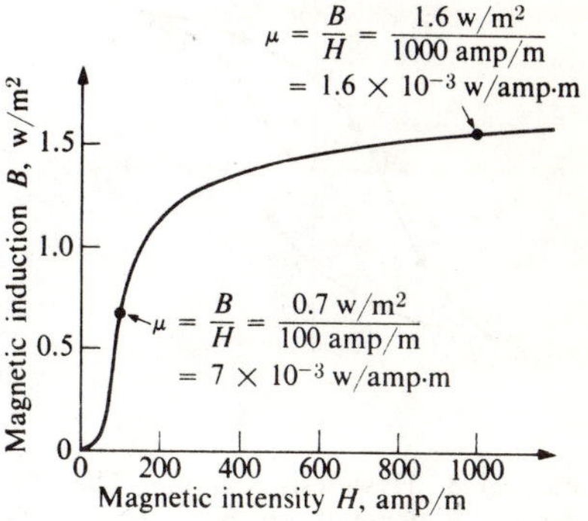

FIG. 24–7. The magnetization curve of annealed iron.

A ferromagnetic material tends to retain a degree of magnetization even when the magnetic field that originally aligned its atomic current loops is removed. Hence its permeability μ does not even have a fixed value at a given H but may take on different values depending on its past history.

Suppose that we place a sample of unmagnetized iron in a coil and vary the current from zero to a maximum in one direction, down through zero to a maximum in the other direction, back through zero to the first maximum, and so on. Figure 24–8 is a plot of B versus H for this cycle. When we first turn on the current, the B-H curve is the same as that of Fig. 24–8. At the point b we reduce the current, so that H drops, but now B does *not* retrace its original path. From b to c the values of B are higher than they were from a to b at corresponding values of H owing to the magnetic "memory" of ferromagnetic materials. At c there is no current in the coil and $H = 0$, yet B nevertheless has the magnitude B_c: the iron sample is now permanently magnetized. Reversing the direction of H does not at first reverse B but merely reduces it, until finally, at d, H is sufficiently negative to bring B to $B = 0$. A further increase in $-H$ takes the B-H curve to e, where B and H are equal in magnitude and opposite in sign to what they were at b. When H is again brought to $H = 0$, B is at f, where $B_f = -B_c$. Increasing H in the positive sense returns H to the point b, but along a curve on which B is always less than it was from a to b. Further cycles simply retrace the curve *bcdefgb*.

From Fig. 24–8 it would seem that, once a ferromagnetic material is magnetized, it cannot be demagnetized, so that $B = 0$ when $H = 0$. There is a procedure that can accomplish this, however. What is done is to carry a magnetized sample through a succession of hysteresis curves, each with a smaller H_{max}. Figure 24–9 shows that, as H_{max} is brought closer and closer to zero, the curve approaches the origin, where $\mathbf{H} = \mathbf{B} = 0$.

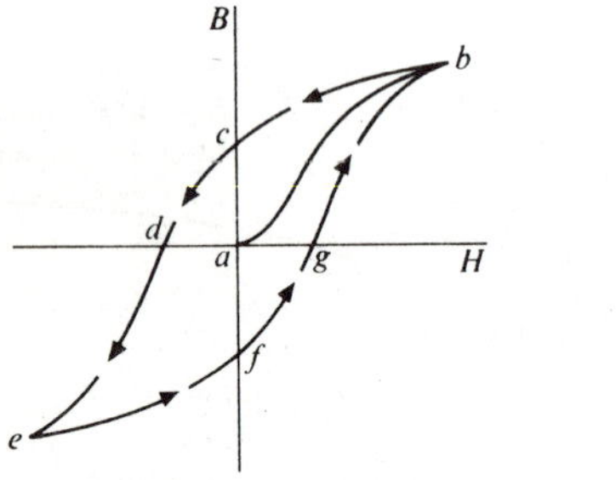

FIG. 24–8. Hysteresis curve. The flux density B depends on both H and the previous history of the sample.

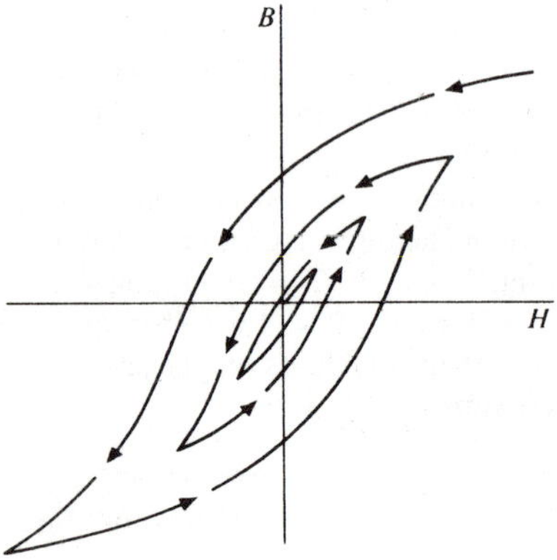

FIG. 24–9. Successive hysteresis loops during the demagnetization of a ferromagnetic sample.

24–10 Einstein-de Haas effect

We mentioned earlier that there is always a mechanical angular momentum associated with a magnetic moment. We have used this fact repeatedly, but have not pointed to any experimental verification. Such a verification was provided by Einstein and de Haas in an effect which bears their name.

If we were to have a cylinder supported longitudinally by a thread and were to apply a magnetic field along the long axis, there would be no torques applied to the cylinder [see Eq. (24–12) with $\theta = 0$]. The cylinder now becomes magnetized. If our hypothesis is correct that the elementary magnetic moments in the cylinder which are now pointing preferentially in one direction have a mechanical angular momentum associated with them, we can conserve angular momentum in the absence of an external torque only if the cylinder as a whole were to rotate. This rotation (which requires great care to observe) is the Einstein-de Haas effect. Careful analysis of this effect yields much information about the magnetic properties of matter. For example, we can learn how many moments contribute to the magnetization of a sample.

Problems

1. Show from the defining equations for various magnetic quantities that the weber is equal to a newton-meter/ampere.

2. Show that the mks units of magnetic moment are either amp·m^2 or joules/weber/m^2.

3. A small magnetic needle has a magnetic moment of 100 amp/m^2, and its moment of inertia is 10^{-5} kg/m^2. It is placed in a region in the earth's magnetic field where $B = 10^{-9}$ weber/m^2. It is then displaced from equilibrium and set into oscillation. What would be the frequency of the oscillations?

4. A small magnet is placed at the center of a solenoid wound with wire at 100 turns/cm. The axis of the coil is perpendicular to the horizontal component of the earth's field. If the horizontal component of the earth's magnetic field is 10^{-9} weber/m^2, what current in the coil will produce a deflection of 10° for the magnet?

5. A magnet has a moment of inertia of 6.25×10^{-5} kg/m^2. When placed 4 m from another magnet which is aligned parallel to the magnetic field **B**, so that its moment is perpendicular to **B**, it experiences a deflection of 10°. When set vibrating in the same field, it vibrates with a frequency of 7.5 vibrations/min. What is the strength of the field?

6. The magnetic moment of an atom is 0.93×10^{-23} joule/weber/m^2. The angular momentum of the electron producing this moment is 1.06×10^{-34} n·m. (a) What is its gyromagnetic ratio? (b) What will its Larmor frequency be in an external field of 2 webers/m^2?

7. The gyromagnetic ratio of an atom is approximately 0.2×10^{12} coul/kg. What must the strength of an external magnetic field be in order for a rotating field, at right angles to the field and of frequency 2×10^{10} cycles/sec, to flip the magnetic moment?

8. The magnetic moment of a silver atom is 0.93×10^{-23} joule/weber/m^2. Silver atoms with a speed of 8×10^2 m/sec are used in a Stern-Gerlach apparatus. The inhomogeneous field is perpendicular to the direction of motion and its gradient is 5 webers/m^2/m. The inhomogeneous field extends over a distance of 0.1 m. The apparatus is 1 m long. (a) What is the force on the silver atoms? (b) What will be the deflection of a silver atom whose moment is parallel to the gradient? (c) What will be the separation of the two traces on the photographic plate? (d) If the beam is defined by slits 0.1 mm wide, will the two traces be distinguishable?

9. The magnetic moment/atom in a certain substance is 10^{-23} amp/m^2. If its susceptibility is 7×10^{-6} in mks units, how many atoms are there per unit volume at 0°C?

10. A long circular rod whose cross section is 5×10^{-3} m^2 has one end placed in a magnetic field of 1 weber/m^2. The other end is far from the field. The field falls to zero in a distance equal to the length of the rod that is placed in the field. The force on the rod is 0.015 newton. What is the volume paramagnetic susceptibility of the rod?

11. What are the mks units of paramagnetic susceptibility?

12. Estimate the mean-square size of the orbit of bismuth if 5 electrons per atom contribute to its diamagnetism and there are 3×10^{22} atoms in 1 cm^3. Is the answer reasonable?

25

Waves

Energy can be transmitted from one place to another in a variety of ways. Suppose, for instance, that we wish to supply energy to a boat in the center of a lake from a position on the shore, with the provision that the precise form in which the energy arrives does not matter. The most obvious thing to do is to throw a stone at the boat. Upon collision, some of the kinetic energy of the stone is transferred to the kinetic energy of the boat.

FIGURE 25–1

A second method requires merely that we drop a stone in the water near the shore, letting the waves that move out transfer energy to the boat by causing it to move up and down (Fig. 25–1). We note in this latter case that what occurs when the stone strikes the water is not a flow of material but, instead, the spreading out of a deformation of the water surface. Energy propagation by means of the motion of a change in a medium, rather than the motion of the medium itself, is called *wave motion*, and it occurs in many forms in nature. We shall subsequently find *electromagnetic waves* of great interest, but in this chapter we shall restrict ourselves to the more familiar water, sound, and other mechanical waves.

25–1 Water waves

If we were somehow to tag individual water molecules and follow them when a series of water waves pass by, we would find that their paths are similar to those shown in Fig. 25–2. Each molecule describes a roughly circular orbit about its original position, with an average velocity of zero, and so never undergoes any

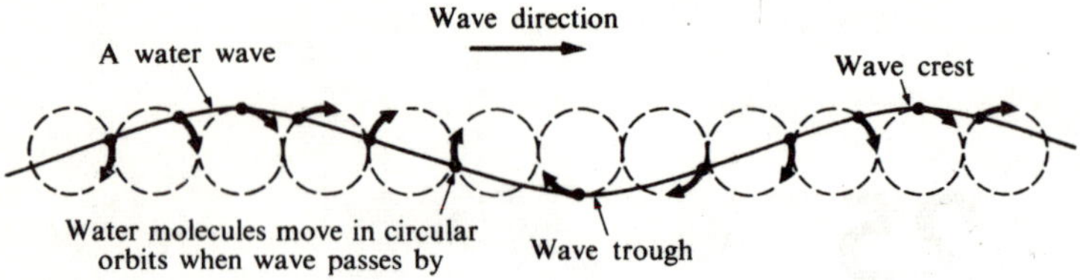

FIG. 25-2. Water molecules move in circular orbits about their original positions when a wave passes by. At the crest of a wave the molecules are moving in the direction the wave is traveling, while in the trough the molecules are moving in the opposite direction. There is no net motion of water involved in the motion of a wave.

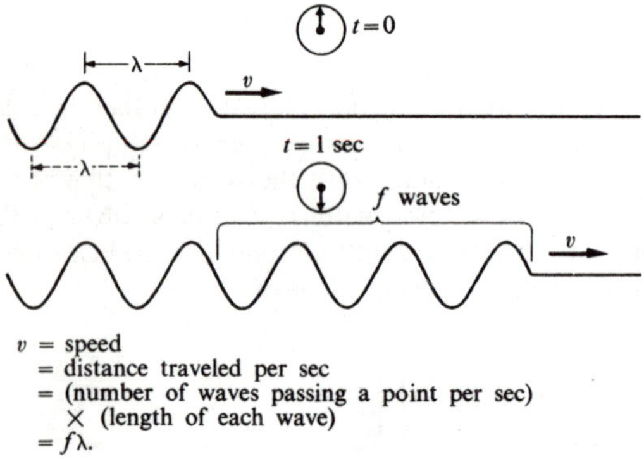

FIGURE 25-3

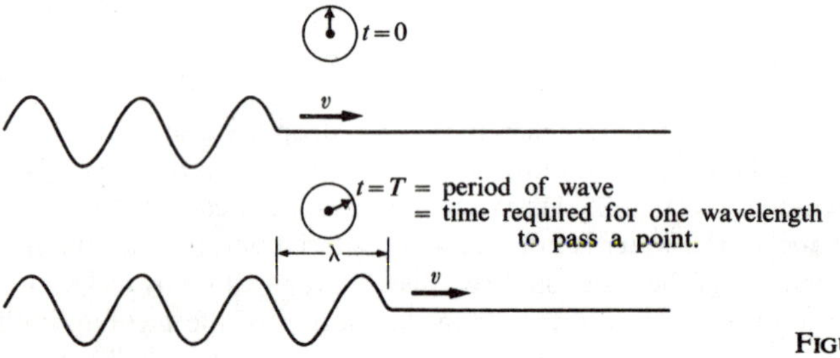

FIGURE 25-4

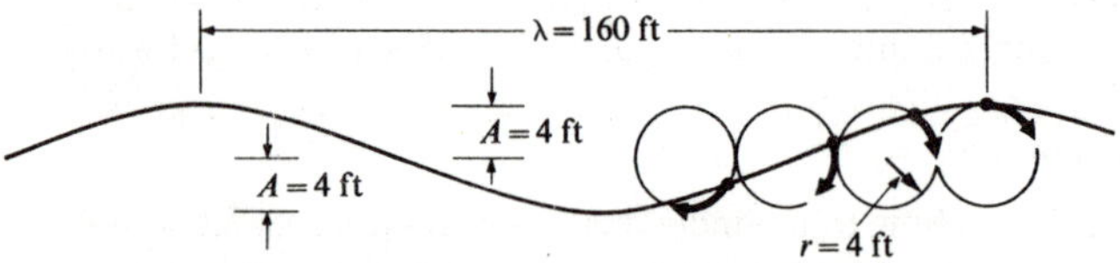

FIG. 25-5. The amplitude of a water wave is the height of its crest above normal sea level, which is equal to the depth its trough lies below sea level.

permanent displacement. The passage of a wave across the surface of a body of water involves the motion of a pattern: energy goes from one place to another by virtue of the changing pattern, but there is no transport of mass.

Three related quantities are useful in describing wave motion (Fig. 25–3). These are the *speed* v with which the wave travels; the *wavelength* λ (Greek letter lambda), which is the distance between adjacent crests, troughs or any other phase of the wave; and the *frequency* f, which is the number of waves that pass a given point per unit time, usually per second. The wave speed is equal to the distance through which a particular wave moves per second. In every second, f waves (by definition) go past a particular point, with each wave representing a distance of λ. Therefore a wave travels a total distance of $f\lambda$ per second, which is the wave speed v. That is,

$$v = f\lambda, \tag{25–1}$$

a basic formula which may be applied to all wave phenomena.

Sometimes it is more useful to consider the *period* T of a wave, which is the time required for one complete wave to pass a given point (Fig. 25–4). Since f waves pass by per second, the period of each wave is

$$T = 1/f. \tag{25–2}$$

If there are five waves passing by per second, for example, each wave has a period of $\frac{1}{5}$ sec. In terms of period T, Eq. (25–1) becomes

$$v = \lambda/T. \tag{25–3}$$

Suppose that we find the distance between adjacent wave crests on the ocean one day to be 160 ft, with a crest passing every 4.5 sec. This means that, since $\lambda = 160$ ft and $T = 4.5$ sec, the wave velocity is

$$v = \frac{\lambda}{T} = \frac{160\text{ ft}}{4.5\text{ sec}} = 36\,\frac{\text{ft}}{\text{sec}}\cdot$$

The frequency of the waves is

$$f = \frac{1}{T} = \frac{1}{4.5\text{ sec}} = 0.22\,\frac{\text{wave}}{\text{sec}}\cdot$$

The motion of the pattern that constitutes a wave in a medium can be much more rapid than the motion of the individual particles of the medium. The waves we have just discussed might have an *amplitude* A of 4 ft, that is, the crests are 4 ft above the normal sea level and the troughs 4 ft below (Fig. 25–5). Thus the water particles are moving in circles whose radii are 4 ft, so that as each wave passes by, the molecules travel a distance d equal to the circumference $2\pi r$ of the circle. Hence

$$d = 2\pi r = 2\pi A = 25\text{ ft}.$$

It takes 4.5 sec for a wave to go past a given point, which means that the particles must cover the 25-ft circumference of their circular orbits in 4.5 sec. The speed v' of the particles is therefore

$$v' = \frac{d}{T} = \frac{25\text{ ft}}{4.5\text{ sec}} = 5.6\,\frac{\text{ft}}{\text{sec}}.$$

The speed of the *wave*, however, is 36 ft/sec, over six times greater. Thus energy can be transported by wave motion at a faster rate than might be possible through direct mass motion.

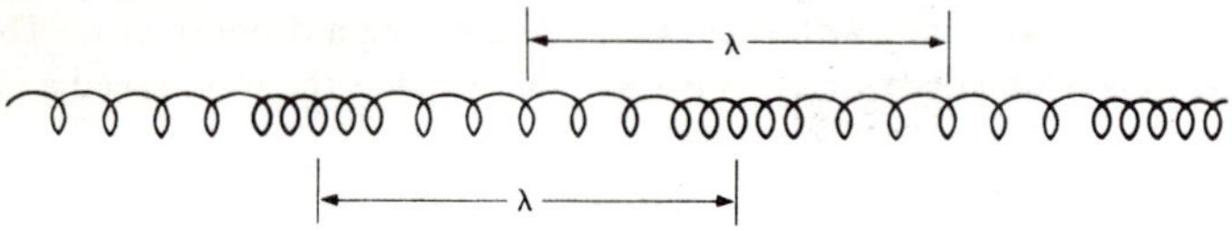

FIG. 25–6. Longitudinal waves in a coil spring.

25–2 Longitudinal and transverse waves

Water waves are actually a combination of two more fundamental types of waves, *longitudinal* and *transverse*. Longitudinal waves occur when the individual particles of a medium vibrate back and forth *in the direction* in which the waves travel. Figure 25–6 shows longitudinal waves in a coil spring; each portion of the spring is alternately compressed and extended as the waves pass by. Longitudinal waves, then, are essentially density fluctuations. Transverse waves occur when the individual particles of a medium vibrate *perpendicular to the direction* in which the waves travel. Figure 25–7 shows transverse waves in a stretched wire; each section of the wire moves from one side to the other and back again as the patterns constituting the waves travel down the wire.

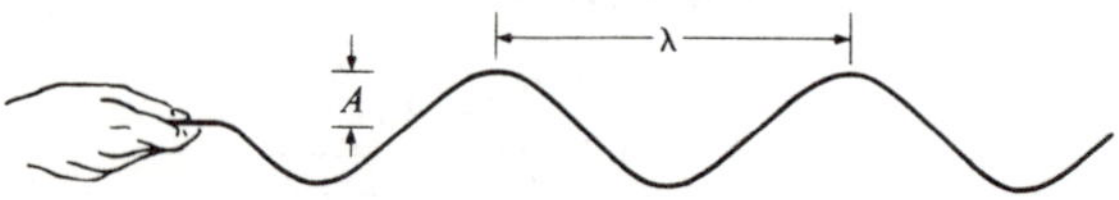

FIG. 25–7. Transverse waves in a stretched wire.

The dynamics of wave motion can best be understood by analyzing how a wave propagates along an elastic string. If we were to displace one end of the string, forces are set up within it which tend to restore it to its equilibrium position. In the diagram (Fig. 25–8), we see that point A will tend to be restored to its equilibrium position. Point A will therefore move downward, while point B, which has an upward force on it, will move up as A moves down. Point B will crest after A. The same thing will happen to point C to the right of B, which will crest after point B is on its way down. Thus the crest originating at A will move down the string as a consequence of its elastic properties. If point A is moved up and down in simple harmonic motion, all points on the string will oscillate with simple

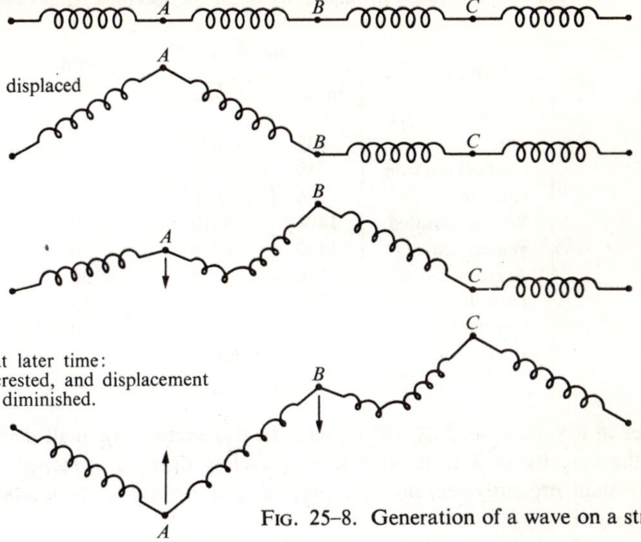

FIG. 25–8. Generation of a wave on a string.

harmonic motion, but the crests will occur at different times. That is the same as saying that the phase of the motion will be different at different points. A similar analysis would hold for longitudinal waves.

The correspondence between harmonic and wave motion permits us to obtain an interesting result. As we found earlier, the work done in extending (or compressing) a spring by a distance A is $\frac{1}{2}kA^2$, where k is the force constant of the spring. If this spring is then released, it oscillates back and forth, with its energy going from potential energy of the spring to the kinetic energy of the particles, and then back to potential energy of the spring, etc., but its total energy at all times equals the $\frac{1}{2}kA^2$ it was originally. In wave terminology, A is the amplitude, and since a wave is a series of harmonic oscillations, its total energy must be proportional to A^2, the square of its amplitude. This dependence of wave energy on A^2 is true for waves of all kinds, longitudinal and transverse, and even applies to electromagnetic waves, which are nonmechanical in character, provided that the amplitude A is properly interpreted. Ocean waves illustrate this relationship: a wave 10 ft high possesses four times the energy of a wave 5 ft high, something no sailor would doubt.

25–3 Wave velocity

The velocity of a wave of any kind is determined by the properties of the medium in which it occurs, and does not depend on the amplitude or frequency of the wave or on the precise manner in which it is generated. Table 25–1 gives the speed of sound waves in various materials; the speed of light or other electromagnetic

TABLE 25–1

The speed of sound in various media at the specified temperatures

Medium	Speed		Temperature, °C
	m/sec	ft/sec	
Air	332	1090	0
Carbon dioxide	259	850	0
Chlorine	206	676	0
Water, distilled	1404	4610	0
Water, sea	1440	4730	0
Acetone	1146	3750	30
Paraffin	1304	4280	15
Copper	3560	11,680	20
Iron	5130	16,830	20

waves in a vacuum is 3×10^8 m/sec. In this section we shall derive a formula for the velocity of a transverse wave traveling down a stretched string, which, as we shall presently see, involves only the tension in the string and its mass per unit length.

Figure 25–9 shows a string under the tension T. If the left-hand end of the string is moved up and down sinusoidally, a sinusoidal wave will go down it at the speed v, as we saw in Section 25–2. If the string is simply jerked upward, a pulse will be propagated with the same speed v; we shall consider the latter case to simplify the calculation. Ideally the "jerk" consists of a constant upward force **F** that acts for a brief time Δt. The bottom of Fig. 25–9 shows the string Δt after the force has been applied. The pulse has gone a distance $v\,\Delta t$, so that the portion of the string to the left of the point P is moving upward with the speed u while the portion of the string to the right of P is still stationary.

A string cannot support shear stresses, and so the resultant of the applied forces **F** and **T** must be parallel to the string where they are exerted. Thus the vector triangle whose vertical and horizontal sides are **F** and **T**, respectively, is

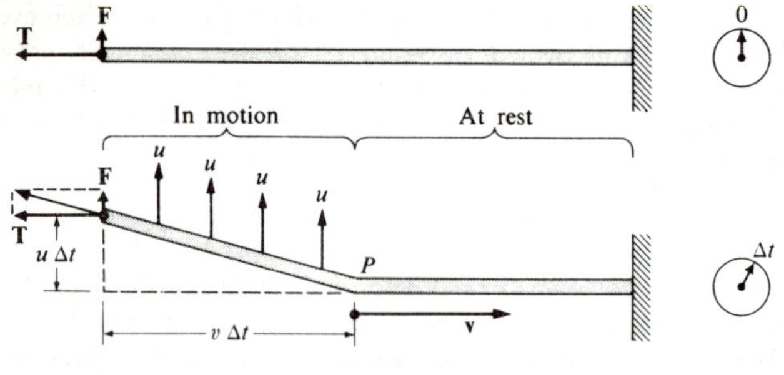

FIG. 25–9. The propagation of a transverse pulse along a stretched string.

similar to the space triangle whose vertical and horizontal sides are $u\,\Delta t$ and $v\,\Delta t$, respectively, and

$$\frac{F}{T} = \frac{u\,\Delta t}{v\,\Delta t} = \frac{u}{v}. \tag{25-4}$$

By jerking the string we have given it an impulse of $F\,\Delta t$. From Eq. (25–4) we find that F can be expressed in terms of the tension T and the speeds v and u, with the result that

$$\text{Impulse} = F\,\Delta t = T\frac{u}{v}\,\Delta t. \tag{25-5}$$

The effect of this impulse has been to set in motion a length $v\,\Delta t$ of the string with the transverse speed u. If the mass per unit length of the string is μ, the mass m of the moving portion of the string is $\mu v\,\Delta t$ and its transverse momentum is

$$\text{Momentum} = mu = (\mu v\,\Delta t)u. \tag{25-6}$$

Since the string was initially at rest, the momentum it has after the force F has acted for the time Δt must be equal to the impulse provided by the force, and so

$$\text{Impulse} = \text{momentum change},$$

$$T\frac{u}{v}\,\Delta t = (\mu v\,\Delta t)u,$$

$$v = \sqrt{T/\mu}. \tag{25-7}$$

The velocity of a pulse (or a wave) in a stretched string is the square root of the string tension T divided by its linear density μ. This conclusion is in accord with our experience: a slack, heavy string means slow waves, while a taut, light string means fast waves.

25–4 Mathematical description of a wave

We shall now look into the mathematical description of wave phenomena, and thereby make more concrete some of the remarks made previously. In doing so we shall assume that the waves are simple harmonic in character—for example, the waves generated when one end of a string is set into simple harmonic motion. This does not constitute any essential limitation to our discussion of wave motion

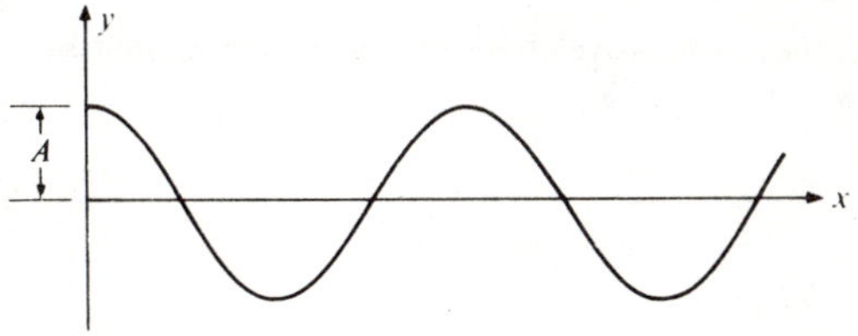

FIGURE 25–10

because, as we shall see, any periodic or even nonperiodic disturbance can be represented as a superposition of simple harmonic waves.

Figure 25-10 shows a string stretched along the x-axis whose vibrations are in the y-direction. If the displacement of the string is $y = A$ at $x = 0$ when $t = 0$, its displacement there at any subsequent time t is given by

$$y = A \cos 2\pi ft. \tag{25-8}$$

This formula gives the displacement of only a single point on the string, namely $x = 0$, for any time t. A complete description of the motion of a wave down a stretched string should, however, tell us what y is at *any* point on the string at *any* time: We require a formula that specifies how y varies with both x and t. To obtain this formula, let us suppose that we begin to shake the end of the string at $x = 0$ when $t = 0$. The resulting wave travels down the string with the speed v in the $+x$-direction, and it has gone the distance $x = vt$ in the time t. The time interval between the departure of the wave from $x = 0$ and its arrival at the point x is therefore $t = x/v$. This is a perfectly general conclusion: The displacement y of the string at x at any time t is the same as the value of y at $x = 0$ *at the earlier time* $t - x/v$.

A formula that gives y in terms of both x and t may thus be found by replacing t in Eq. (25-8) with $t - x/v$, which yields

$$y = A \cos 2\pi f\left(t - \frac{x}{v}\right)\cdot \tag{25-9}$$

This equation describes a wave which is traveling along the string toward the right in Fig. 25-10.

If the wave travels to the left, in the $-x$-direction, the sign of v must be changed, and

$$y = A \cos 2\pi f\left(t + \frac{x}{v}\right)\cdot \tag{25-10}$$

Now a particle of the string at $x = 0$ vibrates in a manner corresponding to that in which a particle at x vibrated at the earlier time $t - x/v$, exactly the reverse of what happens when the wave travels to the right.

Equation (25-9) can be rewritten

$$y = A \cos 2\pi\left(ft - \frac{fx}{v}\right)\cdot$$

The speed, frequency, and wavelength of a wave are related by

$$v = f\lambda,$$

and so

$$y = A \cos 2\pi\left(ft - \frac{x}{\lambda}\right)\cdot \tag{25-11}$$

Since the period T of a wave is the reciprocal of its frequency, $f = 1/T$, we have still another way of describing a traveling wave:

$$y = A \cos 2\pi \left(\frac{t}{T} - \frac{x}{\lambda} \right). \tag{25–12}$$

An alternative method of describing a wave is

$$y = A \cos (\omega t - kx). \tag{25–13}$$

In Eq. (25–13), $\omega = 2\pi f$, and is sometimes called the *circular frequency*. The constant $k = 2\pi/\lambda$ is called the wave number. Of course, in all of our description we have omitted the phase of the wave, which is arbitrary. Thus for an arbitrary wave moving in the $+x$-direction we could write

$$\begin{aligned} y &= A \cos 2\pi \left(ft - \frac{x}{\lambda} + \phi \right) \\ &= A \cos 2\pi \left(\frac{t}{T} - \frac{x}{\lambda} + \phi \right) \\ &= A \cos (\omega t - kx + \phi'), \end{aligned} \tag{25–14}$$

where ϕ or ϕ' is the phase "angle." For waves moving in the $-x$-direction, the minus signs in (25–14) must be replaced by plus signs.

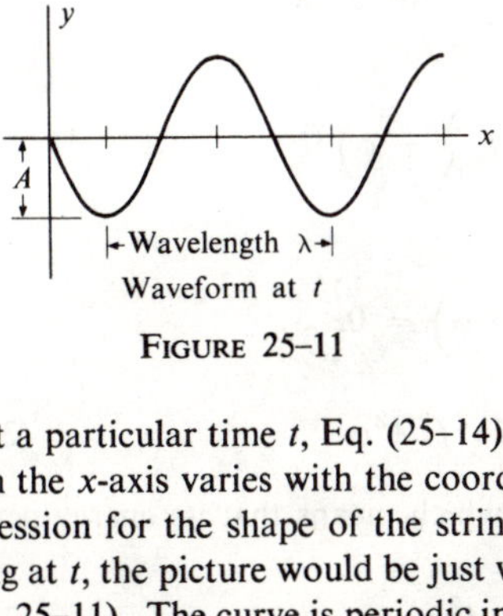

FIGURE 25–11

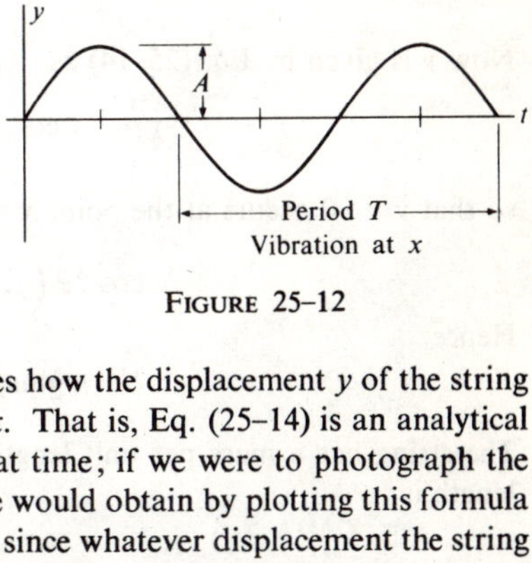

FIGURE 25–12

At a particular time t, Eq. (25–14) specifies how the displacement y of the string from the x-axis varies with the coordinate x. That is, Eq. (25–14) is an analytical expression for the shape of the string at that time; if we were to photograph the string at t, the picture would be just what we would obtain by plotting this formula (Fig. 25–11). The curve is periodic in space, since whatever displacement the string has at some point x is repeated at $x + \lambda$, $x + 2\lambda$, $x + 3\lambda$, and so on.

At a particular position x, Eq. (25–14) specifies how the displacement y varies with time t. Evidently a particle of the string at x undergoes simple harmonic motion of period T and with a phase angle of $\phi = -x/\lambda$ (Fig. 25–12). The displacement of the string at some time t is repeated at $t + T$, $t + 2T$, $t + 3T$, and so on. By successively differentiating Eq. (25–14) with respect to time, we can determine the transverse velocity u and acceleration a of a point on the string:

$$u = \frac{dy}{dt} = -\frac{2\pi A}{T}\sin 2\pi\left(\frac{t}{T} - \frac{x}{\lambda} + \phi\right) = -2\pi f A \sin 2\pi\left(ft - \frac{x}{\lambda} + \phi\right), \tag{25-15}$$

$$a = \frac{du}{dt} = -\frac{4\pi^2 A}{T^2}\cos 2\pi\left(\frac{t}{T} - \frac{x}{\lambda} + \phi\right) = -4\pi^2 f^2 A \cos 2\pi\left(ft - \frac{x}{\lambda} + \phi\right),$$
$$= -4\pi^2 f^2 y. \tag{25-16}$$

Matter in motion possesses energy, and a vibrating string is no exception. As a particle of the string moves up and down during the passage of a wave, its energy is transformed continuously from kinetic energy of the particle to potential energy of the medium, and back again. At $y = 0$, its equilibrium position, the energy is wholly kinetic, while at $y = \pm A$, its maximum displacement on either side of the x-axis, the energy is wholly elastic potential energy of the string. We can readily determine the total energy U of the particle from the fact that its kinetic energy $\frac{1}{2}mu_{\max}^2$ at $y = 0$ is equal to U. In general, if the particle mass is m,

$$\mathrm{KE} = \tfrac{1}{2}mu^2 = 2\pi^2 f^2 A^2 m \sin^2 2\pi\left(ft - \frac{x}{\lambda} + \phi\right),$$

or, since $\sin^2\theta + \cos^2\theta = 1$,

$$\mathrm{KE} = 2\pi^2 f^2 A^2 m\left[1 - \cos^2 2\pi\left(ft - \frac{x}{\lambda} + \phi\right)\right]. \tag{25-17}$$

Now y is given by Eq. (25-14) as

$$y = A\cos 2\pi\left(ft - \frac{x}{\lambda} + \phi\right),$$

so that $y = 0$ occurs at the point x when

$$\cos 2\pi\left(ft - \frac{x}{\lambda} + \phi\right) = 0.$$

Hence

$$U = \tfrac{1}{2}mu_{\max}^2 = 2\pi^2 f^2 A^2 m. \tag{25-18}$$

The string has a mass per unit length of μ, which means that its energy per unit length is

$$U/l = 2\pi^2 f^2 A^2 \mu. \tag{25-19}$$

The wave energy is proportional to the square of the amplitude A and to the square of the frequency f. This result is a general one that applies to all kinds of waves, not just waves in a string, as we said earlier.

As a wave proceeds down a string, portions of the string that were originally at rest are set into motion. Thus energy is being transmitted down the string with the same velocity as the wave velocity v. To find the rate at which the wave transports energy, we first write Eq. (25-19) in the form

$$U = 2\pi^2 f^2 A^2 \mu l,$$

and then differentiate it with respect to time. We obtain

$$\frac{dU}{dt} = 2\pi^2 f^2 A^2 \mu \frac{dl}{dt},$$

or, since $v = dl/dt$,

$$\frac{dU}{dt} = 2\pi^2 f^2 A^2 \mu v. \tag{25-20}$$

The flow of energy down the string occurs at a rate equal to the product of the energy density of the wave and its velocity.

25-5 Composition of waves. Superposition principle

Many interesting phenomena associated with waves occur when two or more waves are transmitted in the same medium simultaneously. An example of this might occur if we drop two rocks into a pond at the same time at two different spots. The wave pattern of the surface of the pond becomes quite complicated, and we need some general principles to treat this situation.

From experience, we find that under most circumstances the response of the medium can be found by adding the amplitudes of each of the undisturbed waves at each instant of time, and this will be the amplitude of the resultant wave. This is known as the *principle of superposition.* If two equal crests approach the same point at the same time, the crest is twice as great. If two equal troughs are there, the trough is twice as deep. If a crest meets a trough of the same amplitude, the surface is undisturbed.

The principle of superposition, while intuitively reasonable, is not always absolutely correct. For example, if the amplitude of each wave is so great that a string would exceed its elastic limit if the amplitudes were added, or if the string would break, the superposition principle would not hold. In most phenomena of interest, however, this principle holds.

Since it is difficult to keep in mind simultaneously the space and time aspects of a wave, we shall first consider the superposition of two harmonic vibrations of the same frequency at a single point. We call the two harmonic vibrations

$$y_1 = A_1 \cos(\omega t + \phi_1), \qquad y_2 = A_2 \cos(\omega t + \phi_2). \tag{25-21}$$

The space factors of the wave are incorporated into the arbitrary phases ϕ_1 and ϕ_2. The resultant of y_1 and y_2 is an harmonic vibration of frequency ω, amplitude A and phase ϕ:

$$y = y_1 + y_2 = A \cos(\omega t + \phi) = A_1 \cos(\omega t + \phi_1) + A_2 \cos(\omega t + \phi_2). \tag{25-22}$$

Expanding the trigonometric functions, we have

$$A(\cos \omega t \cos \phi - \sin \omega t \sin \phi) = A_1(\cos \omega t \cos \phi_1 - \sin \omega t \sin \phi_1) + A_2(\cos \omega t \cos \phi_2 - \sin \omega t \sin \phi_2). \qquad (25\text{–}23)$$

For (25–23) to be valid at all times, the coefficients of cos ωt and sin ωt must be equal, and we have

$$A \cos \phi = A_1 \cos \phi_1 + A_2 \cos \phi_2,$$
$$A \sin \phi = A_2 \sin \phi_1 + A_2 \sin \phi_2. \qquad (25\text{–}24)$$

Equation (25–24) is just the rule for combining vectors of amplitudes A_1 and A_2 that are at angles of ϕ_1 and ϕ_2 with the x-axis. The first equation represents the sum of the x-components, the second the sum of the y-components. From our knowledge of vector analysis, we have that

$$A = \sqrt{A_1^2 + A_2^2 + 2A_1A_2 \cos(\phi_1 - \phi_2)}. \qquad (25\text{–}25)$$

The angle ϕ is also found by vector methods. We thus have the rule:

The resultant of two waves of the same frequency with phases ϕ_1 and ϕ_2 is a wave of that frequency whose amplitude can be found by considering the amplitudes of the initial waves as vectors that make the angles ϕ_1 and ϕ_2, respectively, with a reference direction.

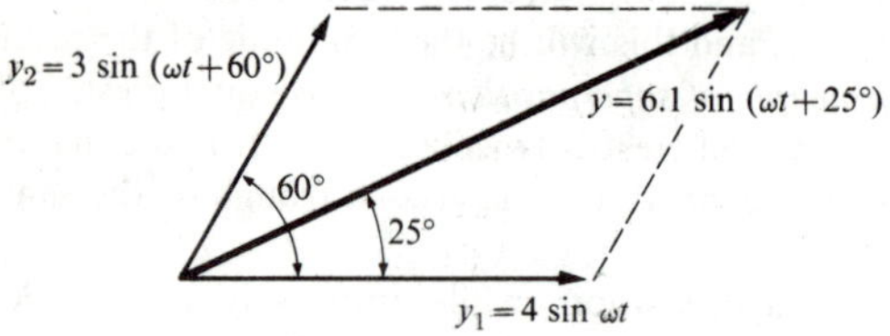

FIG. 25–13. Composition of two simple harmonic motions of the same frequency which are out of phase.

In Fig. 25–13, we illustrate the composition of the waves $y_1 = 4 \sin \omega t$ and $y_2 = 3 \sin(\omega t + 60°)$. We find according to our rules that the resultant is

$$y = y_1 + y_2 = 4 \sin \omega t + 3 \sin(\omega t + 60°)$$
$$= 6.1 \sin(\omega t + 25°),$$

since the vector problem is easy to solve graphically.

The special case where the phase difference is 180° is of interest. In this case, if the amplitudes are the same, the resultant of two waves is zero. This is an important method of distinguishing between wave and particle phenomena. Two particles with kinetic energy at the same point do not disappear into nothingness; two waves, however, can annihilate each other. The energy possessed by these waves does not disappear, of course. It appears at some other point where the phase difference is zero degrees, giving twice the amplitude and four times the energy. The average energy of the resultant wave is then the same as the energy of each separate wave. This phenomenon, known as interference, is characteristic of wave motion and plays an important role in optics, as we shall see in Chapter 29.

25–6 Standing waves

An interesting and practical application of interference phenomena is the analysis of the possible waves (called *standing waves*) that can exist on a string clamped at both ends. The condition that the string be clamped means that there can exist on the string only those vibrations which are compatible with the zero displacement at the ends.

Figure 25–14 shows a wave pulse moving along a stretched string toward a fixed end. When the wave reaches the end, it exerts an upward force on the support. By Newton's third law, the support exerts an equal and opposite force on the string and the wave is reflected with a displacement opposite to the incoming wave. That is, there is a 180° phase change on reflection. The resultant displacement of the string is the superposition of the two waves moving left and right, and is given by

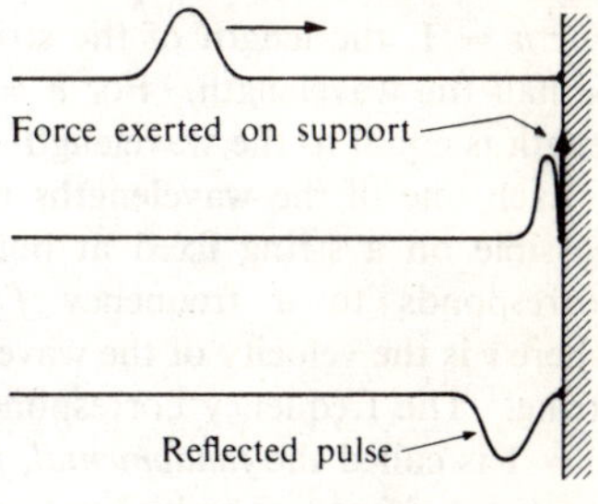

FIG. 25–14. A wave suffers a phase change of 180° when reflected from a fixed support.

$$y = A\cos 2\pi\left(ft - \frac{x}{\lambda}\right) - A\cos 2\pi\left(ft + \frac{x}{\lambda}\right)\cdot \qquad (25\text{–}26)$$

From trigonometry we know that

$$\cos a - \cos b = -2\sin\tfrac{1}{2}(a+b)\sin\tfrac{1}{2}(a-b).$$

Thus

$$y = 2A\sin 2\pi ft \sin\frac{2\pi x}{\lambda}\cdot \qquad (25\text{–}27)$$

The resultant, then, is an harmonic vibration. Those parts of the string for which

$$\sin\frac{2\pi x}{\lambda} = 0 \qquad (25\text{–}28)$$

will always have a zero displacement. These points are called nodes, and between them the string will vibrate in loops. We must now find the values of λ for which Eq. (25–28) is satisfied. At $x = 0$, the displacement will automatically be zero. At the position $x = L$, where L is the length of the string, we require a zero displacement also. The conditions for (25–28) to be satisfied at $x = L$ are

$$\frac{2\pi L}{\lambda} = n\pi, \qquad n = 1, 2, 3, \ldots,$$

or

$$\lambda = \frac{2L}{n}. \qquad (25\text{–}29)$$

Figure 25–15 shows some of the possible ways in which the string can vibrate. For $n = 1$, the length of the string will be half the wavelength. For $n = 2$, the length is equal to the wavelength, etc.

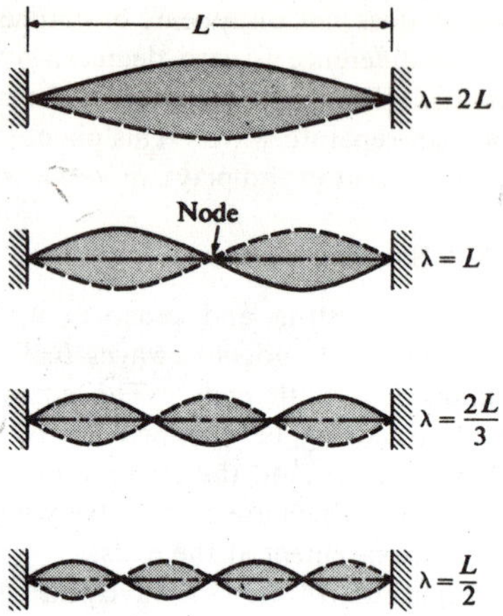

FIG. 25–15. Standing waves on a string.

Each one of the wavelengths that are possible on a string fixed at both ends corresponds to a frequency $f = v/\lambda$, where v is the velocity of the wave on the string. The frequency corresponding to $n = 1$ is called the *fundamental*, f. The frequency corresponding to $n = 2$ is $2f$, to $n = 3$, $3f$, etc. The higher frequencies are called *overtones*, or *harmonics*.

The tone of musical instruments depends on this phenomenon. The frequency of motion of the string sets up a corresponding frequency of a wave in the air. Thus a longer string corresponds to a larger possible λ and a lower f. It should be recognized that the wavelength of the sound is not related directly to the wavelength of the motion of the string, since the velocities of the wave on the string and the sound wave in air differ. Only the frequencies of the two coincide.

25–7 Beats

The superposition of two waves of slightly differing frequencies leads to the phenomenon of beats in sound waves. If two tuning forks (or other sources of single-frequency sound waves) whose frequencies are slightly different are struck at the same time, the sound that we hear fluctuates in intensity. At one instant we hear a loud tone, then virtual silence, then the loud tone again, then virtual silence, and so on. The origin of this behavior is shown schematically in Fig. 25–16; the loud tones occur when the waves from the two forks interfere constructively, thus reinforcing one another, and the quiet periods occur when the waves interfere destructively, thus partially or wholly canceling one another out. These regular pulsations are called *beats*. Observation and calculation both yield the result that the number of beats per second equals the difference between the frequencies of the two sound sources: if tuning forks are used whose frequencies are, say 440 and 445 cycles/sec, five beats per second will be heard.

It is not hard to verify the origin of beats formally. Let us consider a single point in space, corresponding to the position of a listener, and imagine that two sound waves with the same amplitude A but with the different frequencies f_1 and f_2 move past this point. The waves are individually described by the formulas for the displacements they produce:

$$y_1 = A\cos(2\pi f_1 t + \phi_1), \qquad y_2 = A\cos(2\pi f_2 t + \phi_2).$$

The sum of these displacements, which corresponds to what is heard at the point, is

$$y = y_1 + y_2 = A[\cos(2\pi f_1 t + \phi_1) + \cos(2\pi f_2 t + \phi_2)].$$

Again making use of the trigonometric identity

$$\cos a + \cos b = 2[\cos \tfrac{1}{2}(a + b)][\cos \tfrac{1}{2}(a - b)],$$

we obtain

$$y = 2A\cos\left[2\pi\left(\frac{f_1 - f_2}{2}\right)t + \left(\frac{\phi_1 - \phi_2}{2}\right)\right] \times \cos\left[2\pi\left(\frac{f_1 + f_2}{2}\right)t + \left(\frac{\phi_1 + \phi_2}{2}\right)\right]. \quad (25\text{–}30)$$

The composite sound wave may be regarded as having a frequency $\frac{1}{2}(f_1 + f_2)$, which is the *average* of the frequencies of the original waves. The amplitude of the composite wave is given by the quantity in the first set of brackets in Eq. (25–30); we see that the amplitude itself varies with time with a frequency of $\frac{1}{2}(f_1 - f_2)$, which is half the difference between the frequencies of the original waves. In graphic terms, the *envelope* of the waves has a periodic variation. An amplitude maximum occurs whenever $\cos 2\pi[(f_1 - f_2)/2]$ equals $+1$ or -1. Each of these values occurs once per cycle, which means that *two* maxima occur per cycle. The number of beats per second is therefore twice the frequency $\frac{1}{2}(f_1 - f_2)$ of the amplitude variations, and is equal to $f_1 - f_2$, the difference between the original frequencies. Beats are hard to distinguish when the two frequencies differ by more than perhaps 10 cycles/sec.

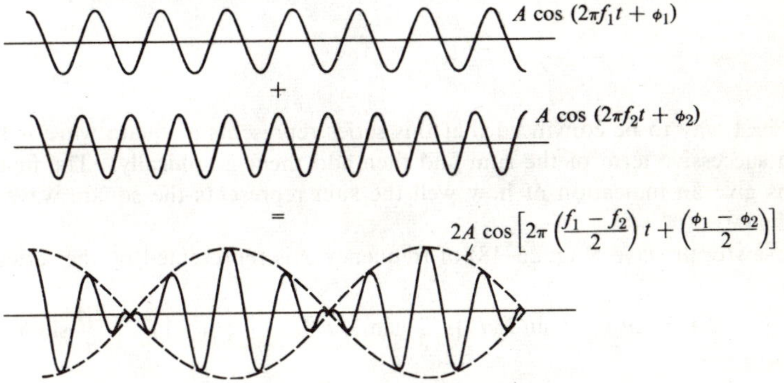

FIG. 25–16. Diagram illustrating the origin of beats.

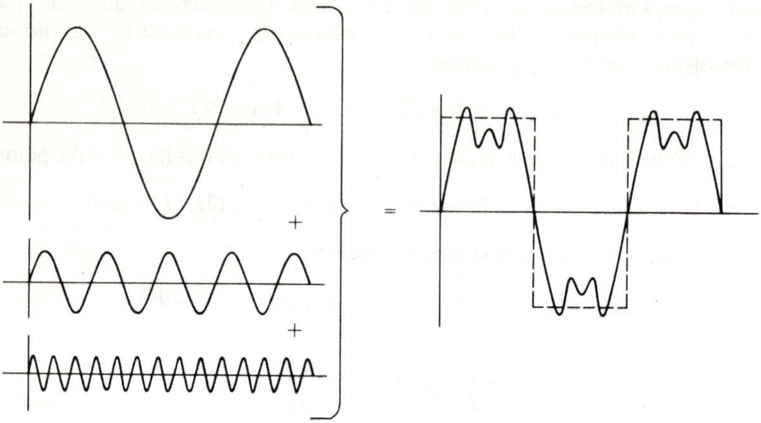

FIG. 25-17. Fourier synthesis of a square wave.

25-8 Fourier series

In Section 25-4 we indicated that there was no limitation on the generality of our treatment of waves, even though we considered only the simplest type. This statement follows from a mathematical theorem due to Fourier which shows that *any* periodic function with frequency ω can be represented as a superposition of harmonic vibrations of frequencies ω, 2ω, 3ω, . . . , with varying amplitudes.

We shall not go into the theory that Fourier developed but shall instead illustrate it by a few examples. A square wave (Fig. 25-17) is about as far from a sinusoidal wave as one can imagine, and yet a square wave of amplitude A can be represented by the superposition of sine waves in the form

$$y = A \sin \omega t + \frac{A}{3} \sin 3\omega t + \frac{A}{5} \sin 5\omega t + \cdots$$

$$\omega = 2\pi f = \frac{2\pi}{T}. \tag{25-31}$$

The best way to be convinced that this series represents a square wave is to plot each successive term of the sum and then add them graphically. The first three terms give an indication of how well the sum represents the square wave, as in the figure.

A sawtooth wave (Fig. 25-18) of frequency ω is represented by the series

$$y = 2A \sin \omega t - A \sin 2\omega t + \frac{2A}{3} \sin 3\omega t + \cdots + (-1)^{n-1} \frac{2A}{n} \sin n\omega t. \tag{25-32}$$

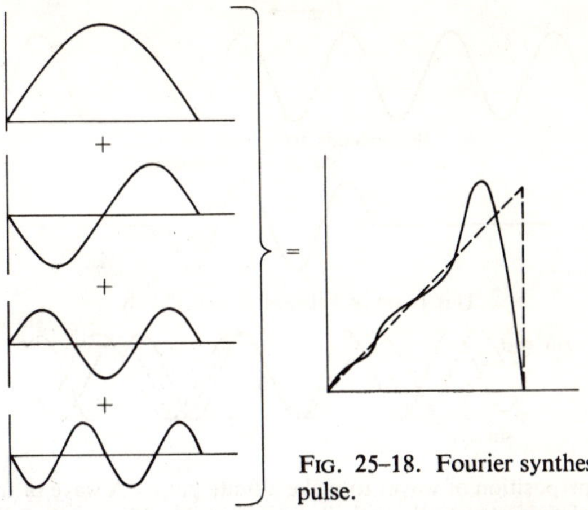

FIG. 25-18. Fourier synthesis of a sawtooth pulse.

Again it is worth plotting the series in order to get some feeling for how the series represents the sawtooth wave form.

The series (25-31) and (25-32) all have a period $T = 1/f$. If we wished to represent a disturbance which is not periodic, we would have to take the period $T \rightarrow \infty$. This would mean that the difference between successive frequencies of $nf - (n - 1)f = f = (1/T) \rightarrow 0$. The sum over successive frequencies then would have to be replaced by an integral over frequencies. Thus to represent an arbitrary shape which is not periodic we must represent it as an integral over frequencies which differ by infinitesimals and whose amplitudes are frequency dependent:

$$y = \int_0^\infty A(\omega) \sin \omega t \, d\omega. \tag{25-33}$$

Thus we see that a wave of any shape can be expressed as a superposition (either a sum or an integral) of simple harmonic vibrations. The harmonic vibration $\sin \omega t$, for example, extends over an infinite time scale $-\infty \leq t \leq \infty$. A vibration like $\sin \omega t$ which extends over a finite time, $0 \leq t \leq T'$ (Fig. 25-19), must be expressed as a superposition of an infinite number of harmonic vibrations. In the latter case we require many frequencies which will interfere constructively in the region $0 \leq t \leq T'$, and which will cancel out the sine wave outside this region. There is an important relationship between the range of frequencies that must be used and the interval T' over which the vibration is different from zero. The relationship is

$$T' \Delta f \simeq 1, \tag{25-34}$$

where Δf = range of frequencies. An exact derivation of (25-34) is beyond the scope of this book. We can, however, make this relationship plausible by the

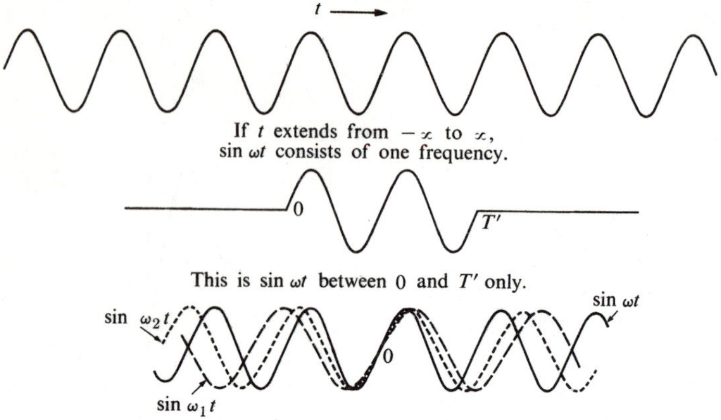

FIG. 25-19. Superposition of waves to make a finite pulse. A wave of frequency greater than ω will interfere destructively with it outside of T'. If $\omega_2 > \omega_1$, the region of destructive interference occurs still farther out.

following argument: It will be recalled from Section 25-7 that if we superpose two waves of frequency ω and ω_1, then the time between nodes is $1/(\omega - \omega_1) = 2\pi/\Delta\omega$. If we wished the vibration to be confined only to a region $T' = 2\pi/\Delta\omega$, we would have to superpose other waves which would interfere with the waves outside the region where the sine wave does not vanish. If we chose a circular frequency ω_2 such that $|\omega - \omega_2| \leq |\omega - \omega_1|$, the node would occur outside the region T' (see Fig. 25-19); i.e., we would get destructive interference outside. In order to make the entire region outside T' disappear, we require that all frequencies from 0 to $\Delta\omega$ be used. Since $\Delta\omega = 2\pi\Delta f = 2\pi/T'$, we have the result (25-34). The meaning of (25-34) is that for a sharp pulse we must superpose many frequencies. For a broad one we need superpose only a few. When we have a single frequency, $\Delta f = 0$ and the pulse is infinitely wide—that is, it extends from $-\infty$ to ∞.

We have considered the Fourier series and Fourier integrals only in the time domain. Similar considerations hold if we wish to expand a function of the space coordinates. In that case we expand in terms of $\sin kx$, $\sin 2kx$, etc., and $\cos kx$, $\cos 2kx$, etc., where $k = 2\pi/\lambda$ = wave number. There is also a relationship, which corresponds to Eq. (25-34), between the extension of a function in space and the distribution of wave numbers required to describe such a function. The relationship is

$$L\,\Delta k \simeq 2\pi, \qquad (25\text{-}35)$$

where L is the extension in space.

The relationships (25-34) and (25-35) are classical mathematical results which hold for any wave phenomenon. We shall see that they play an important physical role in modern physics in the uncertainty relations.

25-9 Doppler effect

If we stand beside a railway track when a train approaches with its whistle blowing, we cannot help noticing that the pitch of the whistle drops suddenly as the train passes by. A careful study would reveal that the pitch of the whistle as the train approaches is *higher* than the normal one when the train is stationary, and its pitch as the train recedes is *lower* than the normal one. We would observe the same phenomenon if the train were at rest with its whistle blowing and we go past it in a rapidly moving car: a higher pitch than usual as we approach the train, a lower one as we recede. The change in frequency of a sound brought about by relative motion between its source and a listener is called the *Doppler effect.*

The origin of the Doppler effect is straightforward. As a moving source emits sound waves, it is tending to overtake those traveling in the same direction. Hence the distance between successive waves is smaller than usual; since this distance is the wavelength of the sound, the corresponding frequency is higher than usual. At the same time the source is moving away from those of its waves that travel in the opposite direction, increasing the distance between successive waves behind it and thereby reducing their frequency. If the source is stationary, a listener moving toward it intercepts more sound waves per unit time than if he were at rest, and accordingly he hears a higher frequency. When the observer moves away from the source, fewer of the waves catch up with him per unit time, and he hears a lower frequency.

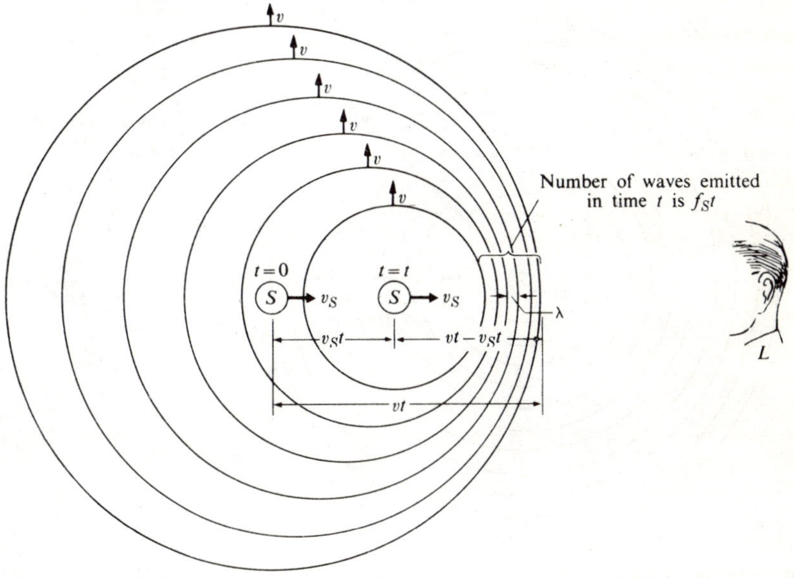

FIG. 25-20. A moving source of sound tends to overtake waves emitted in its direction of motion.

Let us examine the Doppler effect in detail. Figure 25–20 shows a source S that produces sound waves of frequency f_S and is moving with the speed v_S toward a stationary listener L. In the time t a total of $f_S t$ waves is emitted by the source. The first of these waves travels the distance vt in the time t, and when the last of them is emitted the source itself has gone a distance of $v_S t$. (Here v is the speed of sound in air; this speed is a property of the medium in which the sound travels, as we know, and motions of source or observer do not affect its value.) Hence the $f_S t$ waves occupy the distance $vt - v_S t$, and their wavelength is

$$\lambda = \frac{\text{distance}}{\text{number of waves}} = \frac{vt - v_S t}{f_S t} = \frac{v - v_S}{f_S}. \tag{25–36}$$

The corresponding frequency,

$$f_L = \frac{v}{\lambda} = \frac{v f_S}{v - v_S} = f_S\left(\frac{v}{v - v_S}\right), \tag{25–37}$$

is what the listener hears.

When the source is moving *away* from the listener, the $f_S t$ waves are spread through the distance $vt + v_S t$, and the wavelength is accordingly increased to $\lambda = (v + v_S)/f_S$. By adopting the convention that v_S is positive for motion toward the listener and negative for motion away from him, Eqs. (25–36) and (25–37) hold for both situations in their present form.

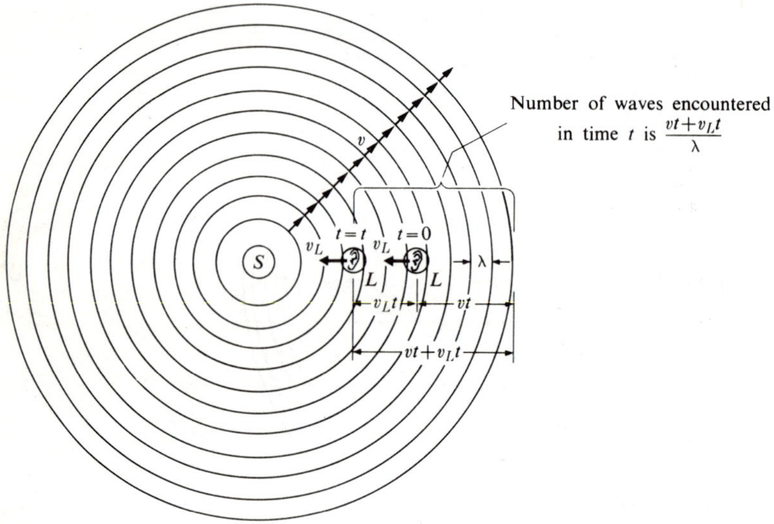

FIG. 25–21. A listener who is moving encounters more waves when approaching a source of sound than a listener who is stationary.

Now we suppose that the listener is moving toward the source of sound at the speed v_L (Fig. 25-21). His speed *relative to the waves* is $v + v_L$, and the number of waves he encounters in the time t is $(v + v_L)t/\lambda$. The frequency he hears is therefore

$$f_L = \frac{\text{number of waves}}{\text{time}} = \frac{(v + v_L)t/\lambda}{t}$$

$$= \frac{v + v_L}{\lambda} = f_S\left(\frac{v + v_L}{v}\right). \tag{25-38}$$

When the listener is moving away from the source, he encounters the smaller number of waves, $(v - v_L)t/\lambda$, in the time t, and the frequency he hears is $(v - v_L)/\lambda$. The convention that v_L is positive for motion toward the source and negative for motion away from the source permits Eq. (25-38) to be used as it is in both cases.

If both source and listener are in motion toward or away from each other, the wavelength λ to be used in Eq. (25-38) is that given by Eq. (25-36) for a moving source. In this event, then,

$$f_L = f_S\left(\frac{v + v_L}{v - v_S}\right). \tag{25-39}$$

The latter formula is the most general one, and reduces to Eq. (25-37) when $v_L = 0$ and to Eq. (25-38) when $v_S = 0$.

The Doppler effect is a very important tool of the astronomer, who is able to determine the speed of approach or recession of stars and galaxies from the shifts in the characteristic wavelengths of light they emit. (We shall discuss these characteristic wavelengths in Chapter 31.) This is the method by which the apparent expansion of the universe was detected. Throughout the sky, characteristic wavelengths in the light from distant galaxies are longer than normal, a phenomenon called the "red shift" because red light has the longest wavelengths. The magnitude of the wavelength shift increases with distance from the earth, which suggests that the entire universe is expanding so that all the bodies in it recede from one another. It is worth noting that the Doppler effect in light is somewhat different from that in sound because light, which does not depend on a material medium for its transmission, has the same relative velocity c to all observers regardless of their state of motion (Chapter 29). The Doppler effect in light waves obeys the formula

$$f_L = f_S\left(\frac{1 + u/c}{\sqrt{1 - u^2/c^2}}\right), \tag{25-40}$$

where f_L is the frequency detected by an observer whose relative speed is u with respect to a source of light of frequency f_S, reckoning u as positive for approach and negative for recession.

25–10 Waves in three dimensions

The description of waves propagating in three dimensions is in many ways similar to the one-dimensional waves we have thus far been considering. Here the equation representing the displacement y of whatever is responding to the wave (air molecules in the case of sound waves, electric and magnetic intensity in the case of electromagnetic waves) may be written as superposition of waves,

$$y = \frac{A}{r} \cos\left[2\pi\left(\frac{t}{T} - \frac{r}{\lambda} + \phi\right)\right], \tag{25–41}$$

where r is the distance from the source. The interpretation of Eq. (25–41) is illustrated in Fig. 25–22: A source of waves is located at $r = 0$ and, if we consider *any* radial line from the source, a wave proceeds with the velocity v along this line as though it were traveling down a string. The amplitude decreases as $1/r$ when we leave $r = 0$ so that, at a distance r from the source, the amplitude is A/r.

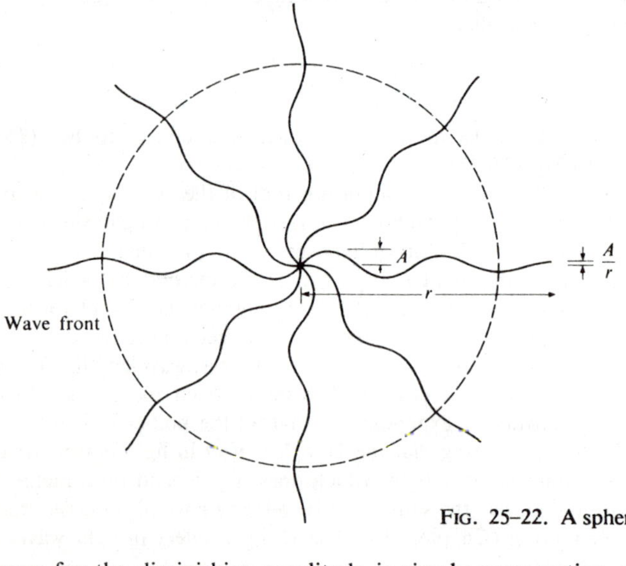

FIG. 25–22. A spherical wave.

The reason for the diminishing amplitude is simply conservation of energy. As the waves spread out in space their disturbances affect a larger and larger volume, and if the source operates at a constant power level this means that the energy density becomes smaller and smaller. In effect the energy is being diluted. As we saw in Eq. (25–18), the energy of the wave is proportional to the square of its amplitude A^2, let us say equal to CA^2. Now consider the amount of energy per unit time passing through an imaginary sphere of radius R_1 (Fig. 25–23). At a time t later this same energy flux passes through an imaginary sphere of radius R_2, where $R_2 = R_1 + vt$. Since the area of a sphere is $4\pi R^2$, the relationship between the amplitudes A_1 and A_2 of the waves at R_1 and R_2 may be found from $E_1 = E_2$ to be

$$\frac{CA_1^2}{4\pi R_1^2} = \frac{CA_2^2}{4\pi R_2^2}, \qquad \frac{A_1}{R_1} = \frac{A_2}{R_2}.$$

Thus the ratio A/r remains constant for all values of r.

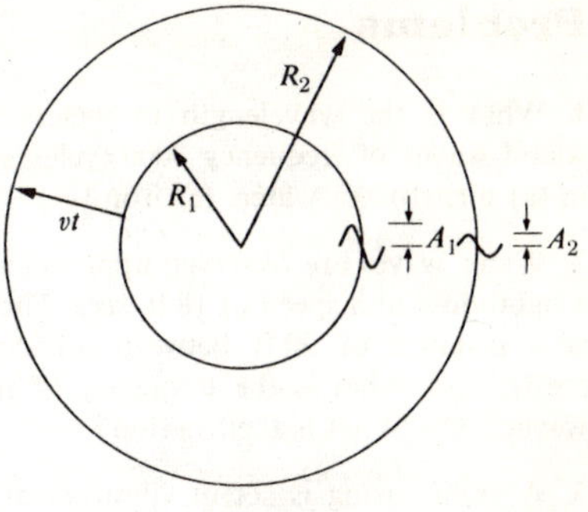

FIG. 25–23. The amplitude of a spherical wave diminishes as $1/R$.

The propagation of a wave in three dimensions is a complicated affair, and to help in analyzing it we shall require some additional concepts. If we trace out the wave motion along all possible radii, it is possible to assign a phase to each point in space at a given time. A surface which joins all points of equal phase is called a *wave front* (Fig. 25–22). In the diagram we have a point source of waves, which propagate unimpeded in all directions. Hence the wave fronts of such a source are all spherical.

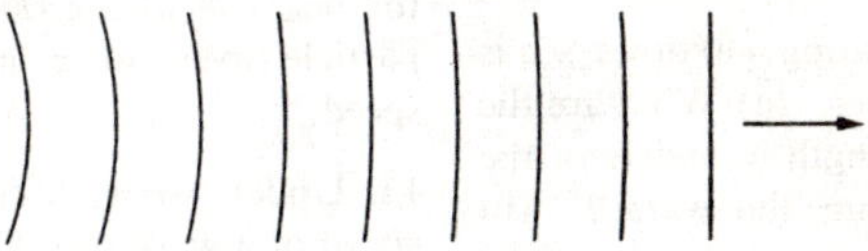

FIG. 25–24. How a spherical wave becomes a plane wave.

As the spherical wave front proceeds farther and farther from its source, the radius of curvature of the front approaches infinity and a small portion of the wave front is very much like a plane. This is shown in Fig. 25–24. Whenever we refer to *plane waves*, then, strictly speaking, we mean waves infinitely distant from their source. [It is also possible to produce plane light waves by using lenses or other optical devices.] Applications of these ideas will be made in Chapter 29.

Problems

1. What is the wavelength in meters of sound waves of frequency 8000 cycles/sec in (a) air, (b) sea water, (c) iron?

2. Water waves are observed approaching a lighthouse at a speed of 18 ft/sec. There is a distance of 20 ft between adjacent crests. (a) What is the frequency of the waves? (b) What is their period?

3. A violin string is set in vibration at a frequency of 440 cycles/sec. How many vibrations does it make while its sound travels 500 ft in air?

4. A particular groove in a phonograph record moves past the needle at a speed of 0.4 m/sec. The sound produced has a frequency of 3000 cycles/sec. What is the wavelength of the wavy indentations of the groove?

5. A tuning fork vibrating 440 times/sec is placed in distilled water. (a) What are the frequency and wavelength in meters of the waves produced within the water? (b) What are the frequency and wavelength in meters of the waves in the surrounding air which the water waves produce when they reach the surface?

6. Show that the right-hand side of Eq. (25–7) has the dimensions of velocity.

7. Show that Eq. (25–9) can be written $y = A \cos [(2\pi/\lambda)(vt - x)]$.

8. The equation of a wave traveling in a string is $y = 0.05 \cos (20t - x)$, where x and y are in feet and t is in seconds. Find the frequency, wavelength, amplitude, and velocity of the wave.

9. The equation of a traveling wave in a string is $y = 0.01 \cos (4t - 0.2x)$, where x and y are in meters and t is in seconds. Find the frequency, wavelength, amplitude, and velocity of the wave.

10. One end of a long, horizontal rope is moved up and down twice per second with a maximum displacement of 3 in. above and 3 in. below its equilibrium position. The string weighs 0.2 lb/ft and is under a tension of 10 lb. (a) Find the amplitude, velocity, frequency, and wavelength of the resulting waves. (b) How much energy per second must be supplied to the rope?

11. Given: the equation of a transverse wave in a certain stretched string is

$$y = 5 \cos \pi(2t - 0.1x).$$

Find: the ratio between the maximum transverse speed of a particle in the string and the speed of the wave.

12. Show that at any point x and time t the slope dy/dx of a stretched string is equal to the negative of the ratio of the transverse particle speed at x and t and the wave speed.

13. Under what circumstances will the speed of a wave traveling down a stretched string be equal to the maximum transverse speed of a particle in the string?

14. A steel wire 1 m long whose mass is 10 gm is under a tension of 400 n, and is tied down at both ends. (a) What is the wavelength of its fundamental mode of vibration? (b) What is the frequency of this mode? (c) What is the wavelength of the sound waves produced in air when the string vibrates in this mode?

15. Two identical steel wires have fundamental frequencies of vibration of 400 cycles/sec. The tension in one of the wires is increased by 2%, and both wires are plucked. How many beats occur per second?

16. Combine graphically two sine waves, whose wavelengths and amplitudes are in

the ratio 2:1, at a time t when they are in phase.

17. The motion of a particle consists of two simple harmonic motions at right angles to each other. The amplitude in the x-direction is A and in the y-direction $2A$. The period of the x-motion is twice that of the y-motion, and it passes through 0 as y reaches its maximum. Find the equation of the path of the particle.

18. Plot the first three terms of the series

$$\frac{2}{\pi} - \frac{4}{\pi} \sum_{n=1}^{\infty} \frac{\cos 2nt}{(2n)^2 - 1},$$

and plot their sum. The series represents the function $\sin t$. Add enough terms so that this becomes clear.

19. Plot the first three terms of the series

$$f(t) = 2 \sum_{n=1}^{\infty} (-1)^{n+1} \frac{\sin nt}{n}.$$

Add enough terms to convince yourself that it represents

$$f(t) = t \qquad (-\pi \leq t \leq \pi).$$

20. The Fourier amplitudes $A(\omega)$ (see Eq. 25–33) which represent $\sin \omega_0 t$, $-T \leq t \leq T$, are given by

$$\sqrt{\frac{2}{\pi}}\, T \frac{\sin (\omega - \omega_0)T}{(\omega - \omega_0)T}.$$

Plot this as a function of ω and show that $\Delta\omega T \simeq 2\pi$.

21. Two simple harmonic motions are in the same straight line. The periods of the motion are 5 and 6 sec, respectively. At the outset the first oscillator is at a maximum displacement and the other is at its minimum. How long will it take for the two amplitudes to be in phase?

22. Two simple harmonic motions are of amplitude 0.2 and 0.3, respectively. The period of each motion is 5 sec. What is the resultant displacement at a given point after 1.2 sec, if at time $t = 0$ the first oscillator is at its maximum and the other is 60° behind?

23. Two tuning forks are vibrating and one hears 3 beats/second. One tuning fork is a standard one, vibrating at 440 vibrations/second. A small piece of wax is placed on the standard fork and the frequency of the beats decreases. What is the frequency of vibration of the second fork?

24. Show that Eqs. (25–37), (25–38), and (25–40) all reduce to

$$f_L = f_S(1 + u/v),$$

where u is the relative speed of approach of source and observer and v the wave speed, when u is small compared with v.

25. Train A is heading east at the speed v_A and train B is heading west at the speed v_B on an adjacent track. The locomotive on train A is blowing its whistle continuously. A passenger on train B observes the frequency of the sound from the whistle to be 307 cycles/sec when the trains approach each other, 256 cycles/sec when they are abreast, and 213 cycles/sec when they recede from each other. From these figures, find the speed of each train relative to the ground, taking the speed of sound as 1100 ft/sec.

26. A motorist goes through a red light and, when he is arrested, claims that the color he actually saw was green ($\lambda = 5.4 \times 10^{-7}$ m) and not red ($\lambda = 6.2 \times 10^{-7}$ m), because of the Doppler effect. The judge accepts this explanation and instead fines him for speeding at the rate of $1.00 for each mi/hr he exceeded the speed limit of 50 mi/hr. If the speed of light is 3×10^8 m/sec, what was his fine?

27. A moving reflector approaches a stationary source of sound of frequency f with the speed u. A listener nearby hears both the original sound waves and the reflected sound waves. Obtain a formula for the number of beats per second the listener hears.

28. A man has two tuning forks, one marked "256 cycles/sec" and the other of unknown frequency. He strikes them simultaneously, and hears 10 beats per second. "Aha," he says, "the other tuning fork has a frequency of 266 cycles/sec." What is wrong with his conclusion?

29. What is the energy crossing unit area per unit time at a distance of 1 m from a point source of 1 watt?

30. The solar constant on earth is 1 cal/cm^2·min. The distance between the earth and the sun is 93×10^6 mi. Estimate the rate of energy output of the sun in watts, assuming that the radiation is emitted isotropically.

26

Alternating-Current Circuits

In the previous chapter we considered energy transmission by means of mechanical waves of various sorts. Now we shall examine at some length *electromagnetic waves*, a remarkable and most important phenomenon for whose understanding we shall have to make use of the basic notions of electricity and magnetism, as well as of wave motion itself.

26–1 Introduction

Even before we discuss the nature of electromagnetic waves it is worth classifying them according to their scale. One way to generate electromagnetic oscillations is merely to rotate a wire loop in a magnetic field, which is how the "alternating current" of ordinary electric power is produced. The frequencies of oscillation range from nearly zero to several hundred cycles per second. Modern electronic circuits can oscillate at very high frequencies, 10^9 cycles/sec or more. And light is an electromagnetic radiation emitted by atoms under suitable circumstances with frequencies in the neighborhood of 10^{15} cycles/sec. In free space all electromagnetic waves travel at a speed of 3×10^8 m/sec, so the wavelengths associated with the above types of sources are, respectively, of orders of magnitude 10^6 m, 10^{-1} m, and 10^{-7} m. This is a useful categorization, since in the first case the wavelength is much larger than the physical size of laboratory apparatus, in the second case it is comparable in size, and in the third case it is much smaller. In each of these cases the method of analysis, like the character of the phenomena, is somewhat different.

When the wavelength is long compared with the dimensions of the apparatus, there are no phase differences between different parts of the apparatus, and, while the current in the circuit varies with time, it has essentially the same value everywhere in the circuit at any instant. This is the situation we shall look into first. In an alternating-current circuit the source of emf varies sinusoidally with time, so that the current direction changes twice per cycle. If the circuit elements are inductors and capacitors as well as resistors, the circuit will behave very differently from the way it would if the emf were constant. In a direct-current (d-c) circuit, current can flow only so long as the capacitors are charging, and then stops when the potential difference due to the stored charge equals the impressed emf. In an a-c circuit, however, the current may reverse itself *before* the capacitors are fully charged, so that an alternating current can continue to flow in the external part of the circuit. The constantly changing current means that an inductive element in a circuit contributes an induced emf, further accenting the difference between a-c and d-c circuits.

We shall begin our mathematical analysis of a-c circuits by treating their mechanical analog, since mechanical systems are closer to our intuition and the formal development is the same.

26–2 The damped harmonic oscillator

As we learned in Chapter 10, when a particle of mass m is attracted to a point by a force proportional to the distance x from that point, in other words by a restoring force given by

$$F = -kx, \tag{26–1}$$

the particle, if displaced, undergoes oscillatory motion of angular frequency

$$\omega_0 = 2\pi f_0 = \sqrt{k/m}. \tag{26–2}$$

This formula was obtained by writing down Newton's second law of motion for the problem,

$$m\frac{d^2x}{dt^2} = -kx, \tag{26–3}$$

and solving it. The solution can be expressed as

$$x = A \sin(\omega_0 t + \phi), \tag{26–4}$$

where A is the amplitude of the motion, ω_0 its angular frequency (equal to $\sqrt{k/m}$), and ϕ the phase angle representing the angular displacement of the particle at the time $t = 0$.

According to Eq. (26–4), the particle when set into motion should continue oscillating indefinitely. Actual harmonic oscillators do not behave in this way; while in some cases they may oscillate for a long period of time, ultimately the amplitude of the motion decreases and comes to a stop. The latter phenomenon is called *damping*. If the amount of damping is very large, as in the situation of an automobile whose springs are damped with good shock absorbers, no oscillation at all occurs—after being displaced, the object gradually returns to its equilibrium position and goes no further.

In general, damping is caused by frictional forces. The potential energy of a stretched spring in a damped oscillator is not completely converted into kinetic energy at $x = 0$, since some of it goes into work done against the frictional forces that are acting. At the end of each vibration the loss in energy is manifested by a shorter extension of the spring, which continues until the amplitude becomes infinitesimal.

The effect of damping, then, is to introduce a resistive force into the equation of motion. This force is usually (though not always) proportional to the instantaneous velocity of the particle. Thus we may write

$$m\frac{d^2x}{dt^2} = -R\frac{dx}{dt} - kx,$$

where R is a constant representing the magnitude of the frictional force. The above equation is usually written in the form

$$m\frac{d^2x}{dt^2} + R\frac{dx}{dt} + kx = 0. \tag{26–5}$$

Solving Eq. (26–5) is more difficult than solving Eq. (26–3) was. Our procedure is to try to guess the general form of the solution and substitute it back into Eq. (26–5) for verification. We choose for the trial solution

$$x = Ae^{-t/T}\sin(\omega t + \phi), \tag{26–6}$$

where ω and T are the constants which must be determined so that Eq. (26–5) is satisfied. In the trial solution the sine function represents the harmonic motion and the exponential function the damping; Fig. 26–1 is a graph of x versus t which illustrates the general form of Eq. (26–6).

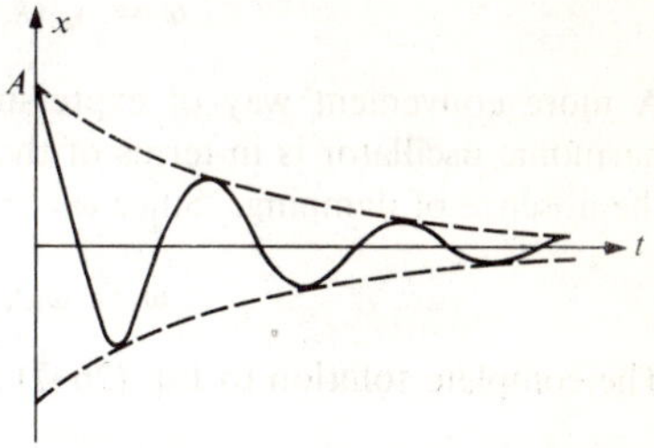

FIG. 26–1. A damped harmonic motion.

When Eq. (26–6) is substituted into the differential equation (26–5), the result is

$$\left\{m\left[-\omega^2\sin(\omega t + \phi) - \frac{2\omega}{T}\cos(\omega t + \phi) + \frac{1}{T^2}\sin(\omega t + \phi)\right]\right.$$
$$+ R\left[\omega\cos(\omega t + \phi) - \frac{1}{T}\sin(\omega t + \phi)\right]$$
$$\left. + k\sin(\omega t + \phi)\right\}Ae^{-t/T} = 0. \tag{26–7}$$

For this equation to be valid, the coefficients of the sine and cosine functions must each equal zero *separately*, since these represent independent oscillations. This yields two equations

$$\left(-m\omega^2 + \frac{m}{T^2} - \frac{R}{T} + k\right)Ae^{-t/T}\sin(\omega t + \phi) = 0,$$

$$\left(-\frac{2\omega m}{T} + R\right)Ae^{-t/T}\cos(\omega t + \phi) = 0,$$

which are satisfied when

$$-m\omega^2 + \frac{m}{T^2} - \frac{R}{T} + k = 0, \tag{26–8}$$

$$-\frac{2\omega m}{T} + R\omega = 0. \tag{26–9}$$

Solving the second of these equations for T, we have $T = 2m/R$. This is the time necessary for the amplitude of the vibration to decay to $1/e$ (which is $1/2.718$ or about 37%) of its initial value. The greater the value of R, which is a measure of the frictional damping forces that are acting, the faster the amplitude drops.

Substituting the value of T into Eq. (26–8) gives

$$-m\omega^2 + \frac{mR^2}{4m^2} - \frac{R^2}{2m^2} + k = 0.$$

This has the solution

$$\omega = \sqrt{(k/m) - (R^2/4m^2)}.$$

A more convenient way of expressing the angular frequency ω of the damped harmonic oscillator is in terms of the frequency ω_0 at which it would oscillate in the absence of damping. Since $\omega_0 = \sqrt{k/m}$, we have that

$$\omega = \omega_0\sqrt{1 - (R^2/4km)}. \tag{26–10}$$

The complete solution to Eq. (26–7) is therefore

$$x = Ae^{-Rt/2m} \sin(\omega t + \phi), \tag{26–11}$$

with ω given by Eq. (26–10).

It is interesting to note that the frequency of oscillation ω is not very much different from ω_0 if the resistive force is small compared with the restoring force. The effect of the resistive force is always to reduce the frequency; the greater R is for a given value of km, the slower the oscillations. When R^2 exceeds $4km$, the value of the square root in Eq. (26–10) becomes imaginary. Under these circumstances no oscillations occur, and the displaced particle returns gradually to the equilibrium position without ever going past it on the other side, as in true harmonic motion. The critical value of R which separates the oscillatory motion from the damped, non-oscillatory case is

$$R = 2\sqrt{km}. \tag{26–12}$$

If R is given just this value, the system is said to be *critically damped*. When critical

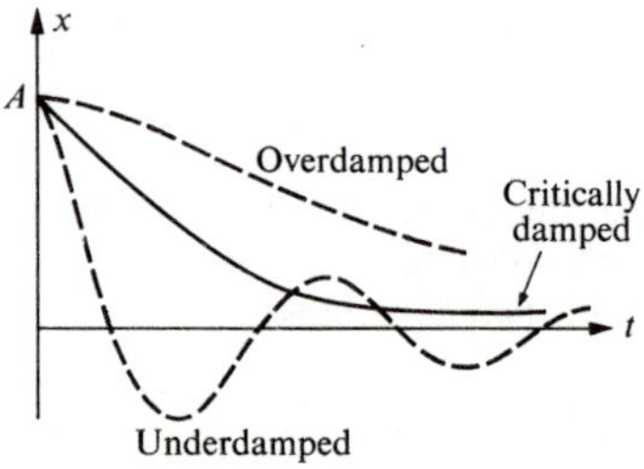

FIG. 26–2. The effect of damping on oscillatory systems.

damping takes place, $\omega = 0$ and x decreases from its initial displacement A exponentially as

$$x = A \exp(-Rt/2m).$$

This is illustrated in Fig. 26-2. Our chief concern in this chapter is with slightly damped motion, with the resistive force well below the condition for critical damping [Eq. (26-12)]. If R exceeds $2\sqrt{km}$, the rate of decay is slower. Critically damped systems return to their equilibrium positions most rapidly.

26-3 Forced oscillations

Let us now suppose that we have a mechanical system which behaves like a damped harmonic oscillator and that we "drive" it by applying an external periodic force. The frequency ω' of this external need not be the same as either ω_0, the natural frequency, or ω, the damped natural frequency of the oscillator. The differential equation of motion is now

$$m\frac{d^2x}{dt^2} + R\frac{dx}{dt} + kx = F_0 \sin \omega' t. \qquad (26\text{-}13)$$

We shall not attempt to describe the method of solving this equation here, but it is not hard to verify that the correct solution is

$$x = A \exp\left(-\frac{Rt}{2m}\right) \sin(\omega t + \phi) + BF_0 \sin(\omega' t + \phi'), \qquad (26\text{-}14)$$

where the quantities in the first term are the same as in Eq. (26-11). Here B is given by

$$B = \frac{1}{\sqrt{m^2(\omega_0^2 - \omega'^2)^2 + R^2\omega'^2}}, \qquad (26\text{-}15)$$

and ϕ' is given by

$$\phi' = \cot^{-1} - \frac{m(\omega_0^2 - \omega'^2)}{R\omega'}; \qquad \phi \text{ is arbitrary.} \qquad (26\text{-}16)$$

The first term of (26-14) represents the damped natural vibrations of the system we are dealing with, and the second term the forced vibrations due to the impressed external force. After a time t sufficiently larger than T the first term becomes negligible compared with the second, and for this reason is called a *transient*. When the transient motion has disappeared, the displacement may then be described by the following equation:

$$x = BF_0 \sin(\omega' t + \phi').$$

We note that the system oscillates with the same frequency as the impressed force, but with a different phase angle. The quantity BF_0, in which F_0 is the maximum value of the applied force, is the amplitude of the oscillations. B itself depends on various parameters, and is a maximum when ω_0, the natural frequency of the system, is equal to ω', the frequency of the driving force. When this occurs, the value of B is $1/R\omega'$. This situation is called *resonance.* In general, resonance occurs whenever a system—mechanical, electrical, atomic, etc.—is subjected to an external excitation whose frequency is that of the natural frequency of the system. Under these circumstances the energy transferred to the oscillating system is a maximum.

The quantity ϕ' is the phase angle between the impressed oscillations and the responding oscillations. At resonance, Eq. (26–16) indicates that $\phi' = \pi/2$.

26–4 Alternating-current circuits

The theory that we have developed may be applied in its entirety to electrical circuits carrying alternating current. Let us connect an inductor, a capacitor, and a resistor in series, as in Fig. 26–3. We assume that the capacitor is initially charged. Since there is no external emf in the circuit, the sum of the voltage drops across each circuit element must be zero. That is,

$$L\frac{di}{dt} + Ri + \frac{q}{C} = 0. \qquad (26\text{–}17)$$

Differentiating with respect to time, we obtain

$$L\frac{d^2i}{dt^2} + R\frac{di}{dt} + \frac{i}{C} = 0, \qquad (26\text{–}18)$$

which is identical in form with Eq. (26–5) for a damped harmonic oscillator. Thus in the circuit of Fig. 26–3 the electric current is analogous to the mechanical displacement of an harmonic oscillator, the inductance L introduces inertial properties analogous to mass, $1/C$ behaves as a spring constant k in providing a restoring force, and the resistance R of the circuit acts to damp out the oscillations.

We are therefore in a position to predict the behavior of the circuit. What we expect is that the current i will oscillate back and forth with a frequency ω, in other words "alternate," with a steadily decreasing amplitude. The variation of the current with time will follow a law like that in Eq. (26–11), which here would be written

$$i = Ie^{-Rt/2L}\sin(\omega t + \phi). \qquad (26\text{–}19)$$

Here, R is analogous to R in Eq. (26–11), L is analogous to m, and C is analogous to $1/k$. The frequency of the oscillations is

$$\omega = \sqrt{(1/LC) - (R^2/4L^2)} = \omega_0\sqrt{1 - (R^2C/4L)}, \qquad (26\text{–}20)$$

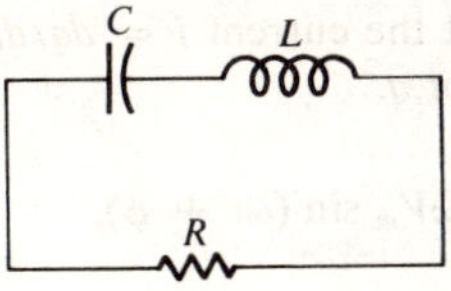

FIG. 26–3. A circuit containing resistance, inductance, and capacitance.

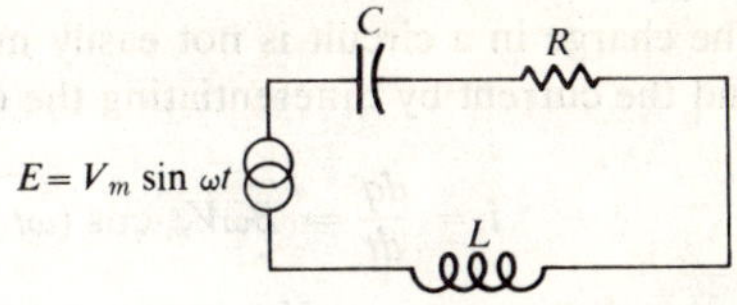

FIG. 26–4. A circuit containing resistance, inductance, and capacitance driven by an external alternating source of emf.

where ω_0 is the frequency in the absence of damping (i.e., when $R = 0$) given by

$$\omega_0 = \sqrt{1/LC} = 2\pi f_0. \tag{26–21}$$

Now let us apply an alternating emf $V_m \sin \omega t$ across the circuit, as in Fig. 26–4. The differential equation that must be satisfied is

$$L\frac{di}{dt} + Ri + \frac{q}{C} = V_m \sin \omega t. \tag{26–22}$$

(Note that we are now using ω to represent the angular frequency of the impressed alternating emf.) Substituting $i = dq/dt$, we have

$$L\frac{d^2q}{dt^2} + R\frac{dq}{dt} + \frac{q}{C} = V_m \sin \omega t. \tag{26–23}$$

Provided that there is not too much damping, the solution to Eq. (26–23) is identical with Eq. (26–14) when the proper electrical quantities are substituted for the mechanical ones. After the transient has died away, the steady state charge q is

$$q = BV_m \sin(\omega t + \phi').$$

The constant of proportionality B is, by reference to Eq. (26–15),

$$B = \frac{1}{\sqrt{L^2(\omega_0^2 - \omega^2)^2 + R^2\omega^2}}$$

$$= \frac{1}{\sqrt{L^2[(1/LC) - \omega^2]^2 + R^2\omega^2}}$$

$$= \frac{1}{\omega\sqrt{R^2 + [\omega L - (1/\omega C)]^2}}.$$

In analogy with Eq. (26–16), the phase angle is

$$\phi' = \cot^{-1} - \left[\frac{\omega L - 1/\omega C}{R}\right]. \tag{26–24}$$

The charge in a circuit is not easily measured, but the current $i = dq/dt$ is. We find the current by differentiating the expression for q,

$$i = \frac{dq}{dt} = B\omega V_m \cos(\omega t + \phi') = B\omega V_m \sin(\omega t + \phi),$$

$$i = \frac{V_m \sin(\omega t + \phi)}{\sqrt{R^2 + [\omega L - (1/\omega C)]^2}}. \tag{26–25}$$

The angle ϕ is the phase difference between the current and the applied voltage. By comparing $\cos(\omega t + \phi')$ and $\sin(\omega t + \phi)$, we have that

$$\phi = \phi' + \pi/2.$$

From Eq. (26–24), we obtain

$$\phi = \tan^{-1}\frac{\omega L - 1/\omega C}{R}. \tag{26–26}$$

For the damped harmonic oscillator, the velocity and the displacement are 90° out of phase. The charge and current in an a-c circuit are also 90° out of phase. This accounts for the difference between Eq. (26–24) and (26–26). The current in a circuit is the important quantity, and we shall use Eq. (26–26) for the phase angle.

26–5 Impedance

Equations (26–25) and (26–26) are the basic ones that describe the current in a series circuit containing inductance, capacitance, and resistance when an alternating emf of frequency ω is applied. These equations, which we have obtained in a rather formal manner by analogy with the behavior of a driven damped harmonic oscillator, can be interpreted in an interesting and useful way.

We begin with the equation for the maximum value i_m of the alternating current,

$$i_m = \frac{V_m}{\sqrt{R^2 + [\omega L - (1/\omega C)]^2}}.$$

Abbreviating the denominator of the expression by Z, we have

$$i_m = V_m/Z, \qquad Z = \sqrt{R^2 + [\omega L - (1/\omega C)]^2}. \tag{26–27}$$

Equation (26–27) resembles Ohm's law in that there is a proportionality between the maximum current and the maximum voltage. The proportionality factor Z is called the *impedance* of the circuit. The impedance is a measure of the response of the circuit to the external emf. For this reason it is not surprising that it depends explicitly on the frequency, since the effects of inductance and capacitance depend on how rapidly the voltage in the circuit is changing. Here the impedance is the ratio of the applied voltage to the current. The mechanical analog of Z is the ratio of the applied force to the velocity of the particle being acted on.

The impedance of an alternating-current circuit consists of three factors, one arising because of the resistance, one because of the capacitance, and one because of the inductance. The various factors are given the following names:

$$\text{Resistance:} \quad R,$$

$$\text{Inductive reactance:} \quad 2\pi fL = \omega L = X_L,$$

$$\text{Capacitance reactance:} \quad \frac{1}{2\pi fC} = \frac{1}{\omega C} = X_C.$$

In terms of these definitions, the impedance may be written

$$Z = \sqrt{R^2 + (X_L - X_C)^2}. \tag{26-28}$$

This is like the formula we obtain when we calculate the magnitude of a resultant vector by using the Pythagorean theorem, which, as we shall see, is essentially the situation here.

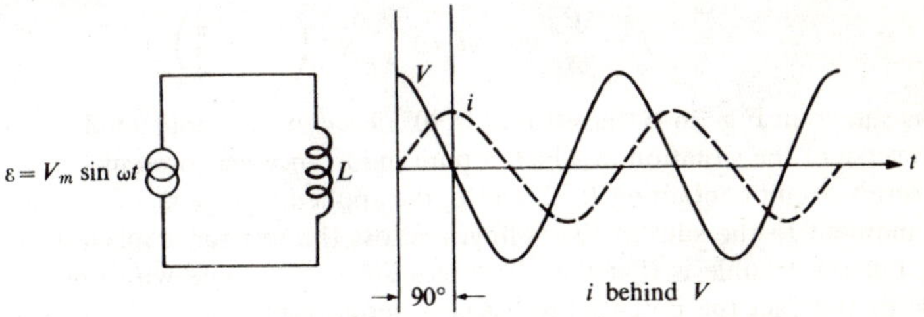

FIG. 26-5. A pure inductive circuit. Note that the emf lags the current.

In order to clarify the meaning of electrical impedance, we shall consider the effects of an inductor, a capacitor, and a resistor separately. If a pure inductor is connected across a generator of alternating emf, as in Fig. 26-5, the equation governing the circuit is

$$L\frac{di}{dt} = V_m \sin \omega t,$$

which has the solution

$$Li = \frac{-V_m}{\omega}\cos \omega t = \frac{V_m}{\omega}\sin(\omega t - \pi/2),$$

$$i = \frac{V_m}{X_L}\sin(\omega t - \pi/2).$$

The current always lags the voltage by $\pi/2$, which is 90°. We note that in this circuit the inductive reactance X_L provides all the impedance.

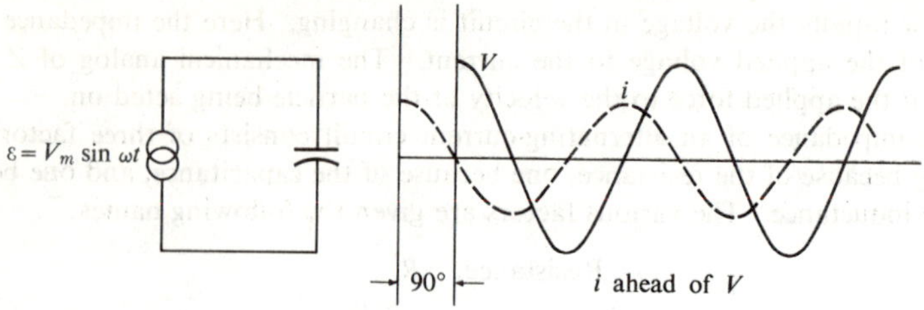

FIG. 26-6. An a-c circuit with a capacitance only. Note that the current leads the emf.

When a pure capacitance is placed across a generator of alternating emf, as in Fig. 26-6, the equation of the circuit is

$$q/C = V_m \sin \omega t.$$

Differentiating, we obtain

$$\frac{1}{C}\frac{dq}{dt} = \frac{i}{C} = V_m\omega \cos \omega t,$$

$$i = \frac{V_m}{X_C} \cos \omega t = \frac{V_m}{X_C} \sin\left(\omega t + \frac{\pi}{2}\right).$$

Here, as shown in Fig. 26-6, the current is 90° *ahead* of the applied voltage, which is the reverse of the situation in which a pure inductance was present.

In a series circuit containing R, C, and L, the applied voltage is, of course, equal at any moment to the sum of the voltages across the resistor, capacitor, and inductor, but the trouble is that these voltages are *out of phase* with one another. Because of this fact the potential differences across each one of them cannot be added arithmetically, and the phase differences must be taken into account. The way in which this is done is to plot the potential differences iR, iX_C, and iX_L on a polar diagram which indicates the phase differences between them as angles. This is done in Fig. 26-7, where we imagine the entire page to be rotating counterclockwise with the angular velocity ω. Here iR is the voltage across the resistor, which is in phase with the current (i.e., reaches its maximum and passes through zero simultaneously with the current); iX_L (which is equal to $i\omega L$) is the voltage across the inductor, which *leads* the current by 90°; and iX_C $(=i/\omega C)$ is the voltage across the capacitor, which *lags* the current by 90°. Adding these voltages, taking due account of their phase differences, is the same as adding them vectorially, and the total potential across the circuit is

$$V = i\sqrt{R^2 + (X_L - X_C)^2} \quad \text{or} \quad i_m = V_m/Z,$$

as in Eqs. (26-28) and (26-27). The phase angle ϕ is

$$\phi = \tan^{-1} \frac{X_L - X_C}{R},$$

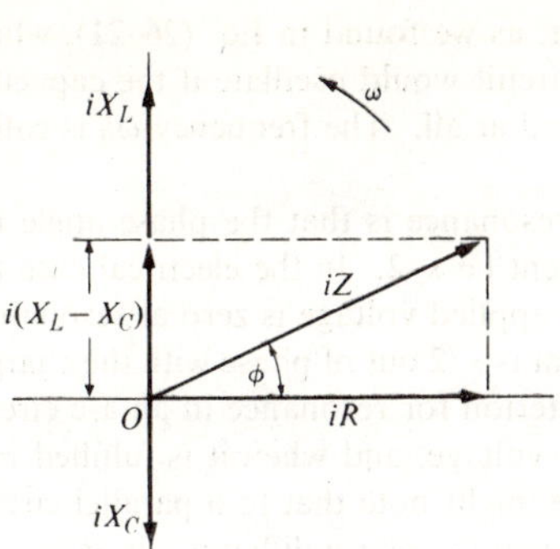

FIG. 26-7. Vector diagram of a resistance, inductance, and capacitance in series in an a-c circuit.

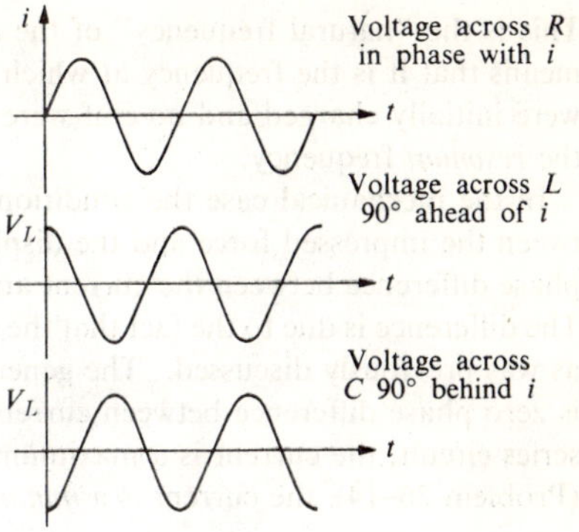

FIG. 26-8. Current and voltage in an LCR a-c circuit.

just as in Eq. (26-26). When $X_L > X_C$, the voltage leads the current, and when $X_C > X_L$, the voltage lags behind the current. Thus this vectorial treatment yields the same results as the solution of the differential equation describing the circuit, with the advantage of being more readily visualized.

Physically, it is not hard to see why there should be phase differences between current and voltage in circuits containing inductance and capacitance when an alternating emf is applied. Figure 26-8 shows the current as a function of time. Since the potential difference across the inductor is given by $L\, di/dt$, it is a maximum when the current is changing most rapidly, which occurs when $i = 0$. Hence the potential difference across L appears to be ahead of the current by 90°. The potential difference across the capacitor depends on the amount of charge stored in it, which is a maximum when $i = 0$, because that is when the current reverses direction and the stored charge is about to disappear only to be subsequently replaced by charge of the opposite sign. Thus the voltage across the capacitor depends on the sign of the current in the previous half-cycle, and therefore appears 90° behind the current.

26-6 Resonance

The maximum current that flows for a given impressed emf occurs when the impedance Z is a minimum. The frequency at which this takes place is specified by the condition that

$$X_L = X_C, \qquad \omega_0 L = \frac{1}{\omega_0 C},$$

so that

$$\omega_0 = \frac{1}{\sqrt{LC}} = 2\pi f_0. \tag{26-29}$$

This is the "natural frequency" of the circuit, as we found in Eq. (26–21), which means that it is the frequency at which the circuit would oscillate if the capacitor were initially charged and no emf were applied at all. The frequency ω_0 is called the *resonant* frequency.

In the mechanical case the condition for resonance is that the phase angle between the impressed force and the displacement be $\pi/2$. In the electrical case the phase difference between the current and the applied voltage is zero at resonance. The difference is due to the fact that the current is $\pi/2$ out of phase with the charge, as was previously discussed. The general criterion for resonance in an a-c circuit is zero phase difference between current and voltage, and when it is fulfilled in a series circuit, the current is a maximum. We might note that in a parallel circuit (Problem 26–14), the current is a *minimum* when the phase difference is zero.

26–7 Effective value of current

It is necessary to adopt some convention as to the way in which the current (and voltage) in an alternating current circuit is specified. A direct current is completely described by saying that its magnitude is 15 amp and its direction is, for instance, clockwise in the circuit (where a diagram must be included for this direction to be unambiguous). In an alternating current, though, both the direction and magnitude vary periodically.

One procedure is to speak of the current in terms of its maximum value $i_{\max}$. The trouble with this procedure is that it does not give a true idea of the rate at which the current is doing work. That is, an alternating current whose maximum value is 5 amp is not equivalent to a direct current of 5 amp.

The conventional method for describing alternating currents is to compare the effectiveness of such currents with direct currents in producing heat in a resistor. The power dissipated into heat by an alternating current is, at any instant,

$$P = i^2R = \frac{V_m^2 R}{Z^2} \sin^2 (\omega t + \phi).$$

What we want is not simply P, but its value averaged over a cycle. (The average value of i over a cycle is 0, but this is not true of i^2.) In the above equation, V_m, R, and Z are all constants, so we need only find the average value of $\sin^2 (\omega t + \phi)$. The average value of $\cos^2 \theta$ and $\sin^2 \theta$ per cycle must be the same, since both curves have the same shape. Furthermore,

$$\sin^2 \theta + \cos^2 \theta = 1,$$

regardless of the value of θ, so the average value of both $\sin^2 \theta$ and $\cos^2 \theta$ is $\frac{1}{2}$. Thus

$$\overline{P} = \overline{i^2}R = \frac{V_m^2 R}{2Z^2},$$

where the bars indicate average values. The direct current that produces the same value of $\overline{P}$ as given above is called the *effective current* (i_{eff}), and is

$$i_{\text{eff}} = \frac{V}{Z\sqrt{2}} = \frac{i_m}{\sqrt{2}} \tag{26-30}$$

since $i_m = V/Z$.

Alternating currents are almost always expressed in terms of their effective values. The same is true for alternating voltages. In terms of decimals,

$$i_{\text{eff}} = 0.707 i_m, \qquad V_{\text{eff}} = 0.707 V_m.$$

Thus conventional "110-volt a-c" power means that the maximum potential difference between the terminals of the generator is actually 155 v.

26-8 Power in a-c circuits

The instantaneous power dissipated in an a-c circuit is

$$P = Vi = [V_m \cos \omega t][i_m \cos (\omega t + \phi)],$$

where ϕ is the phase difference between the current and the voltage. Expanding the second cosine function,

$$P = i_m V_m \cos \omega t\,(\cos \omega t \cos \phi - \sin \omega t \sin \phi).$$

The average power dissipation per cycle is therefore

$$\overline{P} = i_m V_m\,(\overline{\cos^2 \omega t} \cos \phi - \overline{\cos \omega t \sin \omega t} \sin \phi).$$

However, $\overline{\cos^2 \omega t} = \frac{1}{2}$, as we have seen, while it is easy to verify that $\overline{\cos \omega t \sin \omega t} = 0$. Thus

$$\overline{P} = \tfrac{1}{2} i_m V_m \cos \phi = i_{\text{eff}} V_{\text{eff}} \cos \phi. \tag{26-31}$$

Inasmuch as the effective value of current and voltage are normally used in dealing with a-c circuits, the second of Eqs. (26-31) is usually applicable. The quantity $\cos \phi$ is called the *power factor* of the circuit, and is 1 when $\phi = 0$, namely, when the circuit is operated at its resonant frequency. At frequencies greater than and less than the resonant frequency, the power developed is smaller than $i_{\text{eff}} V_{\text{eff}}$ by the factor $\cos \phi$.

Problems

All values are effective values unless otherwise stated.

1. A mass of 1 gm is attached to a spring whose spring constant is 4 n/m. (a) What must be the resistive constant R (in n·m/sec) in order that the natural frequency change by 10%? (b) What must be the value of R for the system to be critically damped? (c) For the value of R under (a), how long will it be before the amplitude of vibration has dropped to 1% of its initial value?

2. An inductance of 10 h and a capacitor of 5 μf are connected in series with a resistor. (a) What must be the value of the resistance in order that the natural frequency of oscillation of the circuit change by 10%? (b) For this value of R, how long will it be before the transient's amplitude drops to 1% of its initial value?

3. A resistance draws 1 amp when connected to a 110-v 60-cycle line. (a) How large a capacitor must be connected in series to drop the voltage to $\frac{1}{2}$ amp? (b) How large an inductance must be connected to accomplish the same purpose?

4. A series circuit consists of a 50-ohm resistance, a 0.1-h inductance, and a capacitor of 10 μf. It is connected across a 100-v 60-cycle line. What would a voltmeter read which is (a) Connected across the resistance? (b) Connected across the resistance and inductance? (c) Connected across the capacitor? (d) Connected across the resistance plus the capacitor? (e) Connected across the inductance? Why is the sum of the readings across the inductance, resistance, and capacitor not equal to 100 v?

5. Repeat the previous question, but change the frequency of the line to 1000 cycles/sec.

6. Find the impedance of the circuits in Fig. 26-9.

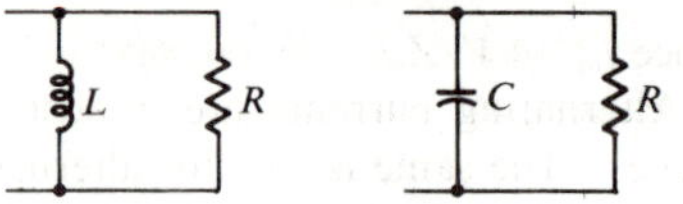

FIGURE 26-9

7. What is the reactance of coils of 1 h, 0.1 h, 1 mh, when connected in series to a 1-μf capacitor at the resonant frequencies?

8. A capacitor of 10 μf is connected across a 110-v 60-cycle line. What is the maximum energy stored in the capacitor?

9. A resistor of 5 ohms is connected in series with an inductance of 22 mh and a capacitor of 100 μf across an a-c source with an effective emf of 260 v and a frequency of 1000 cycles/sec. (a) What is the effective value of the current in the circuit? (b) What is the effective potential difference across the resistor, the inductance, and the capacitor? (c) What is the power factor in the circuit? (d) What is the power dissipated? (e) What is the resonant frequency of the circuit and the current at resonance?

10. A noninductive resistance of 25 ohms is in series with a capacitor and absorbs 968 watts when connected to a 220-v 60-cycle line. What current will this circuit take when connected to a 110-v 25-cycle line, and what power will it absorb?

11. A capacitor 2 μf is in series with a coil whose resistance is 60 ohms and whose inductance can be varied. The combination is connected across a 24-v 60-cycle line. For what value of L will there be a resonance? What will a voltmeter across the capacitor and the inductance read at resonance?

12. The tuning circuit of a radio is effectively resistanceless and has an inductance

of 5 mh. What must be the range of the tuning capacitor to cover the radio broadcast band 550 to 1600 kc/sec?

13. A 1-ohm resistor, 0.04-h inductance, and 30-μf capacitor are connected in series across a line of variable frequency. For what values of the frequencies will the impedance be inductive? For what range will they be capacitive?

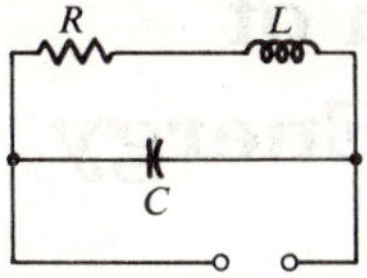

FIGURE 26–10

14. (a) Using the condition that at resonance the current and voltage in the external circuit must be in phase with one another, show that the resonant frequency ω of the parallel circuit in Fig. 26–10 is given by

$$\omega = \sqrt{\frac{1}{LC} - \frac{R^2}{L^2}}.$$

(b) In a parallel resonant circuit is the current in the external circuit a minimum or a maximum?

15. An alternating voltage is connected in series to a resistance of 10 ohms, an inductance of 0.1 h, and a capacitor of 10 μf. (a) For what value of the frequency will the voltage reading across the capacitor be a maximum? (b) For what value of ω will the voltage across the inductance be a maximum? (c) How do these values compare with the resonant frequency?

16. Starting from the formula (26–31) for the power dissipated in a series a-c circuit, show that it is equal to $P \doteq i_{\text{eff}}^2 R$, where R is the resistance in the circuit.

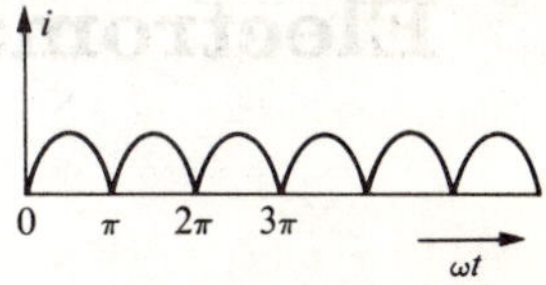

FIGURE 26–11

17. A rectifier supplies a voltage of the shape shown in Fig. 26–11. What is the effective value of the current if the maximum current is 1 amp? What is the average value?

18. A coil has a power factor of 0.1 at 60 cycles. If the voltage is 220 v, the power dissipation is 200 watts. What value of a capacitor must be connected in series with the coil so that the same power is dissipated at 110 v and 60 cycles? What is the power factor in this case?

27

The Transmission of Electromagnetic Energy

We now turn to situations in which the elements used in transmitting an electromagnetic oscillation are comparable in size with the wavelength of the resulting wave. Under these circumstances the entire transmission line will not respond instantaneously to a voltage impressed at one end of the line, and there will be phase lags in both voltage and current along the line. The electrical behavior of a transmission line is entirely analogous to the mechanical behavior of a stretched string when one end is shaken, and in both cases the result is a wave that travels with finite speed.

27–1 Complex analysis

Before we discuss transmission lines it will be useful to develop a notation which makes clear at a glance the phase differences between various parts of the a-c circuit. In the previous chapter the method we used to combine the resistive, inductive, and capacitive impedance vectorially was reminiscent of combining complex numbers. If we relabel the y-axis the i $(=\sqrt{-1})$ axis, then the inductive impedance could be written as $i\omega L$, the capacitive impedance as $-i/\omega C$, and the resistive impedance as R. We now regard the impedance as a complex quantity, and the laws of combination of impedances will be the same as the law of combination of resistances in a d-c circuit. The impedance of a resistance, inductance, and capacitance in series (Fig. 27–1) is then

$$Z = R + i\left(\omega L - \frac{1}{\omega C}\right). \tag{27–1}$$

The advantages of this representation are twofold. First, the circuit analysis is identical with a d-c circuit analysis, except that the quantities we deal with are

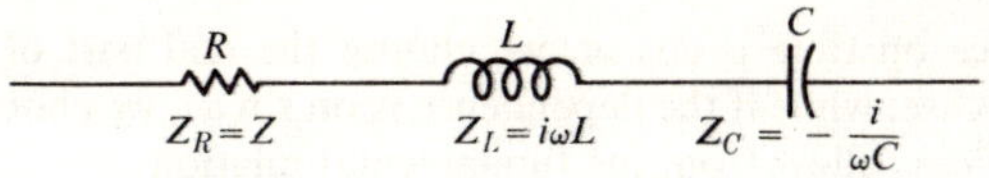

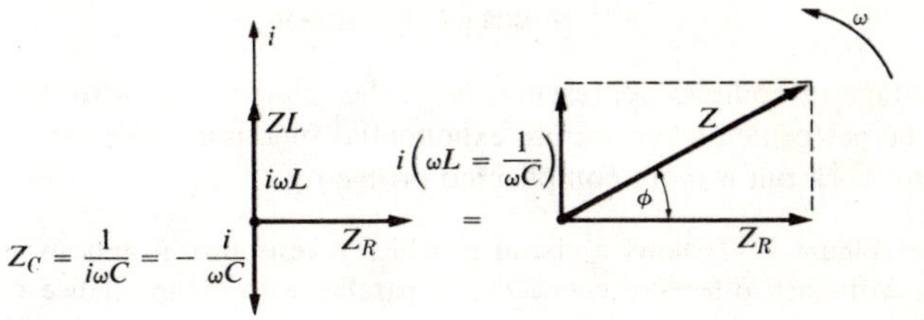

FIG. 27-1. Graphical computation of impedance using complex numbers.

complex. Second, the phase differences are easily calculable. Thus the applied voltage is

$$V = IZ = I\left[R + i\left(\omega L - \frac{1}{\omega C}\right)\right]. \qquad (27\text{–}2)$$

We must remember that V and Z are complex. If we wish to calculate the magnitude of V (or Z), we do this in the same way as we calculate the magnitude of any complex number, so that

$$V = I\sqrt{R^2 + [\omega L - (1/\omega C)]^2}.$$

The phase difference ϕ between current and voltage is, as before,

$$\phi = \tan^{-1}\frac{\omega L - (1/\omega C)}{R}.$$

If the current in the circuit varies according to the equation

$$I = I_0 \cos \omega t, \qquad (27\text{–}3)$$

the voltage varies according to

$$V = V_0 \cos(\omega t + \phi). \qquad (27\text{–}4)$$

In effect, the diagram is rotating with the angular frequency ω. Instead of representing the time dependences of I and V by using trigonometric functions, we can write

$$I = I_0 e^{i\omega t} \qquad (27\text{–}5)$$

and

$$V = V_0 e^{i(\omega t + \phi)}. \qquad (27\text{–}6)$$

If the dependence on time is cos ωt, we choose the real part of the exponential function in each case, while if the dependence is on sin ωt, we choose the imaginary part. These choices follow from the fundamental relation

$$e^{i\omega t} = \cos \omega t + i \sin \omega t. \tag{27–7}$$

The advantage of complex representation is the relative ease with which calculations can be performed when we use exponential functions. We can now use this notation to work out a more complicated problem.

Example: Figure 27–2 shows a circuit in which a resistance R and an inductance L in series with each other are connected in parallel with a capacitance C. Find the resonant frequency of the circuit.

Solution: The relationship between the current and the applied voltage V here is $V = IZ$. The equivalent impedance is found by first finding the impedances of the branches of the circuit, which are

$$Z_1 = R + i\omega L, \qquad Z_2 = \frac{-i}{\omega C} = \frac{1}{i\omega C},$$

and then combining them in a manner similar to the one in which pure resistances are combined in d-c parallel circuits:

$$\frac{1}{Z} = \frac{1}{Z_1} + \frac{1}{Z_2}.$$

We therefore have

$$\frac{1}{Z} = \frac{1}{R + i\omega L} + i\omega C.$$

We now multiply numerator and denominator of the first term on the right by $(R - i\omega L)$, which yields, since $i^2 = -1$,

$$\frac{1}{Z} = \frac{R}{R^2 + \omega^2 L^2} + i\left(\omega C - \frac{\omega L}{R^2 + \omega^2 L^2}\right).$$

Resonance occurs when current and voltage are in phase, which means that at resonance the imaginary part of the above equation must vanish. Hence the condition for resonance is

$$\omega C - \frac{\omega L}{R^2 + \omega^2 L^2} = 0,$$

$$\omega = \sqrt{(1/LC) - (R^2/L^2)}.$$

We note that the resonant frequency of a parallel circuit is different from that of a series circuit and agrees with the result of Problem 26–14. However, when $R^2C/L \ll 1$, the two frequencies are very nearly equal.

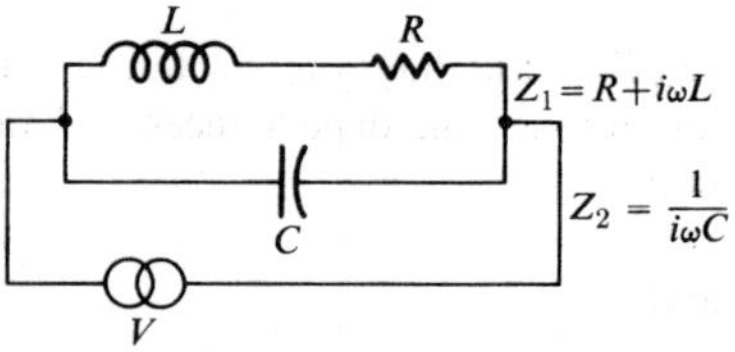

FIG. 27–2. A parallel LCR circuit.

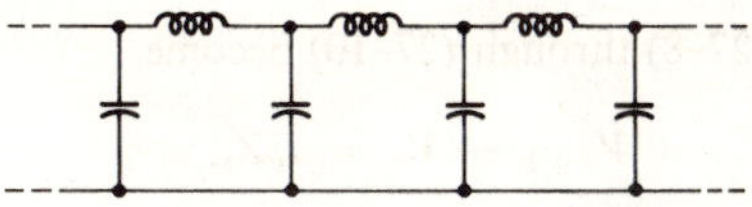

FIG. 27–3. A model of a transmission line. The inductance and capacitance in the line are considered as lumped in separate inductors and capacitors.

27–2 Lumped parameters

In idealized form a transmission line can be thought of as simply two parallel wires. It is impossible to construct such a system without any inductance being present, and though it may be small for short lengths of line, the inductance becomes appreciable for a long line. Similarly the capacitance between the wires cannot be neglected. A model of a transmission line is therefore a series of inductors in series along one side of the line, with capacitors shunted between the wires, as in Fig. 27–3. Here we are considering the reactances as though they were lumped together at definite intervals, which is a reasonable approximation provided that the intervals are small enough for the phase to be essentially the same in each interval.

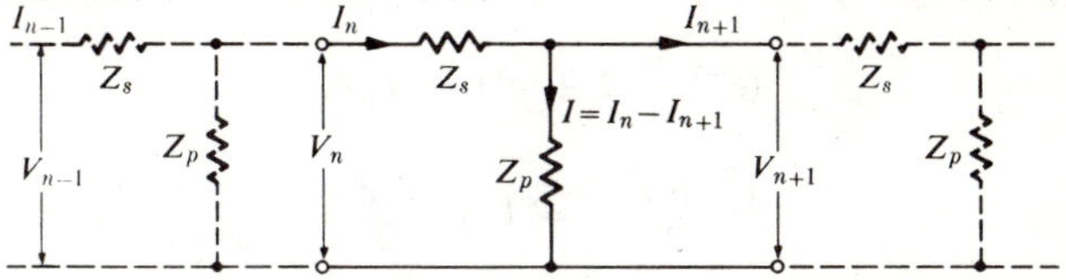

FIG. 27–4. Potential differences in a transmission line. The Z_s are generalizations of the inductances in Fig. 27–3. The Z_p are generalizations of the capacitances.

We shall first consider a uniform transmission line of infinite length. Figure 27–4 shows the nth interval of the line. The potential difference between the two junctions is evidently

$$V_{n+1} - V_n = I_n Z_s, \tag{27–8}$$

the product of the current I_n and the impedance Z_s. The impedance Z_p diverts some of I_n, so that I_{n+1} is smaller. The difference is

$$I_{n+1} - I_n = V_{n+1}/Z_p. \tag{27–9}$$

Similarly,

$$I_n - I_{n-1} = V_n/Z_p. \tag{27–10}$$

For convenience the reciprocal of Z_p is customarily introduced; it is called the *admittance*, Y_p, and is defined by

$$Y_p = 1/Z_p. \tag{27–11}$$

In terms of Y_p, Eqs. (27–8) through (27–10) become

$$V_{n+1} - V_n = I_n Z_s, \tag{27–12}$$

$$I_{n+1} - I_n = V_{n+1} Y_p, \tag{27–13}$$

$$I_n - I_{n-1} = V_n Y_p. \tag{27–14}$$

Eliminating V_{n+1} from Eq. (27–13), we obtain the *difference equation*,

$$(I_{n+1} - 2I_n + I_{n-1}) = Y_p Z_s I_n. \tag{27–15}$$

To solve this equation we shall guess at an expression for I_n and then substitute back in Eq. (27–15) to see whether this expression is correct. Our choice is the exponential

$$I_n = Ae^{-\gamma n}, \tag{27–16}$$

where A and γ are constant. Substituting back in Eq. (27–15), we have

$$Ae^{-\gamma(n+1)} - 2Ae^{-\gamma n} + Ae^{-\gamma(n-1)} = AY_p Z_s e^{-\gamma n}.$$

By dividing through by $Ae^{-\gamma n}$, we obtain an equation that is independent of n, namely

$$(e^{-\gamma} - 2 + e^{\gamma}) = Y_p Z_s,$$

so that

$$\frac{e^{\gamma} + e^{-\gamma}}{2} = 1 + \frac{Y_p Z_s}{2}. \tag{27–17}$$

The left-hand side of this equation is the defining formula of the hyperbolic cosine of γ, $\cosh \gamma$, which means that we might alternatively write

$$\cosh \gamma = 1 + \frac{Y_p Z_s}{2}. \tag{27–18}$$

Any value of γ that satisfies Eq. (27–18) determines the value of the current I_n, through Eq. (27–16). Since Y_p and Z_s in general may be complex, the solution of (27–18) may also be complex. Further, since the hyperbolic cosine is an even function, we shall have two values for γ which satisfy (27–18), one the negative of the other. Hence the general solution of (27–15) is

$$I_n = Ae^{\gamma n} + Be^{-\gamma n}, \tag{27–19}$$

where γ satisfies (27–18). We now let $\gamma = \alpha + i\beta$, and insert the time dependence of I_n through the factor $e^{i\omega t}$. The result is

$$I_n = Ae^{\alpha n} e^{i(\omega t + \beta n)} + Be^{-\alpha n} e^{i(\omega t - \beta n)}. \tag{27–20}$$

Equation (27–20) has an interesting interpretation. We recall from Chapter 25 that waves traveling in the $+x$-direction are represented by functions of the form $f(t - x/v)$, and in the $-x$-direction by functions of the form $f(t + x/v)$. Equation (27–20) can be rewritten

$$I_n = Ae^{\alpha n}e^{i\omega[t+(\beta/\omega)n]} + Be^{-\alpha n}e^{i\omega[t-(\beta/\omega)n]}. \tag{27–21}$$

Now n is a discrete variable which represents positions along the x-axis, which means that Eq. (27–21) describes waves traveling to the right and left along the transmission line with the speed

$$v = \omega/\beta. \tag{27–22}$$

We shall consider the details of wave propagation down a transmission line after we look into the more realistic case of continuously distributed circuit parameters.

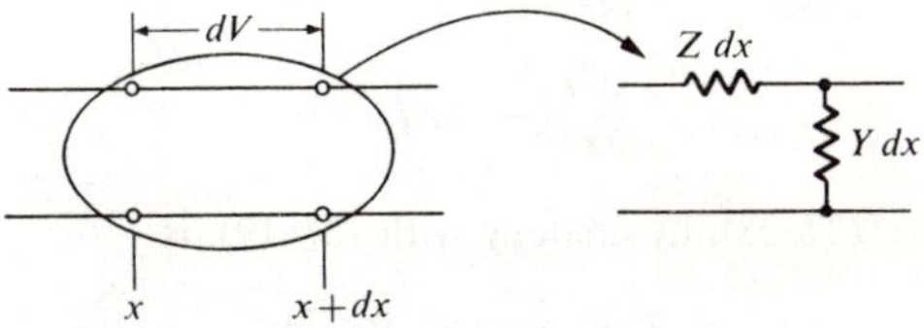

FIG. 27–5. Continuously distributed parameters. The series impedance per unit length is Z, and the parallel admittance per unit is Y.

27–3 Continuous parameters

In an actual transmission line the capacitance, inductance, and resistance present are spread uniformly along the line, and not bunched together at intervals. The analysis of continuously distributed circuit parameters proceeds in much the same way as in the lumped-parameter case. We examine a length dx (Fig. 27–5) of the line and call the series impedance per unit length Z and the parallel admittance per unit length Y. As before, we note that the potential difference between x and $x + dx$ is the product of the series current I through the impedance in this distance and the amount of the impedance, which is $Z\,dx$. That is,

$$dV = -ZI\,dx. \tag{27–23}$$

The diminution of current dI between x and $x + dx$ is the amount diverted by the admittance $Y\,dx$, namely

$$dI = -VY\,dx. \tag{27–24}$$

Hence the variations in V and I with x are

$$\frac{\partial V}{\partial x} = -ZI \tag{27–25}$$

and

$$\frac{\partial I}{\partial x} = -YV. \tag{27–26}$$

It is not difficult to obtain separate equations for V and I from the above two expressions. Our procedure is to differentiate each with respect to x and then eliminate the first derivative by substitution. Thus

$$\frac{\partial^2 V}{\partial x^2} = -Z\frac{\partial I}{\partial x},$$

$$\frac{\partial^2 I}{\partial x^2} = -Y\frac{\partial V}{\partial x}$$

and, from Eqs. (27–25) and (27–26),

$$\frac{\partial^2 V}{\partial x^2} - YZV = 0, \tag{27–27}$$

$$\frac{\partial^2 I}{\partial x^2} - YZI = 0. \tag{27–28}$$

The solution of Eq. (27–28), by analogy with (27–19), is

$$I = Ae^{\gamma x} + Be^{-\gamma x}, \tag{27–29}$$

where $\gamma = \sqrt{YZ}$. A similar solution for V can be obtained from Eq. (27–27). Again inserting the time dependence of I, we obtain

$$I = Ae^{i[\omega t + (\gamma x/i)]} + Be^{i[\omega t - (\gamma x/i)]}, \tag{27–30}$$

which again suggests waves traveling in both directions along the transmission line with the characteristic velocity

$$v = \omega/i\gamma. \tag{27–31}$$

In the case of a real line, $Z = R + i\omega L$, where R is the resistance per unit length and L the inductance per unit length. If the two wires that make up the line are separated by a perfect insulator,

$$Y = \frac{1}{1/i\omega C} = i\omega C,$$

where C is the capacitance per unit length. We therefore obtain for γ the formula

$$\gamma = \sqrt{YZ} = \sqrt{(R + i\omega L)(i\omega C)} = \sqrt{-\omega^2 LC + i\omega RC}. \tag{27–32}$$

Neglecting the resistance R, we have

$$\gamma = i\omega\sqrt{LC}, \tag{27–33}$$

and the velocity of the wave is just

$$v = 1/\sqrt{LC}. \tag{27–34}$$

The effect of a nonvanishing resistance is to make γ complex instead of purely imaginary, and thereby to contribute an attenuation factor. When R is small but not zero,

$$\gamma = \sqrt{-\omega^2 LC + i\omega RC} = \sqrt{-\omega^2 LC[1 - (iR/\omega L)]}$$
$$= i\omega\sqrt{LC}\sqrt{1 - (iR/\omega L)} \approx i\omega\sqrt{LC}\,[1 - \tfrac{1}{2}(iR/\omega L)],$$

and so

$$\gamma \approx i\omega\sqrt{LC} + \tfrac{1}{2}\sqrt{R^2C/L}.$$

Hence we obtain for the current I

$$I = Ae^{-1/2(\sqrt{R^2C/L})x}e^{i\omega(t+\sqrt{LC}x)} + Be^{-1/2(\sqrt{R^2C/L})x}e^{i\omega(t-\sqrt{LC}x)}.$$

In the distance $2\sqrt{L/R^2C}$ the wave amplitude is reduced to $1/e$ of its initial value.

27-4 Coaxial transmission line

Let us apply the above results to a coaxial transmission line. As in Fig. 27-6, the radius of the inner conductor is R_a and that of the outer conductor R_b. In Chapter 22 we found that the inductance per unit length in a coaxial line is

$$L = \frac{\mu_0}{2\pi}\ln\frac{R_2}{R_1},$$

and Problem 23-17 involved verifying that the capacitance per unit length in the line is

$$C = \frac{2\pi\epsilon_0}{\ln(R_2/R_1)}.$$

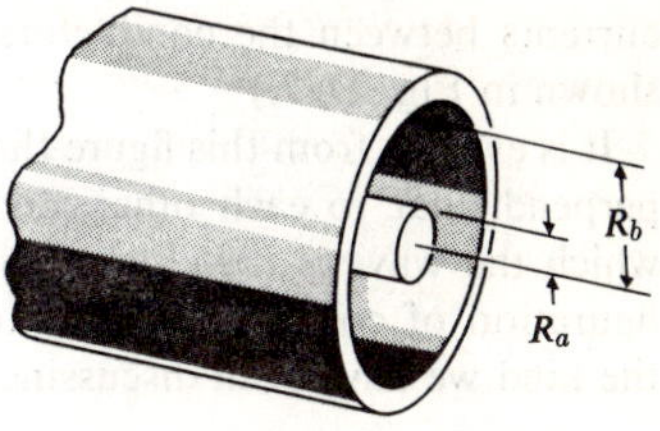

FIG. 27-6. Coaxial transmission line.

If the resistance in the line is negligible, Eq. (27-34) yields for the velocity of propagation

$$v = \frac{1}{\sqrt{\mu_0\epsilon_0}} = \frac{1}{\sqrt{4\pi \times 10^{-7}\,\dfrac{\text{n}\cdot\text{m}^2}{\text{coul}^2} \times \dfrac{1\ \text{n}\cdot\text{sec}^2}{4\pi \times 9 \times 10^9\ \text{coul}^2}}}$$
$$= 3 \times 10^8\ \text{m/sec}. \qquad (27\text{-}35)$$

We might guess that this velocity, since it depends solely on μ_0 and ϵ_0, which are constants characteristic of magnetic and electric phenomena respectively, plays some fundamental role in the physical world. This guess is borne out when we examine the results of measurements of the speed of light, which yield the value 3×10^8 m/sec in free space. And, as we shall discover in the theory of relativity, the velocity of light, c, is one of the most significant quantities in nature.

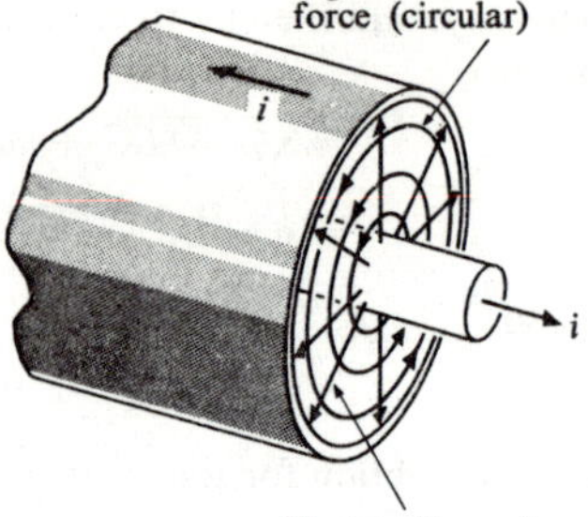

FIG. 27-7. Electric and magnetic lines of force in coaxial transmission line.

When current is flowing in the coaxial cable, electric and magnetic fields are present in the region between the conductors. Since the inner and outer conductors are at different potentials, the electric field at all times (except, of course, when the potentials equal zero twice each cycle) has the radial pattern shown in Fig. 27-7. The direction associated with the lines of force is reversed when the current reverses, but the configuration remains the same. The magnetic field between the conductors, as we saw in Chapter 22, is due to current in the inner wire only, and its lines of force consist of concentric circles, also shown in Fig. 27-7.

(The above picture of the electric and magnetic fields is valid only when the resistance of the wire is neglected. Otherwise there would be a potential difference between two points along the cable arising from the iR drop between them, and the electric field would have a component parallel to the cable. Similarly, leakage currents between the conductors would distort the magnetic field from what is shown in Fig. 27-7.)

It is evident from this figure that the electric and magnetic fields are at all points perpendicular to each other, and both are perpendicular to the direction along which the wave is traveling. This is a general result which is true for any configuration of conductors that can "guide" an *electromagnetic wave*, as waves of the kind we have been discussing are called.

27-5 Poynting's vector

We already know that whenever an electric field exists in space, an energy density of $\frac{1}{2}\epsilon_0 E^2$ is associated with it. Similarly, whenever a magnetic field exists in space, an energy density of $\frac{1}{2}B^2/\mu_0$, which is $\frac{1}{2}\mu_0 H^2$, is involved. Hence the motion of an electromagnetic wave through space must mean that energy flows with the wave. To describe this flow we shall use a vector **S**, called *Poynting's vector*, which points in the direction of the wave (and hence of the energy flow) and whose magnitude is equal to the amount of energy flowing per unit area per unit time in the direction of the vector. Poynting's vector is given by

$$\mathbf{S} = \mathbf{E} \times \mathbf{H}. \tag{27-36}$$

To illustrate Poynting's vector, we shall consider the cases of a current flowing in a wire and the charging of a capacitor in terms of it. In Fig. 27-8 there is an electric field **E** along the wire associated with the current that is flowing. To find the magnitude of this electric field we begin by noting that the potential difference between two points a and b a distance L apart on the wire is

$$V_{ab} = iR,$$

where R is the resistance of the wire between these two points. Since E is given by the potential difference between two points divided by their separation,

$$E = \frac{V_{ab}}{L} = \frac{iR}{L}.$$

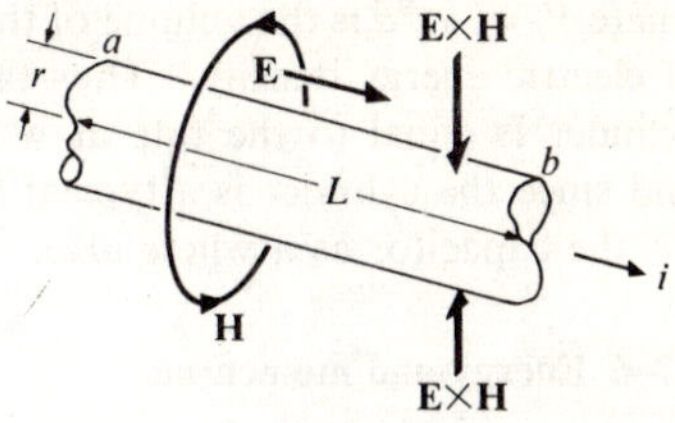

FIG. 27–8. Direction of **E**, **H**, and **E** × **H** around a wire carrying a d-c current.

The magnetic field intensity H at the surface of the wire is, as we know,

$$H = \frac{B}{\mu_0} = \frac{i}{2\pi r}.$$

E and **H** have the directions shown. Since they are perpendicular to each other,

$$S = |\mathbf{E} \times \mathbf{H}| = EH = \frac{i^2R}{2\pi rL} = \frac{i^2R}{A},$$

where A is the surface area of the wire. The Poynting vector is directed radially into the wire, and the total rate of flow of electromagnetic energy into the wire segment is

$$SA = i^2R,$$

which is identical with the power dissipated as heat.

Next, in Fig. 27–9, we have a capacitor which we are in the process of charging. Let us consider an imaginary cylinder whose base has the radius r and whose height is the distance d between the plates of the capacitor. While the capacitor is being charged, the electric field between the plates is varying and, as a result, the displacement current $\epsilon_0\, d\mathbf{E}/dt$ flows. The magnetic field due to the current in the cylinder may be found from the fact that

$$2\pi rH = \pi r^2 \frac{dD}{dt} = \pi r^2 \epsilon_0 \frac{dE}{dt}.$$

The total energy flowing into the capacitor per unit time is the Poynting vector **S** multiplied by the area of the cylinder's sides $A = 2\pi rd$, or

$$SA = EHA = \epsilon_0 E \frac{dE}{dt} \pi r^2 d$$

$$= V \frac{d}{dt}\left(\tfrac{1}{2}\epsilon_0 E^2\right),$$

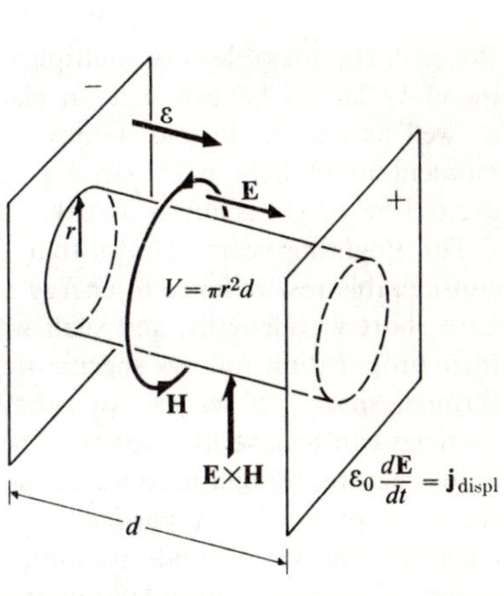

FIG. 27–9. **E**, **H**, and **E** × **H** in a charging capacitor.

where $V = \pi r^2 d$ is the volume of the cylinder and $d(\frac{1}{2}\epsilon_0 E^2)/dt$ is the rate of charge of electric energy density. Thus the rate SA at which energy is flowing into the cylinder is equal to the rate at which electrical energy is being stored up in it, and since the cylinder is a typical region of the capacitor this conclusion follows for the capacitor as a whole also.

27-6 Energy and momentum

Energy is associated with both electric and magnetic fields, and the propagation of an electromagnetic wave has associated with it a flow of electromagnetic energy. In Chapter 30, we shall see that there is an equivalence between mass and energy, with the quantitative relationship being

$$E = mc^2. \qquad (27\text{-}37)$$

Now the Poynting vector representing the energy crossing a unit area per unit time in the direction of propagation of an electromagnetic wave is equal to the energy density u of the field multiplied by the velocity c with which it is transported. That is,

$$S = uc. \qquad (27\text{-}38)$$

If we express Eq. (27-37) in terms of energy and mass densities u and ρ, respectively, we have

$$u = \rho c^2. \qquad (27\text{-}39)$$

Thus, combining (27-38) and (27-39), we obtain

$$S/c^2 = \rho c. \qquad (27\text{-}40)$$

But ρ/c, the mass density multiplied by velocity, is *momentum density*. We are therefore led to believe that an electromagnetic wave carries momentum S/c^2 as well as energy, just as a moving particle has momentum and energy. The momentum of light has been experimentally observed in the form of pressure exerted by a light beam as it strikes an object.

The Poynting vector shows that the wave mode of energy transport bears a considerable resemblance to energy transport by particles. High-frequency waves have short wavelengths, and such waves behave very much like particles in that their propagation follows specific trajectories ("rays") rather than spreading out through space. If waves can behave like particles, we might wonder whether particles can behave like waves. The answer is that particles also appear to have a dual nature, though hardly one that is conspicuous on a macroscopic level, and the wave properties of particles, together with the particle properties of waves, constitute the key to understanding atomic phenomena. We shall return to this aspect of wave motion in later chapters.

Problems

1. A coil whose resistance is 60 ohms and inductance 0.1 h is connected in parallel with a capacitor of 0.1 μf. The combination is connected across a 60-cycle source with an effective voltage of 110 v. What are the effective currents in the external circuit, through the inductance and through the capacitance? Why is the sum of the currents through the inductance and capacitance not equal to the total current?

2. A coaxial transmission line has a resistance of 10^{-3} ohm/m, an inductance of 10^{-6} h/m, and a capacitance of 3×10^{-9} f/m. What is the velocity of propagation of a current pulse along the wire? What will be the percentage of attenuation per meter of the maximum current along the line?

3. A cube 1 cm on a side is located in uniform electric and magnetic fields. If the edges of the cube are parallel to the x-, y-, and z-axes, and the fields have the components

$$H_x = \frac{H}{2}, \qquad H_y = \frac{H}{2}, \qquad H_z = 0,$$

$$E_x = 0, \qquad E_y = 0, \qquad E_z = E,$$

compute the flow of energy into or out of the cube through each face and the net rate of accumulation of energy within the cube.

4. A copper wire with resistivity $\rho = 1.7 \times 10^{-8}$ ohm-m and 4 mm in diameter carries a current of 10 amp. (a) What is the energy dissipated in the wire per meter per second? (b) What are the magnitudes of E and H at the surface of the wire? (c) What are the magnitudes of E and H at a point 1 mm from the axis of the wire? (d) What is the magnitude and direction of **S** at each of the above points? (e) How much power per meter is dissipated in the central cylindrical section of the wire? (f) How much power per meter is supplied by the electromagnetic field to the whole wire and to the central core?

5. A coaxial cable of length 4 m has an inner conductor of radius 2 mm and outer cylinder of radius 2.0 cm. A current of 5 amp goes down the center conductor and returns along the outer conductor. The resistance of the inner conductor is 6×10^{-3} ohm and of the outer conductor 4×10^{-3} ohm. What is the magnitude and the direction of the Poynting vector at the surface of the inner conductor and at the surface of the outer conductor?

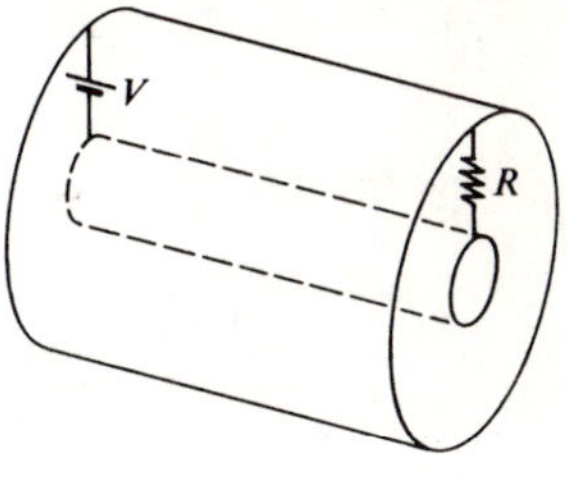

FIGURE 27–10

6. A battery is connected across the inside and outside terminals of one end of a resistanceless coaxial cable and a resistor is connected across the other end (Fig. 27–10). (a) What are the directions of **E** and **H** due to the current flowing in the coaxial cable? What is the direction of **S**? (b) What are the directions of **E** and **H** due to the current flowing in the resistance R? What is the direction of **S**? (c) Are these results physically reasonable?

7. If at a particular time and place the energy density of the electromagnetic field is 3×10^{-20} j/m^3, what is the value of the Poynting vector at that point?

8. Start with Eq. (27–38) and the fact that $u = \frac{1}{2}\epsilon_0 E^2 + \frac{1}{2}\mu_0 H^2$, in a vacuum. Let H be proportional to E and prove that

$$\sqrt{\mu_0}\, H = \sqrt{\epsilon_0}\, E.$$

9. Light is an electromagnetic wave, and the electric and magnetic vectors are perpendicular to the direction of propagation. The intensity of a light pulse is 10^8 watts/m^2. What are the values of the electric and magnetic vectors in the pulse? Use the conditions of Problem 8 for the relation of E and H. What is the momentum density of the wave?

10. The wave of Problem 9 is reflected from a plate whose area is 0.25 m^2. Ninety-five percent of the energy is reflected. What is the force on the plate?

11. Show that in mks units $\sqrt{\mu_0/\epsilon_0}$ is expressed in ohms. Calculate its value.

28

Electromagnetic Waves

Among the most noteworthy achievements of nineteenth-century science was the realization that light consists of electromagnetic waves. Electromagnetic waves themselves were predicted by James Clerk Maxwell in 1864 on the basis of his theory of electric and magnetic fields and their interrelation. The speed of such waves in a vacuum, as obtained from his calculations, was, within the limits of experimental uncertainty, identical with the speed of light in vacuum, and one would surmise that, since both were also transverse waves, they were the same phenomenon. This surmise has since been verified in every detail, and Maxwell's electromagnetic theory of light today occupies a niche in the history of physics alongside Newton's theory of universal gravitation and Einstein's theory of relativity. In this chapter we shall examine some characteristic properties of electromagnetic waves.

28–1 Maxwell's hypothesis and electromagnetic waves

The mechanism by which waves are propagated in an ordinary medium is the constant interchange between the kinetic energy of the particles of the medium and the potential energy of the medium. Thus a moving particle on a string stretches the string. The particle slows down and its kinetic energy is converted to potential energy of the string. Then the stretched string gives up its potential energy to particles further along the string, and thus a wave is propagated.

The electromagnetic waves which proceed down a transmission line, as described in Chapter 27, are also propagated by virtue of a coupling of various forms of energy, although in this case the medium plays a different role. In the electromagnetic case electrical energy produces a current in the transmission line, which in turn produces a magnetic field. The changing magnetic field produces an electric field by induction, which is manifested by an emf. This emf generates a changing current in the transmission line, which in turn produces a magnetic field, thus completing the cycle. We have then a constant interchange between electrical and magnetic energy along a transmission line, and a wave is propagated. We have already noted a certain similarity between magnetic energy and kinetic energy, as well as a similarity between electrical energy and potential energy. The analogy between the waves on a string and electromagnetic waves along a transmission line derives from the interchange of energy between electrical and magnetic forms, which corresponds to the interchange between kinetic and potential energy along the string. The coupling of the electrical and magnetic energy in a transmission line can be seen from Eqs. (27–25) and (27–26). These imply that a changing voltage (or electric field) is equivalent to a current, and that a changing current (or magnetic field) is equivalent to a voltage (or electric field).

There are some essential differences between waves on a string and electromagnetic waves. For one thing, the medium in which the transmission line is immersed plays a different role for electromagnetic and for mechanical waves, and it affects both the energy density of the electromagnetic field and the velocity of propagation of the waves. The medium does not serve as a potential-energy reservoir, as it does in the case of mechanical waves. The function of the transmission line is to provide a current and therefore a changing magnetic field. If no currents were produced, the cycle of interchange between electrical and magnetic energy would seem to be broken and no wave would propagate.

Maxwell considered what might happen in a vacuum in the absence of conductors. Would there be a flow of electromagnetic energy via a wave motion if there were no conductors in which a current could exist? Would the cycle of interchange of electrical and magnetic energy be broken? At first glance this would seem to

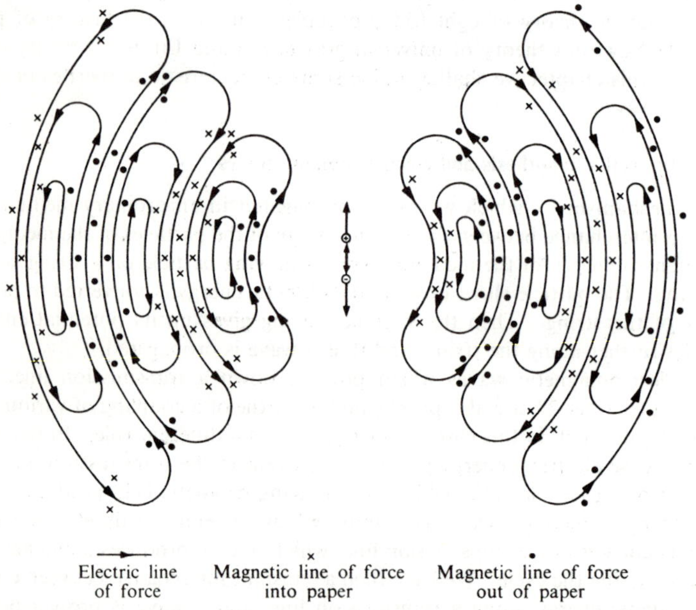

FIG. 28–1. Configuration of electric and magnetic lines of force in the electromagnetic waves that spread out from a pair of oscillating charges.

be true, but it is not. What permits the propagation of electromagnetic waves in a vacuum is the reality of the displacement currents described in Chapter 23. There we saw that a changing electrical field, even in a vacuum, is equivalent to a current, the so-called *displacement current*. This changing displacement current can produce a changing magnetic field, which in turn could produce a changing electric field, thus supporting the necessary cycle.

Since the only purpose that a transmission line has in supporting electromagnetic waves is to supply the current, the existence of a displacement current means that we can dispense with wires, and the electromagnetic waves we would then have in free space would be identical in every respect to those we found in Chapter 27 on transmission lines. The waves would be transverse and their velocity of propagation in free space would be equal to $c = 1/\sqrt{\epsilon_0 \mu_0} = 3 \times 10^8$ m/sec.

In Fig. 28–1 we have drawn the configurations of the electric and magnetic fields that spread outward from a pair of oscillating charges. The actual fields are present in three dimensions, so that the magnetic lines of force form loops in planes perpendicular to the line joining the charges. There are three significant things we should note about these electromagnetic waves, which are as follows:

(1) The variations occur simultaneously in both fields (except close to the oscillating charges), so that the electric and magnetic fields have maxima and minima at the same times and in the same places.

(2) The directions of the electric and magnetic fields are perpendicular to each other and to the direction in which the waves are moving. Light waves are therefore transverse.

(3) The speed of the waves depends only on the electric and magnetic properties of the medium they travel in, and not on the amplitudes of the field variations.

Figure 28–2 is an attempt at portraying (1) and (2) of the above in terms of superimposed graphs showing the electric and magnetic field intensities and directions in an electromagnetic wave at a particular instant. The actual configuration of the fields in space is shown in Fig. 28–1. It is worth keeping in mind that, unlike the other types of wave we have considered—water waves, sound waves, waves in a stretched string—nothing material moves in the path of an electromagnetic wave. The only changes are in field intensities.

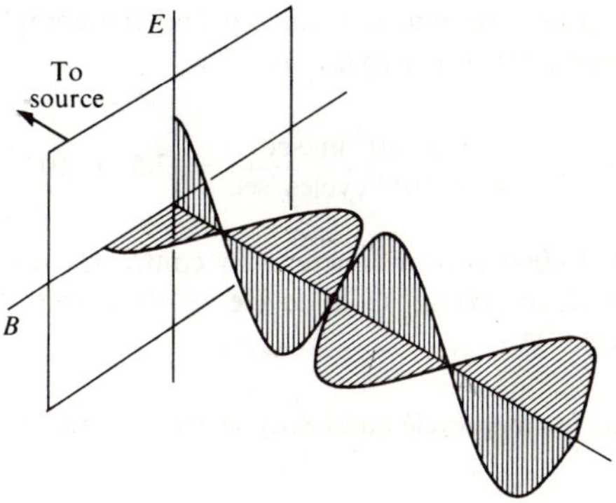

FIGURE 28–2

28–2 Varieties of electromagnetic waves

There is a variety of electromagnetic waves that can propagate in free space without transmission lines or other "wave guides." While all electromagnetic waves share certain basic properties, other features of their behavior depend on their frequencies. Light waves themselves span a brief frequency interval, from 3×10^{14} cycles/sec for red light to 8×10^{14} cycles/sec for blue light. Electromagnetic waves with frequencies between these limits are the only ones that the eye responds to, and specialized instruments of various kinds are required to detect waves with higher and lower frequencies. Figure 28–3 shows the electromagnetic wave *spectrum* from the low frequencies used in radio communication to the high frequencies found in x-rays and gamma rays (about which we shall have more to say in later chapters). The wavelengths corresponding to the various frequencies are also shown.

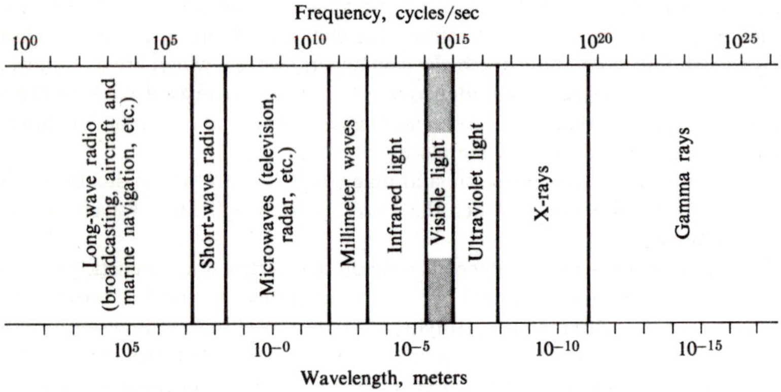

FIG. 28–3. The electromagnetic wave spectrum.

As we recall, the product of the frequency f of a wave and its wavelength λ is just the wave speed, here c. Hence, given the wavelength or frequency of a particular electromagnetic wave, we can immediately find the complementary quantity. The wavelength of red light, for instance, is

$$\lambda = \frac{c}{f} = \frac{3 \times 10^8 \text{ m/sec}}{4 \times 10^{14} \text{ cycles/sec}} = 7.5 \times 10^{-7} \text{ m},$$

which is less than 1/10,000 of a millimeter. By contrast, a 1 megacycle/sec radio wave has a wavelength of 300 m. (1 kilocycle = 10^3 cycles; 1 megacycle = 10^3 kilocycles = 10^6 cycles.)*

* The abbreviations for megacycle and kilocycle are mc and kc.

28-3 Radiation pressure

In the previous chapter we have seen that there is a transport of momentum associated with electromagnetic waves. One of the more interesting consequences of momentum transport by electromagnetic waves is the phenomenon of *radiation pressure*, the pressure these waves exert on any surface they fall upon. We recall from Chapter 14 that the force a stream of particles exerts on a wall is equal to the rate of change of their momenta upon striking the wall. An electromagnetic wave whose Poynting vector has the magnitude S loses the momentum S/c per unit time when it is absorbed by a surface, and so the force it exerts on the wall is S/c per unit area. Since pressure is force per unit area, the pressure p of the wave is

$$p = \frac{S}{c} \text{ (absorbed)} \tag{28-1}$$

as it is absorbed. If the wave is totally reflected, its change in momentum per unit area per unit time is doubled to $2S/c$ (since momentum is a vector quantity), and the pressure it exerts is also doubled to

$$p = \frac{2S}{c} \text{ (reflected).} \tag{28-2}$$

The approximate magnitude of the Poynting vector for solar radiation at the earth is $S = 1400$ watts/m^2. If this radiation is totally absorbed, the resulting pressure on the absorber is

$$p = \frac{S}{c} = \frac{1400 \text{ watts/m}^2}{3 \times 10^8 \text{ m/sec}} = 4.7 \times 10^{-6} \text{ n/m}^2.$$

The seemingly trivial pressure of solar radiation may be responsible for the deflection of comet tails so that they always point away from the sun; it has even been seriously suggested as a means of interplanetary propulsion through the use of large reflecting "sails" on space ships.

If we confine a certain amount of electromagnetic radiation in a container with perfectly reflecting walls, the various waves will bounce around until their directions are random and their average density is uniform. To find the pressure this radiation exerts on the container walls we can proceed almost exactly as we did in the case of randomly moving molecules in the kinetic theory of gases (Chapter 14). There we considered an assembly of N molecules, each with the average kinetic energy $\overline{KE}$, in a box of volume V. The pressure these molecules exert on the walls of the box is given by Eq. (14–10) as $p = \frac{2}{3}(N/V)\,\overline{KE}$, which we can rewrite in terms of the energy density w of the molecules as

$$p = \tfrac{2}{3}w \quad \text{(molecules in a box)} \tag{28-3}$$

since

$$w = \frac{\text{total molecular kinetic energy}}{\text{volume}} = \frac{N\,\overline{KE}}{V}.$$

Equation (28–3) does not hold for trapped radiation because radiation possesses only half as much momentum for a given energy content as molecules do. To verify this statement we note that from the relation $mc = E/c$, the momentum of an electromagnetic wave of energy content E is

$$\text{Momentum} = \frac{E}{\text{speed}} \quad \text{(radiation)},$$

while the momentum of a molecule whose energy is KE is

$$\text{Momentum} = \frac{2\ \text{KE}}{\text{speed}} \quad \text{(molecules)}$$

since its momentum is mv and its kinetic energy $\frac{1}{2}mv^2$. Because pressure results from momentum transfer and radiation of energy density w contains half the momentum of a group of molecules of the same energy density,

$$p = \tfrac{1}{2} \times \tfrac{2}{3}w = \tfrac{1}{3}w \quad \text{(radiation in a box)}. \tag{28–4}$$

The pressure exerted by electromagnetic radiation in a container with reflecting walls is equal to one-third of its energy density.

28–4 Blackbody radiation

All objects give off electromagnetic radiation at a rate that increases with their temperature. The predominant wavelength in the radiation decreases with increasing temperature as an object is heated; the sequence of colors that appears starts with dull red and goes through bright red, orange, yellow, and white to bluish-white. Although a solid must have a temperature of at least 700°C or thereabouts to emit visible light of an intensity sufficient for it to be seen, all bodies at all temperatures radiate electromagnetic waves that can be detected by suitable techniques. The ability of a body to emit radiation is directly proportional to its ability to absorb radiation: a good absorber is a good emitter, and vice versa. This conclusion follows from the fact that a body at the same temperature as its environment must be absorbing and emitting radiation at exactly the same rates. When the temperature of the body exceeds that of its environment, it emits more radiation than it absorbs, and it is this difference we perceive.

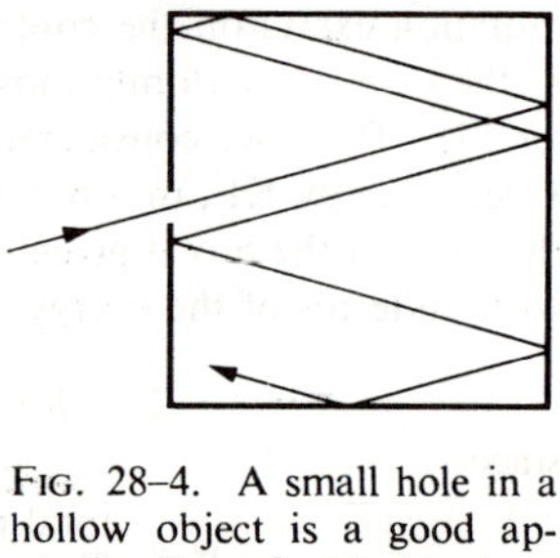

FIG. 28–4. A small hole in a hollow object is a good approximation of a blackbody.

A perfect absorber is one that absorbs all electromagnetic energy that falls upon it, regardless of wavelength. Such an absorber, called a *blackbody* by physicists even though actual objects painted black reflect at least a little radiation, is therefore also the best possible radiator at any temperature. It is convenient to discuss thermal radiation in terms of the behavior of an ideal blackbody in order to avoid such extraneous considerations as the nature of the emitting surface. A good approximation of a blackbody is a small hole in a hollow object (Fig. 28–4), since radiation striking the hole enters the cavity and is trapped there by multiple reflections even if the cavity walls are not perfectly absorbing.

A valuable theoretical approach to blackbody radiation is to consider a blackbody as a thermodynamic system whose working substance is electromagnetic radiation. We can use this approach to show that the energy density of such radiation within a cavity depends only on the temperature of its walls. Let us suppose that this is *not* true. We imagine two cavities, A and B, whose walls are at the same temperature, but cavity A has a higher energy density than cavity B (Fig. 28–5). We join the cavities together, and radiant energy proceeds to flow from A to B until their respective energy densities are equal. Then we separate the cavities and reseal them. Heat now flows from the walls of cavity A to the radiation remaining within to make up for the radiation lost to cavity B, the net result being a lower temperature for cavity A. At the same time some of the additional radiant energy in B is absorbed by its walls, the net result being a higher temperature for cavity B. We may next connect a heat engine between A and B, using the latter as the high-temperature reservoir and the former as the low-temperature one. After the excess heat in A has been converted to work, we can reconnect the cavities, which are again at the same temperature, and then repeat the sequence until no energy is left in both cavities. But the second law of thermodynamics states that it is impossible to construct an engine that, operating in a cycle, produces no effect other than the extraction of heat from a reservoir and the performance of an equivalent amount of work—and the two cavities, initially at the same temperature, constitute a single reservoir from the point of view of thermodynamics. Hence we conclude that the energy density of the radiation within a cavity depends solely on the cavity temperature. Since a blackbody is equivalent to a hole in the wall of a cavity, it follows that the rate at which a blackbody radiates energy is likewise a function solely of the body's temperature.

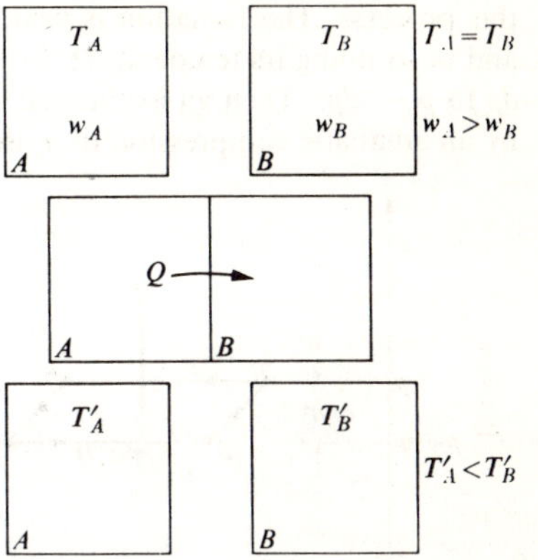

FIG. 28–5. If the energy density w of blackbody radiation were not a function of temperature T only, it would be possible for radiant energy to flow from one body to another when both have the same initial temperature; this violates the second law of thermodynamics.

We can use another thermodynamic argument to find precisely how the rate at which a blackbody radiates energy varies with its temperature. Let us consider a cylinder whose walls are perfectly reflecting which has a movable piston, also a perfect reflector, at one end. The cylinder contains electromagnetic radiation, and we can regard it as a Carnot engine whose working substance is the radiation. Figure 28–6 shows the operating cycle of this engine. The first stage is an isothermal expansion at the temperature T, during which the cylinder volume increases from V_a to V_b. Because the pressure of the radiation is proportional to the energy density, which in turn depends only on temperature, p remains constant during this process. The radiation is next allowed to expand adiabatically from b to c, and in so doing its temperature drops by dT to $T - dT$ and its pressure drops by dp to $p - dp$. Then an isothermal compression from c to d takes place, followed by an adiabatic compression to a, the starting point.

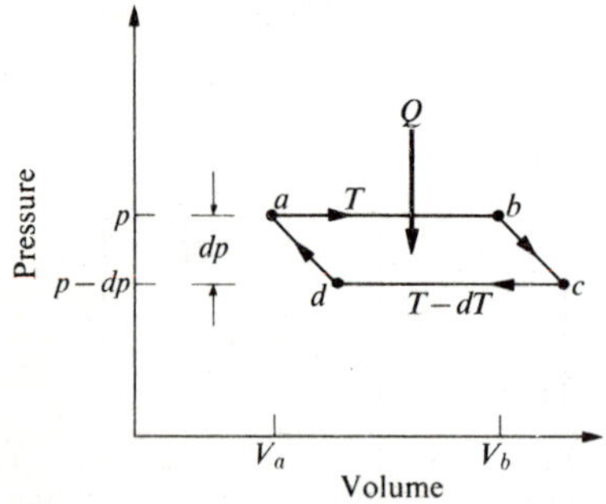

FIG. 28–6. The operating cycle of a Carnot engine whose working substance is radiation.

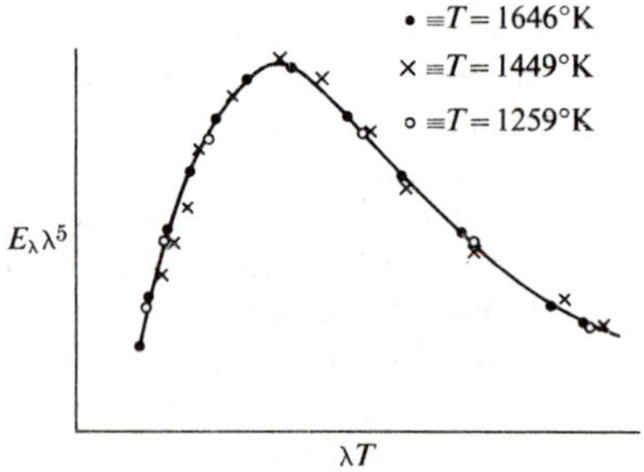

FIG. 28–7. Wien's law [Eq. (28–5)].

The work dW done by the radiation in expanding isothermally by dV is, since $p = \frac{1}{3}w$,

$$dW = p\,dV = \frac{w}{3}\,dV.$$

The internal energy of the radiation is the product of the energy density w and the volume V. Hence the increase dU in internal energy during an isothermal expansion of dV is $dU = w\,dV$. The heat input Q in the expansion from V_a to V_b is, from the first law of thermodynamics,

$$\begin{aligned} Q = \int_a^b dQ &= \int_a^b dU + \int_a^b dW \\ &= \int_{V_a}^{V_b} w\,dV + \int_{V_a}^{V_b} \frac{w}{3}\,dV \\ &= \tfrac{4}{3}w(V_b - V_a). \end{aligned}$$

The work dW done during the entire cycle is the area of the parallelogram $abcd$,

$$dW = dp(V_b - V_a) = \frac{dw}{3}(V_b - V_a).$$

The efficiency Eff of all Carnot engines (see Chapter 16) is

$$\text{Eff} = \frac{dW}{Q} = 1 - \frac{T_2}{T_1}.$$

Here

$$1 - \frac{T_2}{T_1} = 1 - \frac{T - dT}{T} = \frac{dT}{T} \quad \text{and} \quad \frac{dW}{Q} = \frac{dw}{4w}.$$

Hence

$$\frac{dw}{4w} = \frac{dT}{T} \quad \text{and} \quad \int \frac{dw}{w} = 4 \int \frac{dT}{T}.$$

If we let ln b represent the constant of integration, we obtain

$$\ln w = 4 \ln T + \ln b,$$

$$w = bT^4.$$

This result states that the energy density of the radiation trapped in a container with reflecting walls is proportional to T^4, the fourth power of its absolute temperature. If we bore a hole in the container, radiation will leak out at a rate proportional to w and hence to T^4. Since such a hole is in effect a blackbody, the amount of energy R radiated by a blackbody per unit time and per unit area is also proportional to T^4:

$$R = \sigma T^4. \tag{28–5}$$

Equation (28–5) is known as the *Stefan-Boltzmann law* because it was arrived at empirically by Josef Stefan in 1879 and, five years later, was theoretically derived by Ludwig Boltzmann in the above way. The value of the constant σ is

$$\sigma = 5.67 \times 10^{-8}\ \text{watt/m}^2\cdot{}^\circ\text{K}^4.$$

Thus a blackbody whose area is A and whose temperature is T radiates electromagnetic energy at the rate of σAT^4 j/sec. We note that Eq. (28–6) tells us only how much energy is emitted, and nothing about the manner in which the energy is divided among the infinite number of possible wavelengths.

Finally, we can record one other success of the application of thermodynamic reasoning to the problem of blackbody radiation. Wien showed that the energy density must have the form

$$E_\lambda = \frac{1}{\lambda^5} f(\lambda T), \tag{28–6}$$

where f is some function. Figure 28–7 shows an experimental plot of $E_\lambda \lambda^5$ versus λT for a wide range of wavelengths and frequencies, and a plot of Eq. (28–6). A glance will show that the thermodynamic result of Wien is verified. Thermodynamics cannot, however, tell us the form of the function $f(\lambda T)$, which would give the distribution of energy among the various wavelengths of a radiation blackbody.

28-5 The Rayleigh-Jeans formula

An argument almost identical to the one we used above to show that the total radiation rate from a blackbody must depend solely on temperature may be used to show that the distribution of energy among the various wavelengths of the radiation from a blackbody must also depend solely on its temperature. In this case we employ a filter between the cavities when they are joined together which permits only radiation in a very narrow wavelength interval to pass from one to the other, thus restricting the hypothetical energy transfer to that wavelength interval. The rate at which energy is radiated in this interval is the same for all blackbodies at the same temperature, if the second law of thermodynamics is not to be violated.

The empirical data on the spectra of blackbody radiation at various temperatures are shown in Fig. 28-8. The ordinate of the graph is the monochromatic energy density w_λ, where $w_\lambda\, d\lambda$ is the amount of radiant energy per unit volume with wavelengths between λ and $\lambda + d\lambda$. We note that at high temperatures the areas under the curves, which equal the total energy in the radiation, are greater than they are at low temperatures, and the positions of the maxima in the curves, which represent the predominant color, shift to shorter wavelengths as T increases. Both of these effects are entirely in accord with classical physics. As we saw in the previous section, the area under each curve is given by

$$\int_0^\infty w_\lambda\, d\lambda = w = bT^4.$$

The wavelength shift, too, can be derived by thermodynamic reasoning. In agreement with experiment, the wavelength λ_{max} at which w_λ is a maximum turns out

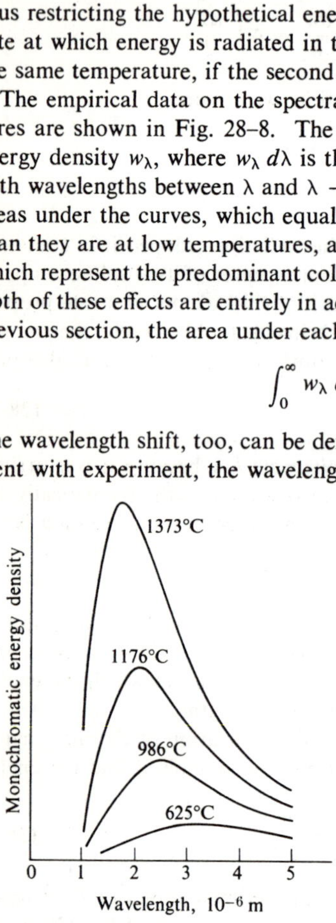

FIG. 28-8. The monochromatic energy density of blackbody radiation as a function of temperature. These curves are the same as those of Fig. 13-4.

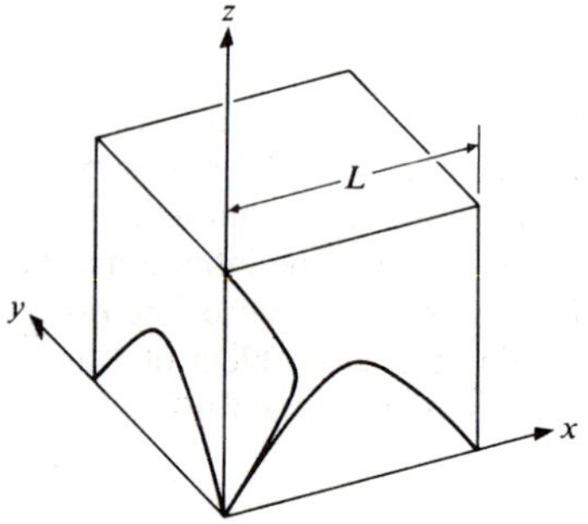

FIG. 28-9. Standing waves of wavelength $\lambda = 2L$ in a cube whose edges are L long. Only one standing wave of this wavelength can exist between reflecting walls L apart, for a total of three.

to be inversely proportional to the absolute temperature of the radiating blackbody. That is,

$$\lambda_{\max} = \frac{C}{T}, \qquad \lambda_{\max} T = C, \tag{28-7}$$

where $C = 2.898 \times 10^{-3}$ m·°K. This relationship is called the *Wien displacement law.*

Still, despite these successes, we have no formula that gives the distribution of wavelengths in the blackbody spectrum at a given temperature.

An attempt to do this, due to Lord Rayleigh and Sir James Jeans in the last century, will be outlined below. Although the Rayleigh-Jeans formula is incorrect, the approach they used can be modified in detail to give the correct theoretical result.

The basic idea of Rayleigh's derivation was as follows. A blackbody consists of radiation in equilibrium with the walls of the enclosure of the blackbody, and these walls are constantly absorbing and re-emitting the radiation. Since equilibrium exists, the energy possessed by the radiation is equal to the energy possessed by the oscillators which emit the radiation. Also, since the energy distribution of all blackbodies is the same experimentally, the result must be independent of the exact nature of the material of the walls, and so it can merely be assumed that the walls consist of charged harmonic oscillators.

The energy density depends on the wavelength because different numbers of oscillators emit the various wavelengths. The reason for this follows from the fact that electromagnetic radiation in a cavity with perfectly reflecting walls must have nodes on the walls. These are the standing waves of Chapter 25. Only three standing waves whose wavelength is $2L$ can exist in a cube of length L and have nodes at the walls (see Fig. 28–9). A shorter wavelength, say $L/2$, can have many more standing waves (see Fig. 28–10). The general criterion is that the standing wave

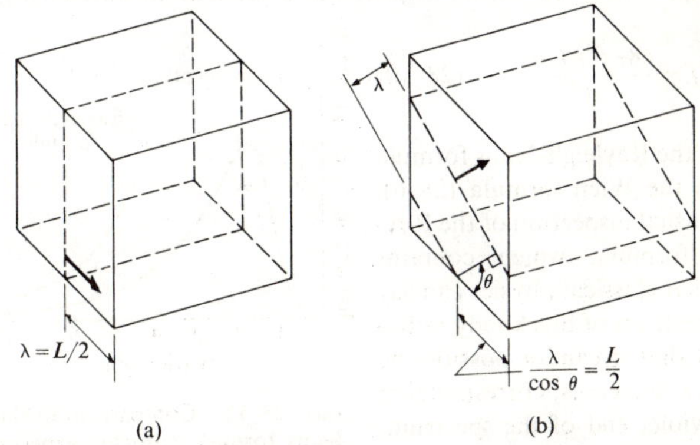

FIG. 28–10. Standing waves in a cube of length L whose wavelength is λ. The general criterion for existence of standing wave is that the projection on each side should be $n\lambda/2$.

must have an integral number of half-waves along each side. (In the one-dimensional case there is only one direction of propagation and $n\lambda/2 = L$.) Using this criterion, one can count (although we shall not do so) the number of standing waves between λ and $\lambda + d\lambda$ that can exist in the cavity. It is given by

$$dN_\lambda = \frac{4\pi}{\lambda^4} L^3\, d\lambda.$$

For each one of these wavelengths there correspond two oscillators. (The observation that there are two is due to Sir James Jeans.) Each one of the oscillators corresponds to the two possible directions of polarization of the electromagnetic waves. Thus the number of oscillators per unit volume is

$$dn_\lambda = \frac{dN_\lambda}{L^3} = \frac{8\pi\, d\lambda}{\lambda^4}. \qquad (28\text{–}8)$$

Next we draw on the principle of equipartition of energy, which states that each degree of freedom has an average energy of $\frac{1}{2}kT$. The one-dimensional oscillator has the equivalent of two degrees of freedom, one for its kinetic and one for its potential energy. The total energy of all the oscillators per unit volume is found by multiplying (28–8) by kT,

$$E_\lambda\, d\lambda = \frac{8\pi}{\lambda^4} kT\, d\lambda, \qquad (28\text{–}9)$$

which is also the amount of energy per unit value in the range of wavelengths between λ and $\lambda + d\lambda$. Equation (28–9) is known as the *Rayleigh-Jeans formula*. To express this formula in terms of frequency f rather than wavelength, we note that

$$f = c/\lambda, \qquad df = -c\, d\lambda/\lambda^2, \qquad E_f\, df = -E_\lambda\, d\lambda,$$

since an increase in frequency corresponds to a decrease in wavelength. Hence

$$E_f\, df = \frac{8\pi f^2 kT\, df}{c^3}. \qquad (28\text{–}10)$$

Note that the Rayleigh-Jeans formula conforms to the Wien formula (28–6). But even a casual inspection of the Rayleigh-Jeans formula, which contains everything that classical physics can say about the spectrum of blackbody radiation, reveals that it cannot possibly be correct. As λ decreases, corresponding to the ultraviolet end of the spectrum, Eq. (28–9) predicts an increase in the radiated energy at a rapidly (as $1/\lambda^4$)

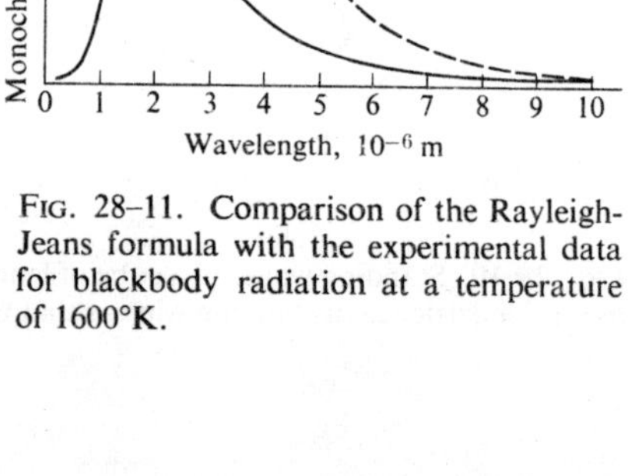

FIG. 28–11. Comparison of the Rayleigh-Jeans formula with the experimental data for blackbody radiation at a temperature of 1600°K.

increasing rate. In fact, in the limit of infinitely short wavelength, $E_\lambda\,d\lambda$ approaches infinity—the "ultraviolet catastrophe" of classical physics, since, of course, $E_\lambda\,d\lambda$ actually goes to zero as λ approaches zero. According to the Rayleigh-Jeans formula, the total energy density in a cavity at *any* temperature, however low, is

$$E = \int_0^\infty E_\lambda\,d\lambda = \infty,$$

which is nonsense, instead of the correct Stefan-Boltzmann result of bT^4.

Oddly enough, considering its hopeless predictions for the short-wavelength end of the radiation spectrum, the Rayleigh-Jeans formula agrees perfectly with observation at long wavelengths. This is evident in Fig. 28–11, where the dashed line was calculated from Eq. (28–9). Because the Rayleigh-Jeans formula contains no arbitrary constants, this partial agreement indicates that it must have some germs of truth in it. The question is, which of the classical laws that went into the Rayleigh-Jeans formula are not really correct?

28–6 Planck's law

The problem of the distribution of energy of blackbody radiation as a function of wavelength was finally solved by Max Planck in 1900. The general approach he used was that of Rayleigh and Jeans, except that he did not take the average energy per oscillator as being equal to kT. In avoiding this latter assumption he was forced to make a revolutionary hypothesis which had repercussions in all phases of submicroscopic physics.

In order better to understand the approach Planck used, let us first see how one could calculate the average energy of an oscillator to be kT. According to the Maxwell-Boltzmann distribution [Eq. (14–14)], the number of oscillators whose energy lies between E and $E + dE$ is

$$dN = N_0 e^{-E/\overline{E}}\,dE, \tag{28–11}$$

where $\overline{E}$ is some number. The average energy, then, of all the oscillators, according to the rules for determining averages, will be

$$E_{\text{av}} = \frac{\int_0^\infty E e^{-E/\overline{E}}\,dE}{\int_0^\infty e^{-E/\overline{E}}\,dE}. \tag{28–12}$$

We can integrate the numerator by parts to obtain

$$E_{\text{av}} = \frac{-\overline{E}Ee^{-E/\overline{E}}\Big|_0^\infty + \overline{E}\int_0^\infty e^{-E/\overline{E}}\,dE}{\int_0^\infty e^{-E/\overline{E}}\,dE}. \tag{28–13}$$

The first term of the numerator of (28–13) is zero, and we obtain finally

$$E_{av} = \bar{E}.$$

In order to obtain our classical result, then, we must choose $\bar{E} = kT$, since $E_{av} = kT$. Equation (28–11) for the number of oscillators in the energy range dE is then

$$dN = N_0 e^{-E/kT}\, dE. \tag{28–14}$$

Equation (28–14) assumes implicitly that E varies continuously from 0 to ∞ in infinitesimally small steps. This is reasonable classically, since the energy of an oscillator depends on the square of its amplitude of motion, and this will vary continuously. Planck assumed, however (with nothing, really, to guide him), that this hypothesis was not correct. Let us assume that the energy of an oscillator varies discontinuously in units of some quantity ϵ. That is, we suppose that the oscillators in the left-hand column have energy as shown on the right,

OSCILLATORS	ENERGY
N_0	0
$N_0 e^{-\epsilon/kT}$	ϵ
$N_0 e^{-2\epsilon/kT}$	2ϵ
$N_0 e^{-3\epsilon/kT}$	3ϵ

and so forth. The average energy of the oscillators would now be

$$E_{av} = \frac{N_0 \sum_{p=0}^{\infty} p\epsilon e^{-p\epsilon/kT}}{N_0 \sum_{p=0}^{\infty} e^{-p\epsilon/kT}} = \frac{1 + \epsilon e^{-\epsilon/kT} + 2\epsilon e^{-2\epsilon/kT} + \cdots}{1 + e^{-\epsilon/kT} + e^{-2\epsilon/kT} + \cdots}. \tag{28–15}$$

The denominator of this expression can be readily summed by noting that it is a geometric series with a ratio $e^{-\epsilon/kT}$, and so the sum is equal to $1/(1 - e^{-\epsilon/kT})$. The numerator can also be summed explicitly by noting that it is equal to

$$-\frac{d}{d(1/kT)}\left(\sum_{p=0}^{\infty} e^{-p\epsilon/kT}\right) = -\frac{d}{d(1/kT)}\,\frac{1}{1 - e^{-\epsilon/kT}} = \frac{\epsilon e^{-\epsilon/kT}}{(1 - e^{-\epsilon/kT})^2}.$$

The average energy, therefore, is

$$E_{av} = \frac{\epsilon e^{-\epsilon/kT}}{1 - e^{-\epsilon/kT}} = \frac{\epsilon}{e^{\epsilon/kT} - 1}, \tag{28–16}$$

where the last result was obtained by multiplying numerator and denominator by $e^{\epsilon/kT}$ The energy distribution is found by multiplying (28–16), the average energy per oscillator, by (28–8), the number of oscillators between λ and $\lambda + d\lambda$:

$$e_\lambda\, d\lambda = \frac{8\pi}{\lambda^4}\,\frac{\epsilon}{e^{\epsilon/kT} - 1}\, d\lambda. \tag{28–17}$$

The functional form of (28–17) is vastly different from the Rayleigh-Jeans formula, but as $\epsilon \to 0$, (28–17) should approach the Rayleigh-Jeans law. That is, as the energy difference of various oscillators varies continuously, we should recover the previous result (28–9). We note that as $\epsilon \to 0$, $e^{\epsilon/kT} \to 1 + \epsilon/kT +$ terms of higher order in ϵ, and the quotient approaches

$$E_\lambda \, d\lambda = \frac{8\pi}{\lambda^4} kT \, d\lambda,$$

the Rayleigh-Jeans formula! But the Planck formula is not at all similar when ϵ does not approach zero.

How is one to determine ϵ? Planck answered this by saying that the functional form of (28–17) should be in accord with Wien's formula (28–6). This can be achieved by making ϵ proportional to $1/\lambda$; thus

$$\epsilon = \frac{\alpha}{\lambda} = \frac{\alpha}{c} f = hf. \tag{28–18}$$

In (28–18), α is a constant; α/c (c = speed of light) is still another constant, h. If we substitute (28–18) into Planck's formula, we obtain

$$E_\lambda \, d\lambda = \frac{8\pi}{\lambda^5} \frac{\alpha}{e^{\alpha/\lambda kT} - 1} d\lambda. \tag{28–19}$$

The energy density is a function of $(1/\lambda^5)f(\lambda T)$, as required by Wien's law. Expressed in terms of frequencies,

$$E_f \, df = \frac{8\pi h f^3}{c^3} \frac{1}{e^{hf/kT} - 1} df. \tag{28–20}$$

We have already seen that as $\epsilon \to 0$, which according to (28–18) implies that $\lambda \to \infty$, Planck's formula reduces to the Rayleigh-Jeans law, which agreed with the experimental data. In this limit the formula is independent of h. In order to see whether it fits experimental data, we must determine h. To do this, we compare it with the Stefan-Boltzmann law (28–6). According to the latter,

$$\int_0^\infty E_f \, df = \frac{8\pi h}{c^3} \int_0^\infty \frac{f^3 \, df}{e^{hf/hT} - 1} = \sigma T^4. \tag{28–21}$$

We introduce a new variable $hf/kT = z$ in (28–21), and it becomes

$$\frac{8\pi h}{c^3} \left(\frac{kT}{h}\right)^4 \int_0^\infty \frac{z^3 \, dz}{e^z - 1} = \sigma T^4.$$

The integral can be evaluated to be $\pi^4/15$. Thus we have that

$$\frac{8\pi}{c^3} \frac{k^4}{h^3} \frac{\pi^4}{15} = \sigma.$$

By inserting the value of $\sigma = 5.67 \times 10^8$ watts/m^2·deg^4, $c = 3 \times 10^8$ m/sec and $k = 1.38 \times 10^{-23}$ j/deg, we obtain a value of $h = 6.52 \times 10^{-34}$ j/sec.

Another method of obtaining a value of h is to note that the Planck distribution formula is consistent with the Wien displacement law. By differentiating (28–19) with respect to λ and setting it equal to zero, we could find the maximum value of E_λ as a function of λ. This yields

$$\frac{-5 \times 8\pi\alpha}{\lambda^6} \frac{1}{e^{\alpha/\lambda kT} - 1} + \frac{8\pi\alpha}{\lambda^5} \frac{e^{\alpha/\lambda kT}}{(e^{\alpha/\lambda kT} - 1)^2} \frac{\alpha}{kT} \frac{1}{\lambda^2} = 0.$$

We can find $\lambda_{\max}$ as the solution of the equation

$$e^{\alpha/\lambda_{\max}kT}\left(1 - \frac{\alpha}{5\lambda_{\max} kT}\right) = 1.$$

The solution of this transcendental equation (other than $\alpha/\lambda_{\max}kT = 0$) is

$$\frac{\alpha}{\lambda_{\max} kT} = 4.965.$$

Since $\alpha = hc$, we have

$$\lambda_{\max} T = \frac{hc}{4.965k}\cdot$$

Wien's displacement constant is experimentally given by 2.898×10^{-3} m·°K. This gives

$$h = \frac{4.965 \times 2.898 \times 10^{-3} \times 1.38 \times 10^{-23}}{3 \times 10^8}$$
$$= 6.58 \times 10^{-34} \text{ j-sec.}$$

The differences in the value of h deduced here are due to the fact that Wien's displacement constant and the Stefan-Boltzmann constant cannot be very accurately determined experimentally.

More accurate experimental determinations give

$$h = (6.62363 \pm 0.00016) \times 10^{-34} \text{ j-sec.}$$

Evidently the Stefan-Boltzmann law as well as the Wien displacement law are satisfied by the Planck distribution, and the values of h agree quite well. With this value of h the Planck distribution law accurately reproduces the empirical data on the spectra of blackbody radiation (Fig. 28–8), with no other adjustable parameters.

This agreement with experiment leaves no doubt about the essential correctness of Planck's formula. But the concept of having oscillators interchange their energy in multiples of the discrete bundles of energy hf is, from a classical point of view, very odd. It is peculiar first because these discrete bundles, or *quanta* as they are called, have no basis in classical physics; and second because the energy of the quanta is proportional to the frequency and not to the square of the amplitude, as one would expect. The smallness of h (of the order of 10^{-34}) makes the quanta for classical oscillators of macroscopic dimensions so small as not to be measurable. If the frequency is 100 cycles/sec, $hf = 6.6 \times 10^{-32}$ j, which is practically zero. When f approaches the frequency of light ($\sim 10^{15}$/sec), the energy is of the order of 10^{-19} j. This energy is several electron volts. In the submicroscopic atomic realm this is an appreciable energy, and thus the quantum hypothesis makes a big difference. We shall see that the quantum concept plays a decisive role in the understanding of all atomic phenomena.

Problems

1. A certain radio station transmits at a frequency of 880 kc/sec. What is the wavelength of these waves?

2. A small marine radar set operates at a wavelength of 3.2 cm and a power of 7 kilowatts per pulse. (a) What is the frequency of the radar waves? (b) How much reaction force is exerted on the antenna whose arc is 0.01 m^2 when each pulse is emitted?

3. The intensity of solar radiation varies inversely with the square of the distance from the sun. The earth is an average of 1.5×10^8 km from the sun. How close to the sun must it be for its radiation pressure to be 1 n/m^2?

4. A radio wave is found to exert a pressure of 10^{-8} n/m^2 on a reflecting surface. (a) How much power per unit area does the wave transport? (b) What are the maximum magnitudes of its electric intensity and magnetic flux density?

5. For good reception a radio wave should have a maximum electric field intensity of at least 10^{-4} v/m when it arrives at the receiving antenna. (a) What is the maximum flux density of the magnetic field of such a wave? (b) What is the magnitude of the Poynting vector of such a wave? (c) What radiation pressure does it exert when it is absorbed?

6. The entire 1000-watt luminous output of a searchlight is focused on a circular spot 2 m in diameter. If half the light is absorbed and half reflected, how much pressure does it exert?

7. Calculate the energy emitted by the earth per unit area per unit time, assuming that it is a blackbody at 10°C.

8. The maximum energy emitted by the sun occurs at a wavelength of 4.65×10^{-5} cm. Calculate the effective temperature of the sun. What is the total energy radiated per second if the diameter of the sun is 8×10^5 mi? Based on this data, what would be the solar constant on the earth? Compare to the correct solar constant of 0.03 cal/cm^2·sec.

9. What is the radiation pressure of a blackbody radiation at a temperature of 6000°K?

10. What is the energy in ev of a quantum whose wavelength is 7×10^{-5} cm? 2×10^{-5} cm?

11. How many quanta per second per square inch fall on the earth from the sun? The solar constant is 0.03 cal/cm^2·sec.

Assume that all the radiation has a wavelength of 5×10^{-7} m.

12. A blackbody is at a temperature of 3000°K. Determine the ratio of the energy density emitted at a wavelength of 10^{-4} cm to the energy density at a wavelength of 5×10^{-5} cm.

13. What must be the temperature of a body in order that the maximum energy density be emitted at 5.50×10^{-5} cm?

14. A classical oscillator has a mass of $\frac{1}{10}$ gm. It oscillates at a frequency of 20 cycles/sec; its amplitude is $\frac{1}{4}$ cm. Find the change of amplitude due to the emission of 1 quantum of radiation. Can this change of amplitude be measured?

15. An oscillator has a frequency of 10^{15} cycles/sec; the amplitude of the motion is $\frac{1}{4} \times 10^{-9}$ cm. The mass of the body is 10^{-26} kg. If this were a classical process, what would be the amplitude change if a quantum of radiation were emitted? Compare the fractional change of amplitude in this problem and the previous one.

16. What is the average density of quanta in a radio wave of which $\lambda = 100$ m and intensity $= 10^{-6}$ watts/m^2? How many quanta in a volume of λ^3?

29
Light

During the nineteenth century, it was established that light is a wave motion which propagates without the benefit of a medium. For other waves a tangible medium is required, so that there can be a constant interchange of energy between the wave and the potential energy of the medium, but in the case of light the constant interchange of electrical and magnetic energy and the propagation of the wave can take place with no medium present.

We shall see later in this chapter that the wavelengths present in light are very small, of the order of 10^{-7} m. A discussion of this end of the electromagnetic spectrum will complete the analysis begun with a-c circuits, where the wavelength was large compared to macroscopic dimensions, and continued with wave guides, where the wavelength was of the same order of magnitude.

29–1 Huygens' principle and reflection

The most important problem in the discussion of light is how light waves propagate in various media and in the presence of various obstacles. Maxwell's electromagnetic theory of light contains all this information, but it is too complicated to apply directly. What we require is a simple, intuitive scheme for determining the propagation of waves. Such a scheme was provided by Christian Huygens in 1678, and it is applicable not only to light waves but to all other waves as well.

Huygens' treatment concerns the propagation of wave fronts. In Chapter 25 we defined a wave front as the locus of all points of a wave having the same phase. Huygens' principle states that: *Every point on a wave front can be considered as a point source of secondary wavelets, which spread out in all directions with the wave speed of the medium. The wave front at any time is the envelope of these secondary wavelets.*

Thus the Huygens construction enables us to predict the wave front at a later time if we know the wave front at any particular time. Figure 29–1 shows how the Huygens principle is applied to the propagation of spherical wave fronts in a uniform medium.

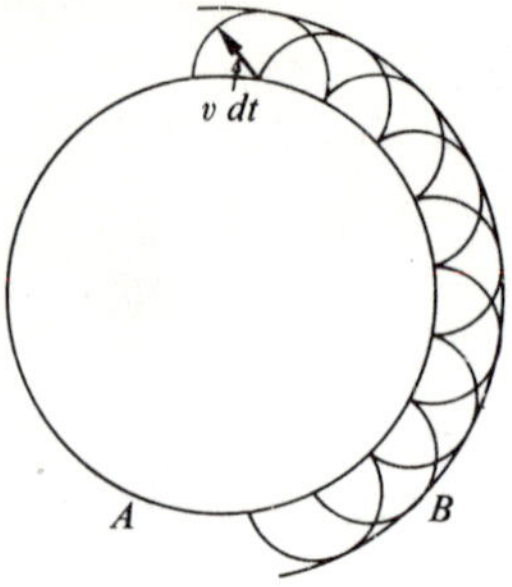

FIG. 29-1. The propagation of a spherical wave in a uniform medium, according to Huygens' principle.

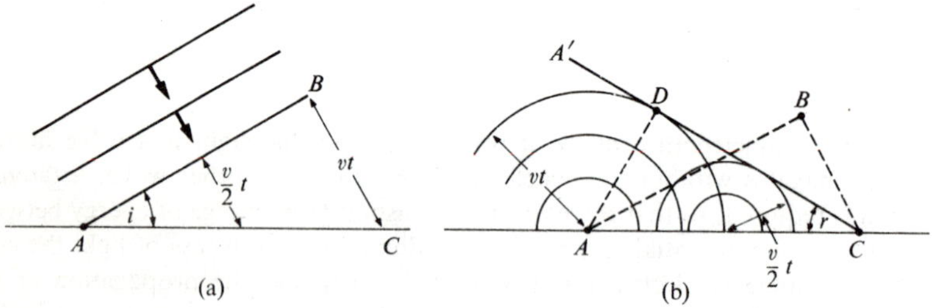

FIG. 29-2. The behavior of successive wave fronts during specular reflection.

Huygens' principle can also be used to determine the laws governing the reflection of a wave front from a surface. We are concerned here only with reflections from a perfectly smooth surface, where the reflection is termed *specular*. We shall consider a "plane wave," as defined in Chapter 25, incident on the reflecting surface (Fig. 29-2). At $t = 0$ the wave front AB just touches the reflecting mirror surface, and secondary wavelets start to spread out from A. After a time t equal to BC/v, the end B of the wave front reaches the mirror at C, and now the secondary wavelets from A are at the point D. The wave front which was AB at $t = 0$ is therefore CD at the later time t. Since

$$AD = vt \quad \text{and} \quad BC = vt,$$

we have

$$AD = BC.$$

The angle i between an approaching wave front and the reflecting surface is called the *angle of incidence;* as before, this angle is equal to that made by an approaching ray with the normal to the surface (Fig. 29-3). The angle r between a receding wave front and the reflecting surface is called the *angle of reflection;* this angle is equal to that made by a receding ray with the normal. Because

$$\sin i = \frac{BC}{AC} \quad \text{and} \quad \sin r = \frac{AD}{AC},$$

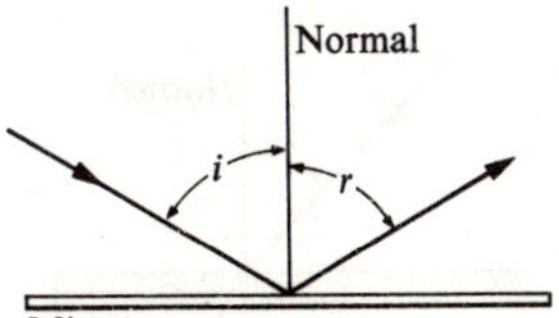

FIG. 29–3. Ray treatment of specular reflection. The angles i and r are the same in both representations.

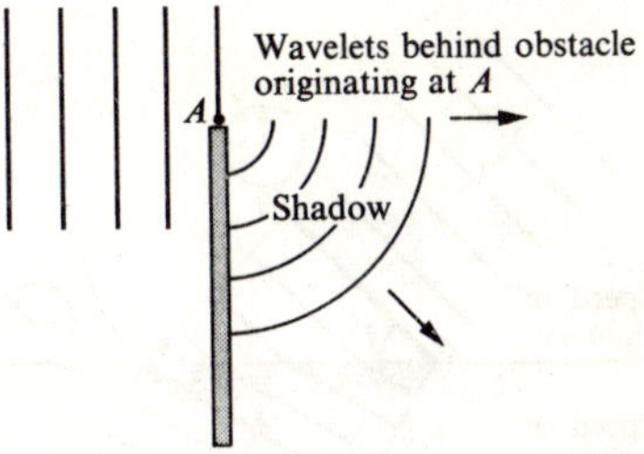

FIG. 29–4. A plane wave incident upon an obstacle is diffracted into the "shadow" region.

the fact that $AD = BC$ means that angles i and r are equal. Thus we have the basic law of reflection: *the angle of reflection of a plane wave front with a plane mirror is equal to the angle of incidence.* The incident and reflected rays and the normal to the surface lie in the same plane.

Finally one can use Huygens' principle to determine the propagation of light around a finite obstacle in its path. Figure 29–4 shows a plane wave incident on this obstacle. A portion of the plane wave is stopped, but since every point on the wave front is a source of wavelets, we can see how the wave can go around the obstacle. This property of waves being able to bend around obstacles is called *diffraction.* It is a property that we are familiar with in connection with sound: An obstacle such as a book placed between a source and a receiver alters the sound, but does not completely prevent it from reaching the receiver.

Particles do not display this property. Bullets fired at an obstacle go past it in straight lines and cast a sharp "shadow." In the case of light the situation is more complicated. Because the wavelength of light is so short, the relative phases of the wavelets play a crucial role in determining the intensity of light in the shadow region. We shall see later that diffraction effects in light phenomena are observable only in distances of the order of magnitude of the wavelength in the vicinity of the shadow boundary, and are therefore difficult to observe. Because of the relatively sharp shadow created by light, Newton and other notable physicists believed that light is a particle phenomenon.

29–2 Refraction of light

Another phenomenon which can be analyzed using the Huygens principle is the refraction of light—that is, the change in direction of a light beam when it passes from one medium to another in which its speed is different. This can be evaluated with the help of the wave-front diagram of Fig. 29–5. At $t = 0$ the wave front AB just comes in contact with the boundary between the two media. We suppose that the speed of light, v_2, in the new medium is less than that in the first one, so that secondary wavelets generated at the interface travel a shorter distance in the same time interval than do wavelets in the first medium. At the moment $t = 0$

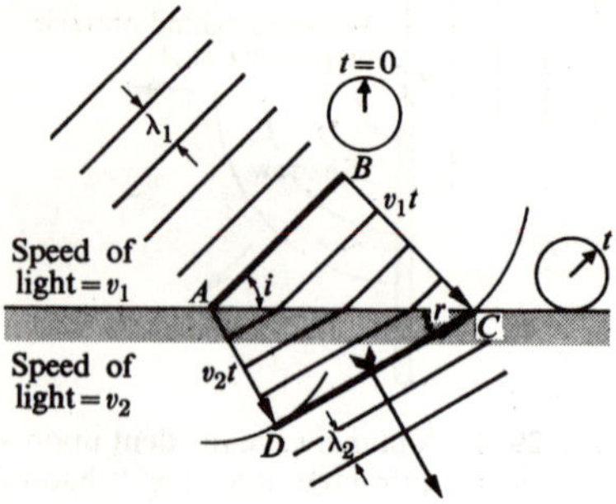

FIG. 29–5. The behavior of successive wave fronts during refraction.

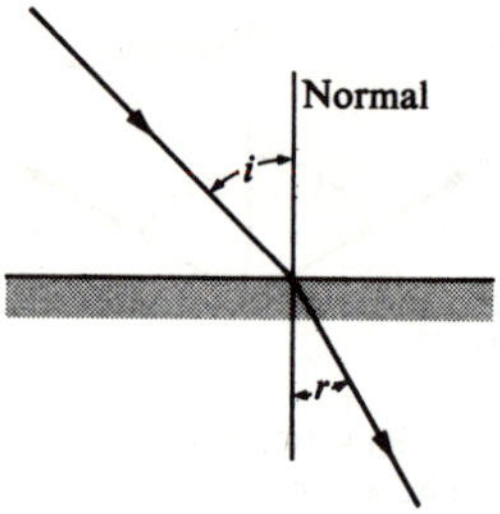

FIG. 29–6. Ray treatment of refraction. The angles i and r are the same in both representations.

when the end A of the wave front AB reaches the boundary, end B is still the distance BC away. Since the speed of light in the first medium is v_1, end B requires the time $t = BC/v_1$ to reach the boundary at C. In this period of time the end A of the wave front proceeds to D, where AD is smaller than BC. Evidently

$$BC = v_1 t \qquad \text{and} \qquad AD = v_2 t.$$

The angle i between an approaching wave front and the boundary between two media is called the *angle of incidence* of the wave front; this angle is equal to that made by an approaching ray (see Section 29–3) with the normal to the boundary, as in Fig. 29–6. The angle r between a receding wave front and the boundary between two media through which it has passed is called the *angle of refraction;* again, this angle is equal to that made by a receding ray with the normal. From Fig. 29–5 we see that

$$\sin i = \frac{BC}{AC} = \frac{v_1 t}{AC}$$

and

$$\sin r = \frac{AD}{AC} = \frac{v_2 t}{AC},$$

so that

$$\frac{\sin i}{\sin r} = \frac{v_1}{v_2}. \qquad (29\text{–}1)$$

The ratio of the sines of the angles of incidence and refraction is equal to the ratio of the speeds of light in the two media. Equation (29–1) is known as *Snell's law* after its discoverer, the seventeenth-century Dutch astronomer Willebrord Snell.

The ratio between the speed of light c in free space and its speed v in a particular medium is called the *index of refraction* of the medium, the symbol for which is n. That is,

$$n = c/v. \qquad (29\text{–}2)$$

TABLE 29–1

Indexes of refraction

Substance	n	Substance	n
Air	1.0003	Glass, flint	1.63
Benzene	1.50	Glycerin	1.47
Carbon disulfide	1.63	Ice	1.31
Diamond	2.42	Quartz	1.46
Ethyl alcohol	1.36	Water	1.34
Glass, crown	1.52		

Table 29–1 is a list of the values of n for a number of substances. The greater the index of refraction, the greater the extent to which a light beam is deflected upon entering or leaving that medium. Since $v_1 = c/n_1$ and $v_2 = c/n_2$, where n_1 and n_2 are the indexes of refraction of the two media, Snell's law, Eq. (29–1), can be written in the alternative form

$$n_1 \sin i = n_2 \sin r. \tag{29–3}$$

29–3 Ray treatment of light

The Huygens construction is cumbersome to use. A simple although less accurate treatment of optical properties involves the definition of rays and the use of rays in the analysis of the propagation of light. A ray is defined as a curve which is perpendicular to the succession of wave fronts of a wave. Thus the ray system which corresponds to a spherical wave expanding in a homogeneous medium is a system of radial lines (Fig. 29–7) originating at the source. The system of rays corresponding to a plane wave is a set of parallel lines perpendicular to the planes.

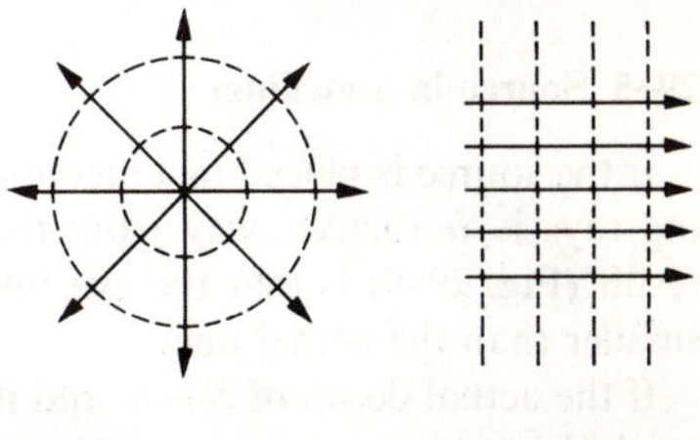

FIG. 29–7. Wave fronts (dashed lines) and rays (solid lines) in the propagation of spherical waves (at left) and plane waves.

The fundamental laws of reflection and refraction derived by Huygens' principle, expressed in ray language, are: (1) the angle of incidence is equal to the angle of reflection. The incident and reflected rays and the normal to the surface lie in the same plane. (2) Snell's law for refraction is obeyed, namely $\sin i/\sin r = n_2/n_1$ = relative index of refraction. Again the incident and refracted rays and the normal to the surface lie in the same plane.

Using the ray treatment, it is possible without undue complications to analyze some simple problems in optics. Several of these will be considered in succeeding sections.

29-4 Reflection from a mirror

If we were to place some luminous object near a reflecting surface, each point on the object would represent a point source of light. Emanating from each point, then, is a uniform distribution of rays. The eye normally intercepts a diverging bundle of these rays. The optics of the eye, plus the brain, interpret any interception of a diverging bundle of rays in terms of a source of light at the intersection of these bundles (Fig. 29-8).

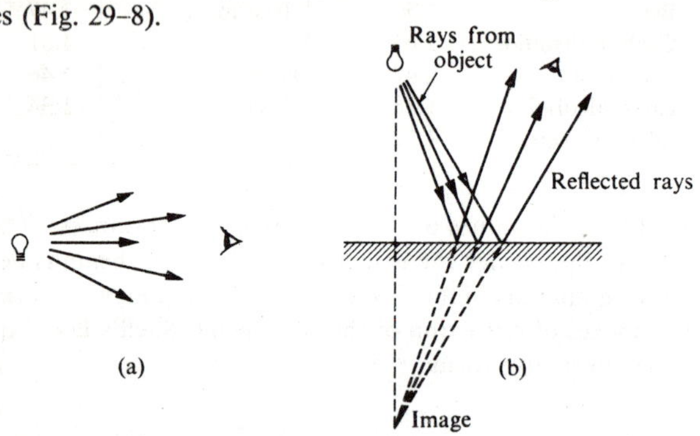

FIG. 29-8. (a) The eye interprets a bundle of diverging light rays in terms of a source located at the intersection of the rays. (b) In reflection, the eye interprets the diverging rays in terms of a source (the *image*) behind the mirror.

Some of these rays are intercepted by the specularly reflecting surface. Each ray obeys the law of reflection, namely, that the angle of incidence is equal to the angle of reflection. When this is done, the eye intercepts the bundle of reflected rays. Since they are a diverging bundle, the brain interprets this as a source arising from the intersection of the rays. Thus the eye "sees" an image of the source at a distance behind the mirror equal to its actual distance from the mirror.

29-5 Source in a medium

If the source is placed in a medium under water, for example, each of the diverging rays is refracted away from the normal, according to Snell's law. The net result (Fig. 29-9) is that the eye interprets a virtual source at an apparent depth smaller than the actual one.

If the actual depth of S is h, and if we restrict ourselves to rays which are nearly vertical, we have

$$\frac{\sin i}{\sin r} \approx \frac{\tan i}{\tan r}, \tag{29-4}$$

since $\sin \theta \approx \tan \theta$ when θ is small. From Fig. 29-9, we see that

$$\tan i = x/h; \qquad \tan r = x/h',$$

where x is the approximate distance between the source and the point O where the ray leaves the surface of the water. Hence

$$\frac{\tan i}{\tan r} = \frac{x/h}{x/h'} = \frac{h'}{h},$$

and so, from Eq. (29-4) and Snell's law, we obtain

$$h'/h = n_2/n_1. \qquad (29\text{-}5)$$

For $n_1 = 1.34$ and $n_2 = 1.00$ (the indices of refraction of water and air, respectively),

$$\frac{h'}{h} = \frac{1.00}{1.34} = 0.745.$$

The apparent depth h' is only about three-fourths of the true depth h.

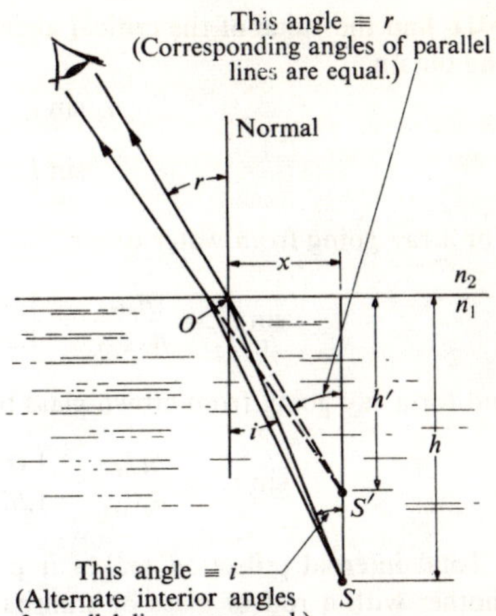

FIG. 29-9. Submerged bodies seem closer to the surface than they actually are.

29-6 Total internal reflection

An interesting phenomenon known as *total internal reflection* can occur when light passes from one medium to another which has a *lower* index of refraction, for instance, from water or glass to air. In this case the angle of refraction is greater than the angle of incidence, and a light ray is bent *away* from the normal (ray 2 in Fig. 29-10). As the angle of incidence is increased, a certain *critical angle* i_c is reached, for which the angle of refraction is 90°. The "refracted" ray now travels along the boundary between the two media and cannot escape (ray 3 in Fig. 29-10). A ray approaching the boundary at an angle of incidence exceeding the critical angle is reflected back inside the medium it came from, with the angle of reflection being equal to the angle of incidence, as in any other instance of reflection (ray 4 in Fig. 29-10).

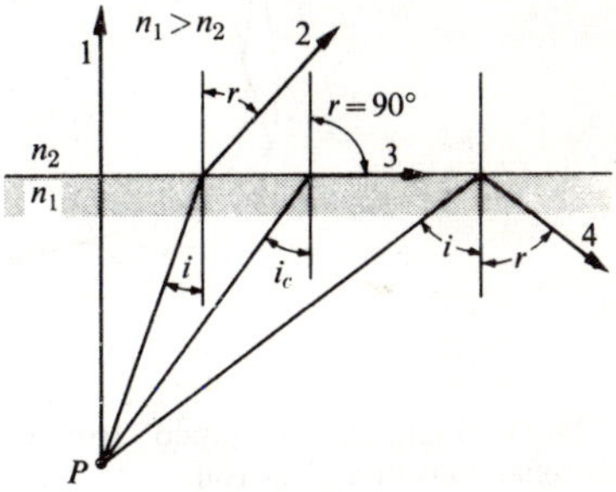

FIG. 29-10. Total internal reflection occurs when the angle of refraction of a light ray going from one medium to another of lower index of refraction equals or exceeds 90°.

To find the value of the critical angle we set $i = i_c$ and $r = 90°$ in Snell's law, and obtain

$$n_1 \sin i_c = n_2 \sin 90°,$$

$$\sin i_c = \frac{n_2}{n_1}. \tag{29–6}$$

For a ray going from water to air,

$$\sin i_c = \frac{n_{\text{air}}}{n_{\text{water}}} = \frac{1.00}{1.34} = 0.746, \qquad i_c = 48°,$$

and for a ray going from crown glass to air,

$$\sin i_c = \frac{n_{\text{air}}}{n_{\text{glass}}} = \frac{1.00}{1.52} = 0.659, \qquad i_c = 41°.$$

Total internal reflection makes it possible to "pipe" light from one place to another with a rod of glass or transparent plastic. As in Fig. 29–11, successive internal reflections occur at the surface of the rod, and nearly all the light entering at one end emerges at the other. If a cluster of narrow glass fibers is used instead of a single thick rod, an image can be transferred from one end to the other, since each fiber carries intact a part of the image.

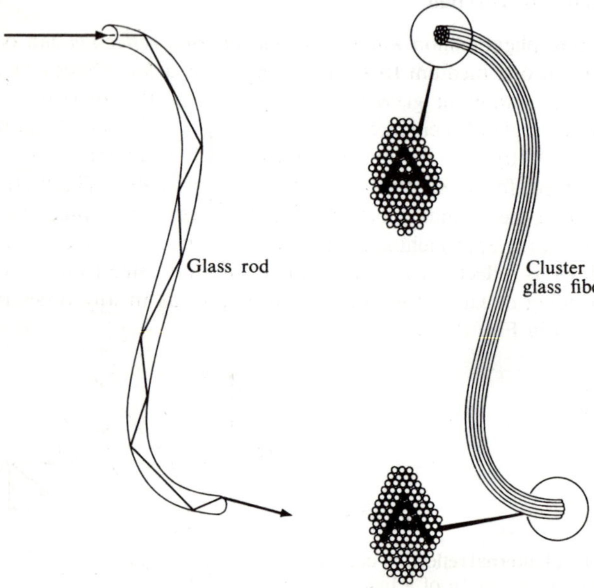

FIG. 29–11. Light can be "piped" from one place to another by means of successive internal reflections in a glass rod.

29-7 Dispersion

The index of refraction of a medium varies to some extent with the wavelength of the light involved, with the shorter wavelengths having the higher values of n. In ordinary glass, the index of refraction for violet light is about 1% greater than that for red light.

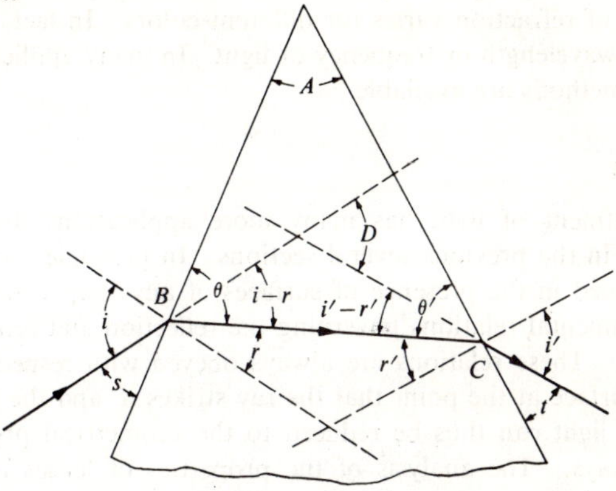

FIG. 29-12. Deviation of a ray which occurs when the ray goes through a prism.

Let us consider a ray of white light (which is a mixture of all wavelengths) that is incident upon a glass prism, as in Fig. 29-12. The angle of deviation D is given by

$$D = (i + i') - (r + r'), \tag{29-7}$$

since the exterior angle of a triangle is equal to the sum of the remote interior angles. Using simple geometry, we have

$$A + \theta + \theta' = 180°,$$

$$\theta = i - r + s, \qquad \theta' = i' - r' + t,$$

$$i + s = 90°, \qquad i' + t = 90°. \tag{29-8}$$

Therefore

$$r + r' = A. \tag{29-9}$$

Thus

$$D = i + i' - A. \tag{29-10}$$

It is helpful to consider all angles small, although the general properties of the result are independent of this assumption. In that case, since $\sin i \approx i$ and $\sin r \approx r$, Snell's law becomes

$$i = nr, \qquad i' = nr',$$

$$i + i' = n(r + r') = nA \tag{29-11}$$

and the angle of deviation is

$$D = A(n - 1). \tag{29–12}$$

Thus the greater the index of refraction, the more a ray would be deviated. A prism can be used to disperse light into its constituent colors or wavelengths, since the index of refraction varies for different colors. In fact, prisms are used to measure the wavelength or frequency of light. In many applications, however, more accurate methods are available.

29–8 Light rays

The ray treatment of light has many more applications than those of the examples given in the previous several sections. In principle the propagation of light can be traced in the presence of surfaces of any shape: All we must do is apply the fundamental relations governing the reflection and refraction of waves at each surface. These relations are always obeyed with respect to the normal drawn to the surface at the point that the ray strikes it, and the problems of the propagation of light can thus be reduced to the geometrical problems involved in tracing the rays. The analysis of the properties of lenses in focusing light belongs in the categories of ray tracing. While this is an interesting and technologically important subject, we shall not consider it in this chapter, but in an appendix.*

The analysis of the propagation of light by ray tracing does not yield the diffraction effects alluded to in the beginning of this chapter. Since the diffraction effects are small (recall that the shadow boundaries seem sharp), we may conclude that the rays include the gross properties of light propagation. It is a good approximation, because the wavelength is so small compared with macroscopic dimensions. The finer effects of diffraction and interference require us to make a more careful analysis, which we shall give later.

If one imagined that the energy of the light wave were confined to the rays, then the propagation of light would be practically indistinguishable from the propagation of particles. The real difference between wave motion and particle motion has to do with the localization of the energy of the latter compared with the former. One can, therefore, conclude that the shorter the wavelength of waves, the more the energy is localized and the greater the resemblance to particle motion.

For many years the nature of light was not understood. Because diffraction effects could not be observed, many people believed that light is a particle phenomenon. The properties of light on reflection would be the same whether it were a particle or a wave motion. The properties of light when refracted could be explained on the basis of a wave phenomenon, provided the speed of light in

---

* A brief introduction to ray tracing is given in Appendix I, which is on the subject of geometrical optics.

a dense medium were less than the speed of light in a vacuum. In order to explain the experimental results of refraction on the basis of particle motion, one would have to say that at the interface the component of the force acting on the "light particle" parallel to the interface would be zero. But the component perpendicular to the interface would be in the direction of the denser medium. This would require the component of velocity in the direction perpendicular to the denser medium to increase, and therefore the net speed in the denser medium would be larger (Fig. 29–13). One could then, in principle, make a crucial experiment to determine whether light was a particle or wave motion by comparing its speed in a dense medium with its speed in vacuum or air. This is the problem we now consider.

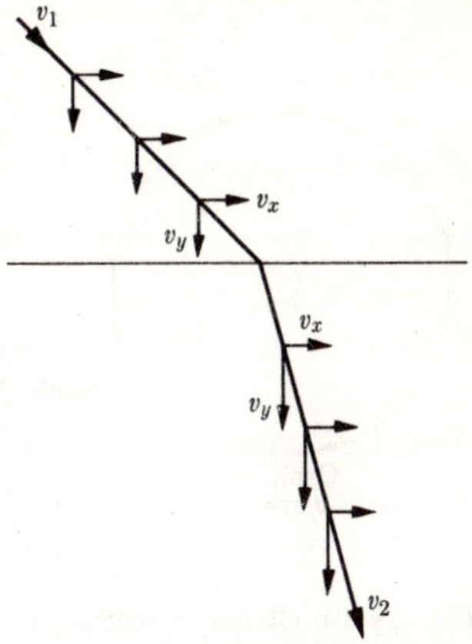

FIG. 29–13. The refraction of light according to the particle picture; v_x remains constant at the interface, but v_y increases.

29–9 The velocity of light

It takes a certain amount of sophistication even to think of light as having a finite velocity. Our senses alone do not indicate otherwise—there is no delay evident between turning on a lamp and seeing the light from it, as there is, for example, between seeing a bat strike a ball and actually hearing the blow. To be sure, even if the velocity of light is not infinite, it must be exceedingly large, and its measurement is not easy.

In determining velocity, a distance and a time interval must be measured in a single experiment. For a high velocity either the distance must be large or the time interval very small. Accurately evaluating small time intervals is a relatively recent accomplishment of science. Earlier ways of finding the velocity of light made use of astronomical methods in which immense distances were involved.

The first observation of the infinite speed of light was performed by Ole Römer in 1675. At the time he was investigating the period of revolution of one of Jupiter's satellites by timing successive eclipses of the satellite. He found that, in the course of the earth's annual motion around the sun (Fig. 29–14)—during which Jupiter effectively remains in the same position owing to its 12-year period—the predicted time of a specific eclipse was about 20 minutes earlier when the earth was nearest to Jupiter than when it was farthest away. Römer interpreted this discrepancy as resulting from the finite time required for light to traverse the earth's orbit; he thus provided the evidence that light travels with a finite speed between 100,000 and 200,000 mi/sec. Today more precise techniques show that the time needed for light to cross the orbit is 16.67 min, and, since this is 10^3 sec and the

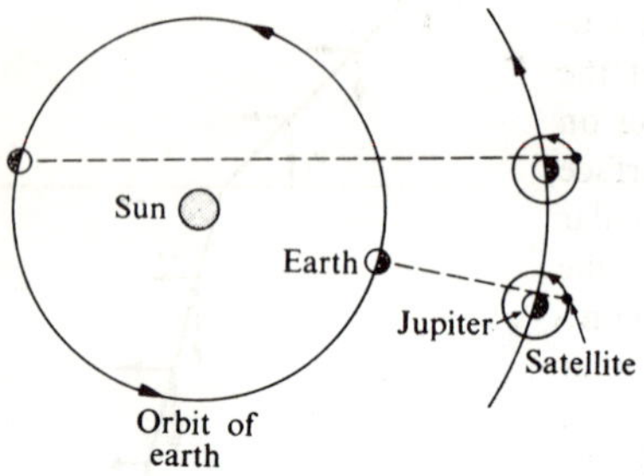

FIG. 29-14. Römer's method for measuring the speed of light.

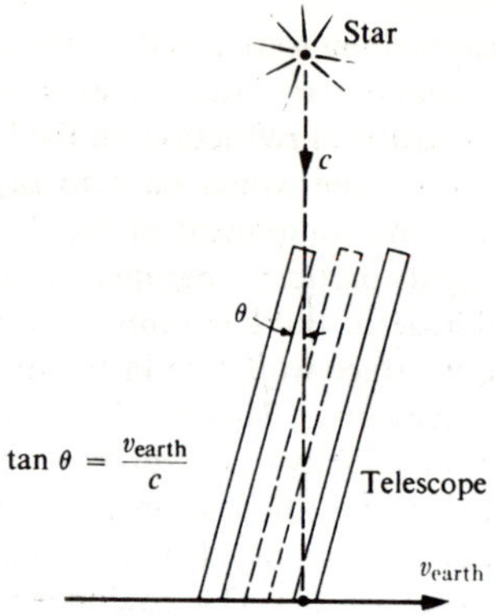

FIG. 29-15. Bradley's method for determining the velocity of light.

orbit has a diameter of 1.86×10^8 mi, the velocity of light becomes 186,000 mi/sec or 3×10^8 m/sec.

Römer's interpretation was not accepted by his contemporaries, who clung to the notion that light has an infinite speed. Not until independent confirmation came fifty years later, as a result of James Bradley's work, was his idea finally believed. What Bradley did was to notice that the apparent direction from earth to any star changes slightly in the course of the year. He was unable to explain this effect, called *aberration*, until one day when, in a boat on the Thames, he observed that the wind's direction relative to the boat varied with the speed of the boat and the direction in which it was headed. What was happening then became clear: as in Fig. 29-15, the telescope is not pointed directly at the star because, as the starlight moves through the telescope tube, the latter is also moving, by virtue of the earth's orbital motion. Just as a shotgun must be aimed a little in front of a moving bird, so must the light reaching the earth from a certain star have originally been directed a little ahead of the moving earth.

As the diagram shows,

$$\tan\theta = \frac{v_{\text{earth}}}{c}, \tag{29-13}$$

where θ is the angle between the true and apparent directions of the star (which is half the total angle through which the star seems to shift during the year), v_{earth} the earth's orbital velocity, and c the velocity of light. The value of θ is about 10^{-4} rad, and so $\tan\theta$ has this value almost exactly also. Therefore

$$c = 10^4\, v_{\text{earth}}.$$

Since the velocity of the earth in its orbit is close to 3×10^4 m/sec, the velocity of light by Bradley's method turns out to be 3×10^8 m/sec.

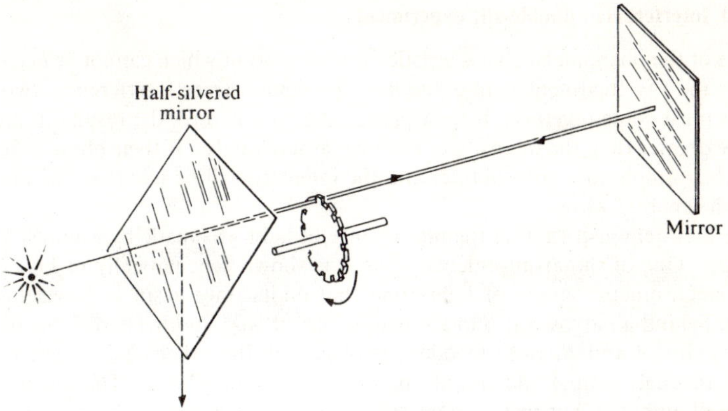

FIG. 29-16. Fizeau's method for measuring the velocity of light.

One thing we may note about Bradley's interpretation is that, by the vector diagram of Fig. 29-15, the velocity of light should be the resultant of **c** and $\mathbf{v}_{\text{earth}}$, or greater than c by itself. According to the theory of relativity, this is not possible. Another discrepancy is that the theory given above holds only for a particle model of light; according to a rigorous analysis by classical wave theory, aberration should not exist. It is only by combining relativity and the wave theory that aberration emerges and Bradley's result becomes really justified.

The astronomical measurements of c do not directly contribute to a distinction between the wave and particle theories of light. A completely terrestial method is required, in order that media of various kinds be traversed by the light, and this, in turn, requires an accurate timing device.

In 1849 Fizeau devised the scheme shown in Fig. 29-16. The notched wheel had approximately 200 teeth along its periphery. Light from the source went through one of the spaces between the teeth and was reflected by a mirror several miles away. If the wheel were stationary, the returning light passed back through the space and was observed by means of a half-silvered mirror. When the wheel was set rotating, at first the returning light was still visible, but at a sufficiently high speed the returning beam disappeared. This meant that, in the time it took the light to travel to the mirror and return, the wheel had turned so that the reflected light was blocked by one of the teeth. Knowing the angular velocity of the wheel, the number of teeth, and the distance of the mirror, the observer could determine c. Modifications of Fizeau's method ultimately made possible the direct evaluation of the velocity of light in material media. The results showed that such velocities were less than the velocity of light in vacuum c, thereby substantiating the wave picture of light. By the time the wave picture was established on this basis, other experiments demonstrating interference phenomena had already made the wave picture the accepted one.

29–10 Interference: double-slit experiment

One of the phenomena characteristic of wave motion which cannot be accounted for by the ray treatment is interference. To demonstrate interference, two light beams must come together. If their phase difference is zero, the resultant intensity will be greater than the intensity of each beam separately. If their phase difference is π, their amplitudes will subtract and the intensity will be less than the intensity of each separate wave.

The first demonstration of the interference of light was made by Thomas Young in 1801. One of the arrangements he used is shown schematically in Fig. 29–17. A monochromatic source of light (one that emits only a single wavelength) is placed behind a narrow slit S in an opaque screen, and another screen having two narrow slits A and B, close together, is placed on the other side. Light from S passes through both A and B, and then to the viewing screen. The screen is not evenly illuminated, but instead consists of a regular pattern of alternate light and dark lines.

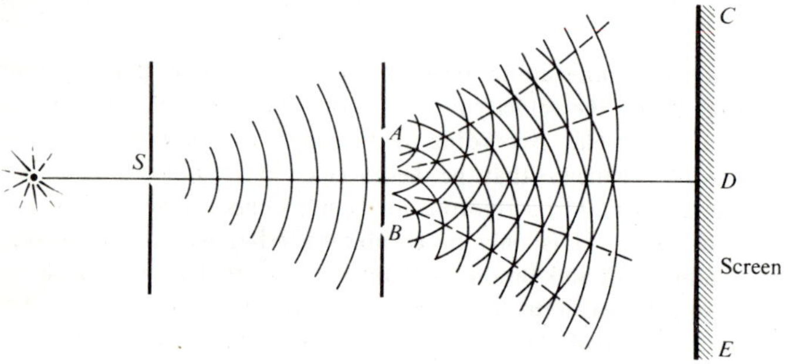

FIG. 29–17. Young's interference experiment.

The interference of light waves readily accounts for the pattern of alternate light and dark lines observed on the screen in Young's double-slit experiment. We see in Fig. 29–18 that light waves from slits A and B travel identical distances in reaching the central position S' on the screen, and consequently they interfere constructively there. The reason for having the light reaching A and B come from the same source S is now evident. Owing to this arrangement a light wave leaving A is always in exactly the same part of its cycle as that leaving B at the same time, so that when a crest leaves A a crest also leaves B, when a trough leaves A a trough also leaves B, and so on. A and B are said to be *coherent* sources of light. When waves from the two slits meet at S' they therefore reinforce one another to produce a bright line. If different sources of light were used to illuminate A and B, the light waves leaving A and B would be independent of one another, and when they met at S' they would usually be in different parts of their cycles. Such sources are called *incoherent*. No interference patterns of light and dark lines would result, although diffraction at the slits would cause the screen to be

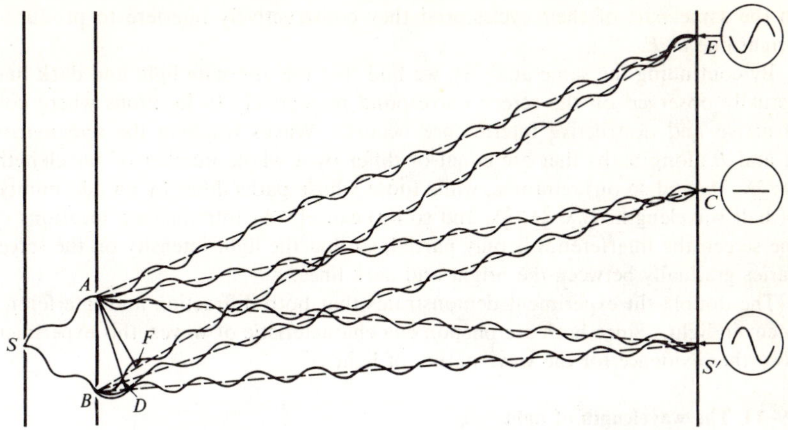

FIG. 29–18. The origin of the double-slit interference pattern.

weakly illuminated. Until recently, interference phenomena were only possible when a single original light beam was split into two or more parts which traveled along different paths and later recombined.

However, the development of a device called a *laser* has made possible a coherent source of light of large aperture. Interference phenomena can now be demonstrated with beams from two different lasers.

Let us examine what happens at the position C on the screen located above S'. The distance BC is longer than the distance AC by the amount BD, as shown in Fig. 29–18. Suppose that BD equals exactly half a wavelength of the light being used, that is, that

$$BD = \tfrac{1}{2}\lambda.$$

When a crest from A reaches C, this difference in path length means that a trough B arrives there at the same time, since $\frac{1}{2}\lambda$ separates a crest and a trough in the same wave. The two cancel each other out, the light intensity at C is zero, and a dark line results on the screen there. At S' the equality of path length gives rise to constructive interference; at C the difference of $\frac{1}{2}\lambda$ in path length gives rise to destructive interference.

If we go past C on the screen we come to a point E such that the distance BE is greater by exactly one wavelength than the distance AE. That is, the difference BF between BE and AE is

$$BF = \lambda.$$

Consequently, when a crest from A reaches E, a crest from B also arrives there, although the latter crest left B earlier than that from A, owing to the longer path it had to cover. Because $BF = \lambda$, waves arriving at E from both slits are always in the same part of their cycles, and they constructively interfere to produce a bright line at E.

By continuing the same analysis, we find that the alternate light and dark lines actually observed on the screen correspond respectively to locations where constructive and destructive interference occurs. Waves reaching the screen from A and B along paths that are equal or differ by a whole number of wavelengths (λ, 2λ, 3λ, and so on) reinforce, while those whose paths differ by an odd number of half wavelengths ($\frac{1}{2}\lambda$, $\frac{3}{2}\lambda$, $\frac{5}{2}\lambda$, and so on) cancel. At intermediate locations on the screen the interference is only partial, so that the light intensity on the screen varies gradually between the bright and dark lines.

The double-slit experiment demonstrates that both diffraction and interference occur in light. Since both are phenomena characteristic of waves, this experiment is further evidence for the wave nature of light.

29-11 The wavelength of light

Interference patterns like those produced by the double slit permit us to determine the wavelength of the light used. To illustrate the procedure, we shall consider the double-slit experiment quantitatively. Figure 29-19 is a diagram of the experiment: d is the separation between the slits, L is the distance from the slits to the screen, and y is the distance from the central point S' on the screen to the point Q whose illumination we are observing. Waves traveling from slit B to Q must travel s further than those from slit A. As we have seen, when the path difference s is

$$s = 0, \lambda, 2\lambda, 3\lambda, \ldots, \tag{29-14}$$

where λ is the wavelength of the light from the source, waves from A and B arrive at Q in the same stage of their cycles and reinforce one another to produce a bright line there. When the path difference s is

$$s = \tfrac{1}{2}\lambda, \tfrac{3}{2}\lambda, \tfrac{5}{2}\lambda, \ldots, \tag{29-15}$$

on the other hand, waves from A and B arrive at Q in the opposite stages of their cycles and cancel one another out to produce a dark line there. The triangles

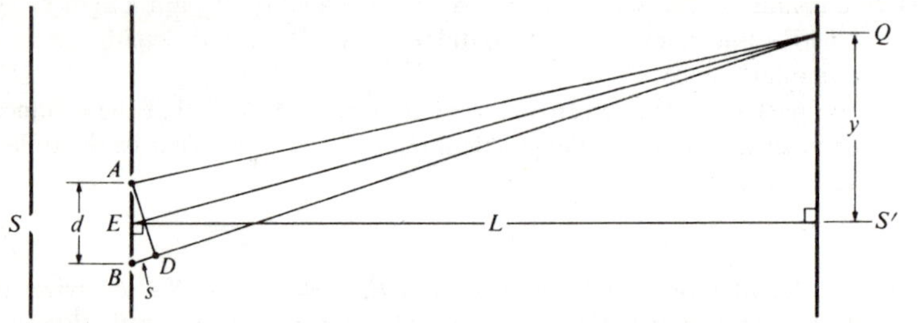

FIG. 29-19. A diagram of the double-slit experiment.

ABD and *QES'* are similar, since each is a right triangle and two sides of each are perpendicular to two sides of the other. Corresponding sides of similar triangles are proportional, and so

$$\frac{s}{y} = \frac{d}{EQ}. \tag{29–16}$$

In an actual experiment, y is very much smaller than L (their ratio is usually less than 1:1000; in Fig. 29–19 y is exaggerated for clarity), which means that EQ is almost exactly equal to L. At the expense of introducing a negligible error, we may substitute L for EQ in Eq. (29–16), with the result that

$$\frac{s}{y} = \frac{d}{L},$$

$$s = \frac{dy}{L}. \tag{29–17}$$

Combining Eqs. (29–14) and (29–15) with Eq. (29–17) yields the conditions for bright and dark lines to occur at Q:

$$\text{Bright lines:}\quad y = 0,\ \frac{L\lambda}{d},\ \frac{2L\lambda}{d},\ \frac{3L\lambda}{d},\ \ldots \tag{29–18}$$

$$\text{Dark lines:}\quad y = \frac{L\lambda}{2d},\ \frac{3L\lambda}{2d},\ \frac{5L\lambda}{2d},\ \ldots \tag{29–19}$$

According to Eq. (29–18), there is a bright line when $y = 0$, corresponding to the center of the screen S', a bright line on either side of S' a distance $L\lambda/d$ from it, another bright line on either side a distance $2L\lambda/d$ from S', and so on (Fig. 29–20). Similarly, Eq. (29–19) states that there is a dark line on either side of S'

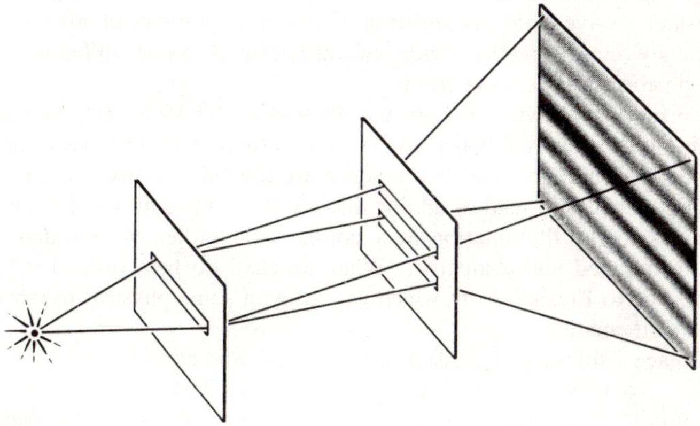

FIG. 29–20. The illumination of the screen in Young's double-slit experiment is not uniform, but consists of alternate bright and dark lines.

a distance $L\lambda/2d$ from it, another dark line on either side a distance $3L\lambda/2d$ from it, and so on.

In an actual experiment, L and d are known initially, and the distance y from the center of the screen to any particular light or dark line can be measured. The wavelength of the light may then be computed by using the equation corresponding to the particular line. As an example, let us consider monochromatic yellow light from a sodium vapor lamp illuminating two narrow slits 1 mm apart. The viewing screen is 1 meter from the slits, and we find that the distance from the central bright line to the bright line nearest it is 0.589 mm. Hence

$$y = \frac{L\lambda}{d},$$

$$\lambda = \frac{yd}{L} = \frac{5.89 \times 10^{-4}\ \text{m} \times 10^{-3}\ \text{m}}{1\ \text{m}} = 5.89 \times 10^{-7}\ \text{m}.$$

The wavelength of the light from the lamp is 5.89×10^{-7} m. Since the wavelength and frequency of a wave are related to its speed by the formula

$$c = f\lambda,$$

the frequency corresponding to the above wavelength is

$$f = \frac{c}{\lambda} = \frac{3 \times 10^{8}\ \text{m/sec}}{5.89 \times 10^{-7}\ \text{m}} = 5.09 \times 10^{14}\ \text{sec}^{-1}.$$

29–12 Fresnel diffraction

In interference phenomena, we consider the superposition of two wave fronts. For diffraction phenomena, we must consider the propagation of a single wave front in the presence of obstacles. The simplest method for analyzing the latter is to consider a wave front as consisting of an infinite number of effective sources, infinitesimally close together with an infinitesimal phase difference between adjacent positions of the wave front.

Because of the behavior of light as represented by Huygens' principle, such a distribution of sources actually results from various parts of a wave front interfering with one another. This is a much more difficult situation to analyze than interference, because instead of adding directly the contributions of a few discrete light sources to the illumination at a point, some rather complicated integrals must be formulated and evaluated. What we shall do here instead is present a treatment, due to Fresnel, from which we can gain some physical insight without elaborate mathematics.

Let us place a source of light at S in Fig. 29–21, and consider the light reaching the point P from the particular wave front shown. The distance of P from the wave front is R. We now draw cones from P to the wave fronts. The slant height of each cone is successively $\lambda/2$ longer than the preceding one. That is, $R + \lambda/2$, $R + 2\lambda/2$, $R + 3\lambda/2$, etc.

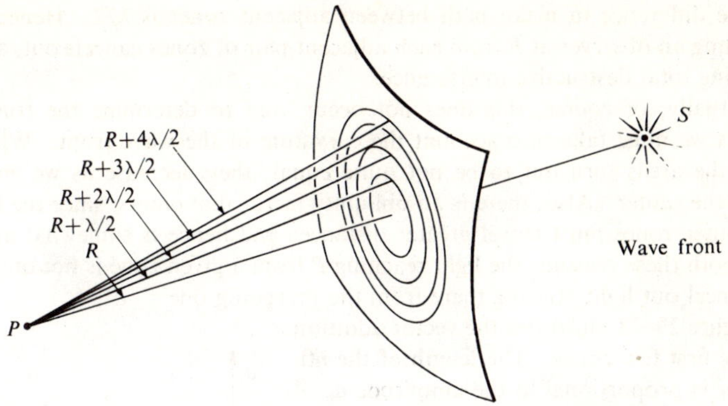

FIG. 29–21. Fresnel zones. The wave front is curved.

For the moment let us suppose that, instead of being curved, the wave front is a plane (Fig. 29–22). We now proceed to calculate the contribution of each ring to the light received at P. This contribution is proportional to the area of the ring. The central circle has the radius x_1, given by

$$x_1^2 = (R + \lambda/2)^2 - R^2,$$

$$x_1 = \sqrt{[R + (\lambda/2)]^2 - R^2} = \sqrt{R\lambda + (\lambda^2/4)}.$$

When the wavelength λ is small compared with R, which we may reasonably assume to be true,

$$x_1 = \sqrt{R\lambda},$$

and the area of the first zone is

$$A_1 = \pi x_1^2 = \pi R\lambda.$$

The second zone has the area

$$A_2 = \pi x_2^2 - \pi x_1^2.$$

Since

$$x_2^2 = \left(R + 2\frac{\lambda}{2}\right)^2 - R^2 \approx 2R\lambda,$$

this area is

$$A_2 = 2\pi R\lambda - \pi R\lambda = \pi R\lambda.$$

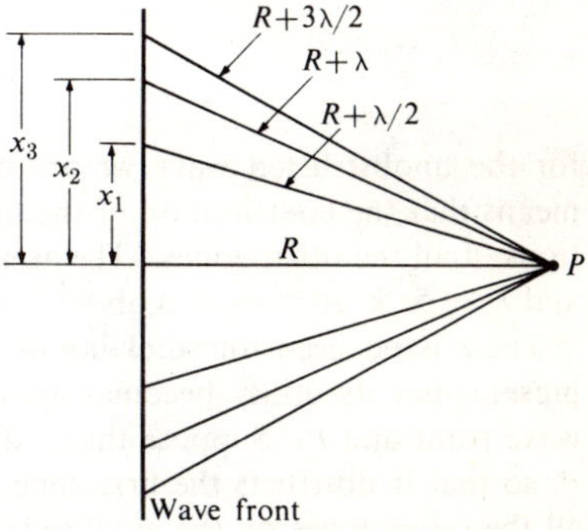

FIG. 29–22. Construction of Fresnel zones for a plane wave front.

In an approximation in which the curvature of the wave front is neglected, then, each zone on the wave front has the same area.

The difference in mean path between adjacent zones is $\lambda/2$. Hence the light reaching an observer at P from each adjacent pair of zones cancels out, apparently yielding total destructive interference.

Actually, of course, this does not occur, and to determine the true state of affairs we must take into account the curvature of the wave front. When we do this, the areas turn out to be not quite equal; they decrease as we go outward from the center. Also, there is an obliquity factor that enters, since the light from the outer zones must travel greater distances and hence is somewhat attenuated. For both these reasons, the light reaching P from a given zone is not quite enough to cancel out light arriving there from the preceding one.

Figure 29–23 illustrates the vector addition of the first few zones. The length of the nth vector is proportional to the amplitude a_n of the light reaching P from the nth zone, and its direction indicates the average phase of light from that zone. (The various vectors should, of course, be superimposed, but they are shown spread out for clarity.) The amplitude of the light reaching P is therefore

$$A = a_1 - a_2 + a_3 - a_4 + a_5 - \cdots \tag{29–20}$$

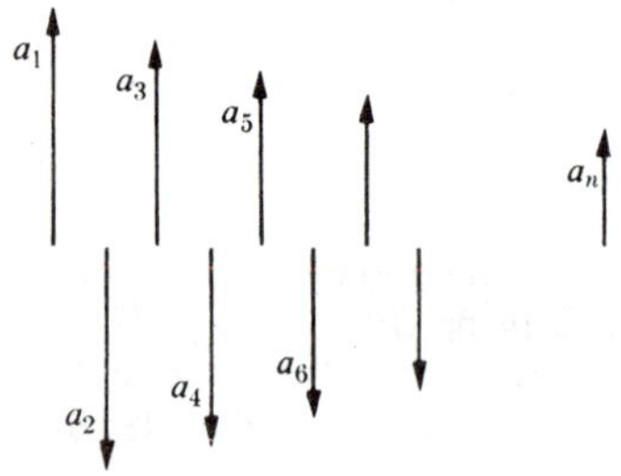

FIG. 29–23. Addition of amplitudes from various Fresnel zones.

The decrease in the size of the vectors a_i is gradual, and does not favor terms of either sign. When it is worked out, the value of A comes out to be exactly

$$A = \frac{a_1}{2}$$

for the unobstructed wave we are dealing with. This perhaps surprising result means that the contribution of the first zone is twice as large as the net contributions of all the other zones. The area of a zone can be computed: when $R = 1$ m and $\lambda = 5 \times 10^{-7}$ m, it is about 5×10^{-7} m^2 (0.5 mm^2), which is rather small.

There is no need for so elaborate a procedure when there are no obstructions present, but its utility becomes apparent when we place something between the wave front and P. Suppose that a disc is interposed between the wave front and P, so that it obstructs the first zone (Fig. 29–24). Now, as we saw, the effect of all the other zones on the amplitude is equal to $-a_1/2$, which, together with the contribution a_1 of the first zone, gave a total amplitude at P of $+a_1/2$ for the amplitude of the light reaching P. (The negative sign is not significant for the illumination of P, since light intensity is proportional to the square of the amplitude.) Thus we have the striking result that introducing an obstruction into the middle of the light path from S to P has no effect on the light intensity reaching P! If light were not a wave and subject to diffraction, no radiation whatever could get from S to P.

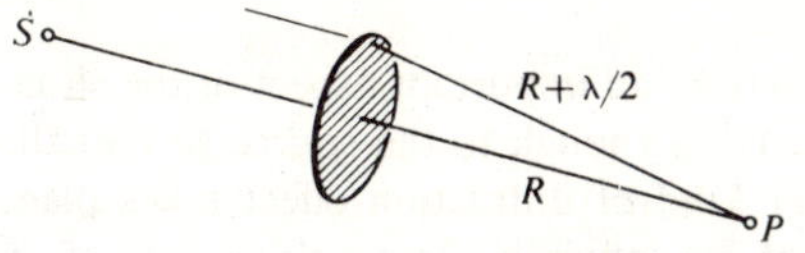

FIG. 29-24. An obstruction of one Fresnel zone keeps light arriving at *P* undiminished.

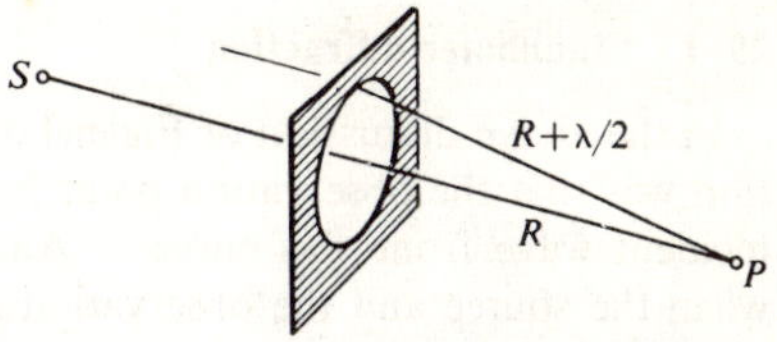

FIG. 29-25. An opening of one Fresnel zone in a screen results in the light intensity at *P* being four times greater than in the absence of any obstruction.

Of course, in order to obtain the above result, we must be sure that the only light removed from the beam is that which comes from the first Fresnel zone. Let us try the inverse procedure, and block off light from all the zones except the first one (Fig. 29-25). Now the amplitude reaching P is simply $+a_1$, so that, since intensity goes as A^2, the illumination at P is four times greater than it was when there were no obstructions at all! It is impossible to reconcile rectilinear propagation of light with these effects, which can be experimentally demonstrated.

Let us continue with these paradoxical diffraction effects. If the opening in the obstruction of Fig. 29-25 were such as to permit light from only the first two zones to reach P, since the amplitudes of light from these zones are very nearly equal but with opposite phase, they would almost completely cancel out and virtually no light at all would get to P. However, by making the hole large enough so that the light from many zones can get through, the amplitude would be very close to $a_1/2$, because the outer zones have relatively little effect when summed up. To see how large an opening is required, 20 zones are sufficient for allowing an amplitude of almost $a_1/2$ to reach P; for a wavelength of 5×10^{-7} m and a distance of 1 m, the area of the opening would have to be about $\frac{1}{3}$ cm^2. Diffraction effects cannot be distinguished with holes of this size or larger, and in fact are almost impossible to discern unless the hole is considerably smaller.

When we analyze the effect of obstacles and openings of any shape on the propagation of light, we find diffraction effects noticeable only in the immediate vicinity of the edges. In a region several wavelengths wide around the edges, rapid variations in intensity occur (Fig. 29-26), indicating the wave nature of light; but elsewhere, because of the smallness of the Fresnel zones, light seems to propagate rectilinearly. (Note that diffraction cannot be observed unless a point or slit source of light is used, since otherwise the finite width of the source means that the various light and dark fringes overlap and are effectively smeared out.)

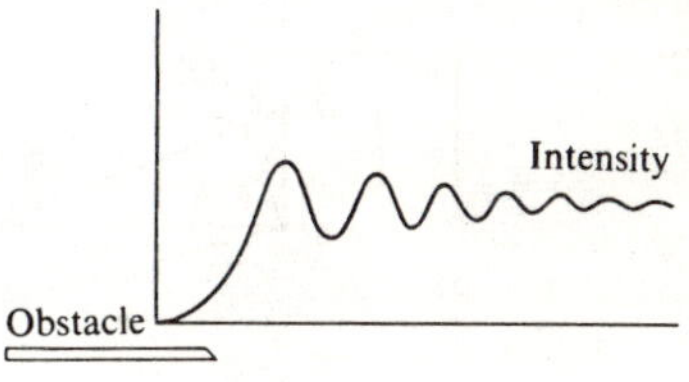

FIG. 29-26. Intensity in the shadow boundary for light incident upon an obstacle.

29–13 Fraunhofer diffraction

In the above discussion of Fresnel diffraction, an important aspect of the situation was that the observation point P was close enough to the source so that the incident wave front was curved. Another kind of diffraction effect takes place when the source and the observation point are infinitely far on either side of an obstacle. Then all the light rays travel the same distance, and only phase differences, not amplitude differences, are involved. In practice the condition that source and observation point be infinitely far from an obstacle is met with the help of lenses, which can make a spherical wave from a plane wave and vice versa (Fig. 29–27). (See Appendix I.)

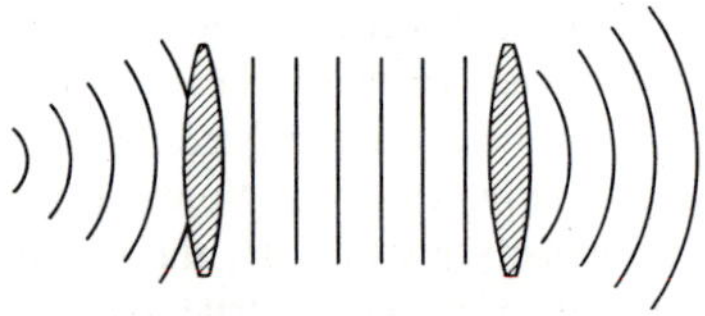

FIG. 29–27. Lenses can convert spherical wave fronts into plane wave fronts, and vice versa.

Let us work out the Fraunhofer diffraction pattern produced by a single slit. In Fig. 29–28, we have a slit of width b upon which plane waves are incident from the left: the emerging light is focused by the lens on a screen some distance away. The light reaching the point P from the slit AB has not all come the same distance from AB. As we go up the slit from the center at O, each succeeding ray must travel a slightly longer distance, while the reverse is true as we go down from O. A wave originating a distance y above O must go the distance

$$B'C' = y \sin\theta \tag{29–21}$$

further than a ray originating at O. The phase angle ϕ at P between rays starting from B' and from O corresponding to the path difference $B'C'$ is

$$\phi = \frac{2\pi B'C'}{\lambda} = \frac{2\pi y \sin\theta}{\lambda}, \tag{29–22}$$

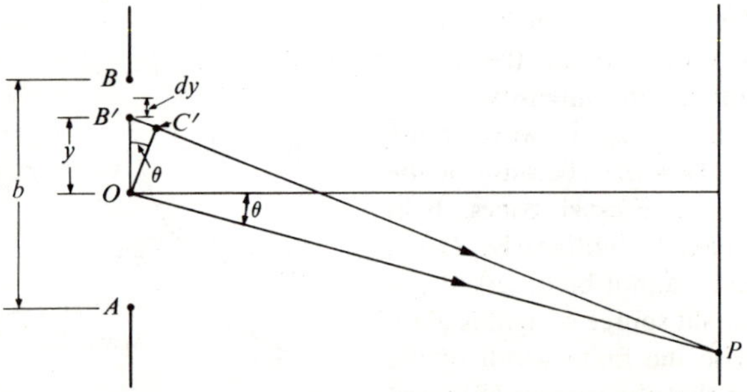

FIG. 29–28. Calculation of Fraunhofer diffraction pattern of single slit.

and the phase difference between rays starting from A and from B is

$$\beta = \frac{2\pi b \sin\theta}{\lambda}. \tag{29-23}$$

What is the amplitude dF of the light reaching P from the infinitesimal portion dy of the slit at B'? If the light has the angular frequency ω corresponding to the wavelength λ,

$$dF = G\,dy \sin(\omega t - \phi) = G\,dy \sin\left(\omega t - \frac{2\pi y \sin\theta}{\lambda}\right), \tag{29-24}$$

where G is a constant appropriate to the particular experiment. In terms of β,

$$dF = G\,dy \sin\left(\omega t - \frac{y\beta}{b}\right). \tag{29-25}$$

In order to find the amplitude F of all of the light from the slit that reaches P, we must integrate dF from $y = -b/2$ to $y = +b/2$, so that

$$F = G\int_{-b/2}^{b/2} \sin\left(\omega t - \frac{y\beta}{b}\right) dy = \frac{Gb}{\beta}\cos\left(\omega t - \frac{y\beta}{b}\right)\Bigg|_{-b/2}^{b/2}$$

$$= \frac{Gb}{\beta}\left[\cos(\omega t - \beta/2) - \cos(\omega t + \beta/2)\right] = Gb\,\frac{\sin\beta/2}{\beta/2}\sin\omega t. \tag{29-26}$$

Equation (29-26) gives the amplitude at P of light reaching there from the slit. As the position of P is varied on the viewing screen, F changes through the factor $(\sin\beta/2)/(\beta/2)$. In order to trace the intensity variation on the screen—in other words, the diffraction pattern—we need consider only the square of this factor,

$$J = \left(\frac{\sin\beta/2}{\beta/2}\right)^2, \tag{29-27}$$

since light intensity is proportional to the square of the wave amplitude.

In Fig. 29-29 we have plotted J against $\beta/2$. The center of the pattern occurs when $\beta/2 = 0$; that is, for

$$\beta/2 = \frac{\pi b \sin\theta}{\lambda} = 0,$$

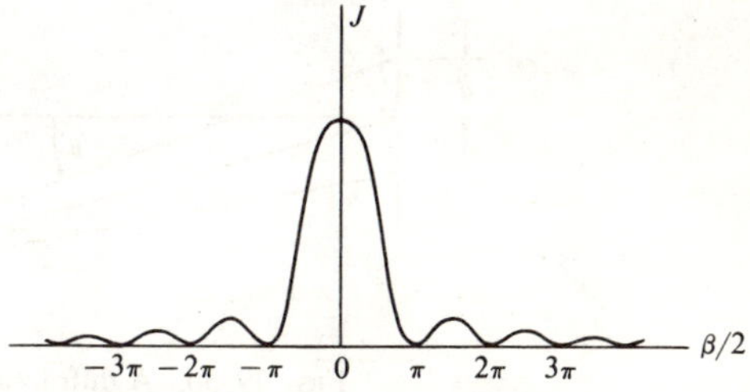

FIG. 29-29. The intensity pattern produced by Fraunhofer diffraction from a single slit. The angle β is the phase difference between rays originating at each edge of the slit.

which corresponds to $\theta = 0$. The various minima on either side of the center occur when $\beta/2 = \pm n\pi$, where n is an integer. That is, we have as minima

$$\beta/2 = \frac{\pi b \sin\theta}{\lambda} = \pm n\pi, \qquad \sin\theta = \pm\frac{n\lambda}{b}. \tag{29-28}$$

This result follows from Eq. (29-27), whose numerator is 0 whenever $\beta/2 = \pm n\pi$. (For $\beta/2 = 0$ the value of J is 1, because $\sin x$ approaches x as x approaches 0.)

The angular width of the bright central maximum is

$$\theta = 2\sin^{-1}\frac{\lambda}{b}.$$

Evidently the width of the maximum depends on b. When b is exactly one wavelength wide, $\theta = 180°$, while if it is a more practical 0.1 mm, θ is approximately $\frac{1}{2}°$.

29-14 The diffraction grating

A diffraction grating essentially consists of a large number of parallel slits, and may operate by reflection (if it is in the form of a mirror with nonreflecting lines ruled across its surface) as well as by transmission of light. To obtain Fraunhofer diffraction, a lens is placed on either side of a grating. Let us consider rays leaving corresponding points in each slit, as in Fig. 29-30. If the path differences between corresponding points on adjacent slits and the viewing point P are integral numbers of wavelengths, all the rays that arrive will be in phase and there will be a bright line on the screen. As the diagram indicates, the condition for this to occur is to have

$$b\sin\theta = n\lambda, \qquad n = 1, 2, 3, \ldots \tag{29-29}$$

as maxima. At intermediate points on the screen there will be darkness, corresponding to the cancellation of out-of-phase light rays. The diffraction pattern from a grating is a somewhat complex affair in its details, although Eq. (29-29) gives the correct positions of the bright lines.

The advantage of using a diffraction grating is that the maxima for a particular wavelength are very sharp. To understand why this should be, we note that a typical grating might have 5,000 to 12,000 lines (which act as slits) per cm. Hence

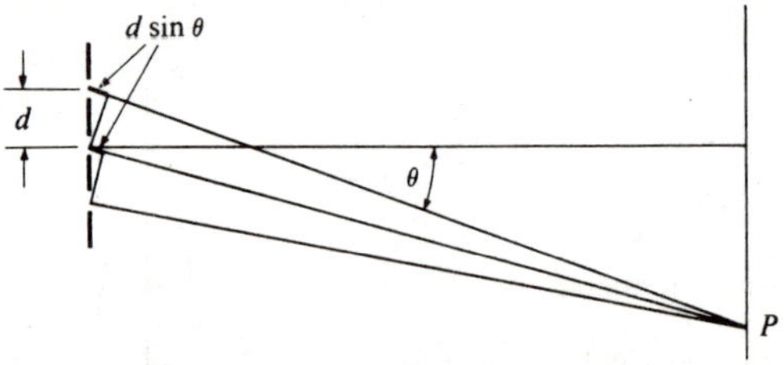

FIG. 29-30. A diffraction grating.

there is plenty of light available to illuminate each maximum well. If we now go to a position on the screen a slight distance away from a maximum, there will be only a small phase difference between rays from adjacent slits. However, since there are a great many slits, even a small phase difference will, when summed over the total number of slits N, decrease the intensity rapidly. In fact, a phase difference of $\phi = 2\pi/N$ brings the intensity to zero. Thus the greater the number of slits, the narrower the width of the bright lines.

The great advantage to having sharp lines in the image produced by a grating is that it permits us to make very precise wavelength measurements. If there are two wavelengths whose values are close together, a sufficiently closely ruled grating permits us to distinguish one from another, while some other optical device whose diffraction maxima are less sharp (for instance a single slit) would not enable us to do so. The ability of an instrument to separate small differences in wavelength is known as its *resolving power*.

29-15 Polarization

While all the experimental evidence we have examined thus far supports the view that light is a wave phenomenon, none of it tells us whether light waves are longitudinal or transverse. These classes of waves were discussed in Chapter 25, and we recall that longitudinal waves are characterized by motion of the particles of the medium parallel to the direction in which the waves travel, while transverse waves are characterized by motion of the particles of the medium perpendicular to the direction in which the waves travel (Fig. 29-31). Sound waves are longitudinal, for example, and the waves that move down a stretched string are transverse. Refraction, diffraction, and interference occur in the propagation of both longitudinal and transverse waves. The evidence that light is a transverse wave must come from other types of experiments.

A definitive distinction between longitudinal and transverse waves may be made based on their capacities to be *polarized.* A polarized beam of transverse waves is one whose vibrations occur in only a single direction perpendicular to

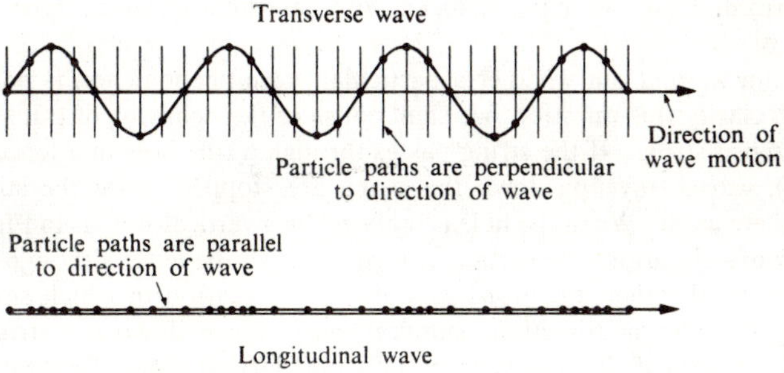

FIG. 29-31. Transverse and longitudinal waves.

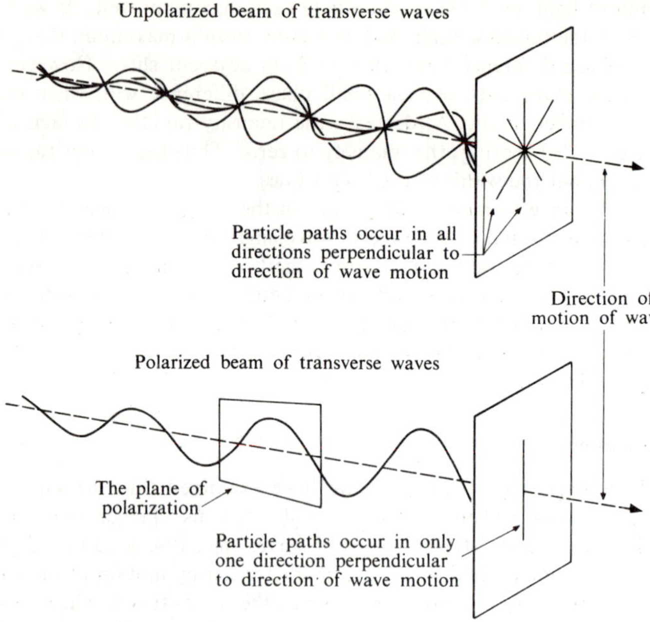

FIG. 29–32. An unpolarized and a polarized beam of transverse waves.

the direction in which the beam travels (Fig. 29–32), so that the entire wave motion is confined to a plane called the *plane of polarization.* When many different directions of polarization are present in a beam of transverse waves, vibrations occur equally often in all directions perpendicular to the direction of motion, and the beam is then said to be *unpolarized.* Since the vibrations constituting longitudinal waves can take place only in one direction, namely that in which the waves travel, longitudinal waves cannot be polarized. By establishing whether or not light can be polarized, then, we have a means for determining which type of wave it consists of.

How can we find out whether a particular wave phenomenon is polarizable or not? To clarify this question, we shall consider the behavior of transverse waves in a stretched string. If the string passes through a tiny hole in a fence, as in Fig. 29–33(a), waves traveling down the string are stopped, since the string cannot vibrate there at all. When the hole is replaced by a vertical slot, as in Fig. 29–33(b), waves whose vibrations are vertical can get through the fence although waves with vibrations in all other directions cannot. In a situation in which several waves vibrating in different directions simultaneously move down the string, the slot stops all but vertical vibrations: an initially unpolarized series of waves has become polarized. Hence we can devise a very effective method for determining whether

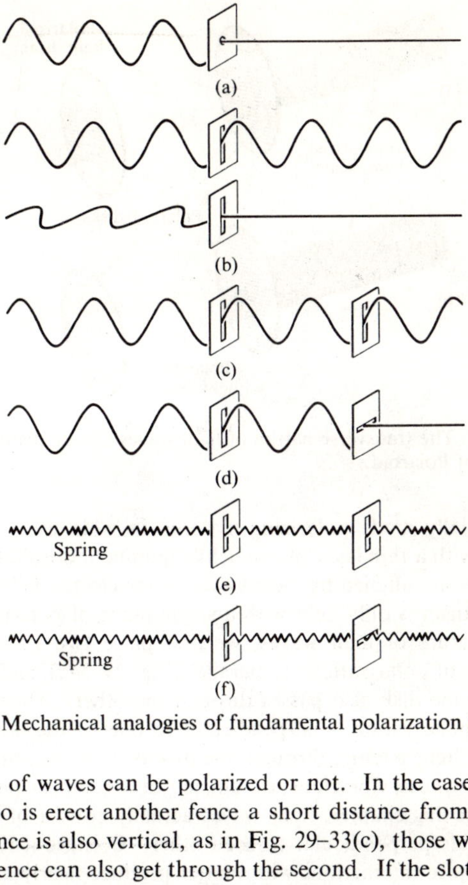

FIG. 29-33. Mechanical analogies of fundamental polarization phenomena.

a particular series of waves can be polarized or not. In the case of the stretched string, what we do is erect another fence a short distance from the first. If the slot in the new fence is also vertical, as in Fig. 29-33(c), those waves that can get through the first fence can also get through the second. If the slot in the new fence is horizontal, however, it will stop all waves that reach it from the first fence (Fig. 29-33d). Should longitudinal rather than transverse waves, say in a spring, go through the fences (Fig. 29-33e and f), it is perfectly possible that their amplitudes might decrease in passing through each slot, but the relative alignments of the slots would not matter, while it is the critical factor in the case of transverse waves.

The above chain of reasoning made it possible for the transverse nature of light waves to be established in the last century. A number of substances are known —for instance, quartz, calcite, and tourmaline—which refract differently light having different planes of polarization. When a beam of unpolarized light is incident on a properly shaped prism of such a substance, only those of its waves whose planes of polarization are parallel to a particular plane in the substance emerge from the other side. The remainder of the waves are absorbed or deflected. *Pola-*

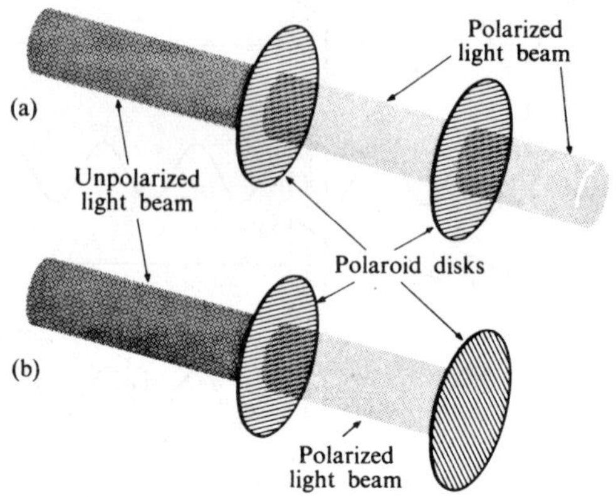

FIG. 29–34. The transverse nature of light waves can be demonstrated with two sheets of Polaroid.

roid is an artificially made polarizing material in wide use. It consists of two sheets of plastic with a thin layer of needlelike quinine iodosulfate crystals between them; the crystals are aligned by means of a strong electric field, and the resulting clear material transmits only light with a single plane of polarization. To exhibit the transverse nature of light waves, we first place two Polaroid disks in line so that their axes of polarization are parallel (Fig. 29–34a), and note that all light passing through one disk also passes through the other. Then we turn one disk until its axis of polarization is perpendicular to that of the other, Fig. 29–34(b), and note that all light passing through one disk is now *stopped* by the other.

Just what is it whose vibrations are aligned in a beam of polarized light? As we said earlier in this chapter, light waves actually consist of oscillating electric and magnetic fields perpendicular to each other. Because it is the electric fields of light waves whose interactions with matter produce nearly all common optical effects, the plane of polarization of a light wave is considered to be that in which both the direction of its electric field and its direction of motion lie. Even though nothing material moves during the passage of a light wave, then, it is possible to establish its transverse nature and to identify its plane of polarization.

It is relatively simple to demonstrate that reflected light should be polarized. Referring to Fig. 29–35, we see that unpolarized light incident on a transparent surface will give rise to a reflected and a refracted beam. The plane of vibration of the waves is perpendicular to the plane of the rays. The vibrations of the unpolarized beam can be resolved in the direction of the plane of the rays and also perpendicular to the plane of the rays (Fig. 29–35). None of the vibrations in the plane can occur in the reflected ray if the angle of the reflection and the angle of refraction are complementary, for in that case we would have a vibration parallel to the direction of motion—an impossibility for a transverse wave.

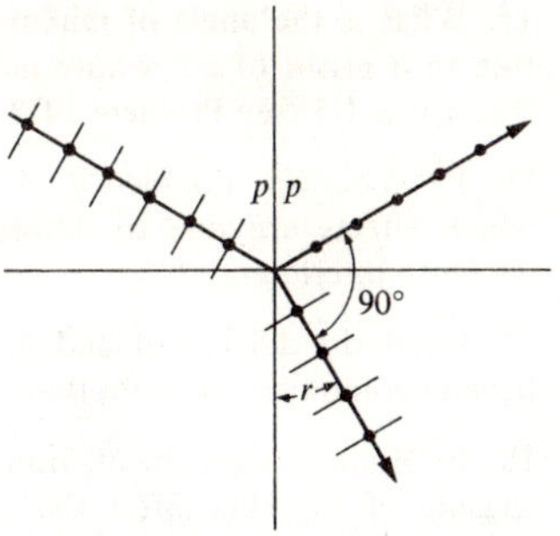

FIG. 29–35. Polarization by reflection.

The angle of incidence p for which the reflected beam is polarized can be found from Snell's law,

$$\frac{\sin p}{\sin r} = n, \tag{29–30}$$

and the fact that $p + r = 90°$. Using this condition we have, from (29–30),

$$\frac{\sin p}{\sin 90 - p} = \frac{\sin p}{\cos p} = \tan p = n. \tag{29–31}$$

The angle p which satisfies (29–31) is called *Brewster's angle*.

Problems

1. What is the index of refraction of a substance in which the speed of light is 2.3×10^8 m/sec?

2. Using the indexes of refraction given in Table 29–1 and taking the speed of light in free space as 3×10^8 m/sec, find the speed of light in (a) air, (b) diamond, (c) crown glass, (d) water.

3. *Fermat's principle* states that light travels between two points in such a manner that the transit time is a minimum. How is the phenomenon of refraction in accord with Fermat's principle?

4. The duration of daylight is extended to a small extent owing to the refraction of sunlight by the earth's atmosphere. Use a diagram to illustrate this effect.

5. The waves used to carry television signals cannot reach receivers beyond the visual horizon of their transmitting antennas, while ordinary radio waves readily travel beyond the visual horizon of their transmitting antennas. Can you think of a reason for this contradictory behavior?

6. Flint glass and carbon disulfide have almost the same index of refraction. How does this explain the fact that a flint glass rod immersed in carbon disulfide is nearly invisible?

7. Why is a beam of white light not dispersed into its component colors when it passes perpendicularly through a pane of glass?

8. Explain why a cut diamond held in white light shows flashes of color. What would happen if it were held in red light?

9. A ray of light passes through a plane boundary separating two media whose indexes of refraction are $n_1 = 1.50$ and $n_2 = 1.30$. (a) If the ray goes from medium 1 to medium 2 at an angle of incidence of 30°, what is the angle of refraction? (b) If the ray goes from medium 2 to medium 1 at the same angle of incidence, what is the angle of refraction?

10. A ray of light is incident at the angle θ on one side of a glass plate of thickness d whose sides are plane and parallel. Show that the ray that emerges from the other side of the plate is parallel to the incident ray but displaced sideways from it by a distance of $\theta d(n - 1)/n$, where n is the index of refraction of the glass. Assume that the index of refraction of air is exactly 1 and that θ is a small angle.

11. Show that when a mirror is rotated through the angle θ, a reflected ray of light is rotated through the angle 2θ.

12. A ray of light strikes a glass plate at an angle of incidence of 60°. If the reflected and refracted rays are perpendicular to each other, find the index of refraction of the glass.

13. The critical angle for total internal reflection in dense flint glass is 37°. Find its index of refraction.

14. The angle of minimum deviation is the angle of deviation of light when it passes through a prism symmetrically. Show that for the angle of minimum deviation D_m

$$n = \frac{\sin [(A + D_m)/2]}{\sin A/2},$$

where A is the angle of the prism and n is the index of refraction.

15. What is the angle of minimum deviation in a prism of 60° whose index of refraction is 1.5 (see Problem 14)?

16. In Young's double-slit experiment, which effects are due to diffraction and which to interference?

17. What do diffraction and interference have in common? How do they differ?

18. In Young's experiment, how does the spacing of the slits affect the pattern of bright and dark lines?

19. Light of unknown wavelength is used to illuminate two parallel slits 1 mm apart. Adjacent bright lines on the interference pattern that results on a screen 1.5 m away are 0.75 mm apart. What is the wavelength of the light?

20. Two parallel slits 0.12 mm apart are illuminated by light of wavelength 6×10^{-7} m. A viewing screen is 1.5 m from the slits. (a) How far from the central bright line is the next bright line? (b) How far is the first dark line? (c) How far is the fifth dark line?

21. Two parallel slits 0.1 mm apart are illuminated by light of wavelength 5.46×10^{-7} m. A viewing screen is 0.8 m from the slits. (a) How far from the central bright line is the second bright line? (b) How far is the third bright line? (c) How far is the third dark line?

22. Two parallel slits 0.2 mm apart are illuminated by light of two wavelengths, 5×10^{-7} m and 6×10^{-7} m. A viewing screen is 2 m away from the slits. How far are the bright lines of one wavelength from the bright lines of the other?

23. Two parallel slits are illuminated by light of two wavelengths, one of which is 5.8×10^{-7} m. On a viewing screen an unknown distance from the slits the fourth dark line of the light of the known wavelength coincides with the fifth bright line

of the light of the unknown wavelength. Find the unknown wavelength.

24. Which of the following can occur in (a) transverse waves and (b) longitudinal waves? Refraction, dispersion, interference, diffraction, polarization.

25. Explain the peculiar appearance of a distant light source when seen through a piece of finely woven cloth.

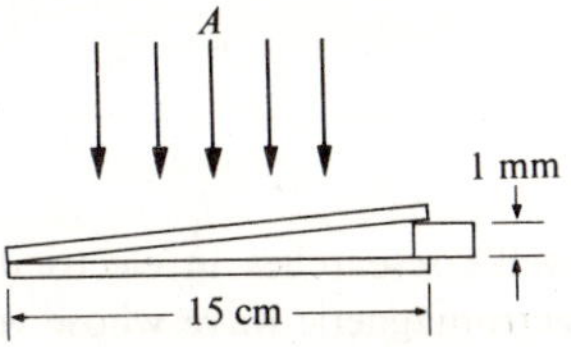

FIGURE 29–36

26. Two flat sheets of glass 15 cm long (Fig. 29–36) are separated at one end by a shim whose thickness is 1 mm. Light of wavelength 5×10^{-5} cm is normally incident. How many fringes per cm will be seen?

27. Light of wavelength 5×10^{-5} cm illuminates a circular obstacle whose diameter is 0.15 cm. The source is infinitely far away. Where along the axis will the first bright spot occur in the shadow region? What is the distance between the first and second bright spot?

28. What is the angular separation of lines whose wavelengths are 5890 A and 5896 A ($A = 10^{-8}$ cm) in second order, if a grating has 1200 lines/cm?

29. A diffraction grating ruled with 6000 lines/cm is illuminated with light of wavelengths 5.890 and 5.896×10^{-5} cm. What is the angular separation between their first order maxima? If the lines are observed 40 m from the grating, what will their separation be?

30. How many lines per cm should a grating have so that in first order the maximum of the 5.890×10^{-5} cm line will fall on the minimum of the 5.896×10^{-5} cm line? The slit width is 10^{-3} mm.

31. An idealized grating which alternates perfectly clear and perfectly opaque spaces gives a spectrum in which all the even orders are missing. What is the ratio of the width of the clear and opaque spaces?

32. Monochromatic light of wavelength 5×10^{-5} cm originates at a distant source and is observed 1 m from a screen. Find the diameter of a circular opening that would expose (a) only one Fresnel zone, (b) three Fresnel zones.

33. Calculate the diameter of a circular opening which will produce an intensity four times the unobstructed intensity at a point 70 cm away along the axis of the opening. Use the wavelength 5.890×10^{-5} cm in your calculation. What must be the diameter of the hole to give complete darkness at this point?

34. How can a single sheet of Polaroid be used to show that light from the sky is partially polarized?

35. What is Brewster's angle for glass whose index of refraction is 1.40?

30

Relativity

One of the principal consequences of Maxwell's researches in electricity and magnetism was the discovery that light is an electromagnetic wave whose velocity in free space is $c = 1/\sqrt{\mu_0 \epsilon_0}$. All velocities must be measured with respect to a specified coordinate system, and the proper one to which the motion of light should be referred is not implicit in the equation governing electromagnetic wave propagation. In the nineteenth century it was assumed that the correct coordinate system for this purpose was one stationary relative to the "fixed stars." An assumption of this kind is equivalent to suggesting that a medium exists which pervades space, and that light travels through this medium with the velocity c with respect to the medium. The hypothetical medium was called the *ether.* The earth travels through space at 3×10^4 m/sec as it rotates about the sun, and so it would seem that the earth must also move through the ether. If so, the actual existence of the ether can be detected by comparing the apparent speeds of light parallel and perpendicular to the earth's motion. This comparison formed the basis of an experiment performed in 1887 by Michelson and Morley.

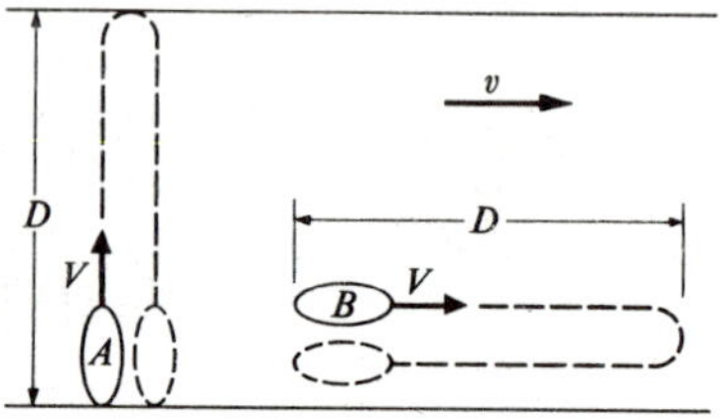

FIGURE 30–1

30–1 The Michelson-Morley experiment

The Michelson-Morley experiment is readily understandable in terms of a simple analogy. Figure 30–1 is a sketch of a river flowing with the speed v. The river's width is D. Two identical boats start out from one bank of the river with the same speed V; boat A is to cross the river to a point on the other bank directly

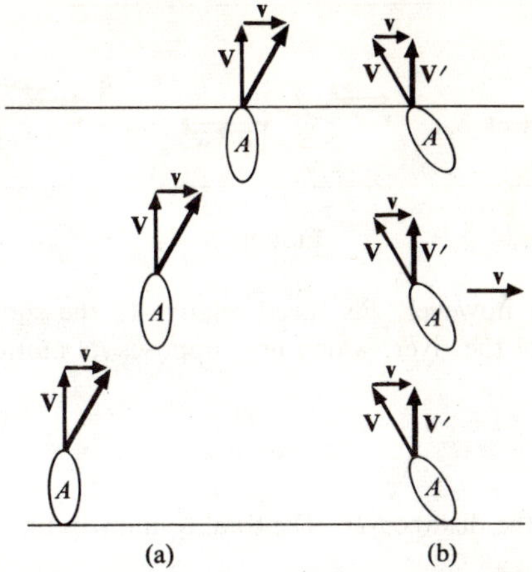

FIGURE 30–2

opposite the starting place and then to return, while boat B is to proceed downstream for the distance D and then to return to the starting place. How much time will each round trip take?

We shall first consider boat A. If A heads directly across the river (as in Fig. 30–2a), the current will carry it downstream from its destination on the other bank opposite where it started, and so it must head somewhat upstream (Fig. 30–2b) in order to compensate for the current. When it does this, it will have an upstream component of its velocity equal to exactly $-v$ in order to cancel out the downstream current v, leaving the component V' as its net speed across the river. From Fig. 30–2 we see that these speeds are related by

$$V^2 = V'^2 + v^2,$$

so that the actual speed with which it crosses the river is

$$V' = \sqrt{V^2 - v^2} = V\sqrt{1 - (v^2/V^2)}.$$

Hence the time for the crossing is the distance D divided by the speed V', and, since the reverse crossing involves exactly the same amount of time, the total round-trip time t_A is twice D/V', or

$$t_A = \frac{2D/V}{\sqrt{1 - (v^2/V^2)}}. \tag{30–1}$$

Boat B is in a different situation. As it proceeds downstream, its speed relative to the shore is its own speed V *plus* the speed v of the river (Fig. 30–3), and it travels the distance D downstream in the time

$$\frac{D}{V + v}.$$

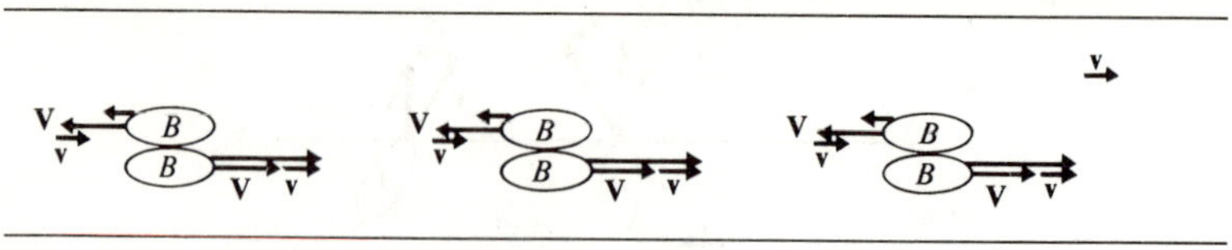

FIGURE 30–3

On the return trip, however, B's speed relative to the shore is its own speed V *minus* the speed v of the river, which now opposes its motion, and it requires the longer time

$$\frac{D}{V - v}$$

to travel upstream the distance D. The total round trip time t_B is the sum of these two, or

$$t_B = \frac{D}{V + v} + \frac{D}{V - v}.$$

Using the common denominator $(V + v)(V - v)$ for both terms, we have

$$t_B = \frac{D(V - v) + D(V + v)}{(V + v)(V - v)}$$

$$= \frac{2DV}{V^2 - v^2}$$

$$= \frac{2D/V}{1 - (v^2/V^2)}, \quad (30\text{–}2)$$

which is longer than t_A. The ratio between the times t_A and t_B is

$$\frac{t_A}{t_B} = \sqrt{1 - (v^2/V^2)}. \quad (30\text{–}3)$$

By sending out a pair of boats in this way, then, a measurement of t_A/t_B together with a knowledge of the boats' speed V enables us to calculate the speed v of the river current.

Now let us restate the above problem in terms of the passage of light waves through the ether. Instead of a current of water relative to the shore we shall have a current of ether relative to the earth as the earth moves around the sun within the conjectured universal sea of ether, and instead of a pair of boats we shall have a pair of light beams from a single source. Figure 30–4 is a schematic representation of the apparatus used in the famous experiment performed by the American physicists Michelson and Morley in 1887. As shown in the figure, parallel light strikes the half-silvered mirror A and is split into two beams, one proceeding in the original direction to the right and the other reflected upward. The former

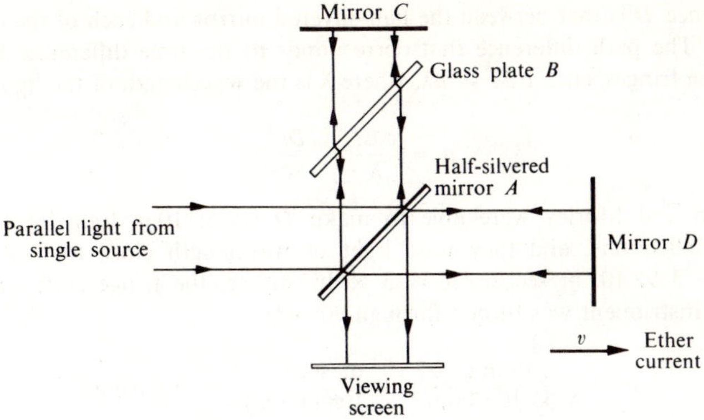

FIG. 30–4. Schematic diagram of the apparatus used by Michelson and Morley to measure the speed of the ether current past the earth.

beam is reflected by the mirror D and returns to A, where some of it is further reflected to a viewing screen. The upward beam is reflected by the mirror C and also returns to A, where part of it is transmitted to reach the viewing screen. The distances from A to C and from A to D are the same, and a glass plate B is inserted in the path of the upward beam, so that each beam passes through identical thicknesses of air and glass between its entry into the apparatus and its arrival at the viewing screen. The presence of an "ether current" as shown would result in the different beams requiring different transit times to go from the source to the viewing screen. Instead of arriving at exactly the same stage in their cycles, the light waves of the two beams would be slightly—but definitely—out of step with each other, and hence would interfere destructively to a corresponding extent.

In the actual experiment the two mirrors are not quite perpendicular, so that a series of bright and dark interference fringes appear at all times on the viewing screen owing to differences in path length between adjacent light waves. When either of the optical paths in the instrument is changed in length, the fringes appear to travel across the screen as the waves successively reinforce and cancel each other at every point. When the instrument is stationary, no information about motion through the ether can be obtained. When it is rotated by 90°, though, the two optical paths exchange orientations with respect to the motion through the ether, and the path that originally required time t_A for the round trip now takes t_B to complete, and vice versa. If t_A and t_B are different, the fringe pattern will move across the screen during the change in orientation.

It is not difficult to estimate the fringe shift to be expected if there is indeed an ether. From Eqs. (30–1) and (30–2), with V now the speed of light c and v the earth's speed relative to the ether, the time difference between the two paths is

$$\Delta t = t_B - t_A = \frac{2D/c}{1 - (v^2/c^2)} - \frac{2D/c}{\sqrt{1 - (v^2/c^2)}} \approx \frac{Dv^2}{c^3}. \qquad (30\text{–}4)$$

The distance D is that between the half-silvered mirror and each of the other two mirrors. The path difference that corresponds to the time difference Δt is $c\,\Delta t$, so that if n fringes shift, $c\,\Delta t = n\lambda$, where λ is the wavelength of the light. Hence

$$n = \frac{c\,\Delta t}{\lambda} = \frac{Dv^2}{\lambda c^2}.$$

Michelson and Morley were able to make D about 10 m long by employing multiple reflections, and they used light of wavelength close to 5×10^{-7} m. Since $v = 3 \times 10^4$ m/sec and $c = 3 \times 10^8$ m/sec, the fringe shift anticipated when the instrument was turned through 90° was

$$n = \frac{10\text{ m }(3 \times 10^4\text{ m/sec})^2}{5 \times 10^{-7}\text{ m }(3 \times 10^8\text{ m/sec})^2} = 0.2\text{ fringe.}$$

Both paths undergo the same fringe shift, but in opposite directions, so the total shift would be 0.4 fringe.

The results of the Michelson-Morley experiment were startling: although their apparatus was capable of measuring the small fringe shift expected, *none was found.* When the experiment was performed at different seasons and in different locations, and when different experiments were tried for the same purpose, the conclusions were always identical—no motion through the ether could be detected.

30–2 Frames of reference

The negative result of the Michelson-Morley experiment deeply affected the conceptual foundations of physics. An immediate conclusion we can draw from the failure of the Michelson-Morley experiment to detect any variations in the velocity of light means that c is the same for all observers, regardless of their motion parallel to or perpendicular to a particular beam of light. This contradicts the customary notions of velocity addition that seemed correct in mechanics: projectiles fired from a moving vehicle in the direction of motion and opposite to the direction of motion have different velocities as measured by a stationary observer.

A second conclusion we can draw is that there can be no absolute coordinate system in the universe, since the velocity of light cannot be referred to such a system. Thus all coordinate systems have the same status: there is no such thing as "absolute motion." If we see an object changing its position with respect to us, we have no way of knowing (even in principle) whether *it* is moving or *we* are moving. In fact, if we were alone in the universe we would have no way of finding out whether we were moving or not. In order to appreciate the magnitude of the revolution in physical thought brought about by the Michelson-Morley result, let us look into the matter of frames of reference in relative motion in some detail.

An event of some kind occurs at a certain moment somewhere and is witnessed by two observers. One of them is in a frame of reference S, and he finds that the event took place at the time t in the position specified by the coordinates x, y, z. The other observer is in a different frame of reference, S', which is identical with S except that it is moving at the constant velocity v in the $+x$-direction (Fig. 30–5). This observer finds that the same event took place at the time t' and has the coordinates x', y', z'. What is the relationship between these sets of measurements?

We have already discussed in Section 5–13 how this problem was analyzed before the advent of relativity theory. Let us now extend that discussion.

Suppose that the clocks in both systems are started at the instant when the origins of S and S' coincide. If this is true, x measurements made in S at the time t exceed similar measurements made in S' by vt, since vt is the distance through which S' has moved relative to S in the x-direction during t. Hence $x = x' + vt$. In the absence of relative motion between S and S' in the y- and z-directions, $y' = y$ and $z' = z$. We assume that the clocks in S and S' are identical, and so there seems to be no reason why t' should not equal t. The set of equations

$$x' = x - vt, \tag{30–5}$$

$$y' = y, \tag{30–6}$$

$$z' = z, \tag{30–7}$$

$$t' = t \tag{30–8}$$

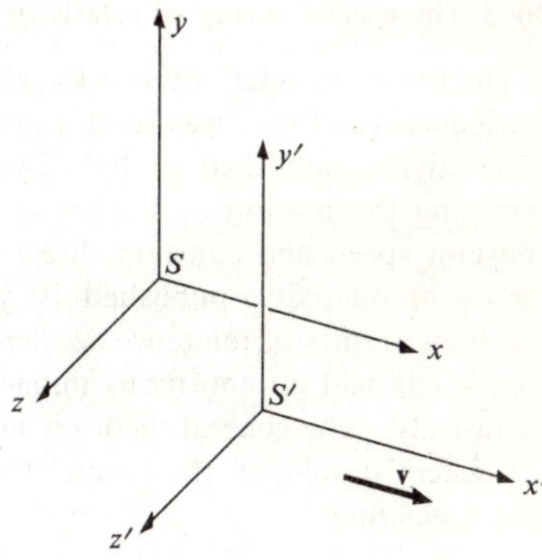

FIGURE 30–5

that relate measurements made in S' with those made in S is called the *Galilean transformation.* To obtain the transformation for velocity components measured in one system corresponding to those measured in the other, we merely differentiate the above expressions for x', y', and z' with respect to t'. This procedure yields

$$v'_x = \frac{dx'}{dt'} = v_x - v, \tag{30–9}$$

$$v'_y = \frac{dy'}{dt'} = v_y, \tag{30–10}$$

$$v'_z = \frac{dz'}{dt'} = v_z. \tag{30–11}$$

The above coordinate and velocity transformations agree completely with our everyday experience, but they do *not* agree with the results of the Michelson-Morley experiment. Because of the absence of a universal "fixed" frame of reference, the formulas of physics must be the same in both S and S'. However, the basic equations of electricity and magnetism are different in S and S' when Eqs.

(30–5) through (30–8) are used to transform measurements made in one frame to their equivalent values in the other, a fact that had caught the attention of physicists even before 1905, although its full significance was not then understood. The second result, namely that the speed of light in free space is the same for all observers, is violated by the Galilean velocity transformation. If a light ray travels in the $+x$-direction at the speed $v_x = c$ in frame S, in the frame S' its speed is given by Eq. (30–9) as $v_x' = c - v$. While at first glance the Galilean transformation seems reasonable, it is inconsistent with experiment and therefore cannot be correct.

30–3 The special theory of relativity

The theory of relativity is a largely successful attempt to work out the physical consequences of the absence of a universal frame of reference. The special theory of relativity, published in 1905 by Albert Einstein, confines itself to problems involving the motion of frames of reference at constant velocity (that is, both constant speed and constant direction) with respect to one another; the general theory of relativity, published 10 years later by Einstein, deals with problems involving frames of reference accelerated with respect to one another. The special theory has had an enormous impact on all of physics, and we shall consider it exclusively. The general theory is beyond the scope of this text.

Einstein developed the special theory of relativity from two postulates. The first states that

The laws of physics may be expressed in the same set of equations for all frames of reference moving at constant velocity with respect to one another.

This postulate follows directly from the absence of a universal frame of reference. If the laws of physics took on different forms for different observers in relative motion, they could infer from these differences which of them were "stationary" in space and which were "moving"; but this distinction does not exist in nature, since there is no universal frame of reference, and the above postulate is an expression of this fact.

The second postulate states that

The speed of light in free space has the same value for all observers, regardless of their state of motion.

This postulate expresses the result of the Michelson-Morley experiment.

At first glance these postulates do not seem so very extraordinary, a misleading impression indeed, as we can show with a simple example. Suppose that we have the two boats A and B once more, this time in a large lake (Fig. 30–6a). Boat A is stationary and boat B is drifting at the constant velocity v. There is a fog present, and the observers in each boat have no idea which boat is moving. At the instant that B is abreast of A (Fig. 30–6b), somebody in one of the boats fires a flare. That light from the flare spreads out uniformly in all directions, something we

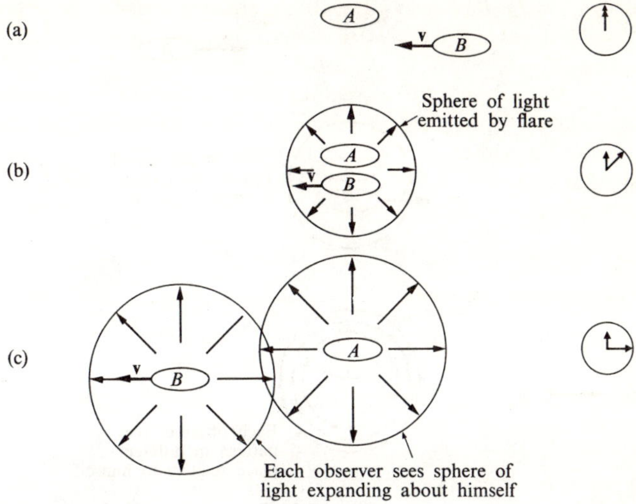

FIGURE 30–6

could verify if our visual responses were fast enough by watching the light scattered by the fog. An observer on either boat must see a sphere of light expanding with *him* at its center (Fig. 30–6c), according to the postulates of special relativity, even though one of them is changing his position with respect to the point where the flare went off. But it is impossible for the observers to tell which of them is undergoing such a change in position, since (owing in this case to the fog) there is no frame of reference other than each boat itself; and so, since the speed of light is the same for both of them, they must both see the identical phenomenon. And, of course, the fog is irrelevant; nothing would be changed in its absence except our ability to visualize what is going on.

What is so remarkable about the situation of Fig. 30–6? Let us consider a more familiar analog. It is a clear day and, instead of firing a flare, somebody drops a stone in the water. A circular pattern of ripples spreads out, as in Fig. 30–7, *which appears different* to observers on each boat. Merely by observing whether or not they are at the center of the pattern of ripples, the observers can tell whether they are stationary with respect to the water or not. Here the water behaves as a frame of reference, and an observer on a boat moving through the water will measure a ripple speed with respect to him different from that measured by an observer on a stationary boat. It is important for us to note that motion and waves *in water* are entirely different from motion and waves *in space;* water is in itself a frame of reference while space is not, and wave speeds in water vary with the observer's motion while wave speeds in space do not.

There is only one way of accounting for the fact that observers in the two boats observe identical expanding spheres of light, and that is to assume that the coor-

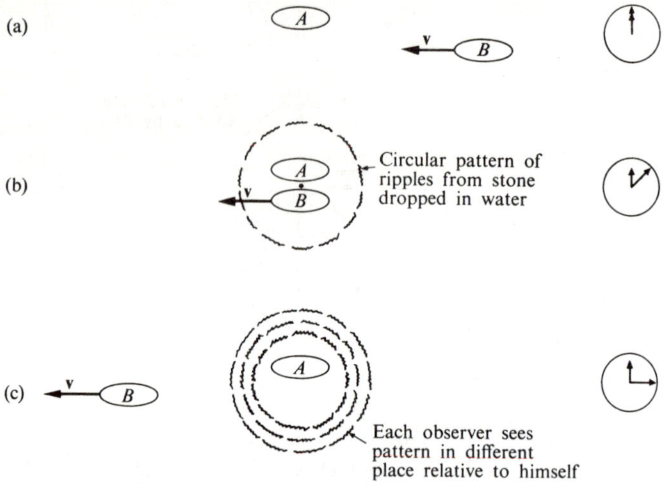

FIGURE 30-7

dinate system of each observer, as seen by the other, is affected by their relative motion. When this idea is worked out, using no more than the ordinary laws of physics and Einstein's postulates, it is found that a host of peculiar effects arise—and all of them, no matter how strange to us, have been confirmed by experiment.

30-4 The Lorentz transformation

Instead of relying on our intuition, as we did in obtaining the Galilean transformation, we can derive a set of transformation equations directly from the postulates of special relativity. Because $x' = x - vt$ is in accord with ordinary mechanics, the new transformation for the x-coordinate must reduce to this formula when the relative velocity v is small. The simplest possibility is

$$x' = k(x - vt), \tag{30-12}$$

where k does not depend on either of the coordinates x or t but may vary with v. This formula has the virtue of being linear in x and x', so that there is automatically one and only one value of x' for a given value of x without our having to impose any subsidiary conditions. From the first postulate of relativity, the inverse transformation from measurements made in S' to those made in S must have precisely the same form as Eq. (30-12) except for a change in the sign of v, and must also have the same constant of proportionality:

$$x = k(x' + vt). \tag{30-13}$$

As before, we have no reason to suppose that there is any difference between y and y' or between z and z', since the relative motion between S and S' is confined to the x-direction:

$$y' = y, \tag{30–14}$$

$$z' = z. \tag{30–15}$$

However, we cannot assume that this equality extends to t and t'. If we substitute the value of x' given by Eq. (30–12) into Eq. (30–13), we obtain

$$x = k^2(x - vt) + kvt',$$

and so

$$t' = kt + \left(\frac{1 - k^2}{kv}\right)x. \tag{30–16}$$

The transformation specified by Eqs. (30–12) and (30–14) through (30–16) satisfies the first postulate of special relativity.

To determine the value of k we return to the situation pictured in Fig. 30–6 of a flare being set off at the instant when two boats in relative motion are directly abreast. Regarding the boats as frames of reference, we can choose their respective coordinates so that the relative velocity v is in the x-direction and, when the flare is fired, $x = x' = 0$ and $t = t' = 0$. An observer in each boat measures the speed with which the pulse of light from the flare spreads out into space. The second postulate of relativity requires that both observers determine the same speed of c, as we said earlier; thus the x-coordinate of the light pulse in the S-system obeys the formula

$$x = ct, \tag{30–17}$$

while in the S'-system the x'-coordinate of the pulse obeys the formula

$$x' = ct'. \tag{30–18}$$

We now use Eqs. (30–13) and (30–16) to convert the x'- and t'-measurements of Eq. (30–18) to their corresponding x- and t-values, which yields

$$k(x - vt) = ckt + \left(\frac{1 - k^2}{kv}\right)cx.$$

Solving for x, we find

$$x = ct\left[\frac{1 + v/c}{1 - (c/v)(1/k^2 - 1)}\right].$$

This formula for x must be the same as that of Eq. (30–17), namely $x = ct$. Equating these expressions for x, we have

$$ct = ct\left[\frac{1 + v/c}{1 - (c/v)(1/k^2 - 1)}\right],$$

from which we obtain

$$k = \frac{1}{\sqrt{1 - (v^2/c^2)}}. \tag{30–19}$$

This result, for all its bizarre character, nevertheless has the reassuring property that, when the relative velocity v is small compared with the velocity of light c, $k \approx 1$ and Eqs. (30–13) and (30–16) reduce to their Galilean equivalents that we know are adequate in ordinary mechanics. Evidently "relativistic" effects occur only at extremely high speeds.

When the above equation for k is inserted in Eqs. (30–13) and (30–16) we obtain the *Lorentz transformation*, which relates measurements of the same event in the S- and S'-frames of reference:

$$x' = \frac{x - vt}{\sqrt{1 - (v^2/c^2)}}, \tag{30–20}$$

$$y' = y, \tag{30–21}$$

$$z' = z, \tag{30–22}$$

$$t' = \frac{t - vx/c^2}{\sqrt{1 - (v^2/c^2)}}. \tag{30–23}$$

This transformation is named in honor of H. A. Lorentz, a Dutch physicist, who proved that it is the only one that permits the laws of electricity and magnetism to have the same forms in frames of reference that are in relative motion. Einstein was able to show its more general significance as the proper coordinate transformation for all measurements, electromagnetic or not.

Equations (30–20) through (30–23) convert measurements of time and space made in the stationary S-frame to their counterparts in the moving S'-frame. As seen from frame S, frame S' is moving with the velocity v along the x-axis, while as seen from frame S', frame S is moving with the velocity $-v$ along the x-axis. Hence the *inverse Lorentz transformation*, which converts measurements of time and space made in the S'-frame to their counterparts in the S-frame, may be obtained from the Lorentz transformation by changing the sign of v. The inverse transformation therefore consists of the equations

$$x = \frac{x' + vt'}{\sqrt{1 - (v^2/c^2)}}, \tag{30–24}$$

$$y = y', \tag{30–25}$$

$$z = z', \tag{30–26}$$

$$t = \frac{t' + vx/c^2}{\sqrt{1 - (v^2/c^2)}}. \tag{30–27}$$

30–5 The Lorentz-Fitzgerald contraction

Measurements of space and time, as we have seen, are not absolute but depend on the relative motion of the observer and that which is being observed. Different observers may obtain different results when they examine the same phenomenon,

FIG. 30-8. (a) An observer at rest relative to a rod measures its length as L_0. (b) An observer moving relative to the same rod measures its length as L.

and all of them may be correct. This situation is a consequence of the fundamental character of the physical universe, and it should not really be surprising that the space-time geometry of nature is not the same as the space-time geometry of Euclid and Galileo. The postulates of special relativity are perfectly reasonable, and if the deductions we are about to obtain from them seem strange, it is only because our personal experience involves frames of reference whose relative motion is on too small a scale to bring out relativistic effects.

Let us begin by considering a stick lying on the ground beside a road, as shown in Fig. 30-8. An observer at rest relative to the stick finds that one end is at $x = x_1$ and the other at $x = x_2$, so that its length L_0 is

$$L_0 = x_2 - x_1.$$

Now a car comes along and, as it passes the stick, a rider measures the length of the stick by simultaneously determining the positions x_1' and x_2' of its ends in his frame of reference. To the rider, who is moving relative to the stick at the constant velocity v, the length of the stick is

$$L = x_2' - x_1'.$$

How are L_0 and L related? From the inverse Lorentz transformation, Eq. (30-24), we have

$$x_1 = \frac{x_1' + vt'}{\sqrt{1 - (v^2/c^2)}} \quad \text{and} \quad x_2 = \frac{x_2' + vt'}{\sqrt{1 - (v^2/c^2)}},$$

so that

$$L_0 = x_2 - x_1 = \frac{x_2' - x_1'}{\sqrt{1 - (v^2/c^2)}} = \frac{L}{\sqrt{1 - (v^2/c^2)}}$$

and

$$L = L_0\sqrt{1 - (v^2/c^2)}. \tag{30-28}$$

The length of an object in motion with respect to an observer is measured by the observer to be shorter than when it is at rest with respect to him. This shortening, known as the *Lorentz-Fitzgerald contraction*, works both ways; to a man in a rocket, measurements indicate that objects on the earth behind him are shorter than they were when he was on the ground by the amount specified in Eq. (30–28). The length of an object is a maximum when measured in a reference frame in which it is stationary, and its length is less when measured in a reference frame in which it is moving. [The ratio L/L_0 is the same in both Eqs. (30–28) and (30–4), but we cannot interpret the Michelson-Morley result solely as evidence for the contraction of their apparatus in the direction of motion because a similar experiment using an interferometer with unequal arms showed no fringe shift either. It is not an instrumental effect; there really is no ether.]

The relativistic length contraction is negligible for ordinary speeds, but is an important effect at speeds close to the speed of light. A speed of 1000 mi/sec seems enormous to us, yet it results in a shortening of the object in the direction of its motion to only

$$\frac{L}{L_0} = \sqrt{1 - (v^2/c^2)} = \sqrt{1 - \frac{(1000\ \text{mi/sec})^2}{(186{,}000\ \text{mi/sec})^2}}$$

$$= 0.999985 = 99.9985\%$$

of its length at rest. On the other hand, something traveling at nine-tenths the speed of light is shortened to

$$\frac{L}{L_0} = \sqrt{1 - \frac{(0.9c)^2}{c^2}} = 0.436 = 43.6\%$$

of its length at rest, a significant change.

Although the Lorentz-Fitzgerald contraction is verified by experimental measurements, the visual appearance of an object in rapid motion (if we could actually see it) would not exhibit such a shortening. This apparent contradiction comes about because light reaching the eye or a camera from the more distant parts of the object left it at an earlier time than that coming from the closer parts. What we see is therefore a composite image, since the object was at different places when the various elements that make up the image left it. This effect lengthens the apparent size of a moving object in its direction of motion, and turns out to exactly cancel the Lorentz-Fitzgerald contraction! A three-dimensional body, for instance a cube, would be seen as though rotated in orientation without any change in shape; in the limit of v approaching c, the angle through which it would be turned would be 180°. A sphere would suffer no distortion whatever. We must distinguish this visual effect from the Lorentz-Fitzgerald contraction itself, which is a physical phenomenon. In the absence of this contraction, moving bodies would appear as though lengthened by a corresponding amount. Interestingly enough, this analysis of the visual appearance of rapidly moving objects was not made until 1959, which was over half a century after Einstein published the special theory of relativity.

30-6 Time dilation

Time intervals, too, are affected by relative motion. A clock moving with respect to an observer appears to tick less rapidly than it does when it is at rest with respect to him; that is, if we observe the length of time some event requires in a frame of reference in motion relative to us, our clock would indicate a longer time interval than a clock in the moving frame.

Let us derive this *time dilation* directly from the Lorentz transformation. Suppose a rider in the car of Fig. 30-8 is located at the point x' in his frame of reference, where, as before, the origins of the stationary frame S (the ground) and the moving frame S' (the car) coincide at $t = t' = 0$. The rider glances at his clock and finds that the time is t'_1. At the same moment a man beside the road finds the time to be t_1, where, according to Eq. (30-27),

$$t_1 = \frac{t'_1 + vx'/c^2}{\sqrt{1 - (v^2/c^2)}}.$$

Later the rider in the car again glances at his clock and notes that the time is now t'_2, so that the time that has elapsed in his frame of reference is

$$t_0 = t'_2 - t'_1.$$

According to the clock on the ground this time interval ended at t_2, where

$$t_2 = \frac{t'_2 + vx'/c^2}{\sqrt{1 - (v^2/c^2)}}.$$

Hence to the man on the ground the period of time involved is

$$t = t_2 - t_1 = \frac{t'_2 - t'_1}{\sqrt{1 - (v^2/c^2)}},$$

or

$$t = \frac{t_0}{\sqrt{1 - (v^2/c^2)}}. \qquad (30\text{-}29)$$

An event occurring in the moving frame (the car) that occupies a time interval of t_0 is longer by the factor $1/\sqrt{1 - (v^2/c^2)}$ to an observer in the stationary frame (the ground).

A good illustration of both the time dilation of Eq. (30-29) and the length contraction of Eq. (30-28) occurs in the decay of unstable particles called mu-mesons, whose properties we shall discuss in greater detail in Chapter 38. For the moment what concerns us is the fact that a mu-meson decays into an electron about 2×10^{-6} sec after it comes into being. Now mu-mesons are created in the atmosphere thousands of meters above the ground by fast cosmic-ray particles incident upon the earth from space, and reach sea level in profusion. Their speeds are observed to be about 2.994×10^8 m/sec, or 0.998 of the velocity of light c.

But in $t_0 = 2 \times 10^{-6}$ sec, the meson lifetime, they can travel a distance of only

$$h = vt_0 = 2.994 \times 10^8 \frac{\text{m}}{\text{sec}} \times 2 \times 10^{-6} \text{ sec} = 599 \text{ m},$$

while they are actually created at elevations 10 or more times greater than this.

We can resolve the meson paradox by using the results of the special theory of relativity. Let us first do this from the frame of reference of the meson, where its lifetime is 2×10^{-6} sec. While the meson lifetime is unaffected by the motion, its distance to the ground appears shortened by the factor,

$$\frac{h}{h_0} = \sqrt{1 - (v^2/c^2)}.$$

That is, while we, on the ground, measure the altitude of the meson as h_0, the meson sees it as h. If we let h be 599 m, the maximum distance the meson can go in its own frame of reference at the speed $0.998c$ before decaying, we find that the corresponding distance h_0 in *our* reference frame is

$$h_0 = \frac{h}{\sqrt{1 - (v^2/c^2)}} = \frac{599 \text{ m}}{\sqrt{1 - [(0.998c)^2/c^2]}}$$

$$= \frac{599 \text{ m}}{\sqrt{1 - 0.966}} = \frac{599 \text{ m}}{0.063} = 9.51 \times 10^3 \text{ m}.$$

Hence, despite its brief life span, the meson is able to reach the ground from quite respectable altitudes.

Now let us apply the special theory of relativity from the point of view of an observer on the ground. On the ground we find the meson altitude to be h_0, but its lifetime is no longer $t_0 = 2 \times 10^{-6}$ sec; its apparent lifetime t *to us* has been extended owing to the relative motion to the value

$$t = \frac{t_0}{\sqrt{1 - (v^2/c^2)}} = \frac{2 \times 10^{-6} \text{ sec}}{\sqrt{1 - [(0.998c)^2/c^2]}}$$

$$= \frac{2 \times 10^{-6} \text{ sec}}{0.063} = 31.7 \times 10^{-6} \text{ sec},$$

almost 16 times longer than when it is at rest with respect to us. In 31.7×10^{-6} sec a meson whose speed is $0.998c$ can travel a distance

$$h_0 = vt = 2.994 \times 10^8 \frac{\text{m}}{\text{sec}} \times 31.7 \times 10^{-6} \text{ sec}$$

$$= 9.51 \times 10^3 \text{ m},$$

the same distance we found previously. The two points of view give consistent results. Time dilation and Lorentz contraction are two different aspects of the same phenomenon.

30-7 The relativity of mass

According to relativity, measurements of the same object or event will be different for observers in relative motion. However, despite these differences, the laws of motion arrived at by the various observers must have exactly the same mathematical form. If we consider in detail a collision between two bodies as witnessed from the reference frames S and S' and impose the requirement that linear momentum be conserved, we would find that mass is no more independent of motion than space and time are. As measured from the earth, a rocket ship in flight is shorter than its twin still on the ground, its clocks tick more slowly, and, in addition, its mass is greater.

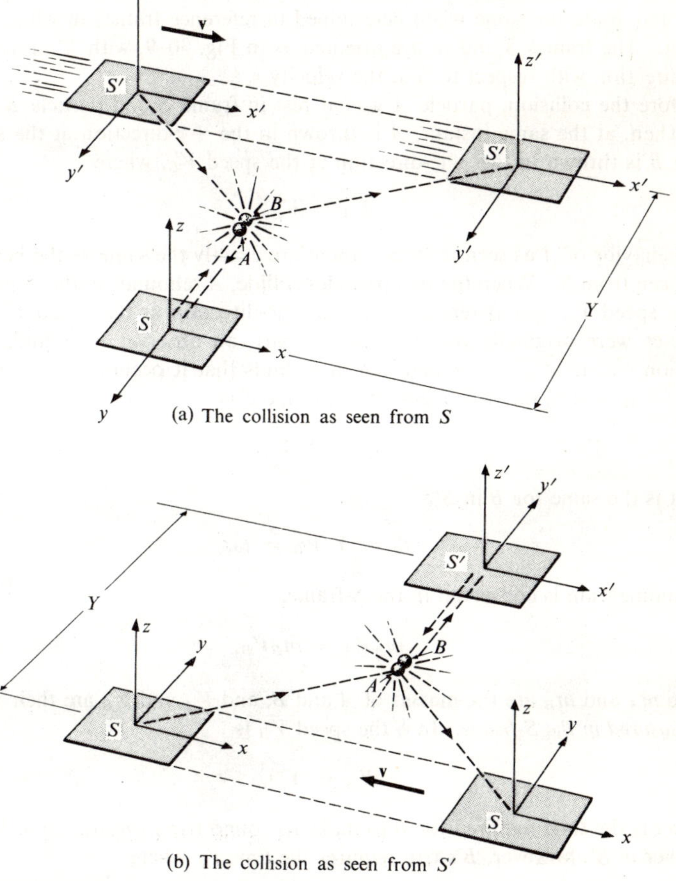

FIG. 30-9. Collision of two identical balls as seen in moving reference frames.

Why do we require, as our point of departure, that momentum be conserved? In order that the equations of electricity and magnetism may be invariant under transformations from one inertial system to another, the Lorentz transformation equations must be used. However, the Galilean transformation is required for similar invariance in classical dynamics, so that a change of some kind must be made in the theoretical structure of dynamics. Einstein found that retaining conservation of momentum as a fundamental law of mechanics permitted leaving the formal structure of mechanics virtually unchanged, though at the expense of mass being a function of velocity instead of being a constant.

Let us consider an elastic collision (that is, a collision in which kinetic energy is conserved) between two particles A and B, as witnessed by observers in the reference frames S and S' which are in uniform relative motion. The properties of A and B are the same when determined in reference frames in which they are at rest. The frames S and S' are oriented as in Fig. 30-9, with S' moving in the $+x$-direction with respect to S at the velocity v.

Before the collision, particle A was at rest in frame S and particle B in frame S'. Then, at the same instant, A is thrown in the $+y$-direction at the speed V_A while B is thrown in the $-y'$-direction at the speed V'_B, where

$$V_A = V'_B. \tag{30-30}$$

The behavior of A as seen from S is therefore exactly the same as the behavior of B as seen from S'. When the two particles collide, A rebounds in the $-y$-direction at the speed V_A and B rebounds in the $-y'$-direction at the speed V'_B. If the particles were originally the distance Y apart, an observer in S finds that the collision occurs at $y = \frac{1}{2}Y$ and one in S' finds that it occurs at $y' = \frac{1}{2}Y$. The round-trip time t_0 for A as measured in frame S is

$$t_0 = Y/V_A, \tag{30-31}$$

and it is the same for B in S',

$$t'_0 = Y/V'_B = t_0. \tag{30-32}$$

If momentum is conserved in the S-frame,

$$m_A V_A = m_B V_B, \tag{30-33}$$

where m_A and m_B are the masses of A and B, and V_A and V_B are their velocities *as measured in the S-frame.* In S the speed V_B is

$$V_B = Y/t, \tag{30-34}$$

where t is the time required for B to make its round trip *as measured in S*. To an observer in S', however, B's trip requires the time t'_0, where

$$t = \frac{t'_0}{\sqrt{1 - (v^2/c^2)}} = \frac{t_0}{\sqrt{1 - (v^2/c^2)}}, \tag{30-35}$$

according to our previous results, Eq. (30–29). Although observers in both frames see the same event, they disagree as to the length of time the particle which is thrown from the other frame requires to make the collision and return.

Replacing t in Eq. (30–34) with its equivalent in terms of t_0, we have

$$V_B = \frac{Y\sqrt{1 - (v^2/c^2)}}{t_0},$$

and, from Eq. (30–31), $V_A = Y/t_0$. Inserting these expressions for V_A and V_B in Eq. (30–33), we see that momentum is conserved if

$$m_A = m_B\sqrt{1 - (v^2/c^2)}. \tag{30–36}$$

We originally supposed that A and B are identical when at rest with respect to an observer, so that the difference between m_A and m_B means that measurements of mass, like those of space and time, depend on the relative speed between an observer and whatever he is observing.

In the above discussion, both A and B are moving in frame S. To obtain a formula giving the mass m of a body measured while in motion in terms of its mass m_0 when measured at rest, we shall consider a similar example in which V_A and V'_B are very small. Now an observer in S will see B approach A with the velocity v, make a glancing collision (since $V_B \ll v$), and then continue on. In the S-frame of reference, then,

$$m_A = m_0 \quad \text{and} \quad m_B = m,$$

and so

$$m = \frac{m_0}{\sqrt{1 - (v^2/c^2)}}. \tag{30–37}$$

The mass of a body moving at the speed v relative to an observer is larger than its mass when at rest relative to the observer by the factor $1/\sqrt{1 - (v^2/c^2)}$. This mass increase is reciprocal; to an observer in the S'-frame of reference,

$$m_A = m \quad \text{and} \quad m_B = m_0.$$

As determined from the earth, a rocket ship in flight is shorter than its twin still on the ground, and its mass is greater. To someone on the rocket ship in flight, the ship on the ground also appears shorter and appears to have a greater mass. (These effects are, of course, unobservably small for today's rocket speeds.)

Provided that momentum p is defined as

$$p = mv = \frac{m_0 v}{\sqrt{1 - (v^2/c^2)}}, \tag{30–38}$$

conservation of momentum is valid in special relativity just as in classical physics.

Relativistic mass increases are significant only at speeds approaching that of light. At a speed one-tenth that of light, the mass increase amounts to only 0.5%, but this increase is over 100% at a speed of nine-tenths that of light. Only atomic particles such as electrons, protons, mesons, and so on have sufficiently high speeds for relativistic effects to take place, and when we deal with such particles, as we have seen, we must be careful to take these effects into account. Historically, the first confirmation of Eq. (30–37) was the discovery of Bucherer in 1908 that the ratio e/m of the electron's charge to its mass was smaller for fast electrons than for slow ones; this equation, like the others of special relativity, has been verified so many times that it is now accepted as one of the basic formulas of physics.

Equation (30–37) has something interesting to say about the greatest speed an object can have. The closer v approaches c, the closer v^2/c^2 approaches 1, and the closer $\sqrt{1 - (v^2/c^2)}$ approaches zero. As the denominator of Eq. (30–37) becomes smaller, the mass m becomes larger, so that if the relative speed v actually were equal to the speed of light, the object's mass would be infinite. The concept of an infinite mass anywhere in the universe is, of course, nonsense on many counts; it would have required an infinite force to have accelerated it to the speed of light, its length in the direction of motion would be zero, by Eq. (30–28), so that its volume would be zero, and it would exert an infinite gravitational force on all other bodies in the universe. Hence we interpret Eq. (30–37) to mean that no material body can ever equal or exceed the speed of light.

We should keep in mind that c in the formulas of relativity refers to the speed of light in free space, 3×10^8 m/sec. In all material media, such as water, glass, and air, light travels more slowly than this, and atomic particles are capable of moving with higher speeds in such media than the speed of light *in them*, though never faster than the speed of light in free space. When an electrically charged particle moves through a substance at a speed exceeding that of light in the substance, a cone of light waves is emitted in a process roughly similar to that in which a ship produces a bow wave as it moves through the water faster than water waves do. The emission of light waves in these circumstances is the *Cerenkov effect*.

30–8 Mass and energy

According to special relativity, Newton's second law of motion must be written

$$\mathbf{F} = \frac{d}{dt}(m\mathbf{v}). \tag{30–39}$$

This is not the same as $\mathbf{F} = m\mathbf{a}$, since

$$\frac{d}{dt}(m\mathbf{v}) = m\frac{d\mathbf{v}}{dt} + \mathbf{v}\frac{dm}{dt} = m\mathbf{a} + \mathbf{v}\frac{dm}{dt},$$

and the time variation of mass dm/dt is not zero if v varies with time. Hence the correct statement of the second law of motion is that the time rate of change of the momentum of an object is equal to the net force acting on it.

In relativistic mechanics, as in ordinary mechanics, the kinetic energy of a moving body is the work needed to bring it from rest to whatever speed v it happens to have. That is,

$$\text{KE} = \int_0^s F\,ds,$$

where F is the component of the applied force in the direction of the displacement ds, and s is the distance through which the force acts. Making use of Eq. (30–39) and the fact that $ds/dt = v$, we obtain

$$\text{KE} = \int_0^s \frac{d}{dt}(mv)\,ds = \int_0^{mv} \frac{ds}{dt}\,d(mv)$$

$$= \int_0^{mv} v\,d(mv) = \int_0^v v\,d\frac{m_0 v}{\sqrt{1 - (v^2/c^2)}}\,.$$

According to the procedure for integrating by parts,

$$\int x\,dy = xy - \int y\,dx,$$

and so, applying this formula here, we have

$$\text{KE} = \frac{m_0 v^2}{\sqrt{1 - (v^2/c^2)}} - m_0 \int_0^v \frac{v\,dv}{\sqrt{1 - (v^2/c^2)}}$$

$$= \frac{m_0 v^2}{\sqrt{1 - (v^2/c^2)}} + m_0 c^2 [\sqrt{1 - (v^2/c^2)}]_0^v$$

$$= \frac{m_0 c^2}{\sqrt{1 - (v^2/c^2)}} - m_0 c^2 = (m - m_0)c^2. \tag{30–40}$$

Equation (30–40) asserts that the kinetic energy of an object is equal to the increase in its mass $m - m_0$ due to its motion multiplied by the square of the speed of light c.

If we rewrite Eq. (30–40) as

$$mc^2 = \text{KE} + m_0 c^2, \tag{30–41}$$

we see that we can interpret mc^2 as the total energy of a body in motion relative to an observer who measures the body's mass as m, provided that the quantity $m_0 c^2$ is the energy the body has when it is at rest relative to this observer. This interpretation has turned out to be correct.

Another useful expression for the total energy of a body is

$$E = mc^2 = \frac{m_0 c^2}{\sqrt{1 - (v^2/c^2)}}\,,$$

$$E = \sqrt{m_0^2 c^4 + p^2 c^2}. \tag{30–42}$$

Equation (30–42) can be verified by substituting the relativistic expression for the momentum p from Eq. (30–38).

We have demonstrated the equivalence of mass and kinetic energy. A more subtle analysis shows that potential energy is also equivalent to mass. The energy

$$E_0 = m_0c^2 \tag{30–43}$$

is called the *rest energy* of the body. Thus a 1-kg mass is equivalent to 9×10^{16} joules. Even a minute bit of matter can yield an immense amount of energy. In fact, the conversion of matter into energy is the energy source of the sun and stars as well as that of atomic and hydrogen bombs, of chemical reactions, and, indeed, of *all* processes in which energy is liberated. The equivalence of mass and energy, a direct consequence of the postulates of special relativity, is the most famous of the various results Einstein obtained early in this century.

When the relative speed v is small compared with c, we expect that the formula

$$\text{KE} = \frac{m_0c^2}{\sqrt{1 - (v^2/c^2)}} - m_0c^2$$

for kinetic energy should become just the $\frac{1}{2}m_0v^2$ we are familiar with and which has been verified by experiment at low speeds. Let us see whether this is true. In general, the binomial theorem of algebra tells us that if some quantity x is much smaller than 1,

$$(1 + x)^n = 1 + nx.$$

Here, with

$$x = \frac{-v^2}{c^2} \quad \text{and} \quad n = -\tfrac{1}{2},$$

we find that the above formula indeed becomes

$$\text{KE} = \left(1 + \frac{\frac{1}{2}v^2}{c^2}\right) m_0c^2 - m_0c^2 = \tfrac{1}{2}m_0v^2.$$

In the foregoing calculation, relativity has met an important test; it has yielded exactly the same results as those of ordinary mechanics at low speeds, where we know by experience that the latter are perfectly valid. It is proper, then, to think of the principles of mechanics that we developed in Chapters 1 through 7 as no more than approximations to the correct relativistic principles of mechanics, valid approximations in nearly all common situations, but requiring modification under certain circumstances.

30–9 Velocity addition

We have already mentioned that the Galilean velocity transformation of Eqs. (30–9) through (30–11) is not consistent with the constancy of the speed of light, since it predicts that different observers will find different values of c, depending on their states of motion. To obtain the correct transformation all that is necessary is to differentiate the appropriate Lorentz transformation equations. For instance, we might wish to relate v'_x, v'_y, v'_z, the components of the velocity of an object as measured in the moving S'-system, to the components v_x, v_y, v_z that would be measured in the stationary S-system. From Eqs. (30–24) through (30–27), noting that both the relative velocity v (assumed in the x-direction) and the speed of light c are constant, we have

$$dx = \frac{dx' + v\,dt'}{\sqrt{1 - (v^2/c^2)}}, \qquad dy = dy',$$

$$dz = dz', \qquad dt = \frac{dt' + v\,dx'/c^2}{\sqrt{1 - (v^2/c^2)}}.$$

Hence

$$v_x = \frac{dx}{dt} = \frac{dx' + v\,dt'}{dt' + v\,dx'/c^2},$$

and dividing through by dt' yields

$$v_x = \frac{dx'/dt' + v}{1 + v/c^2\,dx'/dt'} = \frac{v'_x + v}{1 + vv'_x/c^2}. \tag{30–44}$$

In a similar manner we obtain the other two components of velocity,

$$v_y = \frac{v'_y\sqrt{1 - (v^2/c^2)}}{1 + vv'_x/c^2} \tag{30–45}$$

and

$$v_z = \frac{v'_z\sqrt{1 - (v^2/c^2)}}{1 + vv'_x/c^2}. \tag{30–46}$$

When v is small compared with c, these equations reduce to the Galilean transformation of Eqs. (30–9) through (30–11).

Let us apply the above results to a pulse of light emitted from the moving system S'. If the light is sent out in the $+x$-direction, $v'_x = c$, and

$$v_x = \frac{c + v}{1 + vc/c^2} = \frac{c(c + v)}{c + v} = c;$$

the measured speed of light is exactly the same in both systems. This conclusion further verifies the consistency of the Lorentz transformation with the postulates of relativity. Ordinary mechanics, of course, predicts that the speed of the pulse measured in S would be $c + v$.

Since nothing can move faster than the speed of light, it might seem impossible to pass a space ship whose speed with respect to the earth is $0.9c$ at a relative speed of, say, $0.5c$. Ordinary mechanics indicates that the required speed is

$0.9c + 0.5c = 1.4c$. If we set $v'_x = 0.5c$ and $v = 0.9c$ in Eq. (30–44), however, we obtain

$$v_x = \frac{1.4c}{1 + 0.45} = 0.966c.$$

It is necessary to move less than 8% faster than a space ship whose speed is $0.9c$ in order to pass it at a relative speed of half the speed of light.

Problems

1. The hypothetical speed of the earth through the ether is its orbital speed of 3×10^4 m/sec. (a) If light takes precisely 3×10^{-7} sec to travel up and back through the Michelson-Morley apparatus in the direction parallel to this motion, how long will it take to travel through it in the direction perpendicular to this motion, according to a non-relativistic analysis? (b) If light of wavelength 6×10^{-7} m is used, what fraction of a wavelength does the difference in the above times represent?

2. What is the length of a meter stick moving parallel to its length when its mass is $\frac{3}{2}$ of its rest mass?

3. Show directly from the Lorentz transformation that $x^2 + y^2 + z^2 - c^2t^2 = x'^2 + y'^2 + z'^2 - c^2t'^2$. Interpret the result.

4. Two observers, A on earth and B in a rocket ship whose speed is 2×10^8 m/sec, both set their watches to 1:00 when the ship is abreast of the earth. (a) When A's watch reads 1:30 he looks at B's watch through a telescope. What does it read? (b) When B's watch reads 1:30 he looks at A's watch through a telescope. What does it read?

5. A certain process requires 10^{-6} sec to occur in an atom at rest in the laboratory. How much time will this process require to an observer in the laboratory when the atom is moving at a speed of 5×10^7 m/sec?

6. How fast would a rocket ship have to go for each year on the ship to correspond to two years on the earth?

7. What is the relativistic contraction in the length of an airplane 100 ft long when it moves directly away from an observer at 400 mi/hr?

8. How fast would a rocket ship have to go relative to an observer for its length to be contracted to 99% of its length when at rest?

9. The specific heat of lead is 0.03 kcal/kg·°C. If a bar of lead has a mass of 100 kg at 20°C, how much mass will it gain when heated to 300°C?

10. How much mass is lost by 1 kg of water at 0°C when it turns to ice at 0°C?

11. Compute the rest energy of a proton in joules and in electron volts.

12. Compute the rest energy of an electron in joules and in electron volts.

13. Dynamite liberates 1.3×10^3 kcal/kg when it explodes. What percentage of its total energy content is this?

14. Typical chemical reactions absorb or release energy at the rate of several electron volts per molecular change. What change in mass is associated with the absorption or release of 1 ev of energy?

15. What speed must a body have if its mass is to double?

16. (a) How much energy is required to double the speed of an electron whose initial speed is 7×10^7 m/sec? (b) How much energy is required to double the speed of an electron whose initial speed is 1.4×10^8 m/sec?

17. In Chapter 21 it was stated that the relativistic increase in mass of moving protons limits the ability of a cyclotron to accelerate them. (a) At what speed will a proton's mass exceed its rest mass by 1%? (b) What kinetic energy in electron volts does this speed correspond to?

18. An electron is accelerated through a potential difference of 10^5 volts. (a) What is its speed? (b) What is its mass? (c) Compare the answers to (a) and (b) with the values that classical physics would predict.

19. How much energy is required to accelerate an electron from a speed of: (a) $0.5c$ to $0.9c$? (b) $0.9c$ to $0.99c$?

20. An electron and a proton are accelerated from rest by a potential difference of 10^8 v. (a) What is the increase in the mass of each particle? (b) What is the fractional increase in mass of each particle? (c) What is the final speed of each?

21. The mass of the sun is 2×10^{30} kg. The distance from the earth to the sun is 9×10^7 mi. The solar constant on the earth (the amount of solar energy falling on the earth) is approximately 2 cal/cm^2·min. Estimate the fractional loss in the sun's mass per second.

22. It requires a minimum of 4.48 ev to separate a hydrogen molecule into two hydrogen atoms. What is the fractional change of mass of a hydrogen molecule when it is dissociated into its constituent atoms?

23. It requires a minimum of 2.18×10^6 ev to dissociate a deuteron into a neutron and proton. What is the fractional change of mass of a deuteron when it is dissociated?

24. How many joules of energy per kilogram of rest mass are required to accelerate a space ship to a speed of 2.94×10^8 m/sec?

25. (a) Show that the component of a force in the direction of the velocity of a particle is

$$\mathbf{F}_{\parallel} = \left[\frac{m_0}{(1 - v^2/c^2)^{3/2}}\right] a_{\parallel}.$$

(b) Show that the component of a force perpendicular to its velocity is

$$\mathbf{F}_{\perp} = \left[\frac{m_0}{(1 - v^2/c^2)^{1/2}}\right] a_{\perp}.$$

26. A meter stick at rest makes an angle of 45° with respect to the positive x-axis. If it moves with a velocity of $0.99c$ with respect to the positive x-axis, what is its length and what angle does it make according to an observer at rest with respect to the axis?

27. A lady on Venus sees two space ships approaching her from opposite directions at speeds of $0.8c$ and $0.9c$, respectively. What is the relative speed of the two space ships as measured by an observer on either of them?

31

Quanta

The formulation of the theory of relativity and of the quantum theory of light, both of which took place early in this century, profoundly altered our ways of thinking about the physical world. In the previous chapter we examined some of the remarkable consequences of relativity, consequences so alien to our intuition that our first feeling must be one of astonishment. Now we come to the realm of quanta, which is no less remarkable or astonishing. But these subjective terms are not really justified, because they are based merely on the limits to our imagination that are imposed by our experience; if we were in a world where we were about the size of an electron, relativistic and quantum phenomena would be familiar. We shall find that the ideas presented in this chapter are not only intrinsically of great interest, but are also essential if we are to understand the material that follows on the atom, solids, and the nucleus.

31–1 The photoelectric effect

Toward the end of the nineteenth century a number of experiments were performed that revealed the emission of electrons from a metal surface when light (particularly ultraviolet light) is shined on it (Fig. 31–1). This phenomenon is known as the *photoelectric effect*. It is not, at first glance, anything to surprise us, for light waves carry energy, and some of the energy absorbed by the metal may somehow concentrate on individual electrons and reappear as kinetic energy. Upon closer inspection of the data, however, we find that the photoelectric effect can hardly be explained in so straightforward a manner.

The first peculiarity of the photoelectric effect is that the energy of the emitted electrons (which are called *photoelectrons*) does not depend on the intensity of the light. A bright light yields more electrons than a dim one, but their average energy remains the same. Furthermore, even when the metal surface is faintly illuminated, the photoelectrons leave the surface immediately. This behavior contradicts the electromagnetic theory of light, which predicts that the energy of photoelectrons

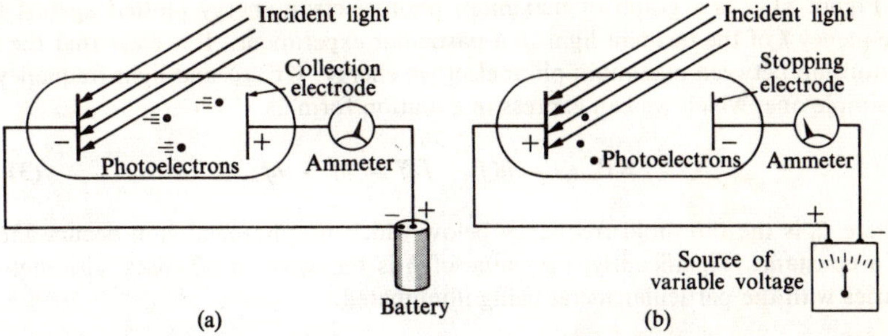

FIG. 31-1. (a) A method of detecting the photoelectric effect. The photoelectrons ejected from the metal plate being irradiated are attracted to the positive collection electrode at the other end of the tube, and the current that results is measured with the ammeter. (b) A method of detecting the maximum energy of the photoelectrons. As the stopping electrode is made more negative, the slower photoelectrons are repelled before they can reach it. Finally a voltage will be reached at which no photoelectrons whatever are received at the stopping electrode, as indicated by the current dropping to zero, and this voltage corresponds to the maximum photoelectron energy.

should depend on the intensity of the light beam responsible for them and that, if the beam has a very low intensity, a certain period of time must elapse on the average before any individual electrons accumulate enough energy to leave the metal.

An example will illustrate why the instantaneous emission of photoelectrons is so remarkable. A detectable photoelectric current results when sodium is irradiated with 10^{-6} watts/m^2 of violet light. There are about 10^{20} atoms in the upper ten layers of sodium atoms per square millimeter, so that, if we arbitrarily assume that the incident light is absorbed in these layers, each atom receives energy at the rate of 10^{-26} watt, somewhat less than 10^{-7} ev/sec (1.6×10^{-19} joule $=$ 1 ev). Energy of the order of 1 ev is needed by an electron in order to make it emerge from a metal surface, so that, on the average, 10^7 sec—several months—would have to elapse before any individual electron would accumulate enough energy to become a photoelectron. Actually, as we said above, there is no detectable time lag.

Still more puzzling is the fact that the photoelectron energy depends on the *frequency* of the light employed. At frequencies below a certain critical one (which is characteristic of the particular metal), no electrons whatever are emitted. Above this threshold frequency, the photoelectrons have a range of energies from zero to a certain maximum value, and *this maximum energy increases with increasing frequency.* High frequencies result in high maximum photoelectron energies, low frequencies in low maximum photoelectron energies. Thus a faint blue light produces electrons with more energy than those produced by a bright red light, although the latter yields a greater number of them.

Figure 31–2 is a graph of maximum photoelectron energy plotted against the frequency f of the incident light in a particular experiment. It is clear that the relationship between maximum photoelectron energy, KE_{max}, and light frequency is a simple one, which we can express in equation form as

$$KE_{max} = h(f - f_0) = hf - hf_0, \tag{31-1}$$

where f_0 is the threshold frequency below which no photoemission occurs and h is a constant. Significantly, the value of h is the same in *all* cases, although f_0 varies with the particular metal being illuminated.

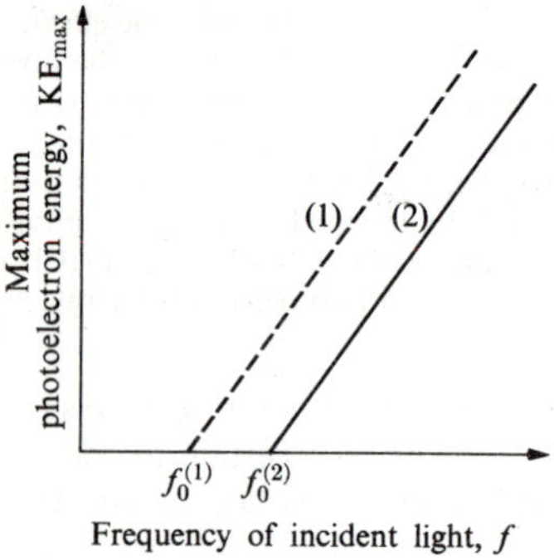

FIG. 31–2. The variation of maximum photoelectron energy with the frequency of the incident light for two target metals. No photoelectrons are emitted for frequencies less than $f_0^{(1)}$ in the case of metal (1) and less than $f_0^{(2)}$ in the case of metal (2). In both cases, however, the angle between the experimental line and either axis is the same. Hence we may write the equation of the lines as $KE_{max} = h(f - f_0)$, where h has the same value in all cases but f_0, the minimum frequency required for photoelectric emission to occur, depends on the nature of the target metal.

Einstein showed that one could understand the result expressed in Eq. (31–1) by adapting Planck's quantum hypothesis to the photoelectric effect. It will be recalled that, in order to derive the correct formula for the energy density of a blackbody, Planck was forced to assume that the oscillators in the walls of a blackbody radiate packets of energy hf, where h was a universal constant, and that

$$E = hf. \tag{31-2}$$

Einstein saw that Planck's idea could be used to interpret the photoelectric effect if light not only is emitted a quantum at a time but also interacts with electrons as quanta. Then the h of Eq. (31–1) is the same as the h of (31–2), and the significance of the former equation becomes clear when it is rewritten

$$hf = KE_{max} + hf_0. \tag{31-3}$$

What this equation states is simply that

$$\text{Quantum energy} = \text{maximum electron energy} + \text{energy required to eject an electron.}$$

The reason for a threshold frequency is clear: it corresponds to the energy required to dislodge an electron from the metal surface, and is called the *work function* of the metal. (There must be such a minimum energy, or electrons would leave metals spontaneously.) And, of course, there are several plausible reasons why not all photoelectrons have the same maximum energy. For instance, the initial energy of all the electrons need not be the same, and an electron may lose some of its initial energy in collisions within the metal before it actually emerges from the surface.

Let us apply Eq. (31–3) to a specific problem. The threshold frequency for copper is $1.1 \times 10^{15}\ \text{sec}^{-1}$. When ultraviolet light of frequency $1.5 \times 10^{15}\ \text{sec}^{-1}$ is shined on a copper surface, what is the maximum energy of the photoelectrons? We find that

$$\begin{aligned} KE_{\max} &= hf - hf_0 = h(f - f_0) \\ &= 6.63 \times 10^{-34}\ \text{j·sec} \times (1.5 - 1.1) \times 10^{15}\ \text{sec}^{-1} \\ &= 2.7 \times 10^{-19}\ \text{j.} \end{aligned}$$

Since 1 ev $= 1.6 \times 10^{-19}$ joule, the maximum photoelectron energy is

$$\frac{2.7 \times 10^{-19}\ \text{j}}{1.6 \times 10^{-19}\ \text{j/coul}} = 1.7\ \text{ev.}$$

Einstein's notion that light travels as a series of little packets of energy (sometimes referred to as quanta, sometimes as *photons*) is in complete contradiction to the wave theory of light (Fig. 31–3). And the wave theory, as we know, has some powerful observational evidence on its side. There is no other way of explaining interference and diffraction effects, for example. According to the wave theory, light spreads out from a source in a manner analogous to the spreading out of ripples on the surface of a lake when a

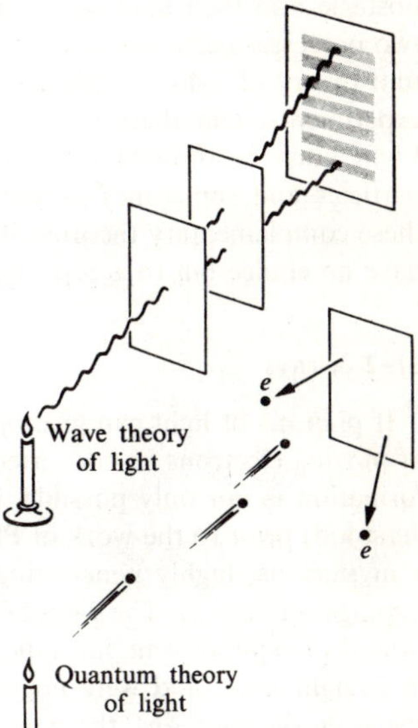

FIG. 31–3. The wave theory of light is necessary to explain diffraction and interference phenomena, which the quantum theory cannot explain. The quantum theory of light is necessary to explain the photoelectric effect, which the wave theory cannot explain.

stone is dropped into it, with the energy of the light distributed continuously throughout the wave pattern. According to the quantum theory, light spreads out from a source as a succession of localized packets of energy, each sufficiently small to permit its being absorbed by a single electron. Yet, despite the particle picture of light that it presents, the quantum theory requires a knowledge of the light frequency f, a wave quantity, in order to determine the energy of each quantum.

On the other hand, the quantum theory of light is able to explain the photoelectric effect. It predicts that the maximum photoelectron energy should depend on the frequency of the incident light and not on its intensity, precisely the opposite of what the wave theory suggests, and it is able to explain why even the feeblest light can lead to the immediate emission of photoelectrons. The wave theory can give no reason why there should be a threshold frequency below which no photoelectrons are observed, no matter how strong the light beam, something that follows naturally from the quantum theory.

Which theory is correct? The history of physics is filled with examples of physical ideas that required revision or even replacement when new empirical data conflicted with them, but this is the first occasion in which two completely different theories are both required to explain a single physical phenomenon. In thinking about this, it is important for us to note that, in a particular situation, light behaves *either* as though it has a wave nature *or* a particle nature, but not both. While light sometimes assumes one guise and sometimes the other, there is no physical process in which both are exhibited. The same light beam can diffract around an obstacle and then impinge on a metal surface to eject photoelectrons, but these two processes occur separately. The electromagnetic theory of light and the quantum theory of light complement each other; each by itself is "correct" in certain experiments, and there are no relevant experiments which neither can explain. Light must be thought of as a phenomenon that sometimes manifests itself as particles and sometimes as waves; while we cannot visualize its "true nature," these complementary theories of light are able to account for its behavior, and we have no choice but to accept them both.

31-2 X-rays

If photons of light can give up their energy to electrons, can the kinetic energy of moving electrons be converted into photons? The answer is that such a transformation is not only possible, but had in fact been discovered (though not understood) prior to the work of Planck and Einstein. In 1895 Roentgen found that a mysterious, highly penetrating radiation is emitted when high-speed electrons impinge on matter. The x-rays (so called because their nature was then unknown) caused phosphorescent substances to glow, exposed photographic plates, traveled in straight lines, and were not affected by electric or magnetic fields. The more energetic the electrons, the more penetrating the x-rays, and the greater the number of electrons, the greater the density of the resulting x-ray beam.

After over 10 years of study, it was finally established that x-rays exhibit, under certain circumstances, both interference and polarization effects, leading to the conclusion that they are electromagnetic waves. From the interference experiments their frequencies were found to be very high, above those in ultraviolet light.

Figure 31–4 is a diagram of an x-ray tube. A battery sends a current through the filament, heating it until it emits electrons. These electrons are then accelerated toward a metallic target by the high potential difference V provided by a voltage source. The tube is evacuated to permit the electrons to reach the target unimpeded. The impact of the electrons causes the evolution of x-rays from the target.

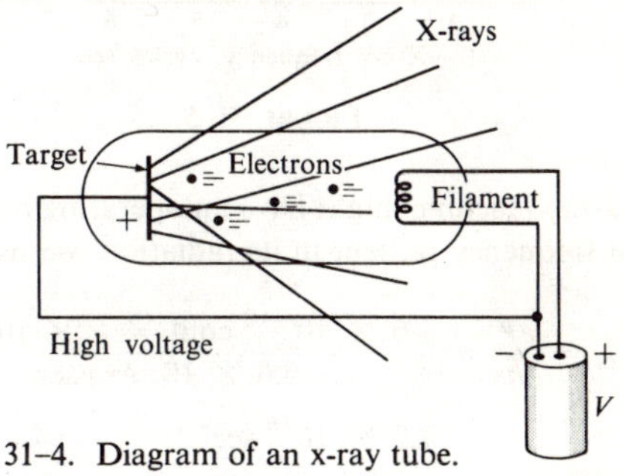

FIG. 31–4. Diagram of an x-ray tube.

What is the physical process involved in the production of x-rays? It is known that charged particles emit electromagnetic waves whenever they are accelerated, and so we may reasonably identify x-rays as the radiation accompanying the slowing down of fast electrons when they strike a metal. The great majority of the incident electrons, to be sure, lose their kinetic energy too gradually for x-rays to be evolved, and merely act to heat the target. (Consequently the targets in x-ray tubes are made of metals with high melting points, and a means for cooling the target is often provided.) A few electrons, however, lose much or all of their energy in single collisions with target atoms, and this is the energy that appears as x-rays. In other words, we may regard x-ray production as an inverse photoelectric effect. The highest frequency f_{max} found in the x-rays emitted from a particular tube should therefore correspond to a quantum energy of hf_{max}, where hf_{max} equals the kinetic energy of the electrons. That is,

$$KE = hf_{max}.$$

If we denote the charge of an electron by e, its kinetic energy after being accelerated through a potential difference V is

$$KE = eV;$$

hence we expect that

$$hf_{max} = eV \tag{31–4}$$

in the operation of an x-ray tube.

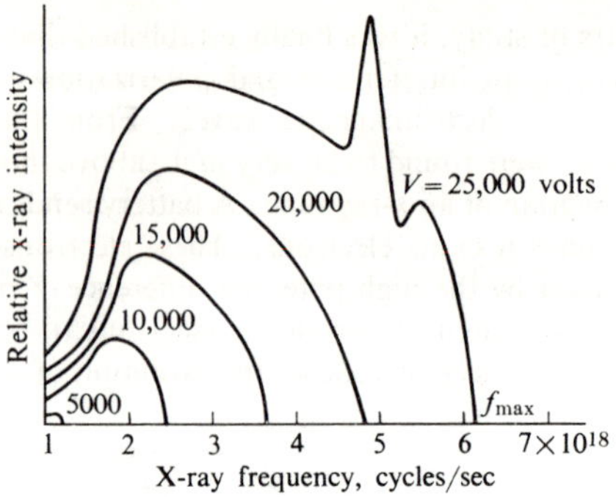

FIGURE 31-5

A conventional x-ray machine might have an operating potential of 50,000 volts. To find the highest frequency present in its radiation, we use Eq. (31-4):

$$f_{\max} = \frac{eV}{h} = \frac{1.6 \times 10^{-19}\ \text{coul} \times 5 \times 10^4\ \text{volts}}{6.6 \times 10^{-34}\ \text{j·sec}}$$
$$= 1.2 \times 10^{19}\ \text{sec}^{-1}.$$

Figure 31-5 shows the x-ray spectra that result when a molybdenum target is bombarded by electrons at several different accelerating voltages. The proportionality between $f_{\max}$ and V predicted by Eq. (31-4) agrees with the measurements, which provides confirmation for the quantum theory of light. The curves show that all frequencies below $f_{\max}$ are emitted, as we expect, and that their intensities vary in a regular manner. In addition, the highest-voltage curve possesses curious, sharp "spikes" that indicate the enhanced production of certain frequencies. These spikes originate in rearrangements of the electron structures of the target atoms that are induced by the bombarding electrons. A number of specific energies may be involved in these rearrangements, depending on the particular target material, so that the presence of discrete x-ray frequencies has a straightforward explanation.

31-3 The Compton effect

In the photoelectric effect, a light photon gives up most or all of its energy to an electron, leading us to attribute particle properties to the photon. How far can we carry the analogy? For instance, can we analyze the case of a photon making a "glancing" collision with an electron just as if both were billiard balls? Let us see what we can expect in such a collision if, indeed, photons behave like particles. Figure 31-6 shows the collision; an x-ray photon strikes an electron, which is initially at rest, and is deflected to one side, while the electron starts to move. As a result of the impact, the photon loses an amount of energy equal to the kinetic

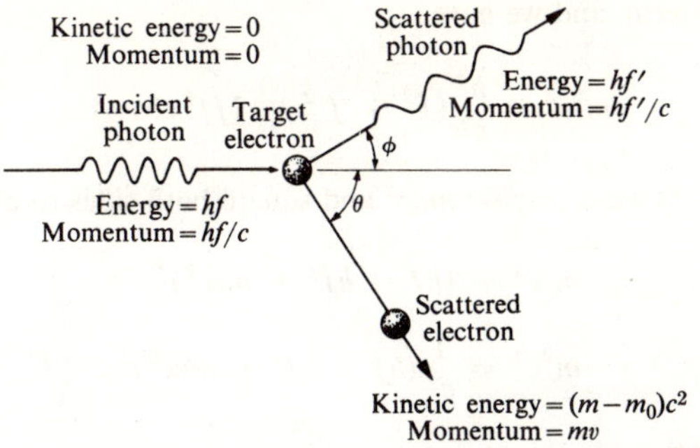

FIG. 31–6. The Compton effect. A vector diagram showing the various momenta and their projections parallel and perpendicular to the original photon direction is included to help in understanding the origin of Eqs. (31–7) and (31–8).

energy $(m - m_0)c^2$ gained by the electron. (The relativistic formula for kinetic energy is the proper one to use here.) If the original frequency associated with the photon is f, its loss of energy means that its subsequent frequency will be the lower value f', where

$$\text{Change in photon energy} = \text{electron kinetic energy},$$

$$hf - hf' = (m - m_0)c^2. \tag{31–5}$$

Momentum as well as energy is conserved in this collision. The velocities involved are so high that it is appropriate to use the relativistic formula $mv = m_0 v/\sqrt{1 - (v^2/c^2)}$ for the electron momentum. What is the momentum of the photon? According to Eq. (27–40), the momentum density of an electromagnetic field is equal to its energy density divided by c. In the photon picture this means that hf/c and hf'/c are, respectively, the initial and final momenta of the photon. The initial and final momenta of the target electron are 0 and mv, respectively. Conservation of the x- and y-components of momentum (see Fig. 31–7) yields

$$\frac{hf}{c} + 0 = \frac{hf'}{c}\cos\phi + mv\cos\theta, \tag{31–6}$$

$$0 = \frac{hf'}{c}\sin\phi - mv\sin\theta. \tag{31–7}$$

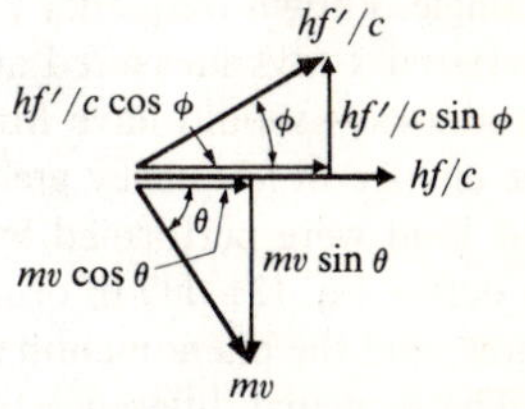

FIG. 31–7. Vector diagram of Compton effect.

We now use Eqs. (31–5), (31–6), and (31–7) to eliminate v and θ. First we transpose the $\cos\theta$ term in Eq. (31–6) and square both sides, and similarly we transpose the $\sin\theta$ term in Eq. (31–7) and square both sides. Adding the resultant equations eliminates the θ term, and we have

$$m^2v^2 = \frac{h^2}{c^2}(f^2 + f'^2 - 2ff' \cos \phi). \tag{31-8}$$

Now in Eq. (31-5) we transpose m_0c^2 and square both sides to obtain

$$m^2c^4 = (hf - hf' + m_0c^2)^2,$$

$$m^2c^2 = \frac{1}{c^2}(hf - hf' + m_0c^2)^2. \tag{31-9}$$

Subtracting (31-8) from (31-9) yields

$$m^2(c^2 - v^2) = \frac{m_0^2}{1 - (v^2/c^2)}(c^2 - v^2) = m_0^2c^2$$

$$= m_0^2c^2 - \frac{2h^2}{c^2} ff'(1 - \cos \phi) + 2hm_0(f - f'). \tag{31-10}$$

Division by ff' gives

$$\frac{1}{f'} = \frac{1}{f} + \frac{h}{m_0c^2}(1 - \cos \phi). \tag{31-11a}$$

Expressed in terms of wavelengths, this result is

$$\lambda' - \lambda = \frac{h}{m_0c}(1 - \cos \phi). \tag{31-11b}$$

The quantity h/m_0c has the dimension of a length and is equal to 0.02426 A. It represents half of the maximum change in the wavelength in Compton scattering. This formula (31-11) can readily be checked by experiment; a beam of x-rays of a single, known frequency f is directed at some target, and the frequencies of the scattered x-rays measured at various angles. According to Eq. (31-11), the scattered x-rays should have lower frequencies than those in the original beam, with the change in frequency greatest for the greatest scattering angle. Experiments of this kind were performed by A. H. Compton in the early 1920's, and he was led to derive Eq. (31-11) in order to account for the results. Theory and experiment agree, and the phenomenon they refer to has become known as the *Compton effect.*

The principal difference between the Compton and photoelectric effects is that in the latter the photon is completely absorbed and momentum conservation plays no role, since the entire metal (which has effectively an infinite mass) takes up the recoil momentum. (For a ball bouncing against a stone wall the conservation of momentum similarly plays no role.) The Compton effect demonstrates further the particle properties of a photon, emphasizing as it does the fact that a photon has momentum.

31-4 Matter waves

As we have seen, electromagnetic waves under certain circumstances have properties indistinguishable from those of particles. It requires no greater stretch of the imagination to speculate whether what we normally think of as particles might not have wave properties, too. This speculation was first made in 1924 by Louis de Broglie. He started with the formula for the momentum of a photon,

$$\text{Momentum} = hf/c,$$

which can be expressed in terms of wavelength λ as

$$\text{Momentum} = h/\lambda$$

since $\lambda f = c$. Hence, for a photon,

$$\lambda = \frac{h}{\text{momentum}}.$$

De Broglie suggested that this equation for wavelength is a perfectly general one, applying to material particles as well as to photons. In the case of particles,

$$\text{Momentum} = mv,$$

and so the *de Broglie wavelength* of a particle is

$$\lambda = h/mv. \tag{31-12}$$

The higher the momentum of the particle, the shorter its wavelength. Also $m_0/\sqrt{1 - (v^2/c^2)}$ must be used for m in Eq. (31-12).

How can de Broglie's hypothesis be verified? Perhaps the most striking exhibition of wave behavior is a diffraction pattern, which depends on the ability of waves both to bend around obstacles and to interfere constructively and destructively with one another. Several years after de Broglie's work, Davisson and Germer in the United States and G. P. Thomson in England independently demonstrated that streams of electrons are diffracted when they are scattered from crystals whose atoms have appropriate spacing (Fig. 31-8). The diffraction patterns they observed were in complete accord with the electron wavelength predicted by Eq. (31-12). As is true for electromagnetic waves, the wave and particle aspects of moving bodies never appear together in the same experiment, so we cannot determine which is the "correct" description. All we can say is that at certain times we must think of a moving body as a particle (in the sense of Newton's laws of motion) and at other times as a wave; whatever the ultimate nature of matter, its manifestations to us exhibit this inescapable duality.

Let us calculate the de Broglie wavelengths of two moving objects, an automobile whose speed is 60 mi/hr and an electron whose speed is 10^7 m/sec. The auto-

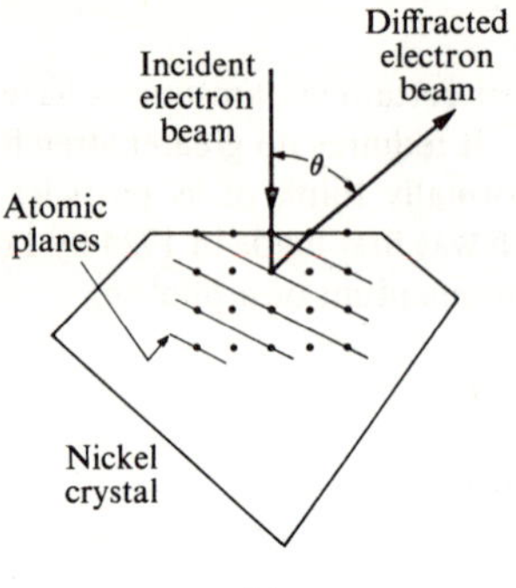

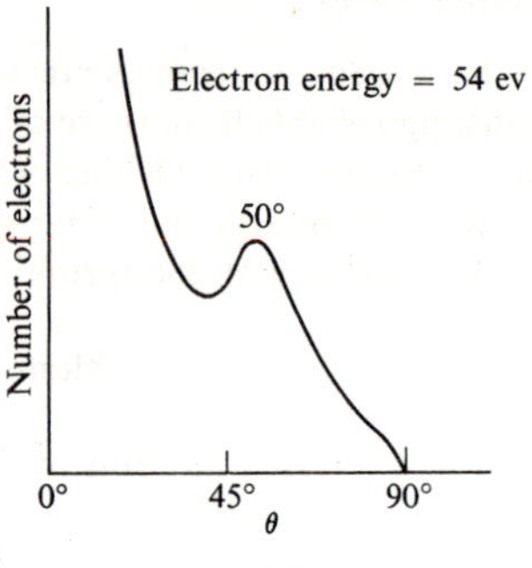

FIG. 31-8. The Davisson-Germer experiment. A peak in the number of diffracted electrons occurs at $\theta = 50°$ in the case of 54-ev electrons, in agreement with de Broglie's wavelength formula.

mobile might weigh 3200 lb, corresponding to a mass of 100 slugs, and 60 mi/hr = 88 ft/sec; hence,

$$\lambda = \frac{h}{mv} = \frac{6.62 \times 10^{-34}\,\text{j·sec} \times 0.738\,\text{ft·lb/j}}{100\,\text{slugs} \times 88\,\text{ft/sec}} = 5.5 \times 10^{-38}\,\text{ft}.$$

The wavelength of the automobile is so small relative to its dimensions that we expect no wave aspects in its behavior. For the electron, however,

$$\lambda = \frac{h}{mv} = \frac{6.62 \times 10^{-34}\,\text{j·sec}}{9.1 \times 10^{-31}\,\text{kg} \times 10^{7}\,\text{m/sec}} = 7.3 \times 10^{-9}\,\text{m},$$

which is of the order of magnitude of atomic dimensions. It is not surprising, then, that electrons with velocities of about 10^7 m/sec exhibit diffraction effects when incident upon crystals.

31-5 The uncertainty principle

The wave nature of a moving particle leads to the important question of what we mean by the position of the particle. If there were only a single wavelength associated with an electron, for example, it would be hard to see what might be meant by its position. We expect, for two reasons, that electrons are actually localized. The first is that experimentally we can focus a beam of electrons on a luminescent screen (for instance that of a television picture tube) into a very minute region. Second, we would expect that there should be a continuous connection between microscopic, atomic phenomena and classical ideas; and classical particles are localized in space. We are therefore led to assume that what is required is a

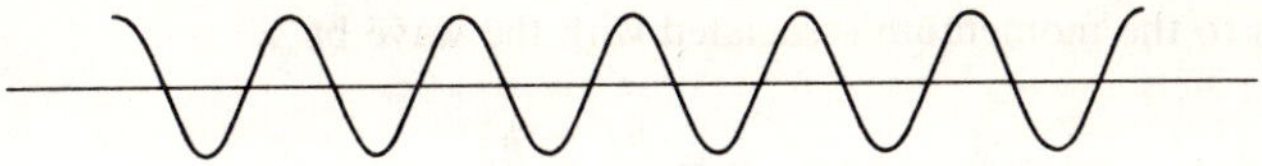

(a)

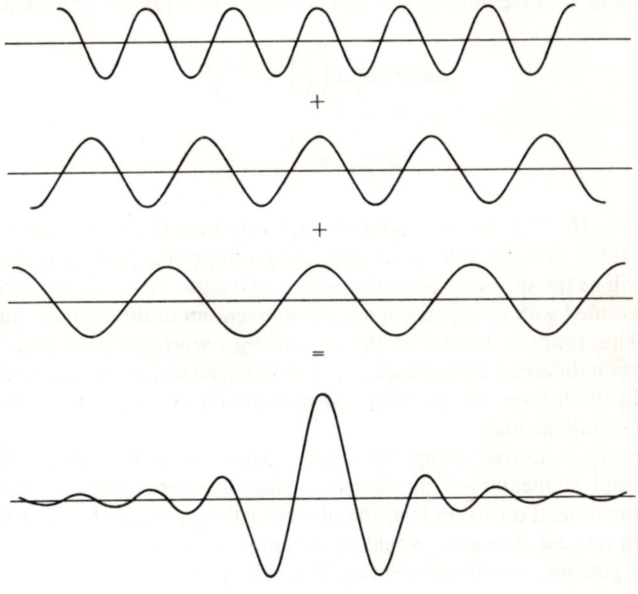

(b)

FIG. 31–9. (a) Representation of an electron by a wave of a single wavelength does not permit us to specify the position of the particle. (b) Superposition of many slightly different wavelengths enables us to form a wave packet which is correlated with electron position.

superposition of various wavelengths to make a packet of waves (Fig. 31–9) which is localized in space, i.e., large in some small region and nearly zero elsewhere, according to the discussion of Chapter 25. The center of this wave packet corresponds to the position of the particle. This picture has some significant consequences. We learned in Chapter 25 that, when a wave packet is formed, the size of the wave packet and the range of wave numbers required to form the packet are related by

$$\Delta x \, \Delta k \approx 1, \tag{31–13}$$

where Δx is the width of the packet and Δk the range of wave numbers required to form the packet. The wavelength, according to the results of the Compton effect, is related to the momentum associated with the wave by

$$p = \frac{hf}{c} = \frac{h}{\lambda}. \tag{31–14}$$

Thus a range of wave numbers Δk can be related to a range of momenta by

$$\Delta k = 2\pi\,\Delta\left(\frac{1}{\lambda}\right) = \frac{2\pi\,\Delta p}{h}$$

and so

$$\Delta x\,\Delta p \approx \frac{h}{2\pi} = \hbar, \tag{31–15}$$

where $\hbar$ ($\approx 10^{-34}$ j · sec) is a symbol frequently used for $h/2\pi$. The meaning of Eq. (31–15) is that the more accurately the position of a particle is specified, the greater will be the spread of possible particle momenta. In the limit where the position is specified with infinite accuracy, the momentum of the particle can have any value. This result is known as the *Heisenberg uncertainty principle*. Evidently an important difference between quantum phenomena and macroscopic phenomena is that, in the former, the position and momentum of a particle cannot both be specified simultaneously.

Heisenberg conceived many experiments which could in principle specify the position and momentum of an electron. One such experiment involved shining one photon of light on an electron and observing the scattered photon with a microscope. In no case, however, would observation permit us to establish simultaneously the position and the momentum of an electron.

The uncertainty principle is now regarded as well established. For macroscopic particles it does not impose any serious limitation on our previous results. If $\Delta x \approx 10^{-10}$ m, then

$$\Delta p \approx \frac{\hbar}{\Delta x} \approx \frac{10^{-34}\ \text{j·sec}}{10^{-10}\ \text{m/sec}} = 10^{-24}\,\frac{\text{kg·m}}{\text{sec}}.$$

For a particle of mass 1 gm, this yields a velocity uncertainty of

$$\Delta p = m\,\Delta v = 10^{-3}\,\Delta v = 10^{-24}\,\frac{\text{kg·m}}{\text{sec}},$$

$$\Delta v = 10^{-21}\ \text{m/sec}. \tag{31–16}$$

This spread in velocities is so small that it could not conceivably be detected. For an electron, however, $m \cong 9 \times 10^{-31}$ kg, and so

$$m\,\Delta v = 9 \times 10^{-31}\,\Delta v = 10^{-24}\,\frac{\text{kg·m}}{\text{sec}},$$

$$\Delta v \cong 10^{6}\ \text{m/sec}. \tag{31–17}$$

This spread in velocities is of the order of actual electronic velocities on the microscopic level. Thus the uncertainty principle is an important one for atomic phenomena, but of no consequence for macroscopic bodies.

A significant application of the uncertainty principle is to the question of whether electrons are present within atomic nuclei. As we shall learn in Chapter 36, nuclei are roughly 10^{-15} m across. If an electron is to be confined within a nucleus, the uncertainty in its position cannot exceed 10^{-15} m. The consequent inherent uncertainty in the electron's momentum is

$$\Delta(mv) \approx \frac{h}{\Delta x} \approx \frac{10^{-34}\ \text{j·sec}}{10^{-15}\ \text{m}} \approx 10^{-19}\ \frac{\text{kg·m}}{\text{sec}}.$$

This momentum uncertainty corresponds to an energy uncertainty of hundreds of millions of electron volts! But there is no evidence that electrons found in atoms ever have much more than 1/1000 of this huge amount of energy, and we therefore conclude that electrons are not present inside atomic nuclei.

The uncertainty principle could also be expressed in the time domain. We saw in Chapter 25, too, that for any wave train,

$$\Delta f\,\Delta t \approx 1. \tag{31–18}$$

Since energy is related to frequency by $E = hf$, we have

$$\Delta E = h\,\Delta f,$$

which gives

$$\Delta E\,\Delta t \approx h. \tag{31–19}$$

The spread in the possible energies of a wave train is related to the time that the system exists with this energy spread by (31–19). Equation (31–19) is, as we shall see, of some importance in estimating the width of spectral lines.

31–6 Causality

Quantum theory and the associated uncertainty principle have not only had a profound effect on our concepts of microscopic physics, but have had important philosophical consequences as well. According to classical notions, if we have a particle and know the forces acting on it, we can determine the subsequent motion of the particle by using Newton's second law of motion, $\mathbf{F} = m\mathbf{a}$, starting from its initial position and velocity. However, these are precisely the two quantities which the uncertainty principle prohibits from being determined simultaneously. The subsequent behavior therefore cannot be precisely determined, and our notions of a precise causal connection between initial forces and subsequent motions must be abandoned on the microscopic level. What the quantum theory proposes instead is a precise prediction about the motion of the waves associated with a particle, but the connection between the wave and the position of the particle is only a statistical one.

Problems

1. If light transfers energy by means of separate quanta, why do we not perceive a faint light as a series of tiny flashes?

2. Yellow light has a wavelength of 6×10^{-7} m. How many photons are emitted per second by a yellow lamp radiating at a power of 10 watts?

3. A radio transmitter operates at a frequency of 880 kc/sec (880×10^3 cycles/sec) and a power of 10,000 watts. How many photons per second does it emit?

4. What is the energy of a quantum for an oscillation of 60 cycles/sec? 10^6 cycles/sec? 10^9 cycles/sec?

5. What is the lowest-frequency light that will cause the emission of photoelectrons from a surface whose nature is such that 1.9 ev is required to eject an electron?

6. Photoelectrons are emitted with a maximum speed of 7×10^5 m/sec from a surface when light of frequency 8×10^{14} cycles/sec is shined on it. What is the threshold frequency for this surface?

7. Light of wavelength 5×10^{-7} m falls on a potassium surface whose nature is such that 2 ev is needed to eject an electron. What is the maximum kinetic energy in ev of the photoelectrons that are emitted?

8. The photoelectric threshold for aluminum is 4770 A, and for potassium 6000 A. What are the work functions for these metals in electron volts?

9. Why do you think the wave aspect of light was discovered earlier than its particle aspect?

10. What potential difference must be applied across an x-ray tube for it to emit x-rays with a minimum wavelength of 10^{-11} m?

11. What potential difference must be applied across an x-ray tube for it to emit x-rays with a maximum frequency of 10^{19} cycles/sec?

12. A target is bombarded by 8×10^4 ev electrons. What is the highest frequency present in the emitted x-rays?

13. Electrons are accelerated in television tubes through potential differences of about 10,000 v. Find the highest frequency of the electromagnetic waves that are emitted when these electrons strike the screen of the tube.

14. Find the energy and momentum of an x-ray photon whose frequency is 5×10^{18} cycles/sec.

15. Find the energy and momentum of an x-ray photon whose wavelength is 2×10^{-11} m.

16. An x-ray photon of frequency 3×10^{19} cycles/sec is scattered through an angle of 90° after colliding with a stationary electron. (a) What is its new frequency? (b) What is the kinetic energy of the electron after the collision?

17. An x-ray photon of frequency 10^{19} cycles/sec is scattered through an angle of 45° after colliding with a stationary electron. (a) What is its new frequency? (b) What is the kinetic energy of the electron after the collision?

18. An x-ray photon of frequency 1.5×10^{19} cycles/sec undergoes Compton scattering. Its new frequency is 1.2×10^{19} cycles/sec. What is the change in the kinetic energy of the electron it has struck?

19. Monochromatic x-rays of wavelength 1 A are scattered from a carbon block. (a) What is the wavelength of the x-rays scattered through 180°? (b) What is the

maximum kinetic energy of the recoil electrons?

20. Determine (a) the direction of the recoil electrons for x-rays of wavelength λ scattered through an angle ϕ, and (b) the energy of the recoil electron.

21. A photon colliding with an electron at rest gives the electron an energy of 10,000 ev. If the photon is scattered at 90°, find the initial and final frequencies of the photon.

22. Give as many reasons as you can to support (a) the wave theory of light, (b) the particle theory of light.

23. Find the de Broglie wavelength of a 10-ton truck whose speed is 60 mi/hr.

24. Calculate the de Broglie wavelength of (a) an electron whose speed is 1×10^8 m/sec, and (b) an electron whose speed is 2×10^8 m/sec. Use relativistic formulas in your calculations.

25. Calculate the de Broglie wavelength of a proton whose kinetic energy is 1 Mev ($=10^6$ ev). This calculation may be made nonrelativistically.

26. (a) How could you experimentally distinguish between a photon whose momentum is 10^{-22} kg·m/sec and an electron whose momentum is 10^{-22} kg·m/sec? (b) What is the energy of each? (c) What is the wavelength of each?

27. How could you experimentally distinguish between an electromagnetic wave whose wavelength is 10^{-11} m and an electron whose de Broglie wavelength is 10^{-11} m?

28. What is the approximate momentum imparted to a proton initially at rest by a measurement which locates its position to 10^{-11} m?

29. (a) An electron is confined in a box 10^{-9} m in length. What is the uncertainty in its velocity? (b) A proton is confined in the same box. What is the uncertainty in its velocity?

30. The electron in a hydrogen atom may be thought of as confined to a region a radius of 5×10^{-11} m from the proton. (a) Use the uncertainty principle to calculate the momentum of the electron, and from this calculate its kinetic energy. (b) What keeps the electron in this region despite its kinetic energy?

31. An harmonic oscillator has a kinetic energy equal to $\frac{1}{2}mv^2 = p^2/2m$, where p is the momentum. Its potential energy is $\frac{1}{2}kx^2 = \frac{1}{2}m\omega^2x^2$. What is the minimum average energy of the harmonic oscillator if the uncertainty in its position times the uncertainty in its momentum is set equal to $\hbar/2$? [*Hint:* The uncertainty in position is the square root of the average value of $(x - x_{av})^2$ which is equal to $(x^2)_{av}$ since the $x_{av} = 0$. The same is true for the momentum.]

32

The Atom

In the previous chapter we saw how Planck's quantum hypothesis accounts for the photoelectric effect, the Compton effect, and the production of x-rays. The most important early application of the quantum hypothesis was made by Niels Bohr to explain the characteristic spectrum of light emitted by atoms of a particular element. Bohr was at Rutherford's laboratory when the latter was doing his important work on atomic structure. We recall from Chapter 21 that, as a result of these investigations, there emerged a picture of the atom characterized by a small region of positive charge $\sim 10^{-15}$ m in radius, which was neutralized by an equal amount of negative charge in the form of electrons at a comparatively great distance away. Bohr adopted this model, and from this and the quantum hypothesis fashioned the first consistent theory of atomic structure. We shall investigate Bohr's theory in this chapter.

32–1 Electron orbits

The negatively charged electrons which surround the positive nucleus in the Rutherford model of the atom cannot be stationary, since the electrostatic force between them and the nucleus would attract them to the latter at once. If the electrons are in motion about the nucleus, however, dynamically stable orbits, like those of the planets about the sun, are possible.

Let us restrict ourselves for the time being to the hydrogen atom, which, having an atomic number of $Z = 1$, is the simplest, with but a single proton as nucleus and one electron around it (Fig. 32–1). Then the centripetal force

$$F_c = \frac{mv^2}{r}$$

holding the electron in an orbit of radius r from the nucleus must be provided by the electrostatic force

$$F_e = \frac{1}{4\pi\epsilon_0}\frac{e^2}{r^2}$$

between them, and

$$\frac{mv^2}{r} = \frac{1}{4\pi\epsilon_0}\frac{e^2}{r^2}. \tag{32–1}$$

The electron speed v is therefore related to its orbit radius r by the formula

$$v = \frac{e}{\sqrt{4\pi\epsilon_0 mr}}. \tag{32–2}$$

The total energy E of the electron in a hydrogen atom is the sum of its kinetic energy

$$\text{KE} = \tfrac{1}{2}mv^2$$

and its electrostatic potential energy

$$\text{PE} = -\frac{1}{4\pi\epsilon_0}\frac{e^2}{r}.$$

(We recall that the latter formula follows from a calculation of the amount of work that would have to be done to bring the electron to an infinite distance from the proton; the potential energy here is a *negative* quantity since, being attracted to the proton, the electron cannot do work on anything outside the atom if it were to be released from its orbital motion.) Hence

$$E = \text{KE} + \text{PE} = \frac{mv^2}{2} - \frac{1}{4\pi\epsilon_0}\frac{e^2}{r}.$$

Substituting for the speed v from Eq. (32–2),

$$E = \frac{1}{4\pi\epsilon_0}\left(\frac{e^2}{2r} - \frac{e^2}{r}\right) = -\frac{1}{8\pi\epsilon_0}\frac{e^2}{r}. \tag{32–3}$$

The total energy of an atomic electron is negative, implying that it is bound to the nucleus. If E were greater than zero, the electron would not be able to remain in a closed orbit about the nucleus. The fact that $E = -\frac{1}{2}V$ is a general result for all orbital motions due to an inverse-square law of force.

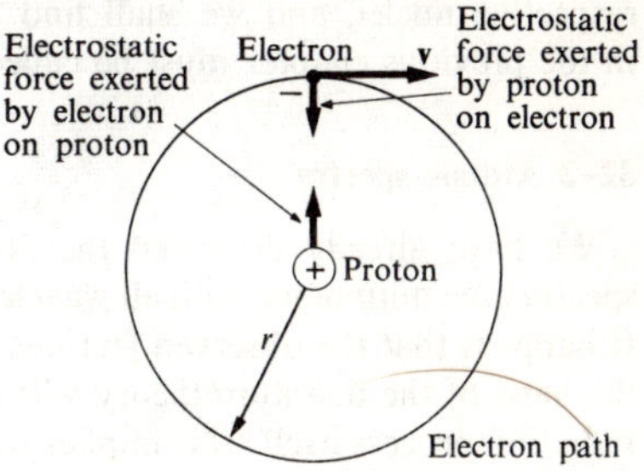

FIG. 32–1. The hydrogen atom in Bohr's theory consists of an electron circling a proton. The electrostatic force exerted by the proton on the electron provides the centripetal force required to hold it in a circular path. The proton is nearly 2000 times heavier than the electron, and so its motion under the influence of the electrostatic force the electron exerts is not significant.

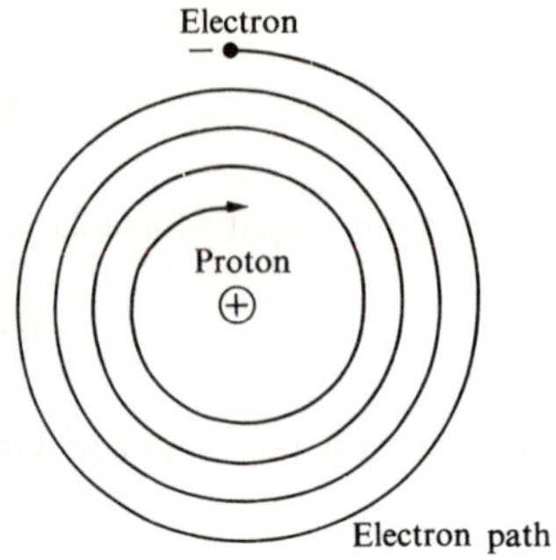

FIG. 32-2. According to electromagnetic theory, an accelerated electron radiates electromagnetic waves. The electron in a hydrogen atom, since it constantly experiences a centripetal acceleration, should therefore radiate energy and spiral into the nucleus.

Experiments indicate that 13.6 ev are required to separate a hydrogen atom into a proton and an electron, which means that its binding energy E is -13.6 ev. Since 13.6 ev $= 2.2 \times 10^{-18}$ joule, we can find the orbital radius of the electron in a hydrogen atom from Eq. (32-3). Since $1/4\pi\epsilon_0 = 9 \times 10^9\ \text{n·m}^2/\text{coul}^2$, this radius is

$$\begin{aligned} r &= -\frac{1}{8\pi\epsilon_0}\frac{e}{E} \\ &= -\frac{9 \times 10^9\ \text{n·m}^2/\text{coul}^2 \times (1.6 \times 10^{-19}\ \text{coul})^2}{2 \times (-2.2 \times 10^{-18}\ \text{j})} \\ &= 5.3 \times 10^{-11}\ \text{m}. \end{aligned}$$

The above analysis is dynamically flawless, but unfortunately it requires that the electron be continuously accelerated owing to its circular motion. And, according to electromagnetic theory, all accelerated electric charges must radiate electromagnetic waves. It is the acceleration experienced by a fast electron when it strikes a target that causes it to emit x-rays, for instance. An atomic electron circling its nucleus to keep from falling into it by electrostatic attraction therefore cannot (according to electromagnetic theory) help radiating away energy, thereby spiraling inward as r decreases until it is swallowed up by the nucleus (Fig. 32-2). Conventional physics is thus completely unable to account for the observed existence of stable atoms whose electrons remain at great distances from their respective nuclei, and we shall find that the once-revolutionary ideas discussed in the previous chapter must be called upon if we are to understand the atom.

32-2 Atomic spectra

We have already discussed the fact that heated solids emit radiation whose spectra are continuous, with all wavelengths present although in varying intensities. It happens that the observed features of blackbody radiation can be explained on the basis of the quantum theory without our having to consider the details of the radiation process itself; this implies that the nature of the heated solid is irrelevant

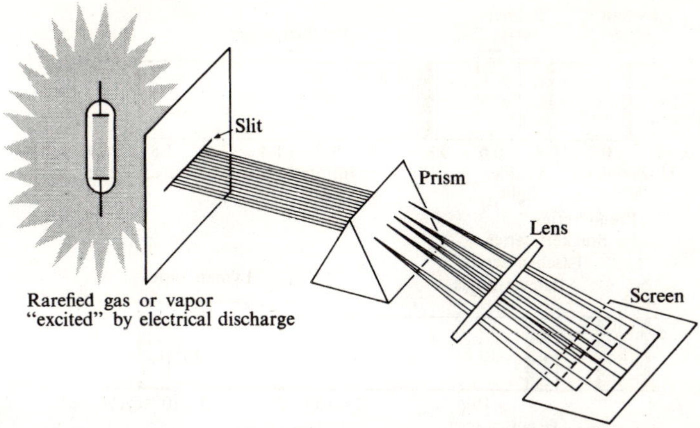

FIG. 32–3. The spectrum of a substance may be obtained by "exciting" a sample of its vapor in a tube by means of an electrical discharge and directing the light it gives off through a prism.

to the process. Since the atoms of a solid are packed closely together, it is reasonable to suppose that their constituent electrons interact with one another. Hence, when the solid is heated to incandescence, any electron behavior characteristic of a particular atomic species becomes subordinate to the collective behavior of a great many interacting electrons.

At the other extreme, the atoms or molecules in a rarefied gas are so far apart on the average that they can be considered as not interacting with one another at all except during brief, relatively infrequent collisions. Here we would expect any emitted radiation to reflect the properties of the individual atoms or molecules that are present, an expectation that is fulfilled experimentally. When an atomic gas or vapor at somewhat less than atmospheric pressure is "excited" by the passage of an electric current through it, light whose spectrum consists of a limited number of discrete wavelengths is emitted. A typical laboratory arrangement for observing such spectra is sketched in Fig. 32–3. Figure 32–4 shows the atomic spectrum of hydrogen; it is called a *line spectrum* from its appearance. Every element exhibits a unique line spectrum when a sample of it is suitably excited, and its presence in a substance of unknown composition can be ascertained by the appearance of its characteristic wavelengths in the spectrum of the substance.

When a molecular gas or vapor at low pressure is excited, "bands" consisting of many individual lines very close together are found in its spectrum (Fig. 32–5). The presence of bands in molecular spectra arises from interactions among the electrons of the atoms in each molecule; hence molecular spectra do not provide as much insight into atomic structure as line spectra do.

It is worth noting that while unexcited gases (for instance, atomic hydrogen at room temperature) do not radiate their characteristic spectral lines, they *absorb*

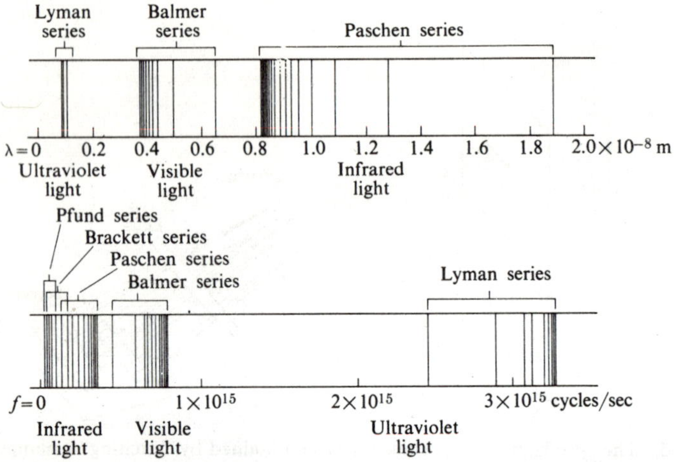

FIG. 32-4. The line spectrum of hydrogen. Note the presence of various series of spectral lines.

FIG. 32-5. A portion of the spectrum of the molecular gas NO showing the presence of bands rather than lines.

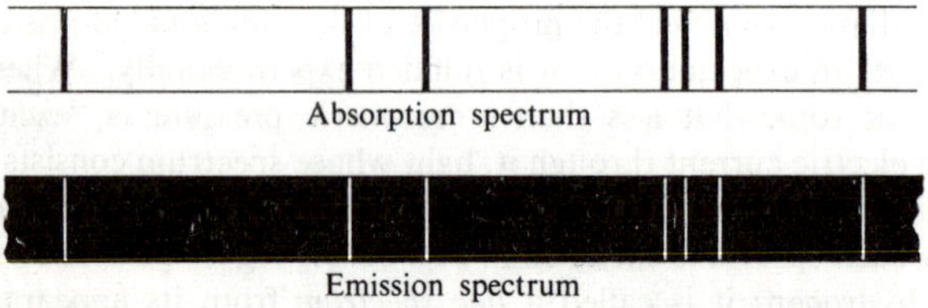

FIG. 32-6. Portions of the emission (bright line) and absorption (dark line) spectra of an element.

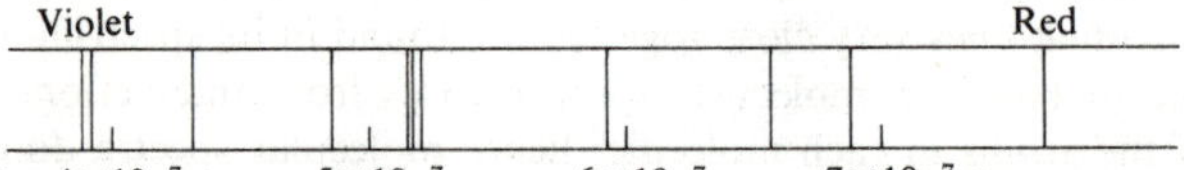

FIG. 32-7. The most prominent Fraunhofer lines in the visible part of the solar spectrum.

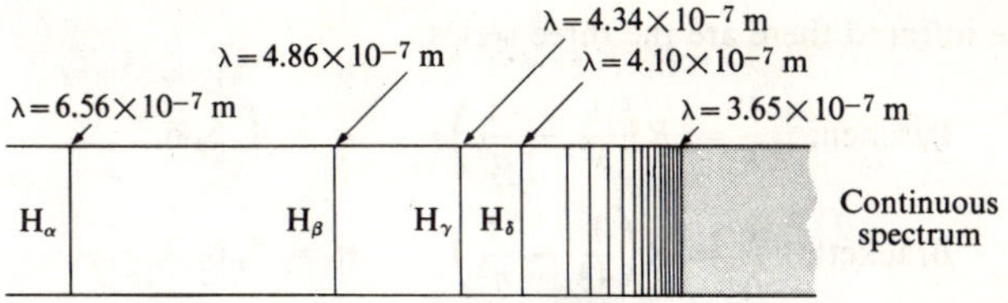

FIG. 32–8. The Balmer series of spectral lines in atomic hydrogen.

light of certain of those particular wavelengths when white light is passed through them. Emission spectra consist of bright lines on a dark background, while absorption spectra consist of dark lines on a bright background (Fig. 32–6). The dark *Fraunhofer lines* in the solar spectrum (Fig. 32–7) occur because the luminous sun, which radiates approximately like a blackbody at 5800°K, is surrounded by an envelope of cooler gas that absorbs light of certain characteristic wavelengths.

The early spectroscopists found that the wavelengths present in atomic spectra fall into definite series, all of which can be expressed by similar simple formulas. The first spectral series was discovered by J. J. Balmer in 1885, who studied the wavelengths emitted by hydrogen in the visible part of its spectrum. As shown in Fig. 32–8, the *Balmer series* consists of a number of lines, the one with the longest wavelength being designated H_α, the next H_β, and so on. As the wavelength decreases, the lines are closer together and weaker in intensity until the *series limit* at a wavelength of 3.65×10^{-7} meter is reached, beyond which there are no more lines but only a faint continuous spectrum. Balmer's empirical formula for this series is

$$\text{Balmer:} \quad \frac{1}{\lambda} = R\left(\frac{1}{2^2} - \frac{1}{n^2}\right), \qquad n = 3, 4, 5, \ldots, \tag{32–4}$$

where R, called the *Rydberg constant*, has the value

$$R = 1.097 \times 10^7 \text{ m}^{-1}.$$

The H_α line corresponds to $n = 3$, the H_β line to $n = 4$, and so on. The series limit corresponds to $n = \infty$, so that its wavelength is $4/R$, in agreement with experiment.

The Balmer series includes only wavelengths in the visible part of the hydrogen spectrum. However, the spectral lines of hydrogen in the ultraviolet and infrared have been found to comprise four other series (Fig. 32–4). In the ultraviolet the *Lyman series* is specified by

$$\text{Lyman:} \quad \frac{1}{\lambda} = R\left(\frac{1}{1^2} - \frac{1}{n^2}\right), \qquad n = 2, 3, 4, \ldots, \tag{32–5}$$

while in the infrared there are the three series

$$\text{Paschen:}\quad \frac{1}{\lambda} = R\left(\frac{1}{3^2} - \frac{1}{n^2}\right), \qquad n = 4, 5, 6, \ldots, \tag{32-6}$$

$$\text{Brackett:}\quad \frac{1}{\lambda} = R\left(\frac{1}{4^2} - \frac{1}{n^2}\right), \qquad n = 5, 6, 7, \ldots, \tag{32-7}$$

$$\text{Pfund:}\quad \frac{1}{\lambda} = R\left(\frac{1}{5^2} - \frac{1}{n^2}\right), \qquad n = 6, 7, 8, \ldots. \tag{32-8}$$

In the above equations the value of R is the same as in Eq. (32-4).

The remarkable regularities in the hydrogen spectrum are not able to lead us directly to an elucidation of the structure of the hydrogen atom, but they do present a definite, quantitative test for any theory of this structure. This test is failed by the classical picture of the atom.

32-3 The Bohr atom

The first theory of the atom to yield results that agree with our observations was proposed by Niels Bohr in 1913. By applying quantum ideas to atomic structure, Bohr arrived at a detailed model which, though subsequently replaced by an abstract description of greater accuracy and usefulness, remains today as the mental image most scientists have of the atom. Our derivation of the Bohr atom differs slightly from that originally given by Bohr, though all the results are identical.

As we saw, the ordinary laws of physics cannot account for the stability of the hydrogen atom, the simplest of all, whose electron must be whirling around the nucleus to keep from being pulled into it and yet must be radiating electromagnetic energy continuously. However, since other phenomena apparently impossible to understand—the photoelectric effect, for instance—found complete explanation in terms of quantum concepts, it is appropriate for us to inquire whether this might also be true for the atom. Let us begin by looking into the wave properties of the electron in a hydrogen atom. The de Broglie wavelength λ of any moving particle of mass m and speed v is

$$\lambda = h/mv,$$

where h is Planck's constant. Now the speed of the electron in a hydrogen atom is, by Eq. (32-2),

$$v = \frac{e}{\sqrt{4\pi\epsilon_0 mr}},$$

and so its wavelength is

$$\lambda = \frac{h}{e}\sqrt{4\pi\epsilon_0 r/m}. \tag{32-9}$$

Substituting 5.3×10^{-11} meter for the radius r of the electron orbit, we find,

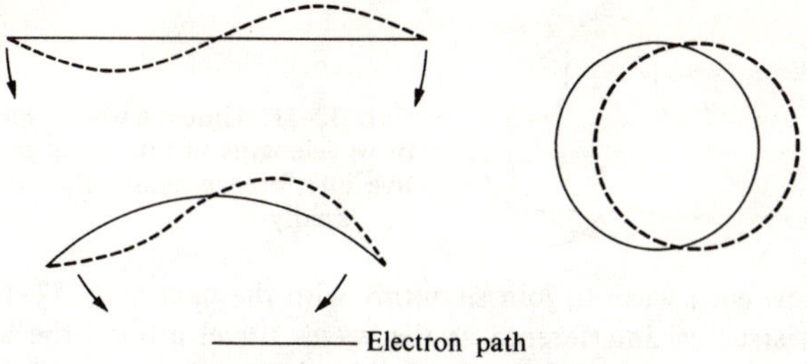

FIG. 32-9. The electron orbit in a hydrogen atom is exactly one de Broglie wavelength in circumference.

again setting $1/4\pi\epsilon_0 = 9 \times 10^9$ n·m²/coul²,

$$\lambda = \frac{6.6 \times 10^{-34}\ \text{j·sec}}{1.6 \times 10^{-19}\ \text{coul}} \sqrt{\frac{5.3 \times 10^{-11}\ \text{m}}{9 \times 10^9\ \text{n·m}^2/\text{coul}^2 \times 9.1 \times 10^{-31}\ \text{kg}}}$$

$$= 33 \times 10^{-11}\ \text{m}.$$

This is a most exciting result, because the electron orbit is exactly

$$2\pi r = 33 \times 10^{-11}\ \text{m}$$

in circumference. Therefore the orbit of the electron in a hydrogen atom corresponds to one complete electron wave joined on itself (Fig. 32-9).

The fact that the electron orbit in a hydrogen atom is one electron wavelength in circumference is just the clue we need to construct a theory of the atom. If we examine the vibrations of a wire loop, as in Fig. 32-10, we see that their wavelengths always fit an integral number of times into the loop's circumference, each wave joining smoothly with the next. In the absence of dissipative effects, such vibrations would persist indefinitely. Why are these the only vibrations possible in a wire loop? A fractional number of wavelengths cannot be fitted into the loop

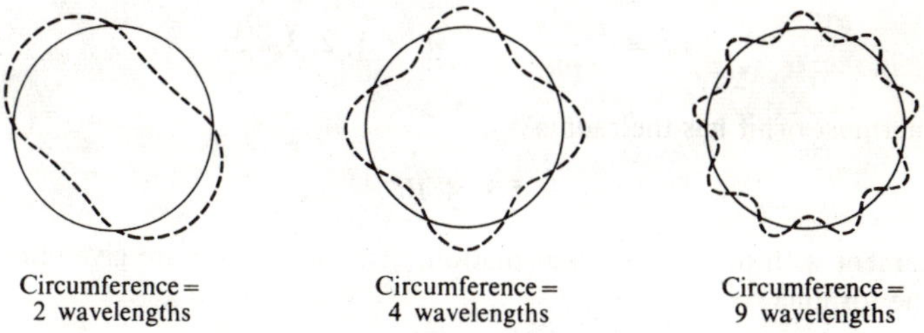

FIG. 32-10. Three possible modes of vibration of a wire loop.

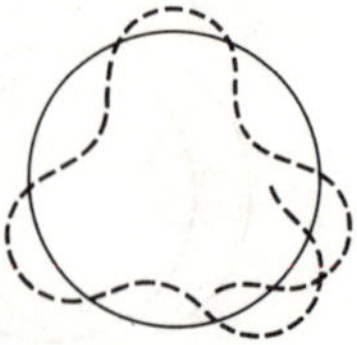

FIG. 32–11. Unless a whole (integral) number of wavelengths fit into the wire loop, destructive interference causes the vibrations to die out rapidly.

and still allow each wave to join smoothly with the next (Fig. 32–11); the result would be destructive interference as the waves travel around the loop, and the vibrations would die out rapidly. By considering the behavior of electron waves in the hydrogen atom analogous to the vibrations of a wire loop, then, we may postulate that

An electron can circle an atomic nucleus indefinitely without radiating energy if its orbit is an integral number of electron wavelengths in circumference.

[The electron speed must be that given by Eq. (32–2), of course, in order that the electrostatic attraction of the nucleus not pull the electron into it.]

The above postulate is the decisive one in our understanding of the atom. We note that it combines both the particle and wave characters of the electron into a single statement; while we can never observe these antithetical characters at the same time, they are inseparable in nature.

It is easy to express in a formula the condition that an integral number of electron wavelengths fit into the electron's orbit. The circumference of a circular orbit of radius r is $2\pi r$, and so the condition for orbit stability is

$$n\lambda = 2\pi r_n, \qquad n = 1, 2, 3, \ldots, \tag{32–10}$$

where r_n designates the radius of the orbit that contains n wavelengths. The quantity n is called the *quantum number* of the orbit. Substituting for λ the electron wavelength given by Eq. (32–9), we have

$$\frac{nh}{e}\sqrt{4\pi\epsilon_0 r_n/m} = 2\pi r_n,$$

and so the stable electron orbits are those whose radii are given by

$$r_n = \frac{\epsilon_0 n^2 h^2}{\pi m e^2}, \qquad n = 1, 2, 3, \ldots \tag{32–11}$$

The innermost orbit has the radius

$$r_1 = 5.3 \times 10^{-11}\ \text{m},$$

in agreement with our previous calculation. The other radii are given in terms of r_1 by the formula

$$r_n = n^2 r_1,$$

so that the spacing between adjacent orbits increases progressively (Fig. 32–12).

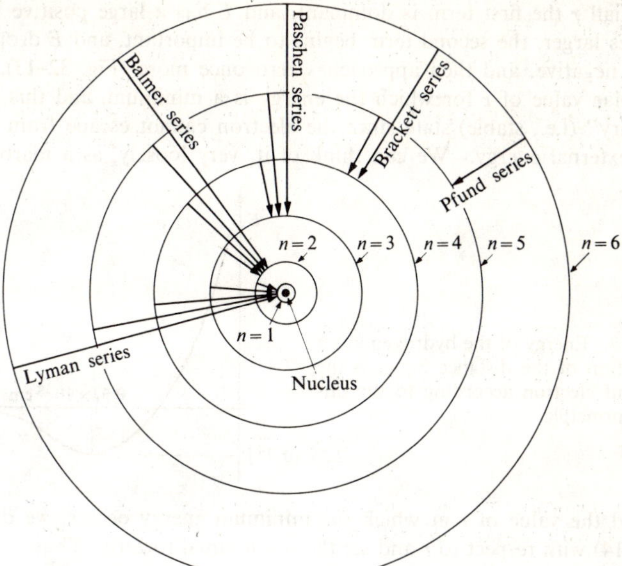

FIG. 32–12. Stable electron orbits in the hydrogen atom according to the Bohr theory. Some of the transitions between orbits that give rise to spectral lines are indicated.

It is interesting to analyze the hydrogen atom from the point of view of the uncertainty principle. The energy of an electron in the vicinity of a proton consists of a potential part, $-e^2/4\pi\epsilon_0 r$, and a kinetic part, $\frac{1}{2}mv^2$, so that the total, as we have already found, is

$$E = \frac{mv^2}{2} - \frac{e^2}{4\pi\epsilon_0 r}. \tag{32–12}$$

What is the lowest possible value that this energy can have? We approach the question in the following way. In view of the wave properties of the electron, a little thought reveals that we cannot state that the electron is the specific distance r from the proton; all that we can say is that the electron wave is localized within the distance r. Corresponding to this uncertainty r in electron position there is an uncertainty mv in the electron momentum; the two are related by

$$mv \approx \frac{h}{2\pi r}. \tag{32–13}$$

We substitute this value of mv into the expression for the total energy,

$$E = \frac{h}{4\pi mr^2} - \frac{e^2}{4\pi\epsilon_0 r}. \tag{32–14}$$

For small r the first term is dominant, and E has a large positive value. As r becomes larger, the second term begins to be important, and E drops to zero, becomes negative, and then approaches zero once more (Fig. 32–13). There is a particular value of r for which the energy is a minimum, and this must be a "stationary" (i.e., stable) state since the electron cannot escape from it without gaining external energy. We can think of it, very loosely, as a marble trapped in a hole.

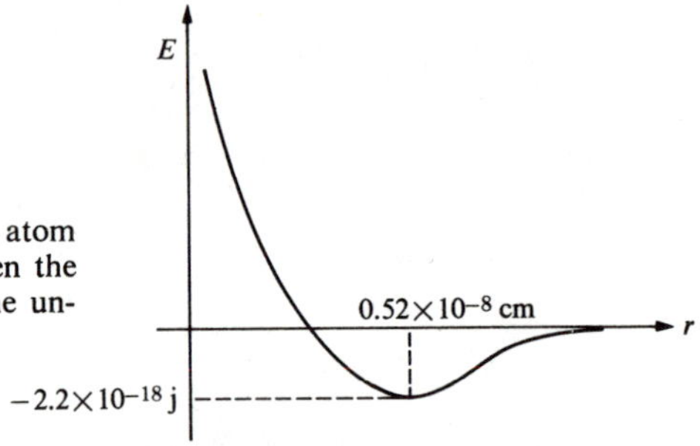

FIG. 32–13. Energy of the hydrogen atom as a function of the distance between the proton and electron according to the uncertainty principle.

To find the value of r at which the minimum energy occurs, we differentiate Eq. (32–14) with respect to r and set the result equal to zero. Thus

$$\frac{dE}{dr} = -\frac{h^2}{4\pi^2 mr^3} + \frac{e^2}{4\pi\epsilon_0 r^2} = 0, \qquad r = \frac{\epsilon_0 h^2}{\pi me^2}.$$

This value of r, the same as in Eq. (32–11), may be thought of as the "radius" of the hydrogen atom. Since the "volume" of the atom really represents the spatial extent of the wave packet required to describe its electron, we cannot think of the atom as a kind of minute hard sphere, but must regard it as more or less of a condensation in space without a specific, tangible boundary.

32–4 Energy levels and spectra

The energy of the electron is not the same in the various permitted orbits. The electron energy E_n is given in terms of the orbit radius r_n by Eq. (32–3) as

$$E_n = -\frac{1}{8\pi\epsilon_0}\frac{e^2}{r_n};$$

substituting for r_n from Eq. (32–11), we see that

$$E_n = -\frac{me^4}{8\epsilon_0^2 n^2 h^2}, \qquad n = 1, 2, 3, \ldots. \tag{32–15}$$

The energies specified by Eq. (32–12), called the *energy levels* of the hydrogen atom, are plotted in Fig. 32–14. These levels are all negative, signifying that the

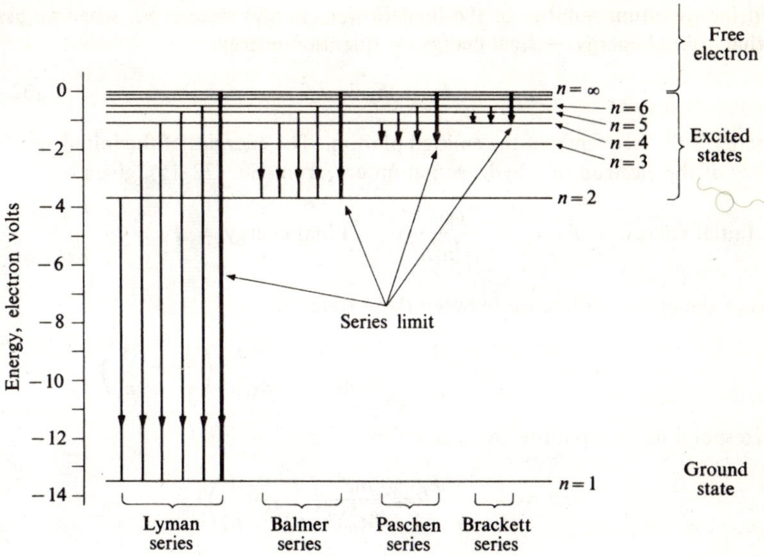

FIG. 32–14. Energy-level diagram of the hydrogen atom. Some of the transitions that give rise to spectral lines are indicated.

electron does not have enough energy to escape from the atom. The lowest energy level E_1, corresponding to quantum number $n = 1$, is called the *ground state* of the atom; the higher levels E_2, E_3, E_4, . . . are called *excited states*. As the quantum number n increases, the corresponding energy E_n approaches closer and closer to zero; in the limit of $n = \infty$, $E_\infty = 0$ and the electron is no longer bound to the nucleus to form an atom. An energy greater than zero signifies an unbound electron, which, since it has no closed orbit that must satisfy quantum conditions, may have any positive energy value whatever.

While everything we have said thus far about the hydrogen atom seems perfectly reasonable, we must nevertheless find a way of directly confronting the equations we have developed with experiment. A particularly striking experimental result is that excited atoms emit line spectra. Can we show that these spectra are a natural consequence of our atomic model?

The presence of a series of definite, discrete energy levels in the hydrogen atom suggests a connection with line spectra. Let us tentatively assert that when an electron in an excited state drops to a lower state, the difference in energy between the states is emitted as a single photon of light. Because electrons cannot, according to our model, exist in an atom except in certain specific energy levels, a rapid "jump" from one level to the other, with the difference in energy being given off all at once in a photon rather than in some more gradual manner, fits in well with this model. If the quantum number of the initial (higher energy) state is n_i

and the quantum number of the final (lower energy) state is n_f, what we assert is that: Initial energy − final energy = quantum energy,

$$E_i - E_f = hf, \tag{32-16}$$

where f is the frequency of the emitted photon. The energies of the initial and final states of the electron in a hydrogen atom are, from Eq. (32-15), given by

$$\text{Initial energy} = E_i = -\frac{me^4}{8\epsilon_0^2 n_i^2 h^2}, \qquad \text{Final energy} = E_f = -\frac{me^4}{8\epsilon_0^2 n_f^2 h^2}.$$

Hence the energy difference between these states is

$$E_i - E_f = \frac{me^4}{8\epsilon_0^2 h^2}\left(\frac{-1}{n_i^2}\right) - \left(\frac{-1}{n_f^2}\right) = \frac{me^4}{8\epsilon_0^2 h^2}\left(\frac{1}{n_f^2} - \frac{1}{n_i^2}\right),$$

corresponding to a photon frequency f of

$$f = \frac{E_i - E_f}{h} = \frac{me^4}{8\epsilon_0^2 h^3}\left(\frac{1}{n_f^2} - \frac{1}{n_i^2}\right).$$

In terms of photon wavelength λ, since $\lambda = c/f$, we have

$$\frac{1}{\lambda} = \frac{f}{c} = \frac{me^4}{8\epsilon_0^2 ch^3}\left(\frac{1}{n_f^2} - \frac{1}{n_i^2}\right). \tag{32-17}$$

Equation (32-17) predicts that the radiation emitted by excited hydrogen atoms should consist of certain wavelengths only. These wavelengths, furthermore, should fall into definite series depending on the quantum number n_f of the final energy level of the electron. Since the initial quantum number n_i must always be greater than the final quantum number n_f, so that there will be an excess of energy to be given off as a photon, the calculated formulas for the first five series are

$$n_f = 1: \quad \frac{1}{\lambda} = \frac{me^4}{8\epsilon_0^2 ch^3}\left(\frac{1}{1^2} - \frac{1}{n^2}\right), \qquad n = 2, 3, 4, \ldots \tag{32-18}$$

$$n_f = 2: \quad \frac{1}{\lambda} = \frac{me^4}{8\epsilon_0^2 ch^3}\left(\frac{1}{2^2} - \frac{1}{n^2}\right), \qquad n = 3, 4, 5, \ldots \tag{32-19}$$

$$n_f = 3: \quad \frac{1}{\lambda} = \frac{me^4}{8\epsilon_0^2 ch^3}\left(\frac{1}{3^2} - \frac{1}{n^2}\right), \qquad n = 4, 5, 6, \ldots \tag{32-20}$$

$$n_f = 4: \quad \frac{1}{\lambda} = \frac{me^4}{8\epsilon_0^2 ch^3}\left(\frac{1}{4^2} - \frac{1}{n^2}\right), \qquad n = 5, 6, 7, \ldots \tag{32-21}$$

$$n_f = 5: \quad \frac{1}{\lambda} = \frac{me^4}{8\epsilon_0^2 ch^3}\left(\frac{1}{5^2} - \frac{1}{n^2}\right), \qquad n = 6, 7, 8, \ldots \tag{32-22}$$

These series agree in form with the empirical spectral series we discussed earlier, with the Lyman series, Eq. (32–5), corresponding to $n_f = 1$, the Balmer series, Eq. (32–4), corresponding to $n_f = 2$, the Paschen series, Eq. (32–6), corresponding to $n_f = 3$, and so on!

Only one thing remains to be determined before we can consider our assertion that line spectra originate in electron transitions from high to low energy states as verified: does the constant term in Eqs. (32–18) through (32–22) have the same value as the Rydberg constant R of the empirical Eqs. (32–4) through (32–8)? Let us compute the value of this constant term. It is

$$\frac{me^4}{8\epsilon_0^2ch^3}$$
$$= \frac{9.1 \times 10^{-31}\,\text{kg} \times (1.6 \times 10^{-19}\,\text{coul})^4}{8 \times (8.85 \times 10^{-12}\,\text{coul}^2/\text{n}\cdot\text{m}^2)^2 \times 3 \times 10^8\,\text{m/sec} \times (6.6 \times 10^{-34}\,\text{j}\cdot\text{sec})^3}$$
$$= 1.097 \times 10^7\,\text{m}^{-1},$$

which is the same as R. Our theory of the hydrogen atom, which is essentially that developed by Bohr in 1913, therefore agrees both qualitatively and quantitatively with experiment. Figure 32–14 shows schematically how spectral lines are related to atomic energy levels; Fig. 32–12 indicates the transitions in terms of Bohr orbits.

At the time Bohr formulated his theory of the atom, the Balmer, Lyman, and Paschen series were known. When techniques were later developed for investigating the far infrared end of the spectrum, the levels terminating at $n = 4$, 5, and 6 were also discovered. In all cases the series agreed exactly with Bohr's predictions.

32–5 Franck-Hertz experiment

The concept of discrete atomic energy levels did not find immediate acceptance among physicists. Since the notion was so foreign to classical thinking, further evidence was necessary to supplement the explanation of the spectrum of hydrogen.

Such evidence was provided by the experiment of Franck and Hertz. To understand their experiment, let us try to predict what happens when electrons impinge on hydrogen atoms. If the energy of the electrons is less than 10.2 ev (the energy difference between the $n = 2$ state and the $n = 1$ state), the electrons cannot excite the hydrogen atoms and just bounce off without losing energy. If the electrons have energies of 10.2 ev, they can excite the target atoms and lose their energy. Thus if we perform an experiment to detect electrons which did not lose energy after passage through a container of hydrogen atoms, we expect to find that after the energy of the electrons is increased to 10.2 ev there will be a diminution in the electrons collected, since some of these have lost energy to the hydrogen atoms. If we continue to increase the energy, the current will drop every time we reach an allowed energy level, although the actual probability of exciting the higher levels is so small that no discernible diminution of electron current is

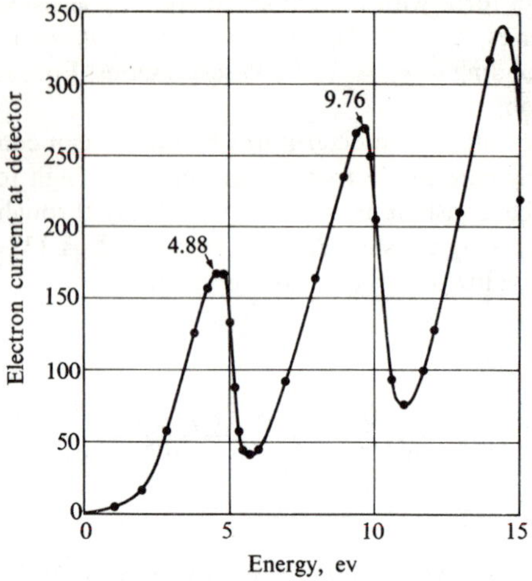

FIG. 32-15. Results of Franck-Hertz experiment with mercury vapor.

likely. However, at 20.4 ev there will be a further conspicuous drop in current due to energy losses of 10.2 ev in two successive collisions.

Franck and Hertz performed the above kind of experiment by passing electrons with carefully controlled energies through mercury vapor. They observed a decrease in the electrons collected when the electrons had initial energies of 4.88 ev. Thereafter drops were observed at various electron energies (Fig. 32-15) which corresponded to multiples of 4.88 ev. This energy divided by Planck's constant h is equal to the frequency of one of the spectral lines of mercury. Thus the Bohr notion of the quantization of energy levels in the atom was further substantiated.

32-6 Atomic excitation

There are two principal mechanisms by which an atom may be excited to an energy level above that of its ground state and thereby become capable of radiating. One of these mechanisms is a collision with another atom during which part of their kinetic energy is transformed into electron energy within either or both of the participating atoms. An atom excited in this way will then lose its excitation energy by emitting one or more photons in the course of returning to its ground state (Fig. 32-16). In an electric discharge in a rarefied gas, an electric field accelerates electrons and charged atoms and molecules (whose charge arises either from an excess or a deficiency in the electrons required to neutralize the positive charge of their nuclei) until their kinetic energies are sufficient to excite atoms they happen to collide with. A neon sign is a familiar example of how applying a strong electric field between electrodes in a gas-filled tube leads to the emission

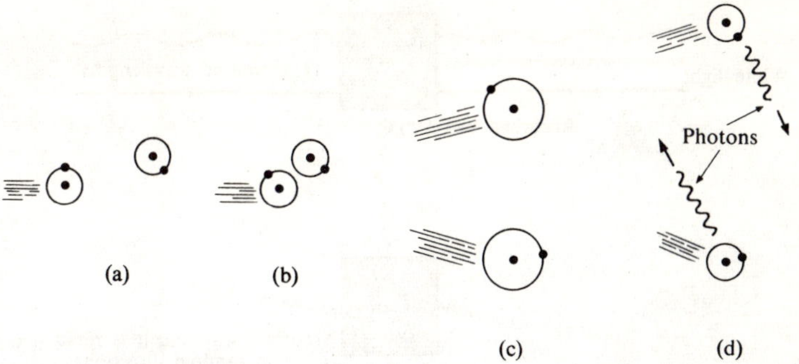

FIG. 32-16. Excitation by collision. In (a) both atoms are in their ground states. During the collision (b) some kinetic energy is transformed into excitation energy, and in (c) the atoms are both in excited states. In (d) they have returned to their ground states by emitting photons.

of the characteristic spectral radiation of that gas, which happens to be orange light in the case of neon.

Another excitation mechanism is the absorption by an atom of a photon of light whose energy is just the right amount to raise it to a higher energy level. A photon of wavelength 6.563×10^{-7} meter is emitted when a hydrogen atom in the $n = 3$ state drops to the $n = 2$ state; hence the absorption of a photon of wavelength 6.563×10^{-7} meter by a hydrogen atom initially in the $n = 2$ state will bring it up to the $n = 3$ state. This process explains the origin of absorption spectra (Fig. 32-17). When white light (in which all wavelengths are present) is passed through hydrogen gas, photons of those wavelengths that correspond to transitions between the ground state of hydrogen and higher energy levels are absorbed. The resulting excited hydrogen atoms reradiate their excitation energy almost at once, but these photons come off in random directions, not all in the same

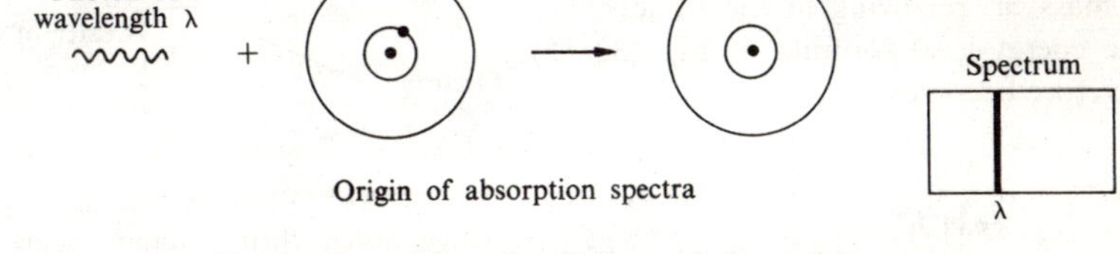

FIG. 32-17. The origins of emission and absorption spectra.

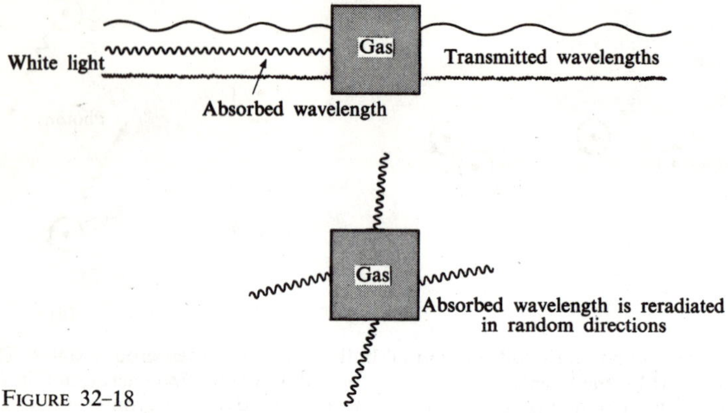

FIGURE 32-18

direction as in the original beam of white light (Fig. 32-18). The dark lines in an absorption spectrum are therefore never totally dark, but only appear so by contrast with the bright background of transmitted light.

32-7 Nuclear motion

In our analysis of the hydrogen atom we have assumed that the nucleus is stationary while the electron revolves around it. This picture cannot be correct because the momentum of a hydrogen atom would have to vary continuously as its electron moves around the nucleus, contrary to the conservation of momentum in an isolated system. What actually happens is that nucleus and electron both revolve about their common center of mass, which, because the nucleus is so much heavier than the electron, is very close to the nucleus (Fig. 32-19). In Chapter 8 we saw that a system of this kind is exactly equivalent to a single particle of mass m' rotating about the position of the heavier particle, where m', the reduced mass of the system, is given by Eq. (8-26) as

$$m' = \frac{m_1 m_2}{m_1 + m_2}.$$

To correct for nuclear motion in the hydrogen atom, then, all we have to do is imagine the electron replaced by a particle of mass m' revolving in the same orbit. The energy-level formula of Eq. (32-15) therefore becomes

$$E'_n = -\frac{m'e^4}{8\epsilon_0^2 n^2 h^2}, \qquad n = 1, 2, 3, \ldots, \tag{32-23}$$

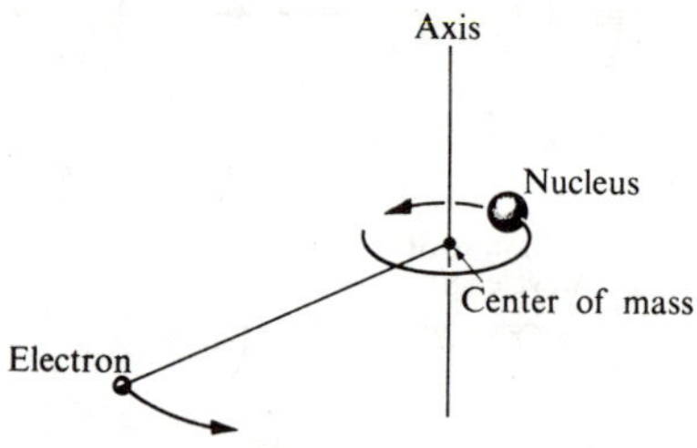

FIG. 32-19. Electron and nucleus revolve about their common center of mass.

where, if m_e is the electron mass and m_n the nuclear mass,

$$m' = \frac{m_e m_n}{m_e + m_n}. \tag{32-24}$$

In the ordinary hydrogen atom this change has the effect of shifting each of the energy levels upward by 0.06%, since each level becomes less negative than it would if the nucleus were stationary according to the formula

$$E'_n = \frac{m'}{m_e} E_n = \frac{m_n}{m_e + m_n} E_n = \frac{1.672 \times 10^{-27}\ \text{kg}}{1.673 \times 10^{-27}\ \text{kg}} E_n = 0.9994 E_n.$$

The correction for nuclear motion affects the hydrogen spectrum to a small but measurable amount. The Rydberg constant R for a stationary nucleus is, to six significant figures,

$$R = 1.09737 \times 10^7\ \text{m}^{-1},$$

while it drops to 1.09678×10^7 m when the reduced mass m' is substituted for the electron mass m in the formula for R. The latter figure is in better agreement with experiment than the former one.

A historically important application of the effect of nuclear motion was the discovery of a type of hydrogen atom whose nucleus has approximately twice the mass of the usual hydrogen nucleus. This variant of hydrogen is known as *deuterium*, or, less formally, as "heavy hydrogen"; about one hydrogen atom in 6000 is a deuterium atom. As we shall learn in Chapter 36, the deuterium nucleus consists of a neutron and a proton instead of a proton alone, which accounts for its greater mass. The spectrum of deuterium is identical with that of ordinary hydrogen except for an increase in the value of R from $1.09678 \times 10^7\ \text{m}^{-1}$ to $1.09707 \times 10^7\ \text{m}^{-1}$ owing to the larger reduced mass in the deuterium atom. The change means that the wavelengths of the spectral lines of deuterium are all slightly reduced. The H_α line, for example, which corresponds to a transition from $n = 3$ to $n = 2$, occurs at a wavelength of 6.56280×10^{-7} m in ordinary hydrogen and at 6.56101×10^{-7} m in deuterium. The existence of this minute difference was responsible for the discovery of deuterium in 1932.

Problems

1. In the Bohr theory of the hydrogen atom the electron is in constant motion. How is it possible for such an electron to have a *negative* amount of energy?

2. Calculate the speed of the electron in the innermost Bohr orbit of a hydrogen atom.

3. (a) Calculate the de Broglie wavelength of the earth. (b) What is the quantum number that characterizes the earth's orbit about the sun? (The earth's mass is 6.0×10^{24} kg, its orbital radius is 1.5×10^{11} m, and its orbital speed is 3×10^4 m/sec.)

4. Explain why the spectrum of hydrogen has many lines, although a hydrogen atom contains only one electron.

5. A beam of electrons whose energy is 13 ev is used to bombard gaseous hydrogen. What wavelengths of light will be emitted?

6. A beam of electrons is used to bombard gaseous hydrogen. What is the minimum energy in ev the electrons must have if the first line of the Balmer series, corresponding to a transition from the $n = 3$ state to the $n = 2$ state, is to be emitted?

7. What is the shortest wavelength present in the Paschen series of spectral lines?

8. What is the shortest wavelength present in the Brackett series of spectral lines?

9. A proton and an electron, both at rest initially at infinity, combine to form a hydrogen atom in the ground state. A single photon is emitted in this process. What is its wavelength?

10. How much energy (in joules and in electron volts) is required to remove the electron from a hydrogen atom when it is in the $n = 5$ state?

11. To what temperature must a hydrogen gas be heated if the average molecular kinetic energy is to equal numerically the binding energy of the hydrogen atom in its ground state?

12. Repeat the derivation of the Bohr theory for a one-electron ion whose nuclear charge is Ze, and show that the energy is proportional to Z^2.

13. Calculate the radius and speed of an electron in the ground state of doubly ionized lithium and compare with the radius and speed of the electron in the ground state of the hydrogen atom.

14. Positronium consists of an electron and a positron, both of the same mass. Compare the wavelength of the radiation arising from an $n = 2$ to $n = 1$ transition in positronium with the H_α line of hydrogen.

15. Show that $e^2/4\pi\epsilon_0\hbar c$ is a dimensionless constant and calculate its value. (It is called the fine-structure constant; $\hbar = h/2\pi$.)

16. Show that the energy of an electron in a hydrogen atom (whose nuclear mass is taken to be infinite) is $E_n = -(1/2n^2)\alpha^2 mc^2$, where $\alpha = e^2/4\pi\epsilon_0\hbar c$ is the fine-structure constant. (mc^2 is the rest energy of an electron and equals 0.511×10^6 ev.)

17. Show that the numerical value of $e^2/4\pi\epsilon_0\hbar c = \alpha = v_1/c$, where v_1 is the speed of an electron in the first Bohr orbit.

18. An electron in hydrogen goes from a state whose binding energy is -0.85 ev to one whose binding energy is -10.2 ev. What is the wavelength of the photon emitted? What transition is this?

19. If the lifetime of the $n = 2$ state is 10^{-9} sec, estimate the relative width $\Delta\lambda/\lambda$ of the line in the $n = 2 \rightarrow n = 1$ transition in hydrogen.

20. An electron spends about 10^{-8} sec in an excited state before it drops to a lower state by giving up energy in the form of a photon. How many revolutions does an electron in the $n = 2$ state of a hydrogen atom make before dropping to the $n = 1$ state?

21. (a) Find the wavelength of the photon emitted when a hydrogen atom goes from the state $n = 4$ to $n = 1$. (b) Find the momentum of this photon. (c) Find the recoil speed of the hydrogen atom after it has emitted this photon.

22. Calculate the average kinetic energy per molecule of a gas at room temperature (20°C), and show that this is much less than the energy required to raise a hydrogen atom from its ground state ($n = 1$) to its first excited state ($n = 2$).

23. When an artificial satellite is about to expire, the radius of its orbit decreases due to the effects of friction with the atmosphere. Does it slow down or speed up when this happens? [*Hint:* Compare this with what happens when an electron in hydrogen makes a transition to a lower state.]

24. *Tritium* is a variety of hydrogen whose nuclei are approximately three times more massive than those of ordinary hydrogen. Show that the H_α line of tritium has a wavelength 2.4×10^{-10} m shorter than the H_α line of ordinary hydrogen.

33

Complex Atoms

33–1 Introduction

The impressive ability of the Bohr theory to account for the wavelengths of the spectral lines of hydrogen means that in some sense it is a correct description of this atom. In other respects, however, the Bohr theory is not so successful. For example, it does not permit us to predict correctly which spectral lines are bright and which ones are faint. Another class of difficulties involves the application of the Bohr theory to atoms more complex than hydrogen atoms. Thus the energy levels of helium, the next most complicated atom, are incorrectly predicted by the theory. Further, many spectral lines in complex atoms, when examined with instruments of high resolution, are *multiplets*; that is, they consist of two or more lines of slightly different wavelengths (Fig. 33–1). These difficulties with the heavier atoms cannot be accounted for by nuclear motion, as they could in the case of deuterium.

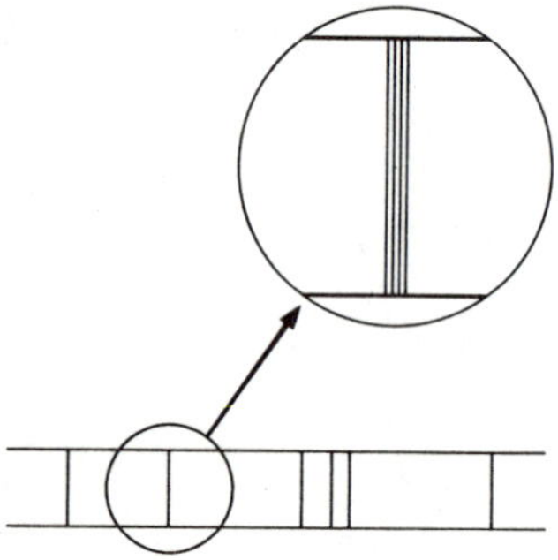

FIGURE 33–1

To modify the Bohr theory it is first necessary to introduce a property of the electron of which Bohr was initially unaware that is known as *electron spin*. Second, there is the *exclusion principle*, discovered by Pauli, that is the key to the structure of more complicated atoms. But even the inclusion of these two effects in the Bohr theory does not entirely resolve its difficulties because the angular momentum properties of the electrons in an atom are incorrectly given by the theory.

The basic trouble with the Bohr theory is that it does not treat the wave properties of the electron consistently. We have seen that the wave properties of the electron imply an uncertainty involving its position and momentum, but the concept of Bohr's orbits violates the uncertainty principle, since the existence of a definite orbit means that we know the position and momentum of the electron at each point in the orbit.

A more satisfactory description of the atom was given forty years ago by Schrödinger and Heisenberg, who consistently took the wave properties of the electron into account. This theory is known as the *wave* or *quantum theory of the atom.* One way of formulating the problem, as was done by Schrödinger, is to write a wave equation for an electron in a given force field. The "standing wave" solutions of this equation give the observed energies of the electron.

Later in this chapter we shall indicate how the Bohr theory can be modified to conform with the accepted ideas of atomic structure derived from quantum theory, although it must be kept in mind that the quantum theory is the correct one. We shall first discuss angular momentum, and then describe electron spin and the exclusion principle. The reason for elaborating the simple Bohr picture by including the correct results of the wave theory is that the result is intuitively accessible, while the Schrödinger and Heisenberg theories are wholly abstract. The Bohr theory turns out to be rigorously correct for those atoms which have large quantum numbers, but its exact agreement with experiment for *all* the hydrogenic states is coincidental.

33–2 Angular momentum

An electron moving in a circular orbit about a positively charged nucleus possesses angular momentum. In fact, the original derivation that Bohr gave of his theory involved the assumption that the angular momentum L of the electron has the magnitude

$$L = mvr = m\omega = \frac{nh}{2\pi} = n\hbar, \tag{33–1}$$

where h is Planck's constant. It is easy to show that (33–1) yields the same value for the radius of a Bohr orbit that we derived in Chapter 32. There we saw that force balance requires

$$\frac{mv^2}{r} = \frac{1}{4\pi\epsilon_0}\frac{e^2}{r^2}, \tag{33–2}$$

and so

$$v^2r^2 = \omega^2 = \frac{1}{4\pi\epsilon_0}\frac{e^2r}{m}, \tag{33–3}$$

while from (33–1),

$$\omega^2 = \frac{n^2\hbar^2}{m^2}. \tag{33–4}$$

Equating (33–2) and (33–3), we obtain

$$r = 4\pi\epsilon_0\frac{n^2\hbar^2}{e^2m},$$

which is identical with (32–11).

According to the original Bohr theory, the value of n which characterizes the orbit and which gives the correct value of the energy in the formula

$$E_n = -\frac{me^4}{8\epsilon_0^2 h^2 n^2}$$

also states that the value of the angular momentum in the orbit is equal to $nh/2\pi = n\hbar$. The ground state of hydrogen thus has an angular momentum of $\hbar$, the first excited state $2\hbar$, and so on. Even before the advent of wave mechanics it was realized that this analysis was not correct, and a different (though still not wholly correct) method of quantization was devised. The results can be summarized as follows: Every state belonging to a quantum number n consists of n substates of identical energy with angular momentum equal to $\hbar, 2\hbar, \ldots, n\hbar$. Each of these orbits is elliptical, and the smaller the angular momentum associated with the orbit the more eccentric it is (see Fig. 33–2). The orbit whose angular momentum is a maximum for a given n, namely $n\hbar$, is a circle. An orbit corresponding to a value of the angular momentum equal to zero would consist of a straight line through the nucleus, a degenerate ellipse. This is an impossible orbit in the Bohr theory, of course. In the Bohr theory, for instance, the ground state of hydrogen (for which $n = 1$) has an angular momentum of $\hbar$.

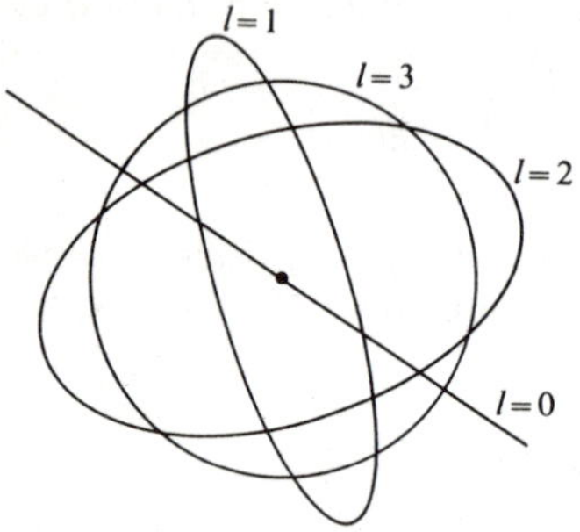

FIG. 33–2. The orbits corresponding to the various possible angular momentum states when $n = 3$. In general, $n = l$ corresponds to a circular orbit; $l = 0$ is not possible classically, but is permitted by wave mechanics.

We shall see in the next section that the angular momentum of the electron in the ground state of silver, due to its orbital motion, can be measured, and it *is* equal to zero. The orbital angular momentum of the electron in the ground state of hydrogen has also been measured, and it too has turned out to be zero. This result is predicted by quantum mechanics. The argument that the zero angular momentum state makes for an impossible orbit is of no consequence in quantum mechanics, since the notion of a mechanical orbit plays no role in this theory; the very concept of an orbit, as was mentioned, violates the uncertainty principle. The correct result is that *every state whose energy quantum number is n consists of at least n states of the same energy whose angular momenta are equal to* $0, \hbar, 2\hbar, \ldots, (n-1)\hbar$.

We see that for very large n, say $n = 100$, the maximum value of l in the modified Bohr theory is $100\hbar$, while it is $99\hbar$ according to quantum mechanics. The eccentricity of the orbit whose angular momentum is $99\hbar$ is so small that it is practically a circle, and in this sense the Bohr theory is valid for large quantum numbers. For the ground state of hydrogen, however, the difference in angular momentum as given by quantum mechanics and as given by the Bohr theory is serious, and we can only regard the accurate agreement between the Bohr theory and experiment as accidental.

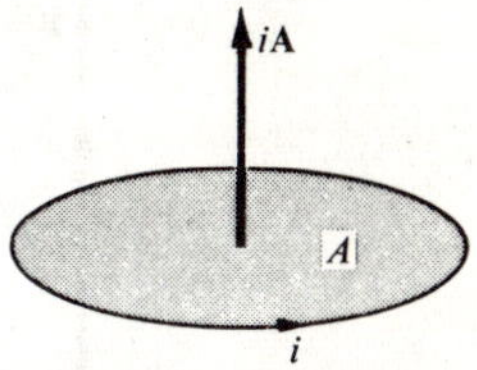

FIG. 33–3. A current i flowing in a plane gives rise to an angular momentum iA in a direction perpendicular to the plane.

The circulating charges as pictured by the Bohr theory give rise to currents, and these currents have magnetic moments whose magnitudes are iA, where A is the area of the orbit. The direction of the magnetic moment is perpendicular to the plane of the orbit (Fig. 33–3). In a random collection of atoms the orbits would point in all possible directions and so would the magnetic moments. The Stern-Gerlach experiment (Chapter 24) showed, however, that only certain directions between the magnetic moment and the magnetic field B were permitted, namely those for which the projection of the angular momentum along the magnetic field is an integral multiple of $\hbar$. If the angular momentum of the atom is $l\hbar$, then the possible values of the projection of the angular momentum in the direction of magnetic field **B** would be $l\hbar$, $(l-1)\hbar$, $(l-2)\hbar, \ldots, -(l-1)\hbar$, $-l\hbar$, which is $2l+1$ possibilities in all (Fig. 33–4). The projection in the direction of **B** is thus limited to $m_l\hbar$, where m_l is called the *magnetic quantum number*. Each one of the states whose projection is different has a different force exerted on it in a magnetic field.

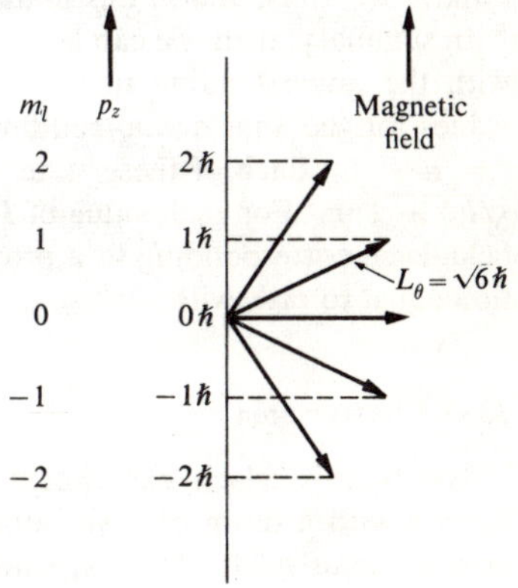

FIG. 33–4. According to the Stern-Gerlach experiment, the possible projections of the angular momentum $\sqrt{l(l+1)}\,\hbar$ in the direction of an external magnetic field are $l\hbar$, $(l-1)\hbar$, $(l-2)\hbar$, $\ldots, -(l-1)\hbar$, $-l\hbar$. The illustration shows the case of $l = 2$, for which m_l can be $+2\hbar$, $+\hbar$, 0, $-\hbar$, and $-2\hbar$.

There is another significant difference between the Bohr theory and the Schrödinger theory. The atom in which the magnetic moment is aligned with an external magnetic field is in the state $m_l = l$, and the corresponding angular momentum is $l\hbar$. This implies that the plane of the orbit is completely defined, and is perpendicular to the magnetic field (Fig. 33–5). Just as a precise orbit violates the uncertainty principle, so does the existence of a definite value of the angular momentum and the direction of the angular momentum violate the same principle.

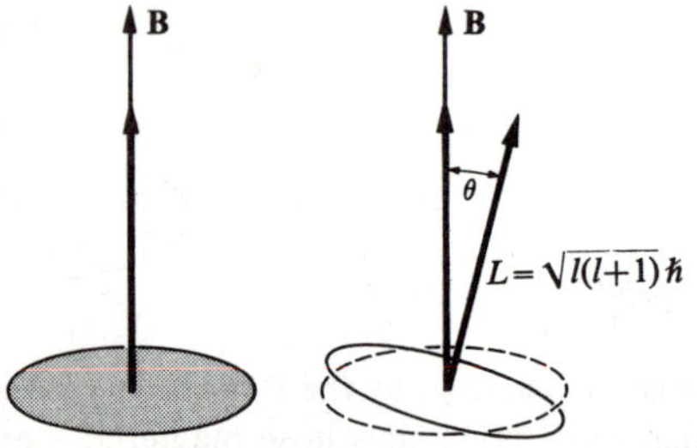

FIG. 33–5. Two versions of the relationship between angular momentum and its projection onto a fixed axis. (a) According to Bohr theory, $L = l\hbar$, and its maximum component in the direction of **B** is $l\hbar$. (b) According to wave mechanics, $L = \sqrt{l(l+1)}\,\hbar$, but its maximum component in the direction of **B** is still only $l\hbar$. Hence there is an uncertainty in the direction of tilt of the plane of the orbit.

The results of the Schrödinger theory overcome this difficulty by showing that the value of the angular momentum is $L = \sqrt{l(l+1)}\,\hbar$, where l can vary from 0 to $n - 1$. The projection of the angular momentum still has a maximum value of $l\hbar$. Since $\sqrt{l(l+1)}$ is greater than l, the plane of the orbit is not quite perpendicular to the magnetic field, and there is an uncertainty in its exact position (Fig. 33–5). In the limit of large values of l there is very little difference between l and $\sqrt{l(l+1)}$, and in this limit the Bohr theory is correct.

In summary, then, we can legitimately use the Bohr picture as an aid to thought, with the understanding that for each quantum number n there are n possible values for the angular momentum quantum number, namely $l = 0$, $l = 1, \ldots,$ $l = n - 1$. Each of these states corresponds to an angular momentum of $L = \sqrt{l(l+1)}\,\hbar$. For each value of l there are $2l + 1$ states in an external magnetic field, each corresponding to a projection of angular momentum in the field direction equal to $m_l\hbar$, with $|m_l| \leq l$.

33–3 Electron spin

The Stern-Gerlach experiment described in Chapter 24 was originally performed with a beam of silver atoms which were split into two beams by the inhomogeneous field. This experiment demonstrated the quantization of the projection of the angular momentum in the direction of the magnetic field, but the two beams were puzzling. If l were equal to zero then there should be no splitting. If l were equal to 1, there should be $2l + 1 = 3$ beams. For a consistent picture of angular momentum there should always be an odd number of components. The difficulty was resolved by the hypothesis that the orbital angular momentum of silver was 0 but that an electron could have a magnetic moment corresponding to an angular momentum of $\frac{1}{2}\hbar$. This phenomenon is now called the *electron spin*, even though it has nothing to do with a spinning electron. It is an intrinsic angular momentum property of an electron, like its charge. Electron spin is described by the spin quantum number s, whose sole possible value is $s = \frac{1}{2}$. The magnitude of the intrinsic spin angular momentum of the electron is given by a formula similar to that describing the orbital angular momentum, namely

$$L_s = \sqrt{s(s+1)}\,\hbar = \frac{\sqrt{3}}{2}\,\hbar. \tag{33-5}$$

The space quantization of electron spin is described by the *spin magnetic quantum number*, m_s, whose values are $+\frac{1}{2}$ and $-\frac{1}{2}$. The component of spin angular momentum along a magnetic field is $m_s\hbar$, so that it can be either $+\frac{1}{2}\hbar$ or $-\frac{1}{2}\hbar$. Because the energy of an electron is different in each orientation of its spin owing to the presence of the magnetic field of its orbital motion, the various energy levels of an atom are split into sublevels, and the presence of these sublevels is responsible for the observed fine structure of spectral lines. (In its own frame of reference the electron "sees" the positively charged nucleus revolving around *it*, and it accordingly experiences a magnetic field like that present at the center of a loop of electric current.)

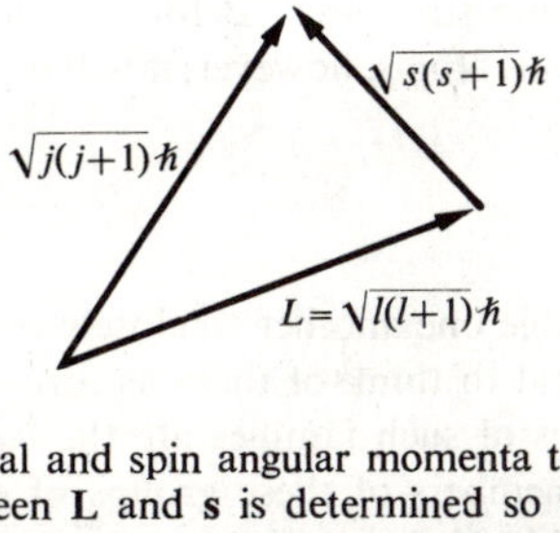

FIG. 33–6. Combining orbital and spin angular momenta to form a total angular momentum J. The angle between **L** and **s** is determined so that the magnitude of J is $\sqrt{j(j+1)}\,\hbar$. Thus if $l = 2$ and $s = \frac{1}{2}$, we can combine angular momentum vectors of magnitudes $\sqrt{6}\,\hbar$ and $\sqrt{\frac{3}{4}}\,\hbar$ to form a vector whose magnitude is $\sqrt{\frac{3}{2}(\frac{5}{2})}\,\hbar = \sqrt{\frac{15}{4}}\,\hbar$. Another possibility, not shown, is $J = \sqrt{\frac{5}{2}\cdot\frac{7}{2}}\,\hbar = \sqrt{\frac{35}{4}}\,\hbar$.

The orbital and spin angular momenta can be combined into a total angular momentum $\mathbf{j} = \mathbf{l} + \mathbf{s}$ with the specification that the total angular momentum is equal to $J = \sqrt{j(j+1)}\,\hbar$, and its projection on an external magnetic field is an integer m_j times $\hbar$; m_j must satisfy the condition $|m_j| \leq j$ (Fig. 33–6).

33-4 The exclusion principle

The introduction of electron spin into the theory of the atom means that a total of four quantum numbers, n, l, m_l, and m_s, is required to describe each possible state of an atomic electron. All atoms other than hydrogen have more than one electron, and it is appropriate to inquire into their ground-state configurations. Do all the electrons in a complex atom have the same sets of quantum numbers, or are the sets all different, or what? This question was answered by Wolfgang Pauli, whose *exclusion principle* states that

No two electrons in an atom can exist in the same quantum state.

That is, each electron in a complex atom must have a different set of the quantum numbers n, l, m_s, and m_l. The exclusion principle can be generalized to refer to the electrons in *any* small region of space, regardless of whether or not they constitute an atom. We shall make use of this generalization in the next chapter.

Pauli was led to the exclusion principle by a study of atomic spectra. It is possible to determine the quantum states of an atom empirically by analyzing its spectrum, since states with different quantum numbers differ in energy (even if only slightly) and the various wavelengths present correspond to transitions between these states. Pauli found several predicted lines missing from the spectra of every element except hydrogen; the missing lines correspond to transitions to and from atomic states having certain sets of quantum numbers. Every one of these absent states has two or more electrons with identical sets of quantum numbers, a result that is expressed in the exclusion principle. Hydrogen, with a single electron, naturally has no absent states.

The Pauli exclusion principle is the final piece of information we require in order to understand atomic structure. Before we actually go into the electron configurations of the various atoms, however, it will be useful to review the periodic law of chemistry.

33-5 The periodic law

Certain elements resemble one another so closely in their chemical and physical properties that it is natural to think of them as forming a "family." Three particularly striking examples of such families are the *halogens*, the *inert gases*, and the *alkali metals*. The members of these groups of elements, with their atomic numbers, are listed in Table 33-1.

The halogens are nonmetallic elements with a high degree of chemical activity. At room temperature fluorine and chlorine are gases, bromine a liquid, and iodine and astatine solids. The halogens have valences of -1, and form diatomic molecules in the vapor state. The inert gases, as their name suggests, are inactive chemically: they form virtually no compounds with other elements, and their atoms do not join together into molecules. The inert gases have valences of zero. The alkali metals, like the halogens, are chemically very active, but they are active

TABLE 33-1

Three groups of elements*

Halogens	Inert gases	Alkali metals
	(2) Helium	(3) Lithium
(9) Fluorine	(10) Neon	(11) Sodium
(17) Chlorine	(18) Argon	(19) Potassium
(35) Bromine	(36) Krypton	(37) Rubidium
(53) Iodine	(54) Xenon	(55) Cesium
(85) Astatine	(86) Radon	(87) Francium

* The atomic numbers of the elements are in parentheses.

as reducing agents rather than as oxidizing agents. They are soft, not very dense, and have low melting points (all except lithium are liquid below 100°C). The alkali metals have valences of +1.

A curious feature of the three groups listed above is that, while the atomic numbers of the member elements of each group bear no obvious relation to one another, each inert gas is preceded in atomic number by a halogen (except in the case of helium) and followed by an alkali metal. Thus fluorine, neon, and sodium have the atomic numbers 9, 10, and 11, respectively, a sequence that persists through astatine (85), radon (86), and francium (87). If we list *all* the elements in order of their atomic number, *elements with similar properties recur at regular intervals.* This observation, first formulated in detail by Dmitri Mendeléev about 1869, is known as the *periodic law.*

A periodic table is a listing of the elements according to atomic number in a series of rows such that elements with similar properties form vertical columns. Table 33-2 is perhaps the most common form of periodic table; the number above the symbol of each element is its atomic weight, and the number below the symbol is its atomic number. The elements whose atomic weights appear in parentheses are radioactive and are not found in nature, but have been prepared in the laboratory using nuclear reactions. The atomic weight in each such case is the mass number of the longest-lived isotope of the element.

The columns in the periodic table are called *groups.* We recognize group I as the alkali metals plus hydrogen, group VII as the halogens, and group 0 as the inert gases. In addition to the elements forming the eight principal groups there are a number of *transition elements* falling between groups II and III. The transition elements are metals which share certain general properties: most are hard and brittle, have high melting points, exhibit several different valences, and form compounds that are paramagnetic. The rows in the periodic table are called *periods.* Each period starts with an active alkali metal and proceeds through less active metals to weakly active nonmetals to an active halogen and an inactive inert gas. The transition elements in each period may be very much alike; the rare earths

TABLE 33–2

Periodic classification of the elements

Group→ Period ↓	I	II											III	IV	V	VI	VII	0
1	1.0080 H 1																	4.003 He 2
2	6.940 Li 3	9.013 Be 4											10.82 B 5	12.011 C 6	14.008 N 7	16.0000 O 8	19.000 F 9	20.183 Ne 10
3	22.991 Na 11	24.32 Mg 12											26.98 Al 13	28.09 Si 14	30.975 P 15	32.066 S 16	35.457 Cl 17	39.944 Ar 18
4	39.100 K 19	40.08 Ca 20	44.96 Sc 21	47.90 Ti 22	50.95 V 23	52.01 Cr 24	54.94 Mn 25	55.85 Fe 26	58.94 Co 27	58.71 Ni 28	63.54 Cu 29	65.38 Zn 30	69.72 Ga 31	72.60 Ge 32	74.91 As 33	78.96 Se 34	79.916 Br 35	83.80 Kr 36
5	85.48 Rb 37	87.63 Sr 38	88.92 Y 39	91.22 Zr 40	92.91 Nb 41	95.95 Mo 42	(99) Tc 43	101.1 Ru 44	102.91 Rh 45	106.4 Pd 46	107.88 Ag 47	112.41 Cd 48	114.82 In 49	118.70 Sn 50	121.76 Sb 51	127.61 Te 52	126.91 I 53	131.3 Xe 54
6	132.91 Cs 55	137.36 Ba 56	* 57–71	178.5 Hf 72	180.95 Ta 73	183.86 W 74	186.22 Re 75	190.2 Os 76	192.2 Ir 77	195.09 Pt 78	197.0 Au 79	200.61 Hg 80	204.39 Tl 81	207.21 Pb 82	209.00 Bi 83	210 Po 84	(210) At 85	222 Rn 86
7	(223) Fr 87	226.05 Ra 88	† 89															

*Rare-earth metals	138.92 La 57	140.13 Ce 58	140.92 Pr 59	144.27 Nd 60	(145) Pm 61	150.35 Sm 62	152.0 Eu 63	157.26 Gd 64	158.93 Tb 65	162.51 Dy 66	164.94 Ho 67	167.27 Er 68	168.94 Tm 69	173.04 Yb 70	174.99 Lu 71
† Actinide metals	227 Ac 89	232.05 Th 90	231 Pa 91	238.07 U 92	(237) Np 93	(244) Pu 94	(243) Am 95	(245) Cm 96	(249) Bk 97	(249) Cf 98	(254) E 99	(252) Fm 100	(256) Mv 101	(253?) No 102	

and actinides are so much alike that they are usually considered as separate categories.

For nearly a century the periodic law has been a mainstay of the chemist because it permits him to predict the behavior of undiscovered elements and provides a framework for organizing his knowledge. One of the triumphs of the quantum theory of the atom is that it enables us to account for the periodic law in complete detail without invoking any new assumptions or postulates.

33–6 Atomic structure

A physical system is stable when its energy is a minimum, which means that in the normal configuration of an atom its constituent electrons are present in the lowest energy states available to them. The electrons in an atom that share the same total quantum number n are said to occupy the same *shell.* This is a useful description, since such electrons average about the same distance from the nucleus and have comparable, though not identical, energies. Electrons in a given shell that also have the same orbital quantum number l are said to occupy the same *subshell.* In complex atoms the various subshells of the same shell vary in energy because electrons with different angular momenta pursue different orbits around the nucleus. (Strictly speaking, electrons do not move in specific orbits. However, the likelihood of their presence varies with position in such a way that a line drawn through the points of greatest probability may be thought of as an orbit. See Fig. 33–7.) In a complex atom the outer electrons are partially shielded from the full nuclear charge of $+Ze$ by the inner electrons, and are therefore less tightly bound than the latter. The extent of this shielding on a particular electron depends on the shape of its orbit, which in turn depends on its orbital quantum number l. When l is large, the corresponding orbit is more or less circular, while when it is small the orbit is pronouncedly elliptical. In an elliptical orbit an electron spends part of its time near the nucleus without being shielded by other electrons, as in Fig. 33–8, and its energy is correspondingly lower than that of an electron in the same shell whose orbit is circular. The general result, then, is that the energies of the various sublevels increase as l increases.

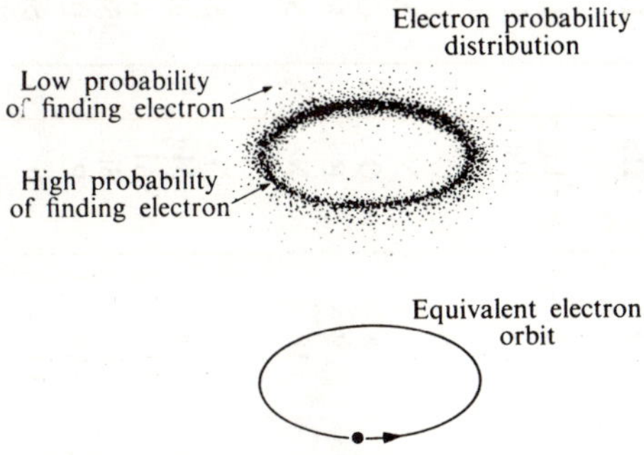

FIGURE 33–7

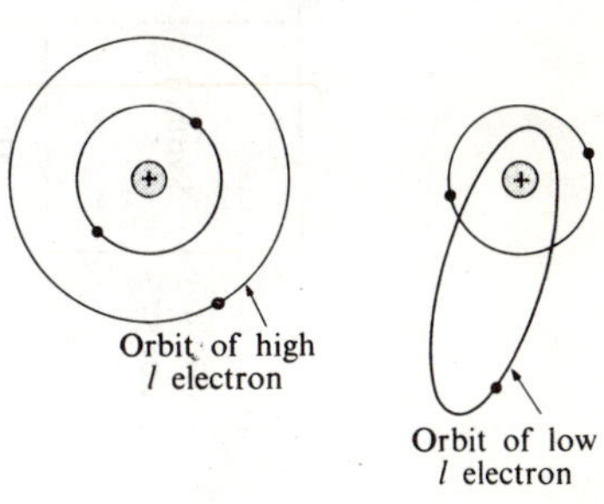

FIGURE 33–8

TABLE 33–3

Electron configurations of the thirty-six lightest elements

Atomic number	Symbol	Element	Electron configuration of atom							
			$n = 1$	$n = 2$		$n = 3$			$n = 4$	
			$l = 0$	$l = 0$	$l = 1$	$l = 0$	$l = 1$	$l = 2$	$l = 0$	$l = 1, \ldots$
1	H	Hydrogen	1							
2	He	Helium	2							
3	Li	Lithium	2	1						
4	Be	Beryllium	2	2						
5	B	Boron	2	2	1					
6	C	Carbon	2	2	2					
7	N	Nitrogen	2	2	3					
8	O	Oxygen	2	2	4					
9	F	Fluorine	2	2	5					
10	Ne	Neon	2	2	6					
11	Na	Sodium	2	2	6	1				
12	Mg	Magnesium	2	2	6	2				
13	Al	Aluminum	2	2	6	2	1			
14	Si	Silicon	2	2	6	2	2			
15	P	Phosphorus	2	2	6	2	3			
16	S	Sulfur	2	2	6	2	4			

17	Cl	Chlorine	2	2	6	2	5			
18	Ar	Argon	2	2	6	2	6			
19	K	Potassium	2	2	6	2	6		1	
20	Ca	Calcium	2	2	6	2	6		2	
21	Sc	Scandium	2	2	6	2	6	1	2	
22	Ti	Titanium	2	2	6	2	6	2	2	
23	V	Vanadium	2	2	6	2	6	3	2	
24	Cr	Chromium	2	2	6	2	6	5	1	
25	Mn	Manganese	2	2	6	2	6	5	2	
26	Fe	Iron	2	2	6	2	6	6	2	
27	Co	Cobalt	2	2	6	2	6	7	2	
28	Ni	Nickel	2	2	6	2	6	8	2	
29	Cu	Copper	2	2	6	2	6	10	1	
30	Zn	Zinc	2	2	6	2	6	10	2	
31	Ga	Gallium	2	2	6	2	6	10	2	1
32	Ge	Germanium	2	2	6	2	6	10	2	2
33	As	Arsenic	2	2	6	2	6	10	2	3
34	Se	Selenium	2	2	6	2	6	10	2	4
35	Br	Bromine	2	2	6	2	6	10	2	5
36	Kr	Krypton	2	2	6	2	6	10	2	6

We shall now construct the periodic table of the elements with the help of the above considerations and the exclusion principle. Our procedure will be to investigate the status of a new electron added to an existing electronic structure. (Of course, the nuclear charge must also increase by $+e$ each time this is done.) In the simplest case we add an electron to a hydrogen atom ($Z = 1$) to give a helium atom ($Z = 2$). Both electrons in helium fall into the same $n = 1$ shell. Since $l = 0$ is the only value l can have when $n = 1$, both electrons have $l = m_l = 0$. The exclusion principle is not violated here, since one electron can have the spin magnetic quantum number $m_s = +\frac{1}{2}$ while the other has $m_s = -\frac{1}{2}$. It is customary to describe this situation by saying that the electrons have *opposite spins*.

Because no more than two electrons can occupy the $n = 1$ shell, helium atoms have *closed shells*. The characteristic properties of closed shells and subshells are that the orbital and spin angular momenta of their constituent electrons cancel out independently and that their effective electric charge distributions are perfectly symmetrical. As a result atoms with closed shells do not tend to interact with other atoms, which we know to be true of helium.

Lithium, with $Z = 3$, has one more electron than helium. There is no room left in the $n = 1$ shell, and so the additional electron goes into the $l = 0$ subshell of the $n = 2$ shell. The outer electron is relatively far from the nucleus in this shell, and is much less tightly bound. The chemical activity of lithium is a consequence of the low binding energy of this electron, which is readily lost to other atoms.

The next element, beryllium, has two electrons of opposite spin in the $l = 0$ subshell of the $n = 2$ shell. Boron, with $Z = 5$, has an electron in the $l = 1$ subshell as well as two in the $l = 0$ one. The $l = 1$ subshell can contain a total of six electrons, corresponding to two electrons of opposite spin in the $m_l = +1$, $m_l = 0$, and $m_l = -1$ states. This subshell is closed (and the $n = 2$ shell also closed) in neon, whose atomic number is 10. We therefore expect neon to be chemically inert, as indeed it is. Fluorine, the element just before neon in the periodic table, lacks but one electron of having a closed outer shell. Just as lithium tends to lose its single outermost electron in interacting with other elements, thereby leaving it with a closed-shell configuration, fluorine tends to gain a single electron to close its outer shell. The very different behavior of the alkali metals, the inert gases, and the halogens which we discussed in the previous section thus find explanation in terms of their respective atomic structures.

Table 33–3 shows the electron configurations of the 36 lightest elements. We note that the sequence of electron addition becomes irregular with potassium, which has an electron in its $n = 4$ shell even though its $n = 3$ shell is incomplete. The origin of this apparent anomaly is that electrons in the $n = 4$, $l = 0$ subshell have less energy, and therefore are more tightly bound, than those in the $n = 3$, $l = 2$ subshell, since the orbits of the former electrons pass closer to the nucleus than the more circular orbits of the latter electrons. The energy difference between the $n = 3$, $l = 2$ and $n = 4$, $l = 0$ subshells is actually quite small, as we can see from the configurations of chromium ($Z = 24$) and copper ($Z = 29$). In both these elements an additional electron is present in the $n = 3$, $l = 2$ subshell at the expense of the $n = 4$, $l = 0$ subshell, leaving a vacancy in the latter that is filled in the succeeding element.

The ferromagnetic properties of iron, cobalt, and nickel are a consequence of their unfilled $n = 3$, $l = 2$ subshells; without violating the exclusion principle, the electrons in these subshells do *not* pair off, and their spins do not cancel out. In the $n = 3$, $l = 2$ subshell of iron, for instance, five of the six electrons have parallel spins, leaving each iron atom with a strongly magnetic character, due to its four unpaired electrons.

As we can see, the quantum theory of the atom is one of the most powerful and fruitful approaches yet devised for understanding the properties of matter.

Problems

1. How many electrons do the elements in group II of the periodic table have in their outermost shells?

2. Find the maximum number of electrons that can be present in each of the subshells of the $n = 4$ shell.

3. What is the electron configuration of the inert gas xenon ($Z = 54$) which contains only closed subshells?

4. What is the electron configuration of the alkali metal francium ($Z = 87$) which has a single electron in its outermost shell? (The inner shells are closed subshells.)

5. Show that the maximum number of electrons in a shell of given n is $2n^2$. [*Hint:* Multiply the average number of electrons in the subshells of the nth shell by the number of subshells in that shell.]

6. How many elements would there be if atoms with occupied electron shells up through $n = 6$ could exist?

7. The energy required to remove an electron from an atom is called its *ionization energy*. Account for the fact that the ionization energies of the elements of atomic numbers 20 through 29 are very nearly the same, although wide variations are found in the ionization energies of other sequences of elements.

8. The valence electron of a lithium atom in the ground state is in an $n = 2$ state. It is approximately a hydrogenic system with a two-electron charge cloud shielding the nucleus. There are two possible angular momentum states corresponding to $l = 0$ (an s-state) and $l = 1$ (a p-state). Which state, the s or p, will have the lower energy, and why?

9. How large must the orbital angular momentum quantum number l be so that the ratio of the difference in magnitude of the orbital angular momentum and its maximum component to its magnitude is no more than one part in 10^6?

10. (a) What is the difference between the maximum value of the total angular momentum of an atom in the modified Bohr theory and that of an atom in quantum mechanics for the state $n = 100$? (b) What is the difference between the maximum values of the z-component of angular momentum in the two theories?

11. An electron is in an orbital angular momentum state corresponding to $l = 3$.

What are the possible values for J and for m_j for each J-value?

12. In a Stern-Gerlach apparatus the oven for the silver atoms is at 1000°K. The gradient of the magnetic field is 50 webers/$m^2 \cdot m$ for a distance of 8.0 cm. It continues for 100 more cm in a field-free region. What is the separation of the atoms on the collector plate? (All atoms should be taken to have the same velocity, given by $\frac{1}{2}mv^2 = \frac{3}{2}kT$.)

13. A beam of free electrons passes perpendicularly into a uniform magnetic field **B** whose magnitude is 2 webers/m^2. What is the difference in energy, in ev, between the electrons whose spins are aligned and those whose spins are anti-aligned with the magnetic field?

34

Molecules and Ions

Molecules exist by virtue of the interactions that occur between atoms under certain circumstances. In this chapter we shall see how the modern theory of the atom is able to account for the properties of molecules, including the variation of the specific heats of molecular gases with temperature that seemed so puzzling in Chapter 15.

34–1 Ionic binding

Atoms are most stable when their electron shells are closed, and tend to gain or lose electrons in order to secure closed shells by joining with other atoms. Two molecular binding mechanisms are known: *ionic* and *covalent*. In ionic binding, electrons are effectively transferred from one atom to another, while in covalent binding, two atoms may share one or more pairs of electrons in order to provide each with a closed outer electron shell. In some molecules one or the other mechanism predominates, while in others both participate to comparable extents in holding molecules together.

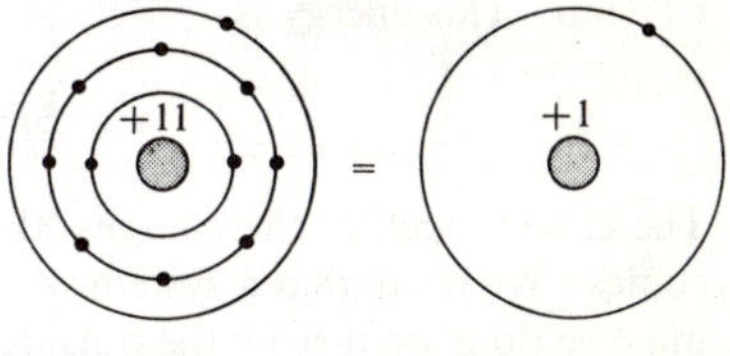

FIG. 34–1. Electron shielding.

Let us see how ionic binding functions in the case of NaCl, the sodium chloride molecule of ordinary table salt. The work that must be done to remove an electron from an atom is called the *ionization energy* of that atom. An atom with just one electron in its outer shell has a relatively low ionization energy which reflects the shielding effect of the inner electrons that cancel out most of the charge of the nucleus (Fig. 34–1). Such an atom tends to lose the outer electron and thereby assume an electron configuration composed solely of closed shells. At the other extreme, an atom lacking just one electron for the completion of its outer shell has a relatively high ionization energy which reflects the greater attractive force exerted by the nucleus when the nuclear charge is less effectively shielded by the intervening electrons. Thus sodium, with filled $n = 1$ and $n = 2$ shells and 1 electron in its $n = 3$ shell, has an ionization energy of 5.1 ev, while chlorine, also with filled $n = 1$ and $n = 2$ shells but with 7 electrons in its $n = 3$ shell, has an ionization energy of 13.0 ev.

Another atomic property relevant to ionic binding is *electron affinity*. The electron affinity of an atom is the amount of energy released when it acquires an additional electron. Since this amount of work must be done to remove the additional electrons after it is in place, electron affinity is a measure of the tendency of an atom to attract electrons in excess of its normal complement. As we would expect, atoms with high ionization energies have high electron affinities, while those with low ionization energies have low electron affinities. Atoms of the former kind, such as chlorine, tend to form negative ions by picking up electrons, and atoms of the latter kind, such as sodium, tend to form positive ions by losing electrons. An ionic bond between an atom of each kind occurs when one or more electrons pass from the atom with the low ionization energy to the one with the high electron affinity, since the resulting ions, both with complete outer shells, then attract each other electrostatically (Fig. 34–2). The bond is stable if more energy is needed to separate the ions than can be supplied by the return of the transferred electron or electrons.

As we said, the ionization energy of sodium is 5.1 ev, so that 5.1 ev of work is required to remove its single outer electron. The electron affinity of chlorine is 3.8 ev, so that 3.8 ev of energy is liberated when an electron is added to fill the vacancy in its outer shell. Transferring an electron from a sodium atom to a chlorine atom to form Na^+ and Cl^- ions therefore involves a net energy expenditure of 5.1 ev − 3.8 ev or 1.3 ev:

$$Na + Cl + 1.3\text{ ev} \rightarrow Na^+ + Cl^-. \tag{34–1}$$

Now we consider the electrostatic potential energy of a charge of $+e$, corresponding to the Na^+ ion, that is a distance r from a charge of $-e$, corresponding to the Cl^- ion. This energy is

$$\text{PE} = -\frac{e^2}{4\pi\epsilon_0 r}. \tag{34–2}$$

The closer together the charges, the more negative their potential energy PE becomes. We recall that a system whose total energy is negative is stable, since work must be done on it from the outside to break it up by making its total energy positive. The PE of a system of a $+e$ and a $-e$ charge is -1.3 ev when the charges are 11×10^{-10} meter apart; hence the total energy of a Na^+ ion and a Cl^- ion is zero when the ions are separated by this distance. When the ions are still closer together, their total energy is negative, and external energy is needed to pull them apart. The actual separation of Na^+ and Cl^- in a NaCl molecule is 2.4×10^{-10} meter, so that this molecule is stable. The work that must be done to break a NaCl molecule into Na and Cl atoms is 4.2 ev; that is,

$$NaCl + 4.2\text{ ev} \rightarrow Na + Cl. \tag{34–3}$$

There are two different repulsive effects that keep the Na^+ and Cl^- ions from being closer than 2.4×10^{-10} meter. Both effects arise because the electron structures of the ions mesh together when they are too near one another (Fig. 34–3). First, when such meshing occurs, the Na and Cl nuclei cease being shielded by their electrons and repel electrostatically. Second, and more important, such meshing means that the Na^+ and Cl^- electrons no longer constitute separate atomic systems but instead constitute a single one. According to the Pauli exclusion principle, no two electrons in the same system can have identical sets of quantum numbers, and so some of the electrons must go into quantum states of higher energy than they would otherwise occupy. The result of both effects is that work must be done to push the ions together to make their electron structures overlap. The various attractive and repulsive forces cancel out at a separation of 2.4×10^{-10} meter, which is why the Na^+ and Cl^- ions in a NaCl molecule are this far apart.

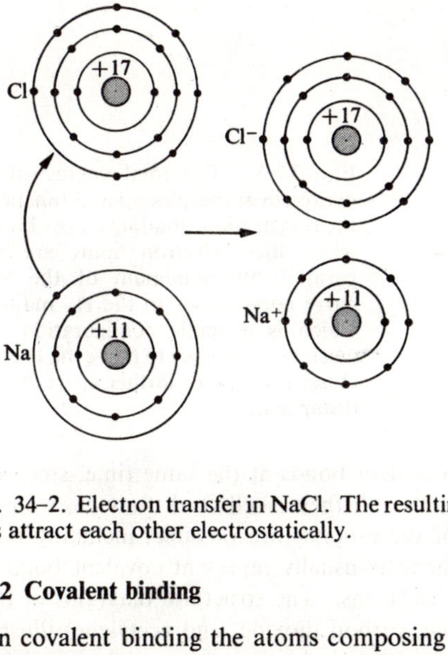

FIG. 34–2. Electron transfer in NaCl. The resulting ions attract each other electrostatically.

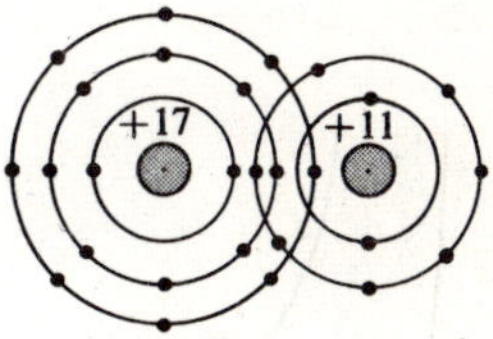

FIGURE 34–3

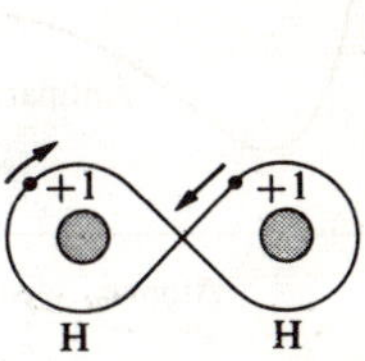

FIGURE 34–4

34–2 Covalent binding

In covalent binding the atoms composing a molecule share one or more pairs of electrons. Let us examine the origin of the covalent bond between the hydrogen atoms in an H_2 molecule, where ionic binding can play no part. If the shared electrons circulate around the H nuclei, as shown in Fig. 34–4, they.spend more time on the average between the nuclei than they do on the outside. The result is an effective negative charge between the positive H nuclei, and the attractive force this charge exerts on the nuclei is more than enough to counterbalance the direct repulsion between them. If the nuclei are too close together, however, their repulsion becomes dominant and the molecule is not stable. The balance between attractive and repulsive forces occurs at a separation of 7.42×10^{-11} meter, where the total energy of the H_2 molecule is -4.5 ev. Hence 4.5 ev of work must be done to break an H_2 molecule into two H atoms:

$$H_2 + 4.5\,\text{ev} \rightarrow H + H. \tag{34–4}$$

According to the Pauli exclusion principle, two electrons can jointly occupy the same region while in the lowest possible energy state when their spin quantum numbers are different, one having $s = +\frac{1}{2}$ and the other $s = -\frac{1}{2}$. In other words, the two electrons must have opposite spins. If the spins were the same, the electrons would be restricted to spending most of their time near the far ends of the H atoms, where they interact least. The result would be both a direct repulsion between the nuclei and a repulsion induced by the outward attraction of the electrons, making a stable molecule impossible (Fig. 34–5).

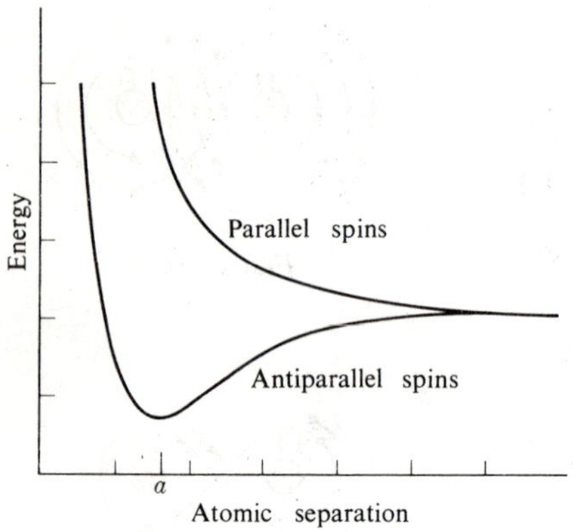

FIG. 34–5. The total energy of two hydrogen atoms plotted as a function of their separation. Binding is possible only when their electron spins are antiparallel; the minimum of the lower curve corresponds to the H_2 molecule, which is a stable configuration since work must be done to move the H atoms closer together or farther apart than the distance a.

Carbon atoms tend to form four covalent bonds at the same time, since they have four electrons in their outer shells and these shells lack four electrons for completion. Various distributions of these bonds are possible, including bonds between adjacent carbon atoms. Chemists usually represent covalent bonds by using a dash for each shared pair of electrons. The structural diagrams of three common covalent molecules (methane, carbon dioxide, and acetylene) illustrate the different bonds in which carbon can participate to form a covalent molecule:

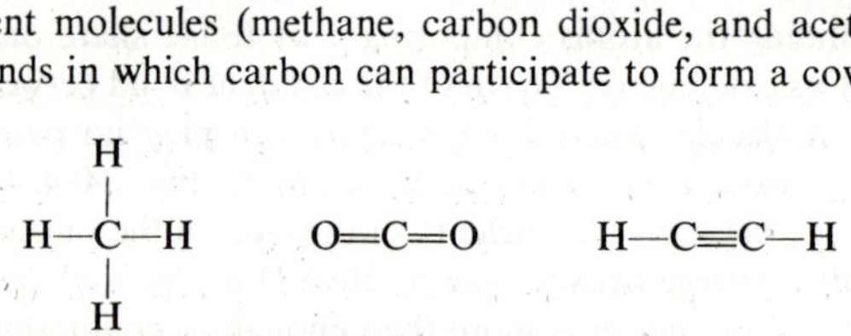

methane carbon dioxide acetylene

Carbon atoms are so versatile in forming covalent bonds with one another, as well as with other atoms, that literally millions of carbon compounds are known, some whose molecules contain tens of thousands of atoms. Such compounds were once thought to originate only in living things, and accordingly their study is known even today as organic chemistry.

34-3 Polar molecules

When atoms join together to form a molecule, their inner, complete electron shells undergo little change. In the limit of pure covalent binding, the atoms in a covalent molecule share their outer electrons evenly, and on the average remain neutral. In the opposite limit of pure ionic binding, however, electrons are transferred from certain atoms to other atoms in the molecule so that all have complete outer electron shells. The resulting molecule contains ions of opposite charge in different locations (Fig. 34-6). A molecule of this kind, as we learned in Chapter 23, is called a *polar molecule*, while a molecule whose charge distribution is symmetric about its center and hence has no dipole moment is called a *nonpolar molecule*. Ionic molecules are polar, as we expect, but so are many covalent molecules: While their component atoms share electrons, the electrons may spend more time in the vicinity of some atoms than in the vicinity of others. Thus the covalent molecule HCl is polar because a chlorine atom has greater attraction for an electron than a hydrogen atom does, while the covalent molecule H_2 is nonpolar because both hydrogen atoms have equal attractions for electrons.

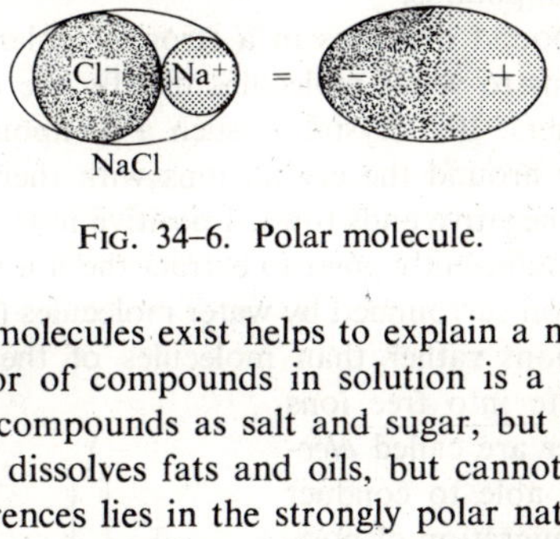

FIG. 34-6. Polar molecule.

The fact that polar molecules exist helps to explain a number of familiar phenomena. The behavior of compounds in solution is a good example. Water readily dissolves such compounds as salt and sugar, but cannot dissolve fats or oils. Gasoline readily dissolves fats and oils, but cannot dissolve salt or sugar. The key to these differences lies in the strongly polar nature of water molecules and the nonpolar nature of gasoline molecules. Water molecules tend to form aggregates under the influence of the electric forces between the ends of adjacent molecules, as shown in Fig. 34-7. Polar molecules of other substances, such as salt and sugar, can join in these aggregates, and are therefore easily dissolved by

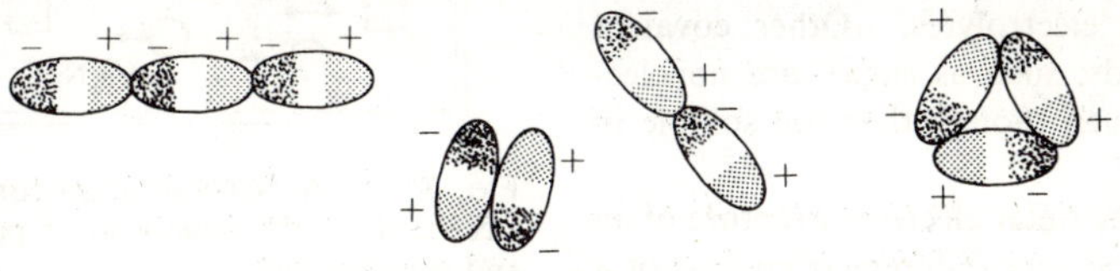

FIG. 34-7. Water molecules.

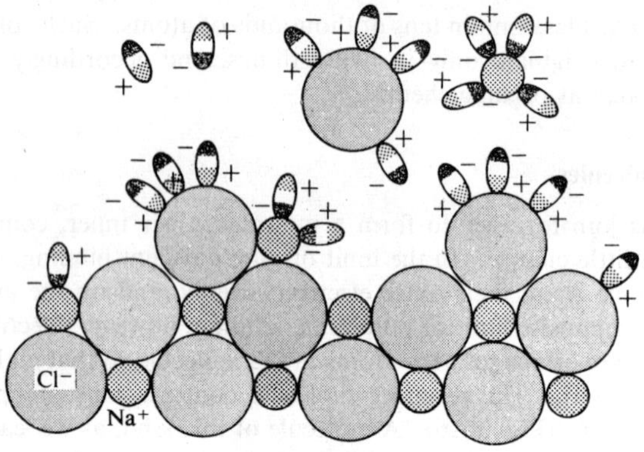

FIG. 34-8. The solution of solid NaCl.

water. The nonpolar molecules of fats and oils, however, do not interact with water molecules. If samples of oil and water are mixed together, the attraction of water molecules for one another acts to squeeze out the oil molecules, and the mixture soon separates into layers of each substance. Fat and oil molecules dissolve only in liquids whose molecules are similar to theirs, which is why gasoline is a solvent for these compounds.

When an ionic compound dissolves in a strongly polar liquid like water, the forces attracting the ions to water molecules usually are greater than the forces holding the ions together. If a crystal of such a compound is placed in water, water molecules cluster around the crystal ions with their positive ends toward negative ions and their negative ends toward positive ions. The attraction of several water molecules is sufficiently great to extract the ion from the crystal lattice, and it moves into solution surrounded by water molecules (Fig. 34-8). The resulting solution contains ions rather than molecules of the dissolved compound. Substances that separate into free ions when dissolved in water are called *electrolytes*, since they are able to conduct electric current by the migration of positive and negative ions. Figure 34-9 shows how a NaCl solution conducts electricity. All ionic compounds soluble in water and some polar covalent compounds, such as HCl, are electrolytes. Other covalent compounds, such as sugar, are nonelectrolytes even though they are soluble in water.

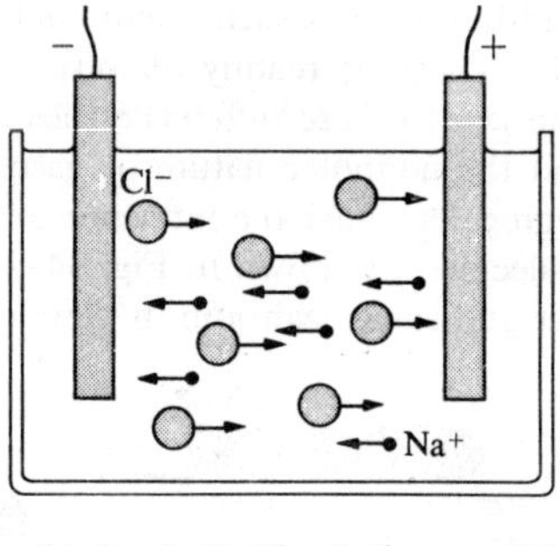

FIG. 34-9. A NaCl solution conducts electricity by the migration of positive and negative ions.

Since the outer electron structure of an ion may be very different from that of a neutral atom of the same species, it is not surprising that the ions of an element may behave very differently from its atoms or molecules. Thus gaseous chlorine is greenish in color, has a strong, irritating taste, and is very active chemically, while a solution of chlorine ions is colorless, has a mild, pleasant taste, and is only feebly active. The contrast in chemical activity follows, of course, from the incomplete outer electron shell in a chlorine atom, a shell which is complete in a chlorine ion.

34-4 Molecular rotation

The energy levels of an atom are the result of interactions between its electrons and nucleus and interactions among the electrons themselves. Electronic energy levels are also present in a molecule but, owing to its composite structure, additional energy states representing rotation and vibration are possible. Molecular energies of rotation and vibration were mentioned in Chapter 15, where they were introduced in order to account for the unusually high specific heats of polyatomic gases. We saw there that the molecules of such gases rotate and vibrate only at relatively high temperatures, although the molecules themselves are free to move about at all temperatures. This paradox is removed by quantum theory, which limits angular momenta and vibrational energies to certain specific values.

We shall begin by examining the rotation of a diatomic molecule about an axis that passes through its center of mass perpendicular to the line joining the two atoms (Fig. 34-10). The moment of inertia about this axis has the value

$$I = m'a^2 \qquad (34\text{-}5)$$

according to Eq. (8-23), where

$$m' = \frac{m_1 m_2}{m_1 + m_2}$$

is the reduced mass. The molecule's angular momentum L when it rotates with the angular velocity ω is

$$L = I\omega.$$

The angular-momentum quantization as discussed in Chapter 33 gives L as

$$L = \sqrt{l(l+1)}\,\hbar, \qquad l = 0, 1, 2, 3, \ldots,$$

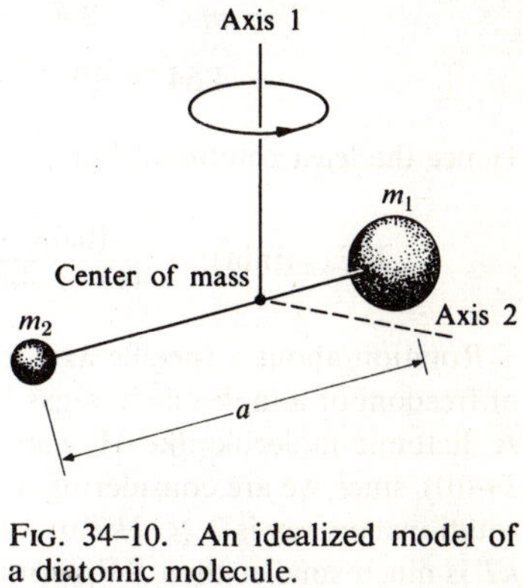

FIG. 34-10. An idealized model of a diatomic molecule.

and is not confined to atomic electrons alone, but turns out to be a general result that applies to all objects and systems that can possess angular momentum. Since Planck's constant h has the minute value of 6.63×10^{-34} j·sec, the quantization of angular momentum is imperceptible for macroscopic bodies. On a microscopic scale, however, this quantization dominates atomic and molecular phenomena. By combining the above expressions for L we find that the permitted angular velocities of a diatomic molecule are

$$\omega = \sqrt{l(l+1)}\,\hbar, \qquad l = 0, 1, 2, 3, \ldots, \tag{34-6}$$

and the corresponding kinetic energies of rotation are

$$KE_{rot} = \tfrac{1}{2}I\omega^2 = l(l+1)\frac{\hbar^2}{2I}, \qquad l = 0, 1, 2, 3, \ldots \tag{34-7}$$

The lowest possible energy of rotation (other than zero) of a molecule whose moment of inertia is I may be found by substituting $l = 1$ in Eq. (34-7), which yields

$$KE_{rot}\,(\text{min}) = \frac{\hbar^2}{I}\cdot \tag{34-8}$$

For the H_2 molecule, where

$$m_1 = m_2 = 1.67 \times 10^{-27}\,\text{kg} \qquad \text{and} \qquad a = 7.42 \times 10^{-11}\,\text{m},$$

the reduced mass is

$$m' = m^2/2m = \tfrac{1}{2}m = 8.4 \times 10^{-28}\,\text{kg},$$

and so the moment of inertia is

$$I = m'a^2 = 8.4 \times 10^{-28}\,\text{kg} \times (7.42 \times 10^{-11}\,\text{m})^2$$
$$= 4.64 \times 10^{-48}\,\text{kg·m}^2.$$

Hence the least rotational kinetic energy a hydrogen molecule can have is

$$KE_{rot}\,(\text{min}) = \frac{(6.63 \times 10^{-34}\,\text{j·sec})^2}{4\pi^2 \times 4.64 \times 10^{-48}\,\text{kg·m}^2} = 2.39 \times 10^{-21}\,\text{j}.$$

Rotation about a specific axis represents a degree of freedom, and each degree of freedom of a molecule in a gas has an average energy of $\frac{1}{2}kT$ associated with it. A diatomic molecule like H_2 can rotate about only two perpendicular axes (Fig. 34-10), since we are considering its atoms as point masses, and so its average total rotational energy is $2 \times \frac{1}{2}kT$ or kT. When the temperature of a gas is so low that kT is much smaller than KE_{rot} (min), virtually none of its molecules rotate, while if kT is much greater than KE_{rot} (min), nearly all of them rotate. When kT is comparable to KE_{rot} (min), some of the molecules rotate and some do not. Thus we expect a gradual transition in the specific heat of a gas with increasing temperature from the value representing pure translational motion to the value representing translation plus rotation. This expectation is fulfilled experimentally, as we can see in Fig. 15-4, where the molar specific heat of H_2 is plotted versus temperature.

The temperature at which kT and KE_{rot} (min) are equal for a diatomic molecule is given by

$$T = \frac{\hbar^2}{kI}. \tag{34–9}$$

Since we have already computed KE_{rot} (min) for H_2, we can find T by dividing it by Boltzmann's constant k:

$$T = \frac{2.39 \times 10^{-21}\ \text{j}}{1.38 \times 10^{-23}\ \text{j/°K}} = 173\text{°K}.$$

The precise value of this temperature is not to be taken too seriously, because the existence of definite rotational energy states means that the manner in which the available average kT of rotational energy per molecule is actually distributed among the molecules in a gas differs from that in which the average of $\frac{3}{2}kT$ of translational energy, not subject to any restriction, is distributed. The molecular speed distribution curves of Fig. 14–3 have quite different rotational counterparts. However, even though we cannot regard the T of Eq. (34–9) as representing, say, the temperature at which half the molecules in a gas rotate and half do not, it is usually not too far away from this temperature and is much easier to calculate. (An exact calculation shows that, at 173°K, a substantial fraction of the H_2 molecules is in the $l = 2$ rotational state and a very few are even in the $l = 3$ state.)

According to Eq. (34–8), the smaller the moment of inertia of a molecule, the greater its minimum kinetic energy of rotation and the greater the temperatures must be if the molecule is to have an appreciable likelihood of rotating. Single atoms can only be excited by electronic transitions. The energies corresponding to these transitions are of the order of electron volts which correspond to temperatures of the order of thousands of degrees. This is why monatomic gases such as helium and argon at moderate temperatures have specific heats corresponding to translational motion only. The first excitation energy of a hydrogen atom is 10.2 ev, which corresponds to a temperature of over 10,000°K. As we saw, the molecules in an H_2 gas are excited in rotational levels in appreciable numbers at 173°K. The "rotation" of a diatomic molecule about its axis of symmetry can be ignored compared with its rotation about axes perpendicular to this one, since I about the latter axis is nearly 2000 times greater.

34–5 Molecular vibration

The bonds that join atoms together to form molecules resist being stretched or compressed beyond their normal lengths, so to speak, and therefore we can crudely picture a molecule as a group of atoms attached to one another by tiny springs. If the usual molecular configuration is disturbed, say by a collision with another molecule or by the absorption of a suitable photon of electromagnetic energy, some or all of the atoms will vibrate. The vibrations of a diatomic molecule are

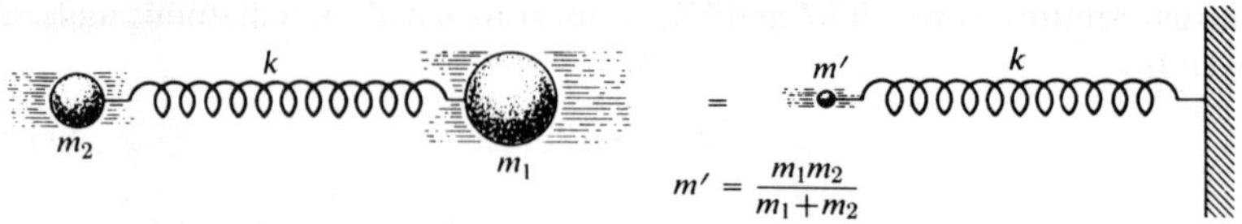

FIGURE 34-11

easy to analyze because only a single bond is present. When the atoms of such a molecule are displaced by a small amount x from their normal separation, the restoring force that acts happens to be very close to the force $F = -kx$ of Hooke's law which is obeyed by ordinary springs. As we learned in Chapter 10, a system of two particles of masses m_1 and m_2 joined by a spring of force constant k vibrates with the same frequency and total energy as a single particle of mass

$$m' = \frac{m_1 m_2}{m_1 + m_2} \tag{34-10}$$

attached to one end of the spring with the other end fixed in place (Fig. 34-11). The frequency of oscillation of a diatomic molecule is therefore

$$f = \frac{1}{2\pi}\sqrt{\frac{k}{m'}}. \tag{34-11}$$

One of the most significant results of quantum theory is that the total energy of a harmonic oscillator is limited to the values

$$E_v = (v + \tfrac{1}{2})hf, \qquad v = 0, 1, 2, 3, \ldots, \tag{34-12}$$

where f is the characteristic frequency of the oscillator and v, the *vibrational quantum number*, may be zero or any integer. The remarkable thing about this formula is not the quantization of energy it embodies, which is hardly surprising in view of the discussions of the previous few chapters, but the fact that the lowest possible energy is $\frac{1}{2}hf$ and not zero. A harmonic oscillator cannot ever be completely at rest! The energy $\frac{1}{2}hf$ is called the *zero-point energy* E_0 of an oscillator with the natural frequency f, and, because Planck's constant h is so small, it is wholly imperceptible in the macroscopic world. The zero-point vibrations of a harmonic oscillator may be thought of as manifestations of the uncertainty principle: An oscillator at rest must have the precisely defined position $x = 0$ and momentum $mv = 0$, while the uncertainty principle does not permit our knowledge of both these quantities to be so exact. We thus have the curious fact that, at absolute zero, polyatomic molecules are still vibrating even though they can neither move nor rotate.

The force constant k of the bond in the H_2 molecule is found to be 5.73×10^2 n/m, and the reduced mass m' of the molecule is, as we found in the previous section, 8.4×10^{-28} kg. The natural frequency of vibration of the H_2 molecule is therefore

$$f = \frac{1}{2\pi}\sqrt{\frac{k}{m'}} = \frac{1}{2\pi}\sqrt{\frac{5.73 \times 10^{2}\ \text{n/m}}{8.4 \times 10^{-28}\ \text{kg}}}$$

$$= 1.32 \times 10^{14}\ \text{cycles/sec.}$$

The molecule's zero-point energy is

$$E_0 = \tfrac{1}{2}hf = \tfrac{1}{2} \times 6.63 \times 10^{-34}\ \text{j·sec} \times 1.32 \times 10^{14}\ \text{cycles/sec}$$

$$= 4.38 \times 10^{-20}\ \text{j},$$

and that of its first excited vibrational state ($v = 1$) is

$$E_1 = \tfrac{3}{2}hf = 13.13 \times 10^{-20}\ \text{j}.$$

The zero-point oscillations do not contribute to the specific heat of gaseous hydrogen because they are always present, regardless of temperature. To go from the zero-point vibrational state to the first excited state requires an energy of

$$E_1 - E_0 = hf = 8.75 \times 10^{-20}\ \text{j}.$$

Vibration involves two degrees of freedom, one for potential energy and one for kinetic energy, and as a result the average vibrational energy per molecule is $2 \times \frac{1}{2}kT$ or kT. As in the case of molecular rotation, we can equate kT with $E_1 - E_0$ to find what we might regard as a typical temperature at which an H_2 molecule will begin to vibrate in its $v = 1$ state. We find that

$$T = \frac{E_1 - E_0}{k} = \frac{8.75 \times 10^{-20}\ \text{j}}{1.38 \times 10^{-23}\ \text{j/°K}} = 6350\text{°K}.$$

An exact calculation shows that, at 6350°K, about half the molecules in gaseous hydrogen are indeed vibrating in excited states, but at this temperature many are in the $v = 2$, $v = 3$, and $v = 4$ states, as well as in the $v = 1$ state.

It is evident that molecular vibration involves considerably higher energies than molecular rotation. Thus the energy of the first excited vibrational state in H_2 is 1.31×10^{-19} joule, while that of its lowest rotational state is 2.39×10^{-21} joule, 55 times smaller. A comparable ratio between vibrational and rotational energies holds for all molecules; hence any molecule that has enough energy to vibrate is certainly rotating as well. Much higher temperatures are required for the vibrational degrees of freedom in a polyatomic gas to absorb thermal energy than for the rotational ones, which is exactly the behavior found experimentally.

It is interesting to compare the frequencies of vibration and rotation in a particular molecule. As we saw, the natural frequency of vibration of the H_2 molecule is 1.32×10^{14} cycles/sec. From Eq. (34–7) we find that the angular velocity of the H_2 molecule when $l = 1$ is

$$\omega = \sqrt{l(l+1)}\frac{\hbar}{I} = \frac{\sqrt{2} \times 6.63 \times 10^{-34}\ \text{j}\cdot\text{sec}}{2\pi \times 4.64 \times 10^{-48}\ \text{kg}\cdot\text{m}^2}$$

$$= 3.22 \times 10^{13}\ \text{rad/sec}.$$

The frequency of revolution f_r that corresponds to this angular velocity is

$$f_r = \frac{\omega}{2\pi} = 5.12 \times 10^{12}\ \frac{\text{rev}}{\text{sec}};$$

an H_2 molecule with $l = 1$ and $v = 1$ rotates 26 times for every vibration.

Our purpose in discussing molecular energy levels in some detail is threefold. First, the crucial role of quantization of energy and angular momentum is demonstrated to hold in the world of molecules as well as in the world of atoms. Second, the mysterious behavior of the specific heats of gases is clarified. And third, the theory of molecular energy levels is what has led to much of our detailed knowledge of molecular structure. Transitions among the various vibrational and rotational energy states in a molecule involve photons of various characteristic wavelengths, and the observed values of these wavelengths permit a great deal of information to be inferred. The exact arrangement of the atoms in a particular molecule can often be established from its rotational spectra, which are generally in the infrared region, since these spectra reveal the moments of inertia of the molecule about three axes. The strengths of the various interatomic bonds in a molecule can be determined from its vibrational spectra. Very complex molecules, such as most organic molecules, can be analyzed indirectly from their spectra. For instance, the bond between the O and H atoms in the OH^- group always has about the same force constant k regardless of the molecule in which OH^- appears, so that the appearance of the wavelengths characteristic of the vibrations of OH^- in the complete spectrum of a compound means that this group is present somewhere in its molecules. Spectroscopy has proved to be one of the most fruitful of our techniques for understanding the structure and behavior of molecules.

Problems

1. How can you account for the fact that lithium and sodium exhibit similar chemical behavior?

2. How can you account for the fact that fluorine and chlorine exhibit similar chemical behavior?

3. Why do lithium atoms not form molecules with each other in the way that fluorine atoms do?

4. The ionization energies of Li, Na, K, Rb, and Cs are, respectively, 5.4, 5.1, 4.3, 4.2, and 3.9 ev. All are in group I of the periodic table. Account for the decrease in ionization energy with increasing atomic number.

5. The ionization energy of potassium is 4.3 ev and the electron affinity of chlorine is 3.8 ev. (a) What is the net amount of

energy required to form a K^+ and Cl^- ion pair from a pair of the same atoms? (b) Considering them as point charges, how close together must a K^+ and a Cl^- ion be if the total energy of the pair is to be zero?

6. The ionization energy of lithium is 5.4 ev and the electron affinity of bromine is 3.5 ev. (a) What is the net amount of energy required to form a Li^+ and Br^- ion pair from a pair of the same atoms? (b) Considering them as point charges, how close together must a Li^+ and a Br^- ion be if the total energy of the pair is to be zero?

7. Why are electrons much more readily liberated from lithium when it is irradiated with ultraviolet light than from fluorine?

8. Why do molecules increase in size with increasing temperature?

9. Why are Na atoms more active chemically than Na^+ ions?

10. Why are Cl atoms more active chemically than Cl^- ions?

11. Give an argument based on the Pauli principle to show that rare gas atoms (He, Ne, Ar, etc.) cannot participate in covalent bonds.

12. Give an argument based on the Pauli principle to show that the molecule H_3 cannot exist.

13. The Hg and Cl atoms in a certain HgCl molecule have masses of 3.32×10^{-25} kg and 5.81×10^{-26} kg, respectively. The molecule emits a photon of wavelength 0.044 m when it undergoes a rotational transition from $l = 1$ to $l = 0$. What is the interatomic distance in this molecule?

14. The moment of inertia of a diatomic molecule increases as its rotational quantum number l increases. Compare this behavior with the equatorial bulge of the earth.

15. The dissociation energy of hydrogen is 4.5 ev (that is, 4.5 ev of work is required to separate the two hydrogen atoms that make up a hydrogen molecule). Assume that the hydrogen molecule behaves exactly like a harmonic oscillator and find the vibrational quantum number corresponding to a vibrational energy as close as possible to the dissociation energy.

16. A pendulum has a zero-point energy of 2.07×10^{-15} ev. Find its length.

17. The H and Cl atoms in a certain HCl molecule have masses of 1.66×10^{-27} kg and 5.81×10^{-26} kg respectively. If the separation of the atoms is 1.29×10^{-10} m, find the number of rotations per second made by the molecule when it is in the $l = 1$ rotational state.

18. The force constant of the bond in the HCl molecule is 470 n/m. (a) Find its natural frequency of oscillation. (b) Is it likely that the molecule will be vibrating in the $v = 1$ state at room temperature?

35

The Solid State

The same physical processes that can hold several atoms together as a molecule can also hold an unlimited number of them together as a solid. In addition, structural regularities in certain solids confer upon them such remarkable properties as the ability to conduct electric current readily. While only an infinitesimal part of the universe is in the solid state, solids constitute much of the physical world of our experience, and modern technology is to a large extent based on the exploitation of the unique characteristics of various solid materials.

35–1 The structure of solids

Most solids are crystalline in nature, with their constituent atoms or molecules arranged in regular, repeated patterns. A crystal is thus characterized by the presence of *long-range order* in its structure. Other solids lack the definite arrangements of atoms and molecules so conspicuous in crystals, and may be regarded in a sense as supercooled liquids whose stiffness is due to an exaggerated viscosity. Examples of such *amorphous* ("without form") solids are pitch, glass, and many plastics. The structures of amorphous solids exhibit *short-range order* only. Some substances, for instance B_2O_3, can exist in either crystalline or amorphous forms. In both cases each boron atom is surrounded by three larger oxygen atoms, which represents a short-range order. In a B_2O_3 crystal a long-range order is also present, as shown in a two-dimensional representation in Fig. 35–1, while amorphous B_2O_3, a glassy material, lacks this additional regularity. The lack of long-range order in amorphous solids means that the various bonds vary in strength. When an amorphous solid is heated, the weakest bonds are ruptured at lower temperatures than the others and the solid softens gradually, while in a crystalline solid the bonds break simultaneously, and melting is a sudden process.

How is it possible to determine the arrangement of atoms in a solid when atomic dimensions are so small? To appreciate this problem, let us calculate the spacing between adjacent atoms in rock salt, which is the crystalline form of NaCl. The density of rock salt is 2.17×10^3 kg/m, and the atomic weights of sodium and chlorine are, respectively, 22.99 and 35.46. The molecular weight of NaCl is

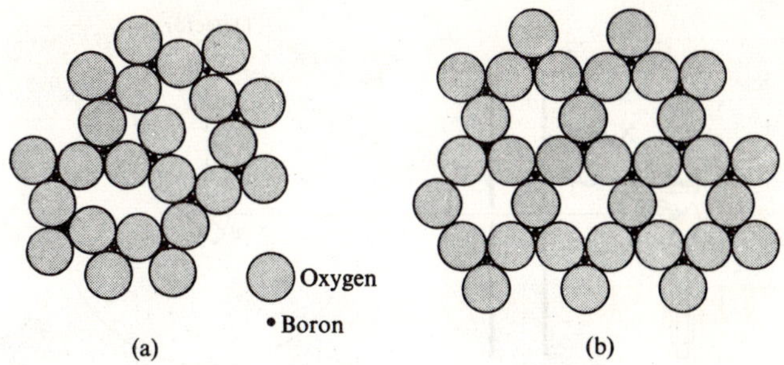

FIG. 35–1. Structure of B_2O_3. (a) Amorphous B_2O_3, a glass, has short-range order only in its structure. (b) Crystalline B_2O_3 has long-range order as well.

therefore 22.99 + 35.46, or 58.45, and each molecule contributes two atoms to the crystal lattice. The number of atoms per unit volume in the crystal is, since Avogadro's number is the number of molecules per kilomole,

$$\text{Atomic density} = 2\,\frac{\text{atoms}}{\text{molecule}} \times 6.025 \times 10^{26}\,\frac{\text{molecules}}{\text{kmole}} \times \frac{1}{58.45\ \text{kg/kmole}} \times 2.17 \times 10^{3}\,\frac{\text{kg}}{\text{m}^3}$$

$$= 4.47 \times 10^{28}\ \text{atoms/m}^3.$$

Let us suppose that each atom occupies a cube a on a side, so that the volume it preempts in the crystal is a^3. Therefore

$$4.47 \times 10^{28}\ \text{atoms} \times a^3 = 1\ \text{m}^3,$$

$$a = 2.82 \times 10^{-10}\ \text{m}.$$

This minute distance represents the spacing between adjacent atoms in a rock-salt crystal.

The atomic structure of a crystal is on too fine a scale to be resolved using visible light, since the shortest wavelengths the eye responds to exceed a by a factor of 10^3. X-rays, however, can be produced with wavelengths well under 10^{-10} meter, which makes them ideal for investigating crystal structures. Several techniques have been developed for this purpose, and we shall examine an especially straightforward one devised in 1912 by W. L. Bragg. More recently, methods employing electron and neutron beams of appropriate de Broglie wavelength have come into use in studying solids, and, in fact, it was the diffraction of electrons by crystals that provided the first experimental proof of de Broglie's hypothesis.

Figure 35–2 shows a beam of x-rays directed at the face of a crystal. Most of them go right through without interacting, but a few are scattered by atoms in their paths. The conditions for a scattered ray to emerge are the same as those for

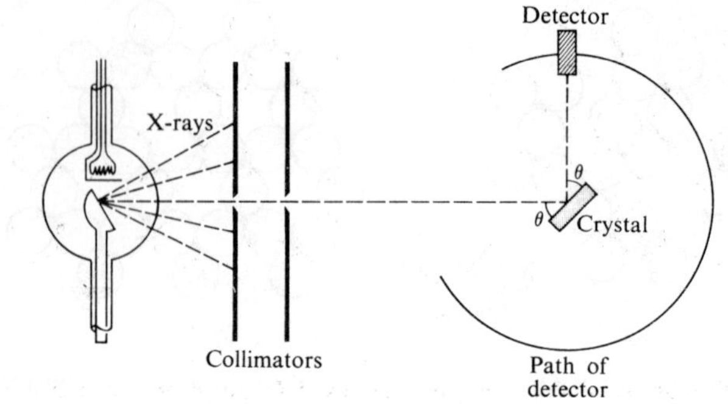

FIG. 35-2. X-ray scattering by a crystal. The angles of incidence and "reflection" are equal.

the reflection of light by a mirror: the scattered and incident rays must lie in a plane perpendicular to the crystal face, and the angle of incidence must equal the angle of scattering. However, x-rays can be scattered by atoms in the interior of the crystal as well as at its surface. Unless all the x-ray waves scattered in a given direction are in phase ("in step"), destructive interference will result and they will cancel one another out. We can derive the condition for constructive interference by considering the scattering of a beam of x-rays from two successive layers of atoms a distance a apart, as in Fig. 35-3. If the beam is incident at the angle θ, both scattered rays also leave at this angle. Ray II must travel the additional distance relative to ray I of $a \sin \theta$ in order to reach its scattering atom, and then a further additional distance of $a \sin \theta$ to emerge from the crystal. Thus ray II travels $2a \sin \theta$ farther than ray I in the course of being scattered. If this path difference is a whole number of x-ray wavelengths, the two rays will reinforce each other. The condition for constructive interference is therefore

$$\text{Path difference} = \text{whole number of wavelengths,}$$

$$2a \sin \theta = n\lambda, \quad n = 1, 2, 3, \ldots \tag{35-1}$$

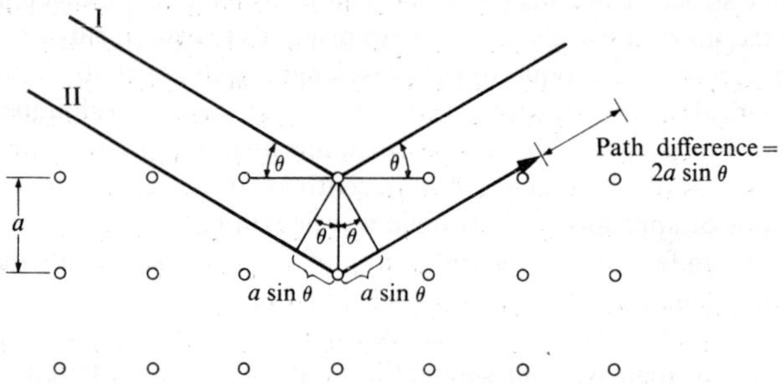

FIGURE 35-3

By directing a beam of x-rays of known wavelength at a crystal and varying the angle of incidence in order to determine the angles at which scattering occurs, the spacing a between adjacent layers of atoms can be ascertained. Of course, the atoms fall into more than one series of parallel layers, as in Fig. 35–4, and a different spacing will be obtained for each of them.

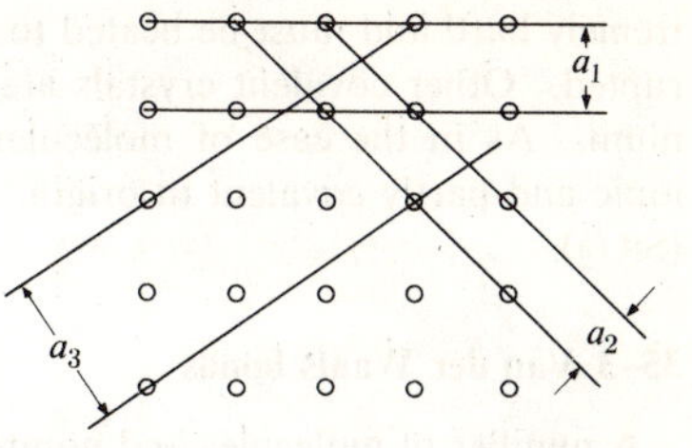

FIG. 35–4. The atoms in a crystal fall into several series of parallel layers.

35–2 Ionic and covalent bonds

The ionic and covalent bonds between atoms that are responsible for the formation of molecules also act to hold many crystalline solids together. Figure 35–5 shows the structure of a NaCl crystal; the small spheres represent Na^+ ions and the large spheres Cl^- ions. Each ion behaves essentially like a point charge, and thus tends to attract to itself as many ions of opposite sign as can fit around it. The latter ions, of course, repel one another, and so the resulting crystal is an equilibrium configuration in which the various attractions and repulsions balance out. In a NaCl crystal each Na^+ ion is surrounded by six Cl^- ions, and vice versa. In crystals having different structures, the number of "nearest neighbors" around each ion may be 3, 4, 6, 8, or 12. Ionic bonds are usually fairly strong, and consequently ionic crystals are strong, hard, and have high melting points.

In covalent crystals the attractive interatomic forces arise from the sharing of electrons between atoms. Figure 35–6 shows the array of carbon atoms in a diamond crystal, with each carbon atom sharing electron pairs with the four other carbon atoms adjacent to it. All the electrons in the outer shells of the carbon atoms participate in the binding, and it is therefore not surprising that diamond is ex-

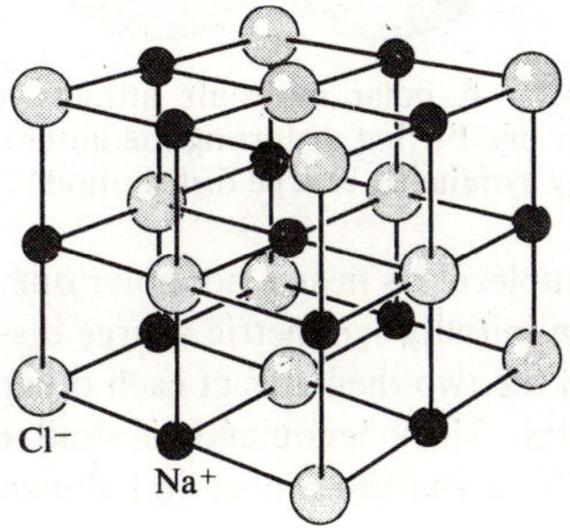

FIG. 35–5. Ionic structure of NaCl crystal.

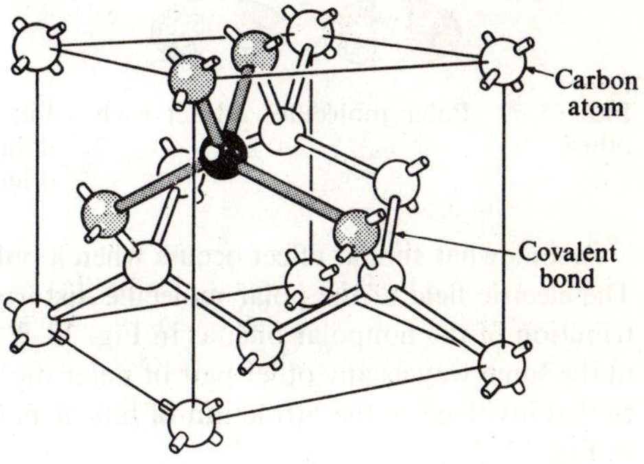

FIG. 35–6. Covalent structure of diamond. The length of each bond is 3.1×10^{-10} m.

tremely hard and must be heated to over 6000°F before its crystal structure is disrupted. Other covalent crystals are those of silicon, silicon carbide, and germanium. As in the case of molecular binding, many crystalline bonds are partly ionic and partly covalent in origin. An example of such mixed binding is quartz (SiO_2).

35–3 Van der Waals bonds

A number of molecules and nonmetallic atoms exist whose electronic structures do not lend themselves to either of the above kinds of binding. The inert gas atoms, which have filled outer electron shells, and organic molecules such as methane,

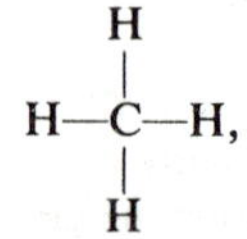

whose valence electrons are fully involved in the molecular bond itself, fall into this category. However, even these virtually noninteracting substances condense into solids and liquids at low enough temperatures through the action of what are known collectively as *van der Waals forces.*

The electrostatic attraction between polar molecules that was discussed in the previous chapter is an example of a van der Waals force. These molecules have asymmetric distributions of charge, with one end positive and the other negative. When one such molecule is near another, the ends of opposite polarity attract each other to hold the molecules together (Fig. 35–7).

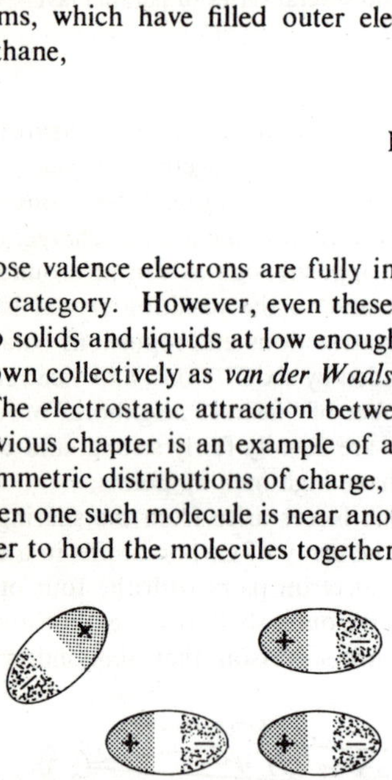

FIG. 35–7. Polar molecules attract each other.

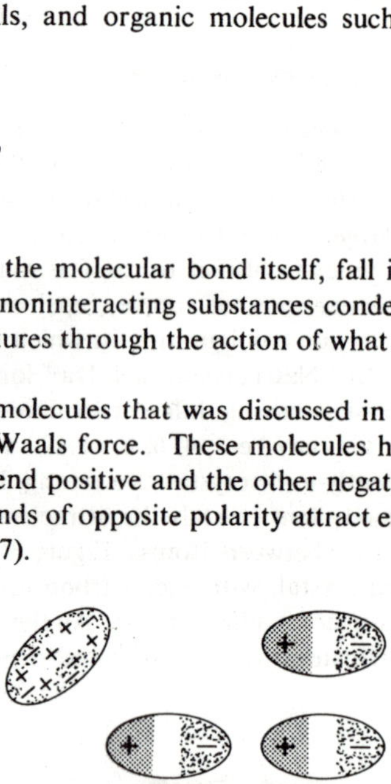

FIG. 35–8. A polar molecule attracts a nonpolar one by first distorting the latter's originally symmetric charge distribution.

A somewhat similar effect occurs when a polar molecule is near a nonpolar one. The electric field of the polar molecule distorts the initially symmetric charge distribution of the nonpolar one, as in Fig. 35–8, and the two then attract each other in the same way as any other pair of polar molecules. This phenomenon is similar to that involved in the attraction of bits of paper by a charged amber rod shown in Fig. 17–3.

Curiously enough, nonpolar molecules also attract one another in the same way that polar molecules attract nonpolar ones. Because the electrons in a molecule are in constant motion, their distribution changes constantly. In a nonpolar molecule the electron distribution is symmetrical *on the average*, but *at any moment* one part of the molecule contains more electrons than usual and the remainder of the molecule contains correspondingly fewer electrons. Thus *every* molecule (and atom as well) behaves as though it is polar, although its polarization may constantly vary in both magnitude and direction. In a true polar molecule the average electron distribution is asymmetrical, while in a nonpolar molecule it is symmetrical; but in either case the molecule is polar in nature at each instant, and so is able to attract other molecules by the mechanism of Fig. 35–8. The fluctuating charge distributions of adjacent "nonpolar" molecules vary almost simultaneously so as to keep in step, as in Fig. 35–9, and the electrostatic forces that result hold the molecules together, provided that their random kinetic energies of thermal origin are not too great.

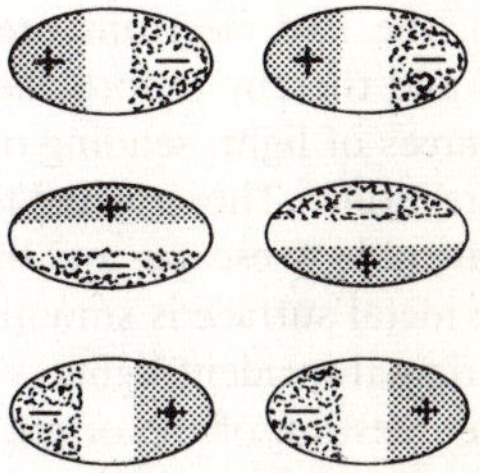

FIG. 35–9. Nonpolar molecules have, on the average, symmetric charge distributions, but at any instant the distributions are not necessarily symmetric. The fluctuations in the charge distributions of adjacent nonpolar molecules keep in step, which leads to an attractive force between them.

In general van der Waals bonds are considerably weaker than ionic, covalent, and metallic bonds; only about 1 percent as much energy is needed to remove an atom or molecule from a van der Waals solid as is required in the case of ionic or covalent crystals. As a result the inert gases and compounds with symmetric molecules liquefy and vaporize at rather low temperatures. Thus the boiling point of argon is $-186°C$ and the boiling point of methane is $-161°C$. Molecular crystals, whose lattices consist of individual molecules held together by van der Waals forces, generally lack the mechanical strength of other kinds of crystals.

35–4 The metallic bond

A fourth important type of cohesive force in crystalline solids is the *metallic bond*, which has no molecular counterpart. A characteristic property of all metal atoms is the presence of only a few electrons in their outer shells, and these electrons can be detached relatively easily to leave behind positive ions. According to the theory of the metallic bond, a metal in the solid state consists of an assembly of atoms that have given up their outermost electrons to a common "gas" of freely moving electrons that pervades the entire metal. The electrostatic interaction between the positive ions and the negative electron gas holds the metal together. This theory has many attractive features. The high electrical and thermal conductivity of metals follows from the ability of the free electrons to migrate through their crystal structures, while all the electrons in ionic and covalent crystals are bound to particular atoms or pairs of atoms. Unlike other crystals, metals may be deformed without fracture, because the electron gas permits the atoms to slide past one another by acting as a kind of lubricant. While certain solids, such as rubber or asphalt, that are amorphous rather than crystalline in structure, can also be deformed readily, they lack the strength and hardness conferred by the metallic bond. Furthermore, since the atoms in a metal interact through the medium of a common electron gas, the properties of mixtures of different metal atoms should not depend critically on the relative proportions of each kind of atom, provided their sizes are similar. This prediction is fulfilled in the observed behavior of alloys, in contrast to the specific atomic proportions characteristic of ionic and covalent solids.

The opacity and metallic luster exhibited by metals may also be traced to the gas of free electrons that pervades them. When light of any frequency shines on a metal, the free electrons are set in vibration by the oscillating electromagnetic fields and thereby absorb the light. The oscillating electrons themselves then act as sources of light, sending out electromagnetic waves of the original frequency in all directions. Those waves that happen to be directed back toward the metal surface are able to escape, and their emergence gives the metal its lustrous appearance. If the metal surface is smooth, the reradiated waves appear to us as a reflection of the original incident light.

The question of the origin of the "electron gas" in a metal is closely related to the question of why hydrogen is the only member of group I of the periodic table which is not a metal. A hydrogen molecule contains two electrons with $n = 1$ and opposite spin, the largest number possible in that state in the same atomic system without violating the Pauli exclusion principle. A hydrogen atom that approaches a H_2 molecule cannot become attached because its electron would have to go to an $n = 2$ state, which has a higher energy than the normal $n = 1$ state. The H_2 molecule exists because H_2 has less energy than $H + H$; the H_3 molecule cannot exist because H_3 has more energy than $H_2 + H$. Solid hydrogen is accordingly an assembly of H_2 molecules held together by van der Waals bonds, bonds that are so feeble that hydrogen melts at $-259°C$.

Lithium, atomic number 3, is the next member of group I of the periodic table and, like all the other members except hydrogen, is indisputably metallic in its behavior. From Table 33-3 we see that each lithium atom has a filled $n = 1$ shell and a single electron in the $n = 2$, $l = 0$ state. This state can accommodate two electrons, just like the $n = 1$, $l = 0$ state of the hydrogen atom. By analogy with hydrogen, then, we expect two Li atoms to join readily to form an Li_2 molecule. However, in a Li atom there are also six vacant $n = 2$, $l = 1$ states whose energies are only a trifle higher than those of the $l = 0$ states of the same shell. A Li atom that approaches a Li_2 molecule can become attached with a covalent bond if its outermost electron goes into an $l = 1$ state, since the energy increment needed for this purpose is less than the energy liberated by the bond formation. There are a total of eight $n = 2$ states in a lithium atom, and so a lithium atom should be able to form covalent bonds with eight other lithium atoms. Lithium atoms do indeed join together in a manner in which each atom has eight nearest neighbors, as shown in Fig. 35-10.

In normal covalent substances each atom contributes one electron per bond, while in solid lithium the single outer electron of each atom must participate in eight separate bonds. To accomplish this task the electron constantly moves from bond to bond, spending only part of its time in each one; a comparable situation occurs in all metals. The electron has no way to remember from which atom it came of the pair it is between at any instant, so it is just as likely to move on to another bond that does not involve its original atom at all. The outer electrons in a lithium crystal therefore wander freely from atom to atom, constituting what is very close to the "gas" of free electrons we spoke of earlier as pervading all metals.

FIG. 35-10. Each lithium atom in solid lithium has eight nearest neighbors.

35-5 Energy bands

When atoms are brought as close together as those in a crystal, they interact with one another to such an extent that their outer electron shells constitute a single system of electrons common to the entire array of atoms. The Pauli exclusion principle prohibits more than two electrons (one with each spin) in any energy level of a system. This principle is obeyed in a crystal because the energy levels of the outer electron shells of the various atoms are all slightly altered by their mutual interaction. (The inner shells do not interact and therefore do not undergo a change.) As a result of the shifts in the energy levels, an *energy band* exists in a crystal in place of each sharply defined energy level of its component atoms (Fig. 35-11). While these bands are actually composed of a multitude of individual energy levels, as many as there are atoms in the crystal, the levels are so near one another as to form a continuous distribution.

The energy bands in a crystal correspond to energy levels in an atom, and an electron in a crystal can only have an energy that falls within one of these bands. The various energy bands in a crystal may or may not overlap, depending on the composition of the crystal (Fig. 35-12). If they do not overlap, the gaps between them represent energy values which electrons in the crystal cannot have. The gaps are accordingly known as *forbidden bands.*

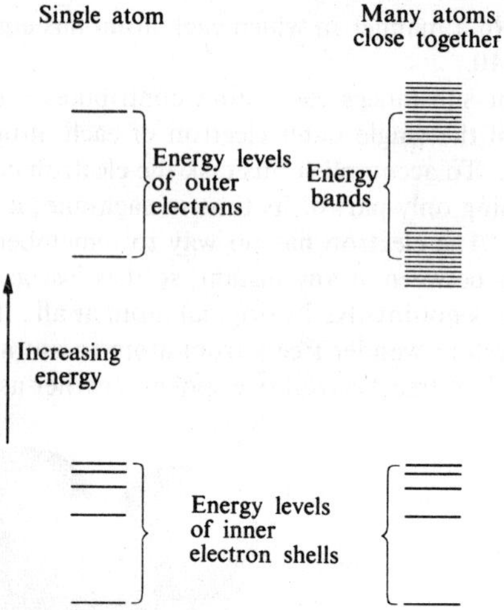

FIG. 35-11. Energy bands replace energy levels of outer electrons in assembly of atoms close together.

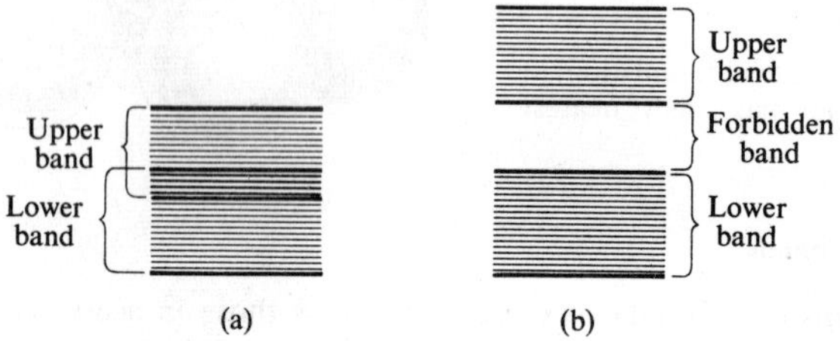

FIG. 35-12. (a) Overlapping energy bands. (b) The gap between energy bands that do not overlap is called a *forbidden band.*

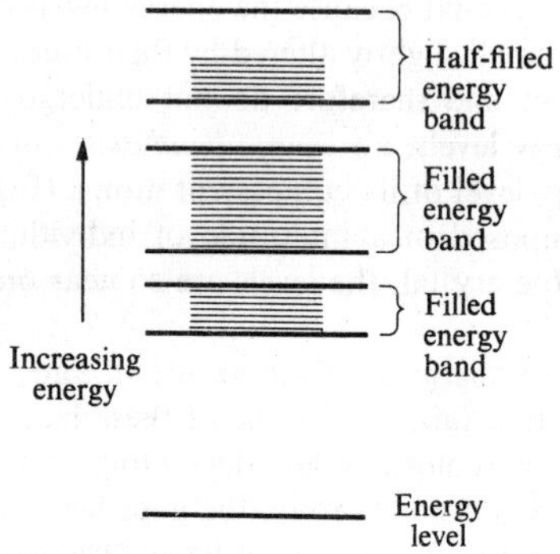

FIG. 35-13. Energy bands in solid sodium.

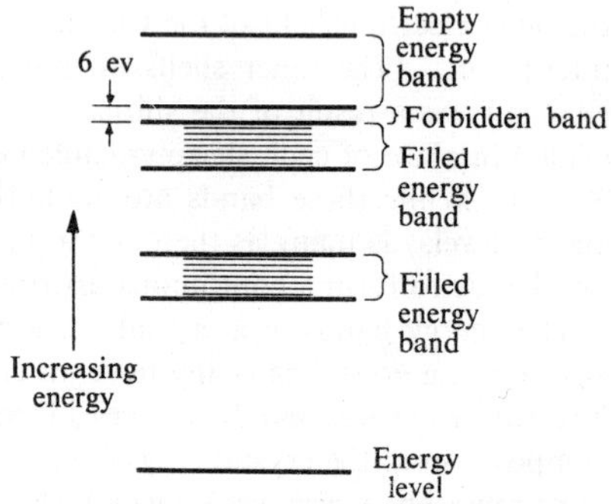

FIG. 35-14. Energy bands in diamond.

The energy bands we have been speaking of contain all the possible energies that can be possessed by electrons. The electrical properties of a crystalline solid depend both on its energy-band structure and on the way in which the bands are normally occupied by electrons. We shall examine a few specific cases to see how the energy-band approach accounts for the observed electrical behavior of such solids.

Figure 35–13 shows the energy bands of solid sodium. Sodium atoms have but one electron in their outer shells. This means that the upper energy band in a sodium crystal is only half filled with electrons, since each level within the band, like each level in the atom, is capable of containing *two* electrons. When an electric field is established in a sodium crystal, electrons readily acquire the small additional energy they need to move up in their energy band. The additional energy is in the form of kinetic energy, and the moving electrons constitute an electric current. Sodium is therefore a good conductor, as are other crystalline solids with energy bands that are only partially filled.

Figure 35–14 shows the energy-band structure of diamond. The two lower energy bands are completely filled, and there is a gap of 6 ev between the top of the higher of these bands and the empty band above it. Hence a minimum of 6 ev of additional energy must be given to an electron in a diamond crystal if it is to be capable of free motion, since it cannot have an energy lying in the forbidden band. Such an energy increment cannot readily be imparted to an electron in a crystal by an electric field. A moving electron in a crystal that is not a member of a free electron gas collides with an atom or another electron after an average of only about 10^{-8} m of travel, whereupon it loses much of the energy it gained from any electric field during its motion. The enormous electric field intensity of 6×10^8 v/m is required if an electron is to gain 6 ev in a path length of 10^{-8} m, well over 10^{10} times greater than the electric intensity needed to cause a current to flow in sodium. Diamond is therefore considered an electrical insulator.

Silicon has a crystal structure similar to that of diamond, and, as in diamond, a gap separates the top of a filled energy band from an empty higher band. However, while the gap is 6 ev wide in diamond, it is only 1.1 ev wide in silicon. At very low temperatures silicon is hardly better than diamond as a conductor, but at room temperature a small proportion of its electrons can possess enough kinetic energy of thermal origin to exist in the higher band. These few electrons are sufficient to permit a limited amount of current to flow when an electric field is applied. Thus silicon has a resistivity intermediate between those of conductors (such as sodium) and those of insulators (such as diamond), and is classified as a *semiconductor*.

35–6 Impurity semiconductors

The conductivity of semiconductors is markedly affected by slight amounts of impurity. Suppose we add several arsenic atoms to a silicon crystal. Arsenic atoms have five electrons in their outermost shells, while silicon atoms have four. When an arsenic atom replaces a silicon atom in a silicon crystal, four of its elec-

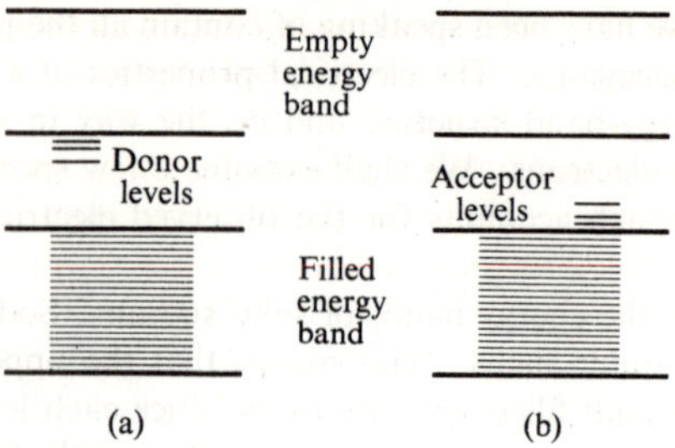

FIG. 35-15. (a) Donor levels due to arsenic atoms in a silicon crystal. (b) Acceptor levels due to gallium atoms in a silicon crystal.

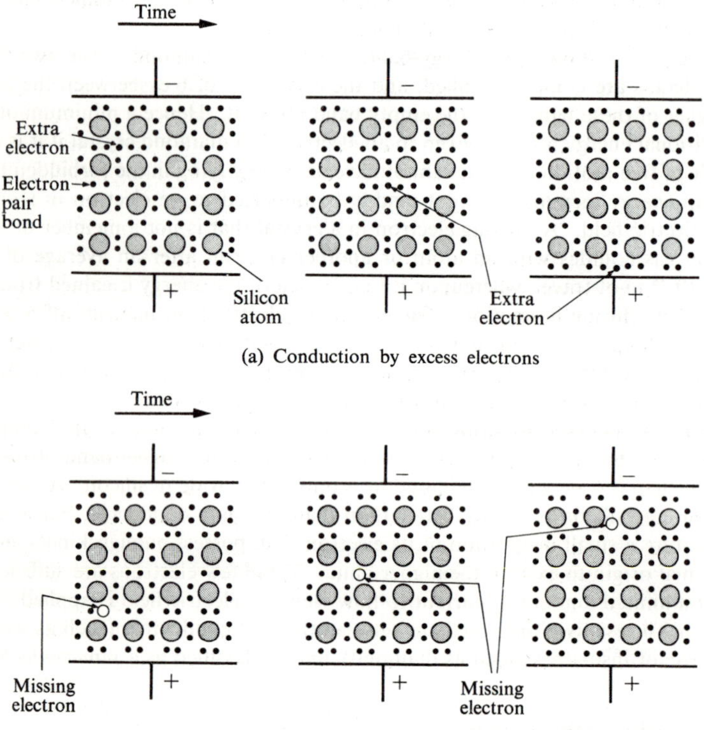

FIG. 35–16. (a) Current in an n-type semiconductor is carried by excess electrons that do not fit into the electron bond structure of the crystal. (b) Current in a p-type semiconductor is carried by the motion of "holes," which are sites of missing electrons. Holes move toward the negative electrode as a succession of electrons move into them.

trons participate in covalent bonds with its nearest neighbors. The remaining electron needs very little energy to be detached and move about freely in the crystal. In an energy-band diagram, as in Fig. 35–15(a), the effect of arsenic as an impurity is to supply energy levels just below the band in which electrons must be present for conduction to take place. These levels are called *donor levels*, and the material is called an *n-type* semiconductor because electric current in it is carried by the motion of negative charges.

If we add gallium atoms to a silicon crystal, a different effect occurs. Gallium atoms have only three electrons in their outer shells, and their presence leaves vacancies called *holes* in the electron structure of the crystal. An electron requires little energy to move into a hole, but as it does so it leaves a new hole in its previous location. When an electric field is applied to a silicon crystal in which gallium is present as an impurity, electrons move toward the positive electrode by successively filling holes. The flow of current here is best described in terms of the motion of the holes, which behave as though they are positive charges, since they move toward the negative electrode (Fig. 35–16). A material of this kind is therefore called a *p-type* semiconductor. In the energy-band diagram of Fig. 35–15(b), we see that the effect of gallium as an impurity is to provide energy levels, called *acceptor levels*, just above the highest filled band; electrons that enter these levels leave behind unoccupied levels in the formerly filled band, which make possible the conduction of current.

The significance of semiconductors in technology arises from the degree of control of electric current that can be accomplished by combining *n*- and *p*-type semiconductors in various ways.

35–7 Ferromagnetism

As a consequence of their spin, electrons have magnetic moments proportional to their spin angular momenta. As we know, the electrons in an atom are arranged in shells about a central nucleus. A complete shell always contains an even number of electrons, half of which are oriented so that their spins point in one direction, while the spins of the other half are in the opposite direction. Hence a complete shell has no net magnetic moment. In the case of all but five elements, inner and outer shells consist of electrons paired off with opposite spins except for a single electron in those shells having an odd number of electrons. In these five elements—iron, cobalt, nickel, gadolinium, and dysprosium—there are vacancies in an inner electron shell, and the electrons there are not paired off with antiparallel spins. In iron, for instance, five of the six electrons in the $n = 3$, $l = 2$ subshell have parallel spins, so that iron atoms have appreciable magnetic moments.

A characteristic property of ferromagnetic materials is that the unpaired electrons in one atom interact strongly with their counterparts in adjacent atoms, causing all their magnetic moments to be aligned. Atoms in these materials group themselves together in assemblies called *domains*, each about 5×10^{-5} m across. In an unmagnetized sample, the directions of magnetization of the domains are randomly oriented, though within each domain all the atomic moments are parallel. When such a sample is placed in an external magnetic field, either the atomic moments within the domains line up with the field or, in pure and homogeneous materials, the domain walls change so that those domains lined up with the field grow at the expense of the others. The former process requires stronger fields to take place than the latter. Hence good "permanent" magnets are inhomogeneous —for instance, made from finely powdered metals sintered together—and, once magnetized, cannot change their magnetization by domain wall motion, and therefore retain their magnetization. Beyond a certain temperature the atomic alignment within the domains disappears, and a ferromagnetic material becomes merely paramagnetic. This temperature, the *Curie temperature*, is 770°C in iron.

35-8 Specific heat

While the specific heat of a substance is a familiar notion and is generally not difficult to determine experimentally, we must use quantum theory to understand why specific heats have their observed values. This was apparent in Chapter 15, where the classical kinetic-molecular theory of gases failed except in the case of a monatomic gas. A similar situation arises in the specific heats of solids.

As before, we shall consider the molar specific heat at constant volume, C_V, which is the heat that must be added to one kmole of a substance, whose volume is held constant, in order to raise its temperature by 1°C. (The specific heat at constant pressure C_p is 3 to 5 percent higher than C_V for solids.) Each of the N_0 atoms in a kmole of a solid can vibrate about its equilibrium position, and, by the principle of equipartition of energy, these vibrations in any direction represent an average of $\frac{1}{2}kT$ of kinetic and $\frac{1}{2}kT$ of potential energy, for a total of kT. Since there are three possible perpendicular directions of vibration, the total thermal energy of each molecule is $3kT$. The total internal energy U of a kmole of any solid is therefore

$$U = 3N_0kT = 3RT,$$

where R is the universal gas constant. From Chapter 15, we recall that the specific heat corresponding to this expression for U is

$$C_V = \frac{dU}{dt} = 3R. \qquad (35\text{-}2)$$

The value of R is 1.99 kcal/kmole·°K, and so we expect that

$$C_V = 5.97 \text{ kcal/kmole·°K}.$$

This result is called the *Dulong-Petit law*. At high temperatures the specific heats of all solids are close to the Dulong-Petit value of $3R$, and even at room temperature the agreement is not bad. At low temperatures, however, the specific heat drops sharply and tends to $C_V = 0$ at absolute zero. Figure 35-17 shows how C_V varies with absolute temperature for gold.

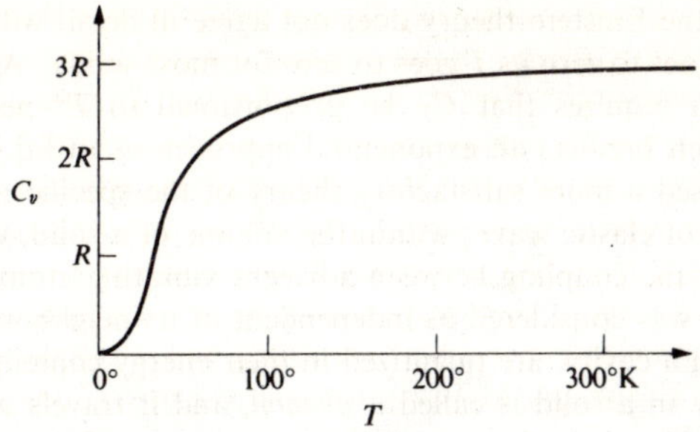

FIG. 35–17. The molar specific heat of gold approaches the Dulong-Petit value of $C_V = 3R$ at high temperatures.

The anomalous behavior of the specific heats of solids was examined by Einstein in 1907, who treated it by analogy with blackbody radiation. Einstein regarded a kmole of a solid as consisting of $3N_0$ harmonic oscillators with the natural frequency f whose energy content is quantized in multiples of hf. The average energy $\overline{E}$ of each oscillator is given by the same formula (Eq. 28–16),

$$\overline{E} = \frac{hf}{e^{hf/kT} - 1} \qquad (35\text{–}3)$$

as that for the average energy per mode of vibration in a cavity (Section 28–6). Hence the total energy of the solid is

$$U = 3N_0\overline{E} = \frac{3N_0 hf}{e^{hf/kT} - 1}, \qquad (35\text{–}4)$$

and, since $N_0 k = R$, its molar heat capacity is

$$C_V = \frac{dU}{dT} = 3R\left(\frac{hf}{kT}\right)^2 \frac{e^{hf/kT}}{(e^{hf/kT} - 1)^2}. \qquad (35\text{–}5)$$

At high temperatures, when $hf \ll kT$, Eq. (35–5) becomes just $C_V = 3R$, the Dulong-Petit law. At low temperatures, when $hf \gg kT$, $e^{hf/kT}$ is much greater than 1, and

$$C_V = 3R\left(\frac{hf}{kT}\right)^2 e^{-hf/kT}. \qquad (35\text{–}6)$$

Thus Einstein's rather simple model of a solid gives a formula for C_V that approaches the Dulong-Petit value of $3R$ at high temperatures and approaches zero at low temperatures. The natural frequency f for a particular solid can be evaluated by fitting Eq. (35–5) to its empirical curve of C_V versus T (Fig. 35–17 is such a curve); a typical frequency is 5×10^{12} cycles/sec, which is reasonable.

Unfortunately the Einstein theory does not agree in detail with the exact manner in which C_V goes to zero as T goes to zero for most solids. A good fit with the experimental data requires that C_V be proportional to T^3 near $T = 0$, while Einstein's approach predicts an exponential approach as in Eq. (35–6). In 1912 Peter Debye devised a more satisfactory theory of the specific heats of solids by thinking in terms of elastic waves within the volume of a solid, which is a way to take into account the coupling between adjacent vibrating atoms. (In Einstein's model each atom was considered as independent of its neighbors.) These waves, like light waves in a cavity, are quantized in their energy content. A quantum of vibrational energy in a solid is called a *phonon*, and it travels with the speed of sound, since sound is an elastic wave. Debye regarded a solid as a volume containing phonons of elastic energy which comprise the internal energy of the solid. When the theory of the specific heat of a solid is developed from this hypothesis, again by analogy with the theory of blackbody radiation, a result is obtained in good agreement with experiment.

The Debye theory is as successful for metals as for nonmetals, which is odd because it completely ignores the presence of free electrons in metals. In a typical metal each atom contributes one electron to the common electron gas, and in one kmole of that metal there are therefore N_0 free electrons. If these electrons behave like the molecules in an ideal gas, each would have an average of $\frac{3}{2}kT$ of kinetic energy and the metal would have

$$U_e = \tfrac{3}{2}N_0kT = \tfrac{3}{2}RT$$

of electron energy per kmole. Hence the molar specific heat of the electrons alone would be

$$C_{Ve} = \frac{dU_e}{dT} = \tfrac{3}{2}R,$$

and the total molar specific heat of a metal should be (at high temperatures where the Dulong-Petit law holds)

$$C_V = 3R + \tfrac{3}{2}R = \tfrac{9}{2}R.$$

In reality $C_V = 3R$ at high temperatures, and so the free electrons do not in fact contribute to the specific heat of a metal. The problem is, why don't they?

The point was made earlier in this chapter that the electrons in a metal sample constitute a single system common to the entire sample. Electrons obey the exclusion principle, and only two of them (one of each spin) can occupy each quantum state. Thus the electron gas in a metal is not comparable with an ideal gas, where no quantum restrictions are present. At absolute zero all the molecules in an ideal gas have zero energy, but in an electron gas the individual electrons occupy states of increasing energy. If N free electrons are present in a particular metal sample, the N lowest energy states will be filled and all those of higher energy will be empty at $T = 0°K$. The uppermost occupied energy level is called the *Fermi level* E_F. In copper, a typical metal, the Fermi level is 7.04 ev, so that in a copper sample at absolute zero electrons with energies of up to 7.04 ev will be present. The average electron energy in the sample will be 4.2 ev; if the free electrons in copper behaved classically, a temperature of nearly 50,000°K would be necessary for their average energy to be this high!

Let us now imagine the copper sample heated above $T = 0$°K. Only electrons with energies very close to E_F can acquire thermal energy, since electrons with less energy must "jump" past the higher filled quantum states in order to reach a vacant one, and to do this they require much more than kT of energy. The Fermi level in copper corresponds to a temperature of 82,000°K, so an electron in the middle of the energy distribution has essentially no chance at all of ever absorbing any thermal energy at actual temperatures below the vaporization temperature of copper. At room temperature kT is about 0.02 ev, and only the very few electrons whose energies are this close to the Fermi level have much chance of absorbing thermal energy and thereby contributing to the specific heat of copper. Figure 35–18 shows the distribution of electron energies in a typical metal at $T = 0$°K, 300°K, and 1000°K; the effect of increasing the temperature is merely to round off the distribution slightly.

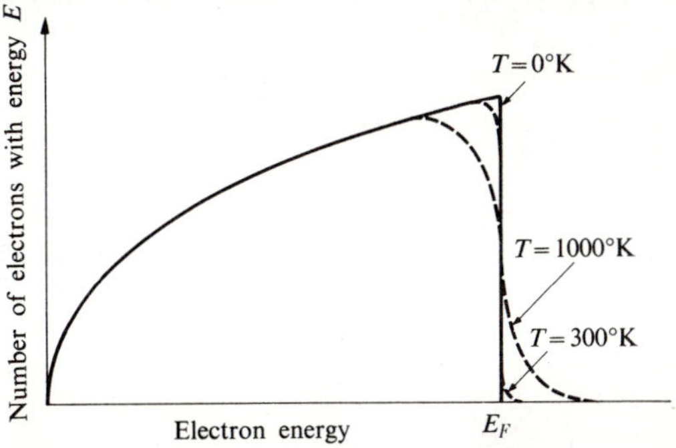

FIG. 35–18. Distribution of electron energies in a metal at three temperatures.

Another puzzling aspect of metallic behavior is answered by the quantum theory of the electron gas. In Chapter 24 we mentioned the phenomenon of paramagnetism, which consists of the alignment of atomic magnetic moments to enhance an external magnetic field. Paramagnetism is a small effect, which is surprising in the case of metals where we would think that the free electrons could all flip over so as to align their spins with the magnetic field. However, all the available quantum states are occupied, and a given electron cannot change its spin direction because another electron in the same energy state already has the opposite spin. Of course, given enough energy to surmount the Fermi level, the electron could align its spin with the field, but we have seen that too much energy is required for more than a minute proportion of the electrons to be able to do this.

Problems

1. What kind of bond holds water molecules together to form ice?

2. Since van der Waals forces can hold inert gas atoms together to form solids, why can they not hold such atoms together to form molecules in the gaseous state?

3. The forbidden energy band in germanium that lies between the highest filled band and the empty band above it has a width of 0.7 ev. Compare the conductivity of germanium with that of silicon at (a) very low temperatures and (b) room temperature.

4. Does the addition of a small amount of indium to germanium result in the formation of an *n*- or a *p*-type semiconductor?

5. What is the upper limit to the contribution the free electrons in a metal can make toward its molar specific heat?

6. Because nearly all energy levels up to the Fermi level are filled in a metal, the free electrons contain a great deal of energy. Can this energy be extracted from the metal?

7. The Fermi level in silver is 5.51 ev. (a) What temperature is necessary for the average molecular energy in an ideal gas to be 5.51 ev? (b) What is the speed of a 5.51-ev electron?

8. Choose the characteristic frequency in Einstein's formula for specific heat as 6.4×10^{12}/sec. Plot the specific heat between $T = 75°K$ and $T = 400°K$.

36

The Nucleus

Until now we have not had to regard the nucleus of an atom as anything but a tiny positively charged lump whose sole function is to provide the atom with most of its mass and to hold its several electrons in place. Since the behavior of atomic electrons is responsible for the behavior of matter in bulk, the properties of matter that we have been exploring, save for mass, have nothing directly to do with atomic nuclei. However, for all its seeming passivity, the nucleus turns out to be of paramount importance in the universe: the chemical elements exist by virtue of the ability of nuclei to possess multiple electric charges, and the energy that is involved in nearly all natural processes has its ultimate origin in nuclear reactions and transformations. Only in recent years have there been tools available for investigating nuclei in any detail, and their many mysteries are still being plumbed.

36–1 Nucleons

What are nuclei composed of? Let us examine a table of the properties of the elements, such as the one in Appendix IV. In this table the elements are listed in the order of their *atomic numbers*, with the atomic number of an element defined as the number of electrons in each of its atoms. Hydrogen, whose atoms have a single electron each, has the atomic number 1; helium, whose atoms have two electrons each, has the atomic number 2; and so on up to lawrencium, whose atoms have 103 electrons each and which therefore has the atomic number 103. We note immediately that atomic mass increases with atomic number, which suggests the hypothesis that all atoms are simply combinations of hydrogen atoms. According to this hypothesis (whose basic idea was put forward about 150 years ago) the helium atom, atomic number 2, should have a nucleus consisting of two protons; the lithium atom, atomic number 3, should have a nucleus consisting of three protons; and so on through the lawrencium atom, atomic number 103, which should have a nucleus consisting of 103 protons. However, atomic masses do *not* increase in steps of one hydrogen atom mass. Helium atoms weigh about four times as much as hydrogen atoms, lithium atoms about seven times as much, and lawrencium atoms more than 260 times as much. So this hypothesis cannot be correct. But it must have some kind of truth in it, because, with few exceptions, atomic masses *are* very close to being exact multiples of the mass of the hydrogen atom.

What we might then suppose is that there are enough protons in each nucleus to provide for the observed atomic mass, with several electrons also present whose negative charges cancel out the positive charges of the "excess" protons. This notion, which had favor for a while, is untenable on several grounds. We discussed one of the objections in Chapter 31: far too much energy is required to localize an electron within a nucleus, according to the uncertainty principle, for it to be a possible nuclear constituent.

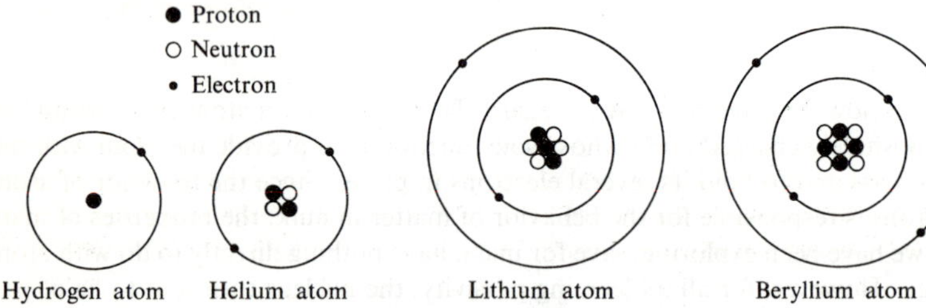

FIG. 36–1. The electronic and nuclear compositions of hydrogen, helium, lithium, and beryllium atoms.

In 1932 the puzzle of the missing nuclear ingredient was solved by Chadwick in England and Joliot and Curie in France. They discovered a very penetrating radiation, which was definitely not electromagnetic in character, as are x-rays and gamma rays, and which readily ejected protons from hydrogen-rich materials like paraffin. The great penetrability suggested a neutral particle, which would not be affected by the strong electric fields in matter. The mass of the particle was revealed by the ease with which it could transfer energy to protons in collisions; efficient energy transfer in collisions can take place only when the participants have identical masses, or very nearly so. This new particle, called the *neutron*, not only could account for the observations of Chadwick, Joliot, and Curie, but also fits perfectly into the nucleus as the uncharged but massive component required to supplement the proton. Present values for the proton and neutron masses are

$$m_{\text{proton}} = 1.6724 \times 10^{-27} \text{ kg},$$
$$m_{\text{neutron}} = 1.6747 \times 10^{-27} \text{ kg}.$$

Neutrons and protons are jointly called *nucleons*.

With the help of the neutron, then, the constitution of nuclei more complex than hydrogen can be figured out readily: helium nuclei contain two protons and two neutrons each, lithium nuclei contain three protons and four neutrons each, and so on (Fig. 36–1).

36-2 Isotopes

We are still faced with the question of how to account for atomic masses, which cannot be exactly expressed in terms of combinations of proton and neutron masses. As it happens, the answer was known even before the composition of nuclei was clearly understood. Experimental measurements showed that what we call chemical elements actually each consist of several different components, called *isotopes*, each having the same nuclear charge (and hence the same atomic number) but with different masses. Chlorine, for example, consists of two isotopes, one whose nucleus contains 17 protons and 18 neutrons and another whose nucleus contains 17 protons and 20 neutrons. In nature, chlorine is composed of about 75% of the former isotope and 25% of the latter, yielding a composite atomic mass of 35.46, the value given in Appendix IV.

Because the chemical properties of an element depend on the distribution of the electrons in its atoms, which in turn depends on the nuclear charge, nuclear structure beyond the number of protons present has little significance for the chemist. The physical properties of an element, however, depend strongly on the nuclear structures of its isotopes, whose behavior may be very different from one another although chemically they are indistinguishable.

The conventional symbols for isotopes all follow the pattern

$$_nX^m,$$

where X is the chemical symbol of the element, n is its atomic number (equal to the number of electrons per atom or, what is equivalent, the number of protons per nucleus), and m is the *mass number* of the isotope in question, which is defined as the total number of nucleons (neutrons plus protons) in each nucleus. Hence ordinary hydrogen is designated

$$_1\mathrm{H}^1,$$

since its atomic number and mass number are both 1, while ordinary helium is designated

$$_2\mathrm{He}^4,$$

since its atomic number is 2 and its mass number is 4. The two isotopes of chlorine mentioned above are designated $_{17}\mathrm{Cl}^{35}$ and $_{17}\mathrm{Cl}^{37}$, respectively.

36-3 Nuclear dimensions

The Rutherford scattering experiment provides information on nuclear dimensions as well as on atomic structure. In that experiment, as we learned in Chapter 23, alpha particles are directed at a thin metal foil and their deviations measured. The observed distribution of scattering angles is consistent with a nucleus of infinitesimal size, provided that the alpha particles are not too energetic. That is, below a certain particle energy the dimensions of the nucleus are small compared with the minimum distance to which the incident alpha particles approach it. At higher energies discrepancies occur between theory and data which suggest that the particles have come so close to the nucleus that it no longer can be represented as a point charge, and from the energy at which these discrepancies appear an estimate can be made of nuclear sizes.

Let us calculate again the distance of closest approach r_0 of the alpha particles in the early scattering experiments. An alpha particle will come closest to a nucleus when it is headed directly at it. At the distance of closest approach the initial kinetic energy KE of the particle is entirely converted to electrostatic potential energy, and so, since the charge of the alpha particle is $2e$ and that of the nucleus is Ze (Z is the atomic number of the nucleus),

$$\mathrm{KE} = \frac{1}{4\pi\epsilon_0}\frac{2Ze^2}{r_0}. \tag{36-1}$$

The most energetic alpha particles of natural origin have KE = 7.7 Mev, which is

$$\mathrm{KE} = 7.7 \times 10^6\ \mathrm{ev} \times 1.6 \times 10^{-19}\ \mathrm{j/ev} = 1.2 \times 10^{-12}\ \mathrm{j}.$$

Since $(1/4\pi\epsilon_0) = 9 \times 10^9\ \mathrm{n \cdot m/coul^2}$, we have from Eq. (36-1),

$$\begin{aligned} r_0 &= \frac{1}{4\pi\epsilon_0}\frac{2Ze^2}{\mathrm{KE}} = \frac{9 \times 10^9\ \mathrm{n\cdot m/coul^2} \times 2Z \times (1.6 \times 10^{-19}\ \mathrm{coul})^2}{1.2 \times 10^{-12}\ \mathrm{j}} \\ &= 3.8 \times 10^{-16}\ Z\ \mathrm{m}. \end{aligned}$$

A typical foil material was gold, $Z = 79$. The distance of closest approach of 7.7-Mev alpha particles to gold nuclei is therefore

$$r_0 = 79 \times 3.8 \times 10^{-16}\ \mathrm{m} = 3.0 \times 10^{-14}\ \mathrm{m}.$$

The scattering of alpha particles by gold foil follows the predictions of Coulomb's law when gold nuclei are considered as point charges, from which we conclude that the radius of the gold nucleus is less than 3.0×10^{-14} m—less than 1/10,000 of the radius of the entire atom.

When higher-energy particles are directed at thin foils, the distribution of scattering angles eventually disagrees with the theoretical expectations. The results show that the volume of a particular nucleus is proportional to its mass number A, the number of nucleons it contains. The volume of a sphere of radius R is $\frac{4}{3}\pi R^3$, so that if we consider a nucleus as spherical in shape, its radius R is proportional to $A^{1/3}$. We can write this relationship in the form

$$R = R_0 A^{1/3}, \tag{36-2}$$

where the experimental evidence yields a value of R_0 of approximately

$$R_0 = 1.2 \times 10^{-15}\ \mathrm{m}.$$

The only isotope of gold found in nature has a mass number of 197, which means that the radius of the gold nucleus is

$$R = (197)^{1/3} \times 1.2 \times 10^{-15}\,\text{m} = 5.82 \times 1.2 \times 10^{-15}\,\text{m} = 7.0 \times 10^{-15}\,\text{m}.$$

Different experiments give different figures for R_0, which is why the above value was called approximate. These differences are primarily due not to experimental uncertainties but to the different ways in which a nucleus can interact with an incident particle. A neutron is acted on exclusively by nuclear forces, for instance, while an electron is acted on exclusively by electrical forces; protons and alpha particles are acted on by both kinds of forces. [Gravitational forces are so small that they play very little role.] Electron scattering leads to a smaller value for R_0 than neutron scattering, which suggests that nuclear matter and nuclear charge are differently distributed in a nucleus.

The density of nuclear matter is about $2 \times 10^{17}\,\text{kg/m}^3$, a billion tons per cubic inch. "White dwarf" stars consist of atoms whose electron structures have collapsed because of immense pressures, and the densities of these stars approach the density of nuclear matter.

36-4 Nuclear binding energy

While the discovery of isotopes permits us to understand why certain elements do not have atomic masses that are multiples of the proton and neutron masses, even those elements that seem to fit the picture well have slight mass variations from simple assemblies of these particles. Table 36-1 is a list of the experimentally determined masses of the neutron and some light isotopes expressed in terms of *atomic mass units* (abbreviated amu), equal to one-twelfth the mass of the $_6C^{12}$ isotope. The value of an amu is

$$1\ \text{amu} = 1.660 \times 10^{-27}\,\text{kg}.$$

The deuteron and triton are nuclei of isotopes of hydrogen with mass numbers 2 and 3, respectively, which implies that they consist of one proton plus, for the deu-

TABLE 36-1

The masses of the five lightest nuclei and the neutron

Name	Symbol	Mass, amu
Proton	$_1H^1$	1.007593
Neutron	$_0n^1$	1.008982
Deuteron	$_1H^2$	2.014186
Triton	$_1H^3$	3.016448
Helium-3	$_2He^3$	3.015779
Alpha particle	$_2He^4$	4.003873

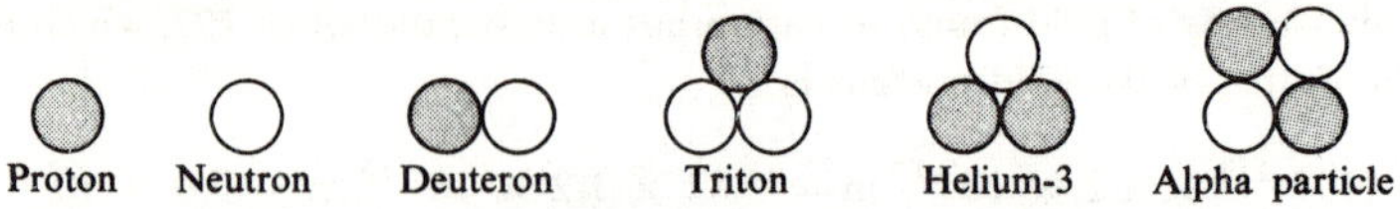

FIG. 36-2. The composition of the six lightest nuclei.

teron, one neutron, and for the triton, two neutrons. Nuclei of $_2He^3$, an isotope of helium, consist of two protons and one neutron each, while ordinary helium nuclei consist of two protons and two neutrons each (Fig. 36-2).

If it is true that a deuteron consists of a neutron plus a proton, having a combined mass of 2.016575 amu, why is this combined mass *more* than the 2.014186 amu mass of the deuteron? The answer is that when nucleons combine to form a nucleus, the total energy of the system (including mass energy) remains constant, although the mass may not. Thus the difference in mass between the constituents of the deuteron and the deuteron itself may appear as additional kinetic energy of the deuteron, energy beyond that which the neutron and proton brought along with them. Correspondingly, this same mass deficiency must be made up by the addition of energy if it is desired to break up a deuteron.

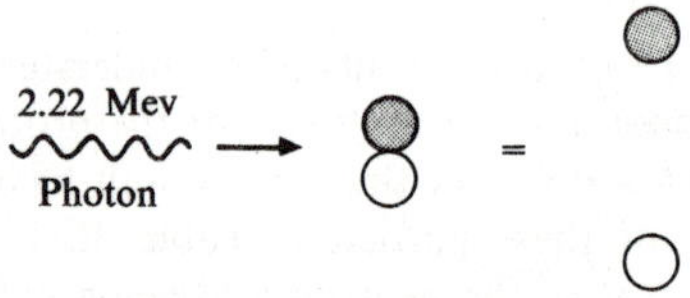

FIG. 36-3. The binding energy of the deuteron is 2.22 Mev, which means that this much energy is required to make up the difference between the deuteron's mass and the combined mass of its constituent neutron and proton. Absorbing 2.22 Mev, for instance by being struck by a 2.22-Mev gamma-ray photon, furnishes a deuteron with enough additional mass to split into a neutron and a proton.

To put this concept on a quantitative basis, we recall that energy E and mass m are related by the formula

$$E = mc^2.$$

One amu is the energy equivalent of

$$1 \text{ amu} = 1.49 \times 10^{-10} \text{ j} = 9.31 \times 10^8 \text{ ev} = 931 \text{ Mev}.$$

The *binding energy* of the deuteron, which is the energy equivalent of the missing mass, is therefore

$$(2.016575 - 2.014186) \text{ amu} \times 931 \frac{\text{Mev}}{\text{amu}} = 2.22 \text{ Mev}.$$

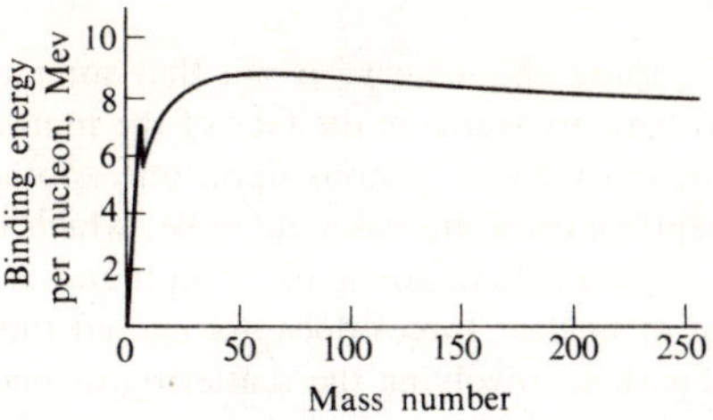

FIG. 36-4. The binding energy per nucleon versus mass number. The higher the binding energy per nucleon, the more stable the nucleus. When a heavy nucleus is split into two lighter ones, a process called *fission*, the greater binding energy of the latter causes the liberation of energy. When two very light nuclei join to form a heavier one, a process called *fusion*, the greater binding energy of the latter again causes the liberation of energy.

This figure is confirmed by experiments that show that the minimum energy a gamma ray must have in order to disrupt a deuteron is 2.22 Mev (Fig. 36–3).

A very interesting curve results when we plot the binding energy per nucleon (that is, the total binding energy of an isotope divided by the number of neutrons and protons that compose it) versus the mass number of the nucleus (Fig. 36–4). Except for an anomalously high peak for ${}_2He^4$, this curve is a surprisingly regular one. The middle range of nuclei have the highest binding energy per nucleon values, about 8.75 Mev/nucleon, which means that they require the greatest amount of energy (per nucleon) to disrupt. Such nuclei are therefore the most stable.

A significant feature of nuclear structure is illustrated by this curve. Suppose that we split the nucleus ${}_{92}U^{235}$, whose binding energy is 7.6 Mev/nucleon, into two fragments. Each fragment will be the nucleus of a much lighter element, and therefore will have a higher binding energy per nucleon than the uranium nucleus. The difference is about 0.8 Mev/nucleon, and so, if such *nuclear fission* were to take place, an energy of

$$0.8 \frac{\text{Mev}}{\text{nucleon}} \times 235 \text{ nucleons} = 190 \text{ Mev}$$

would be given off per splitting. This is a truly immense amount of energy to be produced in a single event. As a comparison, chemical processes involve energies of the order of magnitude of one electron volt per reacting atom, 10^{-8} the energy involved in fission.

Figure 36–4 also shows that if two of the extremely light nuclei are combined to form a heavier one, the higher binding energy of the latter will also result in the evolution of energy. For instance, if two deuterons were to join to make up a ${}_2He^4$ nucleus, over 23 Mev would be released. This process is known as *fusion*, and, with fission, promises to be the source of more and more of the world's power.

36-5 Nuclear forces

The really surprising thing about nuclei is not that some of them are unstable, but rather that *any* of them are stable in the face of the mutually repulsive electrostatic forces exerted by the nuclear protons upon one another. There must be a nuclear force, imperceptible on a macroscopic scale, which is able to hold nuclei together. Because nuclei larger than about 10^{-14} m in radius do not exist, we may assume that the range of nuclear forces does not exceed this figure. Various experiments, in particular those involving the scattering of one nucleon by another in a collision, indicate that nuclear forces are very strong: they give rise to a binding energy between adjacent nucleons on the average over a million times greater than the 13.6 ev binding energy between the proton and electron in a hydrogen atom. Nuclear forces are evidently enormous. It is interesting that there is evidence to indicate that the proton-proton nuclear force is identical in strength and range to the neutron-proton nuclear force.

If we inquire into the "mechanism" by which two electric charges exert forces upon one another, we find that, from a formal mathematical point of view, their interaction can be described in terms of the constant circulation of electromagnetic photons between them. In 1935 the Japanese physicist Yukawa suggested that nuclear forces, too, could be regarded as the result of an interchange of certain particles between nucleons. Today these particles are called *π-mesons*, and may have charges of 0, $+e$, or $-e$. Charged π-mesons have masses of 273 times the electron mass, while neutral π-mesons have masses of 264 times the electron mass.

FIG. 36-5. Particle exchange can lead to attractive or repulsive forces.

The crude analogy illustrated in Fig. 36-5 may help in understanding how meson exchange can lead to both attractive and repulsive forces between nuclei. Each child in the figure has a pillow. When the children exchange pillows by snatching them from each other's grasp, the effect is like that of a mutually attractive force. On the other hand, the children may also exchange pillows by throwing them at each other. Here conservation of momentum requires that the children move apart, just as if a repulsive force were present between them.

According to Yukawa's theory, nearby nucleons constantly exchange mesons without themselves being altered. We note that the emission of a meson by a nucleon at rest which does not lose a corresponding amount of mass violates the law of conservation of energy. However, the law of conservation of energy, like all physical laws, deals only with measurable quantities. Because the uncertainty principle restricts the accuracy with which we can perform certain measurements, it limits the range of application of physical laws such as that of energy conservation. We see from the uncertainty principle in the form

$$\Delta E\, \Delta t \approx h$$

that a process can take place in which energy is not conserved by an amount ΔE provided that the time interval Δt in which the process takes place is not more than $h\,\Delta E$. Thus the creation, transfer, and disappearance of a meson do not conflict with the conservation of energy, provided that the sequence occurs fast enough.

The most attractive aspect of the meson theory of nuclear forces is that the π-mesons predicted by Yukawa actually exist and have the properties expected of them. For a meson to be produced which can emerge from a nucleus and be detected experimentally, enough energy must be supplied to the nucleus, usually by a collision with another particle, for conversion of the energy into meson mass without affecting the invariant masses of its nucleons. When this is done with the help of the energetic particle beams found naturally in cosmic rays or created in laboratory accelerators, mesons can be studied in detail; the results are outlined in Chapter 38. However, despite this success of the meson theory, it has not yet proved useful in understanding nuclear structure. Instead, several very different hypotheses of how nucleons behave within a nucleus seem to be required, depending on which aspect of nuclear phenomena is involved. We shall briefly examine two of them.

Nuclear forces, though strong, have such short ranges that each nucleon in a nucleus effectively interacts with its nearest neighbors only. This situation superficially resembles that of atoms in a solid, but the vibrations of the nucleons about their equilibrium positions would be so great as to disrupt the nucleus if its structure were like that of a crystal. A better analogy is with a liquid, whose molecules move about more or less freely while maintaining a constant intermolecular separation. The *liquid drop model* of the nucleus accounts nicely for the observed variation of nuclear binding energy with mass number A and, as we shall see in the next chapter, it provides a plausible explanation for nuclear fission.

At the opposite extreme from the picture of a nucleus as a drop of liquid is the *independent particle model*, which supposes that nucleons interact with a single force field that pervades the entire nucleus instead of only with their immediate neighbors. Thus the nucleus in this model is analogous to the atom, where only certain quantum states are permitted and, by the Pauli principle, no more than two electrons can occupy each state. Nucleons also obey the Pauli principle, so it is not altogether surprising that the independent-particle model provides an understanding of some nuclear properties.

As we know, the electrons in an atom may be considered as members of various "shells," and the degree to which the outermost shell is filled governs certain aspects of the atom's behavior. Thus atoms with 2, 10, 18, 36, 54, and 86 electrons possess filled electron shells, which are stable structures. These atoms, the inert gases, do not interact with other atoms. A comparable effect is observed in the case of nuclei, where nuclei having 2, 8, 20, 28, 50, 82, and 126 neutrons or protons are exceptionally abundant. Their abundance suggests unusual stability: complex nuclei arose from reactions among lighter ones, and the evolution of larger and larger nuclei was retarded when each stable and hence inert nucleus was formed. Other evidence also suggests that the *magic numbers* 2, 8, 20, 28, 50, 82, and 126 are significant in nuclear structure. The independent-particle model of the nucleus is able to explain the origin of the magic numbers, a strong point in its favor. In recent years attempts have been made to combine the liquid-drop and independent-particle models into a single, consistent scheme, and much progress has been made in this direction.

Problems

1. State the number of neutrons and protons in each of the following nuclei: $_3Li^6$; $_6C^{13}$; $_{15}P^{31}$; $_{40}Zr^{94}$; $_{56}Ba^{137}$.

2. State the number of neutrons and protons in each of the following nuclei: $_5Be^{10}$; $_{10}Ne^{22}$; $_{16}S^{36}$; $_{38}Sr^{88}$; $_{72}Hf^{180}$.

3. The mass of $_{10}Ne^{20}$ is 19.9924 amu. What is its binding energy?

4. The mass of $_{17}Cl^{35}$ is 34.9689 amu. What is its binding energy?

5. Find the binding energy of the triton.

6. What is the minimum energy a gamma-ray photon must have if it is to split an alpha particle into (a) a triton and a proton, and (b) a helium-3 nucleus and a neutron?

7. Compare the electrostatic potential energy of two protons 6×10^{-15} m apart with the binding energy per nucleon of nuclei with $A \approx 60$. (Such nuclei are about 6×10^{-15} m in radius.)

8. The distance between the two protons in a $_2He^3$ nucleus is roughly 1.7×10^{-15} m. (a) Calculate the electrostatic potential energy of these protons. (b) Show that this energy is of the right order of magnitude to account for the difference in binding energy between $_1H^3$ and $_2He^3$. What conclusion can be drawn from this result about the dependence of nuclear forces on electric charge?

9. The deuteron, composed of a neutron and a proton, is a stable nucleus, while no particles consisting of two protons or two neutrons alone have ever been found. Assuming that nuclear forces are charge independent, give an argument based on the Pauli exclusion principle to explain the absence of di-protons and di-neutrons.

37

Nuclear Transformations

While atomic nuclei are the most stable composite bodies known, they may nevertheless undergo transformations in which the identity of a particular nucleus changes. These changes may be spontaneous, as in radioactive decay, or induced, as in reactions between colliding nuclei. Aside from the data they provide on nuclear structure, nuclear transformations are noteworthy because of their role in the evolution of the universe: through them the primeval hydrogen of the universe coalesced into more complex elements, and they are responsible for virtually all of the energy that powers the universe. Today the laboratory of the nuclear physicist is as essential to astronomy as is the telescope.

37–1 Radioactivity

In the previous chapter we discussed only stable nuclei. At the beginning of the twentieth century it became known, as the result of research by Becquerel, the Curies, and others, that some nuclei exist which spontaneously transform themselves into other nuclear species with the emission of radiation. Such nuclei are said to be *radioactive*. The observed radiation is of three types: *alpha particles*, which are the nuclei of ${}_2He^4$ atoms, *beta particles*, which are electrons, and *gamma rays*, which are composed of short wavelength (and hence energetic) photons (Fig. 37–1).

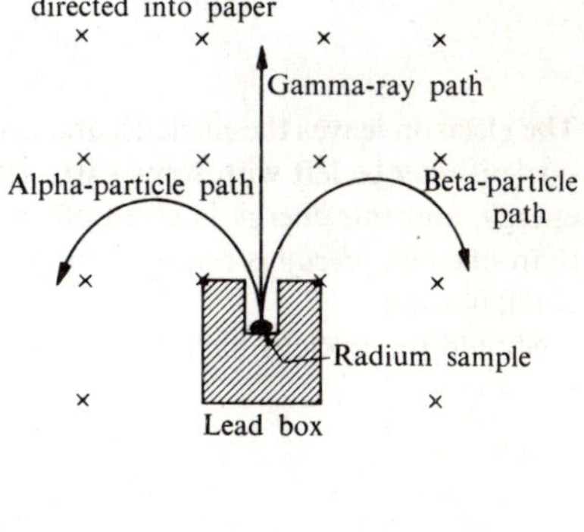

FIG. 37–1. The radiations from a radium sample may be analyzed with the help of a magnetic field. In the figure the direction of the field is into the paper: hence the positively charged alpha particles (actually helium nuclei) are deflected to the left and the negatively charged beta particles (actually electrons) to the right. Gamma rays, which are composed of energetic photons, carry no charge and are not affected by the magnetic field.

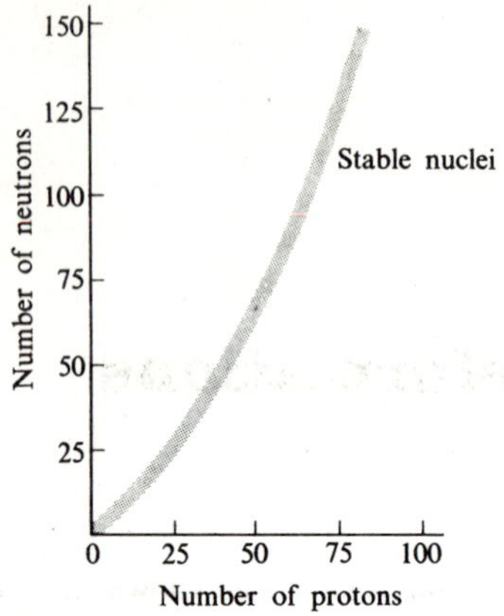

FIG. 37-2. The number of neutrons versus the number of protons in stable nuclei. The larger the nucleus, the greater the proportion of neutrons.

To understand why alpha, beta, and gamma decays take place, let us examine Fig. 37-2, which is a plot showing the number of neutrons versus the number of protons in stable nuclei. For light, stable nuclei the numbers of neutrons and protons are approximately the same, while for heavier nuclei slightly more neutrons than protons are required for stability. (This is a consequence of the strong electrostatic repulsion exerted by protons on one another, which must be balanced by the attractive nuclear forces present between nucleons; to counterbalance the electrostatic repulsion of many protons, a comparatively greater number of neutrons, which produce only attractive forces owing to their electrical neutrality, is necessary for stability.) From Fig. 37-2 it is evident that an element of a given atomic number can have only a very narrow range of possible numbers of neutrons if it is to be stable.

Suppose now that a nucleus exists which has either too many or too few neutrons relative to the number of protons present for stability. We shall consider the former case first. If one of the excess neutrons transforms itself into a proton, this will simultaneously reduce the number of neutrons while increasing the number of protons. To conserve electric charge, such a transformation requires the emission of a negative electron, and we may write it in equation form as

$$n^0 \rightarrow p^+ + e^-. \tag{37-1}$$

The electron leaves the nucleus, and is detectable as a "beta particle." The residual nucleus may be left with some extra energy as a consequence of its shifted binding energy, and this energy is given off in the form of gamma rays. Sometimes more than one *beta decay* is required for a particular unstable nucleus to reach a stable configuration.

Should the nucleus have too few neutrons, the inverse reaction,

$$p^+ \rightarrow n^0 + e^+, \tag{37-2}$$

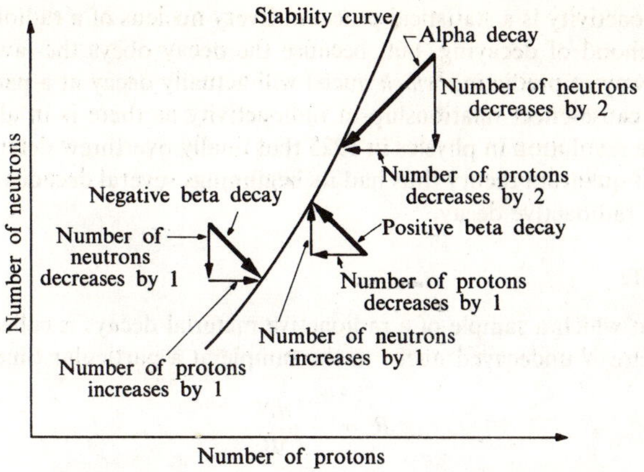

FIG. 37–3. How alpha and beta decays tend to bring an unstable nucleus to a stable configuration.

in which a proton becomes a neutron with the emission of a *positive* electron (known as a *positron*), may take place. This is also called beta decay, since it resembles the emission of negative electrons from an unstable nucleus in every way save for the difference in charge. [Other particles, neutrinos and anti-neutrinos, are emitted in reactions (37–1) and (37–2), but the existence of these particles does not affect the argument.]

Another way of altering its structure to achieve stability may involve a nucleus in *alpha decay*, in which an alpha particle consisting of two neutrons and two protons is emitted. Thus negative beta decay increases the number of protons by one and decreases the number of neutrons by one; positive beta decay decreases the number of protons by one and increases the number of neutrons by one; and alpha decay decreases both the number of protons and the number of neutrons by two. These processes are shown schematically in Fig. 37–3. Very often a succession of alpha and beta decays, with accompanying gamma decays to carry off excess energy, is required before a nucleus reaches stability.

Three aspects of radioactivity are wholly remarkable from the point of view of classical physics. First, the atomic number of a nucleus that undergoes alpha or beta decay changes, so that the nucleus becomes one characteristic of a different element. Elements *can* be transmuted into other elements, though hardly in a manner anticipated by alchemy. Second, radioactive decay liberates energy that can only come from *within* individual nuclei. A gram of radium in a sealed container (to prevent the escape of radon, a gaseous product of its decay that is itself radioactive) evolves energy at the rate of 0.14 kcal/hr; this rate decreases so slowly that after 1600 years it has dropped by only 50 percent. Where does all this energy come from? Not until 1905, when Einstein proposed the equivalence of mass and energy, was this puzzle understood. Third, as we shall see in the next section, radioactivity is a statistical process. Every nucleus of a radioisotope has a certain likelihood of decaying, but, because the decay obeys the laws of chance, we have no way of predicting *which* nuclei will actually decay at a particular time. There is no cause-effect relationship in radioactivity as there is in all of classical physics. The revolution in physics in 1925 that finally overthrew determinism with the advent of quantum theory thus had its beginnings several decades earlier in the discovery of radioactive decay.

37-2 Half-life

The rate at which a sample of a radioactive material decays is called its *activity*, R. If there are N undecayed nuclei in the sample at a particular time,

$$R = -\frac{dN}{dt}. \qquad (37\text{-}3)$$

Because N decreases with time, dN/dt itself is negative, and the minus sign is inserted to make R a more convenient positive quantity. According to Eq. (37-3), R should be expressed in disintegrations/sec, but the activities encountered in practice are usually so high that a larger unit, the *curie*, is more often used. Originally the curie was defined as the activity of 1 gm of radium ($_{88}Ra^{226}$), and its precise value accordingly changed as measuring techniques improved. For this reason the curie is now defined arbitrarily as

$$1 \text{ curie} = 3.70 \times 10^{10} \text{ disintegrations/sec;}$$

the activity of 1 gm of radium is a few percent smaller. The *millicurie* (10^{-3} curie) and *microcurie* (10^{-6} curie) are frequently employed to supplement the curie. A luminous watch dial contains several microcuries of $_{88}Ra^{226}$; ordinary potassium has an activity of about 1 millicurie/kg owing to the presence of the radioactive isotope $_{19}K^{40}$; "cobalt-60" sources of one or more curies are widely used in medicine for radiation therapy, and industrially for the inspection of metal castings and welded joints.

One of the characteristics of radioactivity is that the rate at which the nuclei in a given sample decay always follows a curve whose shape is like that shown in Fig. 37-4. Some isotopes decay faster than others, but in each case a certain definite time is required for half of an original sample to decay. This time is called the

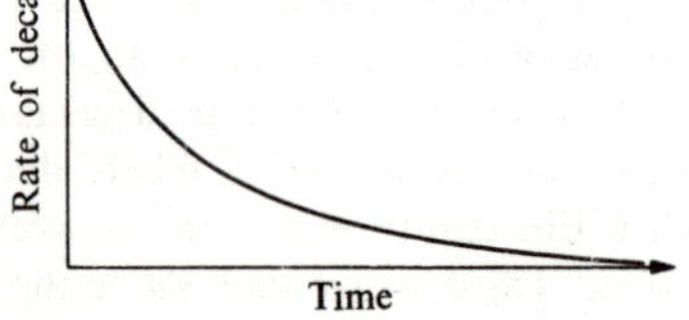

FIG. 37-4. The rate at which a sample of a radioactive substance decays is not constant, but varies with time in the manner shown in the curve. Some isotopes decay faster than others, but all obey curves having the same shape.

half-life, $T_{1/2}$ of the isotope. Observed half-lives range from billionths of a second to billions of years. Radon ($_{86}Rn^{222}$) has a half-life against alpha decay of 3.8 days, for instance: if we start with 1 gm of radon, $\frac{1}{2}$ gm will remain after 3.8 days, $\frac{1}{4}$ gm after 7.6 days, $\frac{1}{8}$ gm after 11.4 days, and so on (Fig. 37–5). This kind of decrease in activity with time is exponential. That is, if the activity of a radioactive sample at a certain moment is R_0, at a time t later it will be

$$R = R_0e^{-\lambda t}. \tag{37–4}$$

The quantity λ, the *decay constant*, has a unique value for each radioisotope.

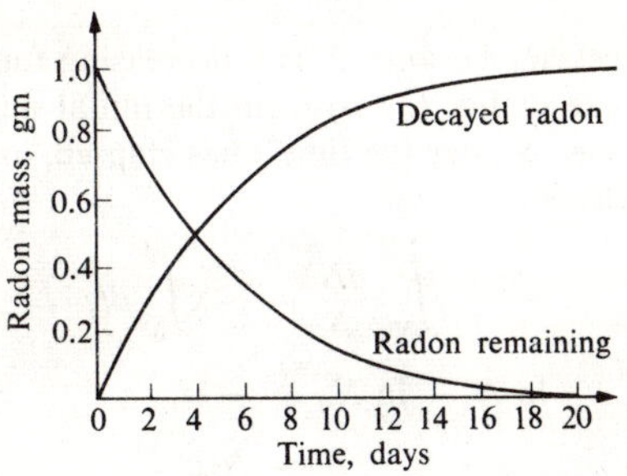

FIG. 37–5. The decay of radon, whose half-life is 3.8 days.

Let us investigate the connection between decay constant λ and half-life $T_{1/2}$. According to the definition of half-life, when $t = T_{1/2}$, $R = \frac{1}{2}R_0$. Hence, substituting these figures into Eq. (37–4), we have

$$\tfrac{1}{2}R_0 = R_0e^{-\lambda T_{1/2}}, \qquad e^{\lambda T_{1/2}} = 2.$$

Taking the natural logarithm of both sides gives

$$\lambda T_{1/2} = \ln 2 = 0.693,$$

$$\lambda = \frac{0.693}{T_{1/2}}. \tag{37–5}$$

The decay constant of radon, which has a half-life of 3.8 days, is therefore

$$\lambda = \frac{0.693}{T_{1/2}} = \frac{0.693}{3.8 \text{ days} \times 86{,}400 \text{ sec/day}}$$

$$= \frac{0.693}{3.3 \times 10^5 \text{ sec}} = 2.1 \times 10^{-6} \text{ sec}^{-1}.$$

Radioactive decay involves individual events taking place within independent nuclei, rather than collective processes that involve more than one nucleus in interaction. This idea is confirmed by experiments which show that the half-life of a particular isotope is absolutely invariant under changes of pressure, temperature, electric and magnetic fields, and so on, which might, if sufficiently strong, influence internuclear phenomena. Samples of radioisotopes decay exponentially because a great many individual nuclei are involved, each having a certain probability of decaying per unit time. If λ is this probability, $\lambda\, dt$ is the probability that one of these nuclei will decay in the time interval dt. Hence the actual number dN of decays that occur in dt is the product of $\lambda\, dt$ and N, the total number of nuclei present in the sample at the start of the time interval:

$$dN = -N\lambda\, dt, \qquad \frac{dN}{N} = -\lambda\, dt. \tag{37–6}$$

(The minus sign is necessary because N is a decreasing function of time.) Integrating the left-hand side of Eq. (37–6) from the initial number of nuclei N_0 at $t = 0$ to the final number N after the time t has elapsed, and the right-hand side from $t = 0$ to $t = t$ yields

$$\int_{N_0}^{N} \frac{dN}{N} = -\lambda \int_0^t dt,$$

$$\ln N - \ln N_0 = -\lambda t,$$

$$N = N_0 e^{-\lambda t}. \tag{37–7}$$

We can go from this formula, which gives the number of undecayed nuclei at the time t in terms of the original number N_0 and the decay probability λ, to the empirical formula of Eq. (37–4) for the activity of a radioactive sample with the help of the definition of activity. From Eq. (37–3),

$$R = -\frac{dN}{dt} = \lambda N_0 e^{-\lambda t}. \tag{37–8}$$

This result is the same as Eq. (37–4) if we identify R_0, the initial activity, with λN_0 and the decay constant λ with the decay probability per nucleus per unit time.

We can use the relationship

$$R = \lambda N \tag{37–9}$$

among activity, decay constant, and number of nuclei to calculate the activity of a given radioactive sample, for instance, that of a milligram (mgm) of radon. A kmole of radon has a mass equal to the atomic mass of radon, 222, expressed in kilograms. One mgm of radon, which is 10^{-6} kg, therefore represents

$$\frac{10^{-6}\text{ kg}}{222\text{ kg/kmole}} = 4.5 \times 10^{-9}\text{ kmole}.$$

Because 1 kmole of any isotope contains an Avogadro's number of nuclei, 1 mgm of radon contains

$$4.5 \times 10^{-9}\text{ kmole} \times 6.02 \times 10^{26}\text{ atoms/kmole} = 2.7 \times 10^{18}\text{ nuclei}.$$

Hence N is 2.7×10^{18} here and, as we found earlier, λ is $2.1 \times 10^{-6}\ \text{sec}^{-1}$ for radon. Substituting these figures in Eq. (37–9), we find that the activity of 1 mgm of radon is

$$\begin{aligned} R = \lambda N &= 2.1 \times 10^{-6}\ \text{sec}^{-1} \times 2.7 \times 10^{18} \\ &= 5.7 \times 10^{12}\ \text{disintegrations/sec} \\ &= 150\ \text{curies}. \end{aligned}$$

The half-life of a radioisotope is not the same as its average lifetime $\overline{T}$, as its activity decreases exponentially and not linearly with time. The average lifetime is simply the reciprocal of the decay probability per unit time λ. In view of the relationship between λ and half-life $T_{1/2}$,

$$\overline{T} = \frac{1}{\lambda} = \frac{T_{1/2}}{0.693} = 1.44\, T_{1/2}.$$

The mean lifetime of radon, whose half-life is 3.8 days, is 5.5 days.

37–3 Alpha decay

Nuclei that contain more than about 210 nucleons are so large that the short-range forces holding them together are barely able to counter-balance the long-range electrostatic repulsive forces of their protons. Such a nucleus can reduce its bulk and thereby achieve greater stability by emitting an alpha particle, which decreases its mass number A by 4. It is appropriate to ask why it is that only alpha particles are given off by excessively heavy nuclei, and not, for example, individual protons or ${}_2\text{He}^3$ nuclei. The reason is a consequence of the high binding energy of the alpha particle, which means that it has significantly less mass than four individual nucleons. Because of this small mass, an alpha particle can be ejected by a heavy nucleus with energy to spare. Thus the alpha particle released in the decay of ${}_{92}\text{U}^{232}$ has a kinetic energy of 5.4 Mev, while 6.1 Mev would have to be supplied from the outside to this nucleus if it is to release a proton, and 9.6 Mev supplied if it is to release a ${}_2\text{He}^3$ nucleus.

Even though alpha decay may be energetically possible in a particular nucleus, it is not obvious just how the alpha particle is able to break away from the nuclear forces that bind it to the rest of the nucleus. Typically an alpha particle has available about 5 Mev of energy with which to escape. However, an alpha particle located at a point near the nucleus but just outside the range of its nuclear forces has an electrostatic potential energy of perhaps 25 Mev; that is, if released from this position it will have a kinetic energy of 25 Mev when it is an infinite distance away as a result of electrostatic repulsion. An alpha particle inside the nucleus therefore should require a minimum of 25 Mev in energy, five times more than is available, in order to break loose (Fig. 37–6). The alpha particle, then, is located in a box whose walls are of such a height that an energy of 25 Mev is needed to surmount them, while the particle itself has only 5 Mev for the purpose.

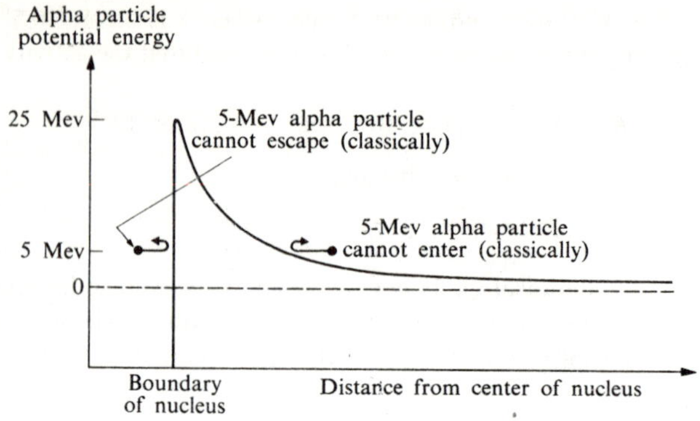

FIG. 37-6. The variation in alpha particle potential energy near a typical heavy nucleus. The alpha particle in this nucleus has 5 Mev available for its escape.

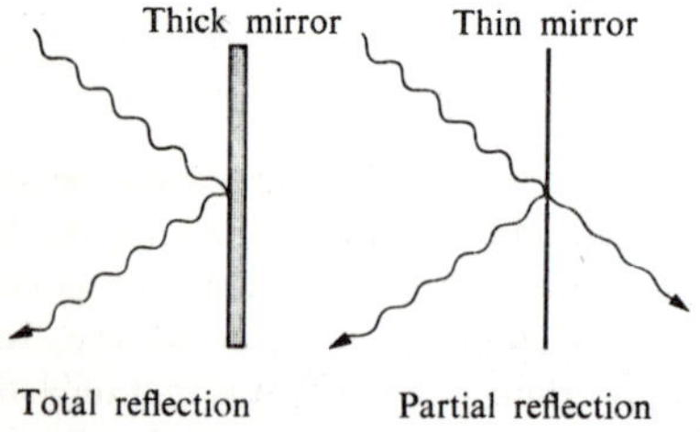

FIG. 37-7. In the process of reflection, a wave penetrates the reflecting surface for a short distance and may pass through it if the mirror is sufficiently thin.

Quantum mechanics provides the answer to the paradox of alpha decay. Two assumptions are needed: (1) an alpha particle can exist as an individual entity within a nucleus, and (2) it is in constant motion there. According to quantum theory, a moving particle has a wave character, so that the proper classical analog of an alpha particle in a nucleus is a light wave trapped between mirrors and not a particle bouncing back and forth between solid walls. Now in order for a light wave to be reflected from a mirror, it must actually penetrate the reflecting surface for a short distance (Fig. 37-7). The intensity of the wave drops off quite rapidly inside the reflecting surface, to be sure, but it *must* penetrate to some extent. If the mirror is thick, all the incident light is reflected. However, if the mirror is very thin, some of the incident light can pass right through the mirror, as shown. The formal theory of this partial transmission, with only minor changes, is able to account quantitatively for alpha decay. The very existence of alpha decay, in fact, is further confirmation of the validity of quantum ideas, since the principles of physics that follow from Newton's laws of motion prohibit such decay.

Of course, a 25-Mev energy barrier is not very "transparent" to a 5-Mev alpha particle. A typical heavy nucleus might be 2×10^{-14} m in diameter, and an alpha particle within it might oscillate back and forth with a speed of 2×10^7 m/sec. Hence the alpha particle strikes the confining nuclear wall 10^{21} times per second, but may nevertheless have to wait as much as 10^{10} years to escape from certain nuclei! The quantum-mechanical phenomenon of barrier penetration is sometimes called the *tunnel effect* because the particle escapes *through* the barrier and not over it.

37-4 Nuclear reactions

When two nuclei approach each other closely enough, it is possible for a rearrangement of their constituent nucleons to occur, with one or more new nuclei formed. These are called *nuclear reactions*, by analogy with chemical reactions in which two or more compounds may combine to form new ones.

An immense number of nuclear reactions are known, all of which obey certain conservation laws. The three most important of these laws are: (1) conservation of electric charge; (2) conservation of number of nucleons; and (3) conservation of mass energy. The reaction

$$_3\mathrm{Li}^7 + {}_1\mathrm{H}^1 \rightarrow 2\,{}_2\mathrm{He}^4 \tag{37-10}$$

involves the fusion of a lithium nucleus of atomic number 3 and mass number 7 with a proton of atomic number 1 and mass number 1 (Fig. 37-8). This reaction results in two alpha particles of atomic number 2 and mass number 4. A total of four protons is present on each side of the arrow, satisfying condition (1). Further, eight nucleons are present on each side of the arrow, taking care of condition (2). To see whether (3) is satisfied, we note that the mass of $_3\mathrm{Li}^7$ is 7.01822, the mass of $_1\mathrm{H}^1$ is 1.00812, and the mass of $_2\mathrm{He}^4$ is 4.00387, all in atomic mass units. Thus

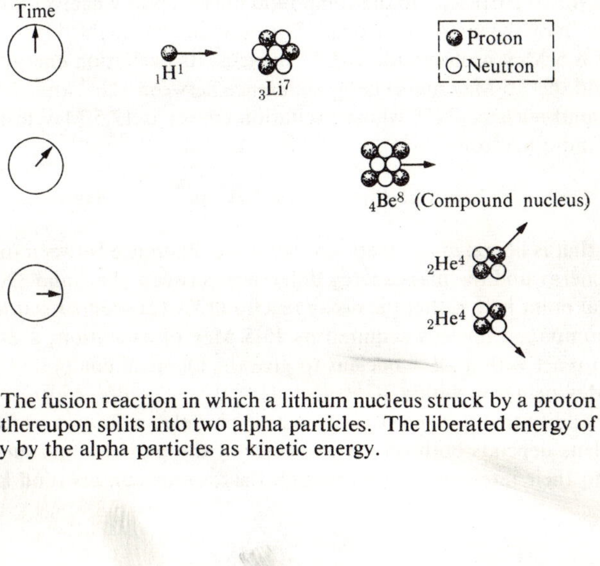

FIG. 37-8. The fusion reaction in which a lithium nucleus struck by a proton absorbs the proton and thereupon splits into two alpha particles. The liberated energy of 17.3 Mev is carried away by the alpha particles as kinetic energy.

the mass excess Q of the left-hand side of Eq. (37-10) is

$$\begin{aligned} Q &= 7.01822 \text{ amu} + 1.00814 \text{ amu} - 2 \times 4.00387 \text{ amu} \\ &= 0.01862 \text{ amu} = 17.3 \text{ Mev}. \end{aligned}$$

This energy is divided between the two alpha particles.

It is important to note that the $_3Li^7$ and proton must have some initial relative velocity, hence initial energy, in order that they approach closely enough for the reaction to take place. This initial energy must be distinguished from the initial energy required by a process whose products have *more* mass than the reactants. Let us examine the reaction

$$_7N^{14} + {}_2He^4 \rightarrow {}_8O^{17} + {}_1H^1. \tag{37-11}$$

Here, since the masses of the nuclei involved are, respectively, 14.00752, 4.00387, 17.00453, and 1.00814 amu, there is a mass *deficiency* of 0.00128 amu, which is 1.2 Mev. This energy must be supplied by the kinetic energy of either the alpha particle or the nitrogen nucleus, or both, if the reaction is to occur at all.

Often nuclear reactions occur in two separate phases. The first phase occurs when a particle (a nucleon or another nucleus) collides with a nucleus and joins with it to form a composite body called a *compound nucleus* (Fig. 37-8). In the second phase the compound nucleus decays into the products of the reaction. The notion of the compound nucleus is a useful one because a particular compound nucleus may be formed in a variety of ways, but it will decay in a characteristic manner that is independent of the way in which it came into being. The amount of excitation energy possessed by the compound nucleus is what determines how it will decay. The excitation energy is the energy equivalent of the mass difference between the initial particles and the compound nucleus plus whatever kinetic energy was brought in by the incident particle. For example, when $_7N^{14}$ is formed by the collision of a 6 Mev proton with a $_6C^{13}$ nucleus, its excitation energy is the sum of 6 Mev and the 7.5 Mev mass energy difference between $_7N^{14}$ and $_6C^{13} + {}_1H^1$. The compound nucleus $_7N^{14}$ whose excitation energy is 13.5 Mev usually decays into $_7N^{13}$ and a neutron:

$$_7N^{14} + 13.5 \text{ Mev} \rightarrow {}_7N^{13} + {}_1n^0 + 3 \text{ Mev}. \tag{37-12}$$

The 3 Mev that is liberated as kinetic energy is the difference between the 13.5 Mev excitation energy and the mass energy difference between $_7N^{14}$ and $_7N^{13} + {}_1n^0$. The essential point here is that the decay reaction (37-12) occurs regardless of how the $_7N^{14}$ compound nucleus acquired its 13.5 Mev of excitation; a 2-Mev alpha particle can react with a $_5B^{10}$ nucleus to give the identical compound nucleus, as can a 3.2-Mev deuteron with a $_6C^{12}$ nucleus.

The probability that a given pair of nuclear particles will react to form a compound nucleus depends both on the identities of the particles and on the energy brought into their interaction, even though the various conservation laws permit the reaction to take place. For example, $_{48}Cd^{113}$ is almost 10,000 times more likely to capture a 0.1-ev neutron than a 10-ev neutron, while it is highly improbable that $_{48}Cd^{112}$ will capture a neutron at all, of any energy.

37–5 Nuclear fission

A particularly significant nuclear reaction that requires a neutron to initiate it is *fission.* In this process, which can take place only in certain very heavy nuclei such as $_{92}U^{235}$, the absorption of an incoming neutron causes the target nucleus to split into two smaller nuclei called *fission fragments* (Fig. 37–9). Because stable light nuclei have proportionately fewer neutrons than do heavy nuclei (Fig. 37–2), the fragments are unbalanced when they are formed and at once release one or two neutrons each. Usually the fragments are still somewhat unstable, and may undergo beta decays (accompanied by gamma decays) to achieve appropriate neutron : proton ratios. The products of fission, such as the fallout from a nuclear bomb burst, are accordingly highly radioactive. Although a variety of nuclear species may appear as fission fragments, we might cite as a typical fission reaction

$$_{92}U^{235} + {_0n^1} \rightarrow {_{92}U^{236}} \rightarrow {_{54}Xe^{140}} + {_{38}Sr^{94}} + {_0n^1} + {_0n^1} + \gamma + 200 \text{ Mev.} \qquad (37\text{–}13)$$

The $_{92}U^{236}$ which is first formed is a compound nucleus, and it is this nucleus that splits in two. The fission fragments $_{54}Xe^{140}$ and $_{38}Sr^{94}$ are both beta radioactive; the former decays four successive times until it becomes the stable isotope

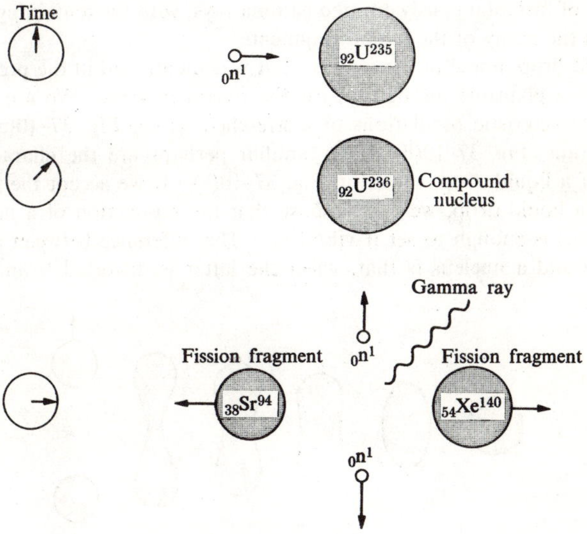

FIGURE 37–9

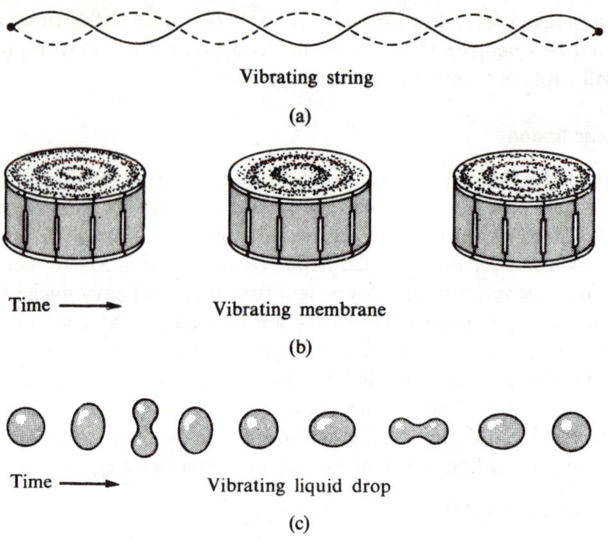

FIGURE 37-10

$_{58}Ce^{140}$, and the latter decays twice in becoming the stable isotope $_{40}Zr^{94}$. About 84% of the total energy liberated during fission appears as kinetic energy of the fission fragments, about 2.5% as kinetic energy of the neutrons, and about 2.5% in the form of instantaneously emitted gamma rays, with the remaining 11% being given off in the decay of the fission fragments.

The liquid drop model of the nucleus that was mentioned in the previous chapter provides a plausible mechanism for the fission process. We are all familiar with the characteristic oscillations of a stretched string, Fig. 37-10(a), and of a taut membrane, Fig. 37-10(b). Less familiar perhaps are the characteristic oscillations of a liquid drop, shown in Fig. 37-10(c). If we accept the picture of a nucleus as a liquid drop, we can suppose that the absorption of a neutron by a heavy nucleus is enough to set it vibrating. The difference between an ordinary liquid drop and a nucleus is that, when the latter is distorted from a spherical

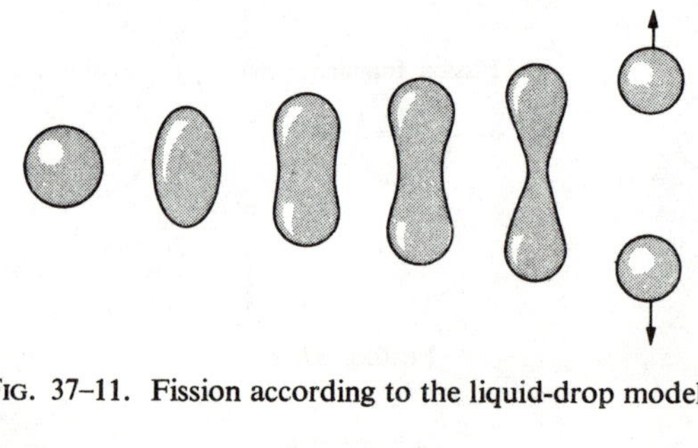

FIG. 37-11. Fission according to the liquid-drop model.

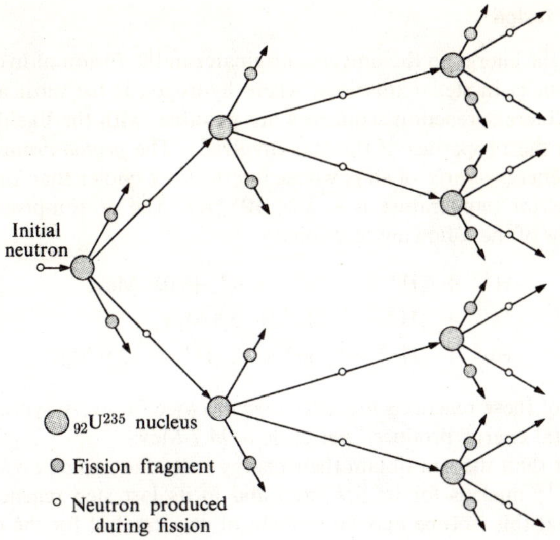

FIG. 37–12. Simplified sketch of a chain reaction in $_{92}U^{235}$.

shape, the short-range nuclear forces holding it together lose much of their effectiveness owing to the larger nuclear surface, where nucleons have fewer bonds. Though the distortion may seriously weaken the attractive forces in the nucleus, the repulsive electrostatic forces of the protons are only slightly affected. If the attractive forces still predominate, the excitation energy of the nucleus is eventually lost through gamma decay: a neutron is captured and a gamma ray is emitted. If the repulsive forces predominate, however, the distortion grows larger and larger until the nucleus splits in two (Fig. 37–11), which is referred to as fission.

Because each fission event liberates two or three neutrons while only one fission is required to initiate it, a rapidly multiplying sequence of fissions can occur in a lump of suitable material (Fig. 37–12). When uncontrolled, a *chain reaction* evolves an immense amount of energy in a short time. If we assume that two neutrons emitted in each fission are able to induce further fissions (the average figure is lower in practice) and that 10^{-8} sec elapses between the emission of a neutron and its subsequent absorption, a chain reaction starting with a single fission will release 2×10^{13} joules of energy in less than 10^{-6} sec! An uncontrolled chain reaction evidently can cause an explosion of exceptional magnitude. When properly controlled so as to assure that exactly one neutron per fission causes another fission, a chain reaction occurs at a constant level of power output. A reaction of this kind makes a very efficient source of power: an output of about 1000 kw is produced by the fission of 1 gm of a suitable isotope per day, as compared with the consumption of over 3 tons of coal per day per 1000 kw in a conventional power plant. A device in which a chain reaction can be initiated and controlled is called a *nuclear reactor*.

37-6 Nuclear fusion

Virtually all the energy in the universe originates in the fusion of hydrogen nuclei into helium nuclei in stellar interiors, where hydrogen is the most abundant element. Two different reaction sequences are possible, with the likelihood of each depending on the properties of the star involved. The *proton-proton cycle* is the predominant energy source of stars whose interiors are cooler than that of the sun (the sun's interior temperature is $\sim 2 \times 10^6$ °K). The proton-proton cycle proceeds by means of the following reactions:

$$\begin{aligned} {}_1H^1 + {}_1H^1 &\rightarrow {}_1H^2 + e^+ + 0.4 \text{ Mev}, \\ {}_1H^1 + {}_1H^2 &\rightarrow {}_2He^3 + 5.5 \text{ Mev}, \\ {}_2He^3 + {}_2He^3 &\rightarrow {}_2He^4 + 2\,{}_1H^1 + 12.9 \text{ Mev}. \end{aligned} \tag{37-14}$$

The first two of these reactions must each occur twice for every synthesis of ${}_2He^4$, so that the total energy produced per cycle is 24.7 Mev.

Stars hotter than the sun obtain their energy from the *carbon cycle*. This cycle requires a ${}_6C^{12}$ nucleus for its first step and in its last step regenerates a ${}_6C^{12}$ nucleus, so that this isotope may be thought of as a catalyst for the process. The carbon cycle proceeds as follows:

$$\begin{aligned} {}_1H^1 + {}_6C^{12} &\rightarrow {}_7N^{13} + 2.0 \text{ Mev}, \\ {}_7N^{13} &\rightarrow {}_6C^{13} + e^+ + 1.2 \text{ Mev}, \\ {}_6C^{13} + {}_1H^1 &\rightarrow {}_7N^{14} + 7.6 \text{ Mev}, \\ {}_7N^{14} + {}_1H^1 &\rightarrow {}_8O^{15} + 7.3 \text{ Mev}, \\ {}_8O^{15} &\rightarrow {}_7N^{15} + e^+ + 1.7 \text{ Mev}, \\ {}_7N^{15} + {}_1H^1 &\rightarrow {}_6C^{12} + {}_2He^4 + 4.9 \text{ Mev}. \end{aligned} \tag{37-15}$$

Here again the net result is the formation of an alpha particle and two positrons from four protons with a total of 24.7 Mev of energy evolved. In the sun both the proton-proton and carbon cycles take place with comparable probabilities.

The energy liberated by nuclear fusion is often called *thermonuclear energy*. High temperatures and densities are necessary for fusion reactions to occur in such quantity that a substantial amount of thermonuclear energy is produced: the high temperature assures that the initial light nuclei have enough thermal energy to overcome their mutual electrostatic repulsion and come together closely enough to react, while the high density assures that such collisions are frequent. A further condition for the proton-proton and carbon cycles is a large reacting mass, such as that of a star, since a number of separate steps are involved in each cycle and much time may elapse between the initial fusion of a particular proton and its ultimate incorporation in an alpha particle. On the earth, where any reacting mass must be very limited in size, an efficient thermonuclear process cannot involve more than a single step. The reactions that appear most promising as sources of commercial power involve the combination of two deuterons to form a triton and a proton,

$$_1H^2 + {}_1H^2 \rightarrow {}_1H^3 + {}_1H^1 + 4.0 \text{ Mev}, \qquad (37\text{–}16)$$

or their combination to form a $_2He^3$ nucleus and a neutron,

$$_1H^2 + {}_1H^2 \rightarrow {}_2He^3 + {}_0n^1 + 3.3 \text{ Mev}. \qquad (37\text{–}17)$$

Both reactions have about equal probabilities. A major advantage of reactions (37–16) and (37–17) is that deuterium (the variety of hydrogen whose atomic nuclei are deuterons, sometimes called "heavy hydrogen") is relatively abundant on the earth, so that there should be no fuel problems in power plants operating on deuteron fusion. While there are many difficulties to surmount in achieving practical thermonuclear power, it will almost certainly become an eventual reality.

Problems

1. Prove that the average lifetime of a radioactive isotope is the reciprocal of the decay probability per unit time.

2. Comment on the similarities and differences between radioactive decay and the death of humans in the population.

3. The half-life of radium is 1600 years. How long will it take for $\frac{15}{16}$ of a given sample of radium to decay?

4. Sixty hours after a sample of the beta emitter $_{11}Na^{24}$ has been prepared, only 6.25% of it remains undecayed. What is the half-life of this isotope?

5. The hydrogen isotope tritium, $_1H^3$, is radioactive and emits an electron with a half-life of 12.5 years. (a) What does $_1H^3$ become after beta decay? (b) What percentage of an original sample of tritium will remain 25 years after its preparation?

6. One millicurie of $_{82}Pb^{214}$ has a mass of 3×10^{-14} kg. Find its decay constant.

7. The half-life of $_{92}U^{238}$ against alpha decay is 4.5×10^9 years. Find the number of disintegrations per second that occur in 1 kg of this isotope.

8. The isotope $_{92}U^{239}$ undergoes two negative beta decays in becoming an isotope of plutonium. State the atomic number and mass number of this isotope.

9. The nuclei $_8O^{14}$ and $_8O^{19}$ both undergo beta decay in order to become stable nuclei. Which would you expect to emit a positron and which an electron?

10. The nucleus $_6C^{11}$ undergoes positive beta decay. What is the atomic number, mass number, and chemical name of the resulting nucleus?

11. The isotope $_{90}Th^{233}$ undergoes two negative beta decays in becoming an isotope of uranium. State the atomic number and mass number of this isotope.

12. The nucleus $_{92}U^{238}$ decays into a lead isotope through the successive emissions of eight alpha particles and six electrons. State the atomic number and mass number of the lead isotope.

13. Radium undergoes spontaneous decay into helium and radon. Why is radium regarded as an element rather than as a chemical compound of helium and radon?

14. How much energy is evolved in the reaction $_1H^2 + {_1H^3} \rightarrow {_2He^4} + {_0n^1}$?

15. The nuclear reaction

$$_3Li^6 + {_1H^2} \rightarrow 2\ {_2He^4}$$

evolves 22.4 Mev. Using the mass values of the deuteron and alpha particle given in Table 36–1, calculate the mass of $_3Li^6$ in amu.

16. In certain stars three alpha particles join in a single reaction to form a $_6C^{12}$ nucleus. The mass of $_6C^{12}$ is 12.0000 amu. How much energy is evolved in this reaction?

17. What are the differences and similarities between nuclear fusion and nuclear fission?

18. A nucleus of $_7N^{15}$ is struck by a proton. A nuclear reaction takes place with the emission of (a) a neutron, or (b) an alpha particle. Give the atomic number, mass number, and chemical name of the remaining nucleus in each of the above cases.

19. A nucleus of $_4Be^9$ is struck by an alpha particle. A nuclear reaction takes place with the emission of a neutron. Give the atomic number, mass number, and chemical name of the resulting nucleus.

20. A reaction often used to detect neutrons occurs when a neutron strikes a $_5B^{10}$ nucleus, with the subsequent emission of an alpha particle. What is the atomic number, mass number, and chemical name of the remaining nucleus?

21. If each fission in $_{92}U^{235}$ releases 200 Mev, how many fissions must occur per second to produce a power of 1 kw?

38
Elementary Particles

The material covered in the previous chapters is so in accord with experiment and observation that it is likely that these ideas will be regarded in the future either in essentially the same way they have been presented here or as limits of as-yet-unformulated theories. Yet the discipline of physics is by no means a static one; on the contrary it is today one of the most active in all of science. A large part of contemporary physical research concerns elementary particles.

We have come a long way from the nineteenth century, when atoms and molecules were considered as the fundamental building blocks of matter. Since that time we have succeeded in explaining the properties of atoms and the nuclei of atoms in terms of more fundamental particles. Each explanation has led, however, to still more complicated problems to be solved. The theory of the structure of the atom left us with the puzzle of explaining the nature of the electron. Research on the properties of nuclei have led in turn to the discovery of the elementary particles. At this moment, approximately 100 of these particles have been discovered, and the physicist is still trying to organize his knowledge about these entities and to discover an all-inclusive theory which will explain their relationships to each other. The task is still not completed, and thus it is likely that present theories of elementary particles will be revised in the near future. In this chapter we shall look into the properties of some of these particles, and into some of the ideas which shed light on their relationships.

38–1 The neutrino

In radioactive decay, as in all other natural processes, energy (including mass energy) is conserved. For this reason the total mass of the products of a particular decay must be less than the mass of the initial nucleus, with the missing mass appearing as photon energy in the case of gamma decay and as kinetic energy in the cases of alpha and beta decay. In gamma and alpha decay the liberated energy is indeed precisely equal to the energy equivalent of the lost mass, but in beta decay a strange effect occurs: instead of all having the same energy, the emitted electrons from a particular isotope exhibit a variety of energies. These energies range from zero up to a maximum figure equal to the energy equivalent of the missing mass in the transformation. This effect is illustrated in Fig. 38-1, which shows the spread in electron energy in the decay of $_{83}Bi^{210}$.

Momentum as well as energy is apparently not conserved in beta decay. As we learned in Chapter 9, when an object at rest disintegrates into two parts, they must move apart in opposite directions in order that the total momentum of the system remain zero. Experiments show, however, that the emitted electron and the residual nucleus do *not* in general travel in opposite directions after beta decay occurs, so that their momenta cannot cancel out to equal the initial momentum of zero.

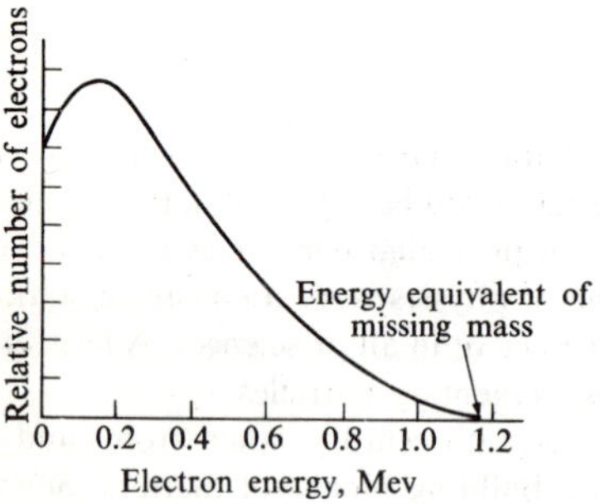

FIG. 38-1. The distribution of electron energies found in the beta decay of $_{83}Bi^{210}$. The maximum electron energy is equal to the energy equivalent of the mass lost by the decaying nucleus minus the electron mass.

A third difficulty concerns angular momentum. The spin quantum number of the neutron, the proton, the electron, and the positron is in every case one-half. The conversion of a neutron into a proton and an electron or of a proton into a neutron and a positron therefore leaves an angular momentum discrepancy of $\frac{1}{2}\hbar$.

The process of beta decay is the first one we have encountered in which the conservation laws of energy, momentum, and angular momentum do not seem to hold. To account for the above discrepancies without abandoning three of the most fundamental and otherwise well-established physical principles, the existence of a new particle was postulated, the *neutrino*, symbol ν (Greek letter nu). The neutrino has no electric charge and no mass, but is able to possess both energy and momentum and has an intrinsic spin of one-half. (Lest this seem unlikely, we might reflect that the photon, also massless, has energy, momentum, and angular momentum. The neutrino is *not* a photon, however, but is an entirely different entity.) According to the neutrino theory, an electron and neutrino are simultaneously emitted in beta decay, as in Fig. 38-2, which permits energy and momentum to be conserved.

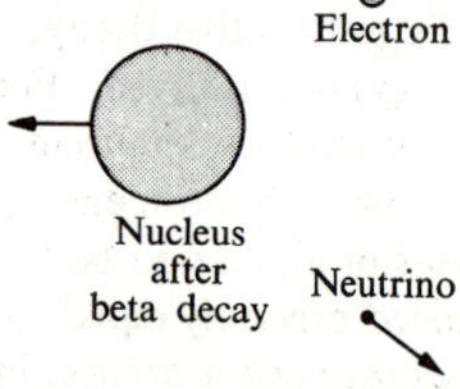

FIGURE 38-2

For a quarter of a century the existence of the neutrino was accepted despite the absence of any direct empirical evidence in support of it. Finally, in 1956, an experiment was performed in which a nuclear reaction that, in theory, could only be caused by a neutrino was actually found to take place. The difficulty to be overcome in detecting neutrinos is their exceedingly feeble interaction with other particles. As we learned in the previous chapter, solar energy originates in nuclear reactions that take place in its interior. Beta decays occur at several stages in the sequences of reactions that participate in the conversion of hydrogen to helium, and a vast number of neutrinos is accordingly produced within the sun. Because neutrinos traverse matter freely, almost all of these neutrinos escape into space and take with them six to eight percent of the total power generated by the sun. The flux of neutrinos from the sun is such that every cubic inch on the earth contains perhaps 100 neutrinos at any instant! In its operation a nuclear reactor also creates a great many neutrinos, and more than 10^{16} neutrinos per second may emerge from each square meter of the shielding around a reactor. A neutrino striking a proton has a small probability of inducing the reaction

$$\nu + \mathrm{p} \rightarrow \mathrm{n} + \mathrm{e}^+ \tag{38-1}$$

in which a neutron and positron are created. By placing a sensitive detecting chamber containing hydrogen near a nuclear reactor, the simultaneous appearance of a neutron and a positron could be registered each time the above reaction occurred. Calculations were made initially of how many such reactions per second should occur, based on the known properties of the detecting apparatus and on the theoretical properties of the neutrino. When this reaction rate was actually found, there was no doubt that neutrinos indeed exist.

As we shall learn in a later section of this chapter, another significant process in which neutrinos are emitted is the decay of charged π-mesons. These particles have masses intermediate between those of electrons and of protons, and may have charges of $+e$ or $-e$. In free space π-mesons decay into μ-mesons and neutrinos; a neutrino is associated with the decay of a positive π-meson and an antineutrino with the decay of a negative π-meson. Are these neutrinos the same particles as those emitted in ordinary beta decay?

The identity of the neutrinos involved in meson decay was finally established in 1962. In this experiment π-mesons were produced by bombarding a metal target with high-energy protons. The meson decays liberated neutrinos, and the interactions of these neutrinos with matter was studied. The only inverse reactions found led to the production of μ-mesons; no electrons whatever were created. Hence the neutrinos set free in meson decay are different from those set free in beta decay. The symbols for the two kinds of neutrino are respectively ν_e and ν_μ.

38-2 Antiparticles

Another recent experimental discovery of a particle whose existence had been predicted theoretically decades ago is that of the negative proton, or *antiproton*, whose symbol is $\bar{\mathrm{p}}$. This is a particle with the same properties as the proton except that it has a negative electric charge. The existence of antiprotons was predicted on the basis that protons and electrons obey very similar fundamental equations. If an antiparticle of the latter exists, so should an antiparticle of the former. Few physicists were surprised when the antiproton was actually found.

Positrons and antiprotons are stable particles. The reason they are so difficult to find is that they are readily *annihilated* upon contact with ordinary matter. When a positron is in the vicinity of an electron, they attract one another electrostatically, come together, and then both vanish simultaneously, with their missing mass appearing in the form of two gamma-ray photons (Fig. 38–3):

$$e^+ + e^- \rightarrow \gamma + \gamma. \tag{38–2}$$

Annihilation with the creation of three photons can also occur but with a much smaller probability. The total mass of the two particles is the equivalent of 1.02 Mev, and so each photon has an energy of 0.51 Mev. (Their energies must be equal and they must be emitted in opposite directions in order that momentum be conserved.) While the similar reaction

$$p + \overline{p} \rightarrow \gamma + \gamma \tag{38–3}$$

can occur when a proton and antiproton undergo annihilation, it is more usual for the vanished mass to reappear in the form of several π-mesons, particles which we shall consider in Section 38–3.

The reverse of annihilation can also take place, with the electromagnetic energy of a photon materializing into a positron and an electron or, if it is energetic

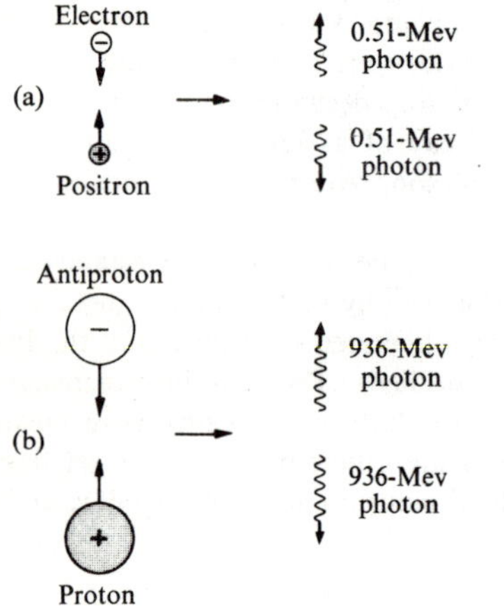

FIG. 38–3. The mutual annihilation of (a) an electron and a positron, and (b) a proton and an antiproton.

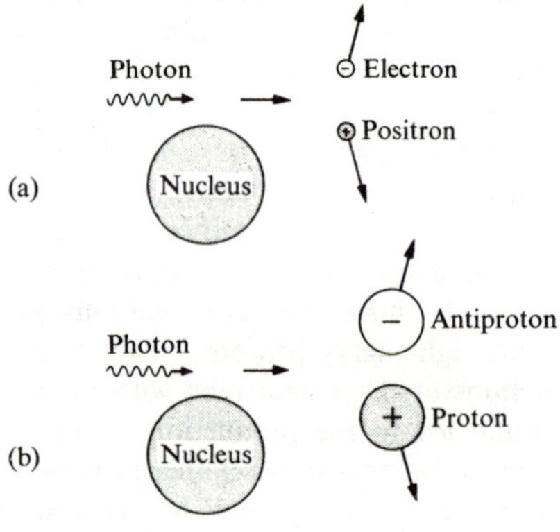

FIG. 38–4. The production of (a) an electron-positron pair and (b) a proton-antiproton pair by the materialization of sufficiently energetic photons. Pair production can occur only in the presence of a nucleus.

enough, into a proton and an antiproton (Fig. 38-4). This phenomenon is known as *pair production,* and requires the presence of a nucleus in order that the laws of the conservation of energy and momentum be satisfied simultaneously. Any photon energy in excess of the amount required to provide the mass of the created particles (1.02 Mev for a positron-electron pair, 1872 Mev for a proton-antiproton pair) appears as kinetic energy.

Antineutrons (symbol $\bar{n}$) and antineutrinos (symbols $\bar{\nu}_e$ and $\bar{\nu}_\mu$) have also been identified. Antineutrons can be detected through their mutual annihilation with neutrons, while more indirect, though equally definite, evidence supports the existence of antineutrinos. The antineutrino differs from the neutrino in that, while the spin axes of both are parallel to their directions of motion, the spin of the former is clockwise and that of the latter is counterclockwise when viewed from behind. A moving neutrino may be thought of as resembling a left-handed screw, and a moving antineutrino as resembling a right-handed screw. An antineutrino is released during a beta decay in which an electron is emitted, and a neutrino is released during a beta decay in which a positron is emitted. Thus the fundamental equations of beta decay are

$$\mathrm{p} \rightarrow \mathrm{n} + \mathrm{e}^+ + \nu_e, \tag{38-4}$$

$$\mathrm{n} \rightarrow \mathrm{p} + \mathrm{e}^- + \bar{\nu}_e. \tag{38-5}$$

Annihilation and pair production are consequences of the facts that matter is a form of energy and that conversions from matter to energy and from energy to matter are no more improbable than conversions from, say, gravitational potential energy to kinetic energy. Relativity and quantum effects underlie the reality of our experience as much as more familiar ones such as the kinetic nature of heat, and within the past fifteen or twenty years the full realization of their significance has profoundly affected the lives of all of us.

38-3 Mesons

If nuclear forces indeed arise from the exchange of mesons between nearby nuclei, as we mentioned in Chapter 36, it should be possible to detect their existence by imparting enough energy to a nucleus for conversion into meson mass without affecting the nuclear mass itself. We can estimate the meson mass from the uncertainty principle,

$$\Delta E \, \Delta t \approx \hbar. \tag{38-6}$$

We assume that the temporary energy discrepancy ΔE is of the same magnitude as the rest energy mc^2 of the meson, and that the meson speed is of the same magnitude as the speed of light c as it travels from one nucleon to another. The time Δt the meson spends between creation and absorption cannot be greater than R/c, where R is the maximum distance that can separate interacting nucleons.

We therefore have

$$\Delta E \, \Delta t \approx \hbar, \qquad mc^2 \times \frac{R}{c} \approx \hbar,$$

$$m \approx \frac{\hbar}{Rc} \cdot \tag{38-7}$$

Nuclear forces have a range of about 1.4×10^{-15} meter. When we substitute

$$R = 1.4 \times 10^{-15} \text{ meter}$$

into Eq. (38-7) we find a predicted meson mass of 2.5×10^{-28} kg, about 275 times the mass of the electron. Of course, the above calculation is hardly a precise one, but, if Yukawa's theory has any validity, the mesons he postulated should have masses between those of electrons and those of nucleons.

Not long after Yukawa's work, charged particles of about the right mass were experimentally discovered in the cosmic radiation. Their discovery was not unexpected because a sufficiently energetic nuclear collision should be able to liberate mesons by providing enough energy to create them without violating conservation of energy, and nuclear collisions between fast cosmic-ray protons from space and oxygen and nitrogen nuclei occur constantly in the atmosphere. However, these particles did not behave at all in the way they were expected to behave. Far from strongly interacting with nuclei, as Yukawa's mesons were supposed to do, they barely interacted at all. Instead of being absorbed in at most a meter of earth, they penetrated thousands of meters into the ground. Finally, in 1947, the explanation for their peculiar behavior was found. The weakly interacting mesons, known as μ-mesons (μ is the Greek letter mu; μ-mesons are often called *muons*) and found so profusely in cosmic rays at sea level, are not the direct products of nuclear collisions, but are secondary particles that result from the decay of the Yukawa mesons. The meson postulated by Yukawa is known as the π-meson (or *pion*), and when outside a nucleus it usually decays in 2.54×10^{-8} sec into a μ-meson and a neutrino plus kinetic energy. That is,

$$\pi^+ \rightarrow \mu^+ + \nu_\mu, \tag{38-8}$$

$$\pi^- \rightarrow \mu^- + \bar{\nu}_\mu. \tag{38-9}$$

(A small percentage of charged π-mesons decay directly into an electron and a neutrino.) The π-meson is observed to have the properties Yukawa attributed to it, and may be loosely thought of as the "glue" that holds nuclei together so strongly. Charged π-mesons have a mass of $273m_e$. Neutral π-mesons also exist; their mass is $264m_e$, a little less than that of their charged counterparts, and they decay in about 10^{-16} sec into two gamma-ray photons:

$$\pi^0 \rightarrow \gamma + \gamma. \tag{38-10}$$

The brief lifetimes of π-mesons are responsible for the delay in their discovery. In the past decade, high-energy accelerators have been built whose particles are able to violently disrupt nuclei with the creation of many π-mesons; considerable progress has been made in our understanding of these mesons and their behavior. The π^--meson is the antiparticle of the π^+-meson, and the neutral π^0-meson is its own antiparticle.

The μ-meson, whose mass is $207m_e$, is less well understood than the π-meson. The relatively long μ-meson lifetime of 2.22×10^{-6} sec and their feeble interaction with matter accounts for the penetrating ability of these mesons. In contrast to the π-meson, which fits in well with the theory of nuclear forces, the μ-meson seems to have no particular function in the scheme of things; why such a particle should exist is one of the current problems of theoretical physics. The μ^+- and μ^--mesons decay into positrons and electrons together with a neutrino and an antineutrino in each case:

$$\mu^+ \rightarrow e^+ + \nu_e + \bar{\nu}_\mu, \tag{38–11}$$

$$\mu^- \rightarrow e^- + \nu_\mu + \bar{\nu}_e. \tag{38–12}$$

The μ^+-meson is the antiparticle of the μ^--meson. There is no neutral μ-meson.

38–4 Strange particles

Not long after the discovery of the π-meson a succession of other, quite unpredicted elementary particles came to light. Even though a good deal is now known about these new particles, they may still be called by their original name of *strange particles* with some justice. It is convenient to divide strange particles into two classes, *K-mesons*, whose masses are all almost exactly $967m_e$, and *hyperons*, whose masses exceed the mass of the proton. Both classes of strange particles, like π-mesons, are found as products of energetic nuclear collisions, and are unstable.

Charged K-mesons may decay in a variety of ways after a mean lifetime of 1.22×10^{-8} sec. The K^+-meson has been observed to have the following six decay schemes, listed in the order of probability:

$$\begin{aligned} K^+ &\rightarrow \mu^+ + \nu_\mu \\ &\rightarrow \pi^+ + \pi^0 \\ &\rightarrow \pi^+ + \pi^+ + \pi^- \\ &\rightarrow \pi^+ + \pi^0 + \pi^0 \\ &\rightarrow e^+ + \pi^0 + \nu_e \\ &\rightarrow \mu^+ + \pi^0 + \nu_\mu. \end{aligned} \tag{38–13}$$

The K^--meson, the antiparticle of the K^+-meson, has similar decay possibilities. Two different versions of the neutral K-meson are known. The K_1^0-meson decays in 10^{-10} sec in either of these ways:

$$\begin{aligned} K_1^0 &\rightarrow \pi^+ + \pi^- \\ &\rightarrow \pi^0 + \pi^0. \end{aligned} \tag{38–14}$$

The K_2^0-meson has the much longer mean lifetime of 6×10^{-8} sec and six possible decay schemes:

$$\begin{aligned} K_2^0 &\rightarrow \pi^+ + \pi^- + \pi^0 \\ &\rightarrow \pi^0 + \pi^0 + \pi^0 \\ &\rightarrow \pi^- + \mu^+ + \nu_\mu \\ &\rightarrow \pi^+ + \mu^- + \bar{\nu}_\mu \\ &\rightarrow \pi^- + e^+ + \nu_e \\ &\rightarrow \pi^+ + e^- + \bar{\nu}_e. \end{aligned} \tag{38-15}$$

Unlike the π^0-meson, which is its own antiparticle, an anti-K^0-meson, the $\overline{K^0}$-meson, is believed to exist.

There are three known families of hyperons, all of whose members are heavier than nucleons and all of which have a nucleon as one of their decay products. There is only a single kind of Λ-hyperon (Λ is the Greek capital letter lambda), a neutral particle whose mass is $2182m_e$ that decays in an average of 2.5×10^{-10} sec into either a proton and π^--meson or into a neutron and a π^0-meson, the latter alternative being the less likely:

$$\begin{aligned} \Lambda^0 &\rightarrow p^+ + \pi^- \\ &\rightarrow n^0 + \pi^0. \end{aligned} \tag{38-16}$$

The $\bar{\Lambda}^0$-hyperon is the antiparticle of the Λ^0-hyperon, and decays into antiparticle counterparts of the above products.

The Σ-hyperons may be neutral or may have either charge (Σ is the Greek capital letter sigma). The Σ^+-hyperon has a mass of $2327m_e$, a mean lifetime of 0.8×10^{-10} sec, and an equal probability of decaying into a proton and a π^0-meson or into a neutron and a π^+-meson:

$$\begin{aligned} \Sigma^+ &\rightarrow p^+ + \pi^0 \\ &\rightarrow n^0 + \pi^+. \end{aligned} \tag{38-17}$$

Only one decay mode is known for the Σ^--hyperon, whose mass is $2341m_e$ and whose mean lifetime is 1.6×10^{-10} sec:

$$\Sigma^- \rightarrow n^0 + \pi^-. \tag{38-18}$$

The Σ^0-hyperon has a mass of $2329m_e$ and a mean lifetime of less than 10^{-12} sec. The Σ^0-hyperon decays into a Λ^0-hyperon and a gamma ray:

$$\Sigma^0 \rightarrow \Lambda^0 + \gamma. \tag{38-19}$$

The $\bar{\Sigma}^+$-, $\bar{\Sigma}^-$-, and $\bar{\Sigma}^0$-hyperons are antiparticles of the Σ^+-, Σ^--, and Σ^0-hyperons.

The third family of hyperons has two members, the Ξ^--hyperon and the Ξ^0-hyperon (Ξ is the Greek capital letter xi). Their masses are about $2583m_e$, and their lifetimes are in the range 10^{-9} to 10^{-10} sec. Both decay into a Λ^0-hyperon and a π-meson,

$$\Xi^- \rightarrow \Lambda^0 + \pi^-, \tag{38–20}$$

$$\Xi^0 \rightarrow \Lambda^0 + \pi^0, \tag{38–21}$$

and both have appropriate antiparticles.

38–5 Categories of elementary particles

In all, about 100 elementary particles have been discovered, of which we have been introduced to no less than 25 in this chapter. Such abundance where scarcity had been expected inspired considerable interest among physicists in the past decade, with the result that partial success has been attained in understanding these entities. But there is still a long way to go before this fundamental puzzle is solved.

Elementary particles fall naturally into four categories, namely, *photons*, *leptons*, *mesons*, and *baryons*. The members of each category share certain basic properties despite differences among themselves in other properties. Photons, as we know, are quanta of electromagnetic energy; they are stable and have zero rest mass. The category of leptons includes two types of neutrinos, electrons, and μ-mesons. Leptons exhibit very weak interactions with nuclei and have a spin of one-half. (We note that in this classification, μ-mesons are considered leptons rather than mesons.) The meson category includes both π- and K-mesons. These particles interact quite strongly with nuclei and all have a spin of zero. Baryons include nucleons and hyperons; they also interact strongly with nuclei, but have spins of one-half. Table 38–1 lists 32 of the longest-lived elementary particles together with some of their characteristic properties; the others exist only for the briefest of instants before decaying.

While it is obvious from Table 38–1 that the various elementary particles fit into the indicated categories on the basis of their masses and spins, experimental evidence of an unusual kind further substantiates this division. Let us assign the quantum number $L = 1$ to the electron and the electron neutrino, and $L = -1$ to their antiparticles. We next assign a muon family quantum number $M = 1$ to the μ-meson and its neutrino, and $M = -1$ to their antiparticles. All other particles are assigned the numbers $L = M = 0$. It is found that in every known process involving leptons, including their production, decay, or interaction with other particles, the total value of L and the total value of M remain constant. Also, let us assign all baryons the special quantum number $B = +1$, all antibaryons $B = -1$, and all other particles $B = 0$. It is found that in every known process involving baryons, the total value of B remains constant. A simple example is the beta decay of the neutron:

$$\begin{array}{lcccc} & n^0 \rightarrow & p^+ + & e^- + & \bar{\nu}_e \\ L = & 0 & 0 & +1 & -1 \\ B = & +1 & +1 & 0 & 0. \end{array} \tag{38–22}$$

TABLE 38-1

Some elementary particles

Name	Particle	Antiparticle	Mass, m_e	Spin	L	M	B	S	Mean lifetime, sec	Category
Photon	γ	(γ)	0	1	0	0	0	0	Stable	Photon
Electron neutrino	ν_e	$\bar{\nu}_e$	0	$\frac{1}{2}$	1	0	0	0	Stable	Lepton
Muon neutrino	ν_μ	$\bar{\nu}_\mu$	0	$\frac{1}{2}$	0	1	0	0	Stable	
Electron	e^-	e^+	1	$\frac{1}{2}$	1	0	0	0	Stable	
Mu-meson	μ^-	μ^+	207	$\frac{1}{2}$	0	1	0	0	2.22×10^{-6}	
Pi-meson	π^+	π^-	273	0	0	0	0	0	2.55×10^{-8}	Meson
	π^0	(π^0)	264	0	0	0	0	0	1.9×10^{-16}	
	π^-	π^+	273	0	0	0	0	0	2.55×10^{-8}	
K-meson	K^+	$\overline{K^+}$	967	0	0	0	0	+1	1.22×10^{-8}	
	K^0	$\overline{K^0}$	974	0	0	0	0	+1	1.00×10^{-10} and 6×10^{-8}	
Proton	p	$\bar{p}$	1836	$\frac{1}{2}$	0	0	1	0	Stable	Baryon
Neutron	n^0	$\overline{n^0}$	1839	$\frac{1}{2}$	0	0	1	0	1.01×10^3	
Lambda-hyperon	Λ^0	$\bar{\Lambda}^0$	2182	$\frac{1}{2}$	0	0	1	−1	2.5×10^{-10}	
Sigma-hyperon	Σ^+	$\bar{\Sigma}^+$	2327	$\frac{1}{2}$	0	0	+1	−1	0.8×10^{-10}	
	Σ^-	$\bar{\Sigma}^-$	2340	$\frac{1}{2}$	0	0	+1	−1	1.6×10^{-10}	
	Σ^0	$\bar{\Sigma}^0$	2332	$\frac{1}{2}$	0	0	+1	−1	10^{-12}	
Xi-hyperon	Ξ^-	$\bar{\Xi}^-$	2580	$\frac{1}{2}$	0	0	1	−2	$\sim 10^{-10}$	
	Ξ^0	$\bar{\Xi}^0$	2570	$\frac{1}{2}$	0	0	1	−2	$\sim 10^{-10}$	

The usual conservation laws of energy, momentum, angular momentum, and charge, plus the new (and still rather mysterious) conservation laws of electron-family, muon-family, and baryon numbers, help in understanding why certain processes involving elementary particles occur and why others do not occur. For instance, the free neutron can decay according to Eq. (38–22) and still obey the above conservation laws, but in no other way. The free proton, which has less mass than the neutron, cannot decay at all: energy conservation prohibits it from decaying with any other baryon as a product, and baryon number conservation prohibits it from decaying into any lighter particles. The μ-mesons are prohibited from decaying into electrons and photons by the reaction

$$\mu^- \rightarrow e^- + \gamma \tag{38–23}$$

by the principle of the conservation of the quantum numbers L and M. The fact that the reaction (38–23) had never been observed led to the successful search for the muon neutrino and the introduction of the electron-family quantum number, L, and the muon-family quantum number, M. Prior to the discovery of the muon neutrino, a single lepton quantum number was used for electrons, μ-mesons, and neutrinos. The reaction (38–23) does not violate the conservation of the lepton quantum number.

Introducing the quantum numbers L, M, and B did not clarify all aspects of elementary particle behavior. Three of the still outstanding problems were the following:

(1) Some heavy particles decay into lighter ones plus gamma rays while others do not. For example, the Σ^0-baryon decays into a Λ^0-baryon and a gamma ray,

$$\Sigma^0 \rightarrow \Lambda^0 + \gamma,$$

while the Σ^+-baryon never decays in the analogous manner

$$\Sigma^+ \not\rightarrow p^+ + \gamma.$$

(2) Physical processes that release large amounts of energy almost always occur faster than processes involving lesser amounts. However, relatively long-lived particles are known whose decays evolve tens or hundreds of Mev's.

(3) Strange particles never come into being singly, but always two or more at a time.

Observations such as these led to the introduction of a further quantum number, the *strangeness number S*. Table 38–1 shows the values of S that are assigned to the various elementary particles. Only the photon and π^0-meson have $L = M = B = S = 0$. These particles are regarded as their own antiparticles. Before we discuss how the strangeness number may be interpreted, it will be necessary to look into another aspect of elementary particle physics.

38-6 Fundamental interactions

Elementary particles can interact with one another in four different ways, and these interactions are responsible for all of the physical processes in the universe. Feeblest is the gravitational interaction, which follows the inverse-square law discovered by Newton. Next is the *weak interaction*, characteristic of leptons, that acts between leptons and either mesons, baryons, or other leptons (besides any electromagnetic forces that may also be present). The weak interaction is responsible for reactions involving neutrinos, such as beta decays. Of greater strength than gravitational and weak interactions are the electromagnetic interactions between all charged particles and also between those with electric or magnetic moments. Most powerful of the four is the *strong interaction* between mesons, between baryons, and between mesons and baryons that gives rise to nuclear forces.

A single elementary particle can interact with others by means of more than one interaction. A proton interacts by means of all four. By virtue of its mass it interacts gravitationally with other masses. By virtue of its charge it participates in the electromagnetic interaction. Since it appears as one of the products of the spontaneous beta decay of the neutron, it is involved in weak interactions. Finally, as a baryon it is one of the particles that has strong interactions.

The relative magnitudes of the strong, electromagnetic, weak, and gravitational interactions are in the ratios $1:10^{-2}:10^{-14}:10^{-39}$. To be sure, the ranges of the forces produced by these interactions are quite different: the nuclear force between two nucleons vastly exceeds the gravitational force between them when they are 10^{-15} m apart, but when they are one meter apart the situation is reversed. Nuclear structure is a consequence of the properties of the strong interaction, while atomic structure is a consequence of those of the electromagnetic interaction. Because bulk matter is electrically neutral and the strong and weak forces are severely limited in their range, gravitational forces, negligible on a small scale, become all-important on a large scale. The role of the weak force in the structure of matter seems to be restricted to causing beta decays that adjust the neutron:proton ratios of nuclei to stable values.

In all processes involving strong and electromagnetic interactions, the strangeness number S is conserved. Thus the decay

$$\begin{array}{lcl} & \Sigma^0 \rightarrow & \Lambda^0 + \gamma \\ S = & -1 & -1 \quad 0 \end{array}$$

that we mentioned earlier conserves S and actually occurs, and the apparently similar decay

$$\begin{array}{lcl} & \Sigma^+ \nrightarrow & p^+ + \gamma \\ S = & -1 & 0 \quad 0 \end{array}$$

does not conserve S and has never been found to occur. The multiple production of strange particles in high-energy nuclear collisions is necessary in order to conserve S, since such collisions involve strong interactions. The relatively long lifetimes of all unstable particles except the π^0-meson and Λ^0-hyperon follow from the assumption that weak interactions are also characteristic of mesons and baryons as well as leptons, though normally dominated by strong or electromagnetic interactions in the former cases. With strong or electromagnetic decay processes impossible except for the π^0-meson and Λ^0-hyperon because S must be conserved in such processes, only the weak interaction is available for particle decay. Events governed by weak interactions are very slow, in agreement with observation. However, even in weak interactions S does not change by more than $+1$ or -1. Thus the Ξ^--hyperon does not decay directly into a neutron in the single step

$$\begin{array}{rccc} & \Xi^- \nrightarrow & n^0 + & \pi^- \\ S = & -2 & 0 & 0 \end{array}$$

but instead requires two separate decays:

$$\begin{array}{rccc} & \Xi^- \rightarrow & \Lambda^0 + & \pi^- \\ S = & -2 & -1 & 0 \end{array} \quad \text{and} \quad \begin{array}{rccc} & \Lambda^0 \rightarrow & n^0 + & \pi^0 \\ S = & -1 & 0 & 0. \end{array}$$

38–7 Conservation theorems

The conservation principles enable the physicist to find relationships among elementary particles and guide him in attempting to find mathematical descriptions of the various laws of force. There are seven conservation laws which apply, so far as we know, to all processes. These are the conservation of

(1) Energy
(2) Linear momentum
(3) Angular momentum (including spin)
(4) Charge
(5) Electron-family number
(6) Muon-family number
(7) Baryon-family number

There are some conservation laws which apply to strong interactions but not to other types. One of these, conservation of strangeness, was discussed in the previous section. Another of these is the *conservation of isotopic spin.*

Isotopic spin is a misnomer, since it is a property of particles that has nothing to do with spin except by analogy. It describes the fact that there are groups of particles with small mass differences that behave identically in strong interactions. For example, the neutron and proton have identical strong interaction properties. The π^+, π^0, and π^- likewise interact in similar ways, and also would be identical except for their charge. The small differences in mass are electromagnetic in origin.

By analogy with splitting of angular momentum states with quantum number l into $(2l + 1)$ different states, an isotopic spin quantum number is assigned to each particle, the number depending on how many particles belong to a single group. The isotopic-spin assignment of some mesons and baryons is given in Table 38–2.

TABLE 38–2

Isotopic-spin assignment of particles

Particle	Isotopic spin
Nucleon (p, n^0)	$\frac{1}{2}$
π-meson (π^+, π^0, π^-)	1
K-meson (K^0, K^+)	$\frac{1}{2}$
η^0-meson*	0
Λ^0-hyperon	0
Σ-hyperon (Σ^+, Σ^0, Σ^-)	1
Ξ-hyperon (Ξ^-, Ξ^0)	$\frac{1}{2}$

* The η^0-meson is a very short-lived one, not discussed in the text. It is important in the eightfold-way scheme for classifying mesons (see Section 38–8).

The isotopic-spin quantum number is conserved in strong interactions; due regard must be given to the fact that isotopic spin combines vectorially in a manner similar to angular momentum. Groups of particles having the same value of isotopic spin are called *multiplets*.

There are still other conservation theorems which are peculiar to strong interactions; we shall not discuss them here. Some, like the conservation of parity, have already played an important part in our understanding of fundamental forces. Some others, related to time reversal, have recently been shown to have been violated in some processes.

38–8 The Eightfold Way

A giant step has recently been taken in organizing our information about elementary particles because of the observation that there are tendencies for these particles to exist in multiplets consisting of eight particles each, called *supermultiplets*. These supermultiplets consist of particles which have the same value of spin but different values of charge, isotopic spin, mass, and strangeness. Two of these groupings are shown in Fig. 38–5.

The same theory that predicts supermultiplets of eight also predicts other groupings of ten and of one. This theory, popularly called the *eightfold way*, has recently scored an impressive success by predicting the existence of an unknown particle

	Baryons				Mesons				
Strangeness	−1	0		1	−1	0		1	Strangeness
Isotopic spin	$\frac{1}{2}$	1	0	$\frac{1}{2}$	$\frac{1}{2}$	1	0	$\frac{1}{2}$	Isotopic spin
Negative charge	Ξ^-	Σ^-			$\overline{K}^-$	π^-			Negative charge
Neutral	Ξ^0	Σ^0	Λ^0	$\cdot$n	$\overline{K}^0$	π^0	η^0	K^0	Neutral
Positive charge		Σ^+		$\cdot$p		π^+		K^+	Positive charge
Mass, Mev	1318	1193	1115	939	496	137	548	496	Mass, Mev

FIG. 38–5. Supermultiplets of mesons and the lightest baryons in the eightfold way.

which belonged in one of the supermultiplets of 10. After the prediction, the particle (called the Ω^--hyperon) was subsequently discovered and had a mass consistent with the predictions of the eightfold way.

All the known elementary particles, including the very short-lived ones (lifetimes $< 10^{-19}$ sec) which have not been discussed here can be classified according to the eightfold way, and the mass differences within each multiplet can be explained approximately.

Physicists feel that they are on the right track to uncover the fundamental puzzle of the force laws and their relationships. Surprises will undoubtedly appear. One new particle is yet to be found experimentally, the meson which carries the weak interaction. (The graviton, which carries the gravitational interaction, exerts too weak a force to be detected.) The role of weak interactions in the scheme of things is yet to be really understood. There are excitement, movement, and almost monthly discoveries in elementary particle physics.

Problems

1. State which of the following are spontaneously emitted by radioactive nuclei, and which are not: electrons, positrons, neutrinos, protons, antiprotons, neutrons, mesons, alpha particles, electromagnetic waves.

2. How much energy must a gamma-ray photon have if it is to materialize into a proton-antiproton pair with each particle having a kinetic energy of 10 Mev?

3. How much energy (in ev) must a gamma-ray photon have if it is to materialize into a neutron-antineutron pair? Is this more or less than that required to form a proton-antiproton pair?

4. A 1-Mev positron collides head on with a 1-Mev electron, and the two are annihilated. What is the energy and wavelength of each of the resulting gamma-ray photons?

5. One proton strikes another, and the reaction

$$p + p \rightarrow n + p + \pi^+$$

takes place. What is the minimum energy the incident proton must have had?

6. Find the energy of each of the gamma-ray photons produced in the decay of the neutral pi-meson. Why must their energies be the same?

7. Why does the neutron not decay into an electron, a positron, and a neutrino?

8. Why does the Λ-hyperon not decay into a π^+- and a π^--meson?

9. A proton can interact with a μ^--meson to form a neutron and another particle. What must this particle be?

10. According to the theory of the continuous creation of matter, the evolution of the universe can be traced to the spontaneous appearance of neutrons in free space. Which conservation laws would this process violate?

11. Show that the laws of conservation of linear momentum and energy prohibit the decay of one particle into a single lighter particle.

12. Show that the laws of conservation of linear momentum and energy prohibit the decay of a lighter particle into a heavier one.

13. Show that the reaction

$$\Lambda^0 \rightarrow p + \pi^-$$

violates the strangeness-conservation law.

14. Show that the reactions

$$\Lambda^0 \rightarrow p + \pi^-$$
$$\Lambda^0 \rightarrow n + \pi^0$$

violate the conservation of isotopic spin.

15. Which conservation theorems are violated by the following reactions?

(a) $e^- \rightarrow \nu + \gamma$

(b) $p + p \rightarrow p + \Lambda^0 + \Sigma^+$

(c) $n \rightarrow \mu^+ + e^- + \gamma$

(d) $\mu^- \rightarrow e^- + e^+ + \nu_\mu$

(e) $e^+ + e^- \rightarrow \mu^+ + \pi^-$.

Appendixes

Appendix I

GEOMETRICAL OPTICS

On the basis of Huygens' principle, or on the basis of the fundamental laws governing the propagation of rays at reflecting or refracting surfaces, it is relatively simple to see how a curved surface can change the shape of plane waves or spherical waves. In Fig. A–1 we show a plane wave ABC incident on a curved reflecting surface. By Huygens' principle, a wavelet that originates at A will propagate to A' while the wavelet from B approaches B'. At this time, the wavelet at C will have reached C'. The wave front $A'B'C'$ will now propagate to the left. If $AB'C$ is a parabolic surface, one can prove geometrically that $A'B'C'$ now propagates as a spherical wave with diminishing curvature. Thus $A''B''C''$ has a smaller curvature than $A'B'C'$. Ultimately all the energy in the original plane wave will pass through F, which is called the *focus* of the parabolic mirror.

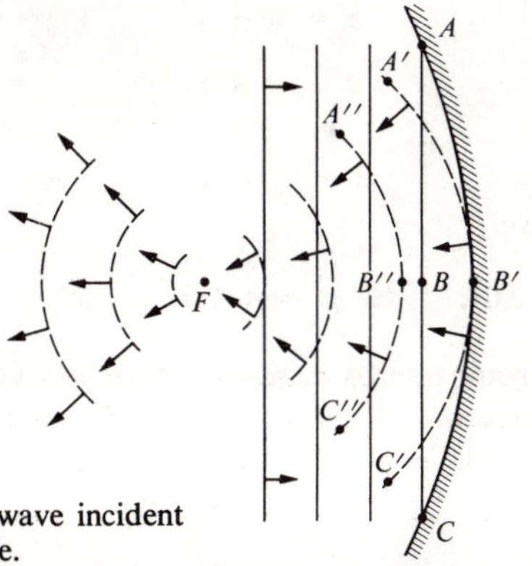

FIG. A–1. A plane wave incident on a parabolic surface.

The parabolic shape, which converts a plane wave into a spherical wave, distorts a spherical wave incident on it and does not produce a spherical one in turn. As a matter of practice, mirror surfaces are usually instead made spherical in contour, and if their apertures are not too large they can be used to focus light which is plane or spherical.

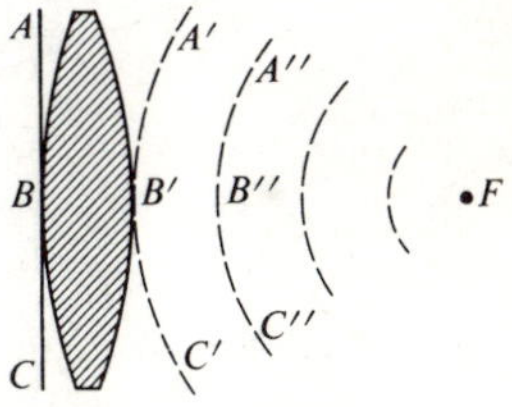

FIG. A–2. A plane wave incident on a lens.

Plane wave fronts are also transformed into spherical ones at a spherical surface between transparent media whose indexes of refraction are different. A plane wave ABC incident on such a surface or combination of surfaces (Fig. A–2) has the middle portion retarded so that when the wavelet originating at B has reached B', the ones originating at A and C have gone a greater distance to A' and C'. Again for small apertures, $A'B'C'$ will be a spherical wave and will propagate so that all the light energy will pass through F, which is again called the focus. Arrangements of dense materials which change the curvature of wave fronts are called *lenses*.

The behavior of curved mirrors and lenses can be analyzed by means of Huygens' principle, but it is much easier to do so in terms of the ray picture, as we shall do.

I. Ray treatment of mirrors

Let us consider a spherical mirror AOB (Fig. A–3) whose center of curvature is at C and whose radius of curvature is R. A point source of light, S, is located on the axis and emits rays in all directions. Rays from the source strike the mirror everywhere; we are interested in the ray SM. CM is normal to the surface AOB, and the angle of incidence i is equal to the angle of reflection CMI. The reflected ray crosses through the axis of the mirror at I. By elementary geometry (the exterior angle is equal to the sum of the remote interior angles), we have

$$\beta = \alpha + i, \tag{A–1}$$

$$\gamma = \beta + i. \tag{A–2}$$

Eliminating i, we have

$$\alpha + \gamma = 2\beta. \tag{A–3}$$

From the figure, we have

$$\tan\alpha = h/SO', \qquad \tan\beta = h/CO', \qquad \tan\gamma = h/IO'. \tag{A–4}$$

We must now make some approximations. If we assume that the angles α, β, and γ are so small that $\tan\alpha \approx \sin\alpha \approx \alpha$, $\tan\beta \approx \sin\beta \approx \beta$, and $\tan\gamma \approx$

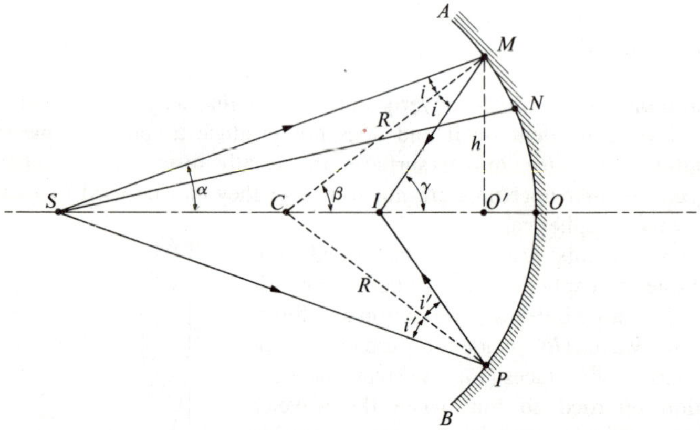

FIG. A–3. Rays from a source reflected from a spherical mirror.

$\sin \gamma \approx \gamma$, then it follows that

$$\alpha \approx \tan \alpha \approx \sin \alpha = \frac{h}{SM} = \frac{h}{SO} = \frac{h}{u}, \tag{A–5}$$

$$\beta \approx \tan \beta \approx \sin \beta = \frac{h}{CM} = \frac{h}{R}, \tag{A–6}$$

$$\gamma \approx \tan \gamma \approx \sin \gamma = \frac{h}{CN} = \frac{h}{CO} = \frac{h}{v}. \tag{A–7}$$

R, we recall, is the radius of curvature of the mirror, and u and v are abbreviations for SO and CO, respectively, the *object distance* and the *image distance*. Substituting (A–5), (A–6), and (A–7) into (A–3), we have

$$\frac{h}{u} + \frac{h}{v} = \frac{2h}{R} = \frac{h}{f},$$

$$\frac{1}{u} + \frac{1}{v} = \frac{2}{R} = \frac{1}{f}. \tag{A–8}$$

At the moment f is just an abbreviation for $R/2$; we shall explore its significance presently. The remarkable result embodied in (A–8) is that it is independent of the angles that the rays make with the axis. All rays obey the same equation, and rays SN and SP will pass through I as will all of the others. Thus all of the light originating at S will pass through the point I if the angle is small enough. These rays are called *paraxial rays*. We thus have a technique for producing images of light sources using mirrors. If the object distance u goes to infinity, Eq. (A–8) reduces to

$$v = f = R/2, \tag{A–9}$$

and rays parallel to the axis pass through the point $v = R/2$. An object at infinity produces a plane wave. Evidently the significance of the distance f is that it is that distance from the mirror where a plane wave will be concentrated. It is called the *focal distance*.

II. Sign conventions for mirrors

Now we consider a source S which is closer to the mirror than the focal distance (Fig. A–4). Here the reflected ray does not cross the axis to the left of O, but instead seems to come from a point I behind the mirror. If the angles i, α, β, and γ are as shown, we obtain

$$\alpha = \beta + i, \tag{A–10}$$

$$2i = \alpha + \gamma. \tag{A–11}$$

Eliminating i again, we have

$$\alpha - \gamma = 2\beta. \tag{A–12}$$

For paraxial rays, letting $SO = u$ and $IO = v$, we have

$$\begin{aligned}\alpha &\approx \sin\alpha \approx \tan\alpha = h/u,\\ \beta &\approx \sin\beta \approx \tan\beta = h/R,\\ \gamma &\approx \sin\gamma \approx \tan\gamma = h/v.\end{aligned} \tag{A–13}$$

Substituting into (A–12),

$$\frac{1}{u} - \frac{1}{v} = \frac{2}{R} = \frac{1}{f}. \tag{A–14}$$

The similarity between (A–8) and (A–14) is not accidental. If we solve (A–8) for $1/v$, we have

$$\frac{1}{v} = \frac{2}{R} - \frac{1}{u}. \tag{A–15}$$

If u were less than $R/2$, as it is in the case we considered in this section, v would turn out to be negative. Thus the result (A–14) is contained in (A–8) with an appropriate sign convention.

The rays which are reflected from the mirror in Fig. A–4 are diverging, and all seem to come from a source I behind the mirror. An eye placed near the mirror would see the diverging rays and would interpret them in terms of a source *I behind* the mirror. This type of image is called a *virtual* one. A negative image distance always signifies a virtual image.

Formula (A–8) is the general mirror formula and must be used with the following sign conventions:

(1) A positive v is a real image distance. The image is a point in real space through which the light actually goes. Negative v's are virtual image distances.

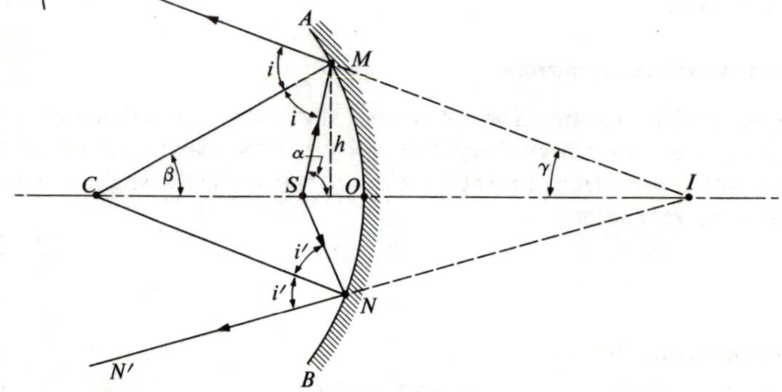

FIG. A–4. Reflection of rays from a source inside the focus of a spherical mirror.

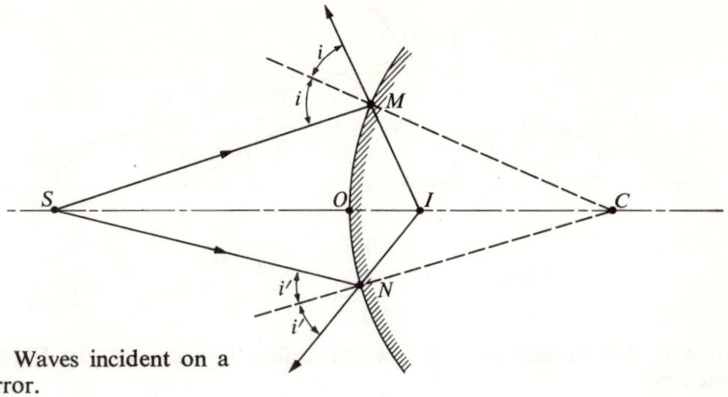

FIG. A–5. Waves incident on a convex mirror.

(2) Concave mirrors have a positive R and positive f associated with them. Convex mirrors have a negative R and a negative f.

(3) Real sources have positive object distances u. There are also virtual sources which have negative object distances. The principal difference between them physically is that the rays from a real source are diverging when they reach the mirror, while they are converging when they reach the mirror from a virtual source. Virtual sources are important when combinations of optical elements are used. For example, if a similar mirror were placed to the left of the mirror in Fig. A–4, the reflected light from the first mirror could be considered a negative source for the second, since the rays are diverging as they are incident on the second.

These sign conventions permit us to locate the image of a source on the axis with any combination of spherical mirrors for paraxial rays. For example, if we were to consider a real source for a convex mirror, Eq. (A–8) would give for the image distance

$$\frac{1}{v} = \frac{1}{f} - \frac{1}{u}.$$

Since f is a negative for a convex mirror and u is positive, v must be negative. Figure A–5 shows the geometrical construction for this case, and bears out the result. Note that, since we have proved that all rays pass through the image point I, we can analyze the problem graphically by considering the intersection of only two rays. In fact, in Fig. A–5, the ray SN is superfluous, since the ray SO is reflected back on itself and, when extended, passes through I.

III. Spherical transmitting surfaces

A spherical interface between materials of different indexes of refraction can also be used for changing the shape of a wave front incident upon it. The altered wave front is transmitted in this case, instead of being reflected. For paraxial rays, all the incident energy will again pass through a single point or will diverge as if it originated at a single point.

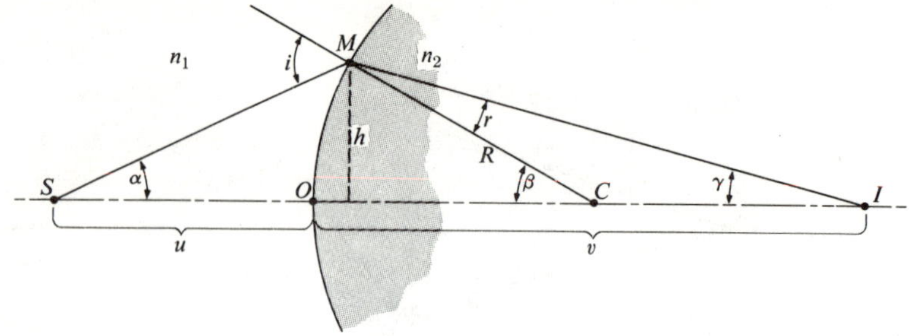

FIG. A–6. Waves incident on a spherical surface between media of different indexes of refraction.

The analysis proceeds in a manner very similar to the one used for mirrors. We consider a source S at a distance u from the mirror on the axis of the spherical surface (Fig. A–6). The only difference between this case and that of the mirror is that the ray is transmitted. The angle of incidence and angle of refraction are related by Snell's law:

$$\frac{\sin i}{\sin r} = \frac{n_2}{n_1}. \tag{A–16}$$

Since we are dealing only with small angles, (A–16) can be replaced by

$$\frac{i}{r} = \frac{n_2}{n_1}. \tag{A–17}$$

The angles are related by

$$\beta = \gamma + r, \tag{A–18}$$

$$i = \alpha + \beta. \tag{A–19}$$

We can eliminate i and r from (A–18) by using (A–17) and (A–19) to obtain

$$\frac{\beta(n_2 - n_1)}{n_2} = \frac{n_1}{n_2}\alpha + \gamma,$$

and finally

$$\beta(n_2 - n_1) = n_1\alpha + n_2\gamma. \tag{A–20}$$

Again we employ the approximations

$$\begin{aligned} h/u &= \tan\alpha \approx \sin\alpha \approx \alpha, \\ h/R &= \tan\beta \approx \sin\beta \approx \beta, \\ h/v &= \tan\gamma \approx \sin\gamma \approx \gamma, \end{aligned} \tag{A–21}$$

and obtain

$$\frac{n_2 - n_1}{R} = \frac{n_1}{u} + \frac{n_2}{v}. \tag{A–22}$$

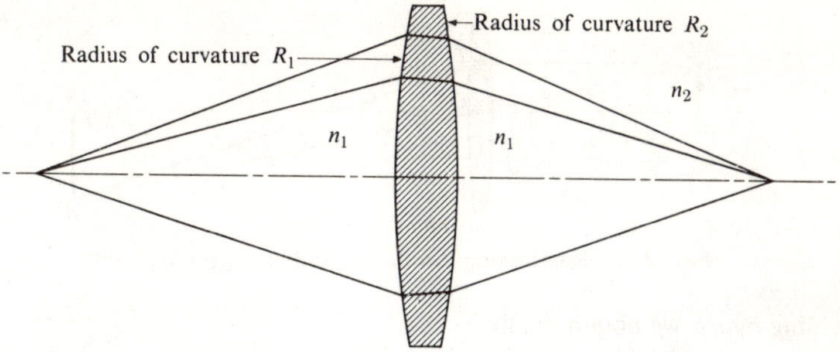

FIG. A–7. A thin lens.

Equation (A–22) for the transmission of waves through a spherical interface is the analog of the mirror equation (A–8). The sign conventions for refraction are similar to the sign conventions for mirrors:

(1) When a convex surface is presented to the incident rays, R is taken as positive; when a concave surface is presented, R is taken as negative.

(2) A positive v corresponds to a real image; a negative v corresponds to a virtual image.

(3) Diverging rays incident on a surface are described by a positive u; converging rays incident on a surface are described by a negative u.

The basic equation, Eq. (A–22), enables us to trace the path of rays through any combination of spherical surfaces. We shall illustrate the use of this formula by applying it to the case of a double convex thin lens (Fig. A–7). According to Eq. (A–22), the image distance v' is given by

$$\frac{n_1}{u} + \frac{n_2}{v'} = \frac{n_2 - n_1}{R_1}. \tag{A–23}$$

Before the rays can converge to the image, they are intercepted by the second spherical surface. If the lens is thin, the object distance for the second surface is equal to the image distance for the first. According to our sign convention the object distance is negative, since the rays are converging toward the second surface. Application of (A–22) gives us

$$-\frac{n_2}{v'} + \frac{n_1}{v} = \frac{n_1 - n_2}{R_2}. \tag{A–24}$$

The first term corresponds to n_1/u in (A–22), with $u = -v'$ and n_2 replacing n_1. The reason for this is that the rays are incident from the medium of index n_2 for the second surface. For the same reason n_1 and n_2 are interchanged in the rest of (A–24) compared with (A–22). Adding Eqs. (A–23) and (A–24), we have

$$\frac{n_1}{u} + \frac{n_1}{v} = (n_2 - n_1)\left(\frac{1}{R_1} - \frac{1}{R_2}\right) = \frac{n_1}{f}. \tag{A–25}$$

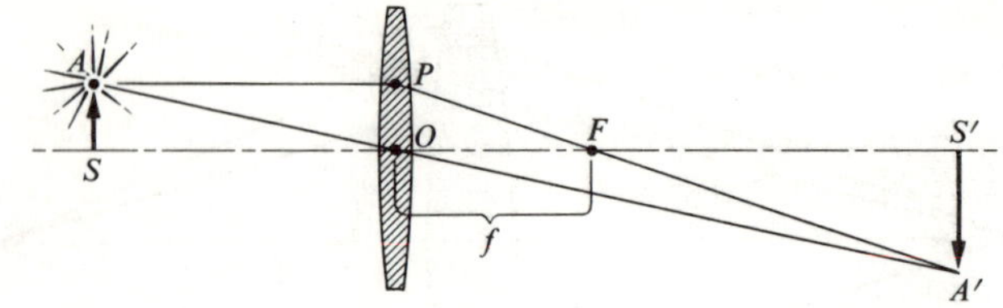

FIG. A–8. Ray tracing through a double convex thin lens.

Dividing by n_1, we obtain finally

$$\frac{1}{u} + \frac{1}{v} = \left(\frac{n_2}{n_1} - 1\right)\left(\frac{1}{R_1} - \frac{1}{R_2}\right) = \frac{1}{f}. \qquad \text{(A–26)}$$

For a double convex lens, R_2 is negative, according to our sign convention. The focal distance is given by

$$\frac{1}{f} = \left(\frac{n_2}{n_1} - 1\right)\left(\frac{1}{R_1} - \frac{1}{R_2}\right); \qquad \text{(A–27)}$$

it is the distance from the lens at which a plane wave will come to a focus ($u = \infty$), and can be computed from a knowledge of the geometrical properties of the lens and its index of refraction.

The basic mirror and thin-lens equations, (A–8) and (A–26), enable us to find the location of the image of sources off the axis as well as on the axis. This is shown graphically for a double convex thin lens in Fig. A–8. A point off the axis A emits rays in all directions. If the rays that pass through the lens are paraxial, they will intersect at a single point, and we need trace only two of them. The two we choose are the ray parallel to the axis AP, which passes through the focus F, and the ray that goes through the center of the lens O, which passes through undeviated. Thus we get an image A' which is below the axis. The image of the arrow SA is an inverted arrow $S'A'$. Other configurations of lenses and mirrors are treated in analogous ways.

Geometrical optics and its application to the design of optical instruments is a well-developed, sophisticated mathematical discipline, and we have scarcely scratched the surface of it here. It is possible, by using suitable combinations of lenses, to eliminate the aberrations that arise from nonparaxial rays as well as many other aberrations which we have not mentioned, and the behavior of lenses which are not thin can be analyzed. The chief limiting element in the design of an optical system is the fact that light is a wave phenomenon and that geometrical optics is an approximate description of the propagation of light. Thus a point source of light can never really produce a point image: Diffraction ultimately causes the image of a point source to have a finite size of the order of λ/D, where λ is the wavelength and D the diameter of the lens. This is the ultimate limitation in the design of optical instruments, and also the ultimate limitation on distinguishing two nearby light sources from one another.

Problems

1. A concave mirror has a radius of curvature of 0.6 m. What is the focal length? Where will an image of an object 1 m, $\frac{1}{2}$ m, $\frac{1}{4}$ m from the mirror be?

2. Repeat Problem 1 for a convex mirror.

3. A glass hemisphere has a diameter of 20 cm and a refractive index of 1.5. It is silvered on its spherical surface. An object is 20 cm from the plane surface on a perpendicular to the plane passing through the center of the sphere. Where will the ultimate image of this object be?

4. Find the focal length of a double convex lens used in air; its index of refraction is 1.48. The radii of curvature of the faces are 1 m and $\frac{1}{2}$ m. Repeat for double concave lens; for convex surface $R_1 = 1$ m, concave surface $R_2 = \frac{1}{2}$ m; and for $R_1 = \frac{1}{2}$ m (convex), $R_2 = 1$ m (concave).

5. Are the optical properties of mirrors changed by using them in media of different indexes of refraction?

6. An object 2 cm high is placed 20 cm from a converging lens whose f is 25 cm. Determine the position of the image graphically, and show whether it is inverted or erect, real or virtual.

7. A lens whose index is 1.60 has a focal length of 30 cm in air. What is its focal length in water? ($n = 1.33$.)

8. Repeat Problem 6 for a diverging lens (f is negative).

9. A concave spherical mirror has a radius of curvature of 20 cm. Find the position of the image of an object placed 5 cm from the mirror.

10. Sketch u as a function of v for a converging lens, a diverging lens, a concave mirror, and a convex mirror.

11. If U is the distance of the object from the focal point to the left of a lens and V the distance from the focal point to the right of the lens, show that

$$UV = f^2.$$

12. A converging lens whose focal length is 9 cm is placed between a source of light and a screen which are 100 cm apart. Where should the lens be placed in order that the image be focused on the screen?

Appendix II

CONVERSION FACTORS

1 meter (m) = 100 cm = 39.4 in. = 3.28 ft

1 centimeter (cm) = 10 millimeters (mm) = 0.394 in.

1 kilometer (km) = 1000 m = 0.621 mi

1 foot (ft) = 12 in. = 0.305 m

1 inch (in.) = 0.0833 ft = 2.54 cm

1 mile (mi) = 5280 ft = 1.61 km

1 m/sec = 3.28 ft/sec = 2.24 mi/hr = 3.60 km/hr

1 ft/sec = 0.305 m/sec = 0.682 mi/hr = 1.10 km/hr
(*Note:* it is often convenient to remember that 88 ft/sec = 60 mi/hr.)

1 mi/hr = 1.47 ft/sec = 0.447 m/sec = 1.61 km/hr

1 radian (rad) = 57.30° = 57°18′

1° = 0.01745 rad

1 revolution/minute (rev/min) = 0.1047 rad/sec

1 kilogram (kg) = 1000 grams (gm) = 0.0685 slug
(*Note:* 1 kg corresponds to 2.21 lb in the sense that the *weight* of 1 kg is 2.21 lb.)

1 slug = 14.6 kg
(*Note:* 1 slug corresponds to 32.2 lb in the sense that the *weight* of 1 slug is 32.2 lb.)

1 atomic mass unit (amu) = 1.66×10^{-27} kg = 1.49×10^{-10} j = 931 Mev

1 newton (n) = 0.225 lb

1 pound (lb) = 4.45 n

1 joule (j) = 0.738 ft·lb = 2.39×10^{-4} kcal = 6.24×10^{18} ev

1 kilocalorie (kcal) = 4,185 j

1 foot·pound (ft·lb) = 1.36 j

1 electron volt (ev) = 10^{-6} Mev = 1.60×10^{-19} j = 1.18×10^{-19} ft·lb
= 3.83×10^{-23} kcal

1 watt = 1 j/sec = 0.738 ft·lb/sec

$\log_e x = 2.3026 \log_{10} x$

Appendix III

USEFUL PHYSICAL CONSTANTS

Quantity	Symbol	Value
Gravitational constant	G	6.67×10^{-11} n·m²/kg²
Acceleration of gravity at earth's surface	g	9.81 m/sec² $= 32.2$ ft/sec²
Atmospheric pressure at sea level	atm	14.7 lb/in² $= 1.01 \times 10^{5}$ n/m²
Absolute zero	0°K	-273.15°C
Boltzmann's constant	k	1.38×10^{-23} j/°K
Avogadro's number	N_0	6.03×10^{26} molecules/kilomole
Universal gas constant	R	8.32 j/mole·°K
Permittivity of free space	ϵ_0	8.85×10^{-12} coul/n·m²
Permeability of free space	μ_0	1.26×10^{-6} weber/amp·m
Electronic charge	e	1.60×10^{-19} coul
Electron rest mass	m_e	9.11×10^{-31} kg
Proton rest mass	m_p	1.67×10^{-27} kg
Neutron rest mass	m_n	1.67×10^{-27} kg
Speed of light	c	3.00×10^{8} m/sec
Planck's constant	h	6.63×10^{-34} j·sec

Solar system constants

Mass, 10^{24} kg	Diameter, 10^6 m	Period of orbit, days	Radius of orbit, 10^9 m
Sun 1.96×10^{6}	1390	——	——
Earth 5.98	12.76	365.3	149
Moon 0.0735	3.48	27.3	0.38

Appendix IV

THE ELEMENTS

Atomic number	Element	Symbol	Atomic weight*	Atomic number	Element	Symbol	Atomic weight*
1	Hydrogen	H	1.008	52	Tellurium	Te	127.6
2	Helium	He	4.003	53	Iodine	I	126.9
3	Lithium	Li	6.939	54	Xenon	Xe	131.3
4	Beryllium	Be	9.012	55	Cesium	Cs	132.9
5	Boron	B	10.81	56	Barium	Ba	137.3
6	Carbon	C	12.01	57	Lanthanum	La	138.9
7	Nitrogen	N	14.01	58	Cerium	Ce	140.1
8	Oxygen	O	16.00	59	Praseodymium	Pr	140.9
9	Fluorine	F	19.00	60	Neodymium	Nd	144.2
10	Neon	Ne	20.18	61	Promethium	Pm	(147)
11	Sodium	Na	22.99	62	Samarium	Sm	150.4
12	Magnesium	Mg	24.31	63	Europium	Eu	152.0
13	Aluminum	Al	26.98	64	Gadolinium	Gd	157.3
14	Silicon	Si	28.09	65	Terbium	Tb	158.9
15	Phosphorus	P	30.97	66	Dysprosium	Dy	162.5
16	Sulfur	S	32.06	67	Holmium	Ho	164.9
17	Chlorine	Cl	35.45	68	Erbium	Er	167.3
18	Argon	A	39.95	69	Thulium	Tm	168.9
19	Potassium	K	39.10	70	Ytterbium	Yb	173.0
20	Calcium	Ca	40.08	71	Lutetium	Lu	175.0
21	Scandium	Sc	44.96	72	Hafnium	Hf	178.5
22	Titanium	Ti	47.90	73	Tantalum	Ta	180.9
23	Vanadium	V	50.94	74	Tungsten	W	183.9

24	Chromium	Cr	52.00	75	Rhenium	Re	186.2
25	Manganese	Mn	54.94	76	Osmium	Os	190.2
26	Iron	Fe	55.85	77	Iridium	Ir	192.2
27	Cobalt	Co	58.93	78	Platinum	Pt	195.1
28	Nickel	Ni	58.71	79	Gold	Au	197.0
29	Copper	Cu	63.54	80	Mercury	Hg	200.6
30	Zinc	Zn	65.37	81	Thallium	Tl	204.4
31	Gallium	Ga	69.72	82	Lead	Pb	207.2
32	Germanium	Ge	72.59	83	Bismuth	Bi	209.0
33	Arsenic	As	74.92	84	Polonium	Po	210
34	Selenium	Se	78.96	85	Astatine	At	(210)
35	Bromine	Br	79.91	86	Radon	Rn	222
36	Krypton	Kr	83.80	87	Francium	Fr	(223)
37	Rubidium	Rb	85.47	88	Radium	Ra	226.1
38	Strontium	Sr	87.62	89	Actinium	Ac	227
39	Yttrium	Y	88.91	90	Thorium	Th	232.0
40	Zirconium	Zr	91.22	91	Protactinium	Pa	231
41	Niobium	Nb	92.91	92	Uranium	U	238.0
42	Molybdenum	Mo	95.94	93	Neptunium	Np	(237)
43	Technetium	Tc	(99)	94	Plutonium	Pu	(242)
44	Ruthenium	Ru	101.1	95	Americium	Am	(243)
45	Rhodium	Rh	102.9	96	Curium	Cm	(245)
46	Palladium	Pd	106.4	97	Berkelium	Bk	(249)
47	Silver	Ag	107.9	98	Californium	Cf	(249)
48	Cadmium	Cd	112.4	99	Einsteinium	Es	(253)
49	Indium	In	114.8	100	Fermium	Fm	(255)
50	Tin	Sn	118.7	101	Mendelevium	Md	(256)
51	Antimony	Sb	121.8	102	Nobelium	No	(253)
				103	Lawrencium	Lw	(257)

*Those in parentheses refer to the most stable isotope of the element.

Appendix V

COMMON LOGARITHMS

N	0	1	2	3	4	5	6	7	8	9
0	....	0000	3010	4771	6021	6990	7782	8451	9031	9542
1	0000	0414	0792	1139	1461	1761	2041	2304	2553	2788
2	3010	3222	3424	3617	3802	3979	4150	4314	4472	4624
3	4771	4914	5051	5185	5315	5441	5563	5682	5798	5911
4	6021	6128	6232	6335	6435	6532	6628	6721	6812	6902
5	6990	7076	7160	7243	7324	7404	7482	7559	7634	7709
6	7782	7853	7924	7993	8062	8129	8195	8261	8325	8388
7	8451	8513	8573	8633	8692	8751	8808	8865	8921	8976
8	9031	9085	9138	9191	9243	9294	9345	9395	9445	9494
9	9542	9590	9638	9685	9731	9777	9823	9868	9912	9956
10	0000	0043	0086	0128	0170	0212	0253	0294	0334	0374
11	0414	0453	0492	0531	0569	0607	0645	0682	0719	0755
12	0792	0828	0864	0899	0934	0969	1004	1038	1072	1106
13	1139	1173	1206	1239	1271	1303	1335	1367	1399	1430
14	1461	1492	1523	1553	1584	1614	1644	1673	1703	1732
15	1761	1790	1818	1847	1875	1903	1931	1959	1987	2014
16	2041	2068	2095	2122	2148	2175	2201	2227	2253	2279
17	2304	2330	2355	2380	2405	2430	2455	2480	2504	2529
18	2553	2577	2601	2625	2648	2672	2695	2718	2742	2765
19	2788	2810	2833	2856	2878	2900	2923	2945	2967	2989
20	3010	3032	3054	3075	3096	3118	3139	3160	3181	3201
21	3222	3243	3263	3284	3304	3324	3345	3365	3385	3404
22	3424	3444	3464	3483	3502	3522	3541	3560	3579	3598
23	3617	3636	3655	3674	3692	3711	3729	3747	3766	3784
24	3802	3820	3838	3856	3874	3892	3909	3927	3945	3962
25	3979	3997	4014	4031	4048	4065	4082	4099	4116	4133
26	4150	4166	4183	4200	4216	4232	4249	4265	4281	4298
27	4314	4330	4346	4362	4378	4393	4409	4425	4440	4456
28	4472	4487	4502	4518	4533	4548	4564	4579	4594	4609
29	4624	4639	4654	4669	4683	4698	4713	4728	4742	4757
30	4771	4786	4800	4814	4829	4843	4857	4871	4886	4900
31	4914	4928	4942	4955	4969	4983	4997	5011	5024	5038
32	5051	5065	5079	5092	5105	5119	5132	5145	5159	5172
33	5185	5198	5211	5224	5237	5250	5263	5276	5289	5302
34	5315	5328	5340	5353	5366	5378	5391	5403	5416	5428
35	5441	5453	5465	5478	5490	5502	5514	5527	5539	5551
36	5563	5575	5587	5599	5611	5623	5635	5647	5658	5670
37	5682	5694	5705	5717	5729	5740	5752	5763	5775	5786
38	5798	5809	5821	5832	5843	5855	5866	5877	5888	5899
39	5911	5922	5933	5944	5955	5966	5977	5988	5999	6010
40	6021	6031	6042	6053	6064	6075	6085	6096	6107	6117
41	6128	6138	6149	6160	6170	6180	6191	6201	6212	6222
42	6232	6243	6253	6263	6274	6284	6294	6304	6314	6325
43	6335	6345	6355	6365	6375	6385	6395	6405	6415	6425
44	6435	6444	6454	6464	6474	6484	6493	6503	6513	6522
45	6532	6542	6551	6561	6571	6580	6590	6599	6609	6618
46	6628	6637	6646	6656	6665	6675	6684	6693	6702	6712
47	6721	6730	6739	6749	6758	6767	6776	6785	6794	6803
48	6812	6821	6830	6839	6848	6857	6866	6875	6884	6893
49	6902	6911	6920	6928	6937	6946	6955	6964	6972	6981
50	6990	6998	7007	7016	7024	7033	7042	7050	7059	7067
N	0	1	2	3	4	5	6	7	8	9

Appendix V. Common Logarithms (*Cont.*)

N	0	1	2	3	4	5	6	7	8	9
50	6990	6998	7007	7016	7024	7033	7042	7050	7059	7067
51	7076	7084	7093	7101	7110	7118	7126	7135	7143	7152
52	7160	7168	7177	7185	7193	7202	7210	7218	7226	7235
53	7243	7251	7259	7267	7275	7284	7292	7300	7308	7316
54	7324	7332	7340	7348	7356	7364	7372	7380	7388	7396
55	7404	7412	7419	7427	7435	7443	7451	7459	7466	7474
56	7482	7490	7497	7505	7513	7520	7528	7536	7543	7551
57	7559	7566	7574	7582	7589	7597	7604	7612	7619	7627
58	7634	7642	7649	7657	7664	7672	7679	7686	7694	7701
59	7709	7716	7723	7731	7738	7745	7752	7760	7767	7774
60	7782	7789	7796	7803	7810	7818	7825	7832	7839	7846
61	7853	7860	7868	7875	7882	7889	7896	7903	7910	7917
62	7924	7931	7938	7945	7952	7959	7966	7973	7980	7987
63	7993	8000	8007	8014	8021	8028	8035	8041	8048	8055
64	8062	8069	8075	8082	8089	8096	8102	8109	8116	8122
65	8129	8136	8142	8149	8156	8162	8169	8176	8182	8189
66	8195	8202	8209	8215	8222	8228	8235	8241	8248	8254
67	8261	8267	8274	8280	8287	8293	8299	8306	8312	8319
68	8325	8331	8338	8344	8351	8357	8363	8370	8376	8382
69	8388	8395	8401	8407	8414	8420	8426	8432	8439	8445
70	8451	8457	8463	8470	8476	8482	8488	8494	8500	8506
71	8513	8519	8525	8531	8537	8543	8549	8555	8561	8567
72	8573	8579	8585	8591	8597	8603	8609	8615	8621	8627
73	8633	8639	8645	8651	8657	8663	8669	8675	8681	8686
74	8692	8698	8704	8710	8716	8722	8727	8733	8739	8745
75	8751	8756	8762	8768	8774	8779	8785	8791	8797	8802
76	8808	8814	8820	8825	8831	8837	8842	8848	8854	8859
77	8865	8871	8876	8882	8887	8893	8899	8904	8910	8915
78	8921	8927	8932	8938	8943	8949	8954	8960	8965	8971
79	8976	8982	8987	8993	8998	9004	9009	9015	9020	9025
80	9031	9036	9042	9047	9053	9058	9063	9069	9074	9079
81	9085	9090	9096	9101	9106	9112	9117	9122	9128	9133
82	9138	9143	9149	9154	9159	9165	9170	9175	9180	9186
83	9191	9196	9201	9206	9212	9217	9222	9227	9232	9238
84	9243	9248	9253	9258	9263	9269	9274	9279	9284	9289
85	9294	9299	9304	9309	9315	9320	9325	9330	9335	9340
86	9345	9350	9355	9360	9365	9370	9375	9380	9385	9390
87	9395	9400	9405	9410	9415	9420	9425	9430	9435	9440
88	9445	9450	9455	9460	9465	9469	9474	9479	9484	9489
89	9494	9499	9504	9509	9513	9518	9523	9528	9533	9538
90	9542	9547	9552	9557	9562	9566	9571	9576	9581	9586
91	9590	9595	9600	9605	9609	9614	9619	9624	9628	9633
92	9638	9643	9647	9652	9657	9661	9666	9671	9675	9680
93	9685	8689	9694	9699	9703	9708	9713	9717	9722	9727
94	9731	9736	9741	9745	9750	9754	9759	9763	9768	9773
95	9777	9782	9786	9791	9795	9800	9805	9809	9814	9818
96	9823	9827	9832	9836	9841	9845	9850	9854	9859	9863
97	9868	9872	9877	9881	9886	9890	9894	9899	9903	9908
98	9912	9917	9921	9926	9930	9934	9939	9943	9948	9952
99	9956	9961	9965	9969	9974	9978	9983	9987	9991	9996
100	0000	0004	0009	0013	0017	0022	0026	0030	0035	0039
N	0	1	2	3	4	5	6	7	8	9

Appendix VI

NATURAL TRIGONOMETRIC FUNCTIONS

Angle					Angle				
Degree	Radian	Sine	Cosine	Tangent	Degree	Radian	Sine	Cosine	Tangent
0°	.000	0.000	1.000	0.000					
1°	.017	.018	1.000	.018	46°	0.803	0.719	0.695	1.036
2°	.035	.035	0.999	.035	47°	.820	.731	.682	1.072
3°	.052	.052	.999	.052	48°	.838	.743	.669	1.111
4°	.070	.070	.998	.070	49°	.855	.755	.656	1.150
5°	.087	.087	.996	.088	50°	.873	.766	.643	1.192
6°	.105	.105	.995	.105	51°	.890	.777	.629	1.235
7°	.122	.122	.993	.123	52°	.908	.788	.616	1.280
8°	.140	.139	.990	.141	53°	.925	.799	.602	1.327
9°	.157	.156	.988	.158	54°	.942	.809	.588	1.376
10°	.175	.174	.985	.176	55°	.960	.819	.574	1.428
11°	.192	.191	.982	.194	56°	.977	.829	.559	1.483
12°	.209	.208	.978	.213	57°	.995	.839	.545	1.540
13°	.227	.225	.974	.231	58°	1.012	.848	.530	1.600
14°	.244	.242	.970	.249	59°	1.030	.857	.515	1.664
15°	.262	.259	.966	.268	60°	1.047	.866	.500	1.732
16°	.279	.276	.961	.287	61°	1.065	.875	.485	1.804
17°	.297	.292	.956	.306	62°	1.082	.883	.470	1.881
18°	.314	.309	.951	.325	63°	1.100	.891	.454	1.963
19°	.332	.326	.946	.344	64°	1.117	.899	.438	2.050
20°	.349	.342	.940	.364	65°	1.134	.906	.423	2.145
21°	.367	.358	.934	.384	66°	1.152	.914	.407	2.246
22°	.384	.375	.927	.404	67°	1.169	.921	.391	2.356
23°	.401	.391	.921	.425	68°	1.187	.927	.375	2.475
24°	.419	.407	.914	.445	69°	1.204	.934	.358	2.605
25°	.436	.423	.906	.466	70°	1.222	.940	.342	2.747
26°	.454	.438	.899	.488	71°	1.239	.946	.326	2.904
27°	.471	.454	.891	.510	72°	1.257	.951	.309	3.078
28°	.489	.470	.883	.532	73°	1.274	.956	.292	3.271
29°	.506	.485	.875	.554	74°	1.292	.961	.276	3.487
30°	.524	.500	.866	.577	75°	1.309	.966	.259	3.732
31°	.541	.515	.857	.601	76°	1.326	.970	.242	4.011
32°	.559	.530	.848	.625	77°	1.344	.974	.225	4.331
33°	.576	.545	.839	.649	78°	1.361	.978	.208	4.705
34°	.593	.559	.829	.675	79°	1.379	.982	.191	5.145
35°	.611	.574	.819	.700	80°	1.396	.985	.174	5.671
36°	.628	.588	.809	.727	81°	1.414	.988	.156	6.314
37°	.646	.602	.799	.754	82°	1.431	.990	.139	7.115
38°	.663	.616	.788	.781	83°	1.449	.993	.122	8.144
39°	.681	.629	.777	.810	84°	1.466	.995	.105	9.514
40°	.698	.643	.766	.839	85°	1.484	.996	.087	11.43
41°	.716	.658	.755	.869	86°	1.501	.998	.070	14.30
42°	.733	.669	.743	.900	87°	1.518	.999	.052	19.08
43°	.751	.682	.731	.933	88°	1.536	.999	.035	28.64
44°	.768	.695	.719	.966	89°	1.553	1.000	.018	57.29
45°	.785	.707	.707	1.000	90°	1.571	1.000	.000	∞

Appendix VII

GREEK ALPHABET

Greek Alphabet

A	α	Alpha	N	ν	Nu
B	β	Beta	Ξ	ξ	Xi
Γ	γ	Gamma	O	o	Omicron
Δ	δ	Delta	Π	π	Pi
E	ϵ	Epsilon	P	ρ	Rho
Z	ζ	Zeta	Σ	σ	Sigma
H	η	Eta	T	τ	Tau
Θ	θ	Theta	Υ	υ	Upsilon
I	ι	Iota	Φ	ϕ	Phi
K	κ	Kappa	X	χ	Chi
Λ	λ	Lambda	Ψ	ψ	Psi
M	μ	Mu	Ω	ω	Omega

Answers to Odd-Numbered Problems

Answers to Odd-Numbered Problems

Chapter 1

1. 6.46×10^9; 1×10^6; 3.516×10^5; 8.4×10^3; 7×10^1; 3.81×10^0; 1.4×10^{-1}; 7.890×10^{-3}; 1.3×10^{-5}; 7.819×10^{-9}

3. 451,000,000; 200,000; 781.9; 5.1; 0.08; 0.000,095,6; 0.000,000,010,03; 0.000,000,000,1

Chapter 2

1. 94.3 mi

3. (a) 8.2 mi/hr, $\theta = 14° 2'$ with horizontal, downstream (b) $11\frac{1}{4}$ min (c) $\frac{3}{8}$ mi

5. 16.7 m, $\alpha = 68° 59'$, $\beta = 53° 17'$, $\gamma = 44° 10'$; α, β, and γ are the angles the resultant makes with the northerly, westerly, and upward directions, respectively.

7. 5.2, 75° south of west

9. 27.6 n, 44° clockwise from the reference direction **11.** 18.8 lb

13. x-component = 9.2 m/sec, y-component = 7.7 m/sec **15.** 41.8°

17. **A** = **B** if cos (**A**, **C**) = cos (**B**, **C**)

21. It is the volume of the parallelepiped whose sides are the vectors **A**, **B**, and **C**.

25. 144

Chapter 3

1. 48 mi/hr **5.** 272.8 mi/hr **7.** 2.25 ft/sec **9.** $(2d/c)/[1 - (v^2/c^2)]$

11. $\theta = 78°41'$ **13.** (a) 2.5 (mi/hr)/sec (b) 3.67 ft/sec^2 **15.** 32 sec

17. 397.2 ft **19.** 2 sec **21.** 64 ft/sec **23.** (a) 2.66 sec (b) 65.12 ft/sec

25. 80 ft/sec, upward **27.** $16(2n - 1)$ **29.** (a) $\frac{1}{9}$ sec, 1 sec (b) $\frac{5}{9}$ sec

31. (a) $v = t^2 + 1$, $x = (t^3/3) + t$ (b) $v = 5/(10t + 1)$, $x = \frac{1}{2}\log(10t + 1) + 1$ (c) $v = v_0 \exp(-kt)$, $x = (v_0/k)[1 - \exp(-kt)]$

33. Acceleration = $ax + b/2$ **35.** 3.05 mi **37.** 62.5 ft/sec

39. $[g(y + \sqrt{x^2 + y^2})]^{1/2}$

41. (a) $y = x - x^2/2$ (b) $y = 1 - 2x^2$ (c) $x^2/A^2 + y^2/B^2 = 1$

43. Average speed = 1 ft/min, average velocity = $\frac{1}{5}$ ft/min

Chapter 4

1. Hour hand, $\pi/21{,}600$ rad/sec; minute hand, $\pi/1800$ rad/sec; second hand, $\pi/30$ rad/sec.

3. (a) $\omega = 40\pi$ rad/sec (b) 5π in/sec

5. 300π in/sec **7.** $\pi/3$ rad/sec **9.** 72×10^5 m/sec

11. (a) $\pi/36 \times 10^{-2}$ m/sec (b) $(\pi^2/648) \times 10^{-4}$ m/sec^2

13. π rad/sec^2; 500 rev **15.** (a) 10 rad/sec^2 (b) 100 rad/sec

17. 1.05×10^3 mi/hr

19. 10.2 rad/sec at an angle of 11° 19′ with the long axis of the football.

21. 8.95×10^3 ft/sec

23. (a) 8 rad/sec^2 (b) 10 rad/sec^2 (c) 10 ft/sec^2

25. (a) $\theta = \omega R/v$ (b) $t = R/v$ (c) $\theta = \arcsin \omega R/v$, $t = R/\sqrt{v^2 - \omega^2 R^2}$ (d) $\omega_c = v/R$

Chapter 5

1. (a) 39.2 n (b) 5 m/sec^2 **3.** (a) 0.82 kg (b) 24.5 m/sec^2

5. 13,200 n **7.** $5\frac{1}{3}$ lb

9. (a) 218.5 lb (b) 181.5 lb (c)–(d) 200 lb (e) 0 **11.** 72 lb

13. (a) 4.9 m/sec^2 (b) 3.92 m/sec **15.** (a) 4.33 sec (b) 9.27 ft/sec

17. 30 lb, to retard motion of crate. **19.** 1.23×10^3 n **21.** 13.4 lb

23. This problem is overdetermined. One cannot specify both the angular velocity and the radius. For the radius given $T = 11.32$ n and $\omega = 4.76$ rad/sec.

25. 384 n **27.** (a) 41.5 ft/sec (b) 1.7 rev/sec **29.** 8 ft/sec

31. He must introduce a force $-mg$ in his coordinate system.

33. Earth exerts force W = weight of book on it. Book exerts force W on table. Table exerts force $-W$ on book. Latter two forces are related as action and reaction. Forces holding book in equilibrium are not related by action and reaction.

35. 1.15 mi **37.** 1.4 hr

39. (a) Plumb line will be deviated south. (b) Ball will fall east of plumb line.

Chapter 6

1. (a) 1040 j (b) 1040 j **3.** 13.6 m **5.** 0

7. 7.5×10^4 j; 20.9 w

9. 1.225 j **11.** 13.6 hp

13. (a) $45k$ (b) $45k$ (c) Yes

15. $(k/4)(a^4 - b^4)$ **17.** 19.5 j

Chapter 7

1. 1.7×10^5 ft-lb **3.** 7.1 m/sec

5. (a) 900 ft-lb (b) 800 ft-lb (c) Work was dissipated in friction.

7. (a) 2000 ft-lb (b) 1600 ft-lb (c) 400 ft-lb
9. At middle of swing; PE maximum at end of swing. **11.** $781\frac{1}{4}$ ft
13. 520 ft-lb **19.** (a) $kr^2/2$ (b) $G \ln r$ (c) $(1/a)e^{-ar}$ (d) $-k/6r^6$

Chapter 8

1. If mass 1 is at origin, mass 2 at $(\sqrt{\frac{3}{2}}, \frac{1}{2})$, mass 3 at $(\sqrt{\frac{3}{2}}, -\frac{1}{2})$, then $R_x = \frac{5}{12}\sqrt{3}$, $R_y = -\frac{1}{12}$ **3.** A distance $3R/8$ toward the pole of the hemisphere
5. $\frac{1}{2}M(R_1^2 + R_2^2)$ **7.** $Ma^2/12$
9. Weight must act along plumb line if total torque about point of support is to be zero.
11. $\sqrt{3g/l}$ **13.** 155.6 n **15.** $3\frac{3}{7}$ ft from center, toward 30-lb child.
17. 3.27 m/sec^2
19. Spool will wind up on thread. Try it!
23. (a) 11.6 rad/sec (b) 5.8 m/sec (c) 50.3 j (d) 100.6 j
25. (a) 4.5 sec (b) 450 j **27.** Solid cylinder **29.** 3.3°

Chapter 9

1. He moves backward at $3\frac{1}{3}$ mi/hr **3.** 2.14 ft/sec
5. (a) $1\frac{1}{3}$ m/sec in same direction (b) 2 j **7.** $\frac{1}{3} \times 10^5$ m/sec
9. 0.82 ft/sec **13.** $3\frac{1}{3}$ ft **15.** 20 m/sec
21. (a) 10 rev/sec (b) 2250 ft-lb initial, 7500 ft-lb final (c) Skater does work changing his moment of inertia.
23. (a) 3.47 rev/sec (b) 30.6 j **25.** $5g/2R\omega$

Chapter 10

1. 5.2 ft/sec **3.** 0.36 m **5.** $T = 1.8$ sec, $f = 0.55$ cycles/sec
7. (a) 24.6 ft-lb (b) 494 ft/sec^2
9. (a) 250 m/sec^2 (b) 1.25 n (c) 125 m/sec^2 (d) 0.625 n **11.** 4.96 sec
13. (a) 1.11 sec (b) 1.11 sec (c) 1.27 sec **15.** 32.3 ft/sec^2
19. $(2k_2 + k_1)l/(k_2 + k_1)$ from end of k_1 spring; $f = (1/2\pi)\sqrt{(k_1 + k_2)/m}$
21. $v = 2\pi(m/(k_1 + k_2))^{1/2}$ **23.** (a) $\frac{1}{2}$/sec (b) -0.01 m
25. $y = 1.84A \cos(\omega t + 22°)$

Chapter 11

1. (a) 40 lb (b) 5 slugs **3.** Because of centrifugal force.
5. The same, by Newton's third law.
7. (a) 0.42×10^{24} n (b) 1.31×10^4 m/sec **9.** 25.9 m/sec^2
11. (a) Force is 597 n in both places. (b) 32.49, 31.47 (c) Tides depend on difference between force on top and bottom of ocean; greater difference for moon.
13. No. Force acting on particle would be along resultant, not along either line of force.

15. $\frac{4\pi}{3} R^3 G M \rho \left[\frac{1}{d^2} - \frac{1}{8}\frac{1}{(d - R/2)^2}\right]$ 17. 1.1×10^4 m/sec^2

Chapter 12

1. 5000 lb/in 3. 7.8×10^{-2} m 5. 0.29 mm
7. (a) 1250 lb (b) 11,250 lb 9. 3.9×10^{-3} in 11. $\frac{1}{2}$ lb/in^2
13. 1500 atm 15. $777 17. $f = (\frac{1}{2}\pi)(\rho_{water}g/\rho a)^{1/2}$
19. 0.1 kg/m^3 greater than at top. 21. 0.26 ft 23. 90% 25. 882 n
27. 0.206 lb
29. Rapid velocity of air behind filter will decrease pressure and cause filter to be pushed against sides.
31. $2A\rho gh$ 33. $2\sqrt{y(H - y)}$
35. (a) 4.43 m/sec (b) 9.8×10^3 n/m^2 37. 23.2 lb/min

Chapter 13

1. 626°F, 2138°F 3. −40° 5. −80°C 7. 138 ft
9. 7.8×10^{-6}/°C 11. 1.25 cm^3 13. $0.10/qt 15. 27.105 m
17. 630 cm^3 19. 28 lb/in^2 gauge
21. Pressure decreases as water cools and condenses. 23. 3.16×10^4 kcal/hr
25. 2.59×10^5 kcal/day 27. Alcohol has smaller specific heat. 29. 20.7°C
31. 0.47°C 33. (a) 2.39×10^{-4} kcal/hr (b) 51.6 kcal (c) 1395 w/m^2
35. 3.38×10^2 m/sec 37. 620 kcal 39. 0.026 kg 41. 0.74 kg
43. Boiling point of water is higher at higher pressure. 45. Liquid

Chapter 14

1. (a) 3.2 kg (b) 6.025×10^{25} (c) 2.24 m^3 3. 43×10^{23} 5. 3.68 kg
7. (a) 5.65×10^{-21} j (b) 7.72×10^{-21} j
9. The average molecular speed increases and the change of momentum per molecule is greater when the molecule rebounds.
11. (a) Remains the same. (b) $(v_{N_2})_{av} = \sqrt{m_{H_2}/m_{N_2}} = 1/3.73 = 0.27$ mi/sec
13. At a given temperature, hydrogen molecules have a higher average speed than any other.
15. 4×10^9/sec 17. 590 m/sec
19. $\lambda = 1.45 \times 10^{-7}$ m, 6.9×10^3 atm
23. At higher pressure the collisions are more frequent, but the energy is transported a shorter distance per collision. The two effects balance out.
25. Total volume occupied by molecules is small compared with container volume. The attractive energy is small compared with kinetic energy.
27. $nRT/V = P$
29. (a) $(16/\pi)^{1/2}\,(m/2kT)^{3/2}$
(b) Av $= (4/\pi)^{1/2}(2kT/m)^{1/2}$; most probable $= \sqrt{2kT/m}$; rms $= (3)^{1/2}(2kT/m)^{1/2}$
(c) $\frac{3}{2}kT$

Chapter 15

1. $-C_V(T_2 - T_1)$ **3.** (a) Isobarically (b) Adiabatically (c) Isobaric
5. (b) $2\pi\sqrt{mV/\gamma pA^2}$ (c) 7.5 cycles/sec
7. It will move 4.7 cm toward warmer side. **9.** 562°C for monatomic gas
11. 12.55×10^{-3} kcal (ether), 3.07×10^{-3} kcal (chlorine), 1.88×10^{-3} kcal (helium)
13. 2.1×10^{-11} n

Chapter 16

1. 25.5% **3.** $\frac{1}{3} \times 10^3$ kcal **5.** *C*; engines *A* and *B* are impossible since their efficiencies are greater than that of a Carnot engine.
7. Room will heat up. More heat is rejected to room than is withdrawn from refrigerator interior.
9. 460 j
13. (a) 1 atm, 6.3 liters; 10 atm, 1 liter (b) 0.425 cal (c) 37% (d) 0.126 j
15. 99 w

Chapter 17

1. Insulating wall. Charge will leak off onto conducting wall.
3. 2.7×10^3 n toward negative charge. **5.** 5.05 m
7. $(9 \times 10^9 q^2)/a^2$ n, 60° out from side between positive and negative charge.
9. 9.6×10^{16} coul $= 6 \times 10^{36}$ electrons
11. 2.47×10^{-3} n **13.** kpQ/R^3
15. 1.7×10^2 j/coul; approx. 1.8×10^2 j/coul
19. (a) $V = 0$ (b) 43.2 j/coul, 2.16 j/coul
21. 0.123 amp, 2.40×10^{-16} amp/m^2 **23.** 3×10^{-9} n/m

Chapter 18

1. Between them, 0.083 m from charge of 1.5×10^{-6} coul
3. 36×10^3 n/coul toward the 4×10^{-6} coul charge
5. (a) 5×10^{-4} n (b) 25×10^{-4} j
7. 2.39×10^4 n/coul at angle of 47° to side; 11.95×10^{-4} n
11. $(1/4\pi\epsilon_0)(Q/R^3)r$ **13.** $(\rho/2\pi\epsilon_0 r)\sin\theta/2$
15. $E_r = 2p\cos\theta/4\pi\epsilon_0 r^3$, $E_\theta = p\sin\theta/4\pi\epsilon_0 r^3$
17. (b) $E_x = 2ax/(x^2 + y^2)$, $E_y = 2ay/(x^2 + y^2)$ (c) $E_r = 2a/r$, $E_\theta = 0$

Chapter 19

1. (a) 0° (b) 90° **3.** 16.3 amp **5.** 2.2×10^{-5} w/m^2 **7.** 10^4 amp
11. 5×10^3 m/sec **13.** (a) 6.9×10^3 m (b) 3.79 m
15. 2×10^{-9} n; attraction is toward wire.
17. $2\sqrt{2}\,\mu_0/\pi l$ **19.** 3.06×10^{-5} w/m^2
21. $(\mu_0/2)ni(\cos\theta_1 - \cos\theta_2)$ **23.** $B = \mu_0 j/2$

Chapter 20

1. (a) 288×10^3 coul (b) 3456×10^3 j 3. 180 coul
5. No. The radius of curvature of a kink or bend is large compared with a mean free path of an electron in the wire.
7. It uses energy. 9. 2.53 hr 11. (a) 100 m (b) 10^{-4}
13. $j = \sigma(4Vr_1r_2)/(r_2^2 - r_1^2)(r_2 + r_1)$; $E = 4Vr_1r_2/(r_2^2 - r_1^2)(r_2 + r_1)$; $R = 1/4\pi\sigma(1/r_1 - 1/r_2)$
15. (a) 0.142 amp (b) 1.69×10^{-2} amp (c) 8.95×10^{-2} amp (d) 0.91 w
17. Series, 7.5 amp; parallel, 7.5 amp
19. Voltmeter, 38×10^3 ohms in series. Ammeter, 5.01 ohms in parallel.
21. 4.8×10^{-3} through 64 ohms, 1.28×10^{-1} through 1 ohm, 6.4×10^{-2} through 3 ohms, 3.2×10^{-2} through 6 ohms.
23. (a) 6.17 ohms (b) 5.48 ohms

Chapter 21

1. 2.9 amp
3. Series: $M_{Cu} = 1.47$ gm, $M_{Al} = 0.42$ gm. Parallel: $M_{Cu} = 2.21$ gm, $M_{Al} = 1.26$ gm.
5. 2.6 ev 7. (a) 131×10^7 j (b) 3.1×10^5 kcal 9. 2.5×10^3 v
11. 1.11×10^{-4} m 13. 2.9×10^{-4} m
15. One should decrease frequency by 0.2% (1.8×10^7cycles/sec) during traversal of 180°

Chapter 22

1. (a) 5×10^{-6} w (b) 1.67×10^{-6} v 3. 10^3 w/m²·sec
5. Flux both increases and decreases in each 180° turn. 7. Lenz's law
9. 160 v; when normal to coil is perpendicular to field.
11. It will always move away from coil. 13. 4×10^{-8} j 15. Downward
17. 60 v 19. (a) $\frac{1}{4}$ j (b) 10^{-2} sec 21. 1.4 h
23. (a) 2.5×10^{-3} sec (b) 0.506 j 25. 0.576×10^{-7} h
27. $N_1N_2(\mu_0\pi/2)(r_2^2/r_1)$ 29. 0.462×10^{-3} sec 31. $\mathcal{E}t/L$
35. 3.96×10^6 j/m³

Chapter 23

1. 25×10^{-3} coul 3. $Q^2/2C$
5. (a) 1.76×10^{-10} farad (b) 7.75×10^{-9} coul (c) 1.75×10^{-7} j
7. $\frac{1}{5}V$; charge is the same. 9. $Q^2/2\epsilon_0 A$
11. (a) $V_f = V_i[1 - \exp(-S/vRC)]$ (b) 4.9×10^2 m/sec
13. (a) 25×10^{-3} sec (b) 0.2 amp (c) 25×10^{-3} sec 17. 8.85×10^{-14} j
19. (a) $$\frac{4\pi\epsilon_1\epsilon_2(a - b)}{(\epsilon_2 - \epsilon_1)\ln(b/a) + (a\epsilon_1 - b\epsilon_2)[(b - a)/ba]} \text{ farad}$$

(b) $$\frac{\epsilon_1\epsilon_2(a - b)}{(\epsilon_2 - \epsilon_1)r + (a\epsilon_1 - b\epsilon_2)[(b - a)/ab]} - \epsilon_0$$

21. (a) 6×10^{-3}, 20×10^{-3} coul (b) 9.3×10^{-3}, 4.7×10^{-3} coul (c) 0.5 j
23. 4 μf **25.** 19.3 Mev **27.** 2.28×10^7 j **29.** $2R$

Chapter 24

3. 0.96/min **5.** 2.61×10^{-3} w/m^2 **7.** 0.63 w/m^2
9. 6.3×10^{26}/m^3 **11.** It is dimensionless.

Chapter 25

1. (a) 0.04 m (b) 0.18 m (c) 0.63 m **3.** 1998 cycles
5. (a) 440/sec, 3.3 m (b) 440/sec, 0.75 m
9. $2/\pi$ cycles/sec, 10π m, 0.01 m, 20 m/sec **11.** $\pi/2$
13. $\lambda = 2\pi \times$ amplitude **15.** 4/sec **17.** $(y/2A) + (2x^2/A^2) = 1$
21. 15 sec **23.** 437 **25.** 93.5 ft/sec, 216 ft/sec **27.** fu/v
29. $\frac{1}{4}\pi$ j/sec·m^2

Chapter 26

1. (a) 0.055 (b) 0.126 (c) 0.17 sec **3.** (a) 13.8 μf (b) 0.505 h
5. (a) 8.8 v (b) $\approx$ 110 v (c) 0.282 v (d) 8.8 v (e) 101 v
7. 10^3 ohms, 316 ohms, 31.6 ohms
9. (a) 1.9 amp
(b) 9.5 v (resistor), 262 v (inductance), 3.03 v (capacitor)
(c) 0.0366 (d) 18.1 w (e) $675/2\pi$, 52 amp **11.** 3.52 h, 530 v, 530 v
13. $>$145 cycles/sec inductive, $<$145 cycles/sec capacitive
15. (a) $\omega = 10^3$/sec (b) 10^3/sec
(c) Less than resonant frequency, but by small amount.
17. Effective value: $(1/\sqrt{2})$ amp. Average value: $(2/\pi)$ amp.

Chapter 27

1. $I = 0.44$ amp, $I_L = 1.56$ amp, $I_C = 0.426$ amp. Phase difference.
3. xy faces 0, $x0z$ $(EH/2)$ in, $x1z$ $(EH/2)$ out, $0yz$ $(EH/2)$ out, $1yz$ $(EH/2)$ in, net flow 0.
5. Inner: $S = 3$ j/m^2·sec toward outside. Outer: $S = \frac{5}{8}\pi$ j/m^2·sec toward inside.
7. 9×10^{-12} j/m^2·sec **9.** 19.4×10^4 v/m, 550 amp/n·m, 0.11×10^{-8} j·sec/m^4
11. 377 ohms

Chapter 28

1. 337 m **3.** 3.3×10^5 km
5. (a) 1.3×10^{-11} amp/m (b) 1.3×10^{-11} w/m^2 (c) 0.43×10^{-19} n/m^2
7. 365 w/m^2 **9.** 0.25 n/m^2 **11.** 2×10^{18}/in^2·sec **13.** 5270°K
15. 1.7×10^{-12} m, much larger (=0.32).

Chapter 29

1. 1.30 **5.** Television waves are shorter in wavelength.
7. The angle of the "prism" is effectively zero. **9.** (a) 25° 40′ (b) 35° 14′

13. 1.66 **15.** 37° 10′

17. In both phenomena there are alternations of intensity. In interference two or more coherent sources are necessary; in diffraction only a source and obstacle.

19. 5×10^{-7} m **21.** (a) 4.4×10^{-3} m (b) 8.7×10^{-3} m (c) 1.09×10^{-2} m

23. 5.1×10^{-7} m

25. Colors are due to interference and diffraction effects of component colors of light.

27. 56.25 cm, 78.8 cm **29.** 1 min, 1.17 cm **31.** $d/2$

33. 1.28 mm, 1.81 mm **35.** 54° 28′

Chapter 30

1. (a) $3 \times 10^{-7}(1 - 5 \times 10^{-9})$sec (b) $\frac{3}{4}$ **5.** 1.014×10^{-6} sec longer

7. 1.8×10^{-11} ft **9.** 3.9×10^{-11} kg **11.** 1.5×10^{-10} j; 940 Mev

13. 6.05×10^{-9}% **15.** $0.866c$ **17.** (a) $0.14c$ (b) 949 Mev

19. (a) 0.58×10^{6} ev (b) 2.45×10^{6} ev **21.** 2.1×10^{-23} **23.** 0.23%

27. $0.988c$

Chapter 31

1. Many photons/sec hit the eye; e.g., 1 microwatt $\sim 10^{13}$ photons/sec.

3. 1.71×10^{31}/sec **5.** 4.6×10^{14} cycles/sec **7.** 0.49 ev

9. The particle aspects of light show up only in its interaction with electrons. Electrons were not discovered until the very end of the nineteenth century.

11. 4.14×10^{10} v **13.** 2.40×10^{18} cycles/sec

15. 6.1×10^{4} ev; 3.32×10^{-23} kg m/sec

17. (a) 9.77×10^{18}/sec (b) 955 ev

19. (a) 1.05 A (b) 5.7×10^{2} ev

21. 1.98×10^{19}/sec; 1.70×10^{19}/sec **23.** 21×10^{-40} m

25. 2.9×10^{-14} m

27. The energy of the photon is 100 times larger. This property can be used to distinguish between the two.

29. (a) 0.1×10^{6} m/sec (b) 0.6×10^{2} m/sec **31.** $\hbar\omega/2$

Chapter 32

1. The potential energy is more negative than the kinetic energy is positive.

3. (a) 0.35×10^{-62} m (b) 28.5×10^{73}

5. Brackett, Lyman, Paschen, Balmer series.

7. 14.8×10^{-7} m **9.** 913 A **11.** 10.5×10^{4} °K

13. $R_{Li^{++}} = \frac{1}{3}R_H = 1.77 \times 10^{-11}$ m; $v_{Li^{++}} = 3\,v_H = 6.6 \times 10^{6}$ m/sec.

19. 0.41×10^{-7}

21. (a) 972 A (b) 0.68×10^{-26} kg m/sec (c) 4.1 m/sec **23.** It speeds up.

Chapter 33

1. 2 **3.** $1s^22s^22p^63s^23p^63d^{10}4s^24p^64d^{10}5s^25p^6$

7. The electrons fill the $n = 3, l = 2$ subshell. **9.** $l > \frac{1}{2} \times 10^6$

11. $J = \frac{7}{2}, \frac{5}{2}$; $-J \leq m_J \leq J$ in steps of one. **13.** 2.32×10^{-4} ev

Chapter 34

1. Both Li and Na atoms have one electron outside a closed shell.

3. Lithium atoms must *lose* an electron each in forming a molecule, and sharing electrons does not permit this to occur.

5. 0.5 ev; 2.9×10^{-9} m

7. Li atoms have one electron outside a closed shell, which is more easily removed than any of the electrons of F atoms whose outer shells lack an electron of completion.

9. Na^+ ions have closed shells, while Na atoms have a single, easily detached outer electron.

11. All rare gas atoms contain only closed electron shells.

13. 2.23×10^{-10} m **15.** 8 **17.** 8.85×10^{11} sec^{-1}

Chapter 35

1. Van der Waals

3. Both are nonconductors at very low temperatures; germanium is the better conductor at room temperature, because its forbidden band is narrower.

5. 1.5 R **7.** 4.26×10^4 °K; 1.39×10^6 m/sec

Chapter 36

1. 3 p, 3 n; 6 p, 7 n; 15 p, 16 n; 40 p, 54 n; 56 p, 81 n

3. 155 Mev **5.** 8.48 Mev

7. Electrostatic potential energy is 0.24 Mev while binding energy per nucleon is 8.75 Mev.

Chapter 37

3. 6400 years **5.** ${}_2He^3$; 25% **7.** 1.24×10^7 sec^{-1}

9. ${}_8O^{14}$ emits a positron, and ${}_8O^{19}$ emits an electron. **11.** 92, 233

13. Helium and radon cannot be combined chemically to form radium nor can radium be broken down into helium and radon by chemical means.

15. 6.015 amu

17. In fission a large nucleus splits into smaller ones, while in fusion small nuclei join together to form a larger one. In both, the product nucleus or nuclei have less mass than the initial one or ones, with the missing mass being evolved as energy.

19. 6, 12, carbon **21.** 3.1×10^{13}

Chapter 38

1. Electrons, positrons, neutrinos, alpha particles, and electromagnetic waves.

3. 1879 Mev; more **5.** 280 Mev

7. Such a decay would not conserve lepton number or baryon number.

9. A neutrino.

13. $S_{\Lambda^0} = -1,\ S_p = 0,\ S_{\pi^-} = 0$

15. (a) Charge.
(b) Baryon number, angular momentum.
(c) Baryon number, angular momentum, electron-family number, muon-family number.
(d) Electron-family number, muon-family number, charge.
(e) Angular momentum, electron-family number, muon-family number.

Appendix I

A-1. $f = 0.3$ m, $2\frac{1}{3}$ m, $1\frac{1}{8}$ m, $-\frac{2}{3}$ m

A-3. 10 cm in front of plane surface for paraxial rays.

A-5. No. **A-7.** 90 cm **A-9.** 10 cm behind mirror.

Index

PRINTED IN JAPAN
TOKYO

BCDE69876